CPSS Power Electronics Series

This series comprises advanced textbooks, research monographs, professional books, and reference works covering different aspects of power electronics, such as Variable Frequency Power Supply, DC Power Supply, Magnetic Technology, New Energy Power Conversion, Electromagnetic Compatibility as well as Wireless Power Transfer Technology and Equipment. The series features leading Chinese scholars and researchers and publishes authored books as well as edited compilations. It aims to provide critical reviews of important subjects in the field, publish new discoveries and significant progress that has been made in development of applications and the advancement of principles, theories and designs, and report cutting-edge research and relevant technologies. The CPSS Power Electronics series has an editorial board with members from the China Power Supply Society and a consulting editor from Springer.

Readership: Research scientists in universities, research institutions and the industry, graduate students, and senior undergraduates.

Indexed by EI Compendex

Yuan Gao · Yan Zhang

Silicon Carbide Power Devices

Characteristics, Test and Application

机械工业出版社
CHINA MACHINE PRESS

Springer

Yuan Gao
Xi'an Jiaotong University
Xi'an, Shaanxi, China

Yan Zhang
Xi'an Jiaotong University
Xi'an, Shaanxi, China

ISSN 2520-8853 ISSN 2520-8861 (electronic)
CPSS Power Electronics Series
ISBN 978-981-96-3479-8 ISBN 978-981-96-3480-4 (eBook)
https://doi.org/10.1007/978-981-96-3480-4

Jointly published with China Machine Press Co.,Ltd.
The print edition is not for sale in China (Mainland). Customers from China (Mainland) please order the print book from: China Machine Press Co.,Ltd.

This Springer imprint is published by the registered company Springer Nature Singapore Pte Ltd.
The registered company address is: 152 Beach Road, #21-01/04 Gateway East, Singapore 189721, Singapore

Preface

Power converters are essential devices for energy utilization, playing a critical role in both industrial production and daily life. At the heart of power converters lie power semiconductor devices, which largely determine the performance of these systems. After decades of development, power semiconductor devices have evolved into a vast family that covers a wide range of voltage (from a few volts to several kilovolts) and current (from a few amps to thousands of amps). Common types of power semiconductor devices include MOSFETs, IGBTs, diodes, GTOs, and thyristors.

Most power semiconductor devices are based on silicon (Si) material, whose characteristics are nearing their theoretical limits, creating a bottleneck for further advancements in power converters. To overcome this limitation, third-generation semiconductor materials such as silicon carbide (SiC) and gallium nitride (GaN) are gaining increasing attention. Compared to Si-based devices, SiC power devices offer higher switching speeds, can operate at higher junction temperatures, and support both high voltage and high current simultaneously. These advantages significantly enhance the performance of power converters, enabling higher energy conversion efficiency, increased power density, and reduced system costs.

SiC power devices are well-suited for applications in automotive traction inverters, onboard chargers for electric vehicles, EV charging stations, photovoltaic systems, uninterruptible power supply (UPS) systems, energy storage systems, and industrial power supplies. Currently, the global SiC supply chain is maturing, and major power semiconductor manufacturers have introduced SiC-based products. As costs continue to decrease, the adoption of SiC power devices is experiencing exponential growth.

The author has dedicated his career to the research and promotion of power semiconductor device testing, assessment, and application technologies, with a particular focus on SiC power devices. His expertise spans device testing equipment and measurement methodologies, along with a deep understanding of SiC device characteristics. Through years of research, the author has recognized the critical importance of mastering the principles of power devices, measurement techniques, and related application technologies. These are essential for advancing research and improving the performance of power converters.

Additionally, the author has observed the growing demand among researchers and engineers for comprehensive knowledge and technical expertise related to SiC devices. This book was written to address this need, aiming to provide readers with an in-depth understanding of SiC device characteristics and testing methods, while identifying potential technical challenges and strategies to address them. It is designed to help researchers quickly grasp the latest developments in the field, offering a solid foundation for further studies, and to support engineers in effectively applying SiC devices, thereby driving industry progress.

The book is divided into 12 chapters, covering the following key topics:

- Chapter 1 is the overview of SiC power semiconductor devices. This chapter begins by discussing the mutual development of power semiconductor devices and power electronics, along with the evolution, characteristics, and limitations of various Si-based power devices. It then introduces the physical properties of SiC semiconductor materials and their advantages in device performance. Additionally, it provides an overview of the development of different stages within the SiC industrial chain, concluding with detailed insights into the technological advancements and product development of SiC diodes, SiC MOSFETs, and SiC power modules.
- Chapter 2 is the key characteristics of SiC diodes. This chapter focuses on the main characteristics of SiC diodes, including maximum ratings, static properties, and limiting parameters.
- Chapter 3 is the key characteristics of SiC MOSFETs. This chapter provides an in-depth exploration of the key features of SiC MOSFETs, covering maximum ratings, static properties, dynamic characteristics, and limiting parameters. It further discusses quality factors, loss calculations, and parameter-based modeling methods for SiC MOSFETs.
- Chapter 4 is the comparison between SiC and Si device characteristics. This chapter compares the performance of SiC devices and Si devices, including a comparison of SiC MOSFETs with Si SJ-MOSFETs and Si IGBTs. It also examines the characteristics of SiC SBDs, SiC MOSFET body diodes, Si FRDs, and Si SJ-MOSFET body diodes.
- Chapter 5 is the double-pulse testing techniques. This chapter begins with the fundamental principles, parameter settings, testing platforms, measurement instruments, and equipment for double-pulse testing. It then discusses the impact of parasitic parameters between voltage measurement points and corresponding mitigation methods. Finally, it provides evaluation criteria for dynamic characteristic test results in different scenarios and key considerations for selecting double-pulse testing equipment.
- Chapter 6 is the testing and failure analysis of SiC devices. This chapter introduces the principles, challenges, and precautions for testing various parameters of SiC devices. It then details the testing projects, principles, and equipment for mass production, system application, and reliability testing. The chapter concludes with an overview of failure analysis techniques for SiC devices.

- Chapter 7 is the impact and mitigation of turn-off voltage spikes. This chapter provides an in-depth discussion of the factors influencing turn-off voltage spikes and outlines three mitigation strategies: loop inductance control, decoupling capacitors, and reducing turn-off speed.
- Chapter 8 is the impact and mitigation of crosstalk. This chapter explains the fundamental principles and key influencing factors of crosstalk, along with two mitigation techniques: Miller clamping and loop inductance control.
- Chapter 9 is the impact and mitigation of common-mode current. This chapter explores the principles and characteristics of signal-path common-mode current, presenting three mitigation strategies: high-CMTI driver chips, high common-mode impedance, and common-mode current diversion. Additionally, it introduces techniques for measuring differential mode interference.
- Chapter 10 is the impact and mitigation of common-source inductance. This chapter examines the impact of common-source inductance on device switching characteristics and crosstalk, highlighting the advantages of Kelvin source packaging. It also presents innovative methods for testing and evaluation.
- Chapter 11 is the driver circuit design. This chapter provides a comprehensive framework for designing driver circuits. It details the selection of gate resistances, driving voltages, and the effects of driver stage characteristics and signal isolation/transmission. Additionally, it introduces various short-circuit detection methods, with a detailed explanation of DESAT short-circuit protection based on measured results. The chapter concludes with reference designs for gate resistors using three commonly used driver ICs.
- Chapter 12 is the key applications of SiC devices. This chapter highlights the performance advantages, technical solutions, and commercial products of SiC devices across various applications. These include main drive inverters, onboard chargers, charging stations, photovoltaics and energy storage systems, motor drives, and uninterruptible power supplies (UPS).

The author has devoted significant effort to structuring the chapters, selecting content, presenting concepts, and illustrating with waveforms to ensure the material is well-organized, easy to understand, and genuinely beneficial for readers' learning and work. First, when explaining knowledge, the author focuses on building a systematic framework for readers, avoiding a "can't see the forest for the trees" approach. Second, in discussing solutions to challenges in SiC device applications, the book emphasizes practical techniques that can address real-world application issues, rather than solely pursuing the latest academic research findings.

Additionally, a considerable portion of the book is dedicated to detailed explanations of testing equipment and methods, addressing a common gap in testing techniques faced by power device and power electronics researchers and engineers. Most of the waveforms in the book are derived from experimental measurements, making the corresponding analyses and conclusions more relevant to real world applications. This approach helps guide application design directly while minimizing the confusion caused by discrepancies between ideal theoretical waveforms and actual test waveforms. Finally, each chapter concludes with a rich selection of references and recommended readings, including industry application manuals and academic papers with strong engineering potential. These resources balance cutting-edge knowledge with practical value, catering to both industry and academia.

Several industry experts contributed materials for this book, providing valuable insights and expertise. Their contributions are acknowledged as follows: Hong Lin (Soitec), Ningyi Liu; Fei Lei, Wansheng Zuo (ALLSIC (Shanghai) Semiconductor Technology Co., Ltd.)—Sect. 1.6; Xiaoyang Zhou, Boqiang Li, Xianlong Zhu (AccoPower Semiconductor Co., Ltd.)—Sects. 1.8; Tiancong Shao, Pei Hu (Beijing Jiaotong University)—Sect. 3.7; Yang Li (Xi'an Jiaotong University)—Sects. 2.1, 2.2, 5.4; Huimin Li (Teledyne LeCroy)—Sect. 5.3.1; Chuan Sun, Fangfang Wang (Tektronix)—Sects. 5.6.1, 5.6.2; Wei Liu, Lei Chen, Shichao Jiang (Hangzhou Firstack Technology Co., Ltd.)—Sect. 5.6.2, 5.6.3; Huipeng Liu, Su Yan (AccoTEST Technology Co., Ltd.)—Sects. 6.1, 6.2.1, 6.2.8; Fenghao Zhong, Haiyang Sun, Jianfei Guo, Chenhuan Wu (Hangzhou Changchuan Technology Co., Ltd.)—Sects. 6.2.2, 6.2.7; Ming Ji (Semight Instruments Co., Ltd.)—Sects. 6.2.3, 6.2.4; Jianming Bu, Shisong Hu, Liujie Wu (HangZhou ZoanRel Electronics Co., Ltd.)—Sects. 6.2.5, 6.3.2; Chunmao Ren, Weiqun Wang, Dongpo Xie (Changchuan Semiconductor (Shenzhen) Co., Ltd.)—Sect. 6.2.6; Xiaoxiao Cao, Weixin Yang, Xiaolei Chu, Chunyang Zhang, Linling Cai (Thermo Fisher Scientific Inc.)—Sects. 6.4.1, 6.4.5, 6.4.6, 6.4.7; Zhengfeng Huang, Xinlei Chen (Tektronix)—Sect. 6.5; Yifan Xiao—Sect. 12.1; Shenhua Zhang—Sects. 12.2, 12.3; Xing Fang, Ye Yuan (Shenzhen Sinexcel Electric Co., Ltd.)—Sect. 12.4; Liangcai Wu, Jiaxing Yu (Shenzhen Growatt New Energy Co., Ltd.)—Sect. 12.5; Dawei Zheng (Eaton Corporation)—Sect. 12.6; Wenbin Xie, Jiaxin Huang (InnoScience (Shenzhen) Semiconductor Technology Co., Ltd.)—Sect. 12.7.

During the preparation of the second edition, Zinan Lou and Ruixi Guo from Xi'an Jiaotong University were responsible for creating some of the illustrations, while Yang Li, Shaopeng Xue, and Jia Shu from Xi'an Jiaotong University handled the manuscript proofreading. Gao Fei from AccoTEST Technology Co., Ltd. conducted several experiments. Additionally, Yanjing He, Xiaofeng Leng, Tao Wang, Bonan Wang, Ting Huang, Wei Shi, Yuheng Wang, Yicong Xie, Bin Zhou, Yutan Zhang, Lihong Chen, Zixiang Tong, Min Luo, Dominic Li, Hua Wang, Wulong Cong, and Jie Liu offered many valuable suggestions and feedback. Further support and assistance were provided by Zengsheng Wang, Bonan Wang, Shiqiang Hao, Jie Dong, Jinshui Zhang, Anping Tong, Guowen Li, Zihao Guo, Yuan Xue, Jun Li, Jie Chen, Qi Dong, and Yang Sun. Heartfelt thanks are extended to all the experts across these fields for their invaluable contributions.

In addition, I would like to thank Prof. Zhaoan Wang from Xi'an Jiaotong University for introducing me to the field of power electronics; Dr. Ming Li for connecting me with power semiconductor devices; and Prof. Jianguo Yang from Xi'an Jiaotong University for making me realize the importance and joy of measurement and testing.

Finally, I must express my special gratitude to my parents for their upbringing, and for the immense understanding and support they provided throughout the process of writing this book.

Due to the author's limited expertise, there may be errors or inadequacies in this book. I sincerely invite readers to offer their criticism and suggestions.

Hanzhong, China Yuan Gao

Contents

List of Figures

List of Tables

Chapter 1
Fundamentals of Power Semiconductor Devices

Electric energy is one of the most crucial forms of energy in human society. To utilize electric energy effectively, one cannot overlook power electronic technology, which converts the form of electric energy. This involves controlling voltage, current, and frequency. Power electronics is a technology that utilizes power semiconductor devices to convert and regulate electrical energy. It amalgamates three disciplines: electricity, electronics, and control theory. As the core of power electronics technology, "one generation power devices, one generation topology and technology". Mastering the key technology of power semiconductor devices is essential to propel the advancement of power electronics.

Si semiconductor-based power devices have over 60 years of development, leading to the emergence of various devices with continuously optimized performance. Presently, Si power devices are nearing their theoretical limits, constraining further enhancements in power converter performance. In order to obtain power devices with superior performance, researchers have redirected their focus towards SiC semiconductor materials and launched new power devices. This provides new vitality for power electronics development.

In this chapter, the developmental trajectory of major Si power semiconductor device technologies is reviewed first. This facilitates an understanding of the device type innovations and structural optimizations made to address the bottlenecks associated with Si devices, building the bridge for the introduction of SiC devices. Subsequently, a comparative analysis is conducted on the distinct physical properties of Si and SiC materials, which verifies the superior characteristics of SiC devices from material level

The SiC power device production has gradually matured after years of development. This chapter presents the development status and relevant technologies in the SiC device production chain, including substrate, epitaxy, devices, packaging, etc. Furthermore, a detailed discussion is conducted on SiC MOSFETs, and SiC power modules, addressing aspects such as technological advancements, industrialization, and the status of products.

Y. Gao and Y. Zhang, *Silicon Carbide Power Devices*, CPSS Power Electronics Series,
https://doi.org/10.1007/978-981-96-3480-4_1

1.1 Power Semiconductor Devices and Power Electronics

Power electronics technology utilizes semiconductor device to achieve efficient conversion and electrical energy control. It involves multiple fields such as power, electronics, control, and computer science. Power devices are the core and foundation of power electronics technology and continuously drive its innovation and application.

In the 1940s, the invention and development of transistors gave birth to power semiconductor devices. The earliest power devices include transistors and diodes, such as power transistor and power diode. These devices are used in fields such as communication, television, and broadcasting. This promotes the quick development of power devices.

In the 1960s and 1970s, thyristors were invented and rapidly advanced. As controllable power semiconductor devices, thyristors can turn on and off using gate drive signals, which promote their widely application in AC power regulation, DC motor driver, inverters, etc. The invention of thyristors brings power electronics technology into the phase control era.

By the end of 1970s, planar power MOSFETs were developed. It can control the drain current through gate voltage. As voltage-controlled unipolar device, this leads to simple drive circuit, low drive power, fast switching speeds, excellent high-frequency properties, and maximum operating frequencies above 1 MHz. They are ideal for high-frequency applications such as switching power supplies and high-frequency induction heating. Furthermore, they possess a wide safe operating area, no secondary breakdown problems, and high robustness. The introduction of power MOSFETs opens up high-frequency applications.

In the late 1980s, trench MOSFETs and IGBTs gradually emerged, marking power devices getting into the electronic application. Trench MOSFETs are an improvement based on planar MOSFETs, featuring a trench structure in the gate region that increases gate capacitance and reduces gate resistance, thus enhancing switching speed and current capacity. IGBTs, a hybrid device that combines the advantages of MOSFETs and bipolar transistors, offer high input impedance, fast switching speeds, and simple drive circuits, as well as high output current density, low on-state voltage drop, and high voltage withstand capability, typically ranging from 600 V to 6.5 kV. IGBTs turn on by applying a forward gate voltage to form a conductive channel and provide base current to the transistor; applying a reverse gate voltage eliminates the channel, turning off the IGBT. Due to their superior performance, IGBTs become the main devices in inverters, UPS systems, frequency converters, motor drives, high-power switching power supplies, and particularly in the electric vehicles and high-speed trains. These new achievements have provided the foundation for the development of high-frequency power electronics technology, driving power electronic equipment towards intelligent and high-frequency advancements.

In the 1990s, superjunction MOSFETs appeared, overcoming the traditional "silicon limit" to meet the needs of high-power and high-frequency applications. They feature a superjunction structure in the drain region, where periodic changes in doping concentration decrease the electric field in the drain region. This increases the device's voltage resistance while maintaining low on-state voltage drops and switching losses. These characteristics make them suitable for high-voltage, high-frequency, and high-efficiency power electronic equipments. The emergence of superjunction MOSFETs marked the entry of power electronics technology into the superjunction era, achieving more efficient energy conversion and control.

In the early twenty-first century, wide bandgap semiconductor devices such as silicon carbide (SiC) and gallium nitride (GaN) power electronics gained attention. However, due to constraints in technological maturity and cost, they are still in the early stages of market development. Compared to silicon-based devices, power electronic devices using wide bandgap semiconductor materials can be smaller, faster, more reliable, and more efficient. This reduces the mass, volume, and lifecycle cost of power electronic components, allowing devices to operate at higher temperatures, voltages, and frequencies, thus achieving higher performance. Given these advantages, wide bandgap semiconductors have extensive application in household appliances, power electronic equipment, new energy vehicles, industrial production equipment, high-voltage direct current transmission equipment, and mobile base stations. The introduction of wide bandgap semiconductor devices has brought power electronics technology into the era of new materials, further enhances the power conversion efficiency.

1.2 Si Power Diode

1.2.1 P–N Junction

Take the formation of N region by doping P-type semiconductor material as an example. The substantial concentration difference prompts holes (majority carriers) in the P region to diffuse towards the N region, while electrons (majority carriers) in the N region diffuse towards the P region. Consequently, near the interface between the N and P regions, the donor element (typically phosphorus) doped in the N region becomes an ionized donor, carrying a positive charge resulting from the loss of electrons. Simultaneously, the doped acceptor element (usually boron) in the P region gains electrons and becomes an ionized acceptor, carrying a negative charge. This region of ionized donors and acceptors forms a space charge region, generating a built-in electric field. The field aligns from the ionized donor to the acceptor, that is, from the N region to the P region (Fig. 1.1).

Holes and electrons drift due to the influence of the built-in electric field. The direction of drift is opposite to the direction of diffusion. Holes drift towards the P region, and electrons drift towards the N region. When the diffusion and drift of electrons and holes reach a dynamic balance, the width of the space charge region stabilizes. The application of an external electric field to the P–N junction results in a change in the space charge region width. When a positive voltage is applied to the P region, the direction of the external electric field is opposite to that of the

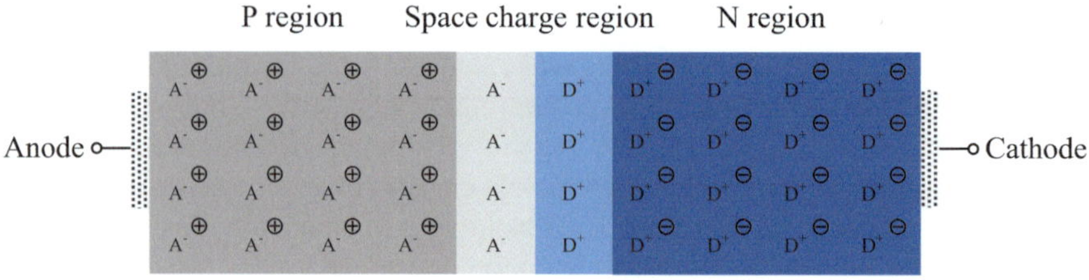

Fig. 1.1 Structure of P–N junction

built-in electric field, hole drift is weakened, diffusion is strengthened, and the space charge region becomes narrower. Similarly, when a negative voltage is applied to the P region, the direction of the external electric field and the built-in electric field are the same, the hole drift is strengthened, the diffusion is weakened, and the space charge region becomes wider.

The main structure of a diode is the P–N junction, with the anode located at the P region and the cathode at the N region. When a reverse voltage is applied across the diode, causing it to be reverse biased, the space charge region widens to withstand the reverse voltage. In this state, diffusion is significantly weakened, and drift dominates, generating the reverse leakage current. The diode is in the off-state. Conversely, when a forward voltage is applied across the diode, P–N junction is forward biased. If the applied forward voltage exceeds the barrier voltage produced by the built-in electric field in the space charge region, diffusion becomes predominant, rising the forward current. In this mode, the diode is in the on state.

1.2.2 P-i-N Diode

The P–N-junction diode has a limited capability to withstand reverse voltage, making it unsuitable for power diode application. To enhance its withstand voltage, a P-i-N diode structure is adopted, incorporating a withstanding voltage layer of thicker lightly doped N^- high-resistance region between the heavily doped P^+ region and N^+ region. This added layer effectively increases the diode's capability to withstand voltage, as depicted in Fig. 1.2. Due to the low doping concentration in the layer, it can be regarded as an intrinsic state compared to high doping, leading to the name "P-i-N diode".

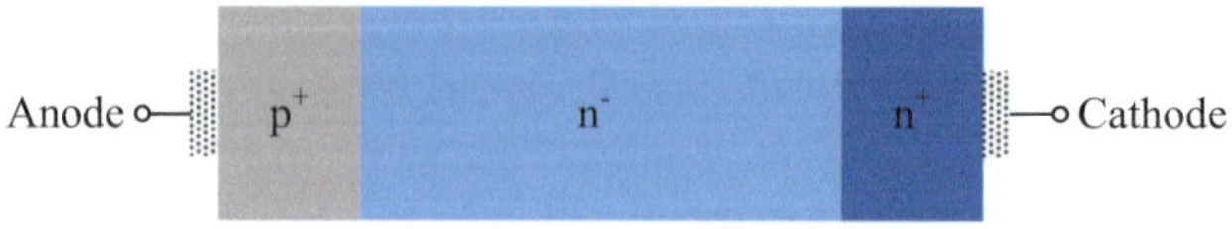

Fig. 1.2 Structure of P-i-N diode

When withstanding the reverse voltage, the P–N-junction barrier in a P-i-N diode increases, causing the space charge region to widen, primarily extending towards the lightly doped N^- region. The wider space charge region, owing to its low doping concentration and increased thickness, enables the N region to endure higher reverse voltages. Conversely, upon the forward voltage, the P–N-junction barrier diminishes, resulting in a narrower space charge region. The P^+ region injects holes into the N^- region, while the N^+ region injects electrons into the N^- region. This process induces conductivity modulation within the N^- region, allowing for a reduced on-state voltage drop under large currents. Combining these characteristics, the P-i-N diode has a small leakage current, exhibits the high withstand voltage capability, and has low on-state voltage drop.

When an on-state P-i-N diode withstands a sudden reverse voltage, it does not turn off immediately but undergoes a reverse recovery process, depicted in Fig. 1.3. During forward conduction, the diode's current is mainly the diffusion of majority carriers, filling the N^- region with both electrons and holes. To turn off the P-i-N diode, it is necessary to extract electrons from the N^- region back to the N^+ region and holes back to the P^+ region. This process results in a reverse recovery current even after the current drops to zero. As the majority carriers continue to decrease, the space charge region forms and expands, gradually reducing the reverse recovery current to zero. Ultimately, the diode withstands the reverse voltage, and the P-i-N diode turns off.

Because the P-i-N diode is a bipolar device, the carrier concentration is high when it turns on, making it difficult to be extracted from the thick N^- region. Consequently, its reverse recovery time is relatively long. Therefore, the P-i-N diode is suitable for medium- and high-voltage applications that are not sensitive to the reverse recovery losses, such as rectifier diode.

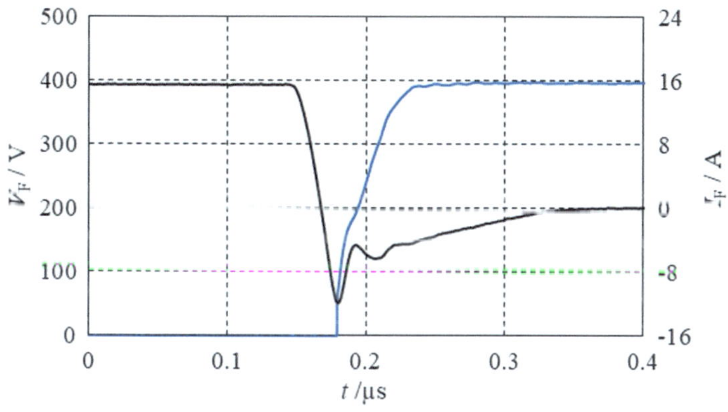

Fig. 1.3 Reverse recovery process of P-i-N diode

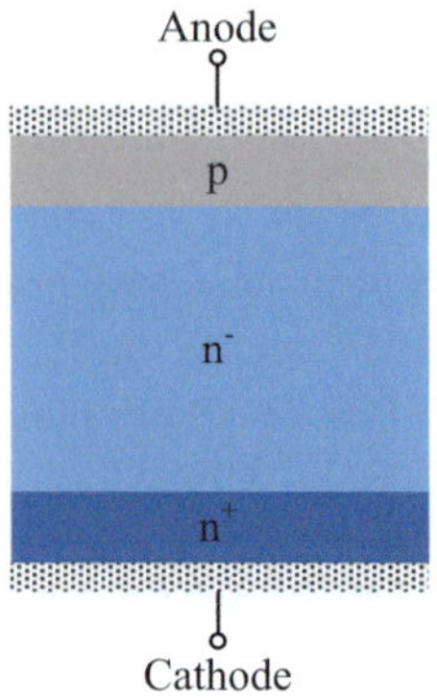

(a) Weak anode structure

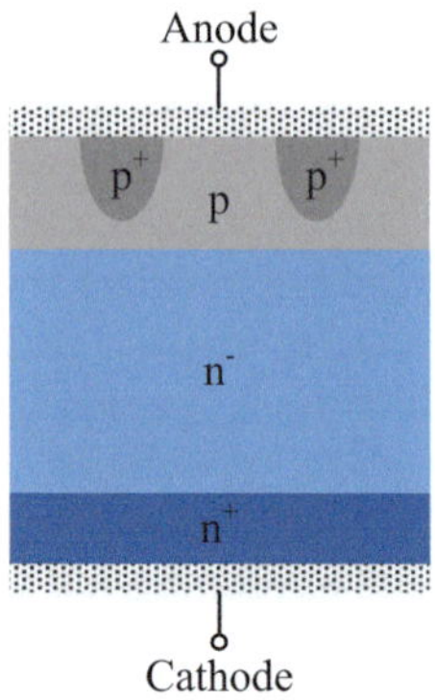

(b) SPEED structure

Fig. 1.4 Structure of fast recovery diode

1.2.3 Fast Recovery Diode

To shorten the P-i-N diode's reverse recovery process, its anode structure is modified to control hole injection efficiency and minimize minority carrier injection during conduction, known as the fast recovery diode (FRD).

The commonly used fast recovery diode has two structures: the weak anode structure and the Self-adjustable P^+ Emitter Efficiency Diode (SPEED) structure, as illustrated in Fig. 1.4.

The weak anode structure reduces anode injection efficiency by lowering the doping concentration of the P region, thus speeding up reverse recovery process. The SPEED structure modifies the anode by adding a heavily doped P^+ region within a lightly doped P region. When the conduction current is low, holes are injected from the P region, resulting in lower injection efficiency, which accelerate the reverse recovery process. As the current increases, holes are injected into the P^+ region, enhancing hole injection efficiency, which improves the device's capability to handle surge currents. Besides changing the anode structure, proton irradiation can also introduce local recombination centers in the anode P^+ region to control carrier lifetime, thereby increasing reverse recovery speed. However, this approach significantly raises leakage current.

The FRD diode is commonly used as freewheeling diode because of its good reverse recovery characteristics. In practical applications, besides a rapid reverse recovery speed, it is important for the reverse recovery characteristics to be "soft", indicating a gradual decrease in reverse recovery current without sudden and rapid drop. Achieving this involves improvement in the diode's cathode structure to control the carrier concentration at the end of reverse recovery.

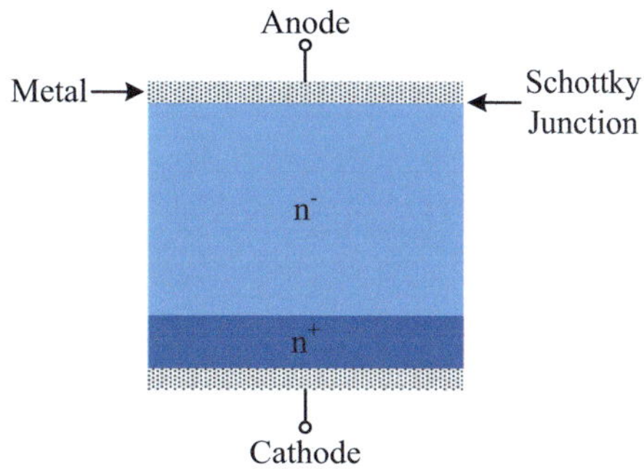

Fig. 1.5 Structure of ordinary Schottky diode

1.2.4 Schottky Barrier Diode

Differing from traditional diode based on P–N junction, the Schottky diode is formed by metal–semiconductor junction and is commonly referred to as Schottky barrier diode (SBD), named after its inventor Walter Schottky. The power Schottky diode includes ordinary power Schottky diode, junction barrier-controlled Schottky diode, Schottky-PiN composite diode, and superjunction Schottky diode.

1. **Ordinary Power Schottky Diode**

Ordinary power Schottky diode is formed by substituting the P^+ region anode of the P-i-N diode with metal. The structure is illustrated in Fig. 1.5. Schottky diode functions as unipolar devices, conducting electricity only with electrons. Consequently, it exhibits faster switching speed and improved reverse recovery characteristics.

Furthermore, the barrier height at the Schottky junction (metal–semiconductor junction) is lower compared to the P–N junction. As a result, the forward voltage of the Schottky diode is approximately 0.3 V, which is lower than the 0.7 V of the P-i-N diode, significantly reducing on-state voltage drop. However, this leads to lower breakdown voltage and higher leakage current, making it more suitable for medium- and low-voltage applications. It's important to note that despite the lower barrier height, the Schottky diode is a unipolar device without the presence of conductivity modulation effects. When the conduction current is large, the forward voltage drop in the drift region remains high. This limitation restricts its application in high-current scenarios.

1. **Junction Barrier-Controlled Schottky Diode**

The junction barrier-controlled Schottky diode, also known as the junction barrier Schottky diode (JBS diode), involves the formation of a P region above the N^- region before establishing the Schottky junction. This P region, together with the N^- and N^+ below, forms a P-i-N diode structure. Therefore, the JBS can be seen as a parallel combination of a typical Schottky diode and a P-i-N diode, as shown in Fig. 1.6.

During forward conduction, the Schottky junction's barrier is low, resulting in a small on-state voltage drop. In reverse blocking, the P region formed by the gap

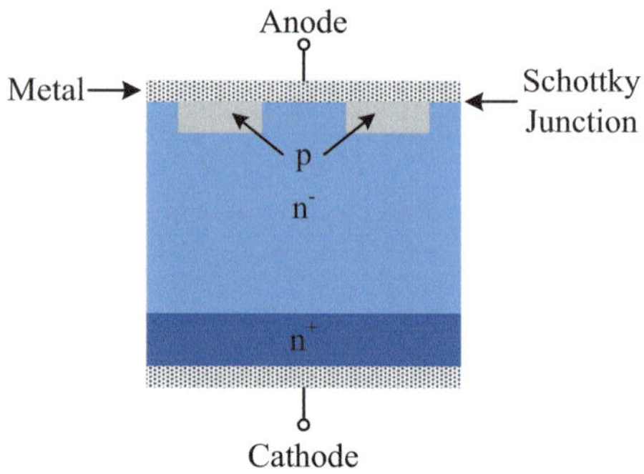

Fig. 1.6 Structure of junction barrier-controlled Schottky diode

and the anode metal form a JFET structure. Under reverse voltage, the width of the space charge region formed by the P region and N region increases and subsequently merging to shield the Schottky junction. At this stage, the P-i-N diode primarily functions, contributing to an increased blocking voltage.

3. **Merged PiN Schottky Diode**

The ordinary Schottky diode and P-i-N diode have characteristics of fast reverse recovery and high power, high withstanding voltage, respectively. The merged PiN Schottky diode, a result of combining these two types, has the advantages of both. Typical structures include the merged PiN Schottky (MPS), trench oxide PiN Schottky (TOPS), and stereotactic field diode (SFD) structure, as depicted in Fig. 1.7.

The structure of MPS is similar to that of JBS diode. By growing multiple P^+ regions within the N^- region of an ordinary SBD, a vertical P-i-N structure is created. The P region depth of MPS Diode is deeper than that of JBS. Under low conduction current, the P-i-N diode turns off. However, with high conduction current, the P^+ region injects holes into the N^- region, inducing a conductivity modulation effect

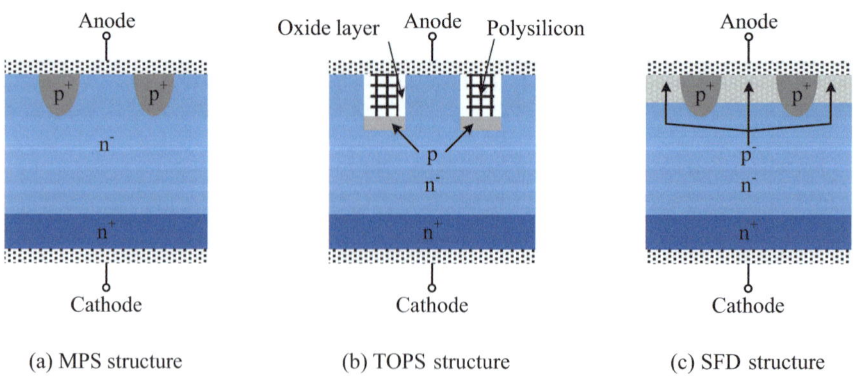

Fig. 1.7 Structure of merged PiN Schottky diode

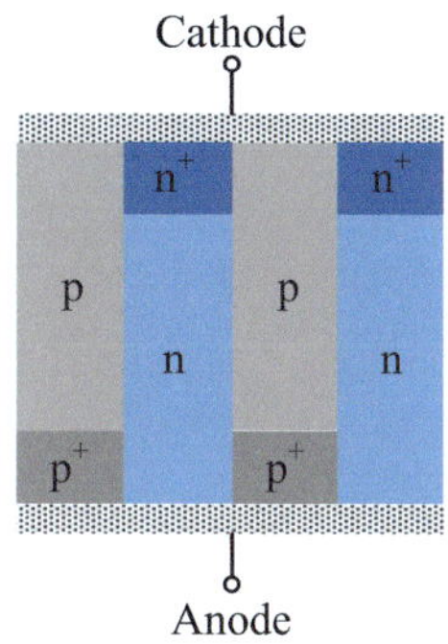

Fig. 1.8 Superjunction Schottky barrier diode

and a small on-state voltage drop. When subjected to a reverse voltage, the P–N junction's space charge area expands, shielding the Schottky junction via the JFET effect, thereby increasing the breakdown voltage.

The TOPS diode initially digs trench, establishes P region through ion implantation, and subsequently fills it with polysilicon and silicon dioxide sequentially to further reduce the hole concentration near the anode side.

The SFD (Stereotactic Field Diode) employs Al-Si instead of the metal anode used in MPS to generate a thin P^- region between the P^+ regions, regulating the anode injection efficiency. This enhances the device's voltage withstanding capability while optimizing its reverse recovery characteristics.

4. **Superjunction Schottky Barrier Diode**

The superjunction Schottky barrier diode (SJ-SBD) features metal electrodes serving as both anode and cathode, with P and N regions positioned in parallel between electrodes. The P region near the anode and the N region near the cathode are heavily doped to establish ohmic contact with the electrode. A Schottky junction is formed between the lightly doped region and the electrode, as shown in Fig. 1.8. During turn-on process, the forward voltage drop is reduced across the vertical Schottky junction. Conversely, during turn-off process, a transverse electric field is generated through the superjunction structure, enhancing the breakdown voltage and reducing the leakage current.

1.3 Si Power MOSFET

1.3.1 Structure and Operating Principle of MOSFET

The Metal–Oxide–Semiconductor Field-Effect Transistor (MOSFET) is a widely used unipolar switching device. Figure 1.9 shows the basic structure of an N-channel MOSFET. It has three terminals: source, drain, and gate. The M–O–S structure comprises the gate metal layer (metal), the SiO_2 layer (oxide) as an insulator, and the P-type substrate semiconductor layer (semiconductor).

Fig. 1.9 Structure of N-channel MOSFET

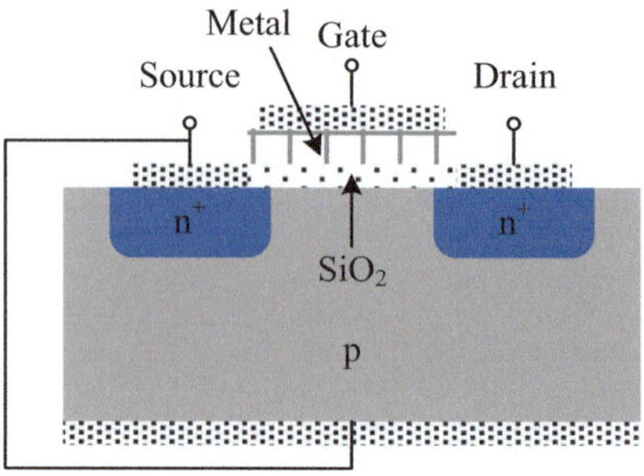

Fig. 1.10 Operating principle of N-channel MOSFET

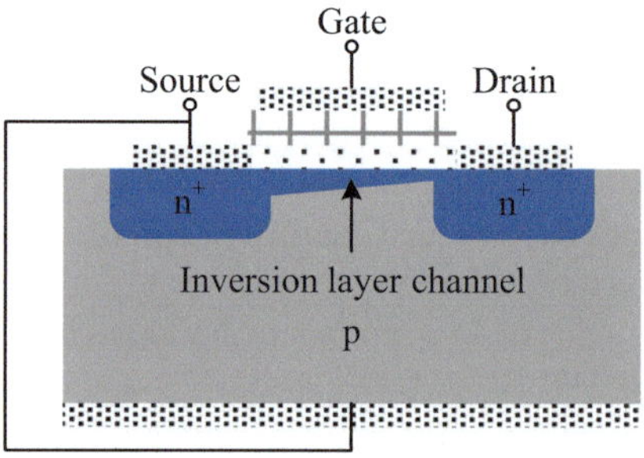

In Fig. 1.10, with a positive gate-source voltage (V_{GS}), the gate is positively charged, inducing a negatively charged inversion layer in the P region under the gate. When V_{GS} exceeds its threshold voltage ($V_{GS(TH)}$), this inversion layer establishes a connection between the source and the N^+ regions of the drain, forming a channel. Upon a forward drain-source voltage (V_{DS}), electrons flow through the channel, resulting in the turn-on process of the MOSFET. If V_{GS} doesn't exceed $V_{GS(TH)}$, a channel cannot form, the MOSFET turns off. During this period, the reverse-biased P–N junction withstands the applied voltage.

1.3.2 Laterally Double-Diffused MOSFET

The turning-on of an N-channel MOSFET relies on the channel in the inversion of P region under the gate. During turn-off process, the widened P–N junction on the drain side primarily withstands voltage. However, a contradiction arises when using the structure in Fig. 1.9 as a power device. If the length of the channel is insufficient, the MOSFET cannot withstand high reverse voltages and thus cannot meet the demands of high-voltage applications. When the channel is too long, it leads to higher channel resistance, resulting in increased losses, and cannot meet the requirements of high-current applications.

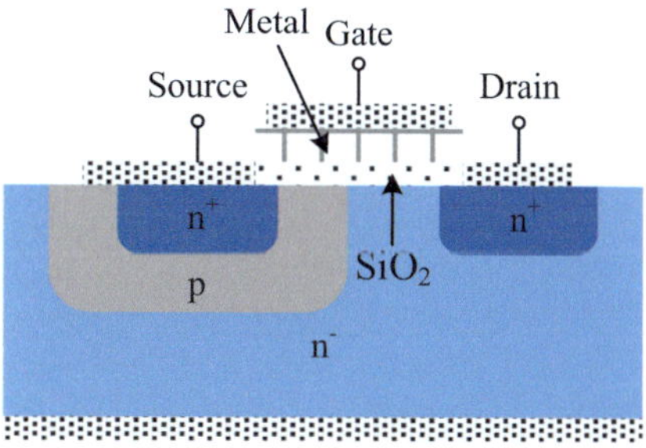

Fig. 1.11 Structure of N-channel LDMOS

To address the contradiction, a lightly doped N^- region is introduced as a drift region between the P region and the drain N^+ region, transforming it into a Laterally Double-Diffused Metal-Oxide Semiconductor (LDMOS), as shown in Fig. 1.11. When a reverse voltage is applied to the LDMOS, the higher doping concentration in the P region leads to the expansion of the depletion region primarily within the drift region. Consequently, the blocking voltage is mainly determined by the length and doping concentration of the drift region. Moreover, the resistivity of the drift region is lower than that of the channel. For LDMOS, a simple method to increase the blocking voltage, while minimizing the relative increase in on-state resistance, is to increase the length of the drift region. This achieves a compromise between blocking voltage and on-resistance.

To enhance capability of withstanding voltage, extending the length of the LDMOS drift region is necessary. However, this extension leads to larger on-resistance in the drift region. At the same time, current flows laterally across the LDMOS surface from the drain to the source, resulting in inefficient use of the substrate materials and occupying a large chip area. This impedes the miniaturization of MOSFETs.

1.3.3 *Vertical Double-Diffused MOSFET*

To address the issues with LDMOS, a strategic modification involves relocating the drain to the backside of the chip, opposite to the source and gate. This is the vertical double-diffused metal-oxide semiconductor (VDMOS), as shown in Fig. 1.12. In VDMOS, the current flows vertically through the MOSFET, maximizing the use of the drift area. This configuration allows for the largest cross-sectional area and the widest channel width, leading to a significant reduction in the on-resistance of the drift area. The blocking voltage of VDMOS is mainly determined by the thickness of the drift region. Increasing the thickness of the drift region can elevate the blocking voltage without affecting the chip area. Moreover, by sitting the source and drain on opposite sides of the chip, VDMOS easily resolves the insulation challenges inherent in high-voltage devices.

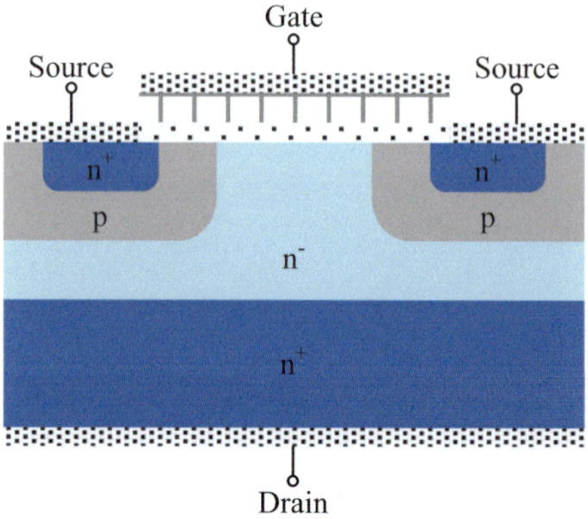

Fig. 1.12 Structure of N-channel VDMOS

According to the direction in which electrons flow through VDMOS, its on-resistance $R_{DS(on)}$ includes: source resistance R_{Source}, channel resistance R_{ch}, and the resistance R_{JFET},, which is caused by narrow current path between the cells (the JFET effect) when electrons leave the VDMOS cell, the drift area resistance R_{drift} and the drain resistance R_{Drain}, as shown in Fig. 1.13.

For VDMOS at different voltage ratings, Table 1.1 shows the ratio of on-resistance of each component to the total $R_{DS(on)}$. In low-voltage VDMOS, R_{ch} and R_{JFET} account for a significant proportion, and in high-voltage VDMOS, R_{drift} comprises the majority of the total $R_{DS(on)}$. To reduce $R_{DS(on)}$,, trench technology and super-junction technology have been developed for low-voltage and high-voltage devices, respectively, which are introduced in the following sections.

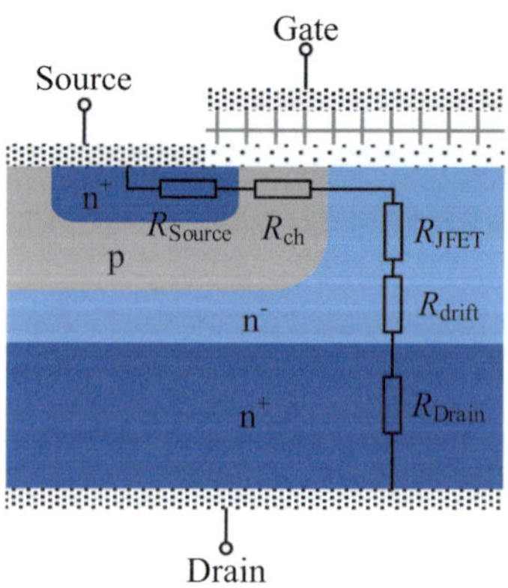

Fig. 1.13 On-resistance composition of N-channel VDMOS

Table 1.1 Proportion of on-resistance of each part in VDMOS [1]

	30 V (%)	600 V (%)
R_{Source}	6	0.5
R_{ch}	30	1.5
R_{JFET}	25	0.5
R_{drift}	31	97
R_{Drain}	8	0.5

1.3.4 Trench-Gate MOSFET

Increasing the cell density of VDMOS can decrease R_{ch}, but it increases R_{JFET}, creating a contradiction between the two. The implementation of trench-gate technology changes the gate structure from traditionally being attached to the surface of the chip to a trench separate the cells. This changes the channel's orientation from horizontal to vertical, resulting in the creation of a trench-gate MOSFET, also known as a Trench MOSFET. As shown in Fig. 1.14, common trench gates are in V-shaped and U-shaped forms.

The JFET effect in VDMOS is mainly due to the P–N–P structure within the cell, creating two reverse-biased dissipation layers. At the same time, due to the low concentration of the N^- region, these dissipation layers are wide, limiting the cross-sectional area of the current path between cells. The trench-gate structure effectively avoids the formation of the P–N–P structure. By directly linking the outlet of the vertical channel to the open drift area, it completely eliminates the JFET effect. Moreover, in contrast to the horizontal direction, the vertical channel occupies a smaller chip area, enabling higher cell density. Consequently, under the same device size, more cells connected in parallel reduce the R_{drift}.

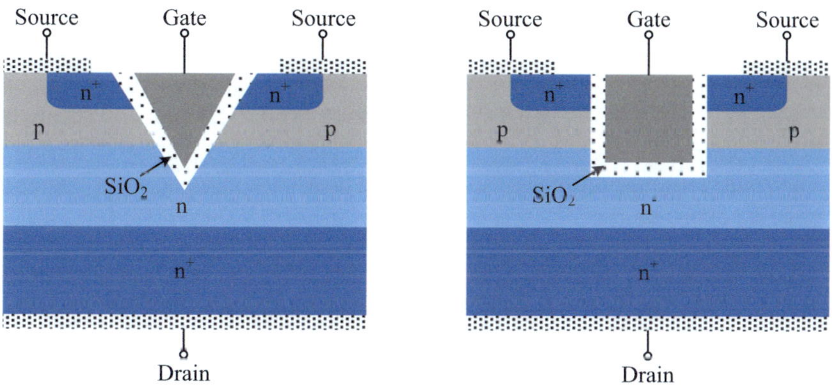

(a) V-shaped trench gate (b) U-shaped trench gate

Fig. 1.14 Structure of N-channel trench-gate MOSFET

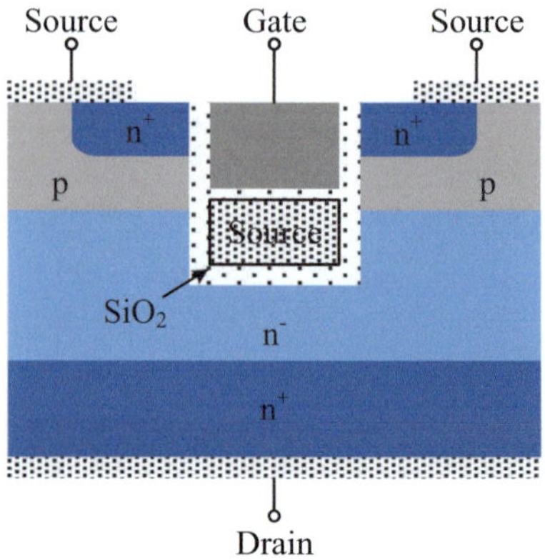

Fig. 1.15 Structure of N-channel shielded trench-gate MOSFET

1.3.5 Shielded Gate Trench MOSFET

When a trench gate is used, the contact area between the gate and the drift region of a single cell increases, and the cell density is higher. This, in turn, leads to an increase in the gate-drain capacitance (C_{GD}), causing the gate charge (Q_G) larger. Consequently, the driving loss increases, and the switching speed slows down. To reduce C_{GD}, a layer of shield gate is added to the trench gate. The lower part of the shield gate is connected to the source, forming a shield gate MOSFET, known as Shielded Gate Trench (SGT) MOSFET, as shown in Fig. 1.15. This modification transforms the original C_{GD} into gate-source capacitance (C_{GS}) and drain-source capacitance (C_{DS}), resulting in a significant reduction in C_{GD}.

1.3.6 Superjunction MOSFET

The $R_{DS(on)}$ of high-voltage VDMOS is mainly dictated by the R_{drift}. At the same time, the thickness of the drift region determines the blocking voltage, and its unit area on-resistance $R_{DS(on),sp}$ is proportional to the 2.5th power of the blocking voltage. As the blocking voltage increases, $R_{DS(on),sp}$ increases rapidly, making VDMOS unable to simultaneously handle high blocking voltage and large current, consequently limiting its potential applications.

Using superjunction technology, the P region extends vertically downward, based on VDMOS, to form a superjunction MOSFET (SJ-MOS), as depicted in Fig. 1.16.

In the turn-off state, the P^- column and N^- column establish a lateral P–N junction, inducing lateral depletion. Maintaining charge balance in the P–N column region enables lateral broadening of the space charge region, leading to complete depletion of the N^- region and the formation of an approximate rectangular electric field, thereby enhancing voltage withstand capability. In the turn-on state, carriers flow from the source through the channel into the N^- region of the superjunction and then

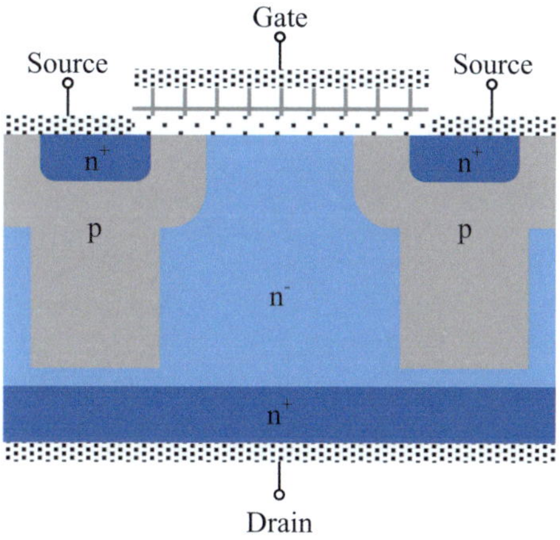

Fig. 1.16 Structure of N-channel superjunction MOSFET

flow into the N^+ substrate to reach the drain. It can be seen that, without compromising the withstanding voltage capability, increasing the doping concentration in the N^- column region notably reduces drift region resistance. Consequently, this reduction significantly decreases the on-resistance, causing $R_{DS(on),sp}$ to be proportional to the 1.4th power of the blocking voltage.

1.4 Si IGBT

1.4.1 Structure and Operating Principle of IGBT

The MOSFET is a unipolar device with a relatively limited current flowing capability. Even when employing superjunction technology to enhance this capability, increasing the chip area still becomes necessary, leading to large junction capacitance, degraded device characteristics, and restricted current flowing capability. For instance, the lowest $R_{DS(on)}$ for a commercial 650 V SJ-MOSFET is approximately 17 mΩ, and its current level is about 130 A. Similarly, the lowest $R_{DS(on)}$ for an 800 V SJ-MOSFET is around 85 mΩ, supporting a current level of about 50 A. On the other hand, the IGBT, a bipolar device with conductivity modulation effects, significantly enhances the device's current flow capability.

The full expansion of IGBT is "Insulated Gate Bipolar Transistor". It is constructed by substituting the N^+ layer of the drain in VDMOS with a P^+ layer, as shown in Fig. 1.17a, including an emitter, gate, and collector. IGBT encompasses three P-N junctions: J_1, the pn junction between the collector P^+ region and the N^- region; J_2, the junction between the N^- region and the P-base region; and J_3, the junction between the P-base region and the emitter N^+ area. Therefore, IGBT has a parasitic P–N–P transistor T_1, an N–P–N transistor T_2, and also possesses a MOSFET structure. Its equivalent circuit is depicted in Fig. 1.17b.

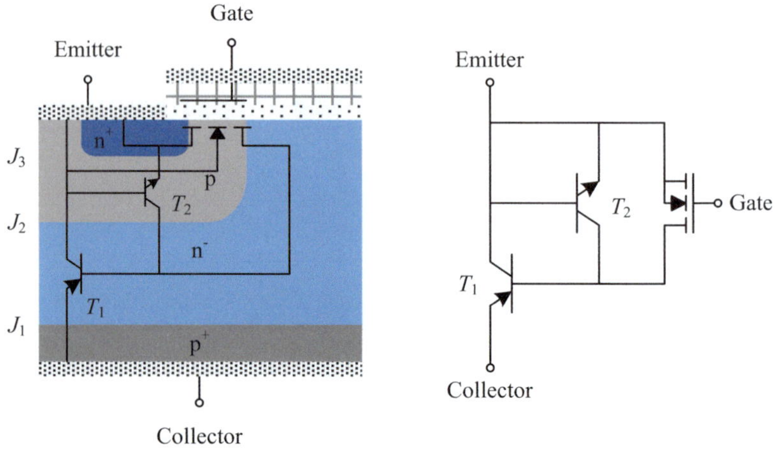

(a) Structure of IGBT (b) Equivalent circuit

Fig. 1.17 Structure and equivalent circuit of IGBT

The working principle of IGBT is shown in Fig. 1.18. When $V_{CE} > 0V$, J_1 is forward biased, and J_2 is reverse biased. An electron channel forms, allowing electrons to flow from the emitter N^+ region into the N^- region. This flow reduces the potential of the N^- region, further enhancing the forward bias of J_1. The P-base region injects numerous holes into the N^- region. Some holes recombine with the electrons moving into the emitter N^+ region, forming a continuous channel for electron current. The remaining holes are driven by the electric field of the reverse-biased J_2 towards the incident electrode N^+ district. When $V_{GE} \leq 0V$, as no electrons flow into the N^- region, the base current required for T_1 cannot be provided. Consequently, the IGBT is in the turn-off state and the reverse-biased J_2 bears the voltage. Therefore, the IGBT can be considered as a transistor controlled by a MOSFET.

IGBT accumulates a large number of non-equilibrium carriers in the N^- region during the turn-on state. It enters into the turn-off state when these carriers completely recombine and dissipate, generating a tail current in the turn-off process. Therefore, its turn-off time is notably longer compared to that of a MOSFET, thus limiting its typical switching frequency less than 100 kHz. Moreover, unlike VDMOS, the emitter junction of the parasitic P–N–P transistor in an IGBT is in reverse bias when $V_{CE} < 0V$, preventing current flow. Therefore, an IGBT cannot conduct in reverse and needs an additional antiparallel diode during operation.

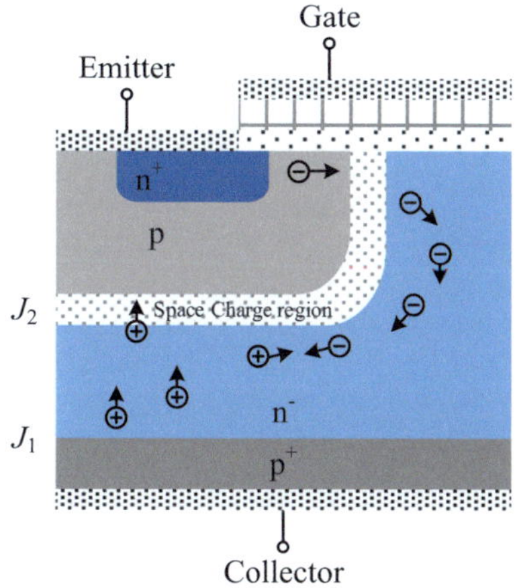

Fig. 1.18 Operating principle of IGBT

1.4.2 *PT-IGBT*

The earliest structure to withstand vertical voltage in IGBTs is the punch-through IGBT (PT-IGBT). When $V_{CE} > 0V$, the J_2 junction becomes reverse biased, causing the depletion layer to expand into the N^- region, extending towards the N-buffer region. This process forms an approximately trapezoidal distribution of electric field intensity. Due to the thicker P^+ region and higher doping concentration, PT-IGBT exhibits high J_1 injection efficiency, thereby reducing the on-state voltage. However, upon turning-on, a significant number of non-equilibrium carriers exist in the N^- region, resulting in an extended turn-off tail current time. A direct approach involves introducing carrier lifetime control technology, which enhances material recombination centers. This control is achieved through high-energy electron irradiation and heavy metal doping in key areas of the device. Despite improving turn-off characteristics, these enhancements often introduce issues such as device parameter drift, characteristic degradation, and reduced stability.

In addition to the contradiction between on-state voltage drop and switching characteristics, PT-IGBT suffers from inherent drawbacks. The on-state voltage drop isn't uniformly distributed along the vertical axis. Furthermore, the carrier lifetime increases with temperature, leading to a negative temperature coefficient in the device, which limits its parallel use. Additionally, this IGBT uses an epitaxial process to create the upper N^- region and MOS structure on a thicker P^+ substrate, adding to its production costs (Fig. 1.19).

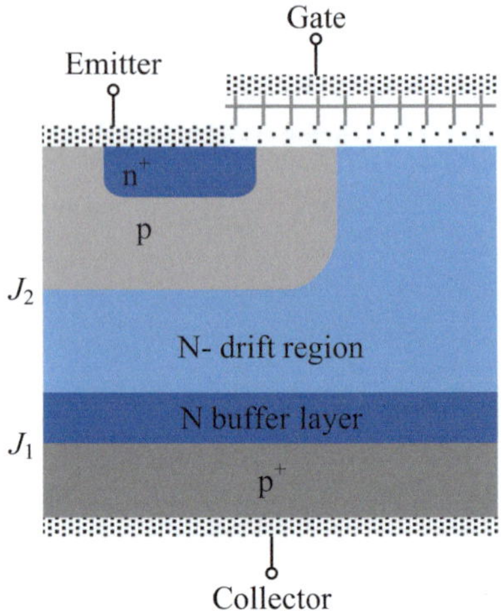

Fig. 1.19 Structure of PT-IGBT

1.4.3 NPT-IGBT

The advancement of transparent collector region technology has enabled the direct processing of non-punch-through IGBT (NPT-IGBT) on a lightly doped N-type region fused silicon single-crystal wafer. On the collector side, ions are implanted to create a P^+ region. When J_2 is reverse biased, the depletion layer extends to the N region. Due to the thickness of the N^- region, even when the peak electric field of J_2 reaches the critical breakdown electric field, it does not penetrate the N^- region. This results in a triangular electric field distribution, ensuring a more uniform vertical distribution of the on-state voltage drop. Using ion implantation to generate the P^+ region helps reduce hole injection efficiency at the J_1, consequently reducing the tailing time. Furthermore, the carrier lifetime of NPT-IGBT remains less affected by temperature changes. In addition, there's a decrease in carrier mobility and a significant increase in contact resistance. The device exhibits a positive temperature coefficient, short circuits, and dynamic avalanche tolerance, promoting its parallel use. However, an obvious drawback of NPT-IGBT is its thick drift region, leading to large turn-off losses (Fig. 1.20).

1.4.4 FS-IGBT

With advancements in thinning technology, the drift region can be reduced in thickness, thereby reducing the on-state voltage drop. However, this improvement raises the issue of minimizing the on-state voltage drop while maintaining the blocking voltage. The Field Stop IGBT (FS-IGBT) offers a potential resolution to this inherent contradiction.

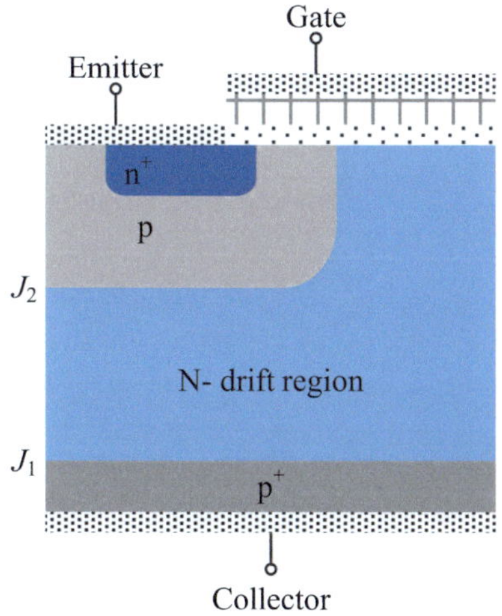

Fig. 1.20 Structure of NPT-IGBT

In Fig. 1.21, considering the advancements in thinning technology and high-energy ion implantation, the FS-IGBT employs a field stopping layer, several microns thick, at the backside of the wafer through ion implantation. Its doping concentration is lower than that of the PT-IGBT's buffer layer. This layer only compresses the electric field without impeding hole injection, allowing for the formation of a trapezoidal electric field distribution like the PT-IGBT. The collector's P^+ region also uses transparent collector technology, resulting in relatively lower electron–hole injection efficiency and a shorter device turn-off tail time. The FS-IGBT demonstrates temperature characteristics and stability comparable to those of the NPT-IGBT. Moreover, the FS-IGBT's N^- region is thinner than that of NPT-IGBT, leading to a smaller on-state voltage drop and requiring fewer carriers for turn-off, consequently reducing turn-off losses. Building upon FS-IGBT, variations like weak punch-through (LPT), soft punch-through (SPT), and other types of IGBTs have emerged.

1.4.5 Trench-Gate IGBT

In addition to enhancing the vertical withstanding voltage structure of the IGBT, the planar gate has been replaced with a trench gate in the lateral aspect, as illustrated in Fig. 1.22. Similar to a trench-gate MOSFET, the trench gate in the IGBT increases channel density, resulting in higher surface utilization efficiency compared to a planar gate. Simultaneously, it avoids the JFET effect, reduces on-state voltage drop, and enhances the device's current-carrying capability. However, the trench-gate structure has some drawbacks. It leads to an increase in the gate-collector capacitance, affecting frequency characteristics and limiting high-frequency applications. Additionally, it reduces the IGBT's avalanche capability and short-circuit tolerance.

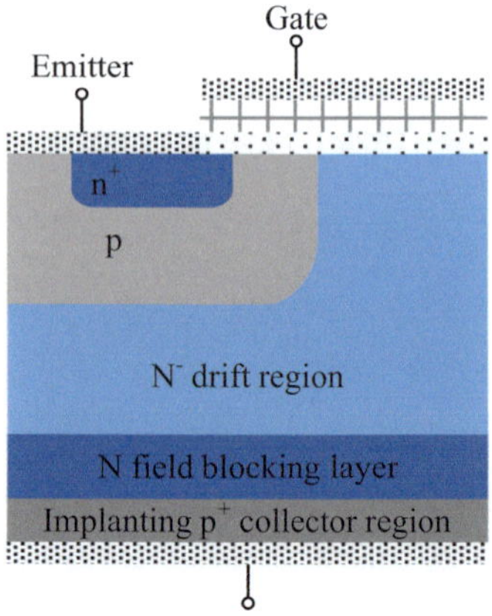

Fig. 1.21 Structure of FS-IGBT

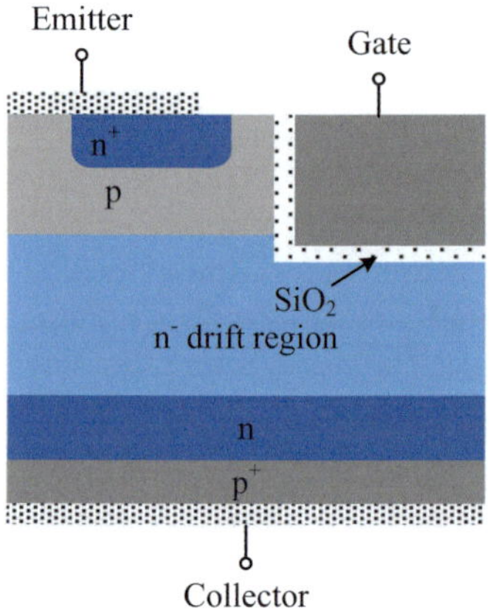

Fig. 1.22 Structure of trench-gate IGBT

1.5 Physical Characteristics of SiC

1.5.1 Crystal Structure

In an atom, electrons move in specific, discrete orbitals under the influence of the potential energy from the atomic nucleus and interactions with other electrons. These orbitals have discrete energy levels, known as energy levels. The ground-state electron configurations of Si and C atoms are Eq. (1.1) and Eq. (1.2), respectively, where the number represents the principal quantum number or layer sequence number, and the letter represents the sublayer or sublevel. The superscript number denotes the number of electrons in that sublevel. It can be observed that both Si and C atoms have four valence electrons in their outermost layer, specifically $3s^2 3p^2$ and $2s^2 2p^2$.

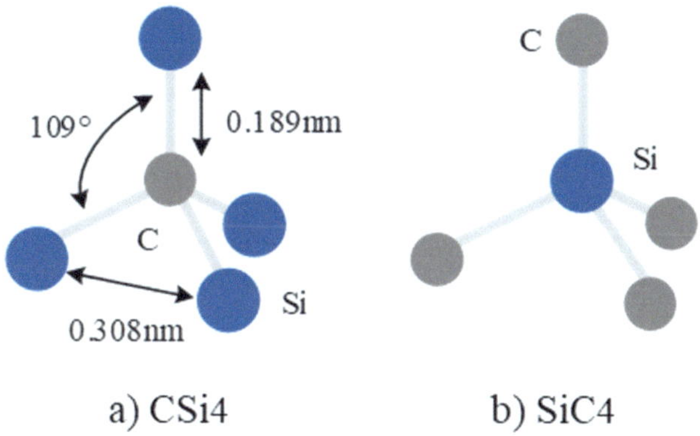

Fig. 1.23 Basic unit of SiC crystal

$$\mathrm{Si{:}1s^2 2s^2 2p^6 3s^2 3p^2} \tag{1.1}$$

$$\mathrm{C{:}1s^2 2s^2 2p^2} \tag{1.2}$$

SiC is a compound semiconductor material composed of Si and C atoms in a 1:1 ratio, forming covalent bonds through sp^3 hybridized electron pairs. The distinctive feature of sp^3 is tetrahedral bonding, resulting in each Si atom being surrounded by 4 C atoms, and each C atom being surrounded by 4 Si atoms. The basic unit of the SiC crystal, as shown in Fig. 1.23, consists of these arrangements.

Although the basic units of SiC crystals are uniform, when they are interconnected and stacked, diverse crystal structures can be generated. Due to the moderate ionicity of SiC bonds and the formation of low-energy stacking faults, SiC exhibits a rich variety of stacking sequences, resulting in over 200 polytypes. In the Ramsdell symbol system, letters denote crystal systems ("C" for cubic, "H" for hexagonal, "R" for rhombohedral), and numbers indicate the number of Si–C double atomic layers in the unit cell along the (001) direction. Common SiC polytypes include 3C–SiC, 4H–SiC, and 6H–SiC, with their stacking ball models and the 1120-face connection schematic diagram illustrated in Figs. 1.24 and 1.25, respectively. Among various crystal structures, 4H–SiC has become the preferred choice for SiC power devices due to its advantageous characteristics.

1.5.2 Energy Band and Bandgap Width

As mentioned above, when atoms are in an isolated state, electrons are located at discrete energy levels. Crystals are composed of a large number of atoms that are closely arranged, and the distance between atoms is very close, so that their electron orbits overlap and affect each other. At the same time, the Pauli exclusion principle

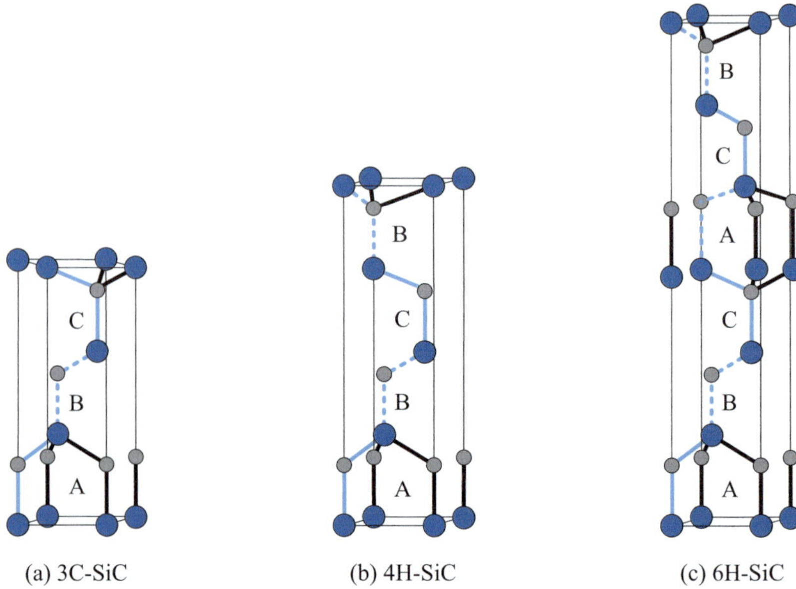

Fig. 1.24 Stacking ball model structure of 24 SiC crystals

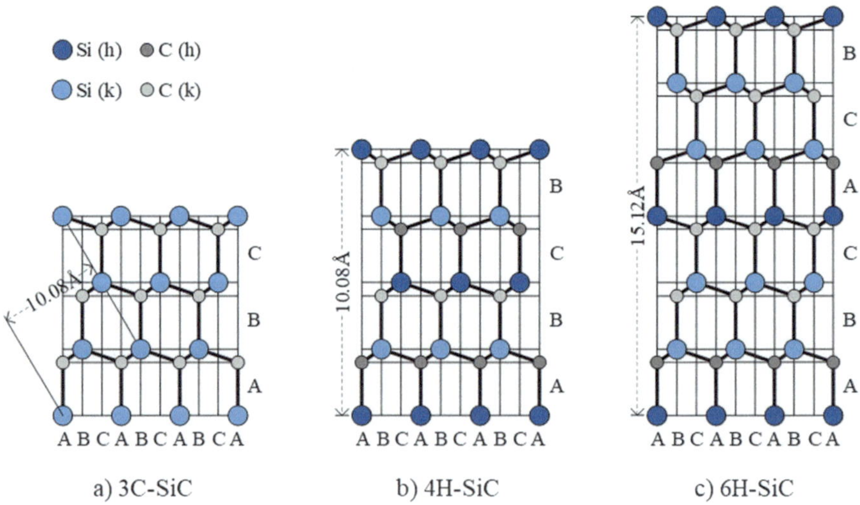

Fig. 1.25 Schematic diagram of (1120) plane connection of 25 SiC crystal

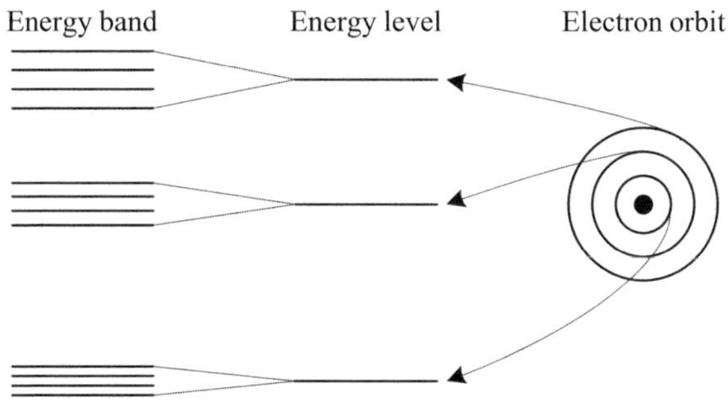

Fig. 1.26 Energy levels split into bands

states that there cannot be two or more particles in the same atomic orbital that are in exactly the same quantum state (principal quantum number, angular quantum number, magnetic quantum number, and spin quantum number). This causes the energy levels with the same energy in each atom to be differentiated into a series of energy levels in a certain energy range to ensure that each electron occupies an independent quantum state. Due to the large number of atoms in the crystal, the energy gap between the split energy levels is very small, and the dense energy levels form energy bands, as shown in Fig. 1.26.

Using silicon (Si) as an example, as the distance between atoms decreases, the splitting of the 3s and 3p energy levels into energy bands is illustrated in Fig. 1.27. For Si atoms that are relatively far apart, the 3s state is fully occupied by 2 electrons, and the 3p state, which can accommodate 6 electrons, contains only 2 electrons. As the distance between atoms decreases, the originally independent energy levels begin to overlap and automatically split into two energy bands at the lattice constant of Si, with 4 quantum states in the lower energy band and the other 4 quantum states in the higher energy band.

As shown in Fig. 1.28, at absolute zero, all 4 electrons occupy the lower energy band. Since the energy band is fully occupied by electrons, it cannot conduct electricity; this energy band is referred to as the valence band. When excited, electrons in the valence band transition to a higher energy band that has all quantum states empty. Due to the limited number of electrons, under the influence of an external electric field, electrons can drift and create an electric current, and this higher energy band is referred to as the conduction band. The electrons in the conduction band and the

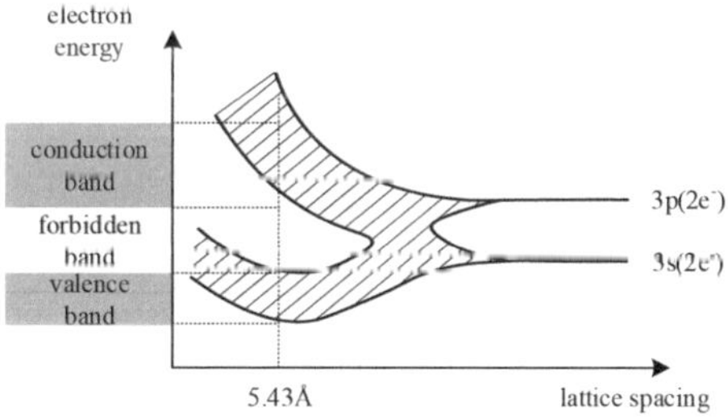

Fig. 1.27 3s and 3p states of Si atoms are split into energy bands

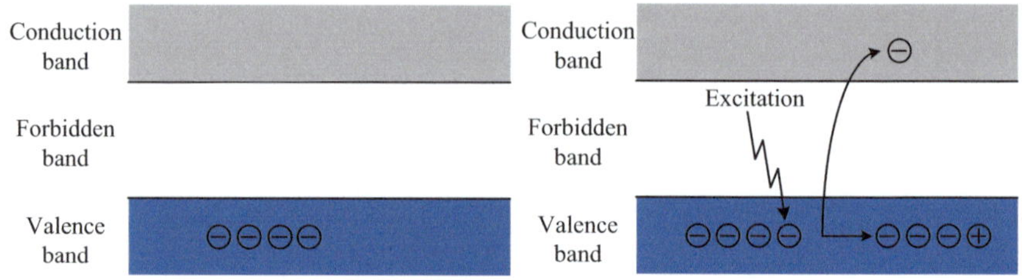

Fig. 1.28 Valence band, conduction band, forbidden band, and excitation

holes created in the valence band due to the absence of electrons are known as charge carriers. The energy range between the valence band and the conduction band, where no electron states exist, is called the band gap. A larger band gap means that more energy is required to excite electrons from the valence band to the conduction band, resulting in better insulating properties of the material.

It's worth noting that in SiC crystals, the process of sp^3 hybridization occurs similarly to Si, where the 3s and 3p states of Si atoms hybridize with the 2s and 2p states of C atoms. However, the band structure in SiC is relatively more complex.

Due to the strong polarity of the Si–C covalent bonds and the tight binding of valence electrons, more energy is required to excite electrons from the valence band to the conduction band in SiC compared to Si. Consequently, the band gap width of SiC is larger than that of Si, with SiC having a band gap width of 3.26 eV, approximately three times that of Si. Hence, SiC is referred to as a wide bandgap semiconductor material.

1.5.3 Breakdown Electric Field Strength

Under the influence of a strong external electric field, electrons in the valence band of semiconductor materials can be excited to the conduction band. When the external electric field reaches a certain strength, the semiconductor material loses its ability to insulate against electricity, resulting in breakdown. The intensity of the external electric field at which this breakdown occurs is referred to as the breakdown field strength.

As mentioned earlier, in high-voltage vertical MOSFETs, the drift region primarily bears the voltage. The thicker the drift region, the higher the breakdown voltage it can withstand. However, this also leads to higher resistance in the conduction path, particularly the drift region resistance R_{drift}, which constitutes the largest proportion of the total on-state resistance $R_{\mathrm{DS(on)}}$, and the specific on-resistance per unit area $R_{\mathrm{DS(on),sp}}$. Additionally, while higher doping concentrations lead to lower resistivity, they also result in steeper electric field gradients, necessitating thicker drift regions. Therefore, simply increasing doping concentration cannot effectively reduce the $R_{\mathrm{DS(on),sp}}$ of Si devices. These factors collectively contribute to the inability of Si MOSFETs

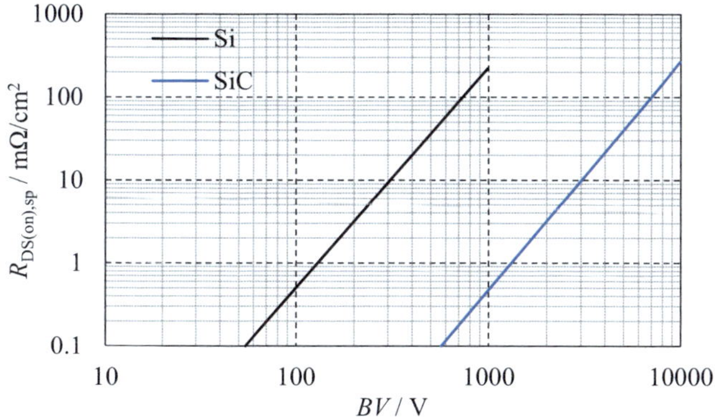

Fig. 1.29 Theoretical values of Si and SiC withstand voltage and $R_{DS(on),sp}$

to achieve high voltage and large current simultaneously within a reasonable chip area.

SiC is a wide bandgap semiconductor material, with a breakdown field strength of 2.5 MV/cm at 25 degrees Celsius, which is 10 times higher than that of Si. Therefore, under the same breakdown voltage conditions, the drift region of a SiC MOSFET can be doped at a higher concentration to reduce resistivity. Simultaneously, it can significantly reduce the thickness of the drift region, leading to a significant reduction in the device's on-state resistance per unit area $R_{DS(on),sp}$. Figure 1.29 illustrates the theoretical values of breakdown voltage and $R_{DS(on),sp}$ for Si and SiC.

Therefore, at the same on-state resistance $R_{DS(on)}$, the chip area of a SiC MOSFET is smaller. For example, at a 900 V voltage rating, a SiC MOSFET can achieve the same $R_{DS(on)}$ with only 1/30 of the chip area compared to a Si MOSFET. A smaller chip area corresponds to smaller junction capacitance, resulting in faster switching speed for SiC MOSFETs and the ability to operate at higher switching frequencies. Similarly, with the same chip area, the on-state resistance $R_{DS(on)}$ of a SiC MOSFET will be lower. Therefore, SiC MOSFETs can simultaneously achieve high voltage and high current. Currently, commercially available SiC MOSFETs at the 1200 V voltage rating have achieved an $R_{DS(on)}$ level of 5 mΩ per chip. Similarly, SiC diodes offer similar advantages compared to Si diodes.

1.5.4 Impurity Doping and Intrinsic Carrier Concentration

A semiconductor without doping foreign atoms is called an intrinsic semiconductor. When the temperature $T > 0$ K, intrinsic excitation occurs, where electrons are excited from the valence band to the conduction band, resulting in intrinsic carriers. To improve the characteristics of semiconductor materials, intrinsic semiconductors are typically doped. In silicon, phosphorus (P) and boron (B) are commonly used

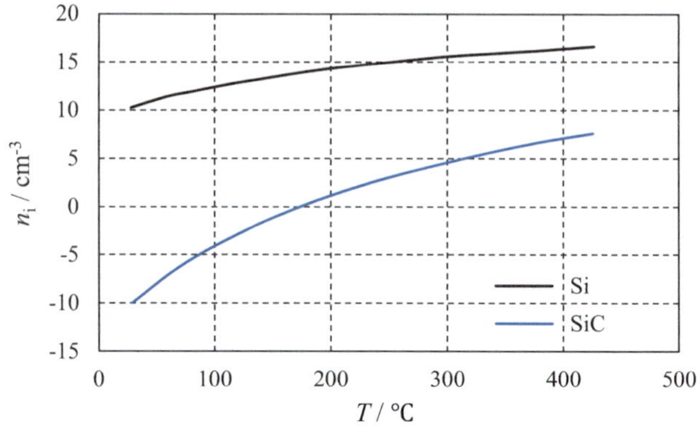

Fig. 1.30 Relationship between intrinsic carrier concentration n_i of Si and SiC and temperature

for doping, while in silicon carbide, aluminum (Al) and nitrogen (N) are commonly employed.

Intrinsic carriers are generated through intrinsic excitation and are related to the bandgap width. The concentration of intrinsic carriers, denoted as n_i, decreases exponentially with the increase in the bandgap width. Due to the larger bandgap width of SiC, its n_i is much smaller than that of Si. At 25 °C, the n_i of SiC is $5 \times 10^{-9}/cm^3$, which is only $1/10^{19}$ of Si's.

Thermal excitation is the primary mechanism for intrinsic excitation, and n_i is temperature-dependent, increasing exponentially with temperature. When the temperature is sufficiently high, n_i becomes too large, turning the semiconductor material into a conductor. This temperature is known as the intrinsic temperature. Figure 1.30 illustrates the relationship between the intrinsic carrier concentration and temperature for Si and SiC. It can be observed that even at high temperatures, the intrinsic carrier concentration of SiC is much smaller than that of Si. This allows SiC devices to operate at higher temperatures, whereas the maximum operating temperature for Si devices does not exceed 200 °C, SiC devices can operate normally at 600 °C.

1.5.5 Carrier Mobility and Saturation Drift Velocity

Carriers undergo drift motion under the influence of an external electric field, forming current. The drift velocity is proportional to the electric field, and the proportionality coefficient is called the mobility. The mobility of Si is 1350 $cm^2/V{\cdot}s$, while the mobility of SiC is 720 $cm^2/V{\cdot}s$. How to address the issue of lower mobility is a key research direction for SiC devices.

Additionally, due to lattice scattering, there is a limit to the drift velocity, known as the saturation drift velocity. The carrier saturation drift velocity of SiC is 2×10^7 cm/s, which is twice that of Si. A higher carrier saturation drift velocity is beneficial for improving the frequency response and switching speed of devices. However, this advantage is more beneficial for radio frequency (RF) devices and offers limited performance improvement for power devices.

1.5.6 Thermal Conductivity

The thermal conductivity of SiC material is 390 W/(m·K), which is 2.6 times that of Si and comparable to the best thermal conductors like silver (419 W/(m·K)) and copper (381 W/(m·K)). With such high thermal conductivity, many people believe that heat dissipation in SiC devices would be easier to manage. However, for devices of the same specifications, the chip area of SiC devices is significantly smaller than that of Si devices. This results in a much larger thermal resistance from the SiC chip to the device casing compared to Si, which actually increases the difficulty of heat dissipation.

1.6 Overview of SiC Industry Chain

1.6.1 Substrate

Based on the type and concentration of doping, SiC substrates can be classified as n-type or semi-insulating. n-type SiC substrates are doped with impurities containing additional free electrons, making them conductive. These wafers serve as the foundational material for producing SiC power devices.

The manufacturing of SiC substrates involves multiple processes, including SiC crystal growth and processing, which includes cutting, grinding, polishing, cleaning, and inspection procedures. Crystal growth requires very high temperatures, typically exceeding 2200 °C. Additionally, its growth rate is much slower than that of Si, with only fractions of a millimeter grown per hour. Furthermore, SiC exists in various stacking configurations or polytypes, with over 200 theoretically feasible arrangements known. These polytypes can easily coexist, transform, or undergo transitions between each other during the growth process, making the SiC crystal growth process very complex.

Currently, the primary method for growing SiC substrates is the physical vapor transport (PVT) method, which has been widely adopted by SiC substrate suppliers. Other growth methods such as high-temperature chemical vapor deposition (HTCVD) and liquid phase epitaxy (LPE) are still under development.

In terms of crystal material synthesis, the traditional Acheson process used for large-scale SiC crystal synthesis cannot produce high-purity SiC grains, thus failing to meet the needs of single-crystal growth. The existing SiC grain synthesis technology generally uses high-temperature (modified from the combustion synthesis) methods, synthesizing in small batches and multiple cycles in high-temperature synthesis furnaces. Due to the small furnace chambers and induction heating methods typically used, the batch synthesis volume is low, usually around 30–60 kg per furnace. Furthermore, due to reaction conditions, the crystal size is generally less than 4 mm. According to current crystal growth research, using large SiC crystal grains for the growth of large-sized SiC single crystals (8 inches and above) can better control the growth rate and maintain the stability of the vaporization rate during the crystal growth process, thereby controlling growth defects, improving yield, and reducing costs. Large-sized SiC crystal grains (0–12 mm) are often obtained through the CVD process abroad, but this process is very expensive and difficult to meet the high demand of the current end market and clients. By conducting fundamental research on the traditional Acheson process, understanding the SiC crystal growth mechanism, and identifying the key factors affecting SiC purity and particle size, new SiC synthesis equipment and processes have been developed to synthesize ultra-large SiC grains that meet crystal growth requirements. Using silicon powder and carbon powder as raw materials and adding appropriate growth aids, the materials are mixed and loaded into a closed furnace. By optimizing the furnace heating method to create a high-temperature and uniform thermal field, high-purity and large-sized SiC can be synthesized through a series of reactions involving Si, C, and vapor-phase reactions. Due to the presence of Si vapor, Si_2C, and SiC_2 at high temperatures, controlling the furnace atmosphere and pressure promotes the growth of crystals to ultra-large grain sizes of 4–8 mm or more. After synthesis, the grown crystal cylindrical particles are classified and further processed. This process enables the large-scale, low-cost synthesis of large-sized SiC grains, thereby meeting the requirements for 8-inch and larger substrate crystal growth processes.

Apart from the slow crystal growth issue, due to the hardness of SiC being close to that of diamond, its cutting and polishing become challenging, resulting in difficulties in subsequent processing and becoming another important factor restricting the development of the silicon carbide device market. Traditional diamond wire cutting cannot achieve thin wafer slicing. The wafer thickness needs to be greater than 350 μm, and the kerf damage exceeds 200 μm, resulting in significant material loss, slow cutting efficiency, and limitations on ingot size. In contrast, laser cutting offers advantages such as ultra-thin wafer slicing (thickness as low as 100 μm), minimal kerf damage, low material loss, high material utilization, fast cutting efficiency, and the ability to cut large ingot sizes. Laser cutting methods such as Disco's KABRA technology and SILTECTRA's cold cutting technology, acquired by Infineon in 2018, are gaining traction. Laser cutting technology will become an important trend in silicon carbide cutting in the future, which can significantly increase the yield rate and further reduce costs.

The production of SiC substrates faces two major challenges: cost and quality. Although the cost of SiC wafers has gradually decreased over the past decade, it still remains significantly higher than that of Si substrates (At the end of 2024, the price of a 150 mm SiC substrate is approximately $280 to $550, while a 300 mm Si substrate is around $100). SiC substrates account for a considerable portion of the cost of SiC devices. Furthermore, compared to Si substrates, the SiC crystal growth or wafer processing exhibits higher densities of defects, directly impacting device performance, reliability, and yield. Therefore, efforts and innovations are continuously made in all production stages to improve the production quality of SiC substrates. In addition to researching single-crystal SiC substrates, some companies are exploring composite SiC substrates achieved by combining single-crystal SiC substrates with carrier wafers. This method aims to reduce wafer costs or improve wafer quality. Companies researching in this direction include Soitec, Sicoxs, and Isabers.

As of the end of 2023, the world's major SiC substrate suppliers are Wolfspeed, Coherent, and SiCrystal. According to Yole's data, in 2022, these three companies accounted for over 70% of the SiC power device substrate market. The rapid growth of the SiC power device market has led many new companies to enter the SiC substrate business, either through internal development or external acquisition. For example, in 2020, SK Siltron acquired DuPont's SiC wafer division. GlobalWafers, one of the world's top five Si substrate suppliers, has also started manufacturing SiC substrates. It is worth noting that Chinese companies are becoming increasingly active in the SiC substrate market, with over 50 companies attempting to produce SiC wafers in 2023. Among them, Shandong Tianyue Advanced Technology Co., Ltd. and Beijing Tiankeheeda Semiconductor Co., Ltd. have successfully entered the global SiC power device supply chain. In 2024, Chinese silicon carbide substrate manufacturers' rankings in the global market continued to rise, not only strengthening their competitiveness but also accelerating the downward trend of global silicon carbide substrate prices.

It is worth noting that, recognizing the critical role of substrates in SiC power devices, major SiC power device manufacturers have attempted vertical integration in this area. For example, ROHM acquired SiCrystal in 2009, STMicroelectronics acquired Norstel AB in 2019, and ON Semiconductor acquired GT Advanced Technologies in 2021. Until the end of 2023, the leading SiC substrate suppliers worldwide are Wolfspeed, Coherent, and SiCrystal. According to data from Yole, these three companies held a combined market share of over 70% in the SiC power device substrate market in 2022. The rapid expansion of the SiC power device market has attracted numerous new entrants into the SiC substrate industry, either through internal development or external acquisitions. For instance, in 2020, SK Siltron acquired DuPont's SiC wafer division. Additionally, GlobalWafers, one of the top five Si substrate suppliers globally, has ventured into SiC substrate manufacturing. Notably, Chinese firms are increasingly active in the SiC substrate sector, with over 50 companies attempting to produce SiC wafers in 2023. Among them, Shandong Tianyue and Tianke Heida have successfully integrated into the global SiC power device supply chain.

The development of 8-inch substrates is a key future trend in the SiC industry. Due to the high difficulty and low yield in SiC substrate preparation, the cost of SiC devices is significantly higher than that of Si products. As substrate sizes increase, the total number of integrated chips per unit also increases. According to Wolfspeed, an 8-inch SiC substrate can yield 89% more 32 mm^2 chips compared to a 6-inch substrate, with edge waste reduced from 14 to 7%. Additionally, according to GTAT, the introduction of 8-inch substrates is expected to reduce SiC device costs by 20–35% compared to the 6-inch platform. Currently, many companies are in the stage of mass production of 6-inch SiC substrates and gradually transitioning to 8-inch production. The latest research directions of n-type SiC crystal growth by domestic and foreign enterprises and institutions mainly include: expanding the crystal diameter, increasing the crystal thickness, controlling crystal defects, improving yield, and optimizing resistivity, etc. In terms of expanding the crystal diameter, by the end of 2024, a total of 31 enterprises worldwide have achieved breakthroughs in 8-inch diameter expansion research, and Chinese enterprises account for 20 of them. 8-inch SiC substrates will be the mainstream product in the coming years. Looking ahead, as 8-inch SiC production processes are optimized and improved, along with updates to production equipment and continuous capacity expansion, the price of SiC substrates is expected to gradually decrease. Finally, it is worth noting that 12-inch silicon carbide wafers were first released at the end of 2024. Although there is still some time before large-scale production is achieved, this breakthrough has opened up new possibilities for the further development of the silicon carbide industry, driving the industry towards more efficient and lower-cost directions. In September 2024, two AR glasses companies released augmented reality glasses using SiC lenses, bringing SiC into the "metaverse" domain and the consumer electronics market; one 6-inch substrate can provide 2 pairs of glasses, one 8-inch substrate can produce 3 to 4 pairs of glasses, and one 12-inch substrate can approximately produce 8 to 9 pairs of glasses. AR glasses have driven the development of SiC towards larger sizes.

1.6.2 Epitaxy

In recent years, there have been improvements in both the size and quality of SiC substrates. However, their quality is still not sufficient to directly fabricate devices. This necessitates the deposition of single-crystal SiC thin films on the substrates to generate epitaxial layers of the required thickness suitable for power devices. As mentioned in the previous section, substrate defects significantly impact device performance and reliability. High-quality SiC epitaxy can help limit the extension of substrate defects into the epitaxial layer or allow benign defect extension from the substrate, known as "performance-degrading" defects.

SiC epitaxy requires precise control by optimizing growth parameters such as temperature, pressure, and gas flow, which can be adjusted accurately and reproducibly through computer control. However, certain parameters are not easily adjustable, such as parasitic growth on reactor walls, cleaning of reactor components, and spare parts replacement. To minimize the impact of these factors, overcoming the learning curve is necessary.

Both chemical vapor deposition (CVD) and liquid phase epitaxy (LPE) can be used for SiC epitaxy fabrication. The CVD method is widely adopted in industrial applications because it can effectively control the flow rate, temperature, and pressure of gas sources during growth, allowing for changes in the layer formation environment and precise control of epitaxial growth parameters with good reproducibility. However, it has a significant issue with polytype mixing. Therefore, the step-flow growth model proposed by Matsunami et al. in 1987 has played a crucial role in the development and quality of epitaxy. Its introduction primarily enabled growth at relatively low temperatures while achieving highly stable control of the 4H polytype—which is more suitable for power devices—thereby solving the polytype mixing issue and triggering the unique development of 4H-SiC in power device applications. Additionally, the introduction of TCS (trichlorosilane) significantly increased the growth rate, achieving more than ten times the traditional growth rate, enhancing not only production speed but also quality control, especially in controlling silicon droplet formation. This is particularly advantageous for thick film epitaxy growth.

There are many defects in SiC epitaxy, and they differ from other crystals due to the unique properties of SiC. The main defects include micropipes, triangular defects, surface carrot defects, and specific issues like step bunching. Many of these defects are directly copied from the substrate, so the quality and processing level of the substrate are crucial for epitaxial growth, especially in controlling defects.

SiC epitaxial defects are generally categorized as fatal and non-fatal. Fatal defects, such as triangular defects and droppings, affect all types of devices, including diodes, MOSFETs, and bipolar devices. They significantly impact the breakdown voltage, which can be reduced by 20% to as much as 90%. Non-fatal defects, such as Threading Screw Dislocations (TSDs) and Threading Edge Dislocations (TEDs), might not affect diodes but can impact the lifespan or cause leakage in MOSFETs and bipolar devices, ultimately affecting the yield of device manufacturing.

In medium- and low-voltage applications, SiC epitaxy technology is relatively mature and can generally meet the requirements of devices like Schottky barrier diodes (SBDs), merged PiN Schottky diodes (MPS), and MOSFETs. However, in high-voltage applications, the development of epitaxy technology is relatively lagging. High-voltage devices require thick films, which currently still have many defects, particularly triangular defects, which mainly affect the fabrication of high-current devices. High-current devices need large chip areas. Moreover, the minority carrier lifetime is relatively low. For high-voltage applications, the device types tend to be bipolar devices, which require high minority carrier lifetimes. We can see that to achieve an ideal forward current, the minority carrier lifetime must be at least 5 μs. Currently, the minority carrier lifetime of epitaxial wafers is about 1–2 μs, which cannot yet meet the demands of high-voltage devices and still requires post-processing technologies.

The key parameters of SiC epitaxial wafers primarily depend on the device design. For instance, epitaxial parameters vary according to the voltage rating of the device. Generally, for low voltage around 600 V, the required epitaxial thickness might be around 6 μm; for medium voltage ranging from 1200 to 1700 V, the thickness needs to be 10–15 μm; and for high voltage of 10 kV and above, the thickness might need to be over 100 μm. Therefore, as the voltage capability increases, the epitaxial thickness increases accordingly. The preparation of high-quality epitaxial wafers becomes very challenging, especially in high-voltage applications, where defect control is crucial and poses a significant challenge.

After decades of development, SiC epitaxy technology has achieved precise control, providing a reliable foundation for manufacturing power devices. Consequently, the landscape of epitaxy suppliers is continuously changing. As of 2023, major companies providing commercial epitaxy include Resonac, Wolfspeed, Coherent, Hantian Tiancheng Electronic Technology (Xiamen) Co., Ltd, and Guangdong Tianyu Semiconductor Co., Ltd.. Meanwhile, an increasing number of SiC substrate companies are incorporating SiC epitaxy as value-added products in their offerings. Considering the significant impact of epitaxial layer quality on device performance and the substantial portion of epitaxy costs in device costs, many SiC device companies also choose to internalize some of the SiC epitaxy processes.

Since 2024, SiC is experiencing fierce competition. The future survival strategy for SiC epitaxial manufacturers is to move towards low cost and deep customer integration.

Low cost has always been the goal for SiC epitaxy. Compared to 6-inch SiC, 8-inch SiC brings significant advantages in production efficiency, cost-effectiveness, and technological applications in wafer fabs. However, due to the large capital investment required for SiC epitaxial equipment and the incompatibility between 6-inch and 8-inch equipment, companies like Hantian Tiansheng and Tianyu have faced many challenges in their transformation. In contrast, some emerging companies can more easily start afresh. Another cost reduction strategy involves replacing traditional induction furnaces with resistance-heated crystal growth furnaces, which can significantly increase ingot thickness, making a solid step towards the low-cost mass production of large-sized SiC substrates and unlocking great potential for continuous substrate cost reduction.

Silicon carbide epitaxial equipment is divided into single-chip type and multiwafer. Multiwafer SiC epitaxial equipment is essentially monopolized by Aixtron, but its equipment is expensive and has long lead times. Promoting China multiwafer epitaxial equipment is an urgent need for cost reduction and efficiency improvement. Additionally, by independently developing key materials such as graphite and quartz for epitaxial furnaces, and possessing parts processing capabilities, critical material support for 8-inch SiC R&D can be provided. This ensures controllable material consistency, improves yield, and further reduces production costs.

Furthermore, the most critical aspect is device validation. Firstly, the validation cycle is long; after epitaxial growth, it takes about 2 months to make MOSFET devices, followed by wafer splitting, packaging and final reliability testing, resulting in an overall validation cycle of approximately 6 months. Therefore, epitaxial factories need to work closely with FAB factories, establishing in-depth technical cooperation at each key stage to provide rapid downstream validation channels for products, ensuring timely feedback and fast iteration of products.

1.6.3 Chip Manufacturing

Due to the different physical properties of Si and SiC, many traditional Si device fabrication techniques cannot be directly applied to the manufacturing of SiC devices.

Incoming inspection is essential when SiC epitaxy reaches the production line. Due to the semitransparent nature of SiC epitaxy, conventional equipment designed for Si devices may inadvertently capture signals from inside the wafer, leading to errors. Therefore, new equipment specifically tailored for SiC epitaxy is required.

In Si device fabrication, doping can be achieved using thermal diffusion and ion implantation according to specific requirements. However, for SiC devices, the low diffusion coefficients of common dopants render thermal diffusion below 1800 °C negligible. Therefore, ion implantation becomes the preferred method. However, performing ion implantation at room temperature results in high-density defects, necessitating the use of high-temperature ion implantation, typically between 400 and 1000 °C.

Furthermore, the post-implantation activation process used to activate the dopants becomes crucial. Achieving dopant activation in SiC requires extremely high temperatures, usually around 1400–1500 °C for N-type and over 1600 °C for P-type. Such temperatures are close to or exceed the growth temperature of the epitaxial layer, which may lead to severe surface degradation, thereby compromising interface features such as Schottky contacts and the channel of field-effect transistors. To mitigate adverse effects, a protective capping layer is commonly employed. However, this approach is costly and may have potential drawbacks, such as reducing the mobility of the MOSFET channel.

Establishing ohmic contacts between the source and drain regions of a MOSFET and between the anode and cathode regions of a diode, involving metal and semiconductor junctions, is crucial for optimizing the efficiency and reliability of SiC devices. However, forming ohmic contacts on SiC faces unique challenges due to its wide bandgap characteristics, which elevate the complexity of the processing technology. Research into achieving ohmic contacts on SiC devices primarily focuses on two main techniques. One approach involves high-dose ion implantation to introduce dopants into the semiconductor material, facilitating the formation of ohmic contacts. The drawback of this method lies in the lattice defects or amorphization tendencies during ion implantation, which may be stable and require high temperature annealing for repair. Another method involves creating an intermediate semiconductor layer with a narrower bandgap or higher carrier density at the interface between the ohmic contact and SiC. This can be achieved through metal deposition and annealing.

The quality and stability of the SiC/SiO_2 interface are crucial for determining the performance and long-term reliability of MOSFET gate oxide structures. Compared to the Si/SiO_2 interface, the SiC/SiO_2 interface requires additional processing to enhance its quality and stability.

The differences in chip manufacturing processes between SiC devices and Si devices are summarized in Table 1.2.

Table 1.2 Differences in chip manufacturing processes between SiC devices and Si devices

Process	Si	SiC	Problems in the process
Epitaxy	Opaque	Opaque	Semioptical process focusing, special defect detection Robots need to adapt
Doping	Diffusion or implantation	Implantation only	Doping area is flat Activation after implantation is required
Annealing	800–1200 °C	> 1700 °C	Potential surface degradation, conventional quartz tubes cannot be used, high-temperature furnace is required
Ohmic contact	Formation temperature 400 °C	Formation temperature 1000 °C	High dose implantation of special composition metal annealing
MOS interface	Si/SiO_2 interface	SiC/SiO_2 interface	Alternative deposition and oxidation post-annealing, channel anti-doping or high-temperature oxidation

Academic institutions and companies have conducted extensive research on various topics related to SiC device chip manufacturing, contributing significantly to the advancement and eventual mass production of SiC devices. The commercialization of the first batch of SiC diodes began in 2001, pioneered by Infineon followed by Cree (now Wolfspeed). In the subsequent years, Infineon and Cree remained major commercial producers of SiC diodes. Discussions regarding SiC-based BJT, JFET, and MOSFET took place before 2010. Significant acceleration in the SiC device industry occurred after ROHM introduced SiC MOSFET in 2010, along with the entry of Cree and numerous other device manufacturers into the market.

Until the end of 2023, there is a wide range of SiC power devices available on the market, covering various voltage levels from 600 to 3300 V, including both discrete devices and power modules. The main focus of development in the SiC power device industry currently lies in improving the durability of SiC MOSFETs, advancing trench-gate technology, and refining packaging techniques to fully leverage the high-temperature performance of SiC.

Due to its outstanding performance and vast market prospects, SiC devices have become a core focus in the power semiconductor industry. Nearly all major suppliers are offering SiC devices, including over 20 global suppliers and over 30 companies in China. The top five ranking companies are STMicroelectronics, Infineon Technologies, ON Semiconductor, Wolfspeed (Cree), and ROHM Semiconductor, all of which are integrated device manufacturers (IDMs), a common business model in the power semiconductor industry. STMicroelectronics has secured a prominent position in the market, particularly by establishing itself in Tesla's supply chain. Since 2024, competition among different participants is fierce, and the industry landscape is continually evolving. Meanwhile, the accelerated entry of Chinese silicon carbide manufacturers is having a profound impact on the global silicon carbide market landscape and increasing the uncertainty in the industry's development.

In addition to IDMs, there are also foundries like X-fab, Xinlian Integrated Circuit Manufacturing Co., Ltd., and Shanghai Jitai Semiconductor Co., Ltd., which provide SiC manufacturing services for design companies that do not have wafer fabs.

1.6.4 Packaging and Testing

Currently, there are various SiC device products available on the market, including discrete devices and power modules. However, most of these are based on traditional packaging technologies inherited from Si devices, influenced by factors such as compatibility, cost, and capacity. To fully leverage the advantages of SiC, optimizing packaging technology is necessary. Specific technical challenges are outlined in Table 1.3.

Currently, industry experts are exploring optimization of packaging designs and layouts, adoption of new materials, and employing various strategies, including but not limited to: high-temperature epoxy resin or silicone encapsulation, silver sintering technology for chip attachment, wire bonding (both wedge and ribbon bonding), direct soldering of the package to heatsinks, module substrates with better thermal conductivity (such as AlN AMB, Si_3N_4 AMB, and IMS with TPG inserts, rather than traditional Al_2O_3 DBC), and integration of heat sinks, among others.

Many of the processes used to validate long-term stability of Si devices can be transferred to SiC devices. However, due to their material characteristics and operating modes, SiC devices require additional and different reliability testing. Key factors affecting reliability include the defect structure of the material, the impact of the larger bandgap on interface traps, higher electric fields generated during operation, and new operating modes combining high voltage and fast switching. The industry has initiated a subcommittee within JEDEC (JC-70.2) dedicated to SiC, aimed at developing identification guidelines and standards for SiC devices.

Table 1.3 Challenges of SiC device packaging technology

SiC versus Si	Challenges faced by traditional packaging	Packaging requirements
High switching speed	High di/dt: higher voltage overshoot and ringing on the device High dV/dt: higher common mode noise Device parallel connection becomes complicated	Low parasitic inductance parameter control
High operating temperature	Packaging materials that can reliably withstand high-temperature operation, traditional packaging materials can withstand up to 175 °C, SiC devices can operate above 200 °C	New packaging materials
High-power density	Due to thermal cycles, the failure rate of bonding wires is higher, and fatigue failures of chip-to-substrate solder joints increase	Eliminate bonding wires Delay or eliminate chip contact failures

In terms of the supply chain, in the automotive industry, some new packaging solutions may directly originate from automotive manufacturers (OEMs) or Tier-1 suppliers. A typical example is Tesla, which has filed for intellectual property rights for its TPAK module design. Additionally, Toyota and Denso possess a significant portfolio of power module technology patents, with over 150 patent families specifically applicable to SiC as of March 2022. Meanwhile, some device companies also supply SiC wafers to packaging firms specialized in power module development. These module manufacturers include Semikron Danfoss, Vincotech, and Guangdong Xinyujun Semiconductor Co., Ltd., among others.

1.6.5 Application

SiC devices offer higher switching frequencies, lower losses, and higher operating temperatures compared to Si devices. These advantages contribute to improved efficiency, increased power density, and reduced system costs across various applications, including electric vehicles, charging infrastructure, solar inverters, battery energy storage systems, industrial and residential motor drives, among others.

By the end of 2023, the market size of SiC devices is estimated to range between $3 billion and $3.5 billion, with specific figures depending on different market and financial analysts. Forecasts indicate a double-digit compound annual growth rate for this market over the next five years, with several top suppliers predicting the market size to exceed $10 billion.

The automotive sector, particularly in electric vehicles, drives the application of SiC devices, especially in main drive inverters, onboard chargers, and DC–DC converters. Market analysts estimate that the automotive industry accounts for 50–70% of the SiC device market. Toyota has been at the forefront of SiC device development for electric vehicles since the early 2010s, attracting significant attention to SiC applications in the automotive sector. However, challenges such as reliability and cost limit the widespread adoption of SiC devices.

A turning point occurred in 2017 when Tesla began mass-producing its popular car, the Model 3. Subsequently, many automotive manufacturers and tier-one suppliers, including BYD, Ford, Volkswagen, Mercedes-Benz, Land Rover, Volvo, General Motors, NIO, and BMW, either adopted SiC devices for electric vehicles or announced related plans. Tesla claimed at its Investor Day in March 2023 to have developed a method capable of reducing the amount of silicon carbide used by 75%, sparking industry discussions. Potential solutions may include using fewer SiC devices, reducing the size of SiC chips, or modifying the topology to achieve hybrid Si–SiC solutions. Despite this proposal, market analysts and device manufacturers believe the impact on the forecast for the SiC power device market will be minimal.

Another noteworthy application is in the field of photovoltaics, where SiC devices demonstrate advantages in various types of solar inverters. SiC diodes are widely used by companies such as ABB, Delta Electronics, SMA, Sungrow Power, Huawei, Fronius, KACO, and others in residential, commercial, and utility-scale solar inverters. The adoption rate of SiC MOSFETs is also increasing due to advantages such as lower on-state resistance, highlighting the growing importance of SiC technology in the renewable energy sector.

Other application areas of SiC devices will be discussed in detail in Chap. 12.

A complete supply chain has formed around the SiC device industry, including SiC substrate suppliers, epitaxy suppliers, chip manufacturers, packaging and testing companies, as well as electronics and automotive companies. Additionally, many equipment suppliers provide specialized machinery for substrate production, chip manufacturing, and testing. Today, a fully developed ecosystem supporting the SiC power device industry has emerged.

1.7 Overview of the Development of SiC Diodes

SiC SBDs replaced Si FRDs in medium-voltage applications due to their superior reverse recovery characteristics. Subsequently, to simultaneously achieve lower forward voltage drop, smaller reverse leakage current, and faster switching speed, SBDs were combined with pin diodes to develop SiC JBS diodes. Then, to address extreme conditions such as surges in high-power applications, the design of the p^+ region in JBS diodes was improved, resulting in the development of MPS diodes, where the p^+ region can be activated under certain conditions to enhance bipolar conduction capability.

1. **JBS Diode**

The first generation of commercial SiC diodes was of the SBD structure, but it had high leakage current and low reverse breakdown voltage, which severely limited its application in the high-voltage field. To address this issue, various manufacturers introduced SiC JBS diodes, whose structure involved integrating multiple pn junctions in the drift region of the SBD diode. By forming barriers through closely spaced p + regions around the Schottky contact and its vicinity, the high electric field on the semiconductor side at the Schottky contact is shielded. The structure is illustrated in Fig. 1.31.

When a reverse voltage is applied, the depletion region formed by the P-N junction extends into the channel region and overlaps with it. As the reverse voltage increases, the depletion layers expand downward into the channel, This way, the Schottky barrier lowering effect can be suppressed, and the reverse leakage current can be significantly reduced, enabling it to be applicable in high-voltage application fields. When the forward voltage of the SiC JBS diode is relatively low, the SBD part within the JBS structure is responsible for conducting the current. When the forward

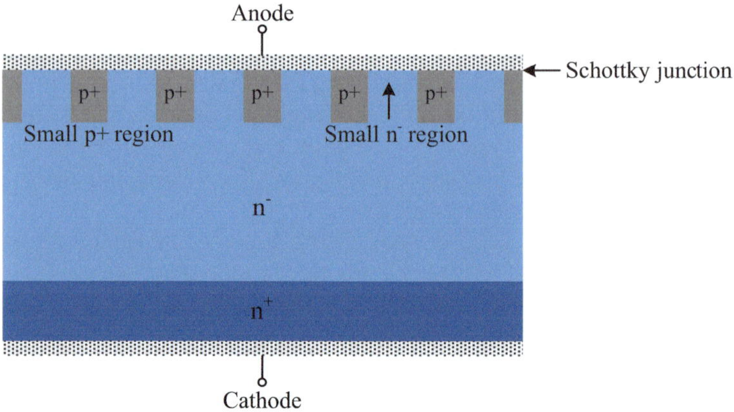

Fig. 1.31 Structure of SiC JBS diode

voltage is higher, the pin diode part within the JBS structure begins to conduct, and the minority carriers in the active region are injected into the drift region, resulting in an electric conductance modulation effect that further reduces its forward conduction voltage drop. It should be noted that although the electric conductance modulation effect enhances the ability of the SiC JBS diode to conduct large currents, in practical applications, under normal conditions, only the SBD part conducts, and the pin part will only be activated when a surge current occurs. Under normal operating conditions, when the SiC JBS diode changes from the conducting state to the off state, since it is also single-polarity carrier conducting like the SBD, it inherits the excellent reverse recovery characteristics of the SBD.

2. **MPS Diode**

During the operation of the diode, it often encounters large current impacts. The heat generated by the large current causes the junction temperature of the diode to rise rapidly within a short period of time. When the temperature becomes too high, it will lead to the failure of the device. Therefore, it is necessary to pay attention to the diode's ability to withstand large current impacts. Generally, this is characterized by surge current.

To enhance the surge resistance capability, the MPS diode structure was developed based on the JBS diode, as shown in Fig. 1.32. The structure of the MPS diode and the JBS diode has no essential difference. The difference lies in the fact that the MPS diode contains both a small-sized p+ region and a large-sized p+ region simultaneously. Among them, the large-sized p+ is used to enhance the diode's ability to conduct large currents. Under high current conditions, the pn junction corresponding to the large-sized p+ will be activated and inject minority carriers into the drift region of the device. The resulting conductance modulation effect will significantly reduce the resistance of the device.

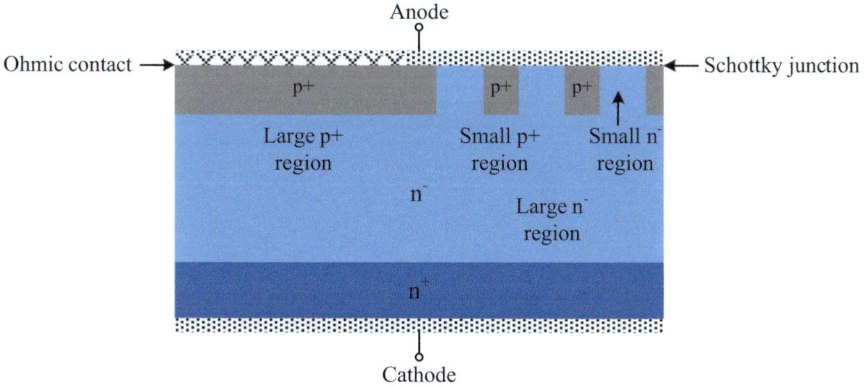

Fig. 1.32 Structure of SiC MPS diode

Compared to SBDs, MPS diodes exhibit lower forward voltage drop. Compared to pin diodes, they store fewer charges in the epitaxial layer, resulting in faster switching speeds. Additionally, the reverse recovery of SiC diodes demonstrates almost constant behavior, with minimal nonlinearity compared to Si FRDs. This eliminates the need to consider variations in temperature and load conditions when designing circuits. Nowadays, MPS diodes have become the mainstream structure solution for SiC diodes.

1.7.1 The structure of commercial SiC MOSFETs

1. **SiC MOSFET ROHM's dual-trench SiC MOSFET**

In 2010, ROHM successfully mass-produced planar SiC MOSFETs. By 2015, ROHM had developed and mass-produced the world's first double-trench SiC MOSFET, as shown in Fig. 1.33 [3]. To address the issue of electric field concentration at the bottom of the gate trench in single-trench structures, which reduces reliability, the double-trench structure introduced trenches in both the source and gate regions. This design mitigates the electric field concentration at the bottom of the gate in single trench structures, thereby improving the reliability of the gate oxide layer in trench MOSFETs. This enhancement in reliability also laid the foundation for large-scale production.

2. **Infineon's half-shielded trench SiC MOSFET**

Infineon's half-shielded trench structure, as shown in Fig. 1.34 [4], features a unique asymmetric trench design where each trench has a channel on one side and is covered by deep p + implantation on the other side. On the left side of the trench wall, the MOS channel is aligned with the plane, optimizing channel mobility. On the right

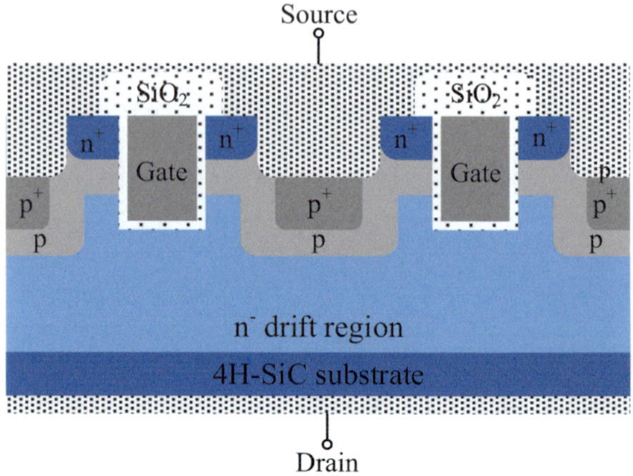

Fig. 1.33 ROHM's dual-trench SiC MOSFET

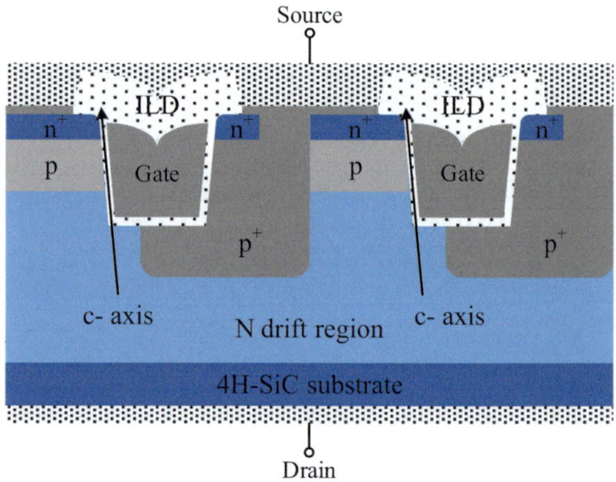

Fig. 1.34 Infineon's half-shielded trench structure SiC MOSFET

side of the trench wall, a significant portion of the trench bottom is embedded into the p + region, which extends below the trench bottom, reducing the critical electric field in the off-state and serving as a body diode. This structure protects the trench corners from peak electric fields, enhancing device reliability, and improving breakdown voltage and switching characteristics. This half-shielded design increases device reliability, allowing for higher system efficiency, switching frequency, and power density.

3. **DENSO's U-shaped trench structure SiC MOSFET**

In March 2023, DENSO announced the development of the first inverter utilizing SiC semiconductors. Their unique trench-type SiC MOSFET structure employs proprietary electric field mitigation technology, enhancing the output capability of each chip and achieving high voltage and low on-resistance, as shown in Fig. 1.35.

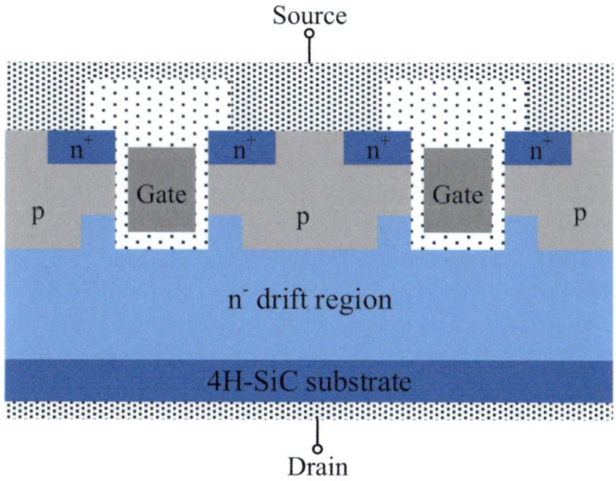

Fig. 1.35 DENSO's U-shaped trench structure SiC MOSFET

4. **Sumitomo's V-Shaped Trench SiC MOSFET**

Sumitomo has developed a V-shaped trench SiC MOSFET using a unique crystal face, as shown in Fig. 1.36. This device offers superior characteristics such as high efficiency, high blocking voltage, and high stability under harsh environments. It supports high current (up to 200 A per single chip), making it suitable for electric vehicles (EVs) and hybrid electric vehicles (HEVs).

5. **Mitsubishi Trench SiC MOSFET**

To address the issue of trench-gate insulation film breakdown under high voltage, Mitsubishi has developed a unique electric field limiting structure, as shown in

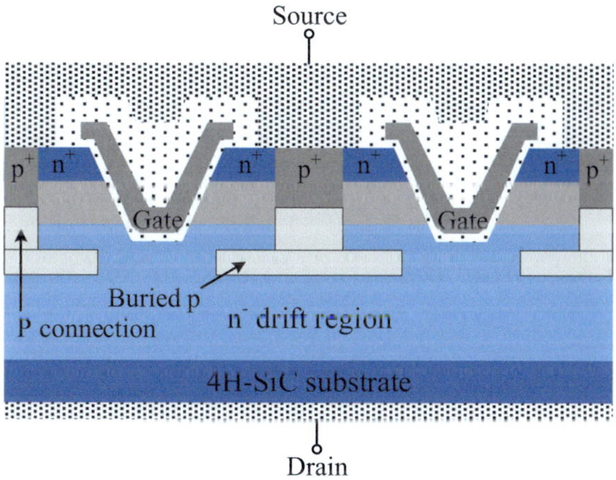

Fig. 1.36 Sumitomo's V-shaped trench SiC MOSFET

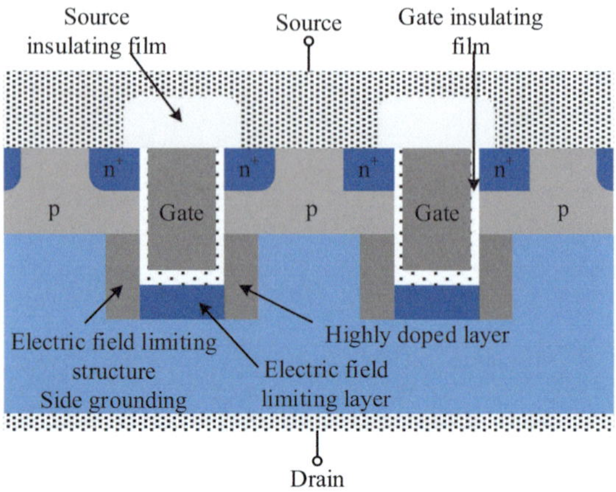

Fig. 1.37 Mitsubishi trench SiC MOSFET

Fig. 1.37 [5]. This design reduces the electric field applied to the gate insulation film to the level of conventional planar-type power semiconductor devices, thereby enhancing the reliability of the insulation film under high-voltage conditions. In the gate trench area, three ion implantation steps form the electric field limiting structure, side grounding electric field limiting layer, and high concentration doped conductive region. This structure reduces the electric field strength applied to the gate insulation film, ensuring device reliability, and facilitates high-speed switching actions through the side grounding layer. Additionally, the high concentration doped layer, created by obliquely implanting nitrogen, reduces the resistance of the current path.

1.8 Development Overview of SiC Power Modules

One generation of devices, one generation of packaging. Due to the superior characteristics of SiC devices, new expectations have been raised for power module packaging. The industry, in response to specific application requirements, has launched a series of power module processing and product solutions tailored for SiC devices.

1.8.1 Manufacturing Process of SiC Power Modules

Unlike discrete devices, SiC power modules integrate multiple SiC chips into the same package according to a specific topology, as depicted in Fig. 1.38. The manufacturing process of SiC power modules is diverse due to the wide range of product types available, such as full-bridge modules encapsulated with silicone gel and cured, or modules formed by injection molding with subsequent trimming and forming.

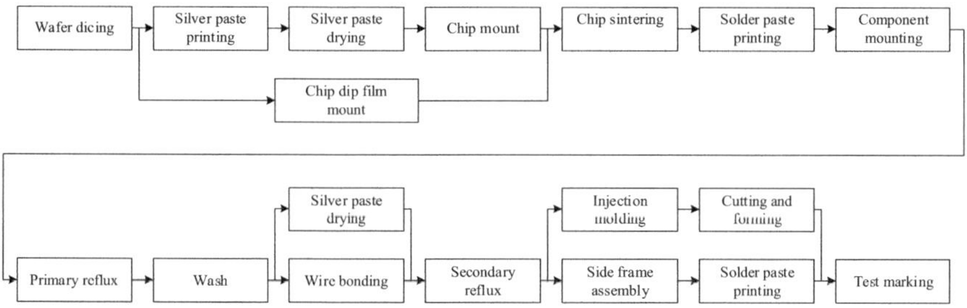

Fig. 1.38 Manufacturing process of SiC power modules

Moreover, there are various implementation methods for specific functional manufacturing steps. For instance, pre-curing SiC chips on ceramic substrates can be achieved through processes like chip dipping in silver paste followed by hot pressing pre-curing, or traditional methods involving printing silver paste and then drying for chip bonding. While the latter involves more steps, it offers higher production yield, making it more suitable for mass production processes. Therefore, the choice of manufacturing route depends not only on the product type but also on factors such as product status and maturity. The manufacturing process illustrated in Fig. 1.38 serves as a general reference.

Compared to traditional Si IGBT power modules, silicon carbide (SiC) power modules offer higher power density, operating junction temperature, and lower stray inductance and thermal resistance. These characteristics place very high technical demands on the mass production lines for SiC power modules. Overseas companies, such as STMicroelectronics and Infineon, which started earlier, have achieved significant economies of scale. Meanwhile, China companies are also catching up rapidly with notable achievements, among which Guangdong Xinjun Semiconductor Co., Ltd. stands out. In 2022, Xinjun Semiconductor's SiC power module manufacturing base officially began operation. This is the first automotive-grade SiC power module production line in China, featuring a fully automated production system and industry-leading manufacturing processes. Overall, the gap between China manufacturers and international companies in terms of manufacturing capability is rapidly narrowing.

Although the manufacturing routes for SiC power modules are diverse and flexible in combination, there are key processes that require special attention.

Firstly, regarding the chip sintering process, the current mainstream view is that the optimal combination for SiC MOSFET chips is silver sintering technology paired with Si_3N_4 ceramic substrates. This combination maximizes the advantages of SiC chips while balancing cost requirements. The silver sintering agent is mainly composed of the following components: silver particles, as the main material, forming the connection layer (nano, micron); binder, as an auxiliary material, to inhibit sintering cracks; diluent, to adjust viscosity and reduce printing difficulty; dispersant, to separate silver particles and prevent agglomeration; and modified fillers, including resin and base metals (optional).

Compared to traditional reflow soldering processes, silver sintering technology offers significant advantages. Firstly, it operates at higher service temperatures; the silver layer formed after sintering requires a temperature of 961 °C to break down again, whereas standard lead-free solder paste typically reflows at temperatures exceeding 220 °C. Secondly, silver sintering technology provides superior heat dissipation. As shown in Fig. 1.39, The theoretical thermal conductivity of the silver sintering layer can reach 200 W/(m·K), approximately four times that of conventional solder paste at around 50 W/(m·K). Utilizing silver sintering technology can significantly reduce the thermal resistance of the module. Under the same chip specifications and quantity, the module can deliver higher currents, thus improving efficiency. Lastly, in terms of process capability, solder paste processes are prone to defects such as voids and chip drifting, whereas silver sintering technology offers enhanced process capabilities (Fig. 1.39).

Next is the wire bonding process, which is a traditional core process in power semiconductor device manufacturing. In the field of power devices, wire bonding primarily refers to the wedge bonding process. Its main function is to achieve the connection between the bonding wire and the bonding area (usually the surface of the chip or the copper surface of the ceramic substrate) under the action of ultrasonic vibration. As shown in Fig. 1.40, The primary function of wire bonding is to realize the electrical connection inside the device (Fig. 1.40).

Fig. 1.39 Microstructure of the silver layer before and after silver sintering [6]

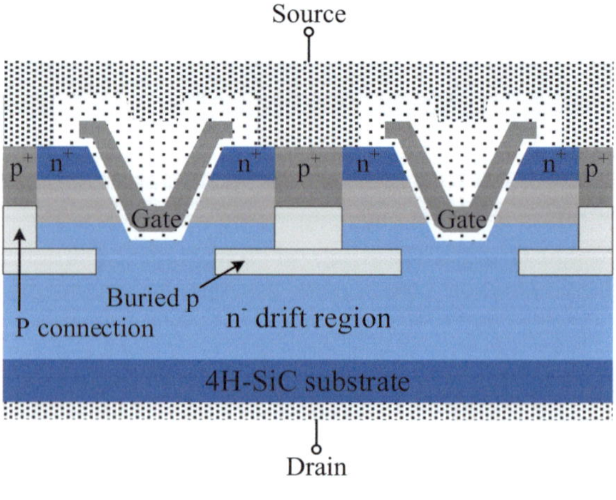

Fig. 1.40 SiC power module aluminum wire lead bond

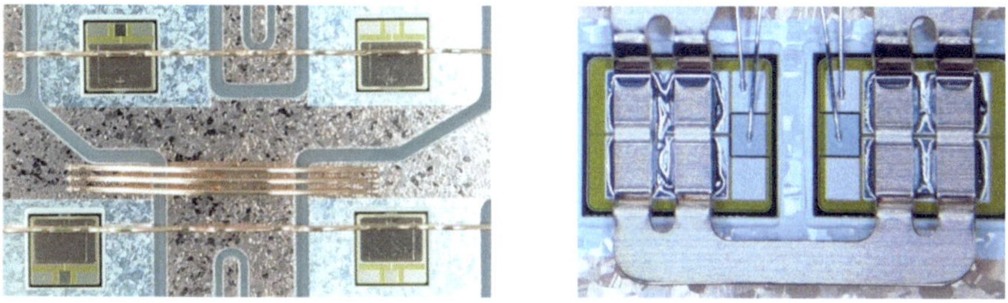

Fig. 1.41 Copper wire bonding and ribbon bonding technology

The quality of the wire bonding process is limited by both materials and process parameters, with the process parameters primarily including the energy, pressure, and time of the bonding. Generally, the wire bonding process exhibits good consistency, providing a solid foundation for determining the process window through Design of Experiments (DOE). Under normal circumstances, for SiC power modules with appropriate aluminum wire selection and reasonable process window settings, the module can withstand tens of thousands of power cycles lasting seconds each.

In recent years, in pursuit of higher reliability in power device electrical interconnection processes, there has been a gradual emergence of Copper Bonding and Clip Bonding technologies in the market. As shown in Fig. 1.41, Compared to aluminum wire bonding, these new technologies generally offer improvements in power cycling lifetimes by a factor of 3–7. However, due to yield limitations and associated costs, these technologies have not yet replaced aluminum wire bonding as the mainstream solution.

For example, the ACCO DRIVE PACK E series SiC power module launched by AccoPower Semiconductor Co., Ltd. in 2022 features a new clip structure for electrical connection of the power devices. This design allows for a three-dimensional spatial structure, enhancing the module's dynamic and static current sharing capabilities and reducing stray inductance. Additionally, thanks to the use of various advanced materials, this series of power modules can withstand 150,000–200,000 cycles in second-level power cycling tests, with a lifespan over three times longer than industry standard products. In future products, a more advanced dual-layer stacked clip structure and full sintering process technology will be adopted to ensure extremely low stray inductance (3 nH) and a very high-power cycling lifespan (over 300,000 cycles) (Fig. 1.41)

1.8.2 Technological Development of SiC Power Modules

Since STMicroelectronics introduced STPAK based on the Tesla Model 3 model in 2018, the technological development of SiC power modules has entered the fast

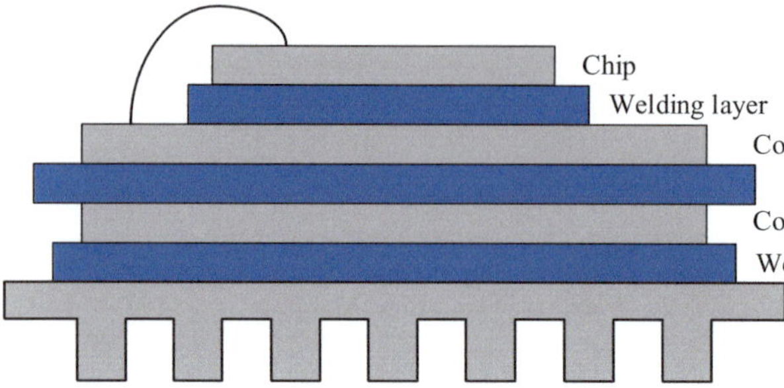

Fig. 1.42 Structure of SiC power module

lane, with various new product schemes and technologies emerging one after another. Fig. 1.42 shows the structure of the SiC power module.

1. **Heat dissipation performance**

Maintaining the same chip specifications, maximizing the output current capability of power modules and reducing their thermal resistance is the most direct and effective approach, which has been the primary focus in the development of SiC power module technology over the past decade. There are three specific dimensions to achieve this (Fig. 1.42).

The first aspect involves maximizing the heat dissipation efficiency of chips on the plane, meaning each chip occupies the maximum area possible. This reduces thermal coupling between chips, preventing localized overheating that could lead to high junction temperatures and waste of overall module performance. However, dispersing chips as much as possible to reduce thermal resistance is advantageous, but unfavorable for parallel current sharing. This is because a scattered chip layout makes it difficult to achieve consistency between power and control circuits among the chips. In this case, it is necessary to optimize the parasitic parameters of the power and control circuits of each chip in a targeted manner to ensure that both static and dynamic current sharing meet requirements.

The second aspect involves optimizing heat transfer in the vertical direction, primarily through structural and material optimization. In terms of structural optimization, adjusting the thickness of copper layers, ceramic layers, and solder layers can achieve lower thermal resistance. However, the implementation of structural optimization schemes is constrained by best practices from the manufacturing process and additional costs associated with non-standard customization. Regarding material optimization, the focus is on using materials with higher thermal conductivity. For example, replacing traditional solder layers with silver sinter layers with a thermal conductivity of up to 200 W/mK, and replacing the currently mainstream Si_3N_4 substrate with Si_3N_4 substrate with a thermal conductivity of up to 120 W/mK.

The third aspect involves optimizing the pin–fin structure of the heatsink. The pin–fin heatsink is a component directly in contact with the cooling liquid, and its structure directly affects the heat dissipation efficiency of the power module and the pressure

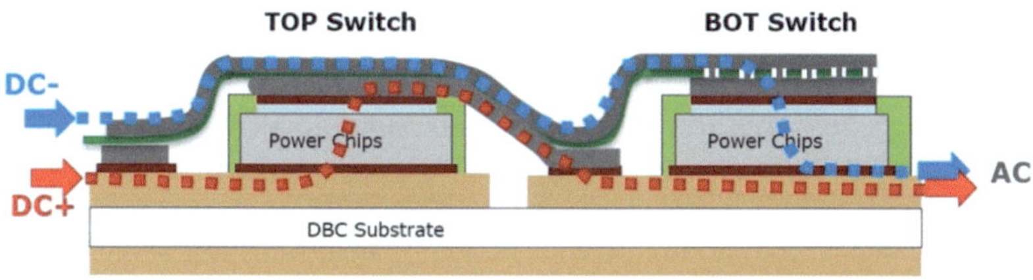

Fig. 1.43 SiC power module's special-shaped heat dissipation pin–fin

drop in the water channel. Generally, there is a trade-off between heat dissipation efficiency and water channel pressure drop. The denser the pin–fin heatsink, the better the heat dissipation, and the lower the thermal resistance to the cooling liquid. However, this also leads to a larger pressure drop in the water channel. However, locally, through special design of the pin–fin structure, a certain reduction in thermal resistance within a certain range can be achieved without a significant increase in water channel pressure drop. A typical example of such a solution is the use of non-cylindrical pin–fin shapes instead of cylindrical ones.

In this direction, China companies like AccoPower Semiconductor Co., Ltd. have developed a unique droplet-shaped pin–fin design. The streamlined droplet shape allows the coolant to flow smoothly, reducing flow resistance. Additionally, by cleverly setting the spacing and size ratios, the cross-sectional area of the coolant passing through the pin-fins remains relatively constant, which reduces pressure loss and flow resistance, significantly lowering the requirements for the cooling system [7] (Fig. 1.43).

2. **Loop Inductance**

With the drive circuit parameters unchanged, the size of the power module's circuit inductance directly determines the voltage spike during turn-off process. As the trend towards high-voltage electrification in new energy vehicles continues, the bus voltage is constantly increasing, from the initial 400 V to the currently popular 800 V. The SiC power modules commonly used now have a rated voltage of 1200 V, which undoubtedly imposes higher requirements on the circuit inductance of the power module.

Reducing the stray inductance of the module can mainly be achieved by reducing the area enclosed by the power circuit. However, due to concerns about supply chain security, automotive customers often adopt a conservative attitude towards customized packaging. With the prerequisite of pin compatibility, the primary approach to reducing stray inductance is to increase the stacking area of the power circuit. Currently, one of the leading solutions on the market is the EMPACK module introduced by Semikron. As shown in Fig. 1.43, By employing a solution similar to flexible circuit boards, it achieves large-area, high-efficiency stacking of the power circuit, significantly reducing the module's stray inductance.

3. **Module reliability**

The advancements primarily manifest in two aspects: module reliability implementation and reliability assessment. In terms of reliability implementation, to achieve better reliability lifespan, silver sintering has gradually replaced solder paste in chip connections, while copper wire bonding and ribbon bonding technologies are gradually replacing aluminum wire bonding. Encapsulation processes using molding compounds or epoxy resin are gradually replacing silicone gel encapsulation. Regarding reliability assessment, the main change lies in evaluation methods becoming more aligned with customer application conditions. For instance, the popularity of online aging tests (such as HTRB, HTGB, and reactive power bench tests), as well as the growing endorsement of dynamic reverse bias (DRB) and dynamic gate spike (DGS) testing methods by both upstream and downstream sectors of SiC modules.

1.8.3 SiC Power Module Solution

Currently, the application of SiC power modules is mainly concentrated in the main drive inverters of new energy vehicles. From the perspective of application technology and product topology, mainstream mass production solutions include single-leg parallel schemes, full-bridge schemes, and half-bridge schemes.

1. **Single MOSFET Parallel Scheme**

The principle of the Single MOSFET parallel scheme involves using multiple single MOSFET devices with the same specifications to form parallel combinations, creating bridge arms. These bridge arms are then connected in series to form a half-bridge, and multiple half-bridges are paralleled to achieve full-bridge topology for inverter functionality. One of the most representative products following this technological path is STMicroelectronics' STPAK device, as shown in Fig. 1.44.

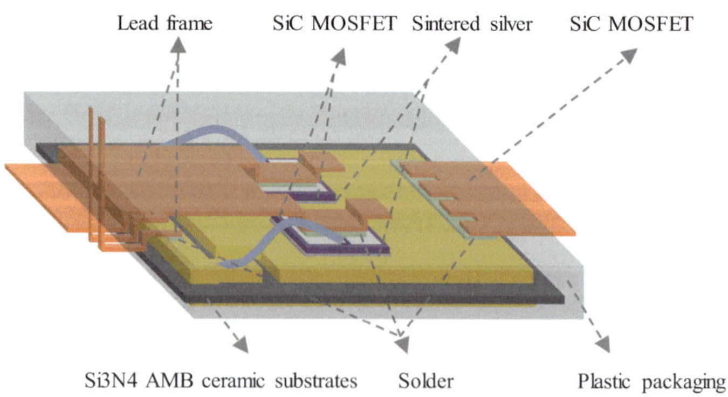

Fig. 1.44 STPAK product of ST

This device integrates 2 SiC MOSFET chips, and the connection between the chips and the Si_3N_4 ceramic substrate utilizes silver sintering technology to reduce thermal resistance and enhance reliability. The exposed side of the ceramic substrate is silver-plated, and a similar silver sintering approach is employed to connect the device to the heatsink water channel. The connection between the busbar and the device is accomplished through laser welding.

The main advantages of the Single MOSFET parallel scheme lie in its design flexibility and relatively low cost. In terms of design flexibility, adjusting the output power of the inverter only requires increasing or decreasing the number of parallel devices, without the need to change the design or specifications of the devices. This greatly shortens the design cycle. In practical applications, the single MOSFET parallel scheme makes it very easy to reuse the same type of device for inverters with different output powers. The cost of SiC is strongly correlated with yield, and device yield is closely related to device production volume. The single MOSFET parallel mode offers cost advantages due to its better cost-effectiveness.

The main drawback of the single MOSFET parallel scheme lies in its high design complexity. The current sharing among parallel MOSFETs is quite intricate, requiring consideration of both static and dynamic current sharing under normal and high-temperature conditions. The effectiveness of current sharing is influenced by device consistency, such as switch voltage and thermal resistance, as well as circuit design factors like parasitic elements in the main circuit and stray inductance in the driving circuit. Typically, the busbar design in the single MOSFET parallel scheme is left to the discretion of the device user, constrained by their understanding of the devices and available resources. This limitation significantly hampers the widespread adoption of the single MOSFET parallel scheme (Fig. 1.44).

2. **Full-Bridge Scheme**

The full-bridge scheme involves power module manufacturers completing the design and manufacturing of three-phase full-bridge topology power modules, while application parties only handle the design and assembly of peripheral circuits. The main representative product of this scheme is the HybridPACK Drive series by Infineon Technologies [8], as shown in Fig. 1.45. In the full-bridge scheme, typically, three

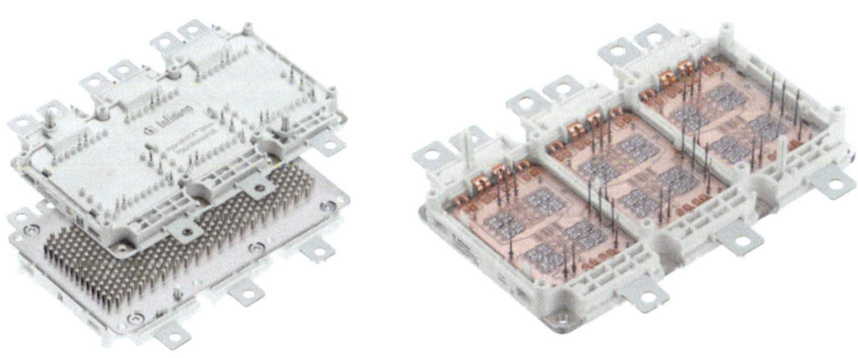

Fig. 1.45 HybridPACK Drive power module of Infineon

half-bridges are integrated within one power module. Each half-bridge is equipped with a temperature sensing resistor. Depending on the required output power, each half-bridge may integrate 6–8 SiC MOSFET chips. The module's DC and AC terminals are located on the upper and lower sides of the module, respectively. The bottom of the module features pin–fin structures for direct liquid cooling. Various signal pins are located on the front side of the module. Internally, the module usually adopts the silicone gel encapsulation technology route.

The advantages of the full-bridge scheme lie in its user-friendly nature for application terminals. There's no need to consider issues like parallel current sharing, and assembly processes are relatively simple. Additionally, there's a wealth of signal sampling ports, making it easy to monitor the module's operating status. As of 2024, the full-bridge scheme is the most widely adopted solution for main drive inverters in new energy vehicles. It's expected that over the next 5–10 years, the full-bridge scheme will continue to thrive.

The drawbacks of the full-bridge scheme include limited design flexibility and potential performance redundancy and waste when the output power is not high and the number of parallel chips is low. Additionally, the overall module size is relatively large, and the fixed terminal positions make it challenging to achieve smaller stray inductance designs.

3. **Half-Bridge Scheme**

The half-bridge scheme involves power module manufacturers completing the design and manufacturing of half-bridge topology power modules, which are then implemented into full-bridge topology by application parties. The main representative product of this scheme is the DCM1000 series by Danfoss, for instance [9, 10]. As shown in Fig. 1.46 These half-bridge modules contain a complete half-bridge and temperature sensing resistor internally. Each bridge arm of the half-bridge may parallel 6–8 SiC MOSFET chips as required. The DC and AC output terminals are placed on both sides, and the device itself includes a liquid-cooled baseplate. The entire device is implemented using encapsulation technology.

The half-bridge scheme serves as a compromise between the single MOSFET parallel scheme and the full-bridge module scheme. Half-bridge power modules achieve better reuse of power devices across different design schemes while also

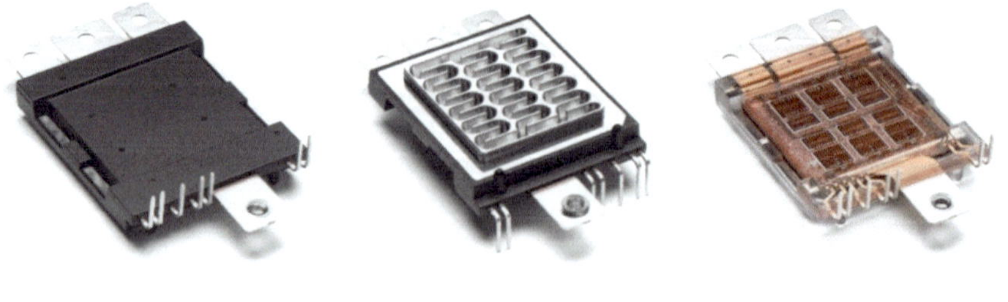

Fig. 1.46 DCM 1000 power module of Danfoss [11]

Table 1.4 SiC power module technology route comparison

Product type	Packaging design difficulty	Application difficulty	Production cost advantage
Single MOSFET	Low	Low	Low
Half bridge	Middle	Middle	Middle
Full bridge	Low	Low	Low

accomplishing parallel current sharing design and providing various sampling interfaces. However, compared to full-bridge modules, the assembly system of half-bridge modules is relatively complex. In comparison to the single MOSFET parallel scheme, its effectiveness in reducing costs and improving power device efficiency is not as significant, and its design adjustments are less flexible (Fig. 1.46).

The comparison of SiC power module products from different technical routes in terms of packaging design, application, and mass production cost is shown in Table 1.4. Each type of product has its own advantages and disadvantages, and users need to choose according to their actual situation. Currently, it is expected that each product type will have a relatively long lifecycle.

Currently, the integration of SiC power modules into the main drive inverters of new energy vehicles has become a mainstream application. For example, Guangdong AccoPower Semiconductor Co., Ltd. has developed a three-phase bridge structure automotive-grade SiC power module for use in new energy vehicle main drive inverters. This module features industry-leading SiC MOSFET chips and advanced silver sintering packaging technology, meeting the high-power density and reliability requirements for new energy vehicle main drive applications. After durability and road tests, the module went into mass production in 2022, filling a domestic gap. As of December 2024, it has been installed in nearly 400,000 new energy vehicles, demonstrating its industrial validation and becoming the most widely used SiC module product in domestic vehicle production and shipments.

References

1. Stefan L (2006) Power semiconductors, 1st edn. EPFL Press, Swiss
2. Casady J, Pala V, Brunt EV et al (2016) Ultra-low (1.25 mΩ) on-resistance 900V SiC 62 mm half-bridge power modules using new 10 mΩ SiC MOSFETs. In: PCIM Europe 2016 international exhibition and conference for power electronics, intelligent motion, renewable energy and energy management, pp 34–41
3. Nakamura T, Nakano Y, Aketa M et al (2011) High performance SiC trench devices with ultra-low Ron. In: 2011 international electron devices meeting 26.5.1–26.5.3
4. Peters D, Basler T, Zippelius B et al (2017) The new CoolSiC™ trench MOSFET technology for low gate oxide stress and high performance. PCIM Europe 2017, international exhibition and conference for power electronics, intelligent motion. Renew Energy Energy Manage 2017:168–174

5. Mitsubishi Electric Corporation (2019) Mitsubishi electric develops trench-type SiC-MOSFET with unique electric-field-limiting structure. News. https://us.mitsubishielectric.com/en/news/releases/global/2019/0930-a/index.html
6. Heraeus (2017) mAgic DA295A non-pressure sinter paste, factsheet
7. Berberich SE, Kasko I, Gross M et al (2017) High efficient approach to utilize SiC MOSFET potential in power modules. In: 2017 29th international symposium on power semiconductor devices and IC's (ISPSD), pp 258–262
8. Jakobi W, Uhlemann A, Schweikert C et al (2018) Benefits of new CoolSiC™ MOSFET in HybridPACK™ drive package for electrical drive train applications. In: CIPS 2018; 10th international conference on integrated power electronics systems, pp 585–593
9. Omid S, Alexander S, Norbert A (2018) DCM™ 1000X—designed to meet the future SiC demand of electric vehicle drive trains. Danfoss Silicon Power GmbH, Bodos Power Magazine, pp 80–83
10. Alexander S (2018) Danfoss silicon power introduces DCM 1000X (Full-SiC) next gen automotive traction inverter power modules. In: WBG power conference
11. DCM™1000 Power Module Technology Platform. https://www.danfoss.com/en/about-danfoss/our-businesses/silicon-power/danfoss-dcm-1000-power-module-technology-platform/

Further Reading

12. Neamen DA (2011) Semiconductor physics and devices: basic principles, 4th edn. McGraw-Hill, New York
13. Baliga BJ (2018) Fundamentals of power semiconductor devices, 2nd edn. Springer, New York
14. van Wyk JD, Lee FC (2013) On a future for power electronics. IEEE J Emerg Selected Topics Power Electron 1(2):59–72
15. Sze SM, Ng KK (2006) Physics of semiconductor devices, 3rd edn. Wiley-Interscience, Hoboken
16. Josef L, Heinrich S, Uwe S (2018) Semiconductor power devices: physics, characteristics, reliability, 2nd edn. Springer, New York
17. Deboy G, Kaindl W, Kirchner U et al (2015) Advanced silicon devices—applications and technology trends. In: Proceedings of IEEE applied power electronics conference (APEC)
18. Yannis T, Colin M (2010) Operation and modeling of the MOS transistor, 3rd edn. Oxford University Press, Oxford
19. Baliga BJ (2011) Advanced high voltage power device concepts. Springer, New York
20. Baliga BJ (2010) Advanced power MOSFET concepts. Springer, New York
21. Jacek K (2011) Low voltage power MOSFETs: design, performance and applications. Springer, New York
22. Williams RK, Darwish MN, Blanchard RA et al (2017) The trench power MOSFET: part I—history, technology, and prospects. IEEE Trans Electron Dev 64(3):674–691
23. Williams RK, Darwish MN, Blanchard RA et al (2017) The trench power MOSFET—part II: application specific VDMOS, LDMOS, packaging, and reliability. IEEE Trans Electron Dev 64(3):692–712
24. Baliga BJ (2015) The IGBT device: physics, design and applications of the insulated gate bipolar transistor, 1st edn. William Andrew, New York
25. Iwamuro N, Laska T (2017) IGBT history, state-of-the-art, and future prospects. IEEE Trans Electron Dev 64(3):741–752
26. Krishna S (2015) The invention and demonstration of the IGBT. IEEE Power Electron Mag 2(2):12–16
27. Baliga BJ (2015) IGBT: the GE story. IEEE Power Electron Mag 2(2):16–23
28. Baliga BJ (2009) Advanced power rectifier concepts. Springer, New York

Chapter 2
SiC Diode Main Characteristics

Current SiC power semiconductor devices include two types: SiC diodes and SiC MOSFETs. Among them, SiC diodes achieved earlier industrial application and serve as the cornerstone driving the SiC device application market.

SiC diodes are uncontrollable devices with fewer parameters compared to fully-controllable devices like SiC MOSFETs. They can be categorized into maximum ratings and static characteristics. Maximum ratings include parameters such as reverse current, breakdown voltage, thermal impedance, power dissipation, continuous forward current, and forward surge current. Static characteristics include forward voltage drop, junction capacitance, junction charge, and junction capacitance energy.

In the following introduction to each parameter of the SiC diode, the data presented is derived from measurements or datasheets without specific differentiation.

2.1 Maximum Ratings

2.1.1 Reverse Current and Breakdown Voltage

When the SiC diode withstands a reverse voltage V_R, where the cathode is at a high potential, a small current flows from the cathode to the anode, known as reverse current I_R. I_R is influenced by the reverse voltage V_R and the junction temperature T_J, as shown in Fig. 2.1. When V_R is low, I_R ranges from several hundred picoamps to tens of nanoamps, indicating the SiC diode is in a reliable off-state. As V_R increases, I_R gradually increases. When V_R exceeds approximately 880 V, I_R increases rapidly and almost vertically, indicating the SiC diode enters avalanche breakdown, unable to effectively block voltage anymore. This 880 V is the clear physical breakdown voltage V_{BR} of the SiC diode. Additionally, during effective off-state, I_R increases

Y. Gao and Y. Zhang, *Silicon Carbide Power Devices*, CPSS Power Electronics Series,
https://doi.org/10.1007/978-981-96-3480-4_2

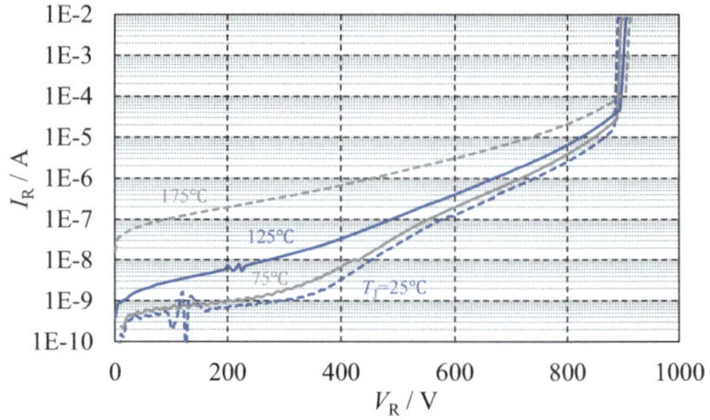

Fig. 2.1 SiC diode I_R–V_R characteristic curve

with T_J. At the critical point of avalanche breakdown, the breakdown voltage V_{BR} increases with T_J.

According to Fig. 2.1, a SiC diode with a rated voltage of 650 V actually has a breakdown voltage V_{BR} around 880 V, allowing for device manufacturing variations and practical application requirements. This explains why the SiC diode does not immediately fail when subjected to short-term voltage spikes slightly above the nominal V_{BR} specified in the datasheet. However, it is still important to avoid exceeding the nominal V_{BR} specified in the datasheet during operation.

As shown in Fig. 2.1 and defined by V_{BR}, V_{BR} is also affected by junction temperature T_J, increasing with T_J as shown in Fig. 2.2. This is because as T_J increases, the carrier mobility decreases due to increased lattice scattering and impurity scattering, making breakdown more difficult and increasing the breakdown voltage.

Obtaining a precise avalanche voltage requires measuring the complete I_R–V_R curve, which is not easy to achieve in mass production or practical applications. Therefore, the breakdown voltage V_{BR} is also defined as the reverse voltage that causes I_R to reach the specified reverse current I_{BR} when the SiC diode withstands it. I_{BR} typically ranges from 100 μA to 5 mA, defined by each manufacturer. Some manufacturers define I_{BR} to increase with device current ratings, while others use a constant I_{BR}.

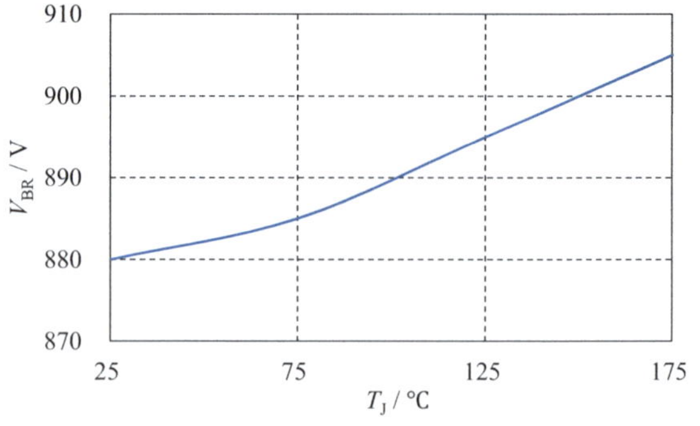

Fig. 2.2 SiC diode V_{BR}–T_J characteristic curve

2.1.2 Thermal Impedance

Thermal impedance consists of thermal capacitance and thermal resistance. Thermal resistance represents the material's ability to impede heat transfer, measured as the reciprocal of the heat transfer coefficient, while thermal capacitance represents the material's ability to store thermal energy. Although semiconductor and power electronics professionals may not be familiar with thermodynamics, analogies between thermal and electrical physics can aid in understanding and memory.

In electrical domain, current and voltage are core physical quantities. Current I represents the flow of charge carriers from point a at potential V_a to point b at potential V_b through a path characterized by resistance R, following Ohm's law:

$$\Delta V_{ab} = V_a - V_b = I \times R \tag{2.1}$$

In thermal domain, heat flux and temperature are core physical quantities. P represents the flow of heat from temperature T_a to temperature T_b through a path characterized by thermal resistance R_{ab}. Thermal physics is governed by Fourier's law, but considering the duality with electrical physics, a similar form to Ohm's law can be generalized:

$$\Delta T_{ab} = T_a - T_b = P \times R_{ab} \tag{2.2}$$

Thus, electrical and thermal physical quantities exhibit symmetric relationships, as shown in Table 2.1.

When a SiC diode operates, it generates losses, and heat transfers sequentially from the chip to the package casing, heatsink, and environment. Based on this, a Cauer thermal impedance model can be established where thermal resistances of each layer of material are connected in series, with all thermal capacitances connected to the ambient reference temperature, as shown in Fig. 2.3. Thus, the nodes in the Cauer model represent the temperatures of each layer of material, providing clear physical meaning.

However, due to the difficulty in obtaining the thermal capacitance and thermal resistance of each layer, datasheets provide thermal impedance curves obtained under single pulse and continuous square wave thermal excitation, as shown in Fig. 2.4. The horizontal axis represents the pulse width t_p, and the vertical axis represents the

Table 2.1 Symmetric relationship between electrical and thermal domain physical quantities

	Electrical domain	Thermal domain
Resistance	Resistance R	Thermal resistance R_{th}
Impedance	Impedance Z	Thermal impedance Z_{th}
Potential	Voltage V	Temperature T
Energy Flow	Current I	Heat flux
Capacitance	Capacitance C	Thermal capacitance C_{th}

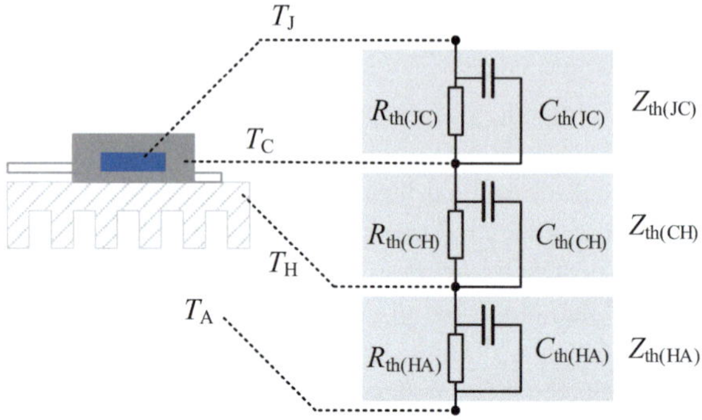

Fig. 2.3 Cauer thermal impedance model

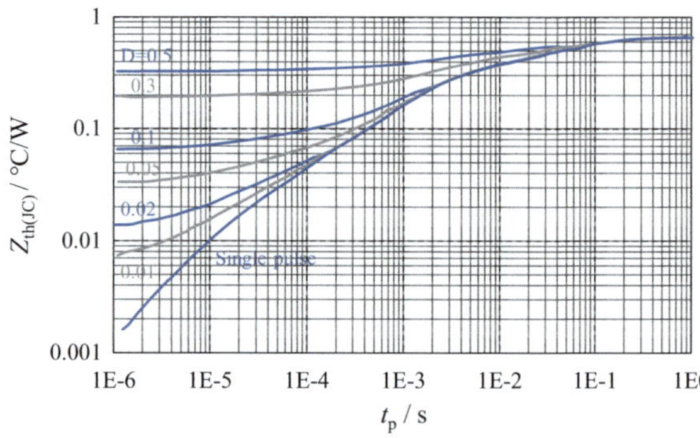

Fig. 2.4 Thermal impedance curve

thermal impedance value $Z_{th(JC)}$. During the pulse heating stage, each layer absorbs heat, leading to a temperature rise. During the pulse off stage, the material dissipates heat to the environment. $Z_{th(JC)}$ is defined as the highest temperature of the material at the end of the temperature rise stage.

The thermal impedance curve under a single pulse is at the bottom, and the shorter the t_p, the shorter the heating time, the less temperature rise, and the smaller $Z_{th(JC)}$. $Z_{th(JC)}$ increases with increasing t_p until it stabilizes, at which point the thermal impedance is the thermal resistance value. The other curves are continuous square wave thermal impedance curves with pulse width t_p, forming a cluster of impedance curves under different duty cycles D. At the same t_p, the higher the D, the higher the proportion of heating time, the higher the material temperature rise, and the higher $Z_{th(JC)}$. Simultaneously, with the increase of t_p, the $Z_{th(JC)}$ curves under different D gradually converge to the thermal resistance value.

Some datasheets from manufacturers also provide Foster model parameters based on thermal impedance curves, as shown in Fig. 2.5.

The expression of the Foster model is given by Eq. (2.3), which can be used for thermal simulation analysis.

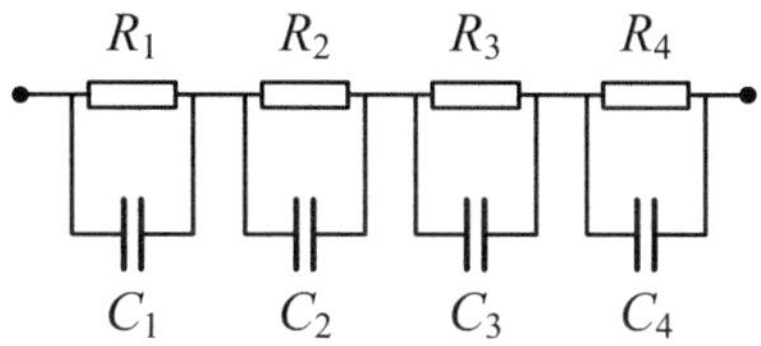

Fig. 2.5 Foster thermal impedance model

$$Z_{th} = \sum_{i=1}^{4} R_i\left(1 - e^{-\frac{t}{\tau_i}}\right),\ \tau_i = R_i C_i \tag{2.3}$$

It is important to note that the Foster model is a numerical fit to the thermal impedance curve. Although it is still represented by a thermal resistance and thermal capacitance network, the capacitance defined between two nodal temperatures does not have physical significance.

2.1.3 *Power Dissipation and Forward Conduction Current*

SiC MOSFETs generate losses during operation. When the losses and heat dissipation conditions are constant, the junction temperature rises and reaches a steady state $T_{J(steady)}$ following Eq. (2.4), where T_C is the case temperature, and $R_{th(JC)}$ is the junction-case thermal resistance.

$$T_J - T_C = P_{tot}(T_C) \cdot R_{th(JC)} \tag{2.4}$$

When $T_{J(steady)}$ equals the device's highest allowable operating junction temperature $T_{J(max)}$, the corresponding power dissipation at the case temperature is the maximum total dissipated power $P_{tot}(T_C)$. According to Eq. (2.4), $P_{tot}(T_C)$ is affected by T_C, and the relationship between the two is provided in the datasheet, as shown in Fig. 2.6, where $P_{tot}(T_C)$ decreases linearly according to Eq. (2.4).

In practical applications, the device dissipates heat to the environment through a heatsink, following Eq. (2.5), where T_A is the ambient temperature, $R_{th(JA)}$ is the junction-ambient thermal resistance, $R_{th(JA)}$ is the case-heatsink thermal resistance, and $R_{th(JA)}$ is the heatsink-ambient thermal resistance. When T_J is at the highest allowable operating junction temperature $T_{J(max)}$, the corresponding power dissipation P_D is the maximum dissipated power $P_{tot}(T_A)$ at the ambient temperature. Datasheets generally do not provide $P_{tot}(T_A)$ data because $R_{th(CH)}$ and $R_{th(HA)}$ need to be determined based on specific heatsink design.

$$T_J - T_A = P_{tot}(T_A) \cdot R_{th(JA)} = P_{tot}(T_A) \cdot (R_{th(JC)} + R_{th(CH)} + R_{th(HA)}) \tag{2.5}$$

The forward conduction current I_F is limited by the maximum dissipation power $P_{tot}(T_C)$ and the duty cycle D, calculated according to (2.6), where $V_F(T_{J(max)}, I_F)$ is

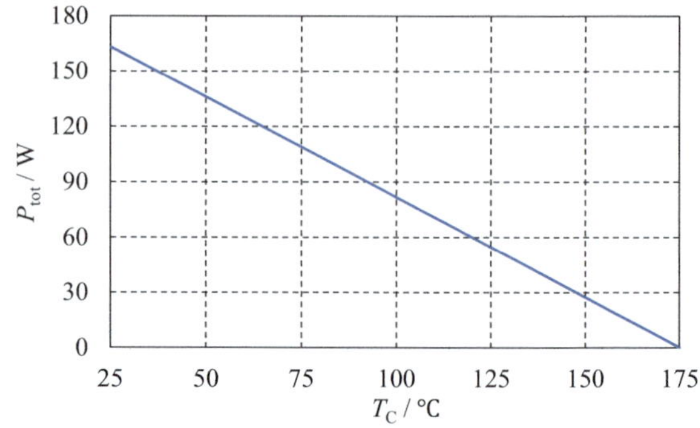

Fig. 2.6 SiC diode $P_{tot}(T_C)$–T_C characteristic curve

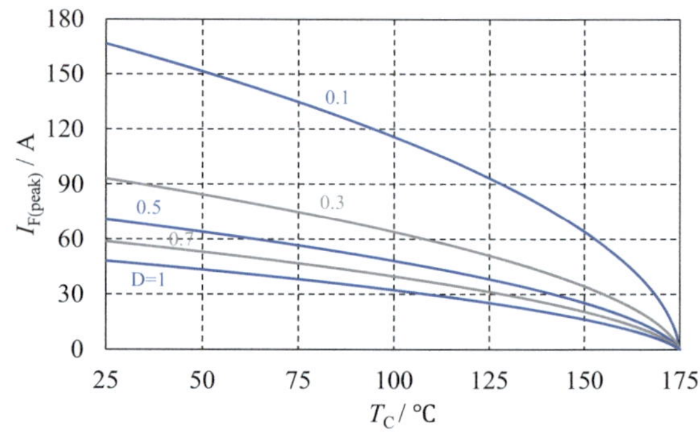

Fig. 2.7 SiC diode I_F–T_C characteristic curve

the conduction voltage drop at maximum allowable junction temperature $T_{J(max)}$ and current I_F.

$$I_F = \frac{P_{tot}(T_C)/D}{V_F(T_{J(max)}, I_F)}. \tag{2.6}$$

Datasheets provide I_F–T_C characteristic curves, as shown in Fig. 2.7, where I_F decreases with T_C according to Eq. (2.6) and increases with decreasing duty cycle D.

2.1.4 Surge Forward Current and i^2t

The maximum peak forward sinusoidal half-wave current that a SiC diode can safely withstand is called the surge forward current. This sinusoidal wave typically has a frequency of 50 Hz or 60 Hz. During the passage of surge current, the SiC diode heats up, and if the junction temperature exceeds maximum rated value, it can lead to device failure.

When the surge current is a single half-sinusoidal wave, it is termed as non-repetitive surge forward current I_{FSM}. When the surge current consists of repeated multiple half-sinusoidal waves, it is termed as repetitive surge forward current I_{FSM}. It's important to note that the surge conditions (number of surges and duty cycle) for I_{FSM} specified in manufacturers' datasheets may vary, so direct comparisons should not be made.

The parameter i^2t also represents the SiC diode's ability to withstand surge currents. It is calculated by integrating the product of I_{FSM} and time:

$$i^2t = \int_0^{10\text{ms}} (I_{FSM} \sin t)^2 \mathrm{d}t = 0.005\, I_{FSM}^2 \tag{2.7}$$

2.2 Static Characteristics

2.2.1 Forward Voltage Drop

When a SiC diode withstands a forward voltage V_F, meaning the anode is at a higher voltage than the cathode, it allows current to flow from anode to cathode, known as forward current I_F. The forward conduction characteristics of a SiC diode are represented by the I_F–V_F curve, as shown in Fig. 2.8.

When V_F is low, I_F is very small, nearly at zero, indicating that the SiC diode is not conducting. As V_F exceeds a certain threshold, the curve begins to rise linearly with V_F. The point where the curve begins to rise sharply is known as the forward voltage drop, marking the onset of normal operation for the SiC diode. As V_F continues to increase, the curve rises again with a steeper slope, indicating operation in the PiN diode of MPS structure dominant region typical during surges or short circuits.

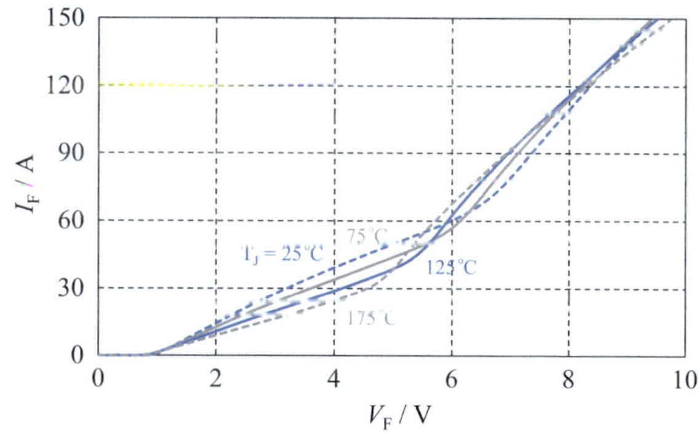

Fig. 2.8 SiC diode I_F–V_F characteristic curve

From Fig. 2.8, it is evident that the forward conduction characteristics of the SiC diode are influenced by T_J, the junction temperature. Firstly, the forward voltage drop decreases with increasing T_J. Secondly, at different temperatures, the I_F–V_F curves intersect near the forward voltage drop point. This intersection point is known as the zero temperature coefficient point. To the left of this point, I_F shows a positive temperature coefficient; to the right, it shows a negative temperature coefficient. Because this zero temperature coefficient point is very close to forward voltage drop, the variation in I_F with temperature is minimal within the operational current range of SiC diodes, allowing them to be safely used in parallel. At the same time, as T_J increases further, the SiC diode enters the PiN region earlier, as the current continues to increase, the curves intersect again at another zero temperature coefficient point.

2.2.2 *Junction Capacitance, Junction Charge, and Junction Capacitance Energy*

When a SiC diode withstands a reverse voltage V_R, a depletion region forms between the p-region and n-region where no charge carriers move. This depletion region can be seen as a dielectric, with the conductive p-region and n-region on either side. This structure behaves like a parallel plate capacitor, where the capacitance is inversely proportional to the width of the depletion region, which increases with V_R. The datasheet of SiC diodes provides a curve showing the relationship between junction capacitance C and reverse voltage V_R, as depicted in Fig. 2.9. C decreases as V_R increases, with a faster rate of change at lower voltages and stabilizing gradually at higher voltages.

Based on the C–V_R curve of the SiC diode, the junction charge Q_C can be obtained by integrating C with respect to V_R, as shown in Fig. 2.10 according to Eq. (2.8).

$$Q_C = \int_0^{V_R} C(V_R)\mathrm{d}V_R \tag{2.8}$$

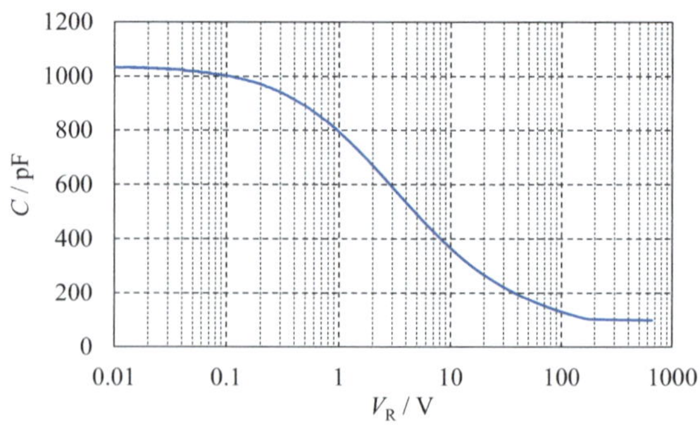

Fig. 2.9 SiC diode C–V_R characteristic curve

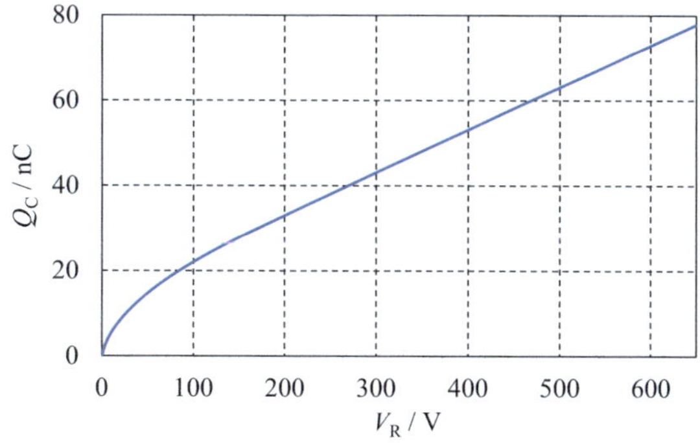

Fig. 2.10 SiC diode Q_C–V_R characteristic curve

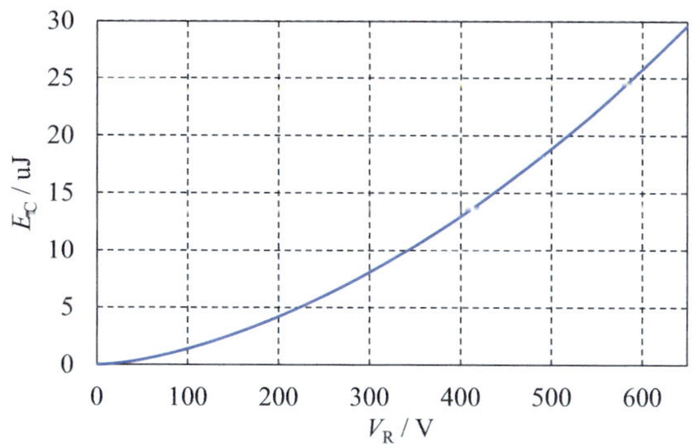

Fig. 2.11 SiC diode E_C–V_R characteristic curve

Further, using the C–V_R curve of the SiC diode and integrating the product of C and V_R with respect to V_R according to Eq. (2.9), the junction capacitance energy E_C curve is obtained, as depicted in Fig. 2.11.

$$E_C = \int_0^{V_R} C(V_R) \cdot V_R \mathrm{d}V_R \tag{2.9}$$

References

1. Pai AP (2018) CoolSiC™ automotive discrete schottky diodes explanation of datasheet nomenclature. Application note, Revision 1.1, Infineon Technologies AG
2. Infineon Technologies AG (2019) CoolSiC™ automotive discrete schottky diodes understanding the benefits of SiC diodes compared to silicon diodes. Application note, AN2019-02_CoolSiC_Automotive_Diode, Revision 1.0
3. Omar H, Vladimir S (2016) 1200V CoolSiC™ Schottky diode generation 5. Application note, Revision. 1.1, Infineon Technologies AG

4. Infineon Technologies AG (2017) Improving PFC efficiency using the CoolSiC™ Schottky diode 650V G6. Application note, AN_201704_PL52_020, Revision 1.0
5. STMicroelectronics (2018) New generation of 650V SiC diodes. Application note, AN4242, Revision 2
6. ROHM Co., Ltd. (2014) SiC power devices and modules. Application note, 14103EBY01

Chapter 3
SiC MOSFET Main Characteristics

SiC MOSFETs have dozens of parameters to characterize their characteristics. These parameters are not only references for device selection but also help engineers to fine-tune transformer designs, such as evaluating and comparing devices based on figures of merit (FOM), modeling and simulating devices, and calculating device losses.

Different manufacturers categorize SiC MOSFET parameters differently. In this chapter, the main parameters are classified into four categories: maximum values, static characteristics, and dynamic characteristics. Maximum values are parameters related to the SiC MOSFET's limit operating points, including breakdown voltage, thermal impedance, maximum power dissipation, maximum drain current, and safe operating area. Static characteristics characterize the voltage and current relationship of the SiC MOSFET's operating point, collectively referred to as I–V parameters, including transfer characteristics, threshold voltage, output characteristics, on-state resistance, diode forward characteristics, and third-quadrant on-state characteristics. Dynamic characteristics refer to the switching process of the SiC MOSFET and the reverse recovery process of the body diode. Since the relationship between junction capacitance, gate charge, and the switching process is close, they are explained as part of the dynamic characteristics, different from some manufacturers' classification methods. It is important to note that although classified as dynamic characteristics, the switching process is still influenced by static parameters, and the two are not completely independent. Limiting characteristics include short-circuit and avalanche capabilities, which represent abnormal operating conditions for SiC MOSFET.

In the following introduction to the various parameters of the SiC MOSFET, the displayed data comes from measurements or datasheets, without special differentiation.

Y. Gao and Y. Zhang, *Silicon Carbide Power Devices*, CPSS Power Electronics Series,
https://doi.org/10.1007/978-981-96-3480-4_3

3.1 Maximum Values

3.1.1 Leakage Current and Breakdown Voltage

When $V_{GS} = 0$ V, the SiC MOSFET is in the off state and bears an applied voltage V_{DS}. A small current, known as the leakage current I_{DSS}, flows from the source to the drain. I_{DSS} is influenced by the applied voltage V_{DS} and the chip junction temperature T_J, as shown in Fig. 3.1. When V_{DS} is low, I_{DSS} is only a few hundred picoamperes to tens of nanoamperes, indicating a reliable off state. As V_{DS} increases, I_{DSS} gradually increases. When V_{DS} exceeds approximately 1650 V, I_{DSS} increases rapidly and almost vertically, indicating that the SiC MOSFET enters avalanche breakdown and can no longer effectively block the voltage. The 1650 V at this point is the breakdown voltage $V_{(BR)DSS}$ with clear physical significance. Additionally, when in effective off state, I_{DSS} increases with the rise of T_J. At the critical point of avalanche breakdown, the transition voltage of I_{DSS} increases with the rise of T_J.

According to Fig. 3.1, the actual $V_{(BR)DSS}$ for a SiC MOSFET with a nominal voltage of 1200 V is approximately 1650 V, which provides a margin considering device manufacturing variability and practical application requirements. This explains why a SiC MOSFET does not immediately fail when subjected to a short-duration voltage spike slightly exceeding the specified off-state voltage $V_{(BR)DSS}$. Nevertheless, it is essential to avoid the terminal voltage exceeding the nominal $V_{(BR)DSS}$ specified in the datasheet.

According to Fig. 3.1 and the definition of $V_{(BR)DSS}$, $V_{(BR)DSS}$ is also influenced by the junction temperature T_J, increasing as T_J rises, as shown in Fig. 3.2. This is because as T_J rises, the carrier mobility decreases due to lattice scattering and impurity scattering, making breakdown more challenging, and the breakdown voltage increases.

To obtain a clear avalanche voltage, it is necessary to measure the complete I_{DSS}–V_{DS} curve, which is not easy to obtain or affects efficiency in mass production or application. Therefore, the breakdown voltage $V_{(BR)DSS}$ is often defined as the reverse voltage at which the drain leakage current I_{DSS} reaches a specified value $I_{(BR)DSS}$ when

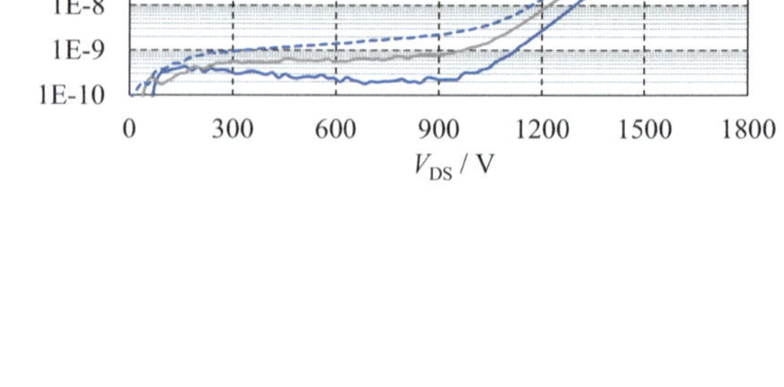

Fig. 3.1 I_{DSS}–V_{DS} characteristics curve of SiC MOSFET

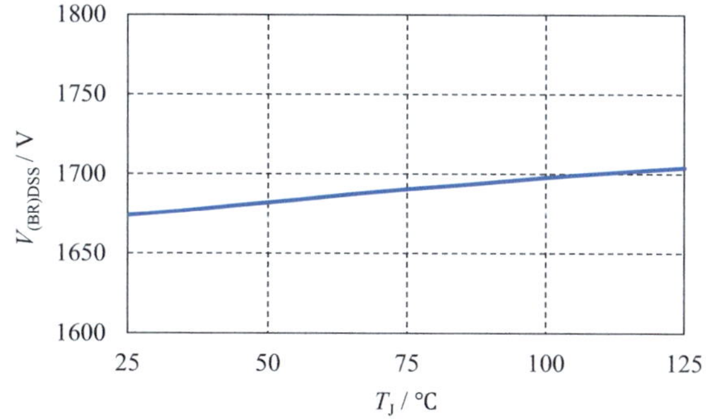

Fig. 3.2 $V_{(BR)DSS}$–T_J characteristics curve of SiC MOSFET

$V_{GS} = 0$ V. The value of $I_{(BR)DSS}$ is generally between 100 uA and 5 mA, defined by each manufacturer. Some manufacturers follow the pattern that the larger the device current rating, the larger $I_{(BR)DSS}$ is, while others choose a constant value for $I_{(BR)DSS}$.

3.1.2 *Power Dissipation and Drain Current*

SiC MOSFETs generate losses during operation. When the losses and heat dissipation conditions are constant, the junction temperature rises and reaches a steady state $T_{J(steady)}$ following Eq. (3.1), where T_C is the case temperature, and $R_{th(JC)}$ is the junction-case thermal resistance.

$$T_J = T_C + P_{tot}(T_C) \cdot R_{th(JC)}. \tag{3.1}$$

When $T_{J(steady)}$ equals the device's highest allowable operating junction temperature $T_{J(Max)}$ the corresponding power dissipation at the case temperature is the maximum total dissipated power $P_{tot}(T_C)$. According to Eq. (3.1), $P_{tot}(T_C)$ is affected by T_C, and the relationship between the two is provided in the datasheet, as shown in Fig. 3.3. When T_C is below 25 °C, $P_{tot}(T_C)$ remains constant. When T_C exceeds 25 °C, $P_{tot}(T_C)$ decreases linearly according to Eq. (3.1).

In practical applications, the device dissipates heat to the environment through a heatsink, following Eq. (3.2), where T_A is the ambient temperature, $R_{th(JA)}$ is the junction-ambient thermal resistance, $R_{th(CH)}$ is the case-heatsink thermal resistance, and $R_{th(HA)}$ is the heatsink-ambient thermal resistance. When T_J is at the highest allowable operating junction temperature $T_{J\,(max)}$, the corresponding power dissipation P_D is the maximum dissipated power $P_{tot}(T_A)$ at the ambient temperature. Datasheets generally do not provide $P_{tot}(T_A)$ data because $R_{th(CH)}$ and $R_{th(HA)}$ need to be determined based on specific heatsink design.

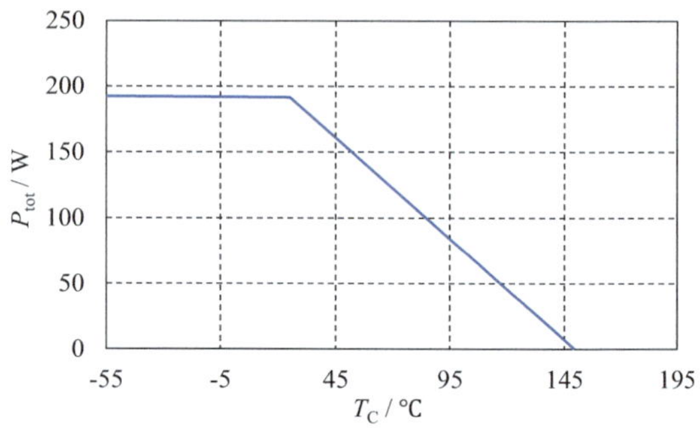

Fig. 3.3 P_{tot} (T_C)–T_C characteristics curve of SiC MOSFET

$$T_J - T_A = P_{tot}(T_A) \cdot R_{th(JA)} = P_{tot}(T_A) \cdot (R_{th(JC)} + R_{th(CH)} + R_{th(HA)}) \tag{3.2}$$

The maximum drain current is divided into the maximum continuous drain current I_D and the maximum pulse drain current $I_{D(pulse)}$, both limited by the maximum power dissipation. This is typically defined based on $P_{tot}(T_C)$.

The device current rating is generally defined by I_D at a specific T_C, calculated using Eq. (3.3), where $R_{DS(on)}\left(T_{J(max)}\right)$ is the on-state resistance at the highest allowable operating junction temperature.

$$I_D = \sqrt{\frac{P_{tot}(T_C)}{R_{DS(on)}(T_{J(max)})}} \tag{3.3}$$

Datasheets provide I_D–T_C characteristic curves, as shown in Fig. 3.4. When T_C is below 25 °C, I_D remains constant; when T_C exceeds 25 °C, I_D decreases according to Eq. (3.3).

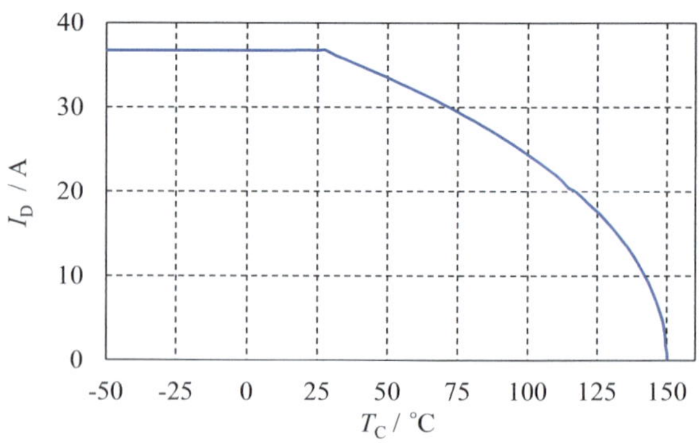

Fig. 3.4 I_D-T_C characteristics curve of SiC MOSFET

As it is a single-pulse current, $I_{D(pulse)}$ is often 2–4 times I_D. $I_{D(pulse)}$ is not only limited by $P_{tot}(T_C)$ but also depends on the pulse width t_p and V_{DS}, calculated using Eq. (3.4).

$$I_{D(pulse)}(V_{DS}, t_p) = \frac{P_{tot}(T_C)}{V_{DS}} = \frac{T_{J(max)} - T_C}{V_{DS} \cdot \leq Z_{th(JC)}(t_p)} \tag{3.4}$$

It is important to note that $I_{D(pulse)}$ does not increase unlimitedly with decreasing t_p and V_{DS} but is always limited by the current-carrying capacity of the bond wire.

3.1.3 Safe Operating Area

The safe operating area (SOA) defines the region within which a SiC MOSFET can operate safely. Its boundaries are determined by the upper limits of drain-source current I_{DS} and drain-source voltage V_{DS} that the SiC MOSFET can withstand during operation. These limits are specifically constrained by $V_{(BR)DSS}$, I_D, and $I_{D(pulse)}$ as introduced earlier. It is imperative for the SiC MOSFET to operate within the safe operating area to avoid premature failure or direct damage. Datasheets generally provide the SOA at T_C is 25 °C, presented in a double-logarithmic coordinate system, as shown in Fig. 3.5.

The boundaries of the SOA consist of four segments: A, B, C, and D.

1. **A $R_{DS(on)}$ Limitation**

Segment A represents the limitation imposed by $R_{DS(on)}$, and the region above A is not feasible for SiC MOSFET operation. It is determined using $R_{DS(on)}$ at T_J (max), expressed in double-logarithmic coordinates as:

$$\log(I_{DS}) = \log(V_{DS}) - \log\left(R_{DS(on)}(V_{DS}, T_J\ (\max))\right) \tag{3.5}$$

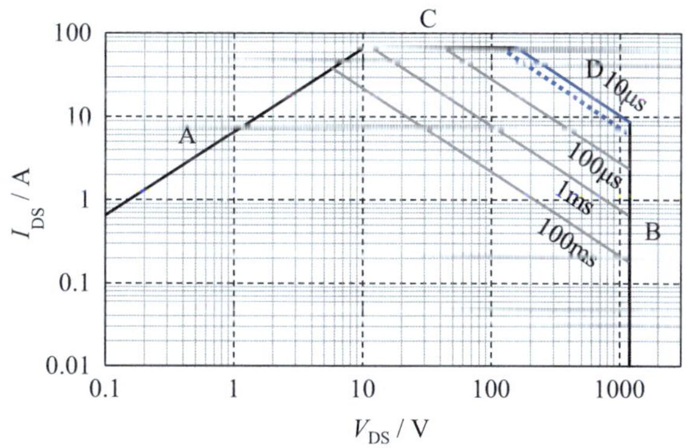

Fig. 3.5 SOA of SiC MOSFET

2. **B $V_{(BR)DSS}$ Limitation**

Segment B represents the limitation imposed by $V_{(BR)DSS}$, the breakdown voltage limit.

3. **C $I_{D(pulse)}$ Limitation**

Segment C represents the limitation imposed by the current-carrying capacity of the bond wire, determining $I_{D(pulse)}$.

4. **D $P_{tot}(T_C)$ Limitation**

Segment D represents the limitation imposed by $P_{tot}(T_C)$, and it is represented by a cluster of $I_{D(pulse)}(V_{DS}, t_p)$ segments obtained from Eq. (3.4) for different t_p, expressed in double-logarithmic coordinates as:

$$\log(I_{DS}) = -\log(V_{DS}) + \log\left(T_{J(max)} - T_C\right) - \log\left(Z_{th(JC)}(t_p)\right) \tag{3.6}$$

In practical applications, where the SiC MOSFET's T_C is significantly higher than 25 °C, the safe operating area presented in Fig. 3.5 may not be applicable directly and requires conversion for design purposes. The T_C does not impact the A, B, and C limitations. However, conversion is necessary for the D limitation. The following example illustrates the conversion of the $t_p = 10$ us condition from $T_C = 25\,°\text{C}$ to $T_C = 100\,°\,\text{C}$.

From Fig. 3.5, $P_{tot}(25°) = 8.5\,\text{A} \cdot 1200\,\text{V} = 10.2\,\text{kW}$. With t_p fixed, $Z_{th(JC)}\,(t_p)$ is not affected by T_C. Hence,

$$P_{tot}(100\,°\text{C}) = P_{tot}(25\,°\text{C})\frac{150\,°\text{C} - 100\,°\text{C}}{150\,°\text{C} - 25\,°\text{C}} = 6.8\,\text{kW}. \tag{3.7}$$

Further, from Eq. (3.7), the intersection points of the $t_p = 10$ us limitation with B and C at $T_C = 100\,°\text{C}$ are (97 V, 70 A) and (1200 V, 5.67 A), respectively. Connecting these two points results in the converted D limitation, represented by a dashed line in Fig. 3.5.

3.2 Static Characteristics

3.2.1 Transfer Characteristics and Threshold Voltage

Transfer characteristics illustrate the impact of V_{GS} on the maximum I_{DS} that a SiC MOSFET can output, represented by the I_{DS}–V_{GS} curve, as shown in Fig. 3.6. As V_{GS} increases, I_{DS} also increases because higher V_{GS} ensures more thorough channel activation, facilitating the flow of electrons. When V_{GS} is less than approximately

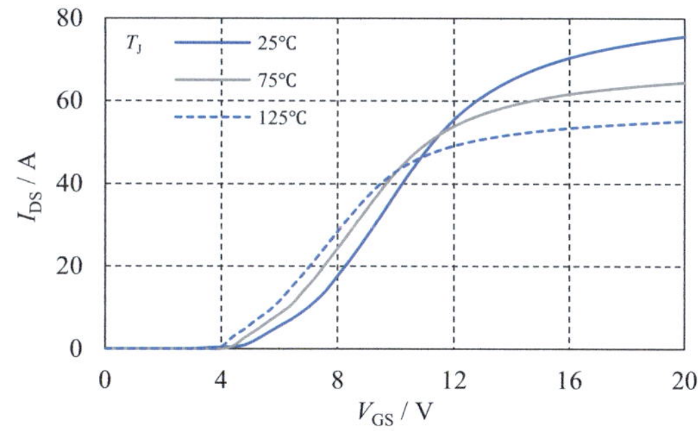

Fig. 3.6 Transfer characteristics' curve of SiC MOSFET

11 V, I_{DS} exhibits a positive temperature coefficient. However, when V_{GS} exceeds around 11 V, I_{DS} displays a negative temperature coefficient. This implies that in parallel operation, the turn-on drive voltage $V_{GS(on)}$ must be sufficiently high to ensure that the SiC MOSFET operates in the region of negative temperature coefficient.

In Fig. 3.6, there is a certain voltage threshold beyond which current output occurs. This voltage is the threshold voltage V_{GS} for the SiC MOSFET. Typically, V_{GS} is defined as the V_{GS} corresponding to I_{DS} greater than a specified threshold current $I_{DS(th)}$. Similar to $I_{(BR)DSS}$, specific standards are provided by individual manufacturers, often following the trend that higher device current ratings correspond to larger $I_{DS(th)}$. The temperature dependence of V_{GS} is illustrated in Fig. 3.7, where higher T_J results in lower V_{GS}.

The SiC MOSFET exhibits a significant Drain-Induced Barrier Lowering (DIBL) effect, causing $V_{GS(th)}$ to increase as V_{DS} decreases, as depicted in Fig. 3.8. At $T_J = 25\,°C$, $V_{GS(th)}$ is 2.04 V at $V_{DS} = 20$ V. At $V_{DS} = 800$ V, $V_{GS(th)}$ decreases to 1.74 V. At $T_J = 175\,°C$, $V_{GS(th)}$ is 1.72 V at $V_{DS} = 20$ V, and it decreases to 1.42 V at $V_{DS} = 800$ V. Due to the fast switching speed of the SiC MOSFET and the further reduction of $V_{GS(th)}$ with increasing junction temperature and V_{DS}, there is a high risk of unintended turn-on and shoot-through faults in the bridge configuration. Therefore, special attention must be paid during the design of the converter.

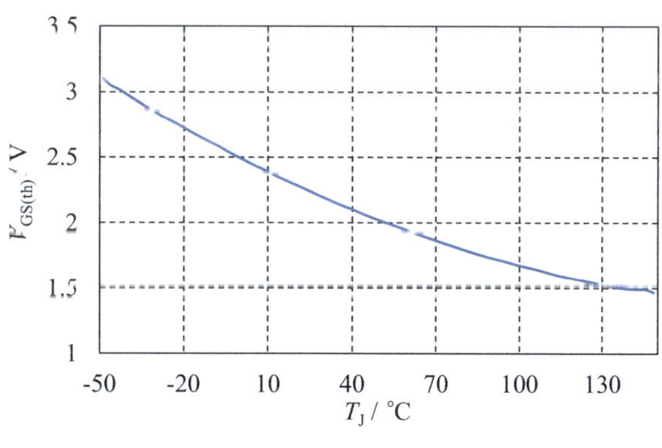

Fig. 3.7 $V_{GS(th)}$–T_J characteristics' curve of SiC MOSFET

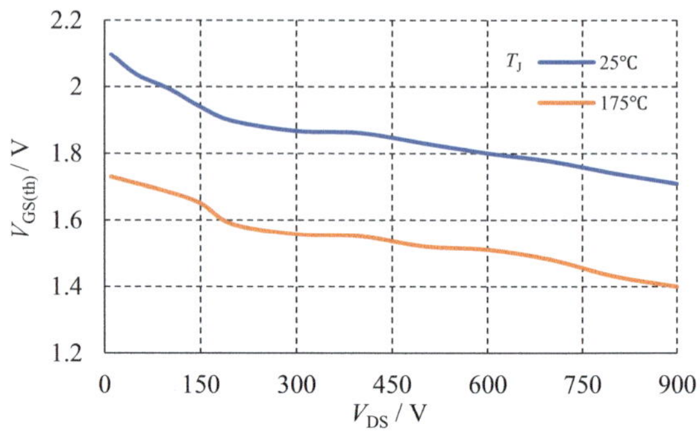

Fig. 3.8 $V_{GS(th)}$–V_{DS} characteristics' curve of SiC MOSFET

Apart from the effects of temperature and Drain-Induced Barrier Lowering (DIBL), SiC MOSFETs also exhibit a threshold voltage hysteresis issue. This means that during the turn-on and turn-off processes, the threshold voltage is not fixed but rather shows a certain degree of historical dependence and nonlinearity. The shape of the I_{DS}–V_{GS} curve resembles a hysteresis loop.

The primary cause of threshold voltage hysteresis in SiC MOSFETs is related to the high interface state density at the SiC/SiO_2 interface. These interface states exchange charges with the semiconductor, leading to threshold voltage drift. Negative gate bias stress increases the number of positively charged oxide traps, resulting in a negative shift in the device's threshold voltage. Conversely, positive gate bias stress causes electrons to be trapped in oxide traps and increases the density of interface traps, leading to a positive shift in the device's threshold voltage.

Threshold voltage hysteresis affects the measurement of $V_{GS(th)}$ and the application of SiC MOSFETs. Different test sequences, test pulse widths, and prestress conditions can lead to deviations in the $V_{GS(th)}$ test results. Additionally, threshold voltage hysteresis impacts the switching characteristics, short-circuit characteristics, gate charge curves, and the risk assessment of false turn-on due to crosstalk in SiC MOSFETs. Many scholars have conducted related research on this topic, so further discussion is not included in this book.

3.2.2 Output Characteristics and On-State Resistance

A set of I_{DS}–V_{DS} curves at different V_{GS} illustrates the output characteristics of the SiC MOSFET, as shown in Fig. 3.9. It can be observed that, at the same V_{DS}, higher V_{GS} results in a larger I_{DS}, consistent with the transfer characteristics. This necessitates that the turn-on drive voltage $V_{GS(on)}$ provided by the driving circuit is being sufficiently high to make full use of chip area and reduce on-state losses.

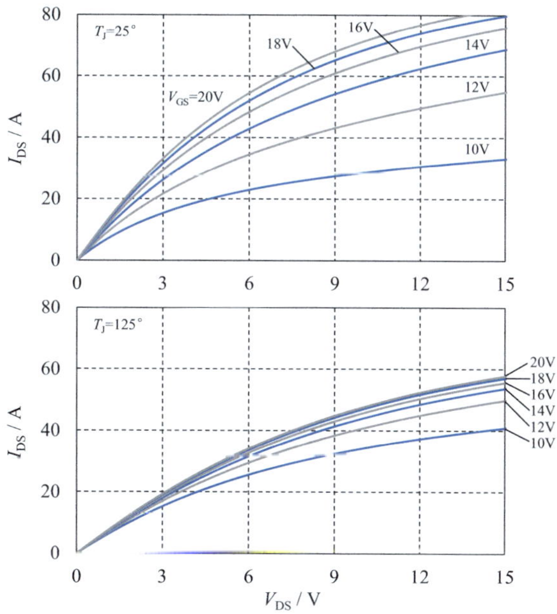

Fig. 3.9 Output characteristic curve of SiC MOSFET

When V_{GS} is less than around 11 V, the I_{DS}–V_{DS} curve at 150°C is higher than junction temperature at 25 °C, and I_{DS} exhibits a negative temperature coefficient. However, when V_{GS} is greater than 11 V, the I_{DS}–V_{DS} curve at $T_J = 150\,°C$ is lower than at $T_J = 25\,°C$, and I_{DS} exhibits a positive temperature coefficient, consistent with the transfer characteristics.

Dividing V_{DS} by I_{DS} gives the on state resistance $R_{DS(on)}$, and the characteristics of $R_{DS(on)}$ can be obtained using the data in Fig. 3.9. At the same V_{GS} and T_J, as I_{DS} increases, $R_{DS(on)}$ also increases. At the same I_{DS} and T_J, higher V_{GS} results in lower $R_{DS(on)}$.

At the same I_{DS}, the temperature characteristics of $R_{DS(on)}$ form a U-shaped curve, as shown in Fig. 3.10. At low temperatures, $R_{DS(on)}$ has a negative temperature coefficient because R_{ch} has a negative temperature coefficient, leading to a decrease in $V_{GS(th)}$ and an increase in channel mobility as temperature rises. On the other hand, R_{JFET} and R_{drift} have positive temperature coefficients, with lattice vibrations becoming more pronounced at higher temperatures, intensifying their hindrance to electrons. At low temperatures, the rate at which R_{ch} decreases is faster than the rate at which R_{JFET} and R_{drift} increase, resulting in an overall decrease in $R_{DS(on)}$. At high temperatures, the rate at which R_{JFET} and R_{drift} increase is faster than the rate at which R_{ch} decreases, leading to an overall increase in $R_{DS(on)}$. Additionally, R_{ch} is influenced by V_{GS}, so the shape of $R_{DS(on)}$ changes with V_{GS}. Smaller V_{GS} results in a larger proportion of R_{ch} in $R_{DS(on)}$, affecting $R_{DS(on)}$ over a wider temperature range. Hence, the trough of $R_{DS(on)}$ corresponds to a higher T_J when V_{GS} is smaller.

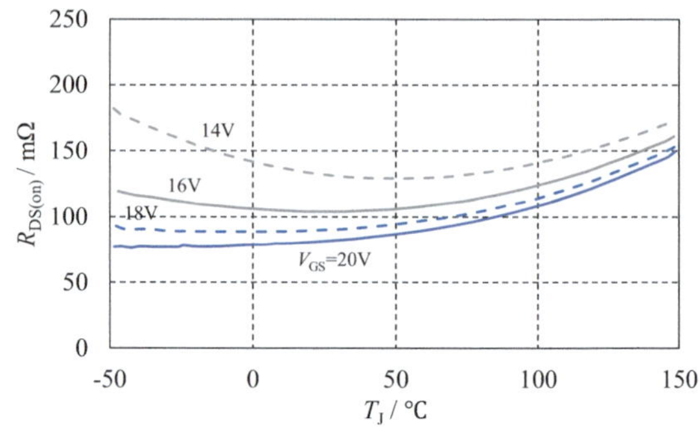

Fig. 3.10 $R_{DS(on)}$–T_J characteristics' curve of SiC MOSFET (I_{DS} = 20 A)

At the same time, R_{ch} is more affected by temperature at lower temperatures, so the decrease in $R_{DS(on)}$ with decreasing T_J is more pronounced.

3.2.3 Body Diode and Third-Quadrant Conduction Characteristics

Similar to Si MOSFETs, SiC MOSFETs also have a body diode, which is a pn-junction diode. When V_{GS} is less than 0 V, the SiC MOSFET is in the off state. Applying a reverse V_{DS} makes the body diode conduct. Figure 3.11 illustrates the conduction characteristics of the body diode, represented by a set of I_{DS}–V_{DS} curves at different V_{GS}. The body diode conducts only when the reverse V_{DS} exceeds a certain voltage, which is the threshold voltage $V_{th(Diode)}$. The more negative V_{GS} is, the larger $V_{th(Diode)}$ $V_{th(Diode)}$ and the higher the conduction voltage drop. Both $V_{th(Diode)}$ and the conduction voltage drop exhibit negative temperature characteristics.

When V_{GS} is higher than $V_{GS(th)}$, applying a reverse V_{DS} to the SiC MOSFET causes it to operate in the third-quadrant conduction state. The conduction characteristics are determined by both the channel and the body diode, as shown in Fig. 3.12. When V_{DS} is less than $V_{th(Diode)}$, the body diode does not conduct, and the current flows entirely through the channel. When V_{DS} is greater than $V_{th(Diode)}$, both the body diode and the channel conduct, and the current is distributed according to the conduction resistance.

From Figs. 3.11 and 3.12, it can be seen that the conduction voltage drop in the third quadrant for SiC MOSFETs is significantly smaller than that of the body diode, especially at low currents. Therefore, synchronous rectification techniques can be employed to operate the SiC MOSFET in the third quadrant, avoiding high losses associated with the conduction voltage drop of the body diode.

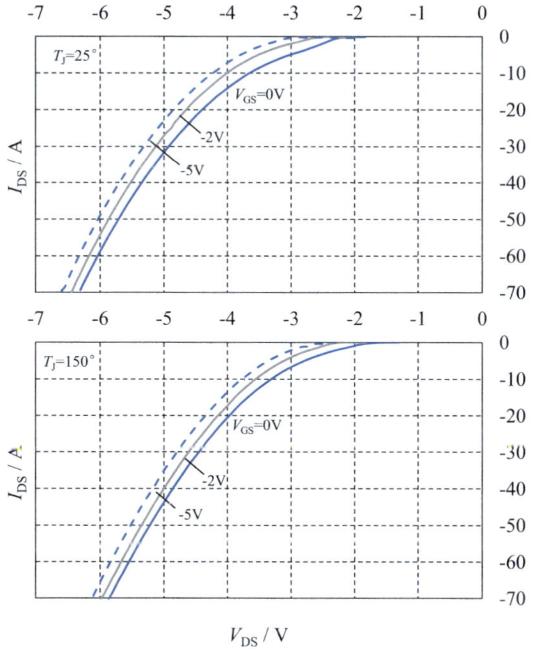

Fig. 3.11 Body diode conduction characteristics curve of SiC MOSFET

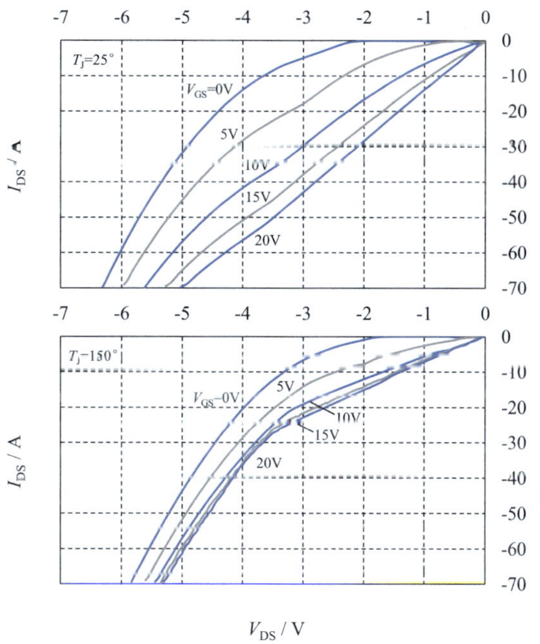

Fig. 3.12 Third-quadrant conduction characteristics curve of SiC MOSFET

3.3 Dynamic Characteristics

3.3.1 *Junction Capacitance*

The structure of the SiC MOSFET is depicted in Fig. 3.13, featuring parasitic capacitances between its gate, drain, and source terminals.

1. **Gate-Source Capacitance C_{GS}**

This capacitance is composed of the channel-gate oxide capacitance $C_{channel}$ and the gate-source parallel-plate capacitance C_{pp} in parallel. C_{pp} is a typical parallel-plate capacitor, unaffected by terminal voltages. $C_{channel}$ is nonlinear and influenced by V_{DS} and channel length. As the depletion region expands with increasing V_{DS}, $C_{channel}$ decreases, but this change is minimal.

2. **Gate-Drain Capacitance C_{GD}**

It consists of the static capacitance of the oxide layer $C_{field\text{-}oxide}$ and the capacitance related to MOS interface depletion $C_{depletion}$ in series. $C_{field\text{-}oxide}$ is constant, while $C_{depletion}$ decreases with rising V_{DS}.

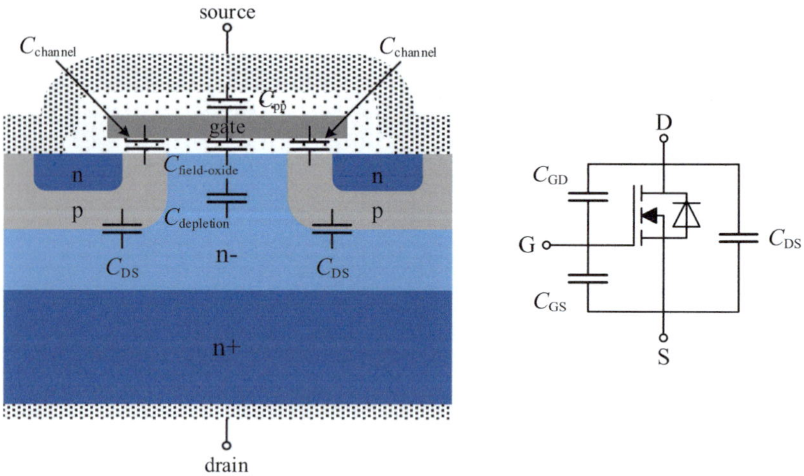

(a) SiC MOSFET structure (b) Equivalent circuit

Fig. 3.13 Junction capacitance of SiC MOSFET

3. 3. Drain-Source Capacitance C_{DS}

It is mainly related to the capacitance of the drain-source pn-junction depletion layer, which decreases with increasing V_{DS}.

The datasheet provides junction capacitance as input capacitance C_{iss}, output capacitance C_{oss}, and reverse output capacitance C_{rss} (also known as the Miller capacitance), defined by Eqs. (3.8), (3.9), and (3.10), respectively.

$$C_{iss} = C_{GS} + C_{GD}, \tag{3.8}$$

$$C_{oss} = C_{DS} + C_{GD}, \tag{3.9}$$

$$C_{rss} = C_{GD}. \tag{3.10}$$

As C_{iss}, C_{oss}, and C_{rss} all contain components affected by V_{DS}, they exhibit nonlinear capacitance characteristics, as shown in Fig. 3.14. When V_{DS} is low, the junction capacitance decreases with increasing V_{DS}; especially C_{oss} and C_{rss} show a

Fig. 3.14 C–V characteristics curve of SiC MOSFET

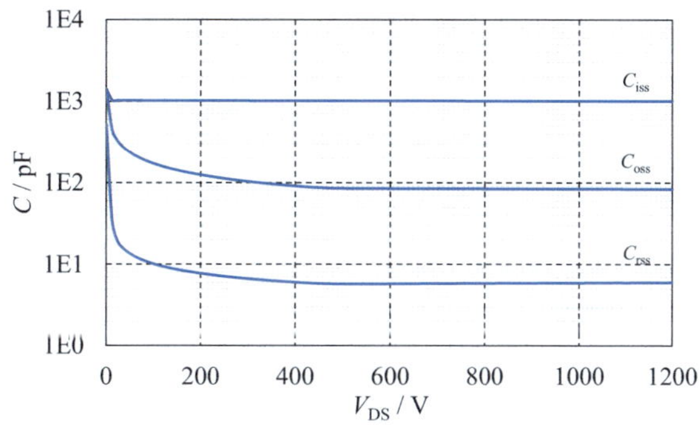

(a) The horizontal axis is a linear axis

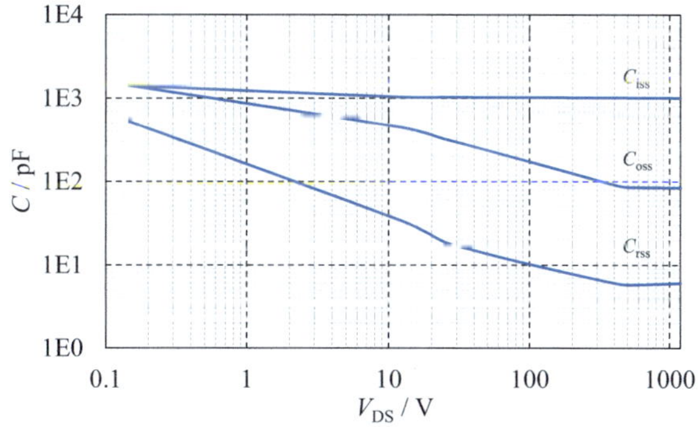

(b) The horizontal axis is a logarithmic axis

more noticeable reduction. As V_{DS} becomes high, the junction capacitance remains relatively constant, with the depletion region showing little change beyond a certain V_{DS}. Simultaneously, $C_{GS} \cdot C_{GD}$, and C_{iss} is dominated by C_{DS}, while $C_{DS} \cdot C_{GD}$, and C_{oss} is dominated by $C_{DS}.C_{DS}$.

It should be noted that the junction capacitance data and C–V characteristic curves provided in datasheets are measured at $V_{GS} = 0$ V, which means that they reflect the C–V characteristics of the SiC MOSFET when it is in the off-state with a gate drive voltage of 0 V. However, when the SiC MOSFET is in the off-state, both C_{GS} and C_{GD} are influenced by V_{GS}. The C–V characteristic curves provided in datasheets do not reflect this feature. Furthermore, these C–V characteristic curves cannot represent the C–V characteristics of the SiC MOSFET when it is in the on-state, which are also influenced by V_{GS} and can be obtained using S-parameter measurements.

During the switching process, the switching state of the SiC MOSFET and V_{GS} both change. This indicates that using only the C-V curves provided in the datasheet cannot accurately describe the switching process of the SiC MOSFET. There have been some research findings on this issue [1, 1].

3.3.2 Switching Characteristics

In most power converters, the switching and freewheeling processes of the SiC MOSFET can be described based on the inductive load circuit, as shown in Fig. 3.15. L is the load inductance, C_{Bus} is the bus capacitance, R_G is the driving resistor, L_{Loop} is the main power loop inductance, L_{DRV} is the driving loop inductance, Q_H and Q_L are SiC MOSFETs housed in a TO-247-4PIN package, and D_H and D_L are the body diodes of Q_H and Q_L, respectively. To simplify the analysis of the switching process, L_{DRV} is ignored. However, to obtain more meaningful waveforms, L_{DRV} is considered in simulations.

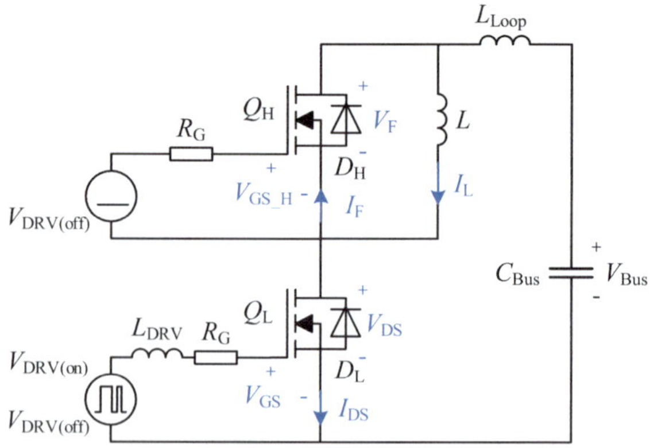

Fig. 3.15 Inductive load circuit

3.3.2.1 Switching Process and Body Diode Reverse Recovery Process

The voltage and current definitions during the turn-on process of the SiC MOSFET are shown in Fig. 3.16. The simulated waveforms of the SiC MOSFET turn-on process and the body diode reverse recovery process are illustrated in Fig. 3.17.

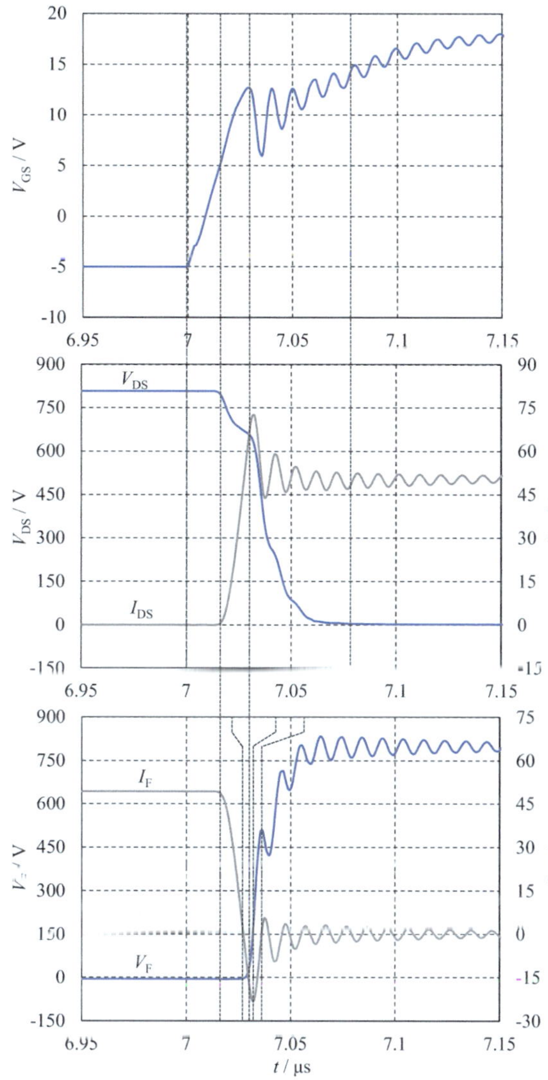

Fig. 3.16 SiC MOSFET turn-on process waveform (simulation waveform)

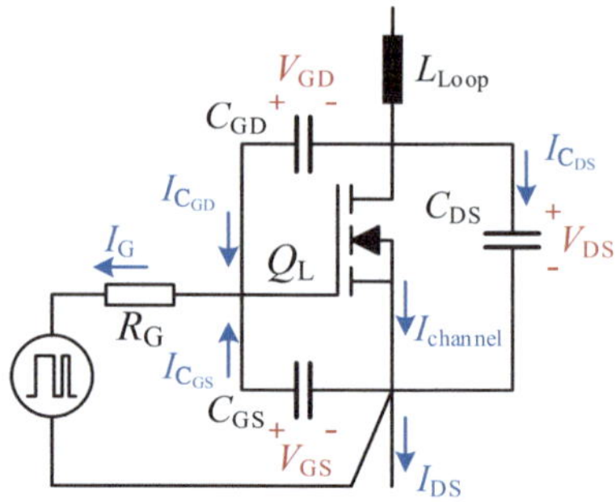

Fig. 3.17 Definition of voltage and current during SiC MOSFET turn-on process

(1) **–t_0**

The V_{GS} of Q_L is the off-state driving voltage $V_{DRV(off)}$, and it is in the off state. I_{DS} is zero, and V_{DS} withstands the bus voltage V_{Bus}. The V_{GS_H} of Q_H is $V_{DRV(off)}$, and the load current I_L continues flowing through D_H and L, resulting in a voltage drop V_F.

(2) **t_0–t_1**

At time t_0, V_{GS} rapidly changes from $V_{DRV(off)}$ to the on-state driving voltage $V_{DRV(off)}$ The driving circuit charges C_{iss} through the driving current I_G, which can be divided into charging C_{GS} with $I_{C_{GS}}$ and charging C_{GD} with $I_{C_{GD}}$.

$$V_{DRV(on)} = V_{GS} + I_G \cdot R_G \tag{3.11}$$

$$I_G = I_{C_{GS}} + I_{C_{GD}} = C_{GS} \cdot dV_{GS}/dt + C_{GD} \cdot dV_{GD}/dt \tag{3.12}$$

Since the SiC MOSFET is still in the off state at this time and V_{DS} remains at V_{Bus}, C_{GS} and C_{GD} are constant. Additionally, dV_{GS}/dt. and dV_{GD}/dt are equal. Combining Eqs. (3.11)and (3.12):

$$V_{DRV(on)} = V_{GS} + C_{iss}.dV_{GS}\big/ dt.R_G \tag{3.13}$$

hence,

$$V_{GS}(t) = V_{DRV(on)} - (V_{DRV(on)} - V_{DRV(off)})e^{-\frac{t-t_0}{R_G \cdot C_{iss}}} \tag{3.14}$$

At t_1, V_{GS} reaches the threshold voltage $V_{GS(th)}$ of the SiC MOSFET. The time duration $t_0 - t_1$ is called the turn-on delay $t_{d(on)}$.

(3) **t_1–t_2**

V_{GS} exceeds $V_{GS(th)}$, and the SiC MOSFET starts conducting. Due to being in the saturation state, $I_{channel}$ is controlled by V_{GS} and can be obtained from Eq. (3.15), where K is a coefficient determined by the device semiconductor parameters.

$$I_{\text{channel}} = K\left(V_{\text{GS}} - V_{\text{GS(th)}}\right)^2 \tag{3.15}$$

It's important to note that I_{channel}. is composed of three parts: I_{DS}; $I_{C_{\text{GD}}}$, and $I_{C_{\text{DS}}}$. Since V_{DS}. is still relatively high at this point, C_{GS} and C_{GD} are small. Simultaneously, $\text{d}V_{\text{DS}}/\text{d}t$ is low. Therefore, in I_{channel}, I_{DS} predominates.

$$I_{\text{DS}} \approx I_{\text{channel}}. \tag{3.16}$$

As I_{L} continuously flows to Q_{L}, I_{DS} rapidly increases. $dI_{\text{DS}}/\text{d}t$ generates a voltage drop on the main power loop inductance L_{Loop}, causing V_{DS} to decrease.

$$\Delta V_{\text{DS}} = L_{\text{Loop}} \cdot dI_{\text{DS}}/\text{d}t. \tag{3.17}$$

Due to $C_{\text{GD}} \ll C_{\text{GS}}$, the impact of ΔV_{DS} on V_{GS} is minimal, and V_{GS} still roughly follows Eq. (3.13).

$$V_{\text{GS}}(t) = (V_{\text{DRV(on)}} - V_{\text{GS(th)}})\left(1 - \text{e}^{-\frac{t-t_1}{\tau}}\right) + V_{\text{GS(th)}}. \tag{3.18}$$

Combining Eqs. (3.15), (3.16), (3.17), and (3.18), it can be observed that $\text{d}I_{\text{DS}}/\text{d}t$ becomes larger, and ΔV_{DS} increases accordingly.

When I_{DS} exceeds the load current, D_{H} undergoes reverse recovery, causing an overshoot in I_{DS}, and V_{GS} continues to rise.

(4) t_2–t_3

At t_2, I_{DS} reaches its peak. D_{H} begins to turn off, and V_{DS} rapidly decreases. I_{channel} still consists of I_{DS}, $I_{C_{\text{GD}}}$, and $I_{C_{\text{DS}}}$. Because $\text{d}V_{\text{DS}}/\text{d}t$ is high, and when V_{DS} is low, C_{GS} and C_{GD} significantly increase, causing $I_{C_{\text{GD}}}$ and $I_{C_{\text{DS}}}$ to approach or even exceed I_{DS}. At this time, the test I_{DS} is obviously smaller than the actual I_{channel}.

$$I_{\text{channel}} = I_{\text{DS}} + C_{\text{GD}} \cdot \text{d}V_{\text{GD}}/\text{d}t + C_{\text{DS}} \cdot |\text{d}V_{\text{DS}}/\text{d}t|. \tag{3.19}$$

Due to the substantial DIBL of the SiC MOSFET, $V_{\text{GS(th)}}$ increases as V_{DS} decreases. At the same time, due to the relatively low channel mobility of the SiC MOSFET, according to (3.15), to ensure I_{channel} flows, C_{GS} needs to be charged to raise V_{GS}. This is different from Si SJ-MOSFET, where V_{GS} remains constant at this stage, and I_{G} only charges C_{GD}, known as the Miller plateau. Due to the slow rise of V_{GS} in the SiC MOSFET, it is called the Miller ramp.

Since $|\text{d}V_{\text{DS}}/\text{d}t| \gg |\text{d}V_{\text{GS}}/\text{d}t|$, then

$$I_{C_{\text{GD}}} = C_{\text{GD}} \cdot \text{d}V_{\text{GD}}/\text{d}t = C_{\text{GD}} \cdot d(V_{\text{GS}} - V_{\text{DS}})/\text{d}t \approx C_{\text{GD}} \cdot |\text{d}V_{\text{DS}}/\text{d}t|. \tag{3.20}$$

It is indicated that dV_{DS}/dt is controlled by the charging speed of C_{GD} due to $I_{C_{GD}}$. When V_{DS} is high, C_{GD} remains almost constant, while when V_{DS} is low, C_{GD} significantly and rapidly increases with decreasing V_{DS}, especially when $V_{GD} < 0$ V. This leads to dV_{DS}/dt being almost constant at high V_{DS} and significantly slowing down at low V_{DS}.

Simultaneously, I_{DS} undergoes decay oscillation, generating corresponding fluctuations in V_{DS}, which, through the coupling effect of C_{GD}, causes oscillations in V_{GS}. The frequencies of the three oscillations are the same.

(5) t_3–

At t_3, V_{DS} drops to the conduction voltage drop $V_{DS(on)}$, marking the end of the switching process. V_{GS} reaches $V_{DRV(on)}$ under the action of the driving circuit.

During the turn-on process of Q_L, when D_H witches from forward conduction to reverse blocking, because it needs to reset carriers to restore the space charge region, the conduction current of the diode will first drop to zero. Then, reverse current will occur, attenuating to zero again. This is the reverse recovery process of the SiC MOSFET, detailed as follows:

(1) –t_1

D_H is forward-conducting, and the conduction current I_F is I_L. The terminal voltage V_F is the conduction voltage drop of D_H.

(2) $t_1 - t_1'$

At t_1, Q_L turns on, and I_F decreases to zero from I_L at the rate of di/dt. V_F increases with the decreasing I_F, and at this point, there are still a large number of carriers in the diode.

(3) $t_1' - t_2$

I_F becomes reverse and increases, playing a role in sweeping out charges. The carrier concentration starts to decrease. Since the carrier concentration is still high, the diode remains in the conducting state.

(4) $t_2 - t_3'$

The carrier concentration continues to decrease, and the diode is not enough to maintain the conducting state, starting to withstand reverse voltage. The rate of change of I_F gradually decreases.

(5) t_3'

I_F reaches the reverse peak I_{rrm}.

(6) $t'_3 - t'_4$.

As the carrier concentration is not sufficient to sustain the reverse current, I_F gradually decreases from I_{rrm} to zero. During this process, V_F oscillates.

The duration from $t_1 - t'_3$ is t_d, and the duration from $t'_3 - t'_4$ is t_f. The reverse recovery time is t_{rr}, and the reverse recovery charge is Q_{rr}.

$$t_{rr} = t_d + t_f, \tag{3.21}$$

$$Q_{rr} = \int_{t_1}^{t'_4} |I_F| dt. \tag{3.22}$$

3.3.2.2 Turn-Off Process

The simulation waveforms of the SiC MOSFET turn-off process, along with the voltage and current definitions during turn-off are shown in Figs. 3.18 and 3.19, respectively.

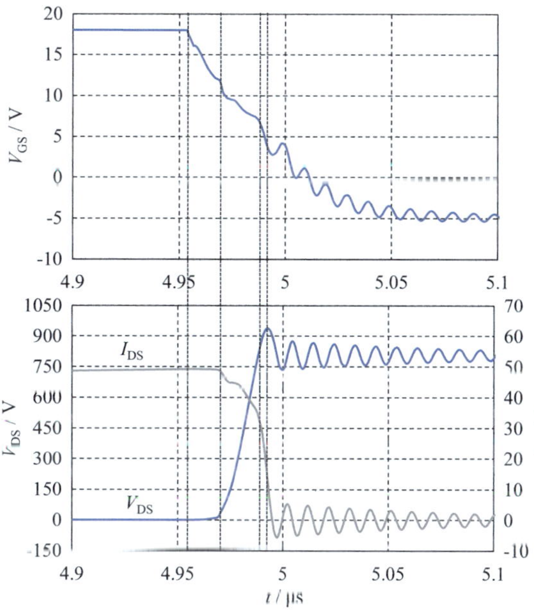

Fig. 3.18 SiC MOSFET turn-off process waveforms (simulation waveforms)

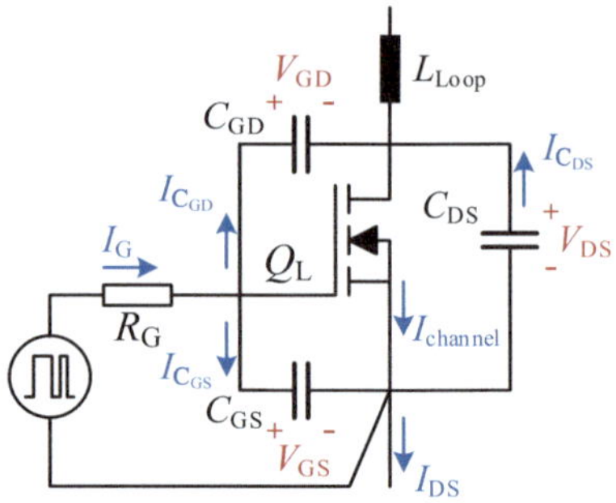

Fig. 3.19 SiC MOSFET turn-off process voltage-current definitions

(1) **$-t_0$**

V_{GS} of Q_L is $V_{DRV(on)}$, indicating that the device is in the conducting state. I_{DS} is the load current I_L and V_{DS} is the conduction voltage drop $V_{DS(on)}$. D_H is in the off state, withstanding the bus voltage V_{Bus}.

(2) **t_0–t_1**

At t_0, V_{GS} quickly transitions from $V_{DRV(on)}$ to $V_{DRV(off)}$. The drive circuit discharges C_{iss} through the drive current I_G, which can be further divided into discharging C_{GS} through $I_{C_{GS}}$ and discharging C_{GD} through $I_{C_{GD}}$.

$$V_{DRV(off)} = V_{GS} - I_G \cdot R_G. \tag{3.23}$$

$$I_G = I_{C_{GS}} + I_{C_{GD}} = C_{GS} \cdot \left| \mathrm{d}V_{GS} / \mathrm{d}t \right| + C_{GD} \cdot \left| \mathrm{d}V_{GD} / \mathrm{d}t \right|. \tag{3.24}$$

Since Q_L is still in the conducting state, I_{DS} remains unchanged. The decrease in V_{GS} leads to an increase in $R_{DS(on)}$ causing a slight increase in $V_{DS(on)}$. As $V_{DS(on)}$ changes very little, C_{GS} and C_{GD} remain constant, and $dv_{GS}/\mathrm{d}t$ and $dv_{GD}/\mathrm{d}t$ are equal.

$$V_{GS}(t) = V_{DRV(off)} - \left(V_{DRV(off)} - V_{DRV(on)}\right) \mathrm{e}^{-\frac{t-t_0}{R'_G C_{iss}}}. \tag{3.25}$$

Before t_1, there was no obvious change in V_{DS} and I_{DS}. Therefore, $t_0 - t_1$ is referred to as the turn-on delay $t_{d(on)}$.

It is important to note that although the forms of (3.25) and (3.14) are similar, C_{GD} in the $t_{d(off)}$ stage is significantly larger than C_{GD} in the $t_{d(on)}$ stage. Additionally, the rate of decrease in V_{GS} in the $t_{d(off)}$ stage is often close to or greater than the rate of increase in V_{GS} in the $t_{d(on)}$ stage. Therefore, $t_{d(off)}$ is longer than $t_{d(on)}$.

(3) **t_1–t_2**

At t_1, Q_L enters saturation, and V_{DS} begins to rise. The voltage at the D_H terminal decreases, and a portion of I_L switches to D_H, causing the discharge of its junction capacitance. Similar to the $t_2 - t_3$ stage in the turn-on process, the rising speed of V_{DS} at this point is also controlled by $I_{C_{GD}}$. As V_{DS} increases, C_{GD} significantly decreases, so dV_{DS}/dt keeps increasing, causing I_{DS} to decrease. During this period, $I_{channel}$ follows (3.15). Due to the joint effect of the decrease in I_{DS} and the DIBL effect, V_{GS} slowly decreases, forming the Miller ramp.

It is noteworthy that I_{DS} is composed of 由 $I_{channel}$, $I_{C_{GD}}$ and $I_{C_{DS}}$. As V_{DS} rises, $I_{C_{GD}}$ discharges C_{GD}, and $I_{C_{DS}}$ charges C_{DS}. At this point, the measured I_{DS} is greater than the actual $I_{channel}$.

$$I_{channel} = I_{DS} - C_{GD} \cdot \left| dV_{GD}/dt \right| - C_{DS} \cdot dV_{DS}/dt. \tag{3.26}$$

(4) t_2–t_3

At t_2, V_{DS} reaches V_{Bus}, and D_H begins to conduct. I_L quickly flows to D_H, I_{DS} rapidly decreases, still composed of $I_{channel}$, $I_{C_{GD}}$ and $I_{C_{DS}}$. follows Eq. (3.15). At t_3, V_{GS} decreases to $V_{GS(th)}$, and $I_{channel}$. decreases to zero. Similar to the t_1-t_2 stage in the turn-on process, the speed of I_{DS} decrease at this point is also controlled by $I_{C_{GS}}$.

Simultaneously, based on Eq. (3.17), the rapid decrease in I_{DS} generates a voltage drop on L_{Loop}, causing an overshoot in V_{DS}.

(5) t_3–

Q_L is completely turned off, and V_{GS} reaches $V_{DRV(off)}$ under the action of the drive circuit. I_{DS} decays oscillation to zero, producing corresponding fluctuations in V_{DS}. Through the coupling effect of C_{GD}, V_{GS} undergoes oscillation, and they exhibit the same oscillation frequency.

3.3.2.3 Switching Energy

Switching energy for SiC MOSFET is generally considered to be generated by the overlap of I_{DS} and V_{DS} during the switching process, calculated by integrating their product. However, only the channel current $I_{channel}$ contributes to losses during the switching process. As explained earlier in the switching process, I_{DS} and $I_{channel}$ are not equal and are given by Eqs. (3.27) and (3.28) during the turn-on and turn off processes, respectively, where $I_{C_{oss}}$ represents the charging and discharging current of C_{oss}.

$$I_{channel} = I_{DS} + I_{C_{oss}}, \tag{3.27}$$

$$I_{channel} = I_{DS} - I_{C_{oss}}. \tag{3.28}$$

This implies that calculating switching energy using I_{DS} would underestimate the turn-on energy E_{on} while overestimating the turn-off energy E_{off}. The deviation current $I_{C_{oss}}$ of $I_{channel}$ and I_{DS} accounts for the charging and discharging of C_{oss}, corresponding to the energy E_{oss}, introducing a correction factor to the switching energy calculation [1]. The switching energy can be calculated using the following formula:

$$E_{oss}(V_{Bus}) = \int_{0}^{V_{Bus}} C_{oss}(V) \cdot V_{DS}(V)\mathrm{d}V, \tag{3.29}$$

$$E_{on} = \int_{t_{E_{on},\,start}}^{t_{E_{on},\,end}} I_{DS}(t) \cdot V_{DS}(t)\mathrm{d}t + E_{oss}(V_{Bus}), \tag{3.30}$$

$$E_{off} = \int_{t_{E_{off},\,start}}^{t_{E_{off},\,end}} I_{DS}(t) \cdot V_{DS}(t)\mathrm{d}t + E_{oss}(V_{Bus}), \tag{3.31}$$

where $t_{E_{on},\,Start}$ and $t_{E_{on},\,end}$ $t_{E_{off},\,Start}$ and $t_{E_{off},\,end}$ are the integration starting points for turn-on and turn-off energy, and the starting points used by different manufacturers are not standardized.

The integration boundaries for E_{on} are 0.1 V_{GS}~$0.03V_{DS}$ and $0.1I_{DS}$~0.1 V_{DS}, while for $E_{off,}$ the boundaries are 0.9 V_{GS}~0.01 I_{DS} and 0.1 V_{DS}~0.1 I_{DS}. The main difference between the two lies in whether the energy during turn-on delay and turn-off delay periods is considered as part of the switching losses. Since the energy during these stages is minimal, the results of the two boundary calculations are very close. It's also important to note that the switching energy values provided in manufacturer datasheets are not corrected using E_{oss}.

3.3.3 Gate Charge

When performing switch control on SiC MOSFETs, the driving circuit needs to charge and discharge their C_{iss}, and the charge amount is the gate charge Q_G. This can be obtained by integrating the gate drive current I_G during the switching process. Q_G represents the demand for driving energy, and the smaller Q_G is, the lower the driving losses.

To improve testing accuracy and simplify integration calculations, a constant current source I_{con} is typically used to drive the tested transistor. Since it is a constant current source drive, multiplying the time axis t of the V_{GS} waveform by I_{con} gives the $V_{GS} - Q_G$ curve. The gate charge Q_G test waveform and the corresponding $V_{GS} - Q_G$ curve for SiC MOSFET with constant current source drive are shown in Fig. 3.20.

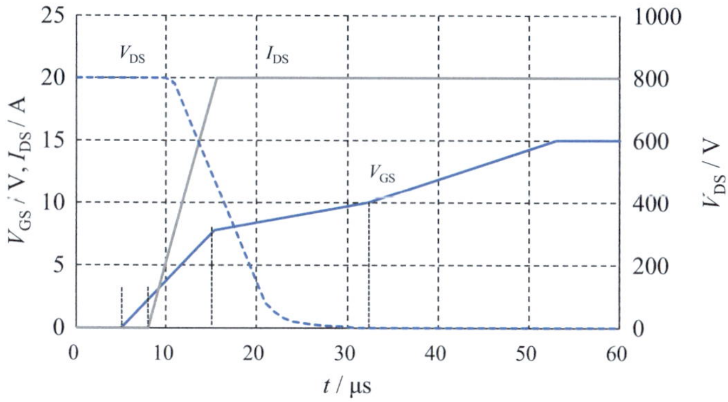

Fig. 3.20 Gate charge test waveform for SiC MOSFET with constant current source drive

(1) t_0–t_1

I_G charges C_{iss} with constant current, C_{iss} remains constant as V_{DS} does not change, and V_{GS} linearly rises from 0 V to $V_{GS(th)}$.

(2) t_1–t_2

V_{GS} exceeds $V_{GS(th)}$, I_{DS} increases from 0A, V_{DS} decreases, C_{iss} remains constant, and V_{GS} maintains a linear rise.

(3) t_2–t_3

When I_{DS} reaches I_{set}, V_{DS} decreases to a low value, due to the DIBL effect and the relatively low channel mobility, SiC MOSFET enters the Miller plateau, and V_{GS} rises slowly. Therefore, in this stage, I_{GS} not only needs to charge C_{rss} but also needs to charge C_{GS} simultaneously.

(4) t_3–

SiC MOSFET is in the on-state, V_{GS} linearly rises to $V_{GS(on)}$. The slope of V_{GS} rise in this stage is smaller than that in the $t_0 - t_2$ stage, due to the significant increase in C_{GS} and C_{GD} with the rise of V_{GD} in the on-state.

3.4 Limiting Characteristics

3.4.1 Short Circuit

Power converters operate in complex environments and conditions, and various situations such as sampling or drive signal interference, operation beyond normal conditions, and component failures can cause power switches to experience short circuits and overcurrent conditions. When a short circuit or overcurrent occurs, the power switch will conduct a current far exceeding normal conditions and will endure high voltage, generating a tremendous amount of heat in a short period. This can potentially lead to device failure and, subsequently, cause the power converter to malfunction or be damaged.

The primary application of SiC MOSFETs is to replace IGBTs in high-voltage, high-power scenarios. In these applications, the short circuit and overcurrent capabilities of IGBTs are of significant concern. Since the chip area of SiC MOSFETs is smaller compared to Si IGBTs, the current density is higher, and during a short circuit, the heat is more concentrated. This thermal aspect results in SiC MOSFETs having weaker short-circuit capabilities. Therefore, more attention needs to be paid to the short circuit and overcurrent capabilities of SiC MOSFETs.

Most applications of SiC MOSFETs use bridge topologies; thus, the short-circuit characteristics are introduced using a half-bridge topology as an example, as shown in Fig. 3.21. When the upper and lower bridge arm SiC MOSFETs form a loop with the bus capacitor and a short circuit occurs, it is referred to as bridge arm shoot-through. This can be further divided into hard-switching fault and fault under load depending on the operating conditions. When a lower bridge arm device, a faulty load, and the bus capacitor form a loop and a short circuit occurs, it is referred to as a load fault short circuit. Moreover, a single-switch short-circuit fault occurs when the high-side switch is shorted through wiring or when a hard-switching failure takes place in a single-switch topology.

1. **Hard-Switching Fault Short Circuit**

A hard-switching fault short circuit occurs when the device is in the off-state before the short circuit happens and immediately experiences a short circuit when it is turned on by the drive signal. The timing sequence is shown in Fig. 3.21. Initially, the V_{GS_H} of Q_H is high, and the V_{GS_L} of Q_L is low. Subsequently, V_{GS_L} becomes high, forming a loop with Q_L, Q_H, and the bus capacitor C_{Bus}, resulting in a short circuit. The waveform of a SiC MOSFET hard-switching fault short circuit is illustrated in Fig. 3.22.

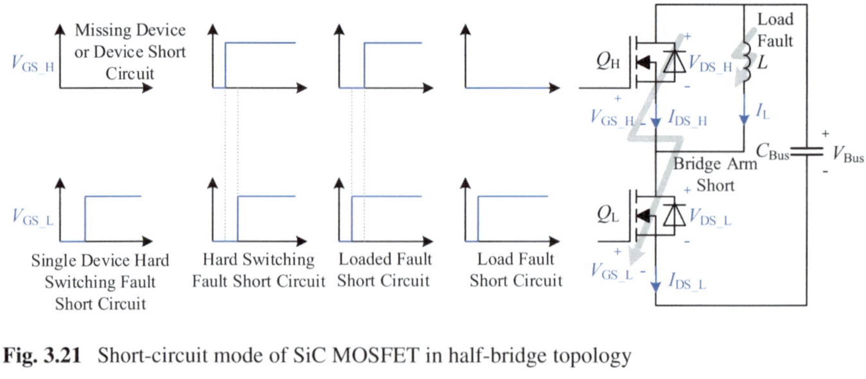

Fig. 3.21 Short-circuit mode of SiC MOSFET in half-bridge topology

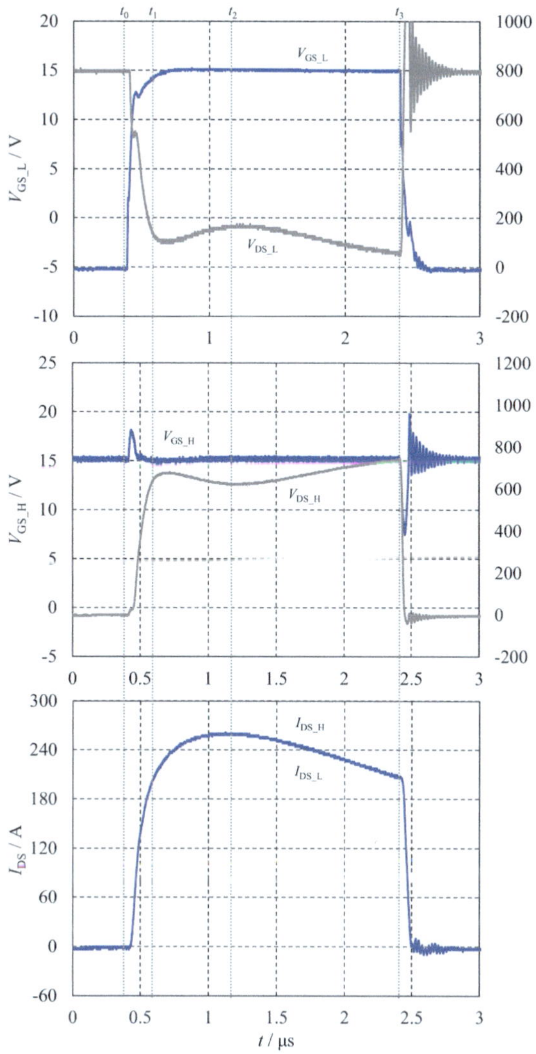

Fig. 3.22 Hard-switching failure short-circuit waveform of SiC MOSFET at 25 °C

(1) $-t_0$

V_{GS_L} of Q_L is at the off-state drive voltage $V_{DRV(off)}$, keeping Q_L in the off-state. I_{DS_L} is zero, and V_{DS_L} withstands the bus voltage V_{Bus}. V_{GS_H} of Q_H is at the on-state drive voltage $V_{DRV(on)}$, keeping Q_H in the on-state, with V_{DS_H} being zero.

(2) t_0–t_1

At t_0 V_{GS_L} quickly changes from $V_{DRV(off)}$ to $V_{DRV(on)}$. Q_L turns on, forming a short-circuit loop with Q_H and C_{Bus}. I_{DS_L} rises rapidly, V_{DS_L} drops quickly, and V_{DS_H}

increases rapidly. The rapid increase in V_{DS_H} creates an upward spike in V_{GS_H} through the Miller capacitance.

(3) t_1–t_2

At t_1, I_{DS_L} reaches its peak and then gradually declines. This is due to the energy from the high voltage and large current increasing the junction temperature of the SiC MOSFET, reducing its current-carrying capacity. V_{DS_L} continues to slowly decrease, approaching zero but still significantly higher than the normal on-state voltage drop. V_{DS_H} continues to slowly decrease, approaching V_{Bus}.

(4) t_2–

At t_2, V_{GS_L} quickly changes from $V_{DRV(on)}$ to $V_{DRV(off)}$. Q_L turns off, rapidly drops to zero, and V_{DS_L} rises quickly, creating an off-state voltage spike and oscillation. V_{DS_H} drops rapidly and oscillates. The rapid decrease in V_{DS_H} creates a downward spike in VGS_H through the Miller capacitance, leading to oscillations.

The waveform of the SiC MOSFET hard-switching fault short circuit at 125 °C is shown in Fig. 3.23. Compared to the waveform at 25 °C in Fig. 3.22, the short-circuit current and the off-state voltage spike of the SiC MOSFET are lower at higher temperatures. Additionally, V_{DS_L} during the short-circuit period is higher at 125 °C.

2. **Single-Switch Hard-Switching Fault Short Circuit**

When the upper switch is bypassed by a wire or a hard-switching fault short circuit occurs in a single-switch topology, only one SiC MOSFET and the bus capacitor form a short-circuit loop. The short-circuit waveform is shown in Fig. 3.24.

(1) –t_0

V_{GS_L} of Q_L is at $V_{DRV(off)}$, keeping it in the off-state. I_{DS} is zero, and V_{DS} withstands the bus voltage V_{Bus}.

(2) t_0–t_1

At t_0, V_{GS_L} quickly changes from $V_{DRV(off)}$ to the on-state drive voltage $V_{DRV(on)}$. Q_L turns on, forming a short-circuit loop with the bus capacitor. I_{DS} rapidly rises from zero, causing a voltage drop across the main power loop inductance, which results in a decrease in V_{DS}. As the rate of rise of I_{DS} slows down, V_{DS} starts to return to V_{Bus}.

(3) t_1–t_2

At t_1, I_{DS} reaches its peak and then gradually decreases. This is due to the energy generated by the high voltage and large current increasing the junction temperature of the SiC MOSFET, which reduces its current-carrying capability. The rate of change of I_{DS} is low, and V_{DS} remains close to V_{Bus}.

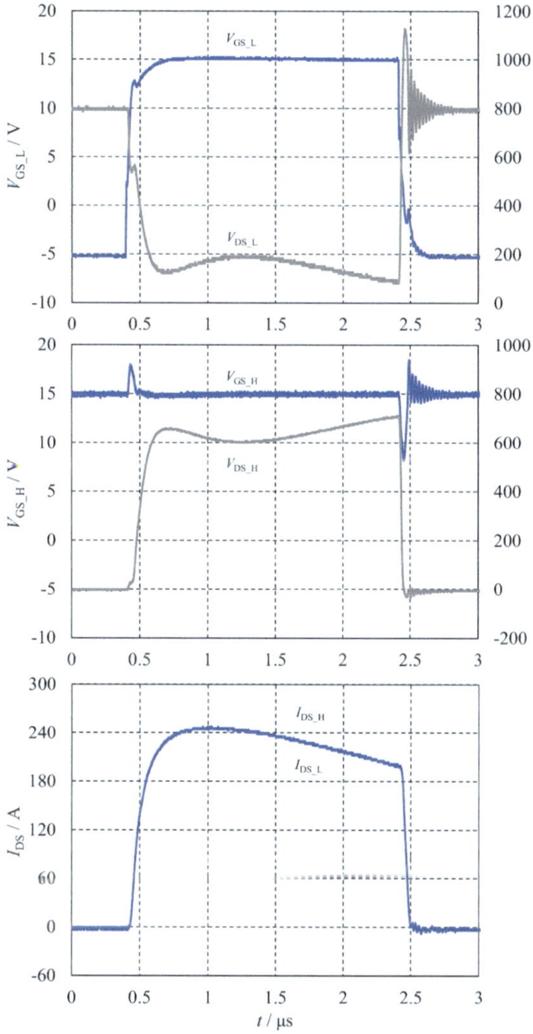

Fig. 3.23 Hard-switching failure short-circuit waveform of SiC MOSFET at 125 °C

(4) t_2–

At t_2, V_{GS_L} quickly changes from $V_{DRV(on)}$. to $V_{DRV(off)}$. Q_L turns off, causing I_{DS} to rapidly drop to zero and generating an off-state voltage spike and oscillation in V_{DS}.

The waveform of a single SiC MOSFET hard-switching fault short circuit at 125 °C is shown in Fig. 3.25. Compared to the waveform at 25 °C in Fig. 3.26, both the short circuit current and the off-state voltage spike of the SiC MOSFET are lower at higher temperatures.

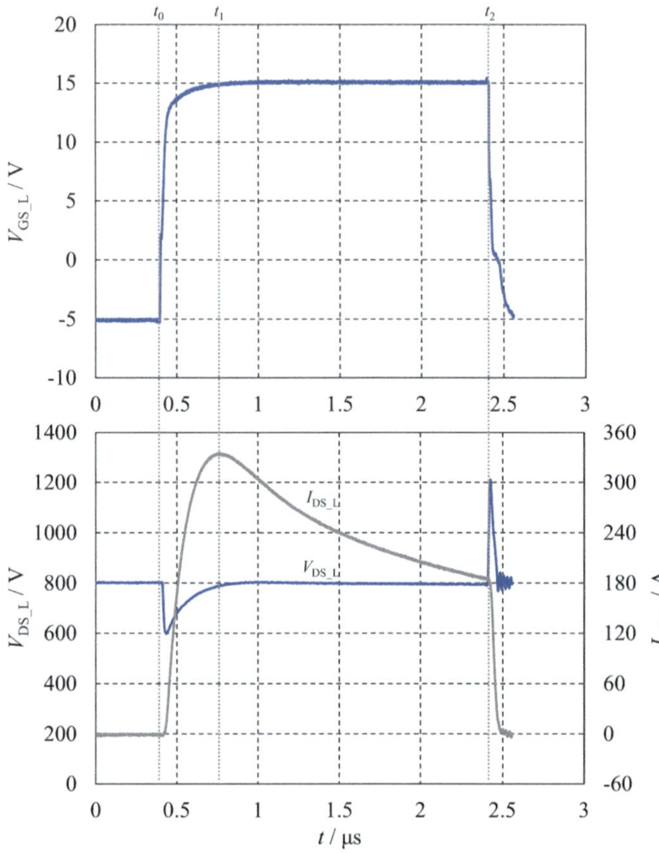

Fig. 3.24 Hard-switching failure short-circuit waveform of single SiC MOSFET at 25 °C

3. **Fault under Load Short Circuit**

A fault under load short circuit occurs when the device is already in the on-state and has load current flowing through it before the short circuit happens. The timing sequence is shown in Fig. 3.21. Initially, V_{GS_H} of Q_H is low, and V_{GS_L} of Q_L is high, allowing the load current to flow through Q_L. Subsequently, V_{GS_H} becomes high, forming a short-circuit loop with Q_L, Q_H, and the bus capacitor. The waveform of a SiC MOSFET fault under load short circuit is illustrated in Fig. 3.26.

(1) **$-t_0$**

V_{GS_L} of Q_L is at the off-state drive voltage $V_{DRV(off)}$, keeping it in the off-state. I_{DS_L} is zero, and V_{DS_L} withstands the bus voltage V_{Bus}. V_{GS_H} of Q_H is also at the off-state drive voltage $V_{DRV(off)}$ keeping it in the off-state. V_{DS_H} is zero.

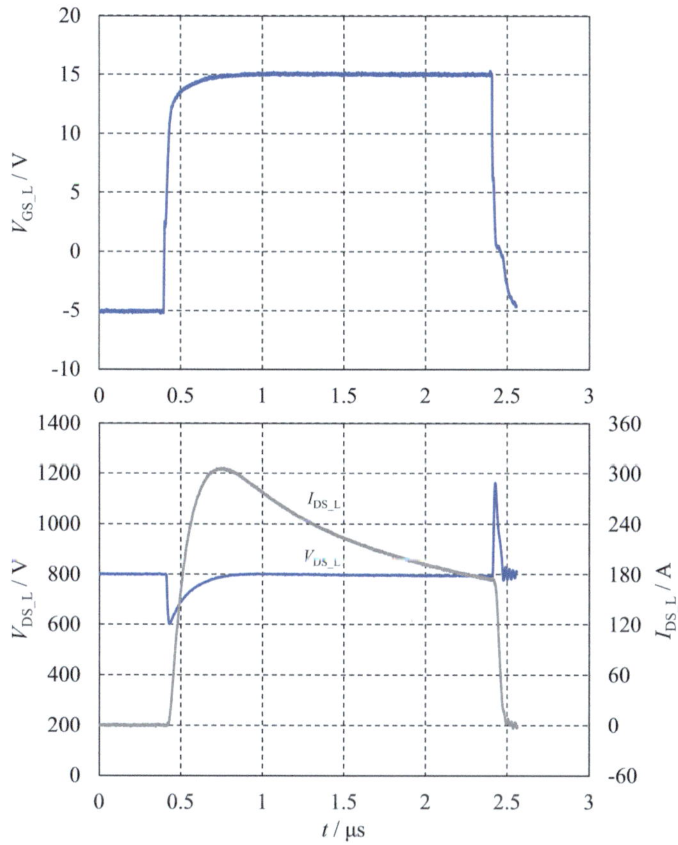

Fig. 3.25 Hard-switching failure short-circuit waveform of single SiC MOSFET at 125 °C

(2) t_0–t_1

At t_0, V_{GS_L} quickly changes from $V_{DRV(off)}$ to the on-state drive voltage $V_{DRV(on)}$. Q_L turns on, forming a loop with the load inductance L and C_{Bus}. V_{DS_L} quickly drops to the on state voltage, and V_{DS_H} rapidly drops to V_{Bus} with oscillations. I_{DS_L} linearly increases.

(3) t_1–t_2

At t_1, V_{GS_H} quickly changes from $V_{DRV(off)}$ to VDRV(on). Q_H turns on, forming a short-circuit loop with Q_L and C_{Bus}. I_{DS_L} rapidly increases, V_{DS_L} quickly rises, and V_{DS_H} rapidly drops. The rapid rise in V_{DS_L} creates an upward spike in V_{DS_L} through the Miller capacitance.

(4) t_2–t_3

At t_2, I_{DS_L} reaches its peak and then gradually declines due to the increased junction temperature of the SiC MOSFET caused by the high voltage and large current,

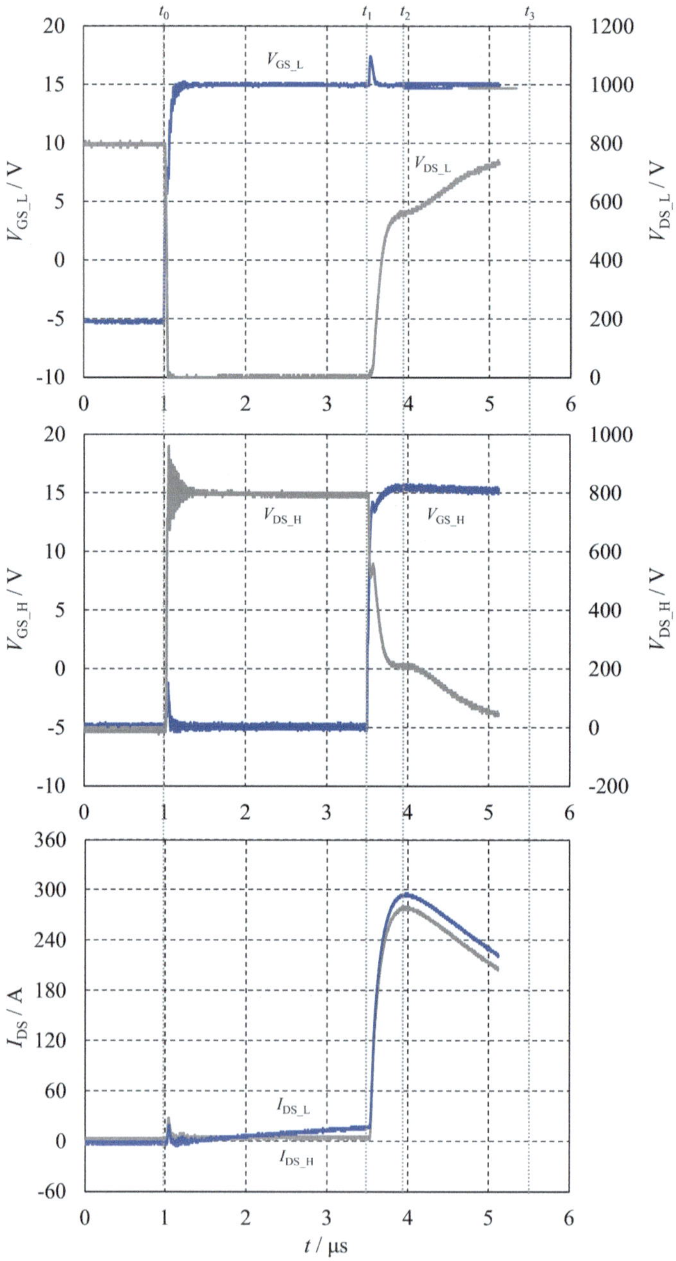

Fig. 3.26 Fault under load short-circuit waveform of SiC MOSFET at 25 °C

reducing its current-carrying capability. V_{DS_L} continues to slowly rise, approaching V_{Bus}, while V_{DS_H} continues to slowly drop, approaching zero but still significantly higher than the normal on-state voltage drop. It is important to note that during the short circuit, I_{SD_L} is the sum of I_{DS_H} and the load inductance current.

(5) t_3–

At t_3, V_{GS_L} quickly changes from $V_{DRV(on)}$ to $V_{DRV(off)}$. Q_L turns off, causing I_{DS_L} to rapidly drop to zero. The load inductance current continues to flow through Q_H in the third quadrant. V_{DS_L} quickly rises, generating an off-state voltage spike and oscillations. V_{DS_H} rapidly drops to the on-state voltage drop. The rapid drop in V_{DS_H} creates a downward spike in V_{GS_H} through the Miller capacitance, leading to oscillations.

The waveform of a SiC MOSFET fault under load short circuit at 125 °C is shown in Fig. 3.27. Compared to the waveform at 25 °C in Fig. 3.2, both the short-circuit current and the off-state voltage spike of the SiC MOSFET are lower at higher temperatures. Additionally, during the short-circuit period, V_{DS_L} is higher and V_{DS_H} is lower at 125 °C.

4. **Load Fault Short Circuit**

Load fault short circuits can occur due to incorrect wiring, dielectric breakdown of the load, motor winding short circuits, and ground faults, etc. In these cases, the load can be considered as a small inductance, reaching extremely high currents in the same on-time. This scenario is depicted in Fig. 3.21. The waveform of a SiC MOSFET load fault short circuit is shown in Fig. 3.28.

(1) –t_0

V_{GS_L} of Q_L is at the off-state drive voltage $V_{DRV(off)}$, keeping it in the off-state. I_{DS_L} is zero, and V_{DS_L} withstands the bus voltage V_{Bus}. The off-state drive voltage $V_{DRV(off)}$ also applies to Q_H, keeping it in the off-state.

(2) t_0–t_1

At t_0, V_{GS_L} quickly changes from $V_{DRV(off)}$ to the on-state drive voltage $V_{DRV(on)}$. turns on, forming a loop with the load inductance L and C_{Bus}. V_{DS_L} quickly drops to the on-state voltage, and I_{DS_L} linearly rises.

(3) t_1–t_2

I_{DS_L} continues to rise, and V_{DS_L} increases continuously. The SiC MOSFET enters saturation, resulting in overcurrent.

(4) t_2–

At t_2, V_{GS_L} quickly changes from $V_{DRV(on)}$ to $V_{DRV(off)}$. Q_L turns off, causing I_{DS_L} to rapidly drop to zero. V_{DS_L} quickly rises, generating an off-state voltage spike and oscillations.

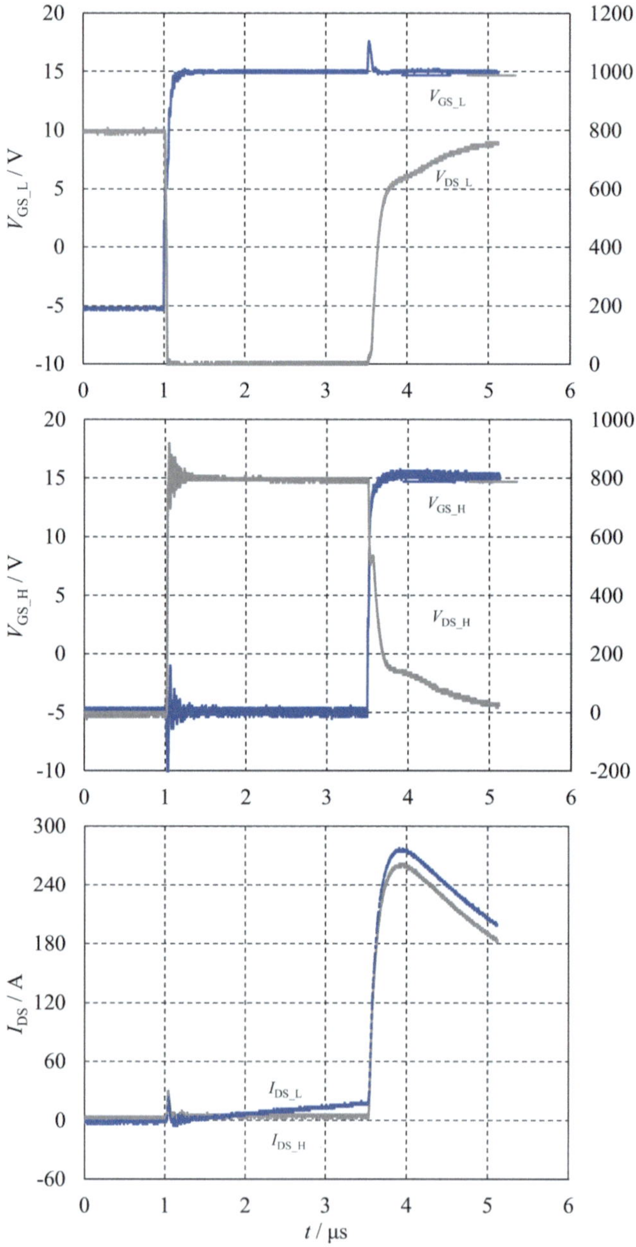

Fig. 3.27 Fault under load short-circuit waveform of SiC MOSFET at 125 °C

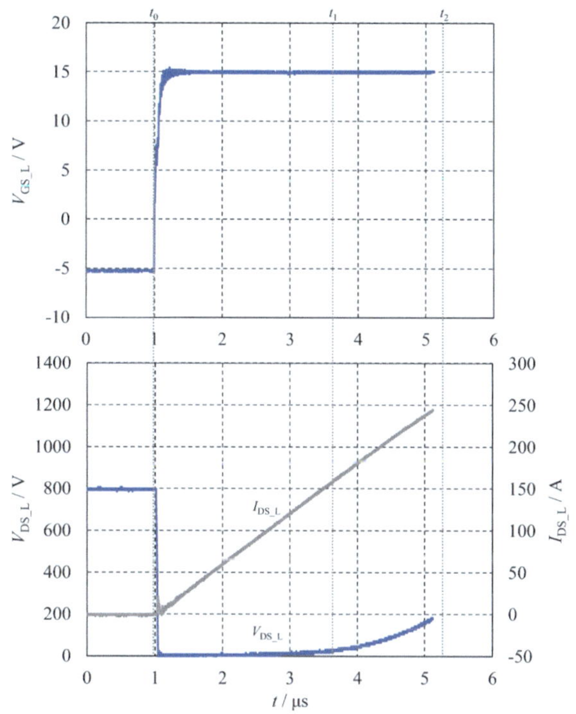

Fig. 3.28 Load fault short-circuit waveform of SiC MOSFET at 25 °C

The waveforms of SiC MOSFET load fault short circuits at different temperatures are shown in Fig. 3.29. Compared to the waveform at 25 °C in Fig. 3.28, both the short-circuit current and the off-state voltage spike of the SiC MOSFET are lower at higher temperatures. Additionally, during the short-circuit period, V_{DS_L} is higher at elevated temperatures.

3.4.2 *Avalanche Breakdown*

The body diode of a SiC MOSFET is a pn-junction diode and is susceptible to avalanche breakdown. During the turn-off process, when the flyback voltage generated by the inductive load or the spike voltage caused by the parasitic inductance of the drain load exceeds the breakdown voltage, the SiC MOSFET will undergo avalanche breakdown. When the pn junction of the body diode is reverse-biased, a strong electric field is generated in the depletion layer, accelerating the free electrons to high kinetic energy. These high-energy electrons can collide with the atoms of the lattice, knocking out other bound electrons and creating electron–hole pairs. This process continues, increasing the number of free electrons and leading to avalanche breakdown.

During avalanche breakdown, the body diode of the SiC MOSFET is subjected to high voltage and large current flow. If the current becomes too large, it can cause the

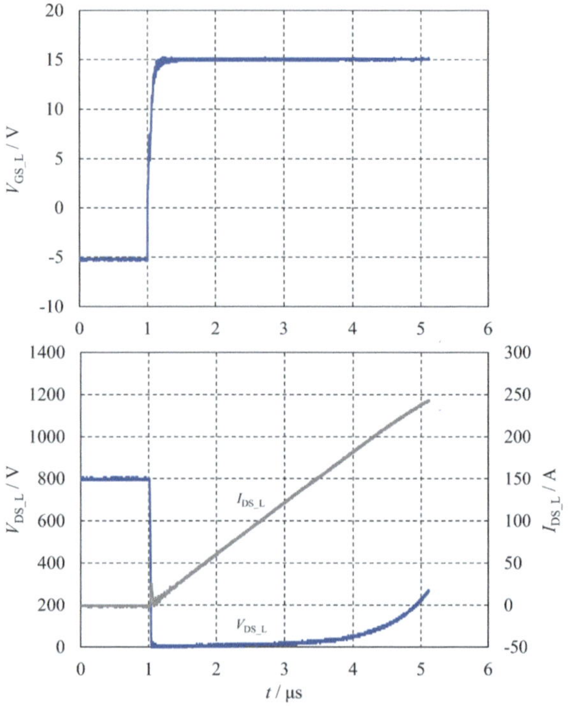

Fig. 3.29 Load Fault short-circuit waveform of SiC MOSFET at 125 °C

parasitic transistor to turn on and enter secondary breakdown, resulting in permanent damage to the SiC MOSFET—this is known as avalanche current breakdown. When the heat generated by the high voltage and large current causes the SiC MOSFET to exceed the chip's temperature limit, the device is damaged, which is termed avalanche energy breakdown. During the SiC MOSFET turn-off process, the V_{DS} rises sharply, and the resulting current can turn on the parasitic transistor, leading to a reduction in its voltage withstand capability, known as dv/dt degradation.

The waveform of SiC MOSFET avalanche breakdown is shown in Fig. 3.30.

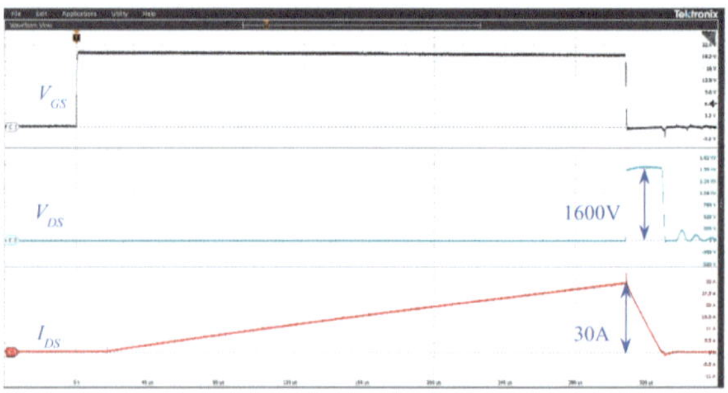

Fig. 3.30 Avalanche waveform of SiC MOSFET

3.5 Figure of Merit

Engineers can gain a basic understanding of device characteristics through device parameters, allowing for device comparisons and assisting in the device selection phase during converter design. In device selection, parameters such as device limits, drive voltage, on-state resistance, junction capacitance, and body diode characteristics need to be considered. Reducing losses and improving efficiency have been the eternal goals pursued in the development of power semiconductor devices and power electronics technology. Therefore, device loss is one of the core concerns in device selection, representing a comprehensive reflection of multiple device parameters. To quickly and simply assess the loss characteristics of devices, the figure of merit (FOM) value has been proposed and widely applied.

1. **BHFFOM** [4]

J. Baliga considered that the on-state loss of the device is influenced by the on-state resistance $R_{DS(on)}$, and the switching loss is generated by charging and discharging the input capacitance C_{iss}. Therefore, the total device loss P_{loss} is represented by Eq. (3.32),

$$P_{loss} = I_{rms}^{2} \frac{R_{DS(on),sp}}{A} + C_{iss,sp} \cdot A \cdot V_{DRV}^{2} \cdot f_{s}, \tag{3.32}$$

Here, I_{rms} is the root mean square current flowing through the device, $R_{DS(on),sp}$ is the specific on-state resistance per unit area, $C_{iss,sp}$ is the specific input capacitance per unit area, A is the device's area, V_{DRV} is the drive voltage, and f_s is the switching frequency. Based on Eq. (3.32), J. Baliga proposed BHFFOM,

$$\text{BHFFOM} = \frac{1}{R_{DS(on),sp} \cdot C_{iss,sp}} \tag{3.33}$$

Since engineers cannot easily obtain the numerical values of $R_{DS(on),sp}$ and $C_{iss,sp}$, BHFFOM is not practical. However, $R_{DS(on),sp}$ has an inversely proportional relationship with $R_{DS(on)}$, and $C_{iss,sp}$ has an inversely proportional relationship with C_{iss}, both related to the device area A

$$R_{DS(on)} = R_{DS(on),sp}/A, \tag{3.34}$$

$$C_{iss} = C_{iss,sp} \cdot A. \tag{3.35}$$

Simultaneously, Q_G comes from C_{iss}, so BHFFOM is extended to the more familiar FOM, FOM,

$$\text{FOM} = R_{DS(on)} \cdot Q_G. \tag{3.36}$$

2. **NHFFOM** [5]

Il-Jung Kim considered that besides C_{iss}, the switching loss caused by the output capacitance C_{oss} should also be taken into account. Therefore, the total device loss is represented by Eq. (3.37),

$$P_{loss} = I_{rms}^2 \cdot R_{DS(on)} + C_{iss,sp} \cdot A \cdot V_{DRV}^2 \cdot f_s + N \cdot C_{oss} \cdot V_{in}^2 \cdot f_s, \tag{3.37}$$

Here, V_{in} is the input voltage of the converter, and N is a coefficient related to the converter topology. Meanwhile, research shows that C_{oss} has a greater impact on losses. Based on Eq. (3.40), Il-Jung Kim proposed NHFFOM,

$$\text{NHFFOM} = \frac{1}{R_{DS(on),sp} \cdot C_{oss,sp}} \tag{3.38}$$

Since the numerical values of $R_{DS(on),sp}$ and $C_{oss,sp}$ cannot be easily obtained, NHFFOM is also not practical. Utilizing the relationship between $C_{oss,sp}$ and C_{oss} and introducing the equivalent output capacitance $C_{oss,eq}$,

$$C_{oss} = C_{oss,sp} \cdot A, \tag{3.39}$$

$$\frac{1}{2} C_{oss,eq} V^2 = \int_0^V C_{oss}(V_{DS}) \mathrm{d}V_{DS} \tag{3.40}$$

In 2009, J. W. Kolar proposed *KFOM*, [6]

$$\text{KFOM} = \frac{1}{R_{DS(on)} \cdot C_{oss,eq}}. \tag{3.41}$$

3. **HDFOM** [7]

Alex Q. Huang divided the MOSFET's turn-on process into several stages, as shown in Fig. 3.31.

(a) $\boldsymbol{t_d}$

The drive current charges C_{iss}, V_{GS} rises from 0 V to the threshold voltage $V_{GS(th)}$, and V_{DS} and I_{DS} remain constant. Define the duration of this stage as t_d and the gate charge as Q_{th}.

(b) $\boldsymbol{t_r}$

The drive current continues to charge C_{iss}, V_{GS} rises from $V_{GS(th)}$ to the Miller plateau voltage V_{plt}, V_{DS} remains constant, and I_{DS} rises from 0A to I_L. Define the duration of this stage as t_r and the gate charge as Q_{gs2}.

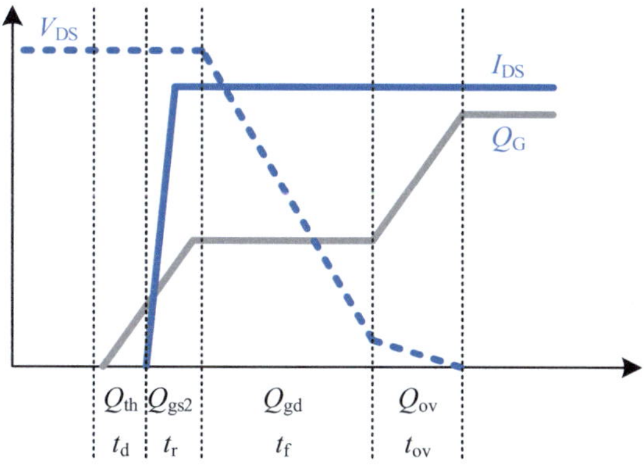

Fig. 3.31 Idealized turn-on process waveform of MOSFET

(c) t_f

V_{DS} decreases from the value of V, the drive current charges the Miller capacitance C_{GD}, and V_{GS} is maintained at V_{plt}. Define the duration of this stage as t_f and the gate charge as Q_{gd}.

(d) t_{ov}

The drive current charges C_{iss}, V_{GS} rises to V_{DRV}, and V_{DS} decreases to the on-state. Define the duration of this stage as t_{ov} and the gate charge as Q_{ov}. The total device loss is then represented by Eq. (3.42),

$$P_{loss} = I_{rms}^2 \cdot R_{DS(on)} + V \cdot I_L \cdot (t_r + t_f) \cdot f_s. \tag{3.42}$$

For high-voltage devices, t_f is much larger than t_r, indicating that t_f plays a dominant role in turn-on loss. Based on (3.42), Alex Q. Huang proposed HDFOM,

$$\text{HDFOM} = \sqrt{R_{DS(on)} \cdot Q_{gd}}. \tag{3.43}$$

4. **YFOM** [8]

HDFOM suggests that t_f plays a dominant role in switch loss, neglecting the loss in the t_r stage. This leads to inaccurate loss evaluation, especially for low-voltage devices and high-voltage devices with increasing speeds, where the loss contribution of t_r cannot be ignored.

Yucheng Ying conducted a more detailed analysis of the device's switching process, as shown in Fig. 3.32.

The $t_{r(on)}$ and $t_{f(on)}$ in the turn-on process are represented by Eqs. (3.44) and (3.45), respectively, and the $t_{r(off)}$ and $t_{f(off)}$ in the turn-off process are represented by Eqs. (3.46) and (3.47), where V_s and I_s are the V_{DS} and I_{DS} during switching.

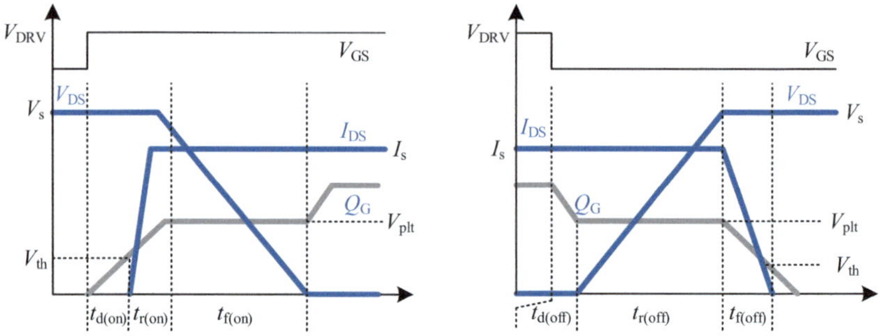

(a) Turn-on process (b) Turn-off process

Fig. 3.32 Idealized switching process waveform for MOSFET

$$t_{\mathrm{r(on)}} = \frac{Q_{\mathrm{gs2}} \cdot R_{\mathrm{G}}}{V_{\mathrm{DRV}} - V_{\mathrm{plt}}}, \tag{3.44}$$

$$t_{f\,(on)} = \frac{Q_{\mathrm{gd}}.R_{\mathrm{G}}}{V_{\mathrm{DRV}} - V_{\mathrm{plt}}}, \tag{3.45}$$

$$t_{\mathrm{r(off)}} = \frac{Q_{\mathrm{gd}}.R_{\mathrm{G}}}{V_{\mathrm{plt}}}, \tag{3.46}$$

$$t_{\mathrm{f(off)}} = \frac{Q_{\mathrm{gs2}}.R_{\mathrm{G}}}{V_{\mathrm{plt}}}. \tag{3.47}$$

This allows the total device loss to be expressed as Eq. (3.48), with the first to fourth terms of the polynomial representing the turn-on loss, turn-off loss, on-state loss, and drive loss, respectively.

$$P_{\mathrm{loss}} = \frac{V_{\mathrm{s}} \cdot I_{\mathrm{s}}}{2} \frac{\left(Q_{\mathrm{gd}} + Q_{\mathrm{gs2}}\right) R_{\mathrm{g}}}{V_{\mathrm{DRV}} - V_{\mathrm{plt}}} f_{\mathrm{s}} + \frac{V_{\mathrm{s}} \cdot I_{\mathrm{s}}}{2} \frac{\left(Q_{\mathrm{gd}} + Q_{\mathrm{gs2}}\right) R_{\mathrm{g}}}{V_{\mathrm{plt}}} f_{\mathrm{s}} + I_{\mathrm{rms}}^2 \cdot R_{\mathrm{DS(on)}} + Q_{\mathrm{G}} \cdot V_{\mathrm{DRV}} \cdot f_{\mathrm{s}}. \tag{3.48}$$

Based on Eq. (3.48), Yucheng Ying proposed YFOM,

$$\mathrm{YFOM} = \left(Q_{\mathrm{gd}} + K_{\mathrm{gs2}} \cdot Q_{\mathrm{gs2}}\right) \cdot R_{\mathrm{DS(on)}} \tag{3.49}$$

where K_{gs2} is,

$$K_{\mathrm{gs2}} = 1 + \frac{V_{\mathrm{DRV}}}{V_{\mathrm{plt}} - V_{\mathrm{th}}} \frac{2V_{\mathrm{plt}}\left(V_{\mathrm{DRV}} - V_{\mathrm{plt}}\right)}{V_{\mathrm{s}} \cdot I_{\mathrm{s}} \cdot R_{\mathrm{g}}} \tag{3.50}$$

5. **FFOM** [9]

YFOM is designed for hard-switching conditions. Under ZVS, the turn-on loss of the device is ignored. Therefore, the total device loss is given by Eq. (3.51), with the first to third terms of the polynomial representing the turn-off loss, on-state loss, and drive loss, respectively.

$$P_{\text{loss}} = \frac{V_s \cdot I_s}{2} \frac{(Q_{\text{gd}} + Q_{\text{gs2}})R_g}{V_{\text{plt}}} f_s + I_{\text{rms}}^2 \cdot R_{\text{DS(on)}} + Q_G \cdot V_{\text{DRV}} \cdot f_s \tag{3.51}$$

Based on this, Dianbo Fu proposed FFOM for ZVS,

$$\text{FFOM} = \left(Q_{\text{gd}} + K_{\text{loss}} \cdot Q_{\text{gs2}}\right) \cdot R_{\text{DS(on)}} \tag{3.52}$$

where K_{loss} is,

$$K_{\text{loss}} = 1 + \frac{V_{\text{DRV}}}{V_{\text{plt}}} \frac{2V_{\text{plt}} V_{\text{DRV}}}{V_{\text{th}} V_s \cdot I_s \cdot R_g} K_{\text{loss}} K_{\text{loss}} \tag{3.53}$$

Based on the various forms of FOM values mentioned above, it is evident that FOM values characterize device losses. The specific expression comes from the loss model and is influenced by the device's operating conditions. Therefore, when comparing and selecting devices, it is necessary to choose an appropriate FOM. If necessary, engineers can establish a loss model for specific topologies and derive corresponding FOM expressions, making the FOM more targeted and obtaining results that are more applicable to real-world applications.

3.6 Calculation of Power Device Losses

Using the figure of merit (FOM) makes it convenient and simple to compare the magnitudes of losses among different devices. However, such comparisons have low accuracy and cannot provide specific numerical values for the impact on converter efficiency. For more precise results, it is necessary to perform power device loss calculations.

Conducting loss analysis during the design phase of a converter helps in making rational decisions for topology selection, device selection, converter parameter design, and thermal design. This, in turn, improves work efficiency and development speed. As power converter design progresses towards finer details, especially with the introduction of Professor J. W. Kolar's multiobjective optimization approach at ETH, loss analysis is gaining more attention. Converter losses mainly include power device losses, drive losses, magnetic device losses, and control circuit losses. Here, we will only discuss power device losses.

3.6.1 Loss Calculation Methods

Device losses comprise conduction losses, switching losses, and reverse recovery losses, represented in the form of power loss.

1. **Conduction Loss**

Conduction loss occurs when the device is conducting, and the instantaneous conduction power loss is the product of the conduction current and the corresponding voltage drop. Since converters generally operate periodically, the average of the instantaneous conduction power loss over one steady-state cycle represents the conduction loss of the device. The conduction current is determined by the converter's operating conditions, and the voltage drop is a function of the conduction current, device junction temperature, and gate drive voltage. The converter's operating cycle is denoted as T_s. The following outlines the calculation methods for various types of conduction losses:

SiC MOSFET forward conduction loss P_C^{MOS} is given by Eq. (3.54), where I_{DS} is the conduction current, T_J is the junction temperature, $V_{DRV(on)}$ is the turn-on drive voltage, and V_{DS} is the voltage drop.

$$P_C^{MOS}\left(I_{DS}(t), T_J, V_{DRV(on)}\right) = \frac{1}{T_s}\int_0^{T_s} V_{DS}\left(I_{DS}(t), T_J, V_{DRV(on)}\right)\cdot I_{DS}(t)\mathrm{d}t. \tag{3.54}$$

Due to SiC MOSFET conduction being analogous to a resistor, P_C^{MOS} can also be calculated using Eq. (3.55), where $R_{DS(on)}$ is the on-state resistance and is a function of I_{DS}, T_J, and $V_{DRV(on)}$.

$$P_C^{MOS}\left(I_{DS}(t), T_J, V_{DRV(on)}\right) = \frac{1}{T_s}\int_0^{T_s} R_{DS(on)}\left(I_{DS}(t), T_J, V_{DRV(on)}\right)\cdot I_{DS}(t)^2\mathrm{d}t \tag{3.55}$$

SiC MOSFET body diode conduction loss $P_C^{MOS,BD}$ is given by Eq. (3.55), where I_F is the forward conduction current, T_J is the junction temperature, $V_{DRV(off)}$ is the turn-off gate drive voltage, and V_F is the voltage drop.

$$P_C^{MOS,BD}\left(I_F(t), T_J, V_{DRV(off)}\right) = \frac{1}{T_s}\int_0^{T_s} V_F\left(I_F(t), T_J, V_{DRV(off)}\right)\cdot I_F(t)\mathrm{d}t \tag{3.56}$$

SiC MOSFET third-quadrant conduction loss $P_{\mathrm{C}}^{\mathrm{MOS,3rd}}$ is given by Eq. (3.57), where I_{SD} is the conduction current, T_{J} is the junction temperature, $V_{\mathrm{DRV(on)}}$ is the turn-on gate drive voltage, and V_{SD} is the voltage drop. Since the third-quadrant characteristic is determined by both the SiC MOSFET channel and body diode, $P_{\mathrm{C}}^{MOS,3rd}$ can only be calculated using V_{SD}.

$$
\begin{aligned}
&P_{\mathrm{C}}^{\mathrm{MOS,3rd}}\left(I_{\mathrm{SD}}(t), T_{\mathrm{J}}, V_{\mathrm{DRV(on)}}\right) \\
&= \frac{1}{T_{\mathrm{s}}}\int_{0}^{T_{\mathrm{s}}} V_{\mathrm{SD}}\left(I_{\mathrm{SD}}(t), T_{\mathrm{J}}, V_{\mathrm{DRV(on)}}\right)\cdot I_{\mathrm{F}}(t)\mathrm{d}t
\end{aligned}
\tag{3.57}
$$

Diode conduction loss $P_{\mathrm{C}}^{\mathrm{Diode}}$ is given by Eq. (3.58), where V_{F} is the voltage drop and is a function of I_{F} and T_{J}.

$$
P_{\mathrm{C}}^{\mathrm{Diode}}(I_{\mathrm{F}}(t), T_{\mathrm{J}}) = \frac{1}{T_{\mathrm{s}}}\int_{0}^{T_{\mathrm{s}}} V_{\mathrm{F}}(I_{\mathrm{F}}(t), T_{\mathrm{J}})\cdot I_{\mathrm{F}}(t)\mathrm{d}t \tag{3.58}
$$

2. **Switching Loss**

Switching losses result from the intersection of voltage and current during the switching process. The energy for each turn-on and turn-off event is represented by E_{on} and E_{off}, respectively. The average values of these energies over one steady-state cycle represent the device's switching losses P_{on} and P_{off}, respectively. The formulas are given by Eqs. (3.59) and (3.60), where $\sum_{i=1}^{N_{\mathrm{on}}} E_{\mathrm{on}}$ and $\sum_{i=1}^{N_{\mathrm{off}}} E_{\mathrm{off}}$ represent the summation of all turn-on and turn-off energies over one steady-state cycle.

$$
\begin{aligned}
&P_{\mathrm{on}}\left(I_{\mathrm{on}}, V_{\mathrm{on}}, T_{\mathrm{J}}, V_{\mathrm{DRV(on)}}, V_{\mathrm{DRV(off)}}, R_{\mathrm{G(on)}}, L_{\mathrm{DRV}}, L_{\mathrm{Loop}}\right) \\
&= \frac{1}{T_{\mathrm{s}}}\left[\sum_{i=1}^{N_{\mathrm{on}}} E_{\mathrm{on}}\left(I_{on,i}, V_{on,i}, T_{\mathrm{J}}, V_{\mathrm{DRV(on)}}, V_{\mathrm{DRV(off)}}, R_{\mathrm{G(on)}}, L_{\mathrm{DRV}}, L_{\mathrm{Loop}}\right)\right],
\end{aligned}
\tag{3.59}
$$

$$
\begin{aligned}
&P_{\mathrm{off}}\left(I_{\mathrm{off}}, V_{\mathrm{off}}, T_{\mathrm{J}}, V_{\mathrm{DRV(on)}}, V_{\mathrm{DRV(off)}}, R_{\mathrm{G(off)}}, L_{\mathrm{DRV}}, L_{\mathrm{Loop}}\right) \\
&= \frac{1}{T_{\mathrm{s}}}\left[\sum_{i=1}^{N_{\mathrm{off}}} E_{\mathrm{off}}\left(I_{off,i}, V_{off,i}, T_{\mathrm{J}}, V_{\mathrm{DRV(on)}}, V_{\mathrm{DRV(off)}}, R_{\mathrm{G(off)}}, L_{\mathrm{DRV}}, L_{\mathrm{Loop}}\right)\right]
\end{aligned}
\tag{3.60}
$$

The energies E_{on} and E_{off} are influenced by various parameters, including I_{on}, V_{on}, I_{off}, V_{off}, T_{J}, $V_{\mathrm{DRV(on)}}$, $V_{\mathrm{DRV(off)}}$, $R_{\mathrm{G(on)}}$, $R_{\mathrm{G(off)}}$, L_{DRV}, and L_{Loop}.

3. **Reverse Recovery Loss**

Similar to switching losses, the average reverse recovery loss P_{rr} is represented by the average value of all reverse recovery energies E_{rr} over one steady-state cycle.

The formula is given by Eq. (3.61), where $\sum_{i=1}^{N_{\mathrm{rr}}} E_{\mathrm{rr}}$ represents the summation of all reverse recovery energies over one steady-state cycle.

$$P_{\mathrm{rr}}(I_{\mathrm{F}}, V_{\mathrm{R}}, T_{\mathrm{J}}, \mathrm{d}i_{\mathrm{F}}/\mathrm{d}t) = \frac{1}{T_{\mathrm{s}}}\left[\sum_{i=1}^{N_{\mathrm{rr}}} E_{\mathrm{rr}}\left(I_{\mathrm{F,i}}, V_{\mathrm{R,i}}, T_{\mathrm{J}}, \mathrm{d}i_{\mathrm{F}}/\mathrm{d}t_{\mathrm{i}}\right)\right] \tag{3.61}$$

E_{rr} is greatly affected by working conditions, including I_{F}, reverse voltage V_{R}, T_{J}, and diode recovery rate $\mathrm{d}i_{\mathrm{F}}/\mathrm{d}t$,. When a diode is part of a switch pair, $\mathrm{d}i_{\mathrm{F}}/\mathrm{d}t$ depends on the switch's turn-on speed, and P_{rr} can also be calculated using Eq. (3.62).

$$\begin{aligned} & P_{\mathrm{rr}}\left(I_{\mathrm{on}}, V_{\mathrm{on}}, T_{\mathrm{j}}, V_{\mathrm{DRV(on)}}, V_{\mathrm{DRV(off)}}, R_{\mathrm{G(on)}}, L_{\mathrm{DRV}}, L_{\mathrm{Loop}}\right) \\ & = \frac{1}{T_{\mathrm{s}}}\left[\sum_{i=1}^{N_{\mathrm{rr}}} E_{\mathrm{rr}}\left(I_{\mathrm{on,i}}, V_{\mathrm{on,i}}, T_{\mathrm{J}}, V_{\mathrm{DRV(on)}}, V_{\mathrm{DRV(off)}}, R_{\mathrm{G(on)}}, L_{\mathrm{DRV}}, L_{\mathrm{Loop}}\right)\right]. \end{aligned} \tag{3.62}$$

SiC MOSFET body diode reverse recovery loss $P_{\mathrm{rr}}^{\mathrm{MOS,BD}}$ and SiC diode reverse recovery loss $P_{\mathrm{rr}}^{\mathrm{Diode}}$ are calculated using the same principles as above.

4. **Total Loss Calculation**

The total loss of the device is the sum of the actual losses occurring during its operation.

$$P_{\mathrm{MOS}} = P_{\mathrm{C}}^{\mathrm{MOS}} + P_{\mathrm{C}}^{\mathrm{MOS,BD}} + P_{\mathrm{C}}^{MOS,3rd} + P_{\mathrm{on}} + P_{\mathrm{off}} + P_{\mathrm{rr}}^{\mathrm{MOS,BD}}, \tag{3.63}$$

$$P_{\mathrm{Diode}} = P_{\mathrm{C}}^{\mathrm{Diode}} + P_{\mathrm{rr}}^{\mathrm{Diode}}. \tag{3.64}$$

Typically, attention needs to be focused on the device losses during steady-state operation of the converter under thermal stability. Therefore, determining the device junction temperature under thermal stability is crucial. The basic method involves using the losses generated in the previous time interval, calculating the new junction temperature through thermal impedance, and using the updated temperature to calculate the losses generated in the next time interval. This process repeats until the junction temperature gradually converges to thermal stability. Once thermal stability is reached, averaging the energy over a steady-state cycle completes the loss calculation.

Taking the buck circuit as an example, Fig. 3.33 illustrates the MOSFET's V_{DS} and I_{DS} waveforms, divided into time intervals:

(1) $\boldsymbol{t_1}$

The junction temperature of the MOSFET is denoted as $T_{J,1}$.

(2) $\boldsymbol{t_1}$–$\boldsymbol{t_2}$

MOSFET is conducting; calculate the conduction losses during $t_1 - t_2$ using $T_{J,1}$ and calculate the junction temperature at t_2 as $T_{J,2}$.

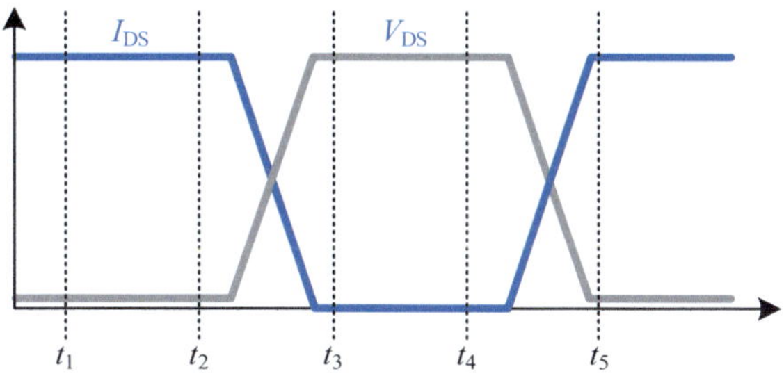

Fig. 3.33 Buck circuit loss calculation

(3) t_2–t_3

MOSFET transitions from conduction to cutoff; calculate conduction and cutoff losses during this interval using junction temperature $T_{J,2}$, resulting in junction temperature $T_{J,3}$ at t_3.

(4) t_3–t_4

MOSFET is in cutoff with no losses; calculate junction temperature at t_4 as $T_{J,4}$.

(5) t_4–t_5

MOSFET transitions from cutoff to conduction; calculate conduction and cutoff losses during this interval using junction temperature $T_{J,4}$, resulting in junction temperature $T_{J,5}$ at t_5.

Device losses are influenced by device characteristics, junction temperature, and external circuitry. Therefore, loss calculations are helpful in completing converter design, providing loss data, comparing differences in heat dissipation schemes for thermal design assistance, evaluating variations in different devices for device selection, and assessing the impact of external circuit parameters to determine suitable drive voltage, drive circuit, and line inductance.

Clearly, device loss calculations require substantial foundational loss data support, including on-state voltage drop vs. on-state current data or on-state resistance data, switch energy data, and reverse recovery energy data. The more dense and comprehensive the data covering various operating points, the higher the accuracy of loss calculations. Generally, the required foundational data can be obtained through manufacturer-provided datasheets or experimental testing methods.

Obtaining foundational loss data through datasheets is cost-effective, straightforward, and fast. However, datasheet data is limited to specific operating points, necessitating interpolation and fitting to supplement the data, introducing some deviation. Moreover, switch energy and reverse recovery energy provided in datasheets are based on the manufacturer's test circuits, but these losses are significantly influenced by external circuitry, further increasing deviation.

Experimental testing provides more loss data at various operating points, with switch energy and reverse recovery energy based on actual circuit parameters, enhancing the accuracy of loss calculations. On-state voltage drop vs. on-state current data or on-state resistance data is measured using a device analyzer, and switch energy and reverse recovery energy are obtained through double-pulse testing, requiring corresponding equipment matching. To obtain typical values, a sufficient number of test samples are needed, resulting in a substantial testing workload, higher costs, and slower speed.

Considering the pros and cons of both methods, it is common to use datasheets for obtaining on-state voltage drop vs. on-state current data or on-state resistance data and conduct experimental testing to obtain switch energy and reverse recovery energy.

Additionally, thermal loss measurement methods can be utilized to acquire loss data.

3.6.2 Simulation Software

The method for calculating losses involves continuously computing losses and updating device junction temperatures based on the device's operating state, using pre-prepared basic loss data. This is precisely the type of task that computer programs excel at. Plexim's PLECS is a system-level power electronics simulation software that integrates device thermal models to facilitate rapid loss calculations.

In PLECS, power devices, magnetic components, capacitors, resistors, and control modules form the electrical circuit. Thermal circuit components include device thermal resistance, heatsink thermal resistance, thermal equivalent networks, and environmental factors. Users input basic loss data for devices into the thermal loss editor, connecting thermal and electrical aspects. During simulation, PLECS solves both the electrical and thermal circuits simultaneously. It records conditions before and after switching instances, retrieves switch losses and reverse recovery energy values from a three-dimensional table, obtains conduction losses through device current and junction temperature, and injects losses into the thermal circuit model to calculate device junction temperatures. The circuit-thermal loss coupled simulation algorithm of PLECS is represented in Fig. 3.34.

PLECS has gained recognition from well-known power semiconductor companies for its excellent performance. ABB, Dynex, GaN System, Infineon, ROHM, and Wolfspeed all provide PLECS device thermal models for their products on the PLECS website, available for free download and direct use in converter designs. Some power semiconductor manufacturers offer web-based simulation tools based on PLECS to make it more convenient for engineers to access PLECS's thermal simulation capabilities.

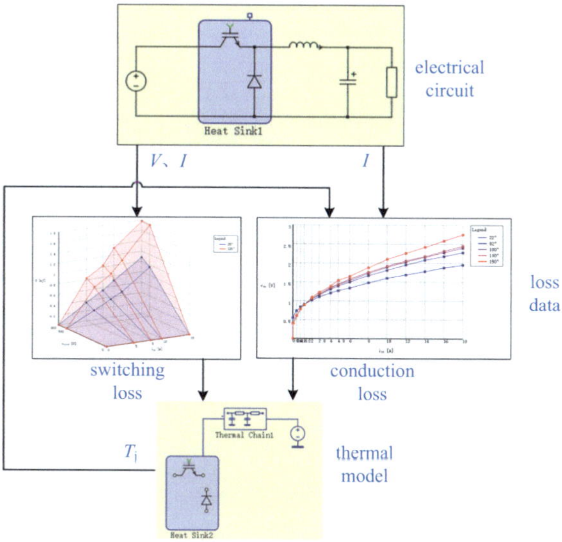

Fig. 3.34 PLECS circuit-thermal loss coupling simulation

1. **Infineon—IPOSIM**

For Infineon IGBT power modules, IPOSIM allows loss analysis for over a dozen common topologies and supports setting the drive resistance. The device model includes Infineon's SiC MOSFET power module and provides examples for single-phase inverters for charging stations and three-phase ANPC inverters for photovoltaic and energy storage applications.

2. **Fuji—Web Simulation Tool**

Designed for Fuji IGBT power modules, this tool enables loss analysis for two-level and T-type three-level inverters.

3. **GaN System—Circuit Simulation Tool**

Tailored for GaN System's GaN HEMT devices, this tool facilitates loss analysis for nearly ten widely used topologies employing GaN devices.

4. **Wolfspeed—SpeedFit Design Simulator**

Focused on Wolfspeed SiC single transistors and power modules, this simulator enables loss analysis for over a dozen commonly used SiC device topologies, including LLC, Totem-Pole PFC, and DNPC circuits. Taking Totem-Pole PFC as an example, the SpeedFit Design Simulator interface is depicted in Fig. 3.35, presenting operational waveforms, device losses, efficiency, and junction temperatures.

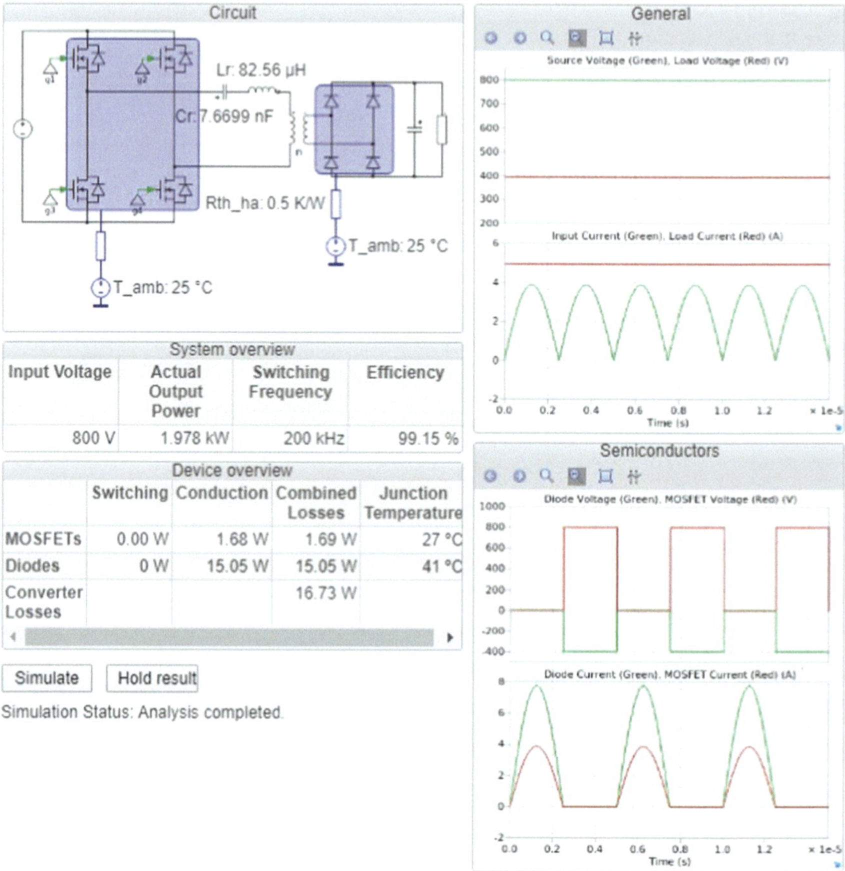

Fig. 3.35 Wolfspeed SpeedFit design simulator [11]

It's important to note that when using PLECS device thermal models or web-based simulation tools provided by various manufacturers, switch loss data is specific to each manufacturer's loss test platform. While the analysis results can serve as a reference for selecting devices and designing converters, they should not be relied upon for accuracy predictions. This is because device switch losses are significantly influenced by circuit parameters, and the losses provided by manufacturers differ from those in converters we design ourselves, often with considerable variations.

In addition to PLECS, Powersim's simulation software PSIM [12] also features device thermal models with similar functionality. The principles and functions of PSIM are comparable to PLECS.

MATLAB/Simulink from MathWorks is another important simulation tool in the field of power electronics. While it doesn't integrate device thermal models, it doesn't mean that MATLAB/Simulink can't achieve device loss assessment. Because the method for loss calculation is explicit, and MATLAB/Simulink has infinite potential for user development, users can build device thermal models using basic modules. Although this incurs development costs, it eliminates the greater time and financial costs associated with changing simulation platforms, and its calculation and analysis of losses are more flexible. MATLAB/Simulink provides an example of loss calculation, such as for a three-level inverter [13], and users can directly use the loss calculation modules or further develop them according to their needs.

3.7 SiC MOSFET Modeling

In the traditional design process of converters, engineers often use simplified circuit simulators along with simplified device models. This requires engineers to have extensive design experience to use simplified models effectively and obtain accurate simulation results. In most cases, engineers tend to directly build prototypes, test them, and then optimize the design. This approach can be time-consuming, requiring multiple iterations before achieving the final design. Additionally, if the prototype circuit board fails, engineers need to spend more time troubleshooting and making corrections.

Although precise device models are not widely used to guide converter design, engineers can still complete product designs according to specifications. This is because the switching speed of traditional power devices does not impose strict requirements on the design, and engineers can solve most problems using their experience. In contrast to traditional devices, SiC MOSFETs have a faster switching speed, and the high dv/dt and di/dt during the switching process impose stringent requirements on circuit design, leading to more pronounced electromagnetic compatibility issues. If the "trial and error" method is still employed, it is likely to consume a significant amount of resources without completing the design. Therefore, there is an urgent need to introduce more accurate models and advanced simulation techniques to predict the transient characteristics of devices in practical circuits, thereby achieving the goal of shortening the design cycle and reducing costs.

Device models fall into two categories: physical models and mathematical models. Mathematical models use a series of equations to describe device parameters and can be considered an application of device parameters. Mathematical models do not require knowledge of some process-related parameters, such as gate oxide thickness, making them the primary choice for power electronics engineers. To ensure that the device model fulfills its intended purpose, two prerequisites must be met: high model accuracy and fast modeling speed. If the model accuracy is not high, simulation results may misguide the design. If the modeling speed is too slow, it severely affects the design progress.

3.7.1 Fundamentals of the SPICE Models

Simulation Program with Integrated Circuit Emphasis (SPICE) is a highly powerful general-purpose circuit simulation tool developed by the Electronics Research Laboratory at the University of California, Berkeley, in 1975. It is primarily used for verifying circuit designs and predicting circuit performance in integrated circuits. Simulation calculations based on SPICE are crucial for the design of power electronic circuits and serve as the primary platform for SiC MOSFET models.

3.7.2 Types of SPICE Models

According to modeling methods, existing SPICE models can generally be categorized into four types: numerical models, physical models, seminumerical models, and equivalent circuit models.

1. **Numerical Models**

Numerical models are based on device process characteristics and are commonly used for simulating power devices and analog circuits. They are primarily developed using simulation software such as Sentaurus TCAD, SILVACO, and MEDICI. The main advantage of numerical models is their ability to provide high simulation accuracy. However, they are complex and less suitable for power electronics circuit simulations. This complexity arises because numerical models require specific material properties and geometric characteristics of device structures, which can be challenging for designers with a background in power electronics. Therefore, numerical models are mainly used for power device design rather than power electronics circuit design.

2. **Physical Models**

Physical models are based on semiconductor physics and accurately reflect the changes in carrier processes within the device. They highly reproduce various device parameters with high modeling accuracy. However, understanding the device mechanisms in depth is necessary during the modeling process due to considerations of semiconductor material physics. Additionally, physical models involve numerous parameters and a complex parameter extraction process, which often results in longer simulation times. These factors collectively limit the application of physical models in power electronics circuit simulations.

3. **Seminumerical Models**

Seminumerical models lie between numerical and physical models. They are developed based on device physics and process parameters. The physical model's parameters and equations are determined using numerical methods, such as using Fourier series or Laplace transforms to obtain solutions to the bipolar diffusion equation, thereby determining carrier distributions in different regions. Despite these efforts, seminumerical models still tend to be complex with many parameters, and their application in power electronics circuit simulations is relatively limited.

4. **Equivalent Circuit Models**

Equivalent circuit models are primarily established based on the external characteristics exhibited by the device. If these models are purely based on mathematical fitting of the device's external behavior without considering internal physical characteristics, they are termed behavioral models. If they partially consider semiconductor physical characteristics based on mathematical fitting of the device's external behavior, they are known as semiphysical models. Equivalent circuit models construct corresponding equivalent circuits based on test data results to characterize device external characteristics. Such models do not involve (or involve very little) device internal material parameters and semiconductor physics knowledge. They are typically constructed based on data from device datasheets. Equivalent circuit models are characterized by their simple structure, ease of understanding, and advantages such as shorter simulation times and better convergence. Therefore, they are more suitable for power electronics circuit design requirements and are widely favored in the power electronics community. As a result, most SPICE models adopt the form of equivalent circuit models.

3.7.3 Levels of Equivalent Circuit Models

The establishment of equivalent circuit models for devices often requires a balance between modeling accuracy and complexity to achieve a balance of result accuracy and simulation convergence. The introduction of model levels allows power electronics engineers to quickly select models based on their intended use and required accuracy. SPICE models are classified into Level 1 to Level 4.

LEVEL 1 model is the classic Shichman-Hodges model, as shown in Fig. 3.36. It is a first-order model for MOSFETs that describes the current–voltage characteristics, taking into account substrate modulation and channel length modulation effects. It is suitable for long-channel MOS transistors with low-precision requirements.

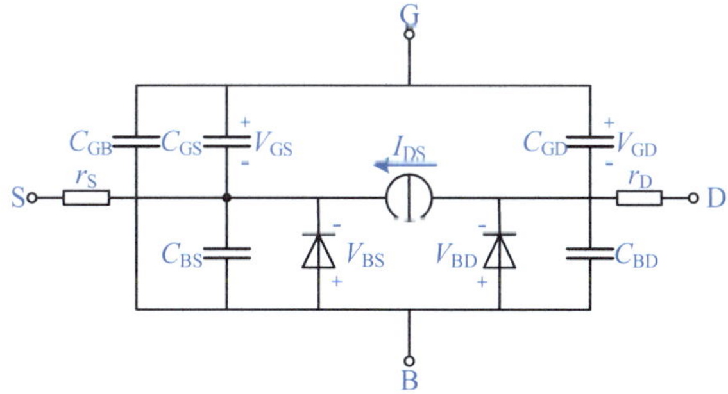

Fig. 3.36 Shichman-Hodges model

Generally, this model is the most appropriate if a simple model is required when the gate length and width of the MOS device are greater than 10 μm, with a low substrate doping concentration.

LEVEL 2 model builds upon the LEVEL 1 model by incorporating second-order effects, making it suitable for short-channel or narrow-channel MOSFETs. Also known as the two-dimensional analytical model, it includes additional parameters that can significantly slow down simulation speed and even lead to non-convergence issues.

LEVEL 3 model is derived from empirical relationships between experimental data and existing theoretical models. It introduces semiempirical formulas for parameters such as threshold voltage, saturation current, channel modulation effects, and source-drain leakage expressions. Additionally, three new model parameters are introduced based on the LEVEL 2 model: empirical parameters for simulating electrostatic feedback effects (EAT), mobility modulation factor (THETA), and saturation field coefficient (KAPPA). This model is an enhanced version of the Shichman-Hodges model, offering higher accuracy and shorter computation times compared to LEVEL 2.

LEVEL 4 model, also known as the Berkeley Short-Channel IGFET Model (BSIM), is physically based. Its parameters are automatically generated from process files using model parameter extraction programs. It is suitable for digital and analog circuits, primarily developed for devices with 1 μm or larger processes, accurately simulating device electrical behavior. Over time, various versions of the BSIM model have been developed, such as BSIM1, BSIM2, BSIM3, BSIM4, BSIM6, BSIM-CMG, BSIM-IMG, each optimized for specific device manufacturing processes and application scenarios with different levels of accuracy, computation speed, and usage scope.

These are the four basic models most commonly used in PSpice. Additionally, other models such as TOM, Curtice Cubic, EKV, Angelov, and EESOF are widely used in device and circuit design. These models were not originally designed for SiC materials and require modifications and improvements to accommodate the characteristics of SiC MOSFET devices.

3.7.4 Modeling Methods

As research on SiC MOSFETs deepens, an increasing number of manufacturers are introducing different types of power devices, leading to their wider application. Device models serve as a crucial bridge for engineers in simulating designs, facilitating communication between system applications and individual components. Establishing a high-precision device model is paramount in the process of device application design. The two most common modeling methods are curve-fitting modeling and subcircuit modeling.

3.7.4.1 Curve-Fitting Modeling Method

Curve-fitting modeling involves using characteristic curves obtained from device datasheets or actual measurements to construct data-point-based models. This method can be implemented using the Model Editor, a built-in tool in PSpice.

Once the basic field-effect transistor model is created using the Model Editor, the system automatically generates interfaces for eight different electrical performances: transconductance, transfer curve, Rds(on) resistance, zero-bias leakage, turn-on charge, output capacitance, switching time, and reverse drain current. By inputting curve data from the device datasheet into these interfaces, the Model Editor can fit various electrical characteristic curves based on the provided data. Each curve corresponds to different model parameters within the model. After fitting all model curves, the necessary parameter values for establishing the model are obtained.

Upon completing the parameter extraction, the model parameter values from the Model Editor are outputted in a ".lib" file format, creating a corresponding MOSFET PSpice library file. This library file is then added to the circuit simulation library files. Additionally, an appropriate package is selected for the model, outputting it in a ".olb" file format, enabling the use of the device model in circuit simulations.

However, due to the utilization of only partial characteristic curve data in this modeling process, although it simplifies modeling steps and reduces the difficulty of model establishment, the accuracy of the established model may not be sufficient. There are inherent differences between the characteristic curves generated by the logical algorithms within the Model Editor and those specified in the device datasheets. Therefore, this method is suitable only for simple circuit simulation applications.

3.7.4.2 Subcircuit Modeling Method

Taking the example of the RF power MOSFET transistor DE150-201N09A, as depicted in Fig. 3.37 of the subcircuit schematic, the core comprises a simple MOSFET used to achieve device switching functionality. When the device is conducting, the drain-source resistance RD is metaphorically described as the conduction resistance of the device. Additionally, the transistor's output capacitance C_{oss} and reverse transfer capacitance C_{rss} are represented using reverse-biased diodes, adjusted by RON and ROFF to control the turn-on and turn-off delays of the transistor. By constructing the above circuit topology and adjusting the corresponding component parameters, the basic operational characteristics of the device can be characterized, completing the modeling of this MOSFET. Naturally, this requires a thorough understanding of circuitry, analog electronics, and semiconductor principles by the development team.

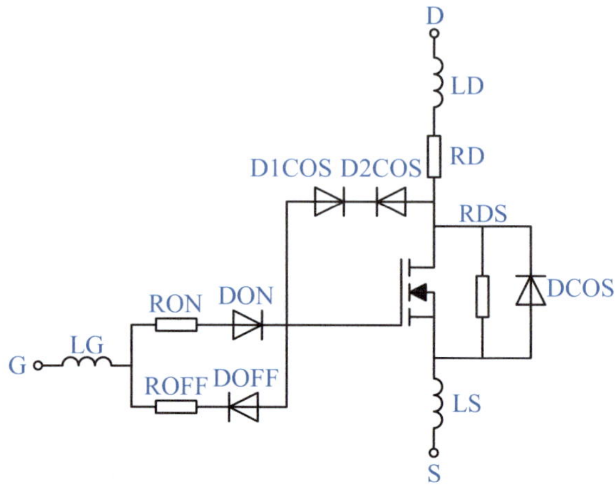

Fig. 3.37 DE150-201N09A subcircuit schematic

Using the method of drawing subcircuit schematics to build models provides a more intuitive understanding of the device's characteristics. However, if the subcircuit constructed uses a large number of components or the circuit connections are complex, it significantly increases the difficulty for developers. Therefore, many manufacturers opt to establish subcircuit models through writing SPICE program statements.

Through programming, complex circuit connections can be cleverly depicted in the form of SPICE program language, greatly enhancing the efficiency of model design. The following mainly introduces the corresponding rules for writing SPICE program statements, facilitating quick understanding of the library files provided by different manufacturers and enabling appropriate improvements based on this foundation.

(1) **Subcircuit and Model Establishment and Invocation**

In the simulation process of power electronic circuits, the device model can be understood as a subcircuit model. Using SPICE language to describe it, we first need to define the subcircuit using the.subckt function and terminate it with.ends. Between these two, the program writing for constructing the subcircuit can be completed based on the relevant characteristics of the device. Generally, the sequence of SPICE program statements for defining a subcircuit is: subcircuit command statement (.subckt), subcircuit name, nodes connected externally to the subcircuit, specific connections within the subcircuit, and subcircuit end command statement (.ends).

Table 3.1 Device model correspondence (switch devices)

Device name	Symbol	Device name	Symbol
Diode	D	MOS field-effect transistor	M
Junction field-effect transistor	J	Bipolar junction transistor	Q
GaAs field-effect transistor	B		

```
.subckt GMOS 1 2 3.
.ends GMOS.
X1 d1 g1 s1 GMOS.
```

The code above defines a subcircuit of an OPAMP, with three external pins labeled 1, 2, and 3. Between the.subckt and.ends functions, the specific subcircuit topology can be described. Additionally, comments can be added using "*" without affecting the circuit's topology structure.

Of course, if this subcircuit needs to be called elsewhere, the subcircuit call command "x" is used, following the rules shown in the last line of the figure: call command statement (x) + name, pins of the subcircuit, name of the subcircuit being called. This reduces a lot of repetitive modeling work.

In addition, during the modeling process, basic device models such as diodes and field-effect transistors are also used. These models often involve numerous parameters, where different parameters represent different meanings and electrical characteristics. Here, we won't go into further details. These models differ from subcircuit models in that they require the.model function for definition. The sequence of SPICE program statements for model definition is: model definition command statement (.model), model name, model type, and relevant parameters of the model. There are also many types of models. Common types used in SPICE are shown in Table 3.1.

Generally, because these device models involve many parameters, they are usually grouped together using "()". If a new line is needed, " + " is used to continue the program, without any other meaning. Below is a typical description statement of an NMOS device model:

```
.MODEL M233 NMOS (LEVEL=3 TOX=0.08U L=2.0U LD=.4U
+ W=1.0 WD=0.5 XJ=1.2U NSUB=4E14 IS=2.1E-14 RB=0
+ RD=0.10 R3=0.03 RDS=1E6 VTO=3.25 UO-450 THETA=0.1
+ ETA=0 VMAX=1.5E6 CBS=1P CBD=3240P PB=0.6 MJ=0.5
+ RG=10 CGSO=2340P CGDO=467P CGB0=0.5P)
```

The above statement defines a simple MOSFET model named M233, with the model type being NMOS, and the parameters inside the parentheses representing various model parameters. Of course, if this model needs to be used elsewhere, similar to the subcircuit call rules, the "X" command statement is used for description.

Table 3.2 Device model correspondence (circuit elements)

Circuit element name	Symbol	Circuit element name	Symbol
Resistor	R	Voltage-controlled voltage source	E
Capacitor	C	Current-controlled current source	F
Inductor	L	Voltage-controlled current source	G
Mutual inductor	K	Current-controlled voltage source	H
Independent current source	I	Independent voltage source	V
Current switch	W	Voltage switch	S

(2) SPICE Description Statements for Circuit Elements

During the process of describing specific subcircuit structures, it is inevitable to use some basic circuit elements such as resistors, inductors, and capacitors. In addition, for some complex, nonlinear elements, they are often equivalently treated as controlled voltage sources or controlled current sources. Here, we need to understand the corresponding SPICE syntax rules for these components.

Generally, the sequence of SPICE description statements for circuit elements is: circuit element symbol + name, nodes connected by the circuit element, and the parameter values of the circuit element. Different symbols correspond to different circuit elements. Below are the commonly used symbols for circuit elements in SPICE program statements (Table 3.2).

It should be noted that there are specific syntax rules for connecting nodes of circuit elements when describing circuit elements. Generally, the "positive node first, negative node second" rule is used. The parameter values of circuit elements can be specific constants. If there is no special indication, these values generally correspond to their respective basic units. Here's an example of a resistor element SPICE description statement:

```
Rdson 12 12 100
```

The above statement names a resistor element called Rdson, connects it between nodes 12 and 13, with current flowing from node 12 to node 13, and its resistance value is 100 Ω.

However, in actual SPICE modeling applications, to improve simulation speed and achieve high-precision SPICE modeling, we often use a semiphysical modeling approach. This approach uses mathematical equations to describe complex physical mechanisms and simplifies the modeling process. These mathematical expressions typically appear in SPICE program statements in the form of controlled sources. Similar to the description rules of circuit elements mentioned above, they are also structured as: circuit element symbol + name, nodes connected by the circuit element, and the parameter values replaced with expressions, formatted as "value = { }".

Inside the braces, the corresponding expression is provided. To facilitate subsequent improvements, some parameters in complex model expressions are often defined using the.param function at the beginning, making the program look more elegant and organized. For nonlinear equations, the IF function is used for description. We will illustrate this through the following example:

```
.param p1 1
.param p2 2
.param p3 3
.param p4 4
Gm d s value {
+ if(V(g,s)<4.5,0,p1+p2*V(g,s))/(1+p4*V(g,s))
+ }
```

Here's what it represents: There is a Gm voltage-controlled current source between nodes d and s. The current magnitude is controlled by voltages Vgs and V(Tj), flowing from node d to node s. Its specific value is: When Vgs is less than 4.5 V, the current of the voltage-controlled current source is 0 A. When Vgs exceeds 4.5 V, it follows the expression inside the curly braces in the program, where parameters $p1$, $p2$, $p3$, and $p4$ are defined using the.param function with values 1, 2, 3, and 4, respectively.

The above covers some basic syntax knowledge of SPICE program statements. It's important to note that SPICE statements are case insensitive—both "D" and "d" mean the same thing. Additionally, pay attention to the placement of spaces and matching parentheses. When writing programs, adhere strictly to the corresponding syntax rules; unnecessary spaces and extra parentheses should be removed to avoid simulation errors. Of course, this is only a part of SPICE program statements; understanding more requires time and effort. If encountering statements or functions that are difficult to understand during research, help documentation in relevant software can assist in comprehension.

Taking the example of Wolfspeed's C3M0016120D SPICE model, its subcircuit schematic can be referenced in the next section on the C3M series models. Its implementation primarily involves writing SPICE program statements, as detailed below.

```
**************************************************************************
******
****Parasitics Included
****Tj = Junction Temperature
****Tc = Case Temperature
****D = Drain
```

```
****G = Gate
****S = Source
************************************************************

.subckt C3M0016120D d g s Tj Tc    //
                                    Establishing the subcircuit model for
                                   C3M0016120D, with external pins d, g, s,
                                   Tc, and Tj
.param Rgint = 2.6                 // Internal gate resistance of 2.6Ω

R1022     Tjc  0    1E6
E1022     Tjc  0    value {limit(v(Tj),-40,260)} // Junction temperature
range: -40°C to 260°C

R100 gk   0    1E6
E100 gk   0    value {limit(v(g1,s1),-8,19)}     // Vgs range: -8V to 19V

e3        NET3 0    Value {Limit(-93.938n*V(Tjc)**3+27.548u*V(Tjc)**2-
+                   5.5921m*V(Tjc)+2.6741,0.8,3.5)}
R_c       NET3 0    1E6

xgmos     d3 d1 gk s1 Tjc NET3 gmos_C3M0016120D  // Referring to gmos
subcircuit model

Ls        s    s1   6.5n  Rser=1.180m   // S pin: inductance, equivalent
series resistance
R_Ls1     s    s1  100                  // S pin: equivalent parallel
resistance

R_g  g1   g2   {Rgint}
Lg        g    g2   6.76n Rser=12.38m   //G pin: inductance, equivalent
series resistance
R_Lg      g    g2   100                 //G pin: equivalent parallel
resistance

Ld        d    d3  3.84n Rser=0.0649m   // D pin: inductance, equivalent
series resistance
R_Ld d    d3   5                        //D pin: equivalent parallel
resistance

vdrain_s  d3   d1   0
Gheat  0  Tj  value{abs((V(d1,s1)*I(Vdrain_s)))+abs((V(g1,g2)**2/Rgint))}

xCGD      d3   g1   cgdmos_C3M0016120D      // Referring to Cgd subcircuit
model
CGS       g1   s1   6120p                   // Cgs fixed value: 6120pF
xCds      d3   s1   cds_C3M0016120D         // Referring to Cds subcircuit
model

R_CGS     g1   s1   10E6                    // Component connections
R_GD g1   d3   10E6
```

```
R_DS d3   s1   10E6

R15  dd12 d1   1E6
e15  dd12 d1   value {
+                   if (V(gk)>V(NET3),
+                        Limit((
+                        (881.9u*V(Tjc)**2+89.38m*V(Tjc)-34.9)*(V(gk)**3)+
+                        (-25.75m*V(Tjc)**2-2.58*V(Tjc)+993.99)*(v(gk)**2)+
+                        (0.16317*V(Tjc)**2+21.0784*V(Tjc)-6210.944)*v(gk)+
+                        (-0.181*V(Tjc)**2-45.93*V(Tjc)+7848))/1000,-4,20)
+                        ,
+                        Limit((
+                        (328u*V(Tjc)**2+0.1011*V(Tjc)-33.982)*(V(gk)**2)+
+                        (1.729m*V(Tjc)**2+0.98748*V(Tjc)-413.27)*v(gk)+
+                        (15.27m*V(Tjc)**2-4.386*V(Tjc)-949.88))/1000,-4,5)
+                        )
+                        }
vdiode1   dd12 dd14 0

Esn1 net1 0    value {I(vdiode1)}
Rsn1 net1 0    1E6

D1         s1   dd14 bodydiode_C3M0016120D temp=25

CDS1       d3   s1   0.1p
CGD1       g1   d3   0.01p
R_DS1      d3   s1   0.5G

.model bodydiode_C3M0016120D d(is=0.88p bv=1590 n=4.80
+    rs=0.015 Tnom=25 tt=2n ibv=500u level=1)    // Body diode
characteristics



R0 N1 Tj 41.15m                // Thermal resistance RC network
R1 N2 N1 90.58m
R2 N3 N2 47.43m
R3 Tc N3 91.1m

C0 Tj 0 2.803m
C1 N1 0 16.49m
C2 N2 0 42.52m
C3 N3 0 129.6m
.ends C3M0016120D

*****************************************

.subckt gmos_C3M0016120D d3 d1 gk s1 Tjc NET3    // Establishing gmos
  subcircuit model

e1         NET1 0    value {0.2}
```

```
R_A        NET1 0    1E6

e2         NET2 0    Value {Limit((
+                    (-139.1u*V(Tjc)**2+10.05m*V(Tjc)+4.896)*v(gk)**2+
+                    (4.183m*V(Tjc)**2-0.4445*V(Tjc)-55.01)*v(gk)+
+                    (-30.04m*V(Tjc)**2+4.052*V(Tjc)+450.7))/
+                    1000,0.01,2)
+                    }
R_B        NET2 0    1E6

e4         NET4 0    Value {Limit((
+                    (968.4u*V(Tjc)**2+32.99m*V(Tjc)-6.2299)*(V(gr)**2)+
+                    (-22.807m*V(Tjc)**2-0.8652*V(Tjc)+111.88)*v(gr)+
+                    (0.12453*V(Tjc)**2+5.5603*V(Tjc)-276.84))/1000,0,5)
+                    }
R_d        NET4 0    1E6

*e99       P99  0    value {0.02}
e9         P9   0    value {Limit((
 +                   (-512.15u*V(Tjc)**2+0.16216*V(Tjc)-5.694)*(V(gr)**2)+
 +                   (10.273m*V(Tjc)**2-3.4705*V(Tjc)+109.74)*v(gr)+
 +                   (-37.963m*V(Tjc)**2+14.846*V(Tjc)-
 245.42))/1200,0.001,3)
 +                   }
 R9         P9   0    1E6

 *e6        NET6   0  Value {1.5}
 e5         NET5   0  Value {Limit((
 +                    (-717.72n*V(Tjc)**2-120.045u*V(Tjc)-18.716m)*V(gk)**4+
 +                    (18.845u*V(Tjc)**2+5.3645m*V(Tjc)+1.6842)*V(gk)**3+
 +                    (48.124u*V(Tjc)**2-56.406m*V(Tjc)-46.154)*V(gk)**2+
 +                    (-3.478m*V(Tjc)**2-0.3262*V(Tjc)+443.821)*v(gk)+
 +                    (11.734m*V(Tjc)**2+5.269*V(Tjc)-760.256))/1000,0.01,3)
 +                    }
 R_e        NET5 0    1E6

 e10        NET10     0    Value {0.2}
 R_K        NET10     0    1E6

 R1001 gr 0     1E6
 E1001 gr 0     value {limit(V(gk),-8,15)}

 e_P8 P8   0    Value {Limit((-613n*V(gk)**4+25.39u*V(gk)**3-
 +              332u*V(gk)**2+1.726m*V(gk)-1.502m)*0.9,0.0001,0.1)
 +                    }
 R_R  P8   0    1E6

 *********************************
 G2 d1 s1 value {              // Drain current
 +    if(V(d3,s1)<=0,
 +         if (V(gk)<V(NET3),
```

```
+         0
+         ,
+         -((((v(NET5)+v(NET4))*(v(gk)-V(NET3))))*
+         (1+v(P9)*v(s1,d3))*(((log(1+exp(v(gk)-V(NET3))))**2)-
+         ((log(1+exp(v(gk)-V(NET3)-(V(NET2)*v(s1,d3)*
+         ((1+exp(-v(NET10)*v(s1,d3)))**v(NET1))))))**2)))*
+         (-2.361u*V(Tjc)**2+272.272u*V(Tjc)+1.0747)
+         )
+         ,
+         if (V(gk)<V(NET3),
+         0
+         ,
+         ((((v(NET5))*(v(gk)-V(NET3))))*
+         (1+v(P8)*v(d3,s1))*(((log(1+exp(v(gk)-V(NET3))))**2)-
+         ((log(1+exp(v(gk)-V(NET3)-(V(NET2)*v(d3,s1)*
+         (1+exp(-v(NET10)*v(d3,s1)))**v(NET1))))))**2)))
+         *(-2.361u*V(Tjc)**2+272.272u*V(Tjc)+1.0747)
+         )
+              )
+              }
R_G2 d1 s1 10E6
.ends gmos_C3M0016120D

****************************************
.subckt cgdmos_C3M0016120D d3 g1       // Cgs subcircuit model
.param k1=2565p
.param k2=0.5
.param ka=90
.param kb=0.3
.param kc=3.0

G11 g1 d11 value {
+         k1*(
+         (1+(limit(v(d3,g1),0,407))*(1+ka*(1+TANH(kb*V(d3,g1)-kc))/2))**-
k2
+         )*ddt(v(g1,d11))
+              }
R_G11 g1 d3 10E6

R_CGD d11 d3 1E-4
.ends cgdmos_C3M0016120D


.subckt cds_C3M0016120D d3 s1     // Cds subcircuit model
.param Cjo = 5179p
.param Vj  = 4.4365
.param M   = 0.643

G12 d12 s1 value {
+         (Cjo/(1+(limit(v(d3,s1),0,580)/Vj)**M))*ddt(v(d12,s1))
+              }
```

```
R_G12 d3 s1 10E6

R_CDS d12 d3 1E-4

.ends cds_C3M0016120D
***************************************
```

3.7.5 Commercial SiC MOSFET Model

To facilitate circuit design for power electronics engineers, many SiC MOSFET manufacturers provide SPICE models tailored to their own devices, each with unique modeling characteristics and approaches.

1. **Infineon Model**

Taking the 1200mV 45mΩ SiC MOSFET as an example, Infineon's approach to modeling SiC MOSFETs is illustrated. Figure 3.38a shows the main structure of the IMZ120R045M1 SiC MOSFET model in a TO-247-4 package. It consists primarily of the integrated SiC MOSFET chip model embedded within the package, parasitic models for the TO-247-4 package, and thermal impedance models describing the thermal resistance Z_{th} between the SiC MOSFET chip junction temperature (T_j) and package case temperature (T_C). The modeling of package parasitics significantly affects switch waveforms and switching energies, especially during fast transient conditions. Therefore, modeling package parasitics requires careful consideration, often involving a combination of finite element simulations and specialized characteristic measurements. This approach aims to clearly distinguish various influences from the chip, package, and measurement environment in actual switch behaviors.

The model of the SiC MOSFET chip is a core component of the entire power device compact model. Figure 3.38b illustrates the main structure of the SiC MOSFET chip

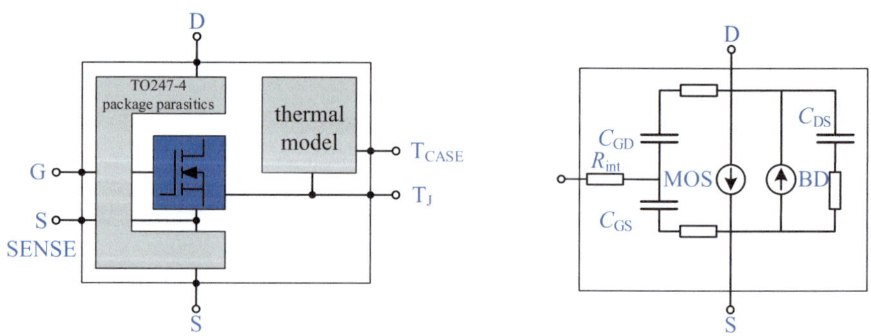

(a) Device Model (b) Chip Model

Fig. 3.38 Infineon SiC MOSFET model equivalent circuit [14] model. It has a relatively simple structure, comprising only two current sources for the MOSFET channel and the body diode, three voltage-dependent capacitors for C_{GS}, C_{DS}, and C_{GD}, and some resistor elements. The capacitance and I–V characteristics of the chip largely determine the overall electrical performance of the device. The chip model contains very few nodes and relies heavily on equations. It does not include implicit or recursive equations, which are common causes of convergence and performance issues. This approach significantly enhances the computational speed of circuit simulations. However, besides the number of nodes, the speed and convergence of compact models also depend on the structure of the implemented equations. Certain types of equations require much larger computational effort than others. To further enhance stability and convergence, it is important to avoid using hard transitions with if, else, and limit statements. Generally, continuous and differentiable functions are preferred over those with poles or discontinuities.

Research indicates that the model shown in Fig. 3.38 has characteristics such as compact structure, minimal elements, explicit equations, continuity, and differentiability, resulting in fast computation speed and strong convergence. Comprehensive comparisons of variations in 960 parameters in dynamic behaviors have revealed overall good consistency between simulations and measurements across all turn-on and turn-off transitions. In most cases, the relative deviations of switching energies, source-drain voltage rates (dv/dt), and drain current rates (di/dt) are below 20%.

2. **Wolfspeed Model**

Wolfspeed's SiC MOSFET SPICE model includes two types: the C2M series model and the C3M series model. This model features a layered code structure, dividing the model into several modules, each represented using subcircuits. Although the equivalent circuit appears relatively simple, the model's complexity arises from the numerous fitting functions and model parameters employed.

The modeling of drain current in the C2M series model is based on improvements to the EKV model to match the specific characteristics of SiC materials. It considers the influence of channel length modulation and introduces additional curve-fitting parameters. Two VCCSs (G1 and G2 in Fig. 3.39a) are used to describe the behavior of forward and reverse drain currents. Additionally, two voltage sources, E2_g and E3_g in the core part, primarily describe transconductance and threshold voltage, both of which are temperature-dependent equations. The VCCS at the drain (G1_d in Fig. 3.39a) describes the on-state resistance of the device, while its reverse bias characteristics are implemented using a standard form of a body diode model. The establishment of this model can utilize Model Editor software to extract relevant parameters.

The C3M series model, on the other hand, is based on the Curtice-Ettenberg FET model. Unlike the C2M series, this model uses only one VCCS (G1 in Fig. 3.39b) to describe the drain current section. Its reverse bias characteristics are implemented using two VCCSs (G2 and G3 in Fig. 3.39b), representing the diode's resistance and current characteristics, respectively.

Fig. 3.39 Wolfspeed SiC MOSFET model equivalent circuit [14]

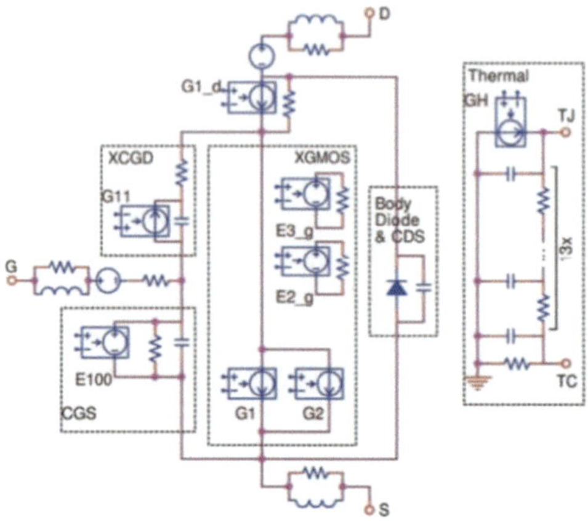

(a) C2M series SiC MOSFET

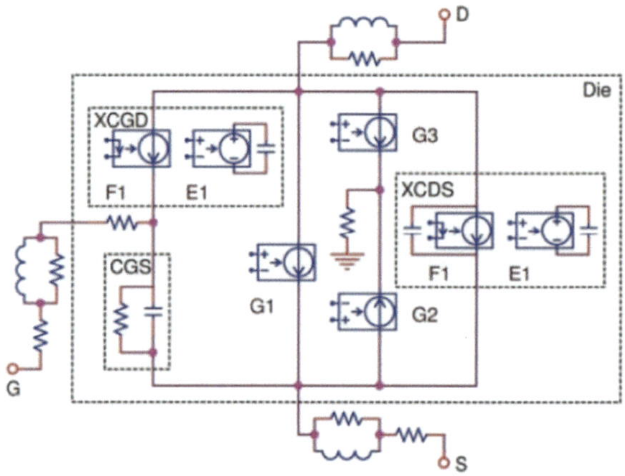

(b) C3M series SiC MOSFET

A key factor affecting the dynamic characteristics of the model is the modeling of junction capacitances. In the C3M series model, the capacitance models are implemented using VCCS (XCGD and XCDS in Fig. 3.39b), where the controlled part includes voltage sources generating voltage derivatives over time, and another part is used to implement capacitor equations with mathematical expressions reflecting their voltage dependencies. The treatment of C_{GS} sets it as a fixed value, assuming that the capacitance does not vary with terminal voltage, a method also employed by ST and ROHM models in this section.

In contrast, the C2M series capacitance model uses a single VCCS (G11 in Fig. 3.39a), directly incorporating voltage time derivatives and capacitance functions, and approximating the latter using hyperbolic tangent functions.

The subcircuit topology of the C2M series model includes a thermal impedance RC network, assuming linear thermal conduction. The thermal conduction path from case temperature (T_C) to junction temperature (T_j) is equivalent to 14 RC pairs, represented in the form of a controlled current source (GH in Fig. 3.39b) generating thermal power, summarizing the heat generated by the gate-source and drain-source circuit sections. The C3M series model does not include thermal components, and thus, it does not account for device self-heating effects and cooling systems.

3. **ST Corporation Model**

ST Corporation's SiC MOSFET model features a relatively more complex equivalent circuit, based on the LEVEL 1 model with two additional temperature coefficients. It has improved the expression of the drain current, described using only one VCCS (Gmos in Fig. 3.40) to capture its behavior. The reverse bias characteristics are primarily implemented through two VCCSs within the body diode model (G_R_didd and G_diode in Fig. 3.40), each used to describe the resistance and current behavior of the diode.

The capacitance section of this model shares a similar topology with Wolfspeed's C3M series model, achieved using VCCS (G_miller and G_cds in Fig. 3.40). However, the capacitance in this model is not merely a function of voltage; it also considers the effect of junction temperature. Therefore, this section includes two additional controlled voltage sources (Edev and Edevc in Fig. 3.40) to reflect temperature dependence.

The model employs 5 RC components to characterize thermal resistance properties. Unlike Wolfspeed's C2M series model, this model incorporates three sources (G_pwl, G_pw2, and G_pw3 in Fig. 3.40), each dedicated to calculating power dissipation related to drain-source current, dissipation in the body diode current, and avalanche breakdown.

4. **ROHM Corporation Model**

ROHM Corporation's SiC MOSFET model is highly similar to ST Corporation's model but includes fewer components. For describing the drain current, this model

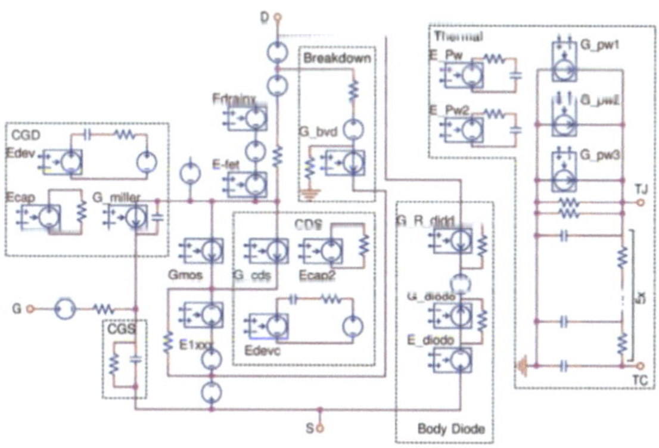

Fig. 3.40 ST SiC MOSFET model equivalent circuit [14]

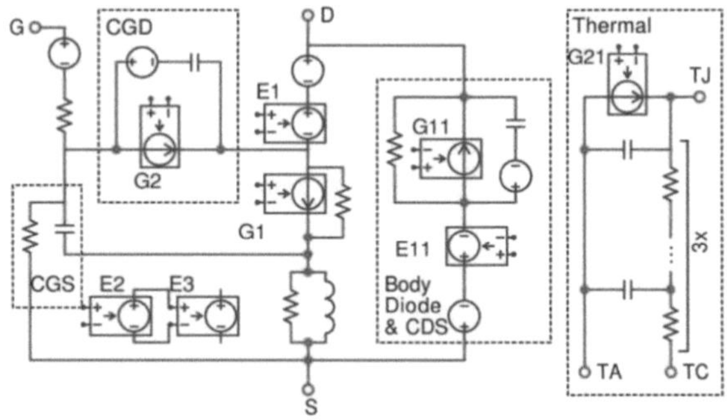

Fig. 3.41 ROHM SiC MOSFET model equivalent circuit [14]

combines methods from the Curtice model and EKV model, incorporating components related to V_{GS} and V_{DS} in its expression. Similar to ST's approach, the model uses a single VCCS (G1 in Fig. 3.41) to describe the drain current. Additionally, E1 in the drain section is used to describe the on-state resistance of the device, influencing both its forward and reverse characteristics.

The body diode model in ROHM's model is also similar, requiring separate descriptions of diode resistance and current characteristics. However, in this model, they are represented using a CCVS (E11 in Fig. 3.41) and a VCCS (G11 in Fig. 3.41). The handling of capacitance models and thermal impedance RC network equivalent models in this model is similar to Wolfspeed's C3M series model, thus not detailed further here.

3.7.6 Challenges in SiC MOSFET Modeling

Due to the structural similarity between SiC MOSFETs and traditional Si MOSFETs, initial SiC MOSFET models heavily borrowed from well-established equivalent circuit models of Si MOSFETs. Despite some variations in SiC MOSFET equivalent circuit models across literature, their basic construction is similar. Common equivalent circuits are depicted in Fig. 3.42, including core components such as the MOS channel M_N, body diode D_N, junction capacitances (C_{GS}, C_{GD}, C_{DS}), and resistances (R_G, R_D, R_S).

Typically, modeling these components requires complex analytic functions derived from semiconductor physics principles, solving Poisson equations, and simplifying formulas. These analytic expressions often depend on device geometries (W, L, NF), temperature (T), and biases (V_{GS}, V_{GD}, V_{DS}),, reflecting real-world device characteristics.

However, due to the unique properties of SiC materials, several reliability issues persist. Currently, SiC MOSFETs face three distinct challenges that need to be addressed:

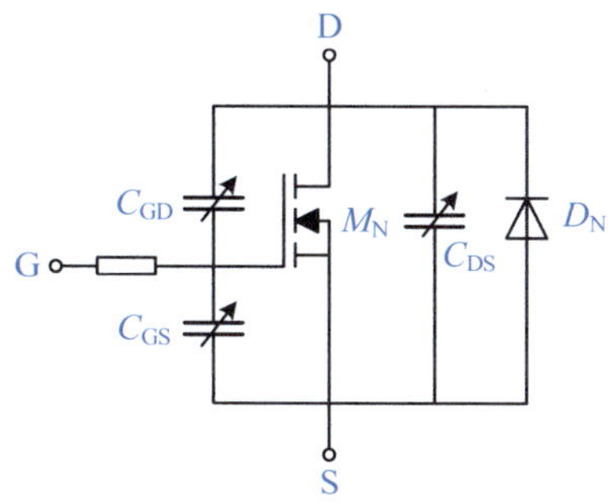

Fig. 3.42 SiC MOSFET equivalent circuit

(1) Lack of High-Voltage High-Current (HVHC) Output Characteristics: Existing SiC MOSFET models typically cover output characteristics for V_{DS} within the range of 20 V to twice the rated current. During switching, SiC MOSFETs pass through the HVHC saturation region, leading to inaccuracies in model fidelity when reproducing device dynamic characteristics. This can be addressed by employing double-pulse tests to simultaneously record V_{GS}, V_{DS}, and I_{DS} of the device under different conditions at slower switching speeds to obtain complete HVHC characteristic curves.

(2) Inadequate Representation of Junction Capacitance: Junction capacitances in SiC MOSFETs play a crucial role in determining their dynamic performance. These capacitances are highly nonlinear and significantly influenced by terminal voltages. Therefore, accurate characterization and modeling of these junction capacitances are essential for correctly reconstructing the SiC MOSFET's transient behavior in the time domain. Traditional modeling methods often only consider junction capacitance when the SiC MOSFET is in the off-state, typically using datasheet provided capacitance curves at $V_{GS} = 0$ V. Unlike capacitance in the off-state, on-state capacitance greatly affects the switch-off transient characteristics from conduction to off-state. Failure to consider on-state capacitance characteristics will result in inaccurate simulation of device switch-off processes. However, on-state capacitance cannot be measured using conventional capacitance meters and requires measurement through S-parameter analysis.

(3) Lack of an Equivalent Circuit Model to Describe Gate Oxide State in SiC MOSFETs: SiC MOSFETs exhibit high trap densities at the SiC/SiO2 interface, leading to significant instability in threshold voltage. Under switching stress, threshold voltage drifts may even become permanent, making changes in this parameter crucial for device characteristic analysis. However, directly describing this characteristic in an equivalent circuit model remains challenging, severely impacting the lifetime assessment of SiC MOSFETs in power electronics circuits and thereby affecting the design progress for users employing SiC MOSFETs.

References

1. Sakairi H, Yanagi T, Otake H, Kuroda N, Tanigawa H (2018) Measurement methodology for accurate modeling of SiC MOSFET switching behavior over wide voltage and current ranges. IEEE Trans Power Electron 33(9):7314–7325
2. Jimenez S, Lemmon A, Nelson B, Deboi B (2021) Comprehensive characterization of MOSFET intrinsic capacitances. IEEE Appl Power Electron Conf Expos (APEC) 2021:1524–1530
3. Li X, Zhang L, Guo S et al (2015) Understanding switching losses in SiC MOSFET: toward lossless switching. In: 2015 IEEE 3rd workshop on wide bandgap power devices and applications (WiPDA). IEEE, pp 257–262
4. Baliga BJ (1989) Power semiconductor device figure of merit for high-frequency applications. IEEE Electron Device Lett 10(10):455–457
5. Kim IJ, Matsumoto S, Sakai T, et al (1995) New power device figure of merit for high-frequency applications. In: Proceedings of international symposium on power semiconductor devices and IC's: ISPSD '95, vol. 104, no 104, pp 309–314
6. Kolar JW, Biela J, Minibock J (2009) Exploring the pareto front of multi-objective single-phase PFC rectifier design optimization—99.2% efficiency vs. 7 kW/dm^3 power density. In: IEEE international power electronics & motion control conference
7. Huang AQ (2004) New unipolar switching power device figures of merit. IEEE Electron Device Lett 25(5):298–301
8. Ying Y (2008) Device selection criteria--based on loss modeling and figure of merit. Virginia Polytechnic Institute and State University
9. Fu D (2010) Topology investigation and system optimization of resonant converters. Virginia Polytechnic Institute and State University
10. Plexim GmbH (2019) PLECS user manual. Rev. 4.3
11. Wolfspeed Speedfit. https://www.wolfspeed.com/speedfit/
12. PowerSim Inc. (2019) IGBT and MOSFET loss calculation in thermal module. Tutorial
13. The MathWorks, Inc. Loss calculation in a three-phase 3-level inverter. Featured Examples. https://ww2.mathworks.cn/help/physmod/sps/examples/loss-calculation-in-a-three-phase-3-level-inverter.html#d117e20378
14. Sochor P, Huerner A, Elpelt R (2019) A fast and accurate SiC MOSFET compact model for virtual prototyping of power electronic circuits. In: International exhibition and conference for power electronics, intelligent motion, renewable energy and energy management (PCIM Europe 2019), pp 1–8
15. Stefanskyi A, Starzak U, Napieralski A (2017) Review of commercial SiC MOSFET models: validity and accuracy. In: International conference mixed design of integrated circuits and systems, pp 488–493

Further Reading

16. ROHM Co., Ltd. (2014) SiC power devices and modules. Application Note, 14103EBY01
17. Infineon Technologies AG (2018) CoolSiC™ 1200 V SiC MOSFET application note. Application Note, AN2017–46, Rev. 1.01
18. Infineon Technologies AG (2018) CoolSiC™ 650 V M1 SiC trench power device Infineon's first 650 V silicon carbide MOSFET for industrial applications. Application note, AN_1907_PL52_1911_144109, Rev. 1.0
19. Basler T, Heer D, Peters D et al (2018) Practical aspects and body diode robustness of a 1200V SiC trench MOSFET. PCIM Europe 2018:536–542
20. Fairchild Semiconductor Corporation (2013) MOSFET basics. Application note, AN-9010, Rev. 1.0.5
21. Fuji Electric Co. Ltd. (2016) Power MOSFET. Application note, AN-080E Rev.1.2

22. TOSHIBA Electronic Devices & Storage Corporation. Power MOSFET electrical characteristics. Application note
23. Infineon Technologies AG (2014) Automotive MOSFETs data sheet explanation. Application note, Rev. 1.2
24. Alan H (2012) Infineon OptiMOSTM power MOSFET datasheet explanation. Application note, AN 2012-03, Rev. 1.1. Infineon Technologies AG
25. Fairchild Semiconductor Corporation (2014) Shielded gate PowerTrench® MOSFET datasheet explanation. Application note, AN-4163, Rev. 1.0.1
26. Björn B, Raffael S, Ulrich S et al (2013) Applying IGBTs. Application note, 5SYA2053, Rev. 04, ABB Switzerland LTD Semiconductors
27. Fairchild Semiconductor Corporation (2001) IGBT basics I. Application note 9016
28. Fairchild Semiconductor Corporation (2002) IGBT basics II. Application note 9020
29. TOSHIBA Electronic Devices & Storage Corporation (2018) IGBTs (insulated gate bipolar transistor). Application note
30. Infineon Technologies AG (2015) Discrete IGBT explanation of discrete IGBTs' datasheets. Application note, AN2015-13, Rev. 1.0
31. STMicroelectronics (2014) IGBT datasheet tutorial. Application note, AN4544, DocID026535, Rev. 1
32. ON Semiconductor Corporation (2012) Reading ON semiconductor IGBT datasheets. Application note, AND9068/D, Rev. 0
33. Infineon Technologies AG (2017) Dynamic thermal behavior of MOSFETs simulation and calculation of high power pulses. Application note, AN_201712_PL11_001, Rev 1.0
34. Maurizio M, Antonino G, Giuseppe S (2015) Thermal effects and junction temperature evaluation of power MOSFETs. Application note, DocID028570 Rev 1, STMicroelectronics
35. TOSHIBA Electronic Devices & Storage Corporation (2018) Power MOSFET maximum ratings. Application note
36. Sanjay H (2009) A practical look at current ratings. Application note. Alpha and Omega Semiconductor Inc.
37. Fei W, Kai L, Anup B (2009) Power MOSFET continuous drain current rating and bonding wire limitation. Application note, Rev. 01, Alpha and Omega Semiconductor Inc.
38. STMicroelectronics (2006) Power dissipation and its linear derating factor, silicon limited drain current and pulsed drain current in MOSFETs. Application note, AN2385, Rev. 1
39. Baschnagel A, Tsyplakov E (2019) IGBT short circuit safe operating area (SOA) capability and testing. Application note, 5SYA2095, Rev. 01, ABB Switzerland LTD semiconductors
40. TOSHIBA Electronic Devices & Storage Corporation (2018) Derating of the MOSFET safe operating area. Application note
41. Schoiswohl J (2017) Linear mode operation and safe operating diagram of power-MOSFETs. Application note, AP99007, Rev. 1.1, Infineon Technologies AG
42. Chen Z (2009) Characterization and modeling of high-switching-speed behavior of SiC active devices. Virginia Polytechnic Institute and State University
43. Bernhard Z (2020) CoolMOS™ gate drive and switching dynamics. Application note, AN_1909_PL52_1911_173913, Rev. 1.0, Infineon Technologies AG
44. Toshiba Electronic Devices & Storage Corporation (2017) Parasitic oscillation and ringing of power MOSFETs. Application note
45. IEC 60747-2:2016 (2016) Semiconductor devices—Part 2: discrete devices—rectifier diodes
46. IEC 60747-8:2010 (2010) Semiconductor devices—discrete devices—Part 8: field-effect transistors
47. IEC 60747-9:2019 (2019) Semiconductor devices—Part 9: discrete devices—insulated-gate bipolar transistors (IGBTs)
48. Xiong Y, Sun S, Jia H et al (2009) New physical insights on power MOSFET switching losses. IEEE Trans Power Electron 24(2):525–531
49. ROHM Co., Ltd. (2020) Calculation of power dissipation in switching circuit. Application note, No.62AN132E Rev.001

50. ROHM Co., Ltd. (2020) Calculating power loss from measured waveforms. Application note, No.62AN134E Rev.001
51. Ulrich N (2015) Determining switching losses of SEMIKRON IGBT modules. Application note AN-1403, Rev.00, SEMIKRON International GmbH
52. Tong Z, Park S (2019) Empirical circuit model for output capacitance losses in silicon carbide power devices. IEEE Appl Power Electron Conf Expos (APEC) 2019:998–1003
53. Nikoo MS, Jafari A, Perera N et al (2020) New insights on output capacitance losses in wide-band-gap transistors. IEEE Trans Power Electron 35(7):6663–6667
54. Tong Z, Zulauf G, Xu JL et al (2019) Output capacitance loss characterization of silicon carbide schottky diodes. IEEE J Emerg Select Top Power Electron 7(2):865–878
55. Lin Z (2019) Study on the intrinsic origin of output capacitor hysteresis in advanced superjunction MOSFETs. IEEE Electron Device Lett 40(99):1297–1300
56. Nikoo MS, Jafari A, Perera N et al (2019) Measurement of large-signal C_{OSS} and C_{OSS} losses of transistors based on nonlinear resonance. IEEE Trans Power Electron 35(3):2242–2246
57. Fedison JB, Harrison MJ (2016) C_{OSS} hysteresis in advanced super junction MOSFETs. IEEE Appl Power Electron Conf Expos (APEC) 2016:247–252
58. Fedison JB, Fornage M, Harrison MJ et al (2014) C_{OSS} related energy loss in power MOSFETs used in zero-voltage-switched applications. IEEE Appl Power Electron Conf Expos (APEC) 2014:150–156
59. ON Semiconductor Corporation (2016) MOSFET gate-charge origin and its applications. Application note, AND9083/D, Rev. 2
60. Jess B (2016) Power MOSFET basics: understanding gate charge and using it to assess switching performance. Application note, AN608, DocID-73217, Vishay Intertechnology, Inc.
61. Hisao K, Ryo T (2014) Selecting best device for power circuit design through gate charge characterization. White paper, 5991–4405. Keysight Technologies, Inc.
62. STMicroelectronics (2012) Power MOSFET: Rg impact on applications. Application note, AN4191, DocID023815, Rev. 01
63. Peter H, Jon H (2007) Understanding diode reverse recovery and its effect on switching losses. Fairchild Semiconductor Corporation, Fairchild Power Seminar
64. STMicroelectronics (2017) Calculation of turn-off power losses generated by an ultrafast diode. Application note, AN5028, DocID030470, Rev 1
65. Kasper M, Burkart RM, Deboy G et al (2016) ZVS of power MOSFETs revisited. IEEE Trans Power Electron 31(12):8063–8067
66. Toshiba Electronic Devices & Storage Corporation (2017) MOSFET avalanche ruggedness. Application note
67. Infineon Technologies AG (2017) Some key facts about avalanche. Application note, AN_201611_PL11_002, Rev. 1.0
68. STMicroelectronics (2006) Power MOSFET avalanche characteristics and ratings. Application note, AN2344, Rev 1
69. STMicroelectronics (2014) The avalanche issue: comparing the impacts of the IAR and EAS parameters. Application note, AN4337, DocID025012, Rev 1
70. Alfio S (2015) Half bridge resonant LLC converters and primary side MOSFET selection. Application note, DocID027986, Rev. 01, STMicroelectronics
71. Infineon Technologies AG (2014) Primary side MOSFET selection for LLC topology. Application note, AN_20105_PL52_001, Rev. 01
72. Infineon Technologies AG (2019) 600 V CoolMOS™ CFD7 latest fast diode technology tailored to soft switching applications. Application note, AN_201708_PL52_024, Rev. 2.1
73. Mantooth HA, Peng K, Santi E et al (2015) Modeling of wide bandgap power semiconductor devices—Part I. IEEE Trans Electron Devices 62(2):423–433
74. Santi E, Peng K, Mantooth HA (2015) Modeling of wide-bandgap power semiconductor devices—Part II. IEEE Trans Electron Devices 62(2):434–442
75. Burkart RM (2016) Advanced modeling and multi-objective optimization of power electronic converter systems. ETH Zurich

76. Feix G, Dieckerhoff S, Allmeling J et al (2009) Simple methods to calculate IGBT and diode conduction and switching losses. In: 2009 13th European conference on power electronics and applications, pp 1–8
77. Dušan G, Marco P, Andreas K (2006) MOSFET power losses calculation using the data-sheet parameters. Application note, Rev. 1.1, Infineon Technologies AG
78. Dušan G, Marco P (2009) IGBT power losses calculation using the data-sheet parameters. Application note, Rev. 1.1, Infineon Technologies AG
79. Anurag A, Acharya S, Bhattacharya S (2020) An accurate calorimetric loss measurement method for SiC MOSFETs. IEEE J Emerg Select Top Power Electron 8(2):1644–1656

Chapter 4
Comparison of SiC and Si Device Characteristics

At 1700 V and below, SiC power devices are available in various voltage ratings. SiC power devices with different voltage ratings are designed for specific application scenarios, replacing different types of Si power devices. The 650–1000 V SiC MOSFETs primarily target Si SJ-MOSFETs and Si IGBTs. The 1200 and 1700 V SiC MOSFETs are mainly designed to replace Si IGBTs. SiC Schottky barrier diode (SiC SBD) is intended to replace Si FRDs. As a newer technology, SiC power devices aim to leverage their inherent advantages to help power converters achieve superior characteristics, such as simpler topology, higher efficiency, and increased power density. In certain application domains, SiC power devices have become the preferred choice, gradually replacing Si power devices with corresponding specifications.

It's crucial to understand the differences between SiC power devices and Si power devices' characteristics. It allows for a clear identification of the advantages of SiC power devices, aiding in their wider adoption. Additionally, this knowledge can assist engineers in explaining or anticipating potential issues that may arise when replacing Si power devices with SiC power devices. The following will compare the characteristics of SiC power devices and Si power devices using datasheets and experimental results. This includes the SiC MOSFET and the Si SJ-MOSFET with a specification of 650 V/20 A, the SiC MOSFET and the Si IGBT with a specification of 1200 V/40 A, and the SiC SBD, the Si FRD, the SiC MOSFET body diode, and the Si SJ-MOSFET body diode, all with a specification of 650 V/20 A.

It is essential to note that the comparative results presented below are based on specific device models. When selecting devices from different manufacturers or employing various technologies, the overall trends in the comparative results remain consistent. However, the specific values and relative proportions of SiC and Si device characteristics will vary.

Y. Gao and Y. Zhang, *Silicon Carbide Power Devices*, CPSS Power Electronics Series,
https://doi.org/10.1007/978-981-96-3480-4_4

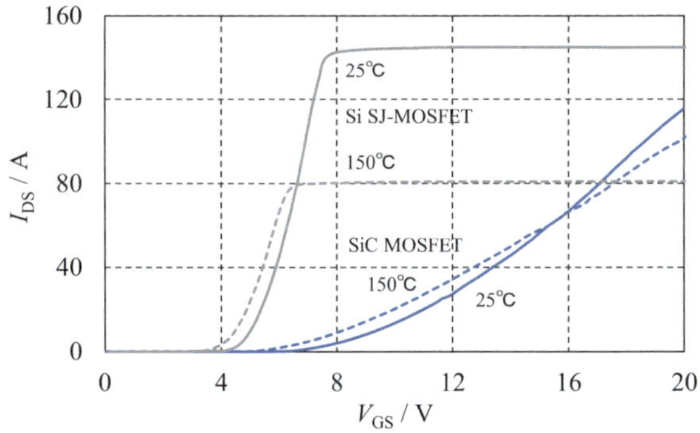

Fig. 4.1 Transfer characteristics curves of Si SJ-MOSFET and SiC MOSFET

4.1 SiC MOSFET and Si SJ-MOSFET

4.1.1 Transfer Characteristics

The transfer characteristic curves of Si SJ-MOSFET and SiC MOSFET are shown in Fig. 4.1.

For SJ-MOSFET, when V_{GS} exceeds $V_{GS(th)}$, the current of the SJ-MOSFET rises rapidly. However, when V_{GS} is higher than a certain value, the output current no longer changes. At this time, the SJ-MOSFET enters the saturation state. This implies that a smaller V_{GS} is sufficient to fully turn on the SJ-MOSFET. Once the SJ-MOSFET enters saturation, a higher V_{GS} won't increase its conduction current capacity. Additionally, with increasing temperature, the output current before saturation exhibits a positive temperature coefficient, while during saturation, both V_{GS} and the saturation current exhibit a negative temperature coefficient.

For SiC MOSFET, when V_{GS} exceeds $V_{GS(th)}$, the rate of current increase for the SiC MOSFET is noticeably slower than the SJ-MOSFET, indicating that the transconductance (g_{fs}) of the SiC MOSFET is smaller. The output current shows a positive temperature coefficient when V_{GS} is below 15 V, and a negative temperature coefficient when V_{GS} is above 15 V. The impact of temperature is significantly smaller than observed in the SJ-MOSFET. It is noteworthy that even when V_{GS} reaches 20 V, the output current of the SiC MOSFET continues to increase, without entering saturation. Due to these characteristics, the turn-on drive voltage for the SiC MOSFET needs to be higher than that of the SJ-MOSFET to ensure full conduction and operation in the negative temperature coefficient region for the SiC MOSFET.

4.1.2 Output Characteristics and On-State Resistance

As shown in Fig. 4.2, when V_{GS} exceeds 8 V, the SJ-MOSFET is fully turned on. Its I_{DS}–V_{DS} curve almost overlaps and is significantly influenced by temperature.

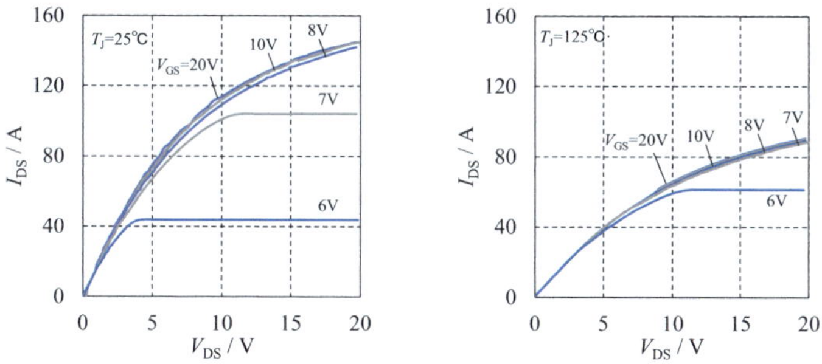

Fig. 4.2 Output characteristics of SJ-MOSFET

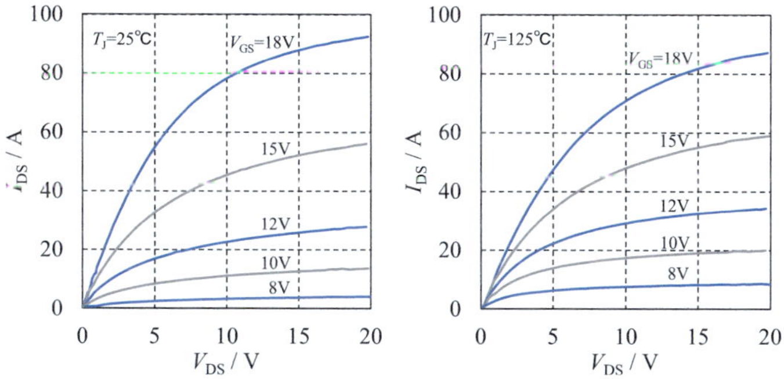

Fig. 4.3 Output characteristics of SiC MOSFET

In contrast, the SiC MOSFET exhibits distinct I_{DS}–V_{DS} curves at different V_{GS} levels, and its temperature sensitivity is relatively mild compared to the SJ-MOSFET, as illustrated in Fig. 4.3. These characteristics align with their respective transfer characteristics.

Figure 4.4 illustrates the $R_{DS(on)}$–T_J characteristics of the SJ-MOSFET and SiC MOSFET. The $R_{DS(on)}$ of the SJ-MOSFET increases with the rise in T_J. This is because the positive temperature coefficients of R_{JFET} and R_{DRIFT} consistently dominate the $R_{DS(on)}$ of SJ-MOSFET. In contrast, the $R_{DS(on)}$-T_J curve of SiC MOSFET exhibits a U-shape. This results from the proportional contributions and temperature characteristics of the various components constituting $R_{DS(on)}$ as explained in detail in this Chap. 4.

Additionally, the impact of temperature on the $R_{DS(on)}$ of SiC MOSFET is significantly milder compared to SJ-MOSFET. At 25 and 150 °C the $R_{DS(on)}$ of SiC MOSFET is 72 and 94 mΩ, respectively, representing a 30% increase. On the other hand, SJ-MOSFET's $R_{DS(on)}$ is 58 at 25 °C and increases to 139 mΩ at 150 °C, indicating a 140% rise. This implies that SiC MOSFET can maintain lower conduction losses even at elevated temperatures, while special attention is needed for the $R_{DS(on)}$ of SJ-MOSFET at high temperatures.

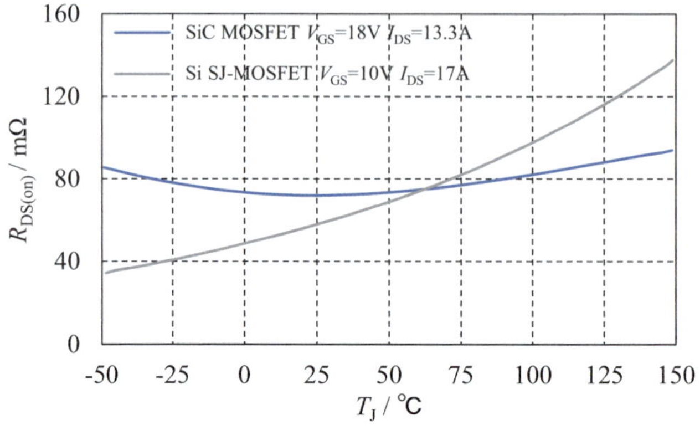

Fig. 4.4 Temperature characteristics of $R_{DS(on)}$ for SJ-MOSFET and SiC MOSFET

4.1.3 C–V Characteristics

The C–V curves for SiC MOSFET and Si SJ-MOSFET are depicted in Fig. 4.5. It is evident that the C_{iss} of SiC MOSFET is significantly smaller than that of Si SJ-MOSFET. The C_{oss} of SiC MOSFET is smaller at low voltages and larger at high voltages compared to Si SJ-MOSFET. The C_{rss} of SiC MOSFET shows a relatively gentle variation with voltage. The C_{iss} of this Si SJ-MOSFET sharply decreases with increasing voltage due to the superjunction structure, followed by a subsequent rise.

4.1.4 Switching Characteristics

Due to the significant impact of the gate drive resistance R_G ($R_G = R_{G(ext)}+R_{G(int)}$) on the switching characteristics of switching devices, $R_{G(int)}$ can be adjusted through chip design and manufacturing processes. $R_{G(ext)}$ can be freely chosen and both are not strongly related to chip materials. Therefore, when comparing the switching characteristics, suitable values of $R_{G(ext)}$ are selected for SiC MOSFET and Si SJ-MOSFET, ensuring that both have the same R_G, and the freewheeling diode is the same SiC SBD. This eliminates the influence of R_G, allowing for a comparison of the impact of SiC and Si materials on MOSFET switching characteristics.

At 25 °C, the turn-on and turn-off processes of the SiC MOSFET and Si SJ-MOSFET are shown in Figs. 4.6 and 4.7. The turn-on delay $t_{d(on)}$ and turn-off delay $t_{d(off)}$ of the SiC MOSFET are 7.7 and 8.1 ns, respectively, which are shorter than the 19.4 and 128.3 ns of the Si SJ-MOSFET. These switching delays significantly impact high-frequency operation. The primary reason for this difference is the larger C_{iss} of the Si SJ-MOSFET mentioned earlier, causing slower rise and fall times of V_{GS} under identical driving circuit and R_G conditions. The switching times of the SiC MOSFET are very close to those of the Si SJ-MOSFET, with the SiC MOSFET

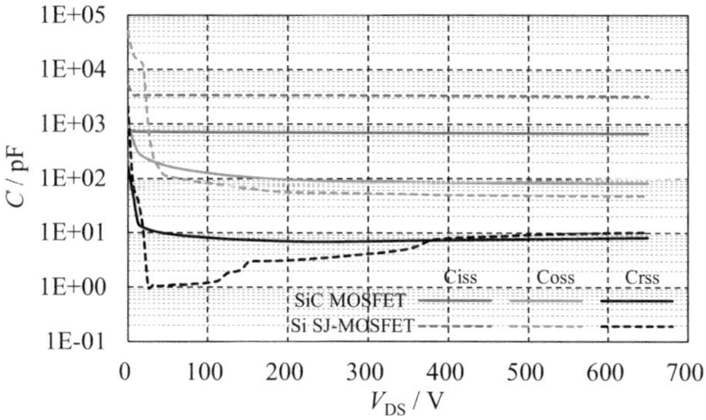

(a) The C axis is a linear coordinate axis

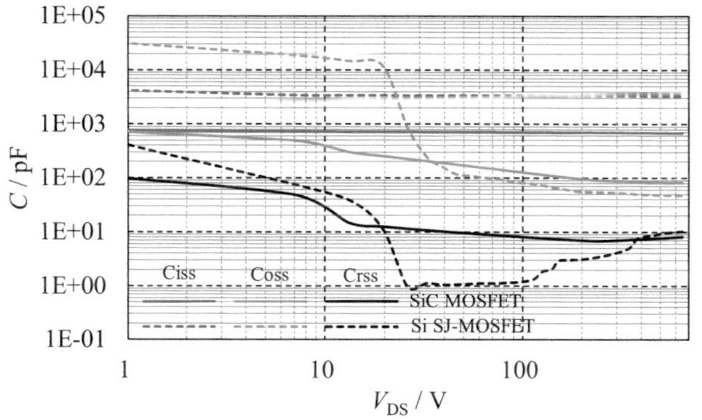

(b) The C axis is a logarithmic axis

Fig. 4.5 C–V characteristics of SiC MOSFET and Si SJ-MOSFET

exhibiting slightly lower switching energy. The SiC MOSFET has higher turn-on current spikes, while the Si SJ-MOSFET has higher turn-off voltage spikes. Overall, the switching speed of the two does not differ greatly. This contrasts with the general perception that SiC devices have much faster switching speeds than Si devices, a view mainly based on comparisons between Si MOSFETs and Si IGBTs. The switching characteristics of SiC MOSFET and Si SJ-MOSFET are shown in Table 4.1.

The turn-on and turn-off processes of the SiC MOSFET and Si SJ-MOSFET at 25, 75, and 125 °C are shown in Figs. 4.8, 4.9, 4.10, and 4.11. As the temperature increases, the turn-on delay of both the SiC MOSFET and Si SJ-MOSFET decreases. However, the switching speed of the SiC MOSFET increases, and its turn-on energy decreases, whereas the opposite is true for the Si SJ-MOSFET. The turn-on current spike of the SiC MOSFET increases, while that of the Si SJ-MOSFET decreases. As the temperature rises, the turn-off delay, turn-off time, and turn-off energy of

Fig. 4.6 Turn-on process of SiC MOSFET and Si SJ-MOSFET

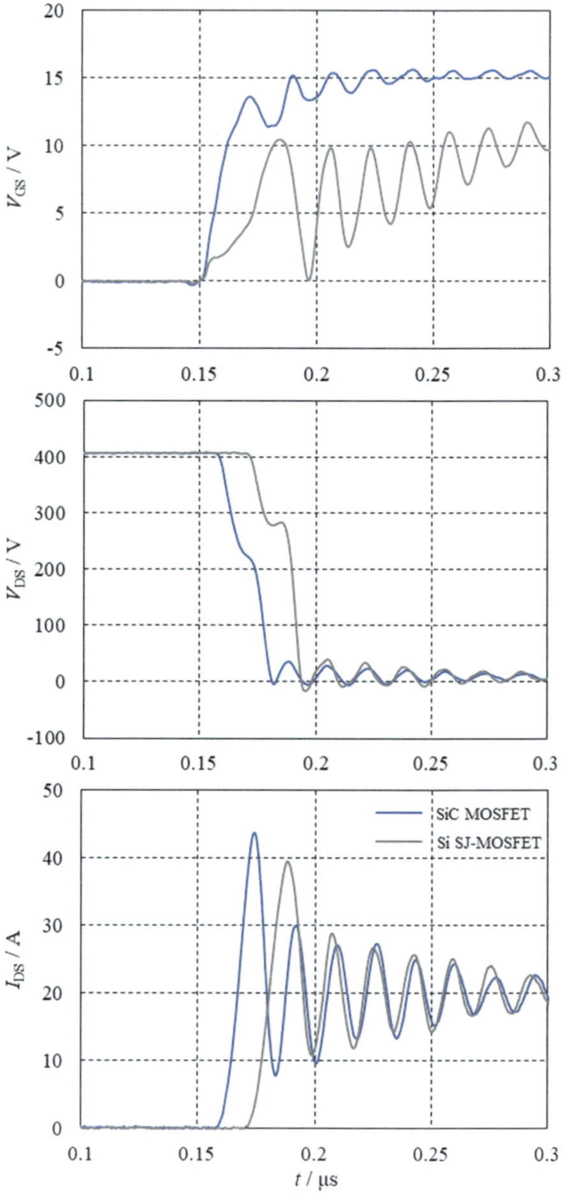

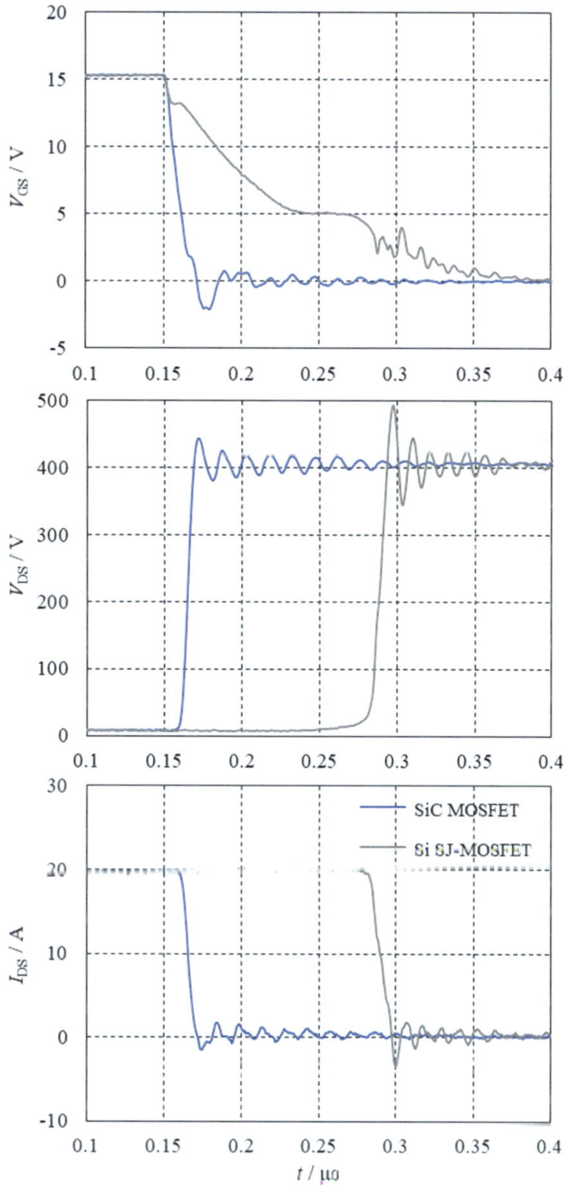

Fig. 4.7 Turn-off process of SiC MOSFET and Si SJ-MOSFET

Table 4.1 Switching characteristics of SiC MOSFET and Si SJ-MOSFET

Devices	SiC MOSFET			Si SJ-MOSFET		
Temperature	25 °C	75 °C	125 °C	25 °C	75 °C	125 °C
$T_{d(on)}$/ns	7.7	7.4	7.0	19.4	18.6	17.4
T_r/ns	18.7	18.0	17.8	18.6	19.7	21.4
I_{spike}/A	43.8	44.4	44.6	39.5	388.5	37.1
E_{on}/μJ	98.4	87.1	84.3	115.6	121.5	133.1
$T_{d(off)}$/ns	8.1	8.7	9.4	128.3	133.4	140.7
T_f/ns	6.5	6.6	6.9	10.2	11.8	13.2
V_{spike}/V	445	446	452	494	495	486
E_{off}/μJ	47.9	49.2	52.4	66.9	82.7	92.5

both the SiC MOSFET and Si SJ-MOSFET increase. The turn-off voltage spike of the SiC MOSFET increases, while that of the Si SJ-MOSFET decreases. Overall, the switching characteristics of the SiC MOSFET are less affected by temperature changes compared to the Si SJ-MOSFET.

4.1.5 Gate Charge

As mentioned earlier, the gate charge Q_G curve is obtained by integrating the C–V curve of the device over the switching process, providing an insight into the switching process from the perspective of charge accumulation. The gate charge characteristics Q_G of SiC MOSFET and Si SJ-MOSFET are illustrated in Fig. 4.12. The gate charges for SiC MOSFET and Si SJ-MOSFET are 19 and 99 nC, respectively. This indicates that driving SiC MOSFET requires less driving energy, Special attention should be paid to the impact of Si SJ-MOSFET operation at ultra-high frequencies on the overall efficiency of the converter. The gate-to-source charges Q_{GS} for SiC MOSFET and Si SJ-MOSFET are 5.5 and 16.5 nC, aligning with the earlier observation that the turn-on time $t_{d(on)}$ of SiC MOSFET is shorter than that of Si SJ-MOSFET. In the Q_{GD} section, Si SJ-MOSFET exhibits the well-known Miller plateau, while SiC MOSFET shows a Miller ramp, reflecting the differences in device characteristics between the two materials.

Fig. 4.8 Temperature influence on SiC MOSFET turn-on process

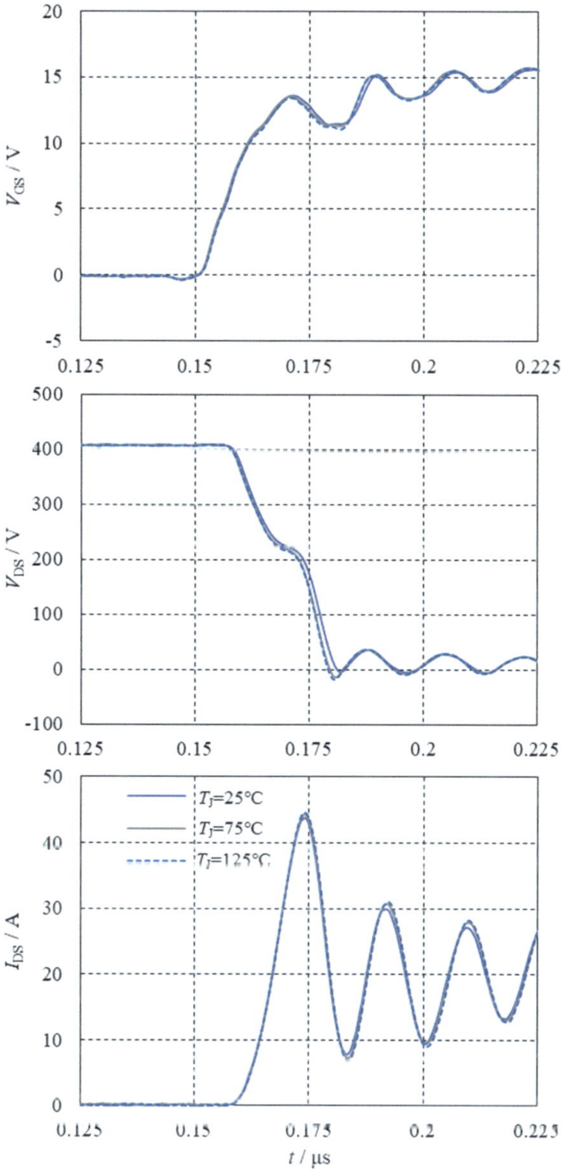

Fig. 4.9 Temperature influence on SiC MOSFET turn-off process

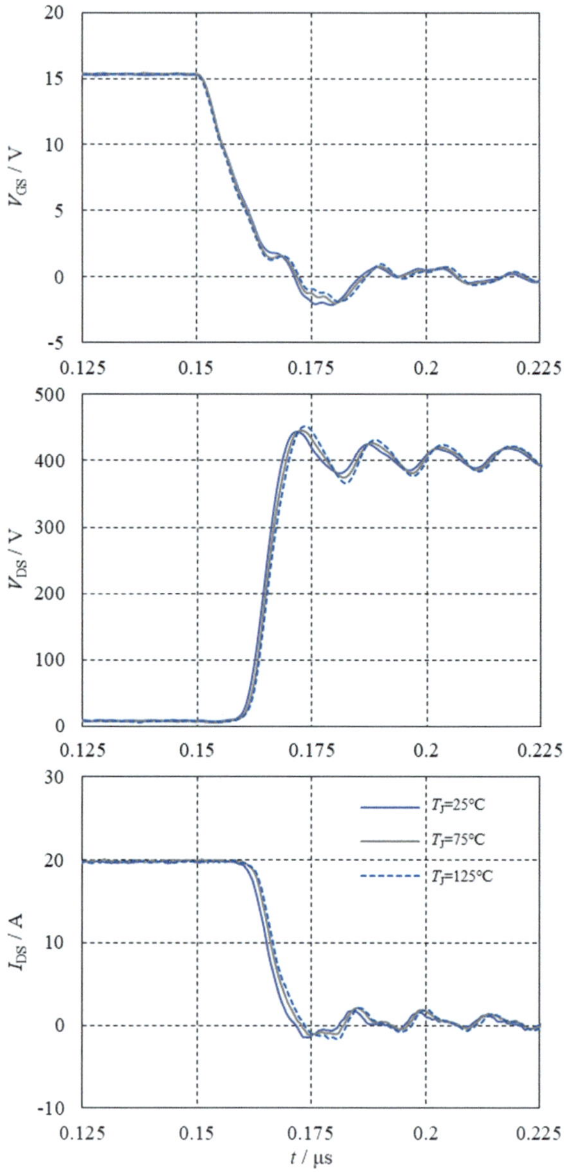

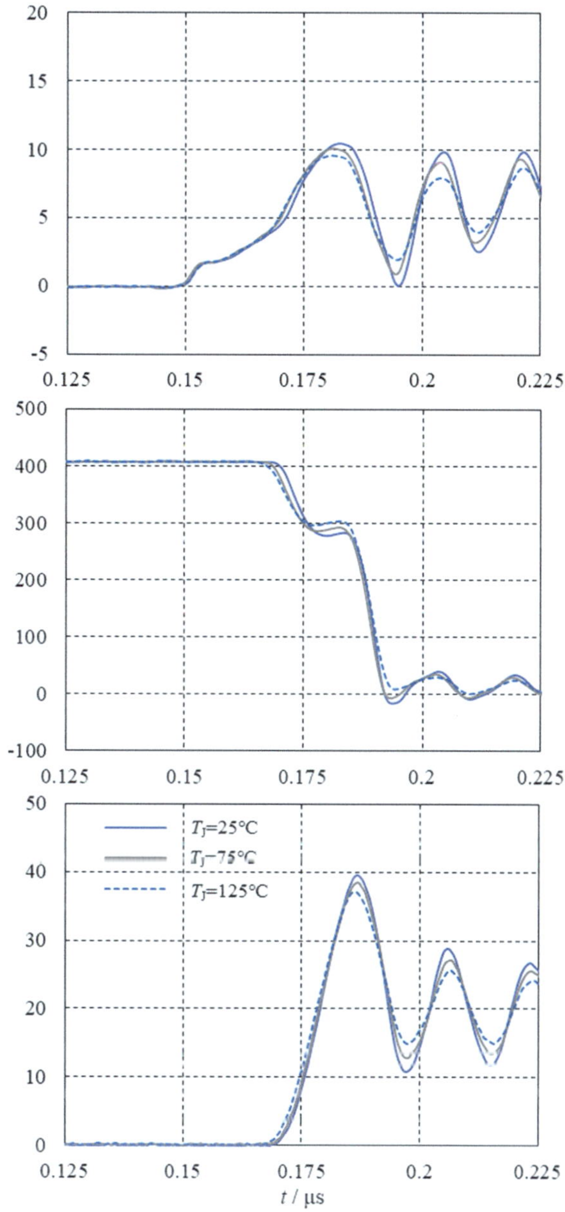

Fig. 4.10 Temperature influence on Si SJ-MOSFET turn-on process

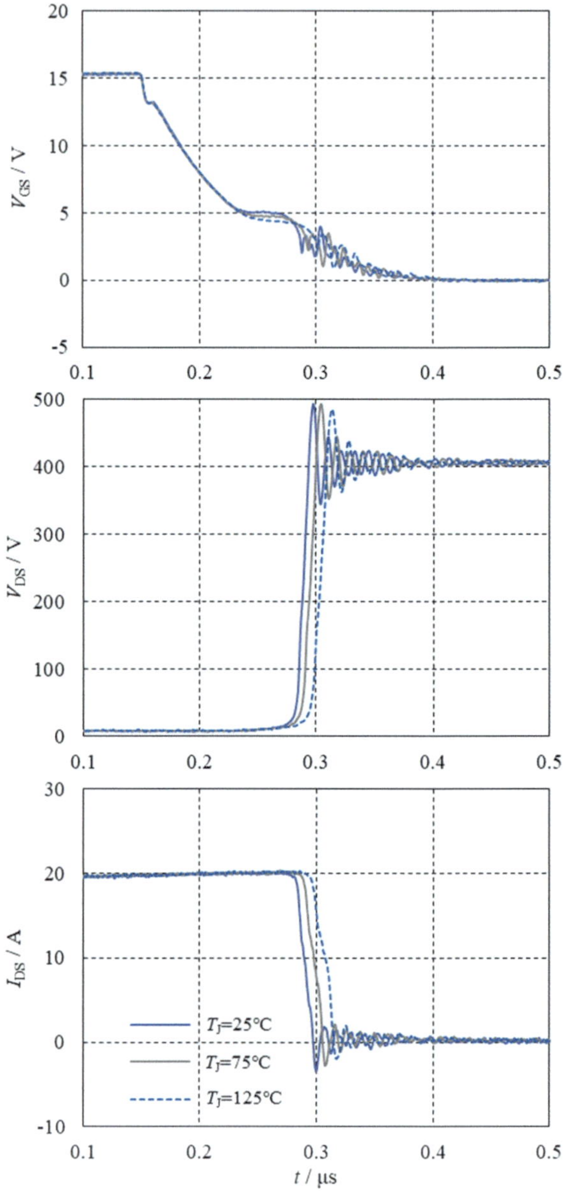

Fig. 4.11 Temperature influence on Si SJ-MOSFET turn-off process

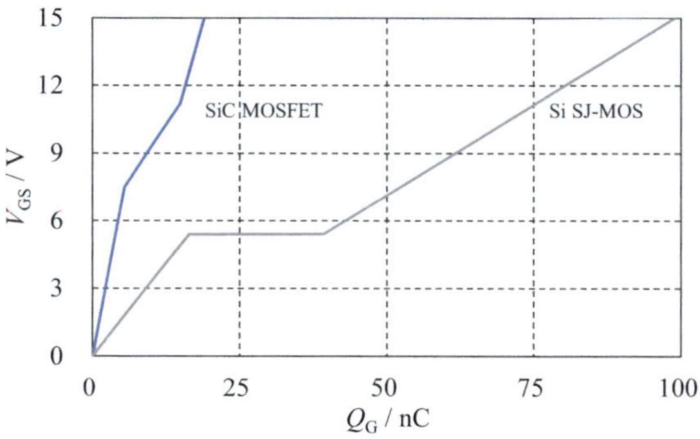

Fig. 4.12 Gate charge characteristics of SiC MOSFET and Si SJ-MOSFET

4.2 SiC MOSFET and Si IGBT

4.2.1 *Transfer Characteristics*

The transfer characteristics of SiC MOSFET and Si IGBT are illustrated in Fig. 4.13. It can be observed that the shapes of their transfer characteristic curves are quite similar. When V_{GS} or V_{GE} exceeds $V_{GS(th)}$ or $V_{GE(th)}$, the output current gradually rises, reaching saturation. At lower V_{GS} or V_{GE}, the current exhibits a positive temperature coefficient. While at higher V_{GS} or V_{GE}, the current shows a negative temperature coefficient. The temperature influence on SiC MOSFET is more significant compared to Si IGBT.

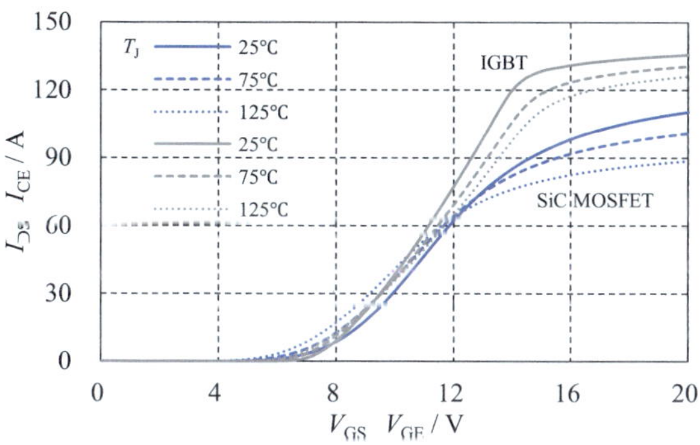

Fig. 4.13 Transfer characteristics of SiC MOSFET and Si IGBT

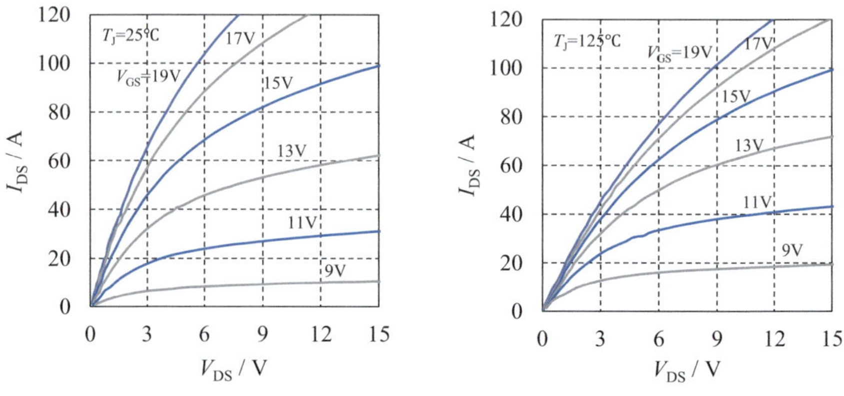

Fig. 4.14 Output characteristics of SiC MOSFET

4.2.2 Output Characteristics

The output characteristics of SiC MOSFET and Si IGBT are depicted in Figs. 4.14 and 4.15. The two curves exhibit significant differences. The I_{DS}–V_{DS} curve of SiC MOSFET starts from the origin due to its resistive characteristics during conduction. In contrast, Si IGBT exhibits current output only when V_{CE} exceeds the saturation voltage drop $V_{CE(sat)}$. This is because IGBT relies on its internally parasitic BJT for conduction. Consequently, at low currents, Si IGBT experiences a larger conduction voltage drop due to the influence of $V_{CE(sat)}$. This can result in lower conduction losses compared to SiC MOSFET. However, at high currents, Si IGBT can carry larger currents with a smaller conduction voltage drop. Because it is a bipolar device with higher transconductance. With temperature variation, the I_{DS}–V_{DS} curve of SiC

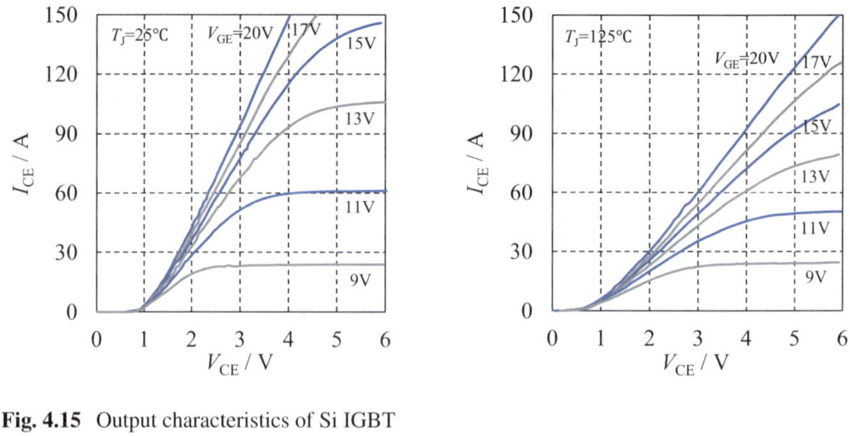

Fig. 4.15 Output characteristics of Si IGBT

Fig. 4.16 C–V curves of SiC MOSFET and Si IGBT

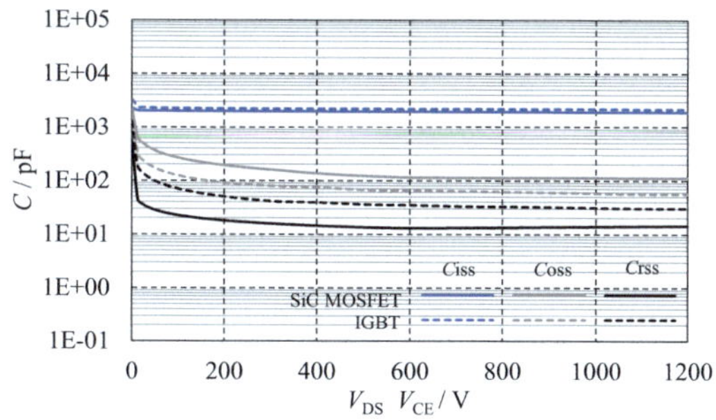

(a) The C axis is a linear axis

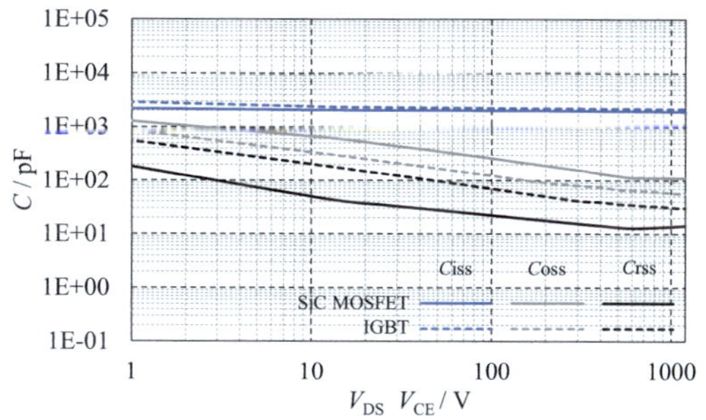

(b) The C axis is a logarithmic axis

MOSFET undergoes a larger amplitude of changes compared to Si IGBT. This is consistent with the characteristics observed in their transfer functions.

4.2.3 C–V Characteristics

The C–V curves of SiC MOSFET and Si IGBT are depicted in Fig. 4.16. The C–V curves of both devices have a generally consistent shape. Additionally, SiC MOSFET exhibits a slightly smaller C_{iss}, larger C_{oss}, and smaller C_{rss} compared to Si IGBT.

4.2.4 Switching Characteristics

Similar to the previous sections, when comparing switching characteristics, appropriate values of $R_{G(ext)}$ are chosen for SiC MOSFET and Si IGBT. This is to ensure that both have the same R_G, and the freewheeling diode is the same SiC SBD.

At 25 °C, the turn-on and turn-off processes of the SiC MOSFET and Si IGBT are shown in Figs. 4.17 and 4.18. The turn-on delay $t_{d(on)}$ and turn-off delay $t_{d(off)}$ of the SiC MOSFET are 11.7 and 17.2 ns, respectively, which are shorter than the 21.31 and 181.2 ns of the Si IGBT. Notably, the $t_{d(off)}$ of the Si IGBT is nearly ten times that of the SiC MOSFET. This discrepancy is mainly due to the larger C_{iss} of the Si IGBT, which results in slower rise and fall times of V_{GS} under identical driving circuit and R_G conditions. Additionally, the SiC MOSFET exhibits significantly faster switching speeds, with higher I_{DS} overshoot during turn-on and higher V_{DS} spikes during turn-off. The slower rise and fall times of V_{CE} in the Si IGBT are primarily due to its larger C_{rss}. Furthermore, the Si IGBT displays a pronounced tail current during turn-off, which is absent in the SiC MOSFET. These factors contribute to the Si IGBT having much higher switching energy compared to the SiC MOSFET. The turn-on energy E_{on} of the SiC MOSFET and Si IGBT are 618.9 and 2565.2 μJ, respectively, with the IGBT's being four times that of the SiC MOSFET. The turn-off energy E_{off} for the SiC MOSFET and Si IGBT are 165.1 and 1627.2 μJ, respectively, with the IGBT's being ten times that of the SiC MOSFET. A comparison of switching characteristics between SiC MOSFETs and Si IGBTs is shown in Table 4.2.

The effects of temperature on the turn-on and turn-off processes of the SiC MOSFET and Si IGBT at 25, 75, and 125 °C are shown in Figs. 4.19, 4.20, 4.21, and 4.22. As temperature increases, the turn-on delay for both the SiC MOSFET and Si IGBT decreases. However, the SiC MOSFET's switching speed increases, and its turn-on energy decreases, while the Si SJ-MOSFET exhibits the opposite trend, with a more significant change compared to the SiC MOSFET. The turn-on current spikes for both the SiC MOSFET and Si IGBT increase slightly as temperature rises. The turn-off delay, turn-off time, and turn-off energy for both the SiC MOSFET and Si IGBT also increase with temperature, but the change is more pronounced for the Si IGBT. Overall, the switching speed of the SiC MOSFET is much faster than that of the Si IGBT, and the characteristics of the SiC MOSFET are less affected by temperature compared to the Si IGBT.

4.2.5 Gate Charge

The gate charge Q_G of SiC MOSFET and Si IGBT is depicted in Fig. 4.23. The Q_G for SiC MOSFET and Si IGBT is 52 and 191 nC, respectively. This indicates that driving SiC MOSFET requires less driving energy. The Q_{GS} for SiC MOSFET and Si IGBT are 14 and 17 nC, respectively. This aligns with the earlier discussion that SiC MOSFET has a shorter $t_{d(on)}$ compared to Si IGBT. The Q_{GD} for SiC MOSFET is 21 nC, significantly smaller than the 105 nC for Si IGBT. This reflects the differences in voltage and current change rates during the switching process between the two. Additionally, Si IGBT exhibits the well-known Miller plateau, while SiC MOSFET shows a Miller slope. This indicates the differences in device characteristics between the two materials.

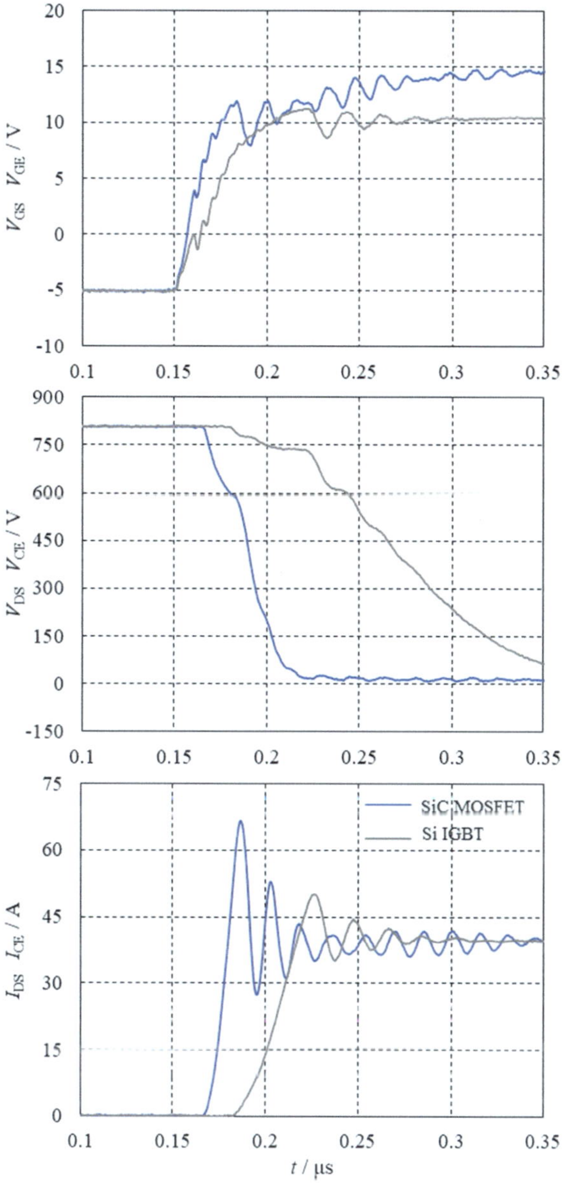

Fig. 4.17 Turn-on process of SiC MOSFET and Si IGBT

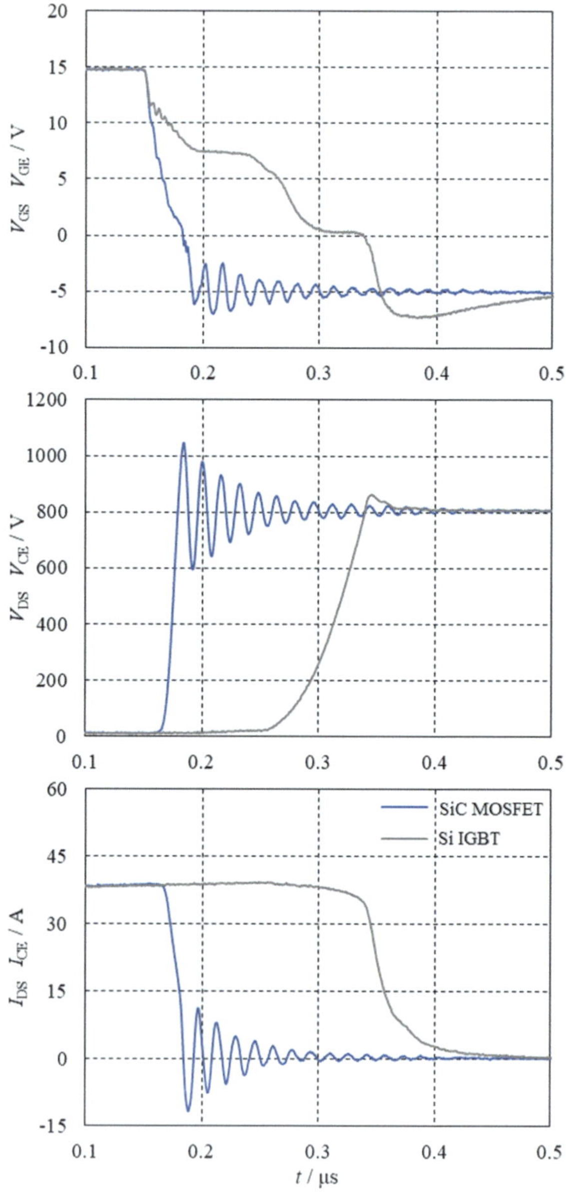

Fig. 4.18 Turn-off process of SiC MOSFET and Si IGBT

Table 4.2 SiC MOSFET and Si IGBT switching characteristics

Device	SiC MOSFET			Si IGBT		
Temperature	25 °C	75 °C	125 °C	25 °C	75 °C	125 °C
$t_{d(on)}$/ns	11.8	11.5	11.5	55.8	56.3	56.0
t_r/ns	37.3	35.4	34.0	117.8	131.8	146.0
I_{spike}/A	66.6	67.0	67.7	50.1	50.6	51.7
E_{on}/μJ	618.9	585.8	584.2	2565.2	2762.0	3027.8
$t_{d(off)}$/ns	16.6	17.4	18.2	122.5	134.6	154.0
t_f/ns	9.5	9.7	9.8	59.9	76.1	100.3
V_{spike}/V	1047	1046	1046	864	859	838
E_{off}/μJ	165.1	193.4	205.4	1627.2	2245.7	3035.5

4.2.6 *Short-Circuit Characteristics*

Figures 4.24, 4.25, 4.26 and 4.27 illustrate the short-circuit test characteristics of SiC MOSFET and Si IGBT. Two scenarios are distinguished: single-device short circuit and hard-switching short circuit. In the single-device short-circuit scenario, the device is directly shorted by connecting both ends to the bus capacitor and applying an on signal. In this case, the short-circuit current and voltage of SiC MOSFET and Si IGBT are relatively close. And Si IGBT exhibits a high peak in V_{GE} during turn-on. The bridge arm short-circuit scenario occurs in the half-bridge topology. The upper and lower switches are sequentially turned on to create a short circuit. In the bridge arm direct connection, there is a notable difference in terminal voltage V_{DS} between SiC MOSFET and Si IGBT. SiC MOSFET exhibits a voltage division state between the upper and lower gate-source voltages, denoted as $V_{DS(H)}$ and $V_{DS(L)}$, while the upper switch of Si IGBT maintains a low voltage denoted as $V_{CE(H)}$.

4.3 SiC SBD and Si Diode

4.3.1 *Conduction Characteristics*

The conduction characteristics of SiC SBD, SiC MOSFET body diode (SiC MOSFET BD), Si FRD, and Si SJ-MOSFET body diode (Si SJ-MOSFET BD) are shown in Fig. 4.28. It can be seen they exhibit noticeable differences.

The turn-on voltage of SiC SBD, Si FRD, and Si SJ-MOSFET BD are very close, around 0.7 V. In contrast, the turn-on voltage of SiC MOSFET BD is approximately 1.8 V. This is attributed to its nature as a SiC PN-junction diode, which has a higher barrier voltage. As the terminal voltage V_F exceeds the turn-on voltage, the conduction current I_F for each diode increases at different rates with the growing V_F. Among them, Si SJ-MOSFET BD exhibits the fastest rise in conduction current, followed

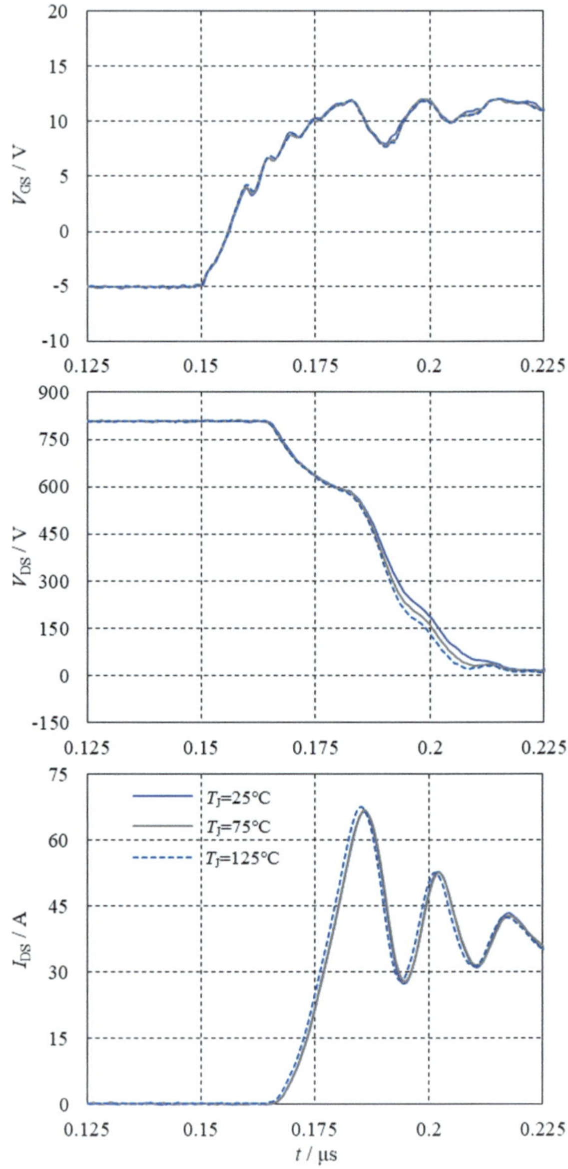

Fig. 4.19 Temperature influence on turn-on process of SiC MOSFET

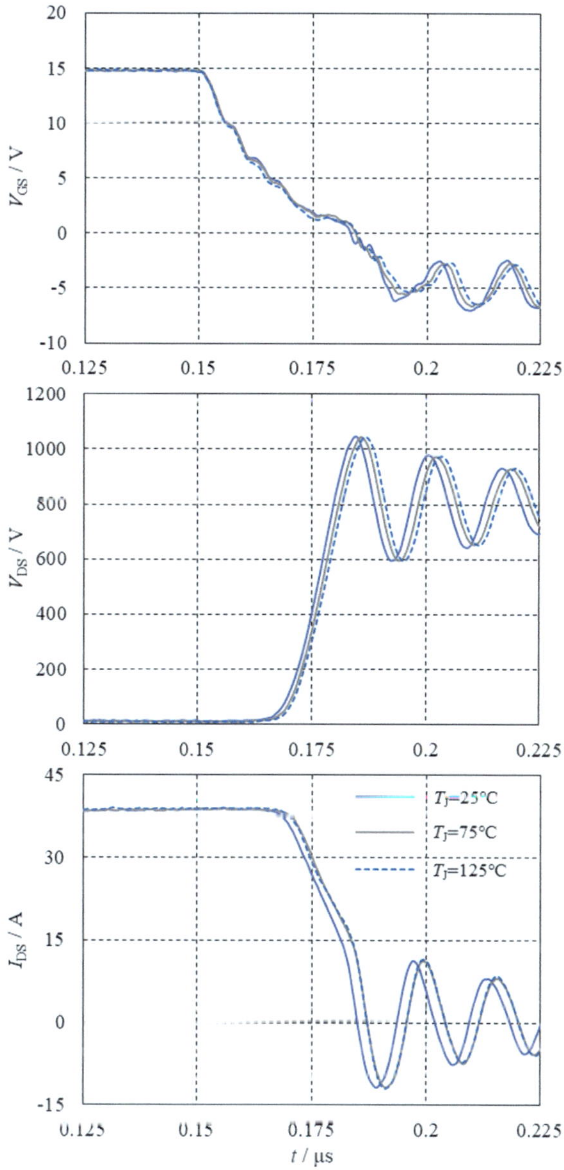

Fig. 4.20 Temperature influence on turn-off process of SiC MOSFET

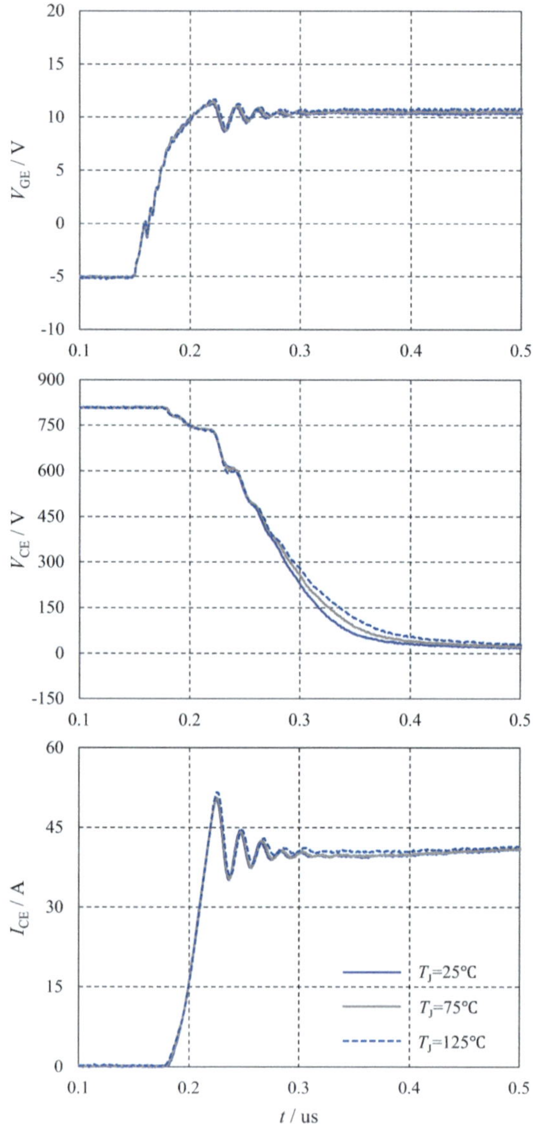

Fig. 4.21 Temperature influence on turn-on process of Si IGBT

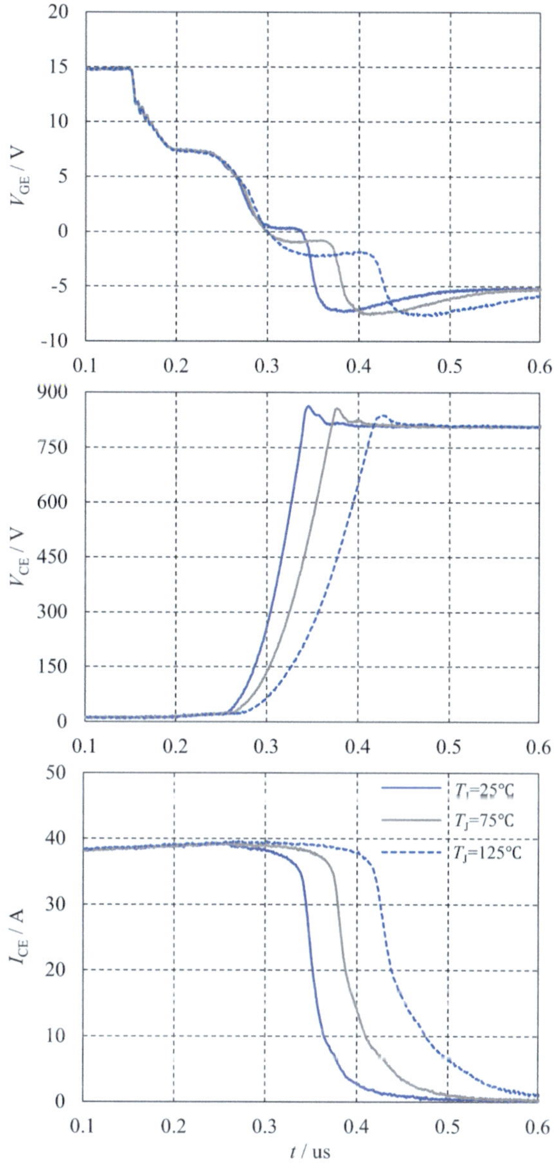

Fig. 4.22 Temperature influence on turn-off process of Si IGBT

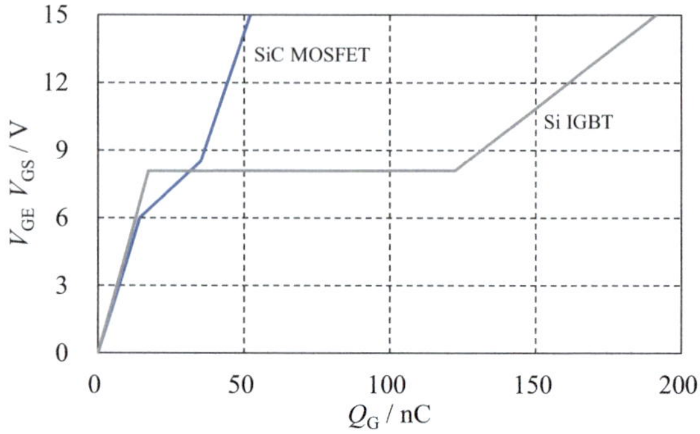

Fig. 4.23 Gate charge of SiC MOSFET and Si IGBT

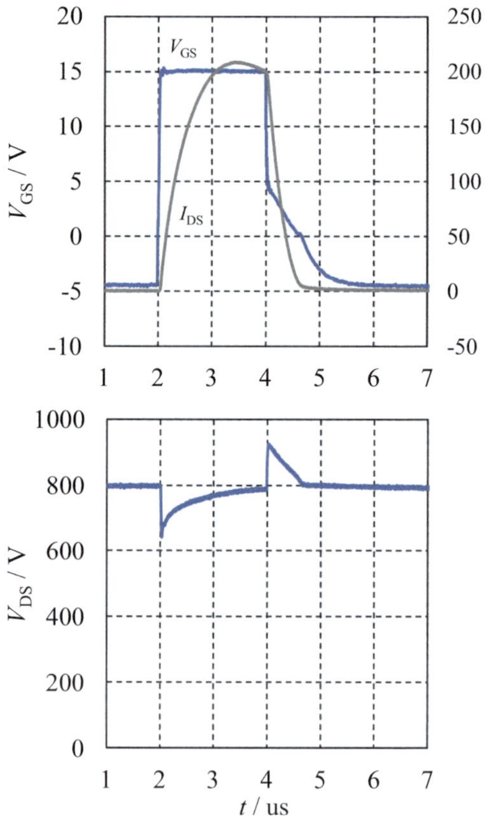

Fig. 4.24 SiC MOSFET single-device hard-switching short circuit

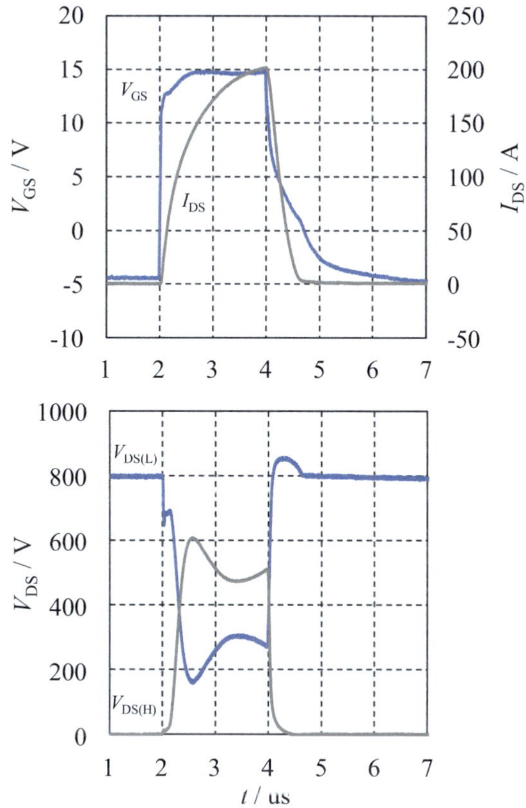

Fig. 4.25 SiC MOSFET hard-switching short circuit

by SiC SBD and Si FRD, while SiC MOSFET BD shows the slowest rise. Therefore, at the same conduction current, Si SJ-MOSFET BD incurs the least conduction losses, while SiC MOSFET BD experiences the greatest conduction losses. Consequently, when employing SiC MOSFET in applications requiring its body diode for freewheeling, synchronous rectification techniques are often utilized to operate SiC MOSFET in the third quadrant, minimizing conduction losses.

The conduction characteristics of various diodes under the influence of temperature and gate voltage are shown in Fig. 4.29. On the one hand, all four diodes exhibit positive temperature coefficients at low V_F and negative temperature coefficients at higher V_F. On the other hand, the zero temperature coefficient points differ, with SiC SBD and Si SJ-MOSFET BD at around V_F is 1 V, Si FRD at around V_F is 2 V, and SiC MOSFET BD at around V_F is 3 V. Additionally, in the negative temperature coefficient region, the impact of temperature is more pronounced for SiC SBD and SiC MOSFET BD compared to Si SJ-MOSFET BD and Si FRD.

Furthermore, the influence of gate voltage on the output characteristics of the body diode varies. The output characteristics of Si SJ-MOSFET BD are unaffected by gate voltage, while negative gate voltage significantly affects the output characteristics

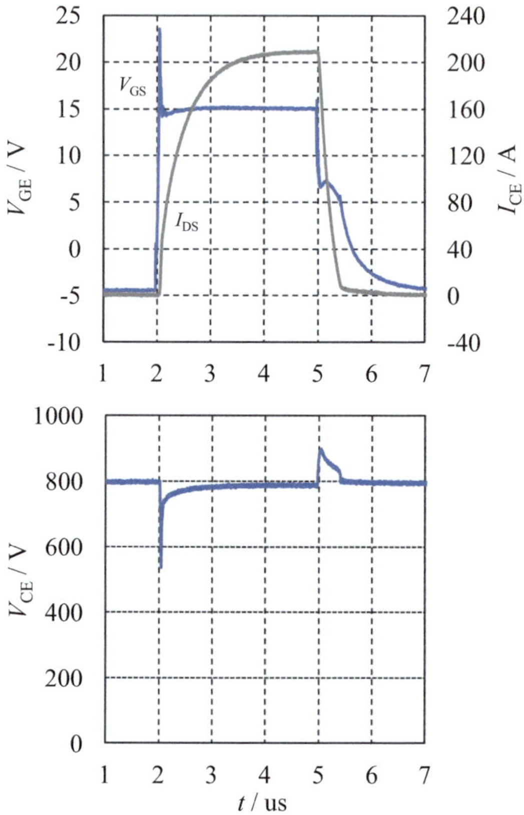

Fig. 4.26 Si IGBT single-device hard-switching short circuit

of SiC MOSFET BD. When applying negative gate voltage to SiC MOSFET, its turn-on voltage increases, and the output current decreases.

4.3.2 Reverse Recovery Characteristics

Reverse recovery is a critical diode characteristic. The shorter reverse recovery time with smaller reverse recovery charge can enhance the performance of converters. Si FRD and Si SJ-MOSFET FR-BD, optimized for the reverse recovery of the body diode, are introduced to achieve better reverse recovery characteristics. In the following discussion, Si SJ-MOSFET FR-BD will be added to the four diodes mentioned earlier.

The reverse recovery waveforms of SiC SBD, SiC MOSFET BD, Si FRD, Si SJ-MOSFET BD, and Si SJ-MOSFET FR-BD are shown in Fig. 4.30. The test condition is 15 A current and a reverse recovery speed dI_F/dt of 1000 A/μs for the first three and 100 A/μs for the latter two. Si SJ-MOSFET BD and Si SJ-MOSFET FR-BD are

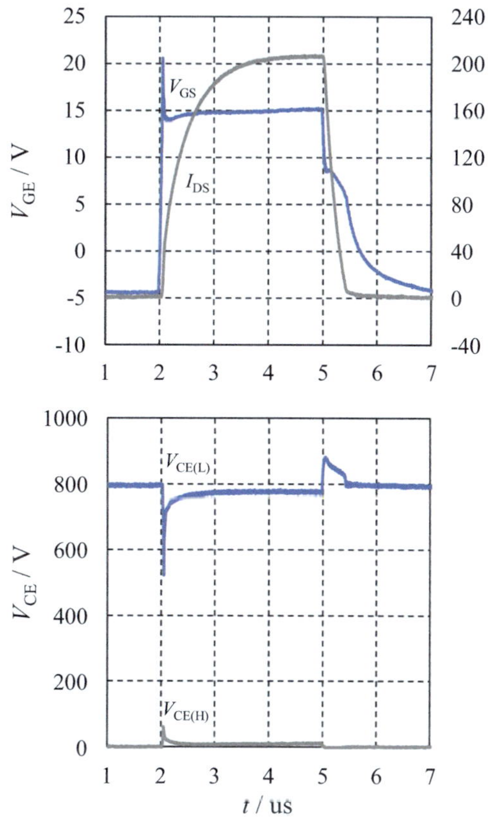

Fig. 4.27 Si IGBT hard-switching short circuit

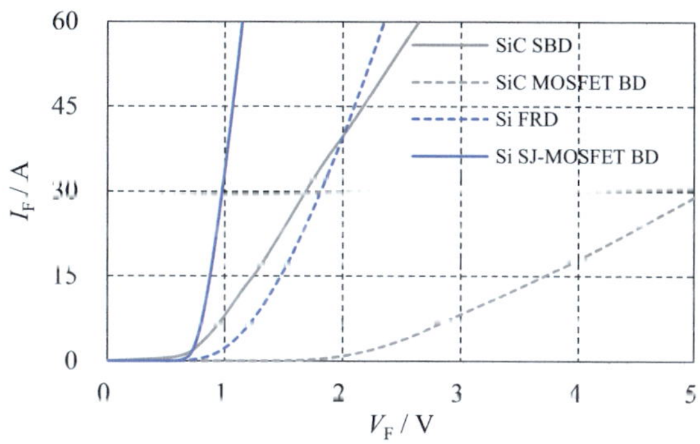

Fig. 4.28 Conduction characteristics of various diodes

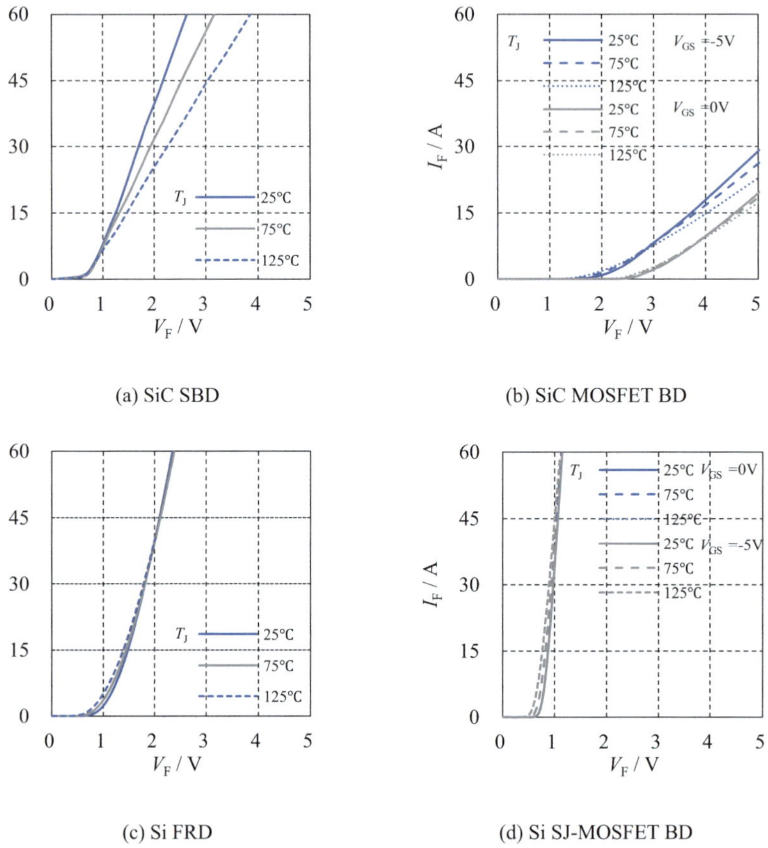

Fig. 4.29 Conduction characteristics of various diodes under the influence of temperature and gate voltage

tested at 100 A/μs due to their lower safe reverse recovery speed. Excessively high $\mathrm{d}I_F/\mathrm{d}t$ may lead to damage.

Based on the waveforms in Fig. 4.30, the reverse recovery characteristics of the five diodes are summarized in Table 4.3. The table provides specific values for reverse recovery time t_{rr}, reverse recovery charge Q_{rr}, and reverse recovery peak current I_{rrm}. It also presents normalized values using SiC SBD and SiC MOSFET BD as references respectively, this facilitates clearer observation of the relative ratio relationships in reverse recovery characteristics across different diode types. The parameters t_{rr}, Q_{rr}, I_{rrm} for SiC SBD, SiC MOSFET BD, Si FRD, Si SJ-MOSFET BD, and Si SJ-MOSFET FR-BD are as follows: 9.2 ns/26.0 nC/4.6 A, 12.8 ns/45.9 nC/5.9 A, 139.8 ns/614.1 nC/12.0 A, 404.1 ns/6137.1 nC/31.1 A, and 118.6 ns/722.7 nC/10.1 A, respectively.

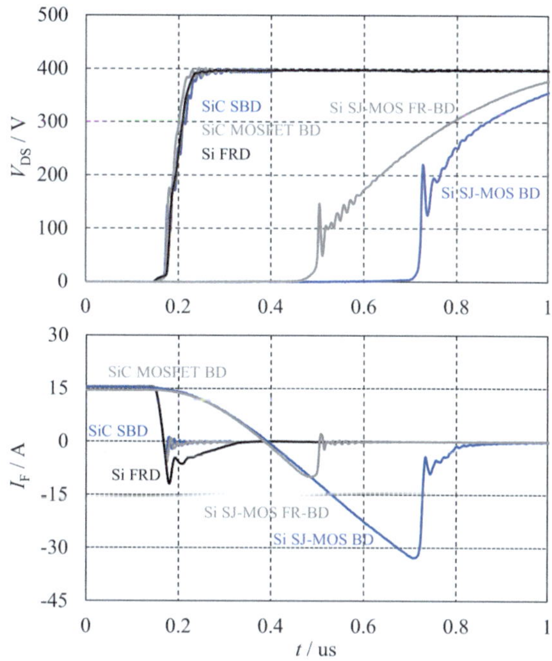

Fig. 4.30 Reverse recovery waveforms of various diodes

Table 4.3 Reverse recovery characteristics of various diodes

	t_{rr}/ns	t_{rr}/p.u.	t_{rr}/p.u.	Q_{rr}/nC	Q_{rr}/p.u.	Q_{rr}/p.u.	I_{rrm}/A	I_{rrm}/p.u.	I_{rrm}/p.u.
SiC SBD	9.2	1.0	0.7	26.0	1.0	0.6	4.6	1.0	0.8
SiC MOSFET BD	12.8	1.4	1.0	45.9	1.8	1.0	5.9	1.3	1.0
Si FRD	139.8	15.2	10.9	614.1	23.6	13.4	12.0	2.6	2.0
Si SJ-MOSFET BD	404.1	43.9	31.6	6137.1	236.0	133.7	31.1	6.8	5.3

SiC SBD shows significantly better parameters in all aspects, demonstrating the best reverse recovery characteristics. Despite Si FRD being specifically optimized for reverse recovery, its t_{rr}, Q_{rr}, and I_{rrm} are 15.2 times, 23.6 times, and 2.6 times those of the SiC SBD. This is the reason for the widespread replacement of Si FRD with SiC SBD in certain applications.

While the parameters of SiC MOSFET BD are slightly inferior to those of the SiC SBD, they still outperform the other three Si diodes significantly. Si SJ-MOSFET BD exhibits the worst reverse recovery characteristics among the five diodes, with t_{rr}, Q_{rr}, and I_{rrm} being 31.6 times, 133.7 times, and 5.3 times those of SiC MOSFET BD. The Si SJ-MOSFET FR-BD, introduced to enhance reverse recovery characteristics,

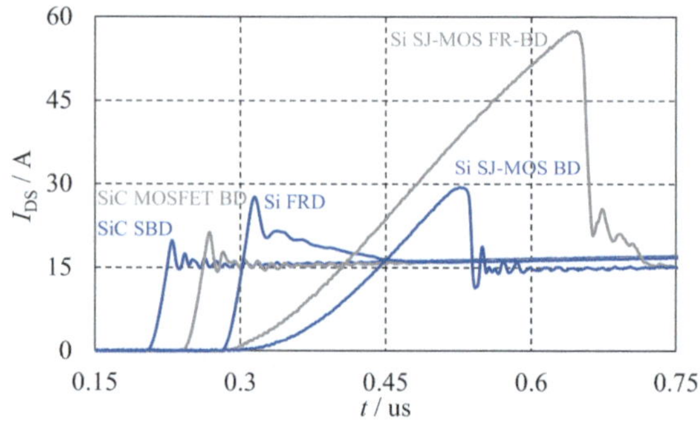

Fig. 4.31 Influence of various diode reverse recovery characteristics on the switching process

shows improvements compared to Si SJ-MOSFET BD. But it still has t_{rr}, Q_{rr}, and I_{rrm} being 9.3 times, 15.7 times, and 21.7 times those of SiC MOSFET BD, indicating a considerable gap. It is noteworthy that generally higher dI_F/dt leads to longer reverse recovery time and larger reverse recovery charge. Even at 100 A/μs, Si SJ-MOSFET BD and Si SJ-MOSFET FR-BD exhibit significant differences compared to SiC SBD, let alone at higher values. Therefore, SiC MOSFET BD has made significant progress in reverse recovery characteristics compared to Si MOSFET and even outperforms Si FRD.

As discussed in Chap. 2, in the half-bridge circuit with an inductive load, the current spike during the turn-on of the switch is caused by the reverse recovery of the diode. Therefore, diodes with different reverse recovery characteristics will have varying effects on the turn-on current waveform of the switch. When the switch is the same SiC MOSFET, using the aforementioned five diodes as the freewheeling diode. The turn-on speed is consistent with that shown in Fig. 4.30, the turn-on waveform of the SiC MOSFET is depicted in Fig. 4.31. It can be observed that the turn-on current waveform of the SiC MOSFET matches the shape and values of the diode reverse recovery current waveform.

The influence of current on the reverse recovery characteristics of various diodes is shown in figures from Figs. 4.32, 4.33, 4.34, 4.35 and 4.36. The reverse recovery characteristics based on the waveform are given in Table 4.4. The test conditions are: T_J is 25 °C, dI_F/dt is 1000 A/μs or 100 A/μs, and I_F gradually increasing from 5 to 25 A. As I_F increases, the reverse recovery characteristics of SiC SBD and SiC MOSFET BD remained almost unchanged. The reverse recovery time, charge, and peak current of Si FRD, Si SJ-MOSFET BD, and Si SJ-MOSFET FR-BD gradually increased. Among them, the change in Si SJ-MOSFET BD is the most significant.

Figures 4.37, 4.38, 4.39, 4.40 and 4.41 illustrate the impact of temperature on the reverse recovery characteristics of various diodes. The reverse recovery characteristics based on waveforms are provided in Table 4.5. The test conditions include: I_F is 15 A, dI_F/dt is 1000 A/μs or 100 A/μs, and T_J is from 25 to 125 °C. As T_J increases, the reverse recovery characteristics of SiC SBD remain almost unchanged. This shows that its reverse recovery characteristics are almost unaffected by temperature.

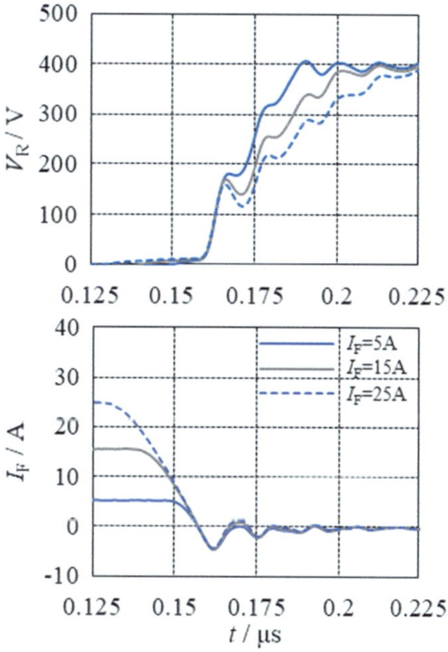

Fig. 4.32 Effect of I_F on reverse recovery of SiC SBD

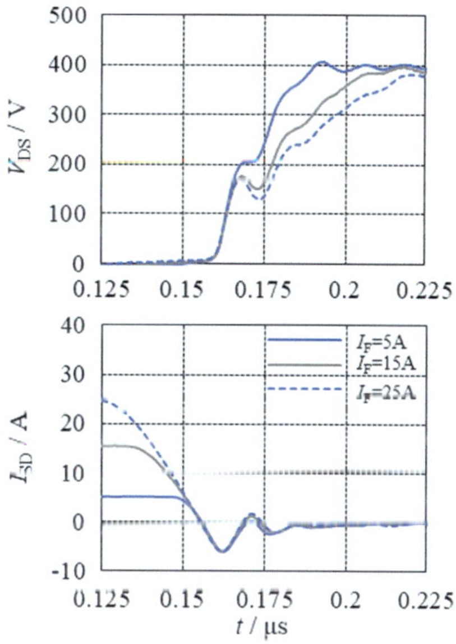

Fig. 4.33 Effect of I_F on reverse recovery of SiC MOSFET BD

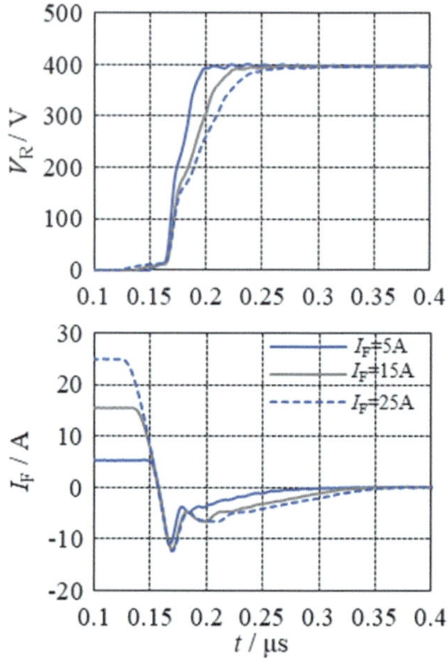

Fig. 4.34 Effect of I_F on reverse recovery of Si FRD

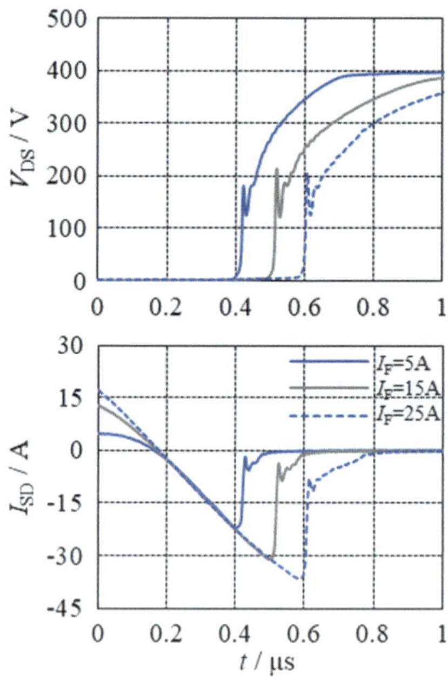

Fig. 4.35 Effect of I_F on reverse recovery of Si SJ-MOSFET BD

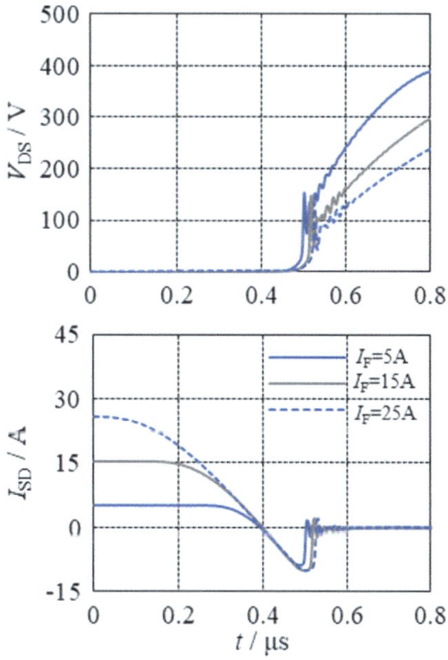

Fig. 4.36 Effect of I_F on reverse recovery of Si SJ-MOSFET FR-BD

Table 4.4 Reverse recovery waveforms of various diodes at different currents

	t_{rr}/ns			Q_{rr}/nC			I_{rrm}/A		
I_F/A	5	15	20	5	15	20	5	15	20
SiC SBD	10.3	9.2	8.8	27.0	26.0	23.8	4.5	4.6	4.3
SiC MOSFET BD	13.3	12.8	12.6	47.3	45.9	46.2	6.0	5.9	6.0
Si FRD	85.9	139.8	160.1	320.6	614.1	746.6	10.9	12.0	12.3
Si SJ-MOSFET BD	263.2	404.1	547.8	2943.3	6137.1	9686.6	22.0	31.1	36.6
Si SJ-MOSFET FR-BD	107.8	118.6	128.6	556.4	722.7	788.2	8.8	10.1	10.1

While the reverse recovery time, charge, and peak current of other diodes gradually increase. Si FRD and Si SJ-MOSFET FR-BD show the most significant changes.

Figures 4.42, 4.43, 4.44, 4.45 and 4.46 illustrate the impact of reverse recovery speed on the reverse recovery characteristics of various diodes. Table 4.6 provides the reverse recovery characteristics based on waveforms. The test condition is: I_F at 15 A, T_J at 25 °C, and dI_F/dt increases from 100 to 1000 A/us or from 50 to 200 A/μs. As dI_F/dt increases, the reverse recovery time, charge, and peak current for each diode type significantly increase.

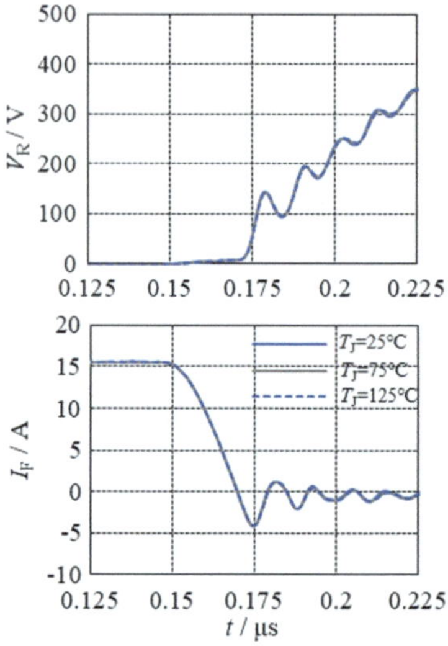

Fig. 4.37 Effect of T_J on reverse recovery of SiC SBD

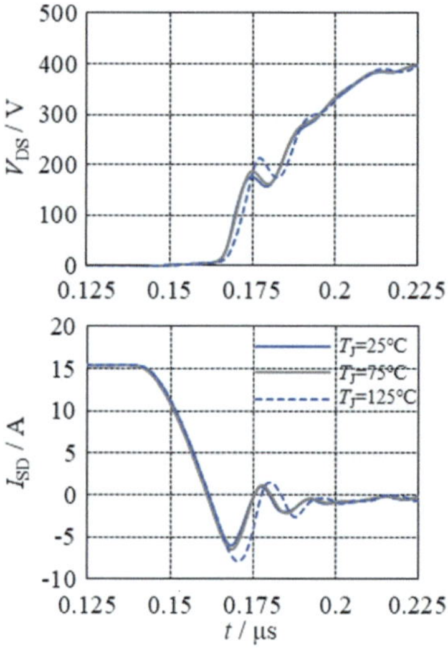

Fig. 4.38 Effect of T_J on reverse recovery of SiC MOSFET BD

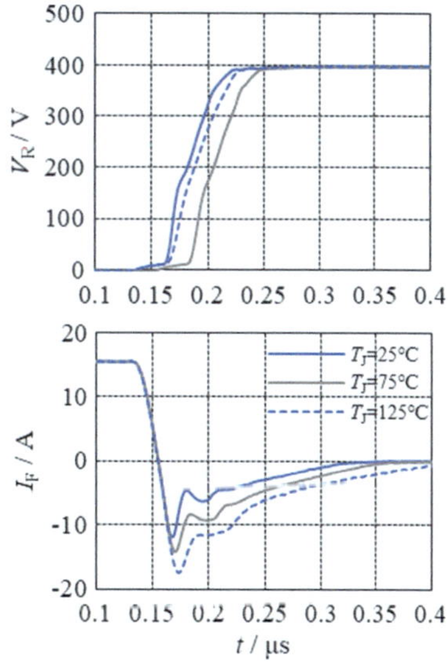

Fig. 4.39 Effect of T_J on reverse recovery of Si FRD

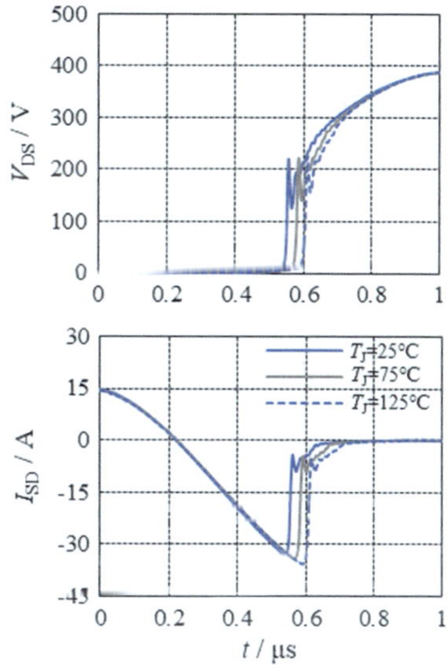

Fig. 4.40 Effect of T_J on reverse recovery of Si SJ-MOSFET BD

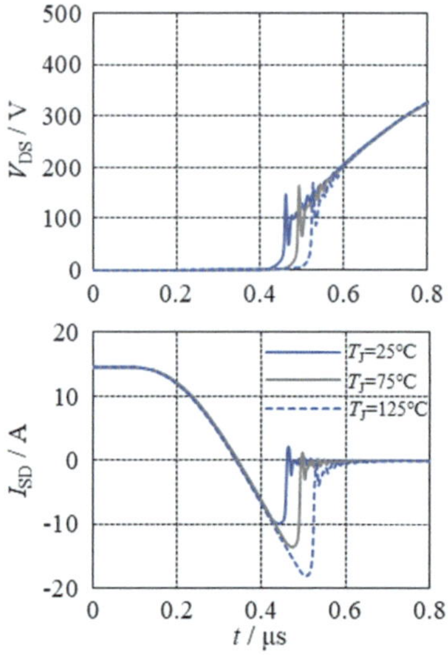

Fig. 4.41 Effect of T_J on reverse recovery of Si SJ-MOSFET FR-BD

Table 4.5 Reverse recovery waveforms of various diodes under different temperatures

	t_{rr}/ns			Q_{rr}/nC			I_{rrm}/A		
T_J/°C	5	15	20	5	15	20	5	15	20
SiC SBD	8.7	8.9	9.0	22.7	23.0	22.8	4.2	4.1	4.1
SiC MOSFET BD	13.1	13.7	15.4	47.5	54.5	76.1	6.0	6.5	7.9
Si FRD	139.7	167.0	206.0	597.5	963.3	1401.9	11.9	14.2	17.4
Si SJ-MOSFET BD	391.4	429.1	456.2	6245.6	7081.9	8038.8	32.8	34.1	35.8
Si SJ-MOSFET FR-BD	118.4	148.5	188.9	709.0	1159.4	1892.6	9.9	13.6	18.0

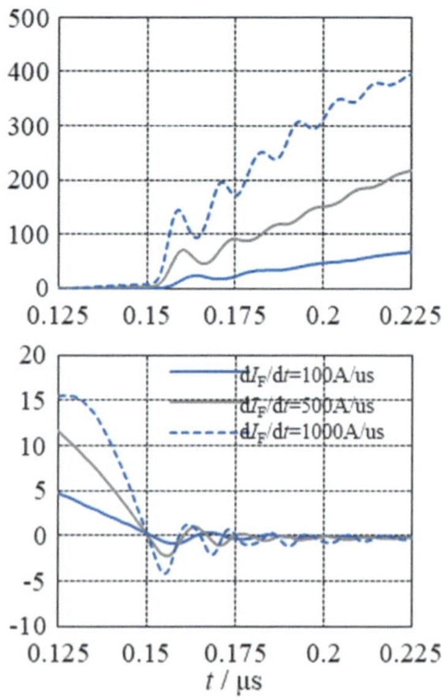

Fig. 4.42 Effect of dI_F/dt on reverse recovery of SiC SBD

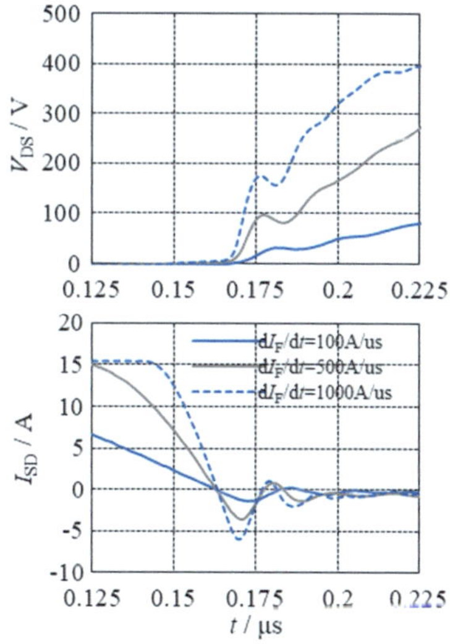

Fig. 4.43 Effect of dI_F/dt on reverse recovery of SiC MOSFET BD

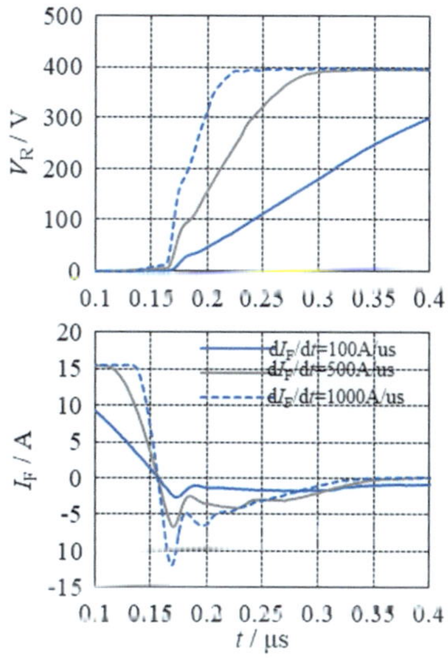

Fig. 4.44 Effect of dI_F/dt on reverse recovery of Si FRD

Fig. 4.45 Effect of dI_F/dt on reverse recovery of Si SJ-MOSFET BD

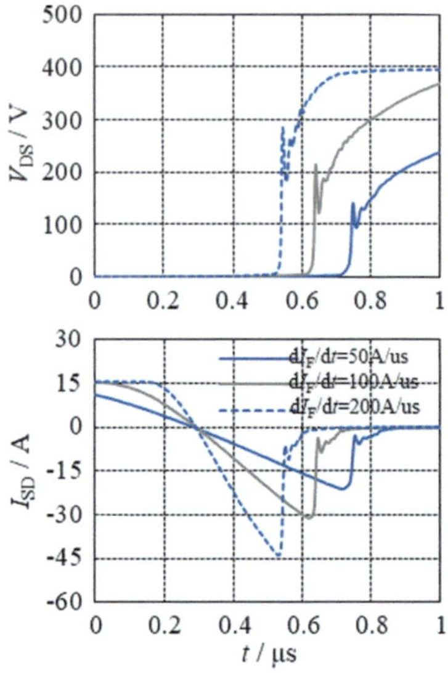

Fig. 4.46 Effect of dI_F/dt on reverse recovery of Si SJ-MOSFET FR-BD

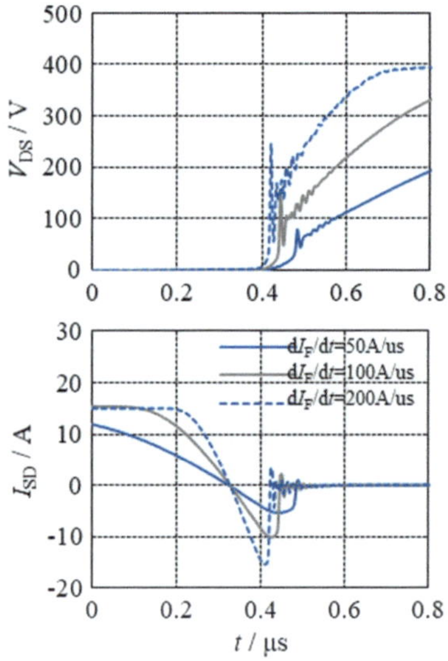

Table 4.6 Reverse recovery characteristics of various diodes at different dI_F/dt

	t_{rr}/ns			Q_{rr}/nC			I_{rrm}/A		
dI_F/dt/A/μs	100	500	1000	100	500	1000	100	500	1000
SiC SBD	12.1	10.0	8.7	7.1	14.2	22.7	0.9	2.2	4.2
SiC MOSFET BD	18.5	14.1	13.0	16.0	30.8	47.5	1.4	3.6	6.0
Si FRD	297.3	175.6	139.8	363.9	525.2	614.1	2.6	6.7	12.0
dI_F/dt/A/μs	50	100	200	50	100	200	50	100	200
Si SJ-MOSFET BD	5017.5	404.1	294.5	29,026.0	6137.1	6484.4	21.2	31.1	43.8
Si SJ-MOSFET FR-BD	161.0	118.6	95.2	560.3	722.7	844.3	5.4	10.1	15.4

References

1. ROHM Co., Ltd. (2014) SiC power devices and modules. Application note, 14103EBY01
2. Infineon Technologies AG (2018) CoolSiC™ 1200 V SiC MOSFET application note. Application note, AN2017-46, Rev. 1.01
3. Infineon Technologies AG (2018) CoolSiC™ 650 V M1 SiC Trench power device Infineon's first 650 V silicon carbide MOSFET for industrial applications. Application note, AN_1907_PL52_1911_144109, Rev. 1.0
4. Infineon Technologies AG (2019) CoolSiC™ automotive discrete Schottky diodes understanding the benefits of SiC diodes compared to silicon diodes. Application note, AN 2019–02_CoolSiC_Automotive_Diode, Rev. 1.0
5. Omar H, Vladimir S (2016) 1200V CoolSiC ™ Schottky diode generation 5. Application note, Rev 1.1, Infineon Technologies AG
6. Infineon Technologies AG (2017) Improving PFC efficiency using the CoolSiC™ Schottky diode 650V G6. Application note, AN_201704_PL52_020, Rev 1.0
7. STMicroelectronics (2018) New generation of 650 V SiC diodes. Application note, AN4242, Rev 2
8. STMicroelectronics (2015) Monolithic Schottky diode in ST F7 LV MOSFET technology: improving application performance. application note, AN4789, DocID028669, Rev. 01
9. ON Semiconductor (2023) onsemi EliteSiC Gen 2 1200V SiC MOSFET M3S Series. Application Note, AND90204/D
10. Liang M, Zheng TQ, Li Y (2014) Performance evaluation of SiC MOSFET, Si CoolMOS and IGBT. Int Power Electron Appl Conf Expos 2014:1369–1373

Chapter 5
Double-Pulse Test

Switching characteristic test and reverse recovery characteristic test are both based on double-pulse test, a method that has been used for over 20 years, although not exclusive to SiC devices. However, due to the faster switching speed of SiC devices, higher demands are placed on the skills of test personnel and the performance of test equipment. This necessitates enhancements to the test techniques and equipment originally designed for Si devices. Consequently, these two tests stand out as one of the differences between SiC and Si devices in terms of test, garnering more attention. Therefore, this chapter will first discuss switching characteristic test and reverse recovery characteristic test separately, highlighting their importance, before introducing other test items in the next chapter.

In this chapter, we will first discuss the rationality of characterizing device switching characteristics and reverse recovery characteristics during power converter operation using double-pulse test. We will then introduce the basic principles, test procedures, and test platforms of double-pulse test. We will focus on discussing the principles of setting test parameters, which serve as the foundation for conducting double-pulse test but have not been fully discussed before. Finally, we will discuss the measurement instruments based on measured waveforms and provide reasonable measurement recommendations for respectively. The new requirements of SiC devices for double-pulse test mainly manifest in this aspect. This chapter also innovatively proposes that crosstalk is an important dynamic characteristic of SiC MOSFET, and thus the crosstalk test is also discussed based on actual measurements.

The parasitic parameters between voltage measurement points are one of the important factors affecting measurement results, but they have not received enough attention and discussion in the past. During previous test and research, engineers often overlooked the parasitic parameters between voltage measurement points, confusing the measured waveforms with deviations and the true waveforms intended for further analysis. This can lead to incorrect analysis of device switching processes, erroneous research results, and incorrect evaluation of converter designs. This chapter will discuss this issue and propose methods to eliminate the influence of parasitic parameters between voltage measurement points on measurement results.

Y. Gao and Y. Zhang, *Silicon Carbide Power Devices*, CPSS Power Electronics Series,
https://doi.org/10.1007/978-981-96-3480-4_5

Double-pulse test covers the entire process of device research, development, production, and application, with different requirements at each stage. Applying the highest standards to all stages would severely sacrifice efficiency and cost-effectiveness. Therefore, it's necessary to analyze and discern specific requirements for each stage and propose corresponding rational demands. Once the requirements are clarified, judgment criteria are needed. In the past, the evaluation of double-pulse test results has been based on characteristic values obtained from test waveforms, which is convenient and easy to implement. However, waveform accuracy is the foundation of the value of these characteristic values. Therefore, analyzing the switching and reverse recovery waveforms obtained from measurements is essential to correctly evaluate double-pulse test results. From this perspective, while test techniques are indeed crucial, the requirements for test and judgment criteria are equally important. Unfortunately, this aspect has not received the attention it deserves.

5.1 Power Converter Working Mode

According to the form of energy conversion, power converters are divided into four major categories: DC–DC converters, AC–DC converters, DC–AC converters, and AC–AC converters. Each category of converters includes various circuit topologies, forming a vast topology system. Just for the commonly used DC–DC converter topologies, there are Buck, Boost, Buck-Boost, Cuk, Flyback, Forward, Push–Pull, and Half-Bridge, among others. In addition, power converters can be classified into various types such as multi-level and soft-switching types. In practical applications, suitable topologies are chosen for converter design based on different application scenarios, power levels, specifications, and cost requirements.

Taking Buck circuits, three-phase full-bridge inverter circuits, DNPC circuits, and LLC circuits as representatives of DC–DC converters, DC–AC converters, multi-level converters, and soft-switching converters, respectively, the commutation process during device switching is as follows.

1. **Buck Converter**

The topology structure and working principle of the Buck circuit are illustrated in. During the time interval t_0 to t_1, the switch S is turned on, allowing the inductor current I_L to flow through S and the inductor L towards the load, causing I_L to linearly increase. At time t_1, S is turned off. Due to the inability of the inductor I_L to change instantaneously, I_L commutates to the diode D for freewheeling. During the time interval t_1 to t_2, I_L flows through D and L to provide current to the load, causing I_L to linearly decrease. At time t_2, S is turned on again, causing I_L to commutate to S and D to turn off. The Buck circuit can be viewed as a half-bridge circuit composed of S and D, with the inductor L tapped from the midpoint of the bridge. Commutation occurs between S and D, and L acts to stabilize the current (Fig. 5.1).

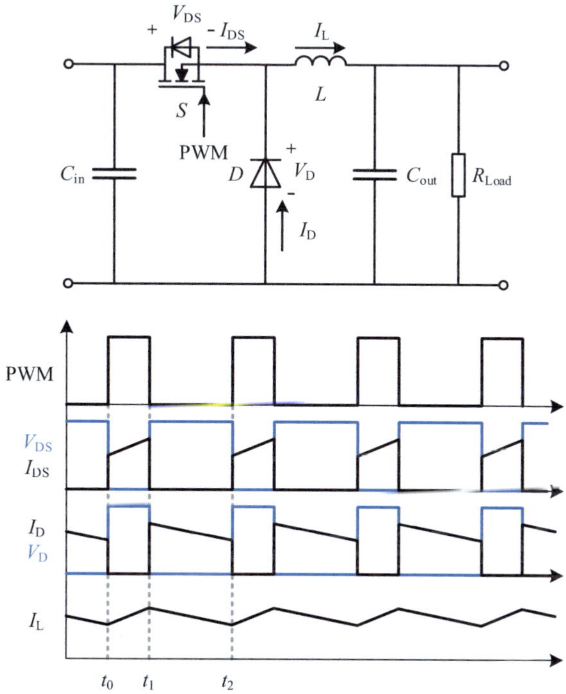

Fig. 5.1 BUCK circuit and working principle

2. **Three-Phase Full Bridge Inverter Circuit**

The three-phase full-bridge inverter circuit generates three phase three-level line voltages at the AC output side by switching different conduction combinations, and utilizes PWM principle to obtain sinusoidal voltage. As shown in Fig. 5.2, under an inductive load with an impedance angle $\varphi < 60$ degrees, at time t_0, the switch combination is (S_5, S_6, S_1), where $V_{AB} = V_d$, $V_{BC} = -V_d$, and $V_{CA} = 0$. At time t_1, switching to (S_6, S_1, S_2) occurs, S_5 is turned off, S_2 is turned on, resulting in $V_{BC} = 0$ and $V_{CA} = -V_d$. Since I_C cannot change instantaneously, I_C first commutates to the reverse parallel diode D_2 of S_2 for freewheeling until after time t_2, I_C reverses, S_2 is forward biased, and D_2 is turned off. Thus, although there are many switching devices, the commutation still takes place within the devices in the same bridge arm.

3. **DNPC Inverter Circuit**

The DNPC circuit can operate four -quadrant, with four types of flow replacement modes, as shown in Fig. 5.3.

When I_{out} is positive and V_{out} is positive, S_2 remains conductive while S_1 undergoes switching action, and commutation occurs between S_1 and D_n.

When I_{out} is positive and V_{out} is negative, S_3 remains conductive while S_2 undergoes switching action, and commutation occurs between S_2 and D_4.

When I_{out} is negative and V_{out} is negative, S_3 remains conductive while S_4 undergoes switching action, and commutation occurs between S_4 and D_p.

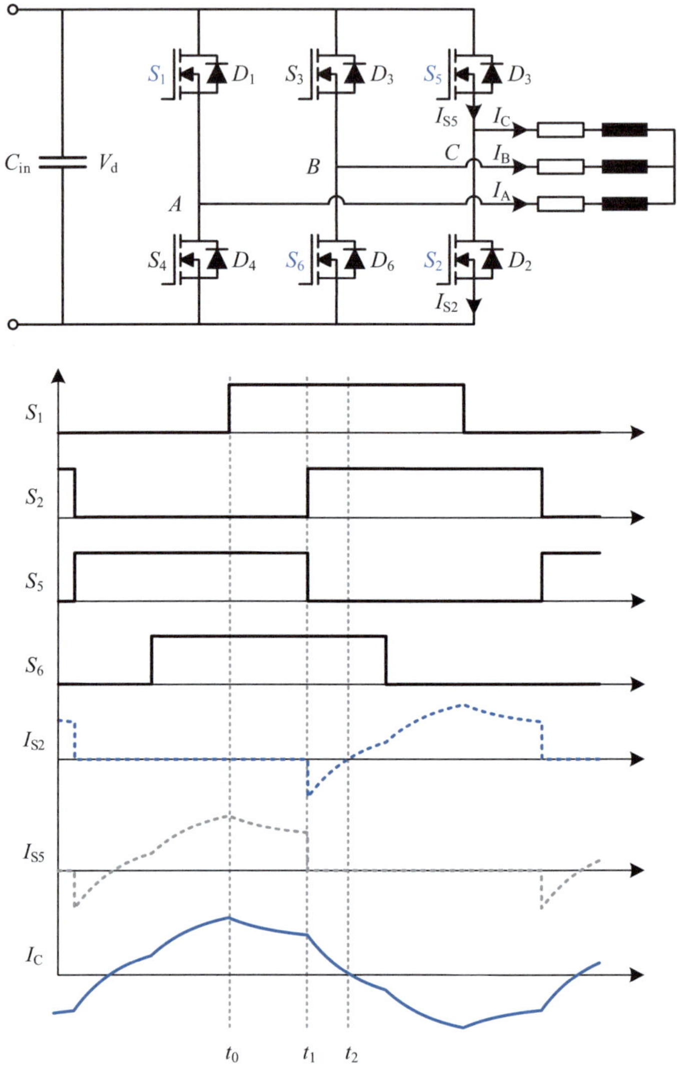

Fig. 5.2 Three-phase full-bridge inverter circuit and working principle

When I_{out} is negative and V_{out} is positive, S_2 remains conductive while S_3 undergoes switching action, and commutation occurs between S_3 and D_1.

4. **LLC Resonance Converter**

The LLC resonance converter changes the transmission power by regulating the frequency of the switch. Its optimal working point is that its switching frequency is equal to the resonance frequency, as shown in Fig. 5.4. At time t_0, the switch transistors S_1 and S_4 are turned off. Since the resonant current I_r is positive, I_r commutates

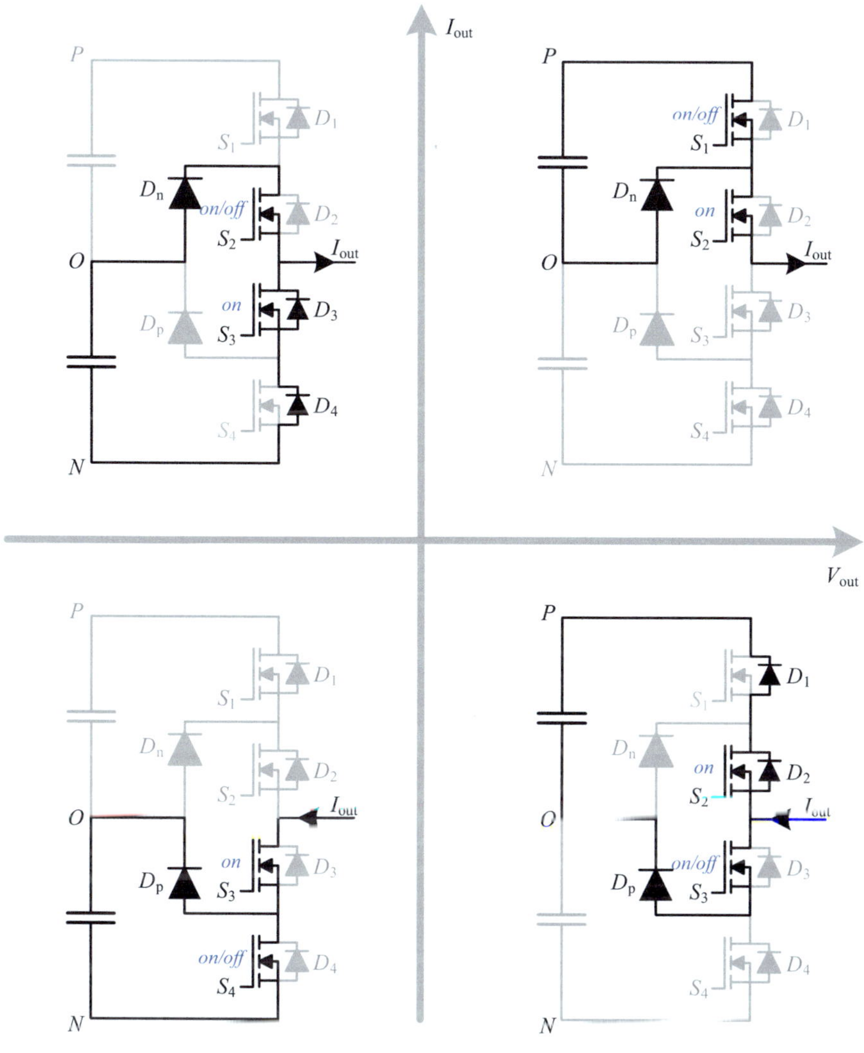

Fig. 5.3 DNPC circuit and working principle

to the body diodes D_2 and D_3 for freewheeling. At time t_1, after the dead time has ended, S_2 and S_3 receive the turn-on signal and operate in the third quadrant conduction state. At time t_2, as I_r changes from positive to negative, S_2 and S_3 are forward biased, achieving ZVS.

From the above introduction of the above four typical converters, it can be observed that although the topology types and operating principles differ, the commutation modes of the switch transistors during the switching process have commonalities. That is, the commutation process occurs between switch transistors or between switch transistors and diodes in the half-bridge structure. Therefore, test, evaluating,

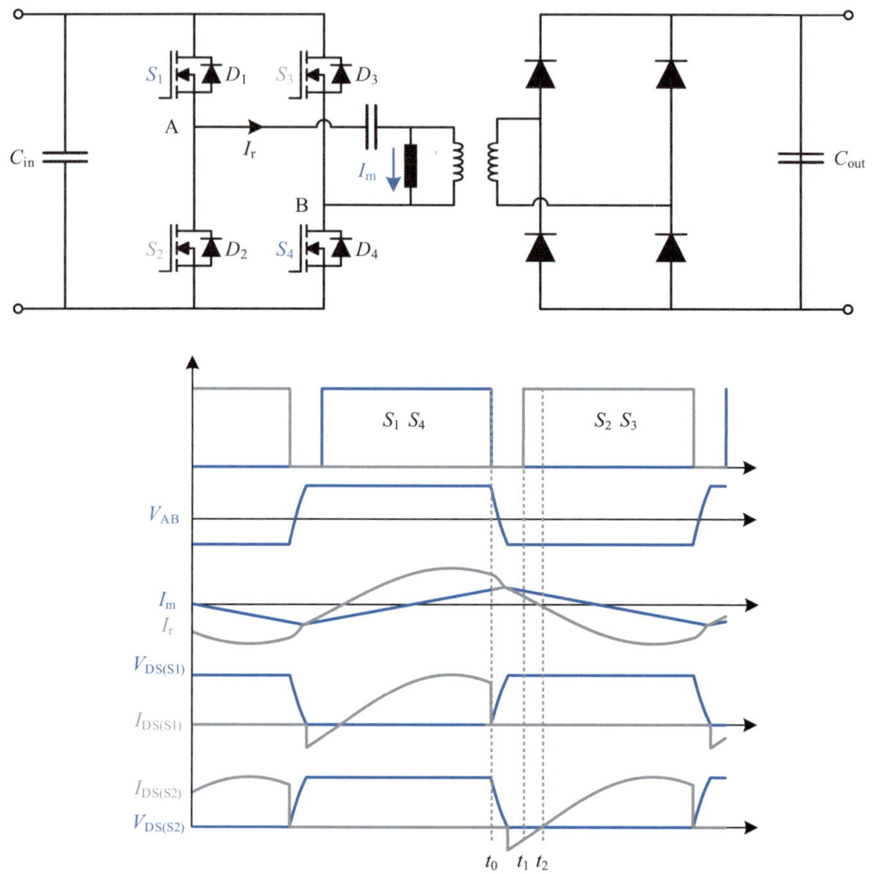

Fig. 5.4 LLC resonant circuit and its working principle

and comparing the switching characteristics of switch transistors can all be based on the inductive load half-bridge circuit. Moreover, when the parameters of the half-bridge circuit match the actual parameters of the converter, the results based on the inductive load half-bridge circuit can be directly applied to converter design.

5.2 Basis of Double-Pulse Test

5.2.1 Basic Principles of Double-Pulse Test

Based on the inductive load half-bridge circuit, a comprehensive evaluation method for switch transistor switching characteristics has been developed, known as double-pulse test. The double-pulse test circuit consists of a bus capacitor C_{Bus}, the tested

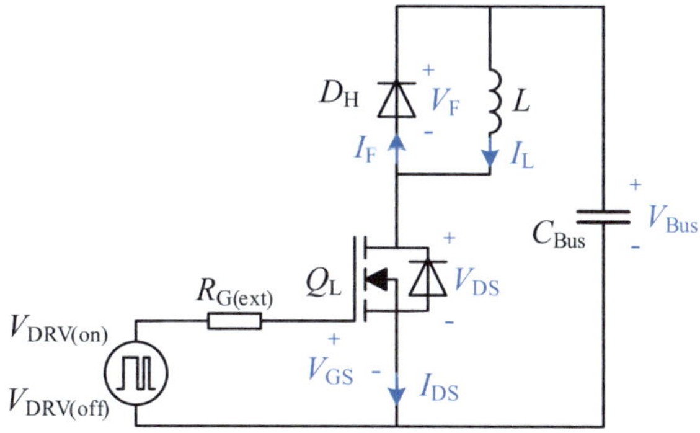

Fig. 5.5 Double-pulse test circuit

switch transistor Q_L, the companion diode D_H, drive circuitry, and load inductance L, as shown in Fig. 5.5.

During the test, sending a double-pulse drive signal with appropriate pulse width to Q_L allows obtaining the switching characteristics of Q_L at the specified voltage V_{set} and specified current I_{set}. The entire test process is illustrated in Fig. 5.6.

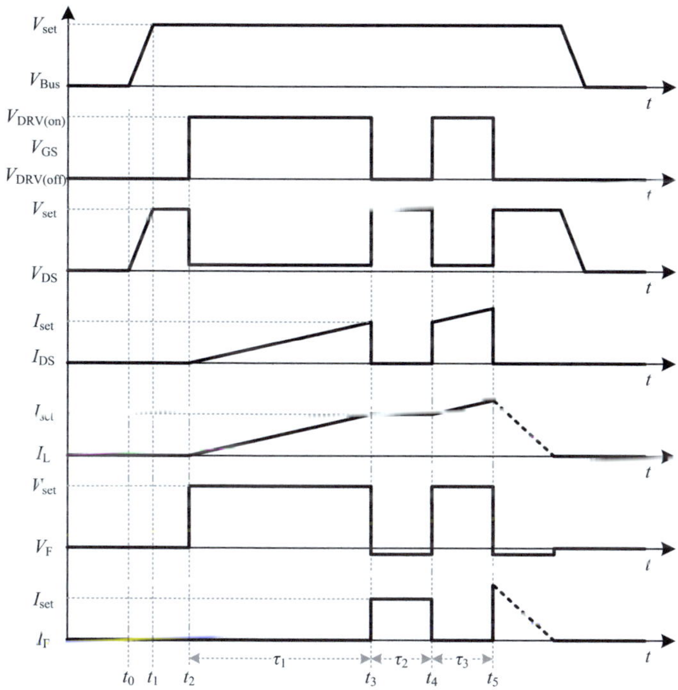

Fig. 5.6 Complete double-pulse test process

(1) $-t_0$

The bus voltage $V_{\text{Bus}} = 0\,\text{V}$, the drive circuit outputs the turn-off drive voltage $V_{\text{DRV(off)}}$, and Q_L is in the off state.

(2) $t_0 - t_1$

The DC power supply starts to charge C_{Bus}, and V_{Bus} gradually reaches the set voltage V_{set}. Since D_{H} is shorted by L, Q_L bears the bus voltage.

(3) $t_1 - t_2$

No action;

(4) t_2

The drive circuit of Q_L is triggered to send an turn on signal, causing the drive voltage to transition from $V_{\text{DRV(off)}}$ to the turn on drive voltage $V_{\text{DRV(on)}}$. This results is in the turn on of Q_L in the beginning.

(5) $t_2 - t_3$

The time duration is denoted as τ_1. Q_L, L, and C_{Bus} form a loop, with the load current I_L beginning to rise according to (5.1).

$$\mathrm{d}I_L(t)/\mathrm{d}t = \frac{\left(V_{\text{Bus}} - I_L(t)\cdot R_{\text{es}(L)} - I_L(t)\cdot R_{\text{DS(on)}}\right)}{L}, \tag{5.1}$$

where $R_{\text{es(L)}}$ is the equivalent series resistance of the load inductance. In the actual test circuit, C_{Bus} is sufficiently large, maintaining V_{Bus} at a constant V_{set}. Meanwhile, both $R_{\text{es(L)}}$ and the $R_{\text{DS(on)}}$ of the SiC MOSFET are small enough to simplify (5.1) to (5.2), ensuring that I_L increases linearly.

$$DI_L(t)/\mathrm{d}t \approx V_{\text{set}}/L \tag{5.2}$$

(6) t_3

When I_L reaches the specified current I_{set},

$$I_L(t_3) = I_{\text{set}} = \tau_1 \cdot V_{\text{set}}/L \tag{5.3}$$

At this point, a turn off signal is sent to the drive circuit of Q_L, causing the drive voltage to transition from $V_{\text{DRV(on)}}$ to $V_{\text{DRV(off)}}$. Q_L switches off at I_{set}, and I_L continues to flow through D_{H}.

(7) $t_3 - t_4$

The time duration is denoted as τ_2. D_{H} and L form a loop, where I_L slowly decreases according to (5.4). Here, V_{F} represents the forward voltage drop across D_{H}.

$$\mathrm{d}I_L(t)/\mathrm{d}t = \frac{V_F(I_L(t)) + I_L(t)\cdot R_{\text{es}(L)}}{L} \tag{5.4}$$

(8) t_4

I_L decreases until it reaches $I_L(t_4)$, and when I_L decreases significantly,

$$I_L(t_4) = I_{\text{set}} - \int_{t_3}^{t_4} \frac{V_F(I_L(t)) + I_L(t) \cdot R_{\text{es}(L)}}{L} \mathrm{d}t \quad (5.5)$$

At this point, an activation signal is sent to the drive circuit of Q_L, causing Q_L to turn on at $I_L(t_4)$. D_H undergoes reverse recovery.

(9) $t_4 - t_5$

The time duration is denoted as τ_3. It shares the same current loop as the stage $t_2 - t_3$, where I_L increases linearly according to (5.2).

(10) t_5

I_L rises to $I_L(t_5)$,

$$I_L(t_5) = I_L(t_4) + \tau_3 \cdot V_{\text{set}}/L \quad (5.6)$$

At this moment, a turn-off signal is sent to the drive circuit of Q_L, causing Q_L to turn off at $I_L(t_5)$. I_L continues to flow through D_H.

(11) $t_5 -$

The current loop during the stage $t_3 - t_4$ is identical. I_L decreases slowly according to (5.4) until it reaches 0 A, taking a longer time, represented by dashed lines. Following this, the DC power supply is turned off, discharging C_{Bus}, causing V_{Bus} to drop to 0 V.

Throughout the entire test process, the device under test Q_L undergoes two cycles of activation and deactivation, forming two pulses. The first pulse occurs from t_2 to t_3, while the second pulse occurs from t_4 to t_5. The interval between t_3 and t_4 represents the pulse gap, hence the name double-pulse test. The focus lies on the moments t_3 and t_4, corresponding to the deactivation and activation processes of Q_L at specified voltage V_{set} and current I_{set}, respectively. By measuring and recording the waveforms of Q_L's V_{GS}, V_{DS}, and I_{DS}, its switching characteristics can be analyzed and evaluated.

According to Sect. 5.1, it is known that power converters have two types of commutation modes: MOS-diode and MOS-MOS configurations. When conducting double-pulse test, it's necessary to select device models and configurations that match those of the actual converter. For the MOS-MOS configuration, it only requires replacing the diode D_H with the SiC MOSFET Q_H, and continuously applying the turn-off signal during test, as shown in Fig. 5.7.

Additionally, at t_4, it's not only the turn-on process of Q_L, but also the reverse recovery process of the body diode of D_H or Q_H. Directly measuring the upper transistor involves floating ground measurements, which can lead to inaccurate results due to the abrupt common-mode voltage changes. Therefore, when test the reverse

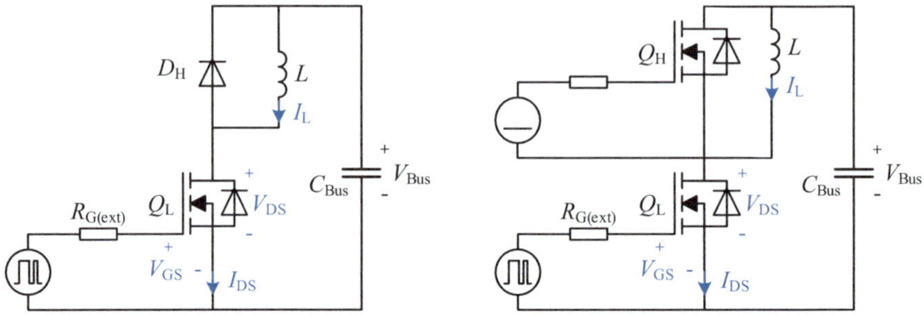

(a) MOS-Diode (b) MOS-MOS

Fig. 5.7 SIC MOSFET's switching characteristics test circuit

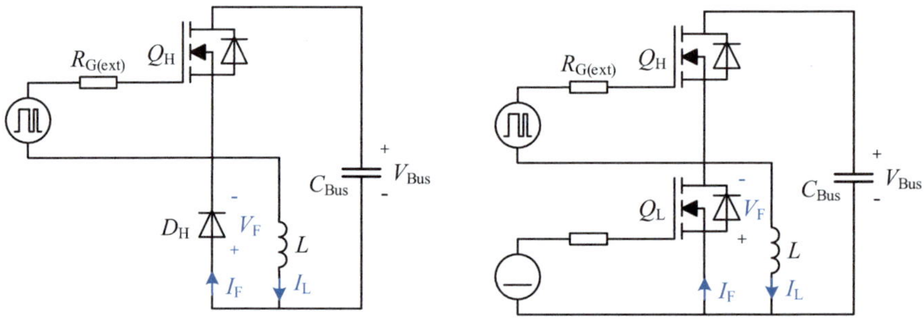

(a) MOS-diode (b) MOS-MOS

Fig. 5.8 Test circuit with reverse recovery characteristics of diode

recovery characteristics of the diode, it's often preferable to use a circuit as shown in Fig. 5.8. The device under test is the lower transistor, with the load inductor connected in parallel across its terminals. The companion device is the upper transistor, which undergoes on/off actions during test.

5.2.2 Double-Pulse Test Parameter Settings

5.2.2.1 Pulse Width Time

1. **Width of First Pulse** τ_1

The purpose of the double-pulse test is to evaluate the switching characteristics of the device under specified voltage and current conditions. The voltage is set by the DC power supply output, while the current is established by the first pulse. To achieve the specified current I_{set} more accurately, the value of τ_1 can be calculated using (5.7). It's important to note that (5.7) is derived from the simplified version of (5.1), leading to some deviation in the actual current achieved. However, by appropriately selecting C_{Bus} and controlling $R_{es(L)}$, the error can be minimized, and adjustments to τ_1 can be made based on empirical values.

$$T_1 = I_{set} \cdot L/V_{set} \tag{5.7}$$

At time t_2, Q_L turns on. Due to the effects of the equivalent parallel capacitance $C_{ep(L)}$ of L, reverse recovery of the companion diode, and the junction capacitance C_{DH}, current spikes and oscillations occur. Therefore, it is necessary to wait for the current oscillations to settle before turning off Q_L, to avoid affecting the turn-off process of Q_L at time t_3. This is why there is a requirement for the lower limit of the duration τ_1, typically greater than 1us-2us.

On the other hand, an excessively long τ_1 can cause significant temperature rise in the device, resulting in the test results not reflecting the switching characteristics of the device under specified temperature conditions. This is why there is a requirement for the upper limit of the duration τ_1. For individual devices, τ_1 generally should not exceed 15us, while for high-power modules, τ_1 generally should not exceed 50us.

2. **Pulse Interval τ_2**

At time t_3, Q_L turns off. It's crucial to observe the entire process from when VGS of Q_L starts to decrease until the oscillation of V_{DS} ends. This is why there is a requirement for the lower limit of the duration τ_2, typically greater than 1us-2us.

On the other hand, during the pulse gap, I_L decreases slowly according to (5.4). If the decrease is too large, then at t_4, I_L will deviate significantly from I_{set}, which fails to meet the requirement for Q_L to turn on at t_4 while maintaining I_{set}. This is why there is a requirement for the upper limit of the duration τ_2, which is determined by V_F, $R_{es(L)}$, L, and the allowable current drop amplitude.

3. **Width of the Second Pulse τ_3**

At time t_4, Q_L turns on. It's essential to observe the entire process from when V_{GS} of Q_L starts to rise until the oscillation of I_{DS} ends. This is why there is a requirement for the lower limit of the duration τ_3, typically greater than 1us-2us.

On the other hand, during the second pulse period, I_L increases according to (5.2), reaching $I_L(t_5)$ at time t_5. When $I_L(t_4)$ is not significantly different from I_{set}, $I_L(t_5)$ is

$$I_L(t_5) = (\tau_1 + \tau_3) \cdot V_{set}/L \tag{5.8}$$

An excessively high $I_L(t_5)$ can lead to excessively high turn-off voltage spikes. When these spikes exceed the device's voltage tolerance, it can result in device damage. This is why there is a requirement for the upper limit of the duration τ_3. It can be requested that τ_3 is less than 0.5 times τ_1, which means $I_L(t_5)$ is less than 1.5 times I_{set}. The shorter the duration, the better, but this should be determined based on specific circumstances.

5.2.2.2 Load Inductor

The inductance of the load inductor is limited by the following factors:

1. **Commutation Path**

In the converter, the inductance L of the load inductor is sufficiently large, much greater than the inductance L_{Loop} of the main power commutation loop. This ensures that during the switching process, the current I_L remains essentially unchanged, and the high-frequency current for commutation mostly passes through the main power commutation loop. Similarly, in the double-pulse test, L also needs to be much greater than L_{Loop} to achieve the same effect. L_{Loop} generally ranges from a few nanohenries to two hundred nanohenries, while L typically ranges from tens of microhenries to several hundred microhenries.

2. **Width of the First Pulse τ_1**

The first pulse width τ_1 is determined by (5.7). When V_{set} and I_{set} are determined, a larger L leads to a larger τ_1, while a smaller L results in a smaller τ_1. The value of L needs to meet the requirements of the upper and lower limits of τ_1.

3. **The Turn on Current ($I_{L(t4)}$) at Time t_4**

$I_L(t_3) = I_{\text{set}}$, and between pulses,$I_L$ gradually decreases according to (5.4). At time t_4, $I_L(t_4)$ is given by (5.5). In the double-pulse test, it is required that $I_L(t_4)$ is not significantly different from I_{set}. Therefore, under this premise, the minimum value of the load inductance $L_{\min}$ be calculated using (5.9).

$$L \geq \tau_2 \cdot \frac{V_{\text{F}}(I_{\text{set}}) + I_{\text{set}} \cdot R_{\text{es}(L)}}{K_i \cdot I_{\text{set}}} \tag{5.9}$$

4. **The Turn off Current $I_L(t_5)$ at Time t_5**

In order to avoid excessive turn-off current, $I_L(t_5)$ is required to be less than 1.5 times of I_{set}.

$$(\tau_1 + \tau_3) \cdot V_{\text{set}}/L \leq 1.5 I_{\text{set}} \tag{5.10}$$

5.2.2.3 Bus Capacitance

During test, it's necessary to ensure that V_{Bus} remains constant at V_{set}. During the first pulse period, the response speed of the DC power supply is slow, and I_L is provided by C_{Bus}, causing V_{Bus} to experience a certain degree of decrease. To avoid excessive voltage drop in V_{Bus}, C_{Bus} needs to satisfy (5.11), where K_{v} represents the allowable voltage drop ratio, typically ranging from 0.5 to 2%.

$$C_{\mathrm{Bus}} \geq \frac{LI_{\mathrm{set}}^2}{2K_{\mathrm{v}}V_{\mathrm{set}}^2} \tag{5.11}$$

From (5.11), it's evident that the larger the L, the greater the requirement for C_{Bus}. However, an excessively large C_{Bus} can result in prolonged charging and discharging times, leading to more severe consequences in case of circuit failure during test. Therefore, there's a tendency to choose a smaller load inductance to reduce the demand on C_{Bus}.

5.2.2.4 Parameter Setting Method

Based on the above research, there are mutual influences and constraints among pulse width, load inductance, and bus capacitance. For instance, to shorten τ_1 and reduce C_{Bus}, choosing a smaller L is preferable. However, to ensure that $I_L(t_4)$ closely matches I_{set}, L cannot be too small. Therefore, in designing parameters for double-pulse test, careful calculations are needed to determine the ranges of values for each parameter.

The method for determining these three parameters is as follows:

(1) **Step 1**

Specify the minimum values for τ_1, τ_2, and τ_3 as $\tau_{1,\min}$, $\tau_{2,\min}$, and $\tau_{3,\min}$ respectively, ensuring that

$$T_{1,\min} \geq 2\tau_{3,\min} \tag{5.12}$$

(2) **Step 2**

The preset alternative range for τ_1 is $\tau_{1,\min} - \tau_{1,\max}$. Calculate the corresponding range of inductance values using (5.13).

$$L_{\tau_1,\min} = \frac{\tau_{1,\min} \cdot V_{\mathrm{set}}}{I_{\mathrm{set}}} \leq L \leq \frac{\tau_{1,\max} \cdot V_{\mathrm{set}}}{I_{\mathrm{set}}} = L_{\tau_1,\max} \tag{5.13}$$

Based on the above range of L, calculate the range of the equivalent series resistance $\mathrm{R}_{\mathrm{es}(L)}$ as $R_{\mathrm{es}(L),\min} - R_{\mathrm{es}(L),\max}$.

(3) **Step 3**

Prescribe the value of τ_2, and substitute $R_{\mathrm{es}(L),\min} - R_{\mathrm{es}(L),\max}$ into (5.14) to obtain the range of L values as $L_{\tau_2,\min} - L_{\tau_2,\max}$.

$$L = \tau_2 \cdot \frac{V_{\mathrm{F}}(I_{\mathrm{set}}) + I_{\mathrm{set}} \cdot R_{\mathrm{es}(L)}}{K_i \cdot I_{\mathrm{set}}} \tag{5.14}$$

(4) **Step 4**

According to the corresponding relationship between $L_{\tau_1,\min}-L_{\tau_1,\max}$, $L_{\tau_2,\min}-L_{\tau_2,\max}$ and $R_{\mathrm{es}(L),\min}-R_{\mathrm{es}(L),\max}$, determine the value range of L and select its minimum value as L to proceed to the fifth step; if $L_{\tau_1,\max} \leq L_{\tau_2,\min}$, return to the third step and reduce the τ_2 preset until L is determined; if τ_2 is reduced to $\tau_{2,\min}$ and L is not determined yet, then return to step 2 and increase the $\tau_{1,\max}$ preset.

(5) **Step 5**

Determine the value of τ_3 to satisfy (5.15),

$$\tau_{3,\min} \leq \tau_3 \leq 0.5\tau_1 \tag{5.15}$$

(6) **Step 6**

Calculate C_{Bus} according to (5.11).

5.2.3 *Dynamic Process of SiC Devices*

Traditionally, the dynamic processes of power devices refer to their transitions between on-state and off-state, including the switching processes of switch devices and the forward and reverse recovery processes of diodes or MOSFET body diodes. The former belongs to active processes, while the latter belongs to passive processes. These dynamic processes are mentioned in various sources such as device datasheets, textbooks, and materials explaining device characteristics and test standards. It's important to note that avalanche breakdown, short circuits, and surges, while they change the operating state of power devices, are abnormal operating conditions to be avoided, demonstrating the device's ability to withstand extreme conditions rather than being considered dynamic processes.

Based on the introduction to the double-pluse test process in Sect. 5.2.1, it is evident that in the double-pulse test circuit, the reverse recovery process of the diode or SiC MOSFET body diode corresponds to the turn-on process of its counterpart device. Specifically, as shown in Fig. 5.8, the reverse recovery process of the bottom device is observed during the second turn-on process of the top device.

In the topology of the half-bridge circuit or during the commutation process equivalent to the half-bridge circuit topology, the switching actions of the switching devices (MOSFETs or IGBTs with anti-parallel diodes) will cause fluctuations in the drive waveform of their counterpart devices, resulting in crosstalk. This will be discussed in detail in Chap. 8. It's evident that crosstalk is an unavoidable byproduct when switching devices are turned on or off in a half-bridge circuit. It's worth noting that the switching devices experiencing crosstalk undergo transitions in the conduction and turn-off states of their body diodes or anti-parallel diodes. Compared to Si power switching devices, SiC MOSFETs have faster switching speeds, lower threshold voltage $V_{GS(th)}$, poorer gate oxide reliability, and incompletely resolved gate oxide reliability issues. Therefore, crosstalk has a more severe impact on the reliability and application safety of SiC MOSFETs. Extra attention should be paid to this issue when applying SiC MOSFETs. The crosstalk test circuit is the same as the one used for measuring reverse recovery, as shown in Fig. 5.8. It observes the turn-off crosstalk and turn-on crosstalk of the bottom device during the first turn-off process of the top device and the second turn-on process, respectively. It can be seen that the switching process, the reverse recovery process and the crosstalk process can all be studied using the dual-pulse test. Moreover, the latter two processes occur simultaneously with the switching process.

From this, it can be seen that the switching process, reverse recovery process, and crosstalk process can all be studied using double-pulse test, with the latter two occurring alongside the switching process. In summary, considering the inherent characteristics of SiC power devices and the problems encountered in practical applications, the dynamic processes worthy of attention for SiC devices should include: switching, reverse recovery, and crosstalk. Based on the test results of the switching process and reverse recovery process, device characteristic analysis, loss calculation, driver circuit validation, circuit debugging, and other tasks can be performed. Similarly, based on the test results of the crosstalk process, crosstalk characteristic analysis, crosstalk suppression method research, driver circuit validation, circuit debugging, and other tasks can be conducted.

5.2.4 Double-Pulse Test Platform

The double-pulse test platform comprises a test board, load inductor, high-voltage DC power supply, low-voltage DC power supply, signal generator, oscilloscope, voltage probe, and current sensor, as shown in Fig. 5.9.

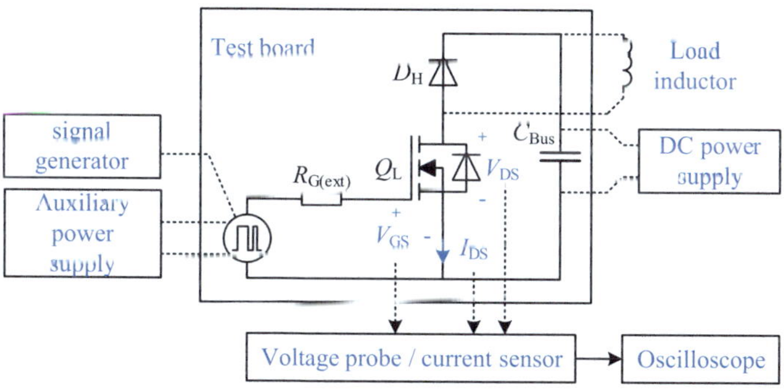

Fig. 5.9 The structure of the double-pulse test platform

5.2.4.1 Test Board

The test board generally includes the main power circuit, driver circuit, decoupling capacitors, bus capacitors, and external interfaces, typically designed by the tester. The basic requirement for the test board is to ensure normal characteristics of the device under test and accurate measurements, without abnormal waveforms occurring under the majority of test conditions. Abnormal waveforms may include unintentional turn-on or turn-off, voltage spikes exceeding device limits, and significant discrepancies between switch waveforms and theoretical expectations. If certain characteristics of the device under test exceed the limits of the test board and cause abnormal waveforms, targeted optimization and improvements to the test board are necessary.

From the analysis of the switching process of SiC MOSFETs in Sect. 3.3.2, it's evident that their switching characteristics are greatly influenced by external circuit parameters such as main power loop inductance, driver loop inductance, and driver circuit drive capability. Therefore, even with the use of completely identical measurement instruments, significant differences in test results can occur simply due to differences in test boards. Thus, comparing test results obtained from different test boards or test equipment is meaningless. Furthermore, the switching times and switching energies stated in datasheets are based on tests conducted by the device manufacturers using their own test boards. Therefore, it's normal for deviations to exist between the results obtained from other test circuits or converter products and those stated in the datasheet, and these differences can be quite significant. In other words, whether the measured values match or approximate those in the datasheet cannot be used as a standard to evaluate the performance or accuracy of double-pulse test equipment. At this point, many engineers have a misconception.

Ensuring completion of double-pulse test for SiC MOSFETs requires two main aspects: a test board that meets the requirements of SiC MOSFET characteristics and the correct measurement methods. Measurement methods primarily involve selecting appropriate test equipment and connection methods for measurement points to ensure accuracy. On the other hand, engineers typically need to design the test board according to specific requirements.

Compared to Si MOSFETs and Si IGBTs, SiC MOSFETs exhibit significant differences in characteristics, which impose more stringent requirements on test boards. These differences also present application challenges when using SiC MOSFETs. Main challenges include issues such as turn-off voltage spikes, crosstalk, common-mode currents, and driver circuit design, which will be discussed in detail in subsequent chapters. Therefore, I won't delve further into these issues here.

To facilitate engineers in assessing the switching characteristics of SiC MOSFETs, many device manufacturers have introduced evaluation boards. These evaluation boards are not complete power converters; typically, they only include half-bridge circuits and corresponding driver circuits. When adding inductors, double-pulse test can be conducted to comprehensively evaluate the device's switching characteristics. When adding inductors and loads, the evaluation board can function as an open-loop operated converter for assessing device characteristics such as Buck/synchronous

(a) Cree KIT8020-CRD-5FF0917P-2 [1]

(b) Cree KIT-CRD-CIL12N-XM3 [2]

(c) Littelfuse EVAL_DCP_01 [3]

(d) Littelfuse EVAL_GDEP_01 [4]

Fig. 5.10 Evaluation board with coaxial resistors

Buck, Boost/synchronous Boost, single-phase inverters, and bidirectional DC–DC converters. This allows for a comprehensive evaluation of the device's conduction, switching, and thermal performance characteristics.

On evaluation boards from Wolfspeed and Littelfuse, coaxial resistors are used for current sensing, enabling complete double-pulse test, as depicted in Fig. 5.10. Littelfuse's EVALGDEP01 driver board, functioning as a satellite board, connects to the main power motherboard via connectors, offering up to 7 types of driver circuits that can be freely replaced. Engineers can also design their own driver boards to compare different driver circuits and study SiC driver characteristics. Unlike evaluation boards designed for discrete package devices, Wolfspeed's KIT-CRD-CIL12N-XM3 is used to evaluate 1200 V/450 A SiC MOSFET power modules. Due to the significantly increased current rating, larger coaxial resistors are chosen to prevent overheating damage.

Infineon and ROHM evaluation boards do not use coaxial resistors for current sampling. Instead, they mainly evaluate devices by running them in an open loop with a load and an inductor, as shown in Fig. 5.11.

Unlike typical evaluation boards, the ROHM P02SCT3040KR-EVK-001 offers a wide range of features, with the following key characteristics [7], allowing for more comprehensive device evaluation:

(a) Infineon EVAL-1EDC20H12AH-SIC [5]

(b) Infineon EVAL-PS-E1BF12-SiC [6]

(c) ROHM P02SCT3040KR-EVK-001 [7]

Fig. 5.11 Evaluation board without shunt resistor

(1) Supports evaluation of TO-247-4L and TO-247-3L packages
(2) Operates with a single power supply (+ 12 V)
(3) Double-pulse testing up to 150 A
(4) Switching operation up to 500 kHz
(5) Supports various power topologies (Buck, Boost, Half-Bridge) Built-in isolated gate driver power supply, adjustable with a variable resistor (+ 12 V to + 23 V)
(6) Selectable gate drive negative bias and zero bias via jumper pins
(7) Prevents simultaneous conduction of the high and low sides
(8) Built-in overcurrent protection features (DESAT, OCP)

Manufacturers provide customers with access to test circuit schematics, PCB files, and BOM, allowing engineers to create their own test boards. This serves as excellent material for learning and mastering the characteristics and usage of SiC MOSFETs.

5.2.4.2 Load Inductor

Inductors used in the primary power circuit of converters typically feature a magnetic core, with common core materials being ferrite and powdered iron, often in toroidal or E-shaped configurations. When designing inductors, it's crucial to select the appropriate core based on the actual operating conditions to avoid core saturation and maintain a relatively constant inductance value. The same requirements apply to the

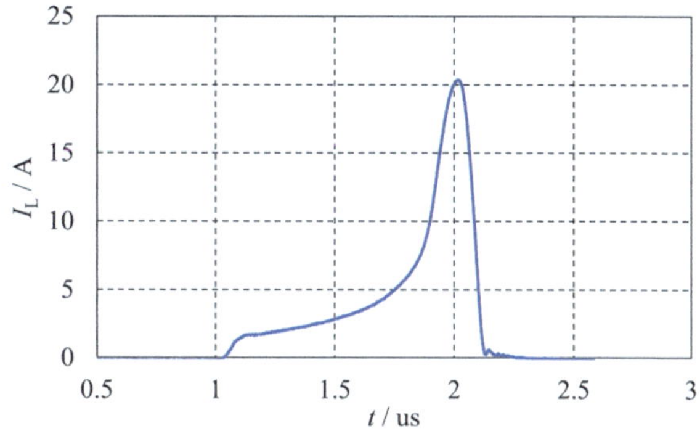

Fig. 5.12 The current waveform is abnormal caused by the inductive saturation

load inductors used in double-pulse test; otherwise, abnormal current waveforms may occur during test, as illustrated in Fig. 5.12.

To address the saturation issue associated with magnetic core inductors, one option is to use air-core inductors as load inductors. With air as the magnetic medium, saturation is not a concern. However, since there is no magnetic core, the inductance will significantly increase, necessitating the selection of thicker wire to avoid heating and voltage drop issues caused by excessive internal resistance of the inductor. However, the hollow inductor has the disadvantages of excessive volume, heavy weight and high cost. Moreover, the equivalent parallel capacitance of the hollow inductor is larger than that of the magnetic-coated inductor, which has a negative impact on the dynamic characteristic test. Regarding this issue, some relevant scholars have conducted research [8]. Therefore, for engineers with certain technical capabilities, choosing a well-designed magnetic-coated inductor offers better performance and cost advantages.

5.2.4.3 High-Voltage DC Power Supply

In double-pulse test, the high-voltage DC power supply charges the bus capacitor, providing the bus voltage required for the test. The fundamental requirement for this power supply is that its output voltage meets the test requirements. For SiC MOSFETs with voltage ratings of 1700 V and below, considering the typical bus voltage range in practical applications, it is recommended to select the voltage output capability of the high-voltage DC power supply as shown in Table 5.1.

In addition to output voltage, the output current capability is the second important indicator of a DC power supply. In double-pulse test, the speed of charging the bus capacitor doesn't need to be very fast, and both the current establishment and the commutation process during switching are completed through the bus capacitor. Therefore, the current output capability requirement for the DC power supply is not high. This also brings additional benefits: low-power supplies are more affordable, and they generate minimal noise during operation.

Table 5.1 Double-pulse test requirements for high-voltage power output voltage requirements

Device voltage level (V)	High-voltage DC power output voltage (V)
650	500
1200	1000
1700	1500
3300	3000
6500	6000
10,000	9000

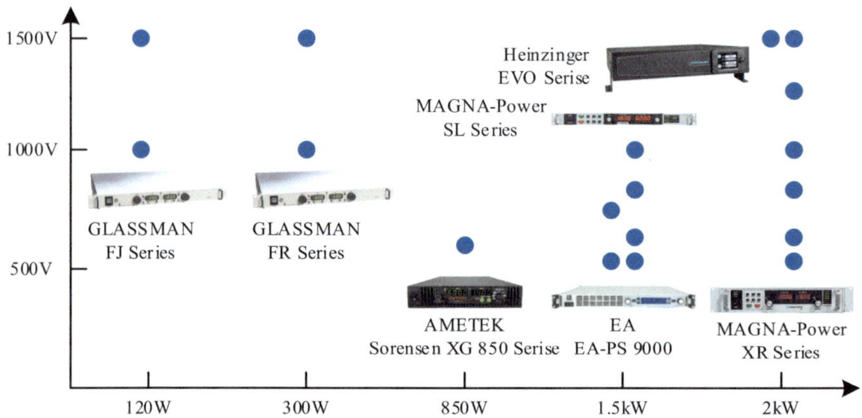

Fig. 5.13 High-voltage dc power supply for double-pulse test

Based on the requirements for output voltage and output current, selecting a DC power supply is straightforward. Figure 5.13 shows some partial examples of high-voltage DC power supplies suitable for double-pulse test of SiC MOSFETs [9–14].

5.2.4.4 Low Voltage DC Power Supply

The low-voltage DC power supply is used to supply power on the drive circuit. Generally, the desktop power supply is generally selected, as shown in Fig. 5.14.

5.2.4.5 Signal Generator

During the double-pulse test process, it's necessary to send double-pulse commands to the driver circuit. Since the signals sent are straightforward and have no special requirements, they can be generated by the tester using an MCU development board,

Fig. 5.14 Tektronix 2220 series low voltage dc power supply [15]

which is cost-effective and flexible to operate. However, it's crucial to pay extra attention to its reliability and stability, especially considering the strong electromagnetic interference generated during device switching. Improper design of the development board or signal output wiring can easily lead to interference and output errors.

To avoid these issues, a signal generator can be used, as shown in Fig. 5.15. The Tektronix AFG31000 series signal generators are specifically equipped with

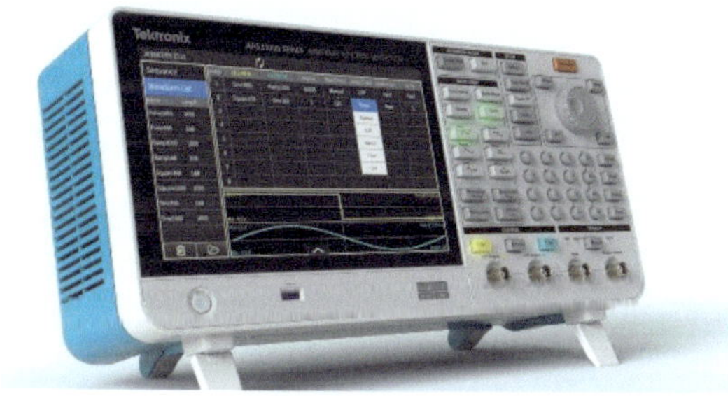

(a) Tektronix AFG31000 series signal generator [15]

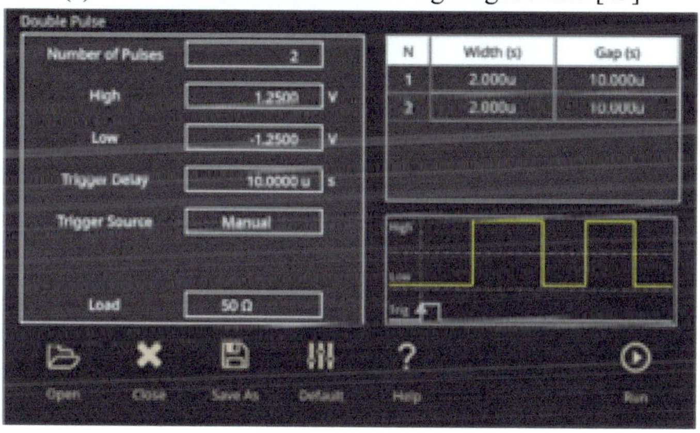

(b) Tektronix double pulse test application [16]

Fig. 5.15 Signal generator

double-pulse test applications, eliminating the need for external PC applications or manual editing. Users can directly generate two waveforms with variable pulse widths (ranging from 20 ns to 150 μs) on the touchscreen display, significantly simplifying the double-pulse pulse width settings.

Furthermore, in double-pulse test, it's essential to measure the drain-source voltage V_{DS}, gate-source voltage V_{GS}, and drain-source current I_{DS} of the device under test. Choosing the appropriate oscilloscope, voltage probes, and current probes is crucial for achieving accurate measurements. We will now delve into a detailed discussion on this topic.

5.3 Measurement Instrument

5.3.1 Oscilloscope

5.3.1.1 Frequency Characteristics

1. **Bandwidth**

Bandwidth is the most commonly mentioned crucial parameter of an oscilloscope, as it determines the oscilloscope's ability to measure high-frequency signals. It is primarily determined by the analog front end of the oscilloscope. Figure 5.16 illustrates a frequency-gain curve for an oscilloscope, where the gain decreases with increasing frequency. The frequency at which the gain drops to − 3 dB is defined as the bandwidth, corresponding to the half-power point.

When the oscilloscope's bandwidth is insufficient, the high-frequency components of the measured signal are attenuated, reducing measurement accuracy. For example, when using an oscilloscope with a bandwidth of 100 MHz to measure a 150 MHz sine wave with a peak-to-peak voltage V_{pp} of 1 V, the displayed V_{pp} on the

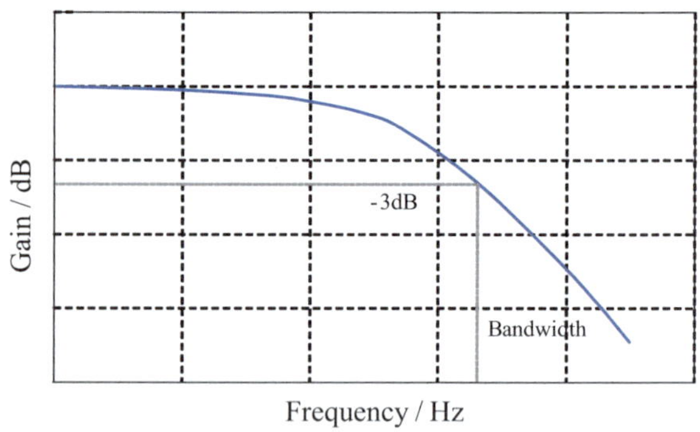

Fig. 5.16 Oscilloscope frequency-gain curve schematic diagram

oscilloscope is only 0.62 V. However, when using an oscilloscope with a bandwidth of 500 MHz, the measured V_{pp} is 1 V. This illustrates that the oscilloscope's bandwidth does not necessarily represent the highest frequency it can accurately measure. To ensure measurement accuracy, it's essential to ensure that the frequency range of the signal under test falls within the region where the oscilloscope's gain is close to 0 dB.

2. **Rising Time**

Rise time is another oscilloscope parameter closely related to bandwidth, with an approximate conversion relationship between them. The rise time of an oscilloscope refers to the time it takes for the oscilloscope to measure a waveform's amplitude to rise from 10 to 90% when measuring an ideal step signal. It's important to note that rise time does not represent the fastest edge speed that the oscilloscope can accurately measure. Since ideal step signals do not exist in reality, it's common practice to use signals with rise times 3–5 times faster than the oscilloscope's rise time to measure the oscilloscope's rise time. The oscilloscope's rise time can be found in the manual and generally doesn't need to be verified. Rise time is another oscilloscope parameter closely related to bandwidth, with an approximate conversion relationship between them. The rise time of an oscilloscope refers to the time it takes for the oscilloscope to measure a waveform's amplitude to rise from 10 to 90% when measuring an ideal step signal. It's important to note that rise time does not represent the fastest edge speed that the oscilloscope can accurately measure. Since ideal step signals do not exist in reality, it's common practice to use signals with rise times 3–5 times faster than the oscilloscope's rise time to measure the oscilloscope's rise time. The oscilloscope's rise time can be found in the manual and generally doesn't need to be verified

The error in amplitude will inevitably lead to errors in rise time measurement. Therefore, when the bandwidth is insufficient, the measured signal appears slower, and the measured rise time of the signal is larger than the true value. To accurately measure fast rise time signals, it's crucial for the oscilloscope to have a sufficiently high bandwidth and a short enough rise time.

3. **Frequency Response Method**

Based on the above analysis, both the accuracy of amplitude measurements and the accuracy of rise time measurements of fast rise time signals require the oscilloscope to have sufficient bandwidth. This is the foundation for accurate measurements and is a crucial factor to consider when selecting an oscilloscope. To determine the bandwidth requirements for measurements, it's also necessary to consider the characteristics of the oscilloscope's frequency response curve, known as its frequency response mode. There are two types of frequency response modes for oscilloscopes: Gaussian response and maximum flatness response, as shown in Fig. 5.17.

Named after its response resembling the characteristics of a low-pass Gaussian filter, the Gaussian response is commonly adopted by oscilloscopes with bandwidths below 1 GHz. It features minimal rise time and no overshoot when measuring step signals. However, due to its gradual gain reduction starting far below the bandwidth,

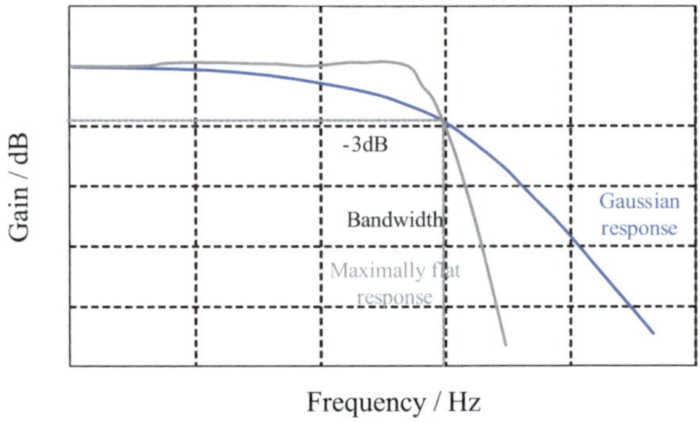

Fig. 5.17 Oscilloscope frequency response method

signals are severely attenuated within the bandwidth range. Hence, achieving high measurement accuracy requires oscilloscopes with significantly higher bandwidth than the signal being measured. The relationship between the rise time RT_{scope} and the bandwidth BW_{scope} of a Gaussian response oscilloscope can be expressed as:

$$RT_{scope} \approx 0.35/BW_{scope} \tag{5.16}$$

Named for its maximum gain flatness within the bandwidth range, the maximum flatness response is typically adopted by oscilloscopes with bandwidths of 1 GHz and above. The maximum flatness response differs significantly from the Gaussian response in that it exhibits a broader flatness range within the bandwidth, with the gain beginning to decrease only near the bandwidth. This results in higher measurement accuracy within the bandwidth. Additionally, it features a faster roll-off rate outside the bandwidth, providing better out-of-band suppression. The rise time of a maximum flatness response oscilloscope is given by:

$$RT_{scope} \approx 0.4 \sim 0.5/BW_{scope} \tag{5.17}$$

It's important to note that while oscilloscopes have two types of response modes, it doesn't imply that oscilloscopes of the same response type and bandwidth exhibit identical frequency response curves. Frequency response curves vary noticeably between different manufacturers and series of oscilloscopes.

The maximum signal frequency f_{max} can be estimated using Eqs. (5.18) and (5.19), where $RT_{10-90\%}$ and $RT_{20-80\%}$ represent the rise times defined by the signal's rise from 10 to 90% and from 20 to 80%, respectively.

$$F_{max} \approx 0.4/RT_{10-90\%} \tag{5.18}$$

$$f_{max} \approx 0.5/RT_{20-80\%} \tag{5.19}$$

Table 5.2 Relationship between rising time measurement accuracy and oscilloscope bandwidth

Rising time measurement error (%)	Gaussian response BW_{scope}	Maximum flat response BW_{scope}
~ 20	$1.0 f_{max}$	$1.0 f_{max}$
~ 10	$1.3 f_{max}$	$1.2 f_{max}$
~ 3	$1.9 f_{max}$	$1.4 f_{max}$

The greater the difference between the oscilloscope bandwidth and the maximum signal frequency, the higher the accuracy of the signal rise time. The specific values are shown in Table 5.2.

The discussion so far has only considered the bandwidth of the oscilloscope. However, the measurement system consists of both the oscilloscope and the probe, and it's important to consider the overall bandwidth of the measurement system. For Gaussian response oscilloscopes, the system bandwidth and rise time are calculated using Eqs. (5.20) and (5.21). When using a maximum flatness response oscilloscope, the system bandwidth and rise time of the oscilloscope-probe combination are typically provided by the oscilloscope manufacturer.

$$\frac{1}{BW_{sys}} = \frac{1}{\sqrt{BW_{scope}^2 + BW_{probe}^2}} \tag{5.20}$$

$$RT_{sys} = \sqrt{RT_{scope}^2 + RT_{probe}^2} \tag{5.21}$$

Since the frequency response curve of the oscilloscope probe combination is not available in the datasheet, it's challenging to accurately calculate the error in amplitude measurement. When selecting oscilloscopes and probes for analog signal test, it's prudent to choose ones with bandwidths at least five times higher than the maximum frequency of the signal under test. Considering the switching speed of SiC MOSFETs, it is advisable to ensure that both the oscilloscope and the probe have bandwidths of at least 500 MHz.

5.3.1.2 Vertical Features

1. **ADC Quantitative Position**

In digital oscilloscopes, the Analog-to-Digital Converter (ADC) is the core component that converts continuous analog signals into discrete digital signals. The most crucial parameter of an ADC is its resolution, which determines the vertical resolution of the ADC. The resolution of an ADC with N bits divides the full scale into 2^N equal parts, resulting in a resolution of $1/2^N$ of the full scale. Therefore, the higher the resolution of the ADC, the higher its vertical resolution.

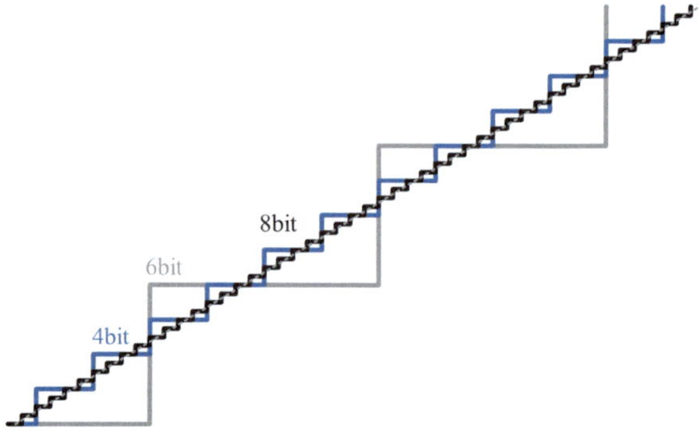

Fig. 5.18 ADC quantification position on measurement resolution

Indeed, you can think of an ADC as a special type of balance scale where each weight it has is of equal mass. When the total mass of all weights is constant (i.e., the ADC's full scale range is fixed), the smaller the mass of each individual weight (i.e., the higher the ADC's resolution), the smaller the measurement error and the higher the accuracy of the balance scale (ADC). The resolution of an 8-bit ADC is respectively 4 times and 16 times higher than that of a 6-bit ADC and a 4-bit ADC. As shown in Fig. 5.18, as the number of bits in the ADC increases, the measurement results become more precise.

To ensure the measurement capability of high-frequency signals, digital oscilloscopes typically employ FLASH-type ADCs with a quantization bit depth of 8 bits. Taking the test of a 1200 V SiC MOSFET as an example, if the vertical scale of the oscilloscope is set to 150 V/div, then the full scale is 1200 V. In this scenario, the resolution of an 8-bit ADC is 4.69 V, while that of a 12-bit ADC is 0.29 V.

2. **Noise**

The waveform displayed on the oscilloscope is the result of the signal under test being superimposed with noise, making the waveform appear fuzzy and fat. The sources of noise include the analog front end, ADC, probe, cables, etc. In terms of the overall noise of the oscilloscope, the contribution of ADC quantization error is usually small, while the noise introduced by the analog front end is typically more significant. When no probe is connected, the noise of the oscilloscope can be observed, exhibiting random characteristics as shown in Fig. 5.19.

The same series of oscilloscope has 7 products from 500 MHz to 8 GHz. Figure 5.20 shows the RMS noise value when the input impedance is 50Ω [17].

For a specific oscilloscope, the noise increases with the increase in vertical scale. This is because a part of the separation of the oscilloscope noise is based on the relative noise of the oscilloscope range, which is determined by the vertical scale. When the vertical scale is small, this part of the noise can be ignored; but when the vertical scale is large, the noise will occupy the dominant position.

Under the same vertical scale, the higher the bandwidth of the oscilloscope, the greater the noise. This is because the noise at the front end of the oscilloscope is randomly distributed in Gauss. The higher the bandwidth, and the wider the noise

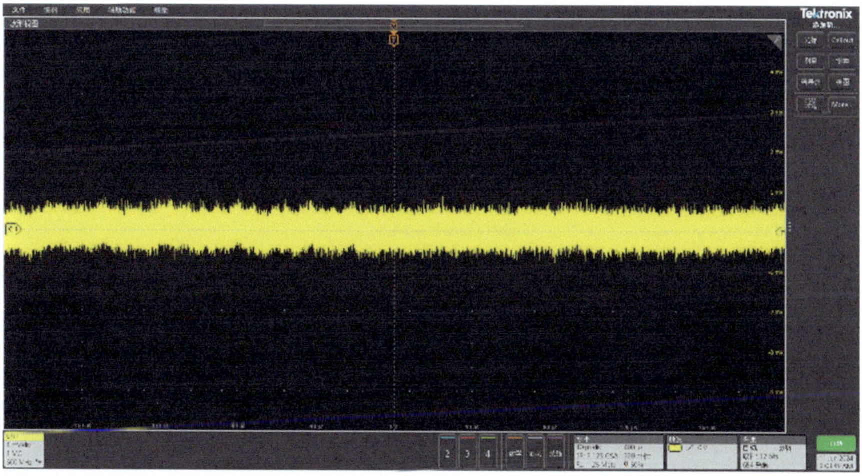

Fig. 5.19 The noise of the oscilloscope

Fig. 5.20 Same series 500 MHz to 8 GHz bandwidth oscilloscope 50 Ω input impedance noise

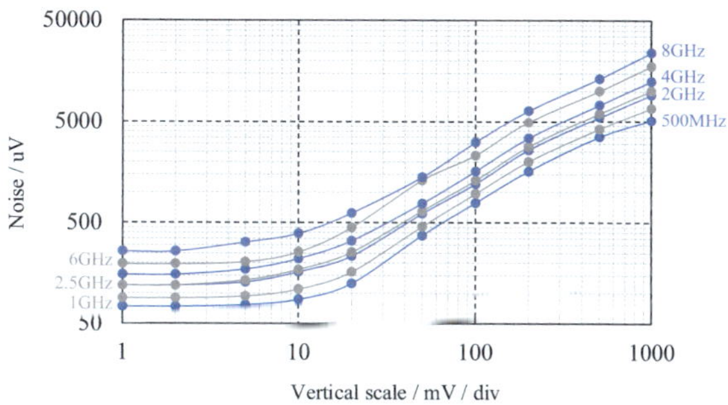

spectrum, the greater the noise. Low-bandwidth oscilloscope can effectively filter high-frequency noise, while high -frequency noise will enter the bandwidth of the high bandwidth oscilloscope.

During double-pulse test, some probes used require the oscilloscope's input impedance to be set to 1 MΩ. Taking a 500 MHz oscilloscope as an example, when the input impedance is set to 1MΩ, the noise becomes significantly larger than when set to 50 Ω, as shown in Fig. 5.21. Additionally, the noise of the oscilloscope is also influenced by the amplification factor. The higher the amplification factor, the greater the noise. This is because the noise of the entire measurement system has two sources: the probe and the oscilloscope's analog front end. A probe with an attenuation ratio of N:1 attenuates the signal to 1/N before sending it to the oscilloscope. Then, the oscilloscope amplifies the signal by a factor of N before displaying it. While amplifying the signal, the oscilloscope also amplifies its own noise, exacerbating the impact of oscilloscope noise.

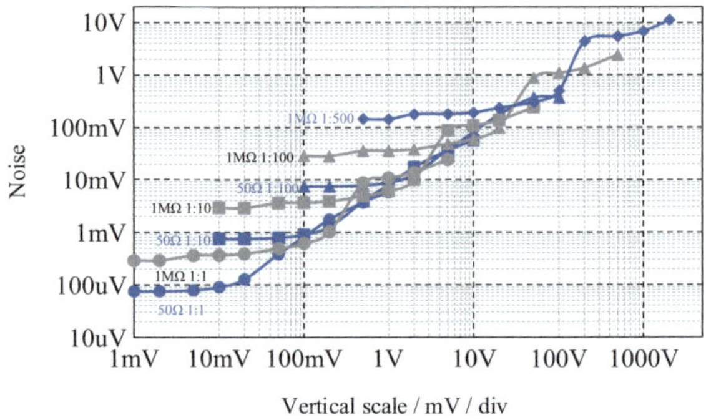

Fig. 5.21 The effect of the oscilloscope input impedance and amplification multiple on noise

3. **Effective Number of Bits**

Due to noise and other non-ideal factors, the actual resolution of an oscilloscope is often lower than the ideal resolution corresponding to the quantization bits of the ADC. The quantization bits calculated based on the actual resolution achievable by the oscilloscope are referred to as the Effective Number of Bits (ENOB).

Oscilloscopes with 8-bit ADCs typically have ENOBs around 6 bits. Therefore, in the previous example of measuring the V_{DS} of a 1200 V SiC MOSFET, the actual resolution achieved is not 4.6875 V as expected from an ideal 8-bit ADC, but only 18.75 V. This illustrates how noise significantly reduces the vertical resolution of the oscilloscope. Thus, ENOB provides a more accurate characterization of the oscilloscope's vertical resolution compared to the quantization bits of the ADC, serving as a basis for oscilloscope selection and error analysis.

ENOB is determined by sweeping a sine wave of fixed amplitude and analyzing it in the time domain or frequency domain. It is depicted by an ENOB curve. The ENOB of an oscilloscope is influenced by factors such as the sampling rate, vertical scale, and the proportion of the measured signal in the vertical direction of the screen. Therefore, when specifying ENOB, the corresponding test conditions need to be indicated.

Figure 5.22 illustrates the ENOB of oscilloscopes from the same series with bandwidths ranging from 1 to 8 GHz [18]. These values were obtained under the following conditions: a sampling rate of 20 GS/s, vertical scale of 100 mV/div, and the measured signal occupying 90% of the vertical direction of the screen.

The ENOB curve of the same series oscilloscope is not flat, indicating that ENOB is not a fixed value but fluctuates with frequency. Oscilloscope manufacturers typically do not provide ENOB curves; instead, they often only specify the average ENOB value in the datasheet.

For oscilloscopes from the same series, the ENOB decreases as the bandwidth of the oscilloscope increases. For oscilloscopes with bandwidths of 1 GHz, 2 GHz, 4 GHz, and 8 GHz, the average ENOB values are 7.8 bits, 7.5 bits, 7.2 bits, and 6.4 bits respectively. This trend correlates with the increase in noise with higher bandwidth, as shown in Fig. 5.20. Therefore, when selecting an oscilloscope, blindly pursuing higher bandwidth is not appropriate. Excessively high bandwidth can lead to excessive noise and a significant decrease in ENOB.

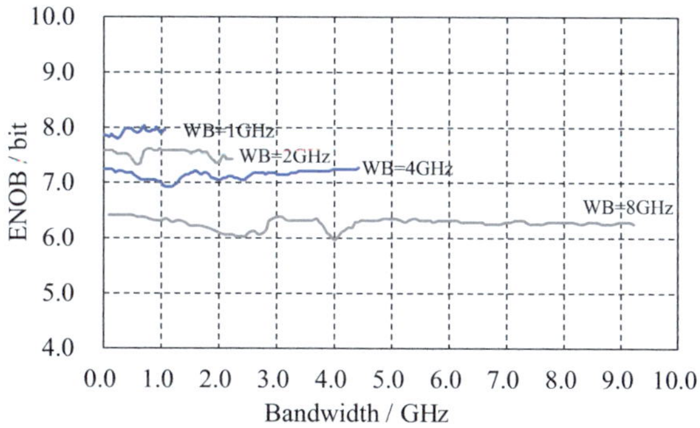

Fig. 5.22 The same series 1 GHz to 8 GH bandwidth oscilloscope ENOB

The noise and ENOB described above only pertain to the oscilloscope itself and do not consider the influence of the probe. Similar to bandwidth, what needs attention in measurement is the ENOB of the entire measurement system. As seen from the previous discussion on the impact of amplification factor on noise, using probes with lower attenuation ratios helps control system noise and improve system ENOB.

4. **High-Resolution Oscilloscope**

Since the introduction of the first digital oscilloscope in the mid-1980s, the 8-bit ADC has been a constant feature of oscilloscopes. Increasing the bandwidth and sampling rate of ADCs has always been a major concern for high-speed oscilloscopes. However, after 2010, oscilloscopes reached a bottleneck in terms of bandwidth and sampling rate. Meanwhile, the requirements for test equipment in power systems, embedded systems, and EMC/EMI test demanded higher precision and a wider dynamic range. As a result, manufacturers began to focus more on the research and development of high-resolution oscilloscopes and introduced a variety of models, significantly enhancing the vertical resolution of oscilloscopes.

The improvement of high-resolution oscilloscopes compared to 8-bit oscilloscopes mainly lies in the acquisition chips. On one hand, it requires ADCs with higher bit depths, and on the other hand, it necessitates low-noise analog front-ends, as shown in Fig. 5.23.

As shown in Fig. 5.24, the noise of three oscilloscopes with a bandwidth of 2.5 GHz is depicted. It's evident that the oscilloscope with a 10-bit resolution has significantly lower noise compared to the other two 8-bit oscilloscopes. A lower noise level in the oscilloscope enables more effective utilization of the ADC's bit depth, thereby enhancing the ENOB.

Using the same Tektronix MSO64B BW-1000 oscilloscope and identical probes, two double pulse tests were conducted, one with the oscilloscope set to 8-bit resolution and the other to 12-bit resolution. Other settings remained unchanged. The test results are shown in Fig. 5.25. It can be observed that in the 8-bit resolution mode, the waveform exhibits significant aliasing, while in the 12-bit resolution mode, the waveform appears much smoother. This demonstrates the benefits of high resolution.

Fig. 5.23 Tektronix's low noise simulation front end of high-resolution oscilloscope

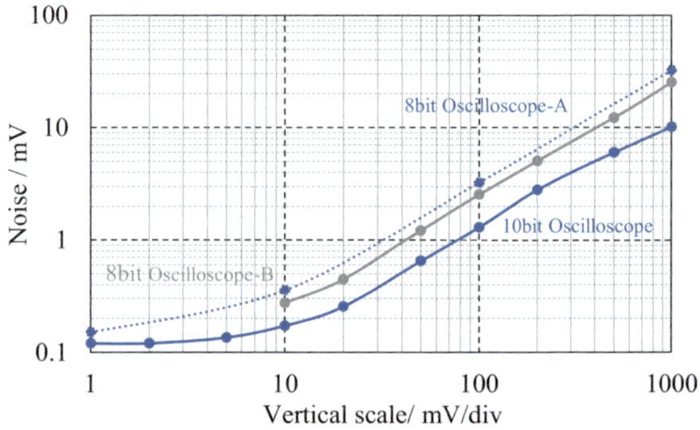

Fig. 5.24 Comparison of oscilloscope noise

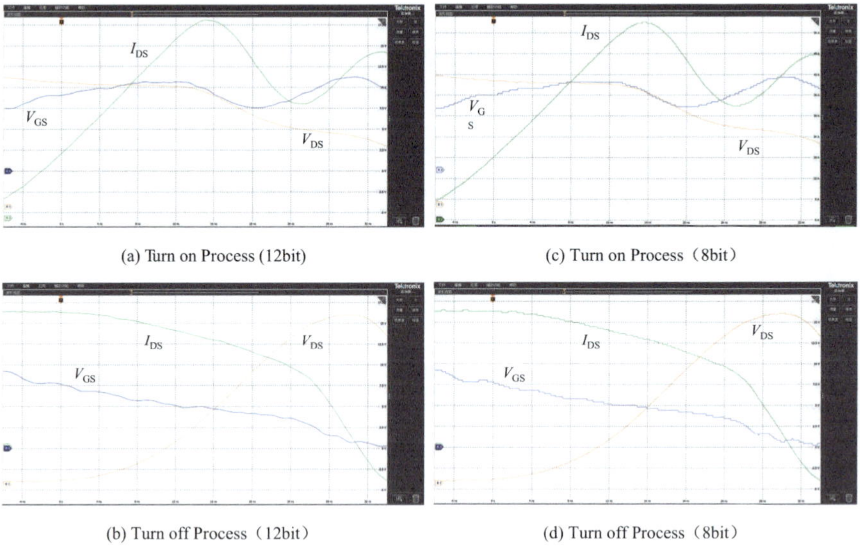

Fig. 5.25 Effect of the resolution of the oscilloscope on measurement

5.3.1.3 Horizontal Characteristics

1. **Sampling Rate**

The sampling rate refers to the number of samples taken by an oscilloscope from the measured signal per unit time, typically measured in GS/s (Gigasamples per second). Ensuring that the oscilloscope's sampling rate satisfies the Nyquist-Shannon sampling theorem is a fundamental requirement. The Nyquist-Shannon sampling theorem states that if $x(t)$ is a band-limited signal, meaning that $X(j\omega) = 0$ for $|\omega| > \omega_M$, and if the sampling frequency $\omega_s > 2\omega_M$ is greater than twice the maximum frequency, then $x(t)$ can be uniquely determined by its samples $x(nT)$, where $n = 0, \pm 1, \pm 2 \ldots$. In other words, to accurately reconstruct the analog signal without distortion, the sampling rate should be at least twice the highest frequency component in the analog signal.

Since the frequency response of an oscilloscope is not an ideal low-pass filter, there are still many high-frequency components beyond the bandwidth, as shown in Fig. 5.26. When the bandwidth is 1 GHz, the Gaussian response has a slow roll-off outside the bandwidth, with high-frequency components attenuated only until around 3 GHz. On the other hand, the maximum flat response has a faster roll-off, with high-frequency components attenuated by around 2 GHz. Therefore, sampling at twice the bandwidth does not satisfy the Nyquist-Shannon sampling theorem. Typically, for Gaussian response oscilloscopes, the sampling rate should be greater than four times the bandwidth, while for maximum flat response oscilloscopes, the sampling rate should be greater than 2.5 times the bandwidth.

In addition to the requirements of the Nyquist-Shannon sampling theorem, the sampling rate also needs to be sufficiently high to capture the details of the switching process of SiC MOSFET and obtain enough sampling points for calculation and analysis. Figure 5.27 illustrates the effect of sampling rate on the number of sampling points at different switching times. To achieve the required number of sampling points, a higher sampling rate is needed for shorter switching times. Nowadays, mid-range oscilloscopes can easily achieve sampling rates of 10–20 GS/s, which can fully meet the requirements of SiC MOSFET double-pulse test.

The sampling rate and resolution of regular oscilloscopes are noticeably different from those of high-resolution oscilloscopes. Figure 5.28 illustrates waveforms

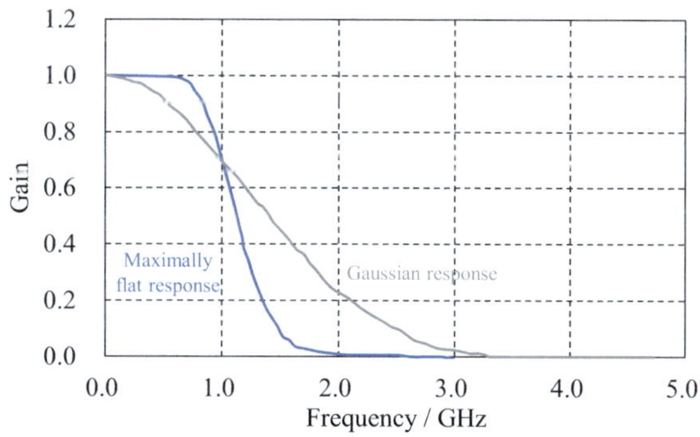

Fig. 5.26 1 GHz bandwidth oscilloscope frequency response curve

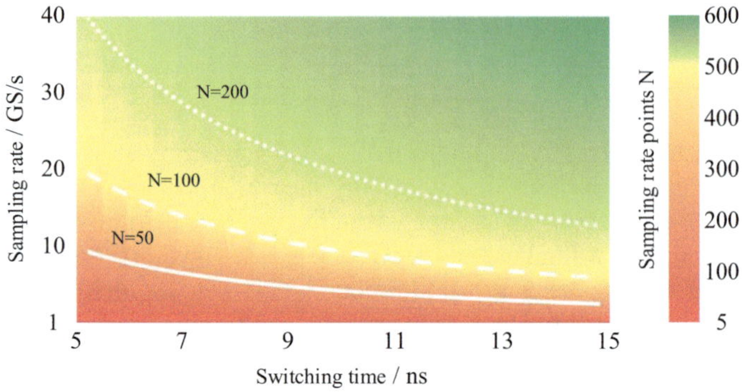

Fig. 5.27 The relationship between sampling points and sampling rates

Fig. 5.28 The measurement results of ordinary oscilloscope

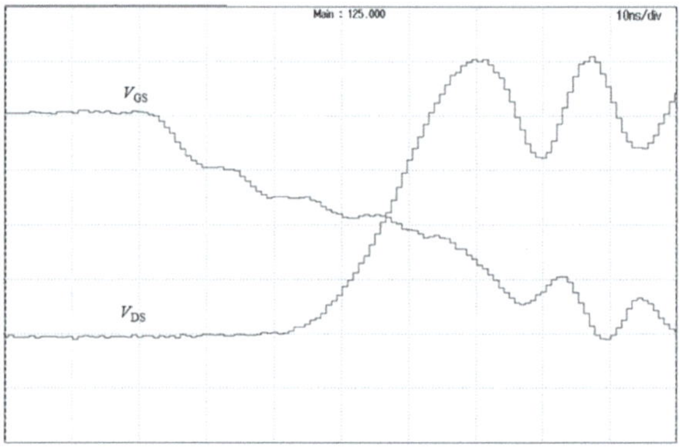

(a) Turn on Process

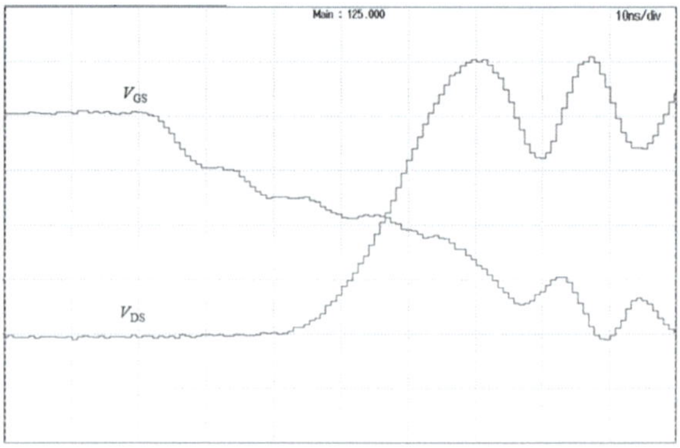

(b) Turn off Process

obtained using a regular oscilloscope with an 8-bit ADC and a sampling rate of 1.25 GS/s, showing more pronounced aliasing artifacts. It is evident that measuring the switching characteristics of SiC MOSFETs requires oscilloscopes with sufficiently high sampling rates and ADC bit depths.

2. **Inherent Jitter**

The internal circuit defects of an oscilloscope can cause a horizontal offset of the ADC sampling points from their ideal positions, known as the inherent baseline jitter of the oscilloscope. The sources of oscilloscope jitter are diverse and may include errors introduced by multiple ADCs operating in tandem, jitter in the ADC sampling clock input signal, and other internal sources of jitter. The collective effect of these sources of horizontal offset error constitutes an overall horizontal timing error, referred to as the equivalent sampling clock jitter (or simply sampling jitter) or inherent source jitter clock (SJC).

We often focus on the requirement of the Nyquist theorem for the sampling rate to be at least twice the highest frequency of the signal being measured, but we may overlook the prerequisite of uniformly spaced sampling. When inherent jitter is significant, the signal reconstructed by the oscilloscope using the sinc(x)/x function will deviate noticeably from the actual measured signal, as shown in Fig. 5.29.

High-sampling-rate oscilloscopes are often achieved through interleaved multiple ADCs, as illustrated in Fig. 5.30. The input signal is fed to two ADCs, each driven

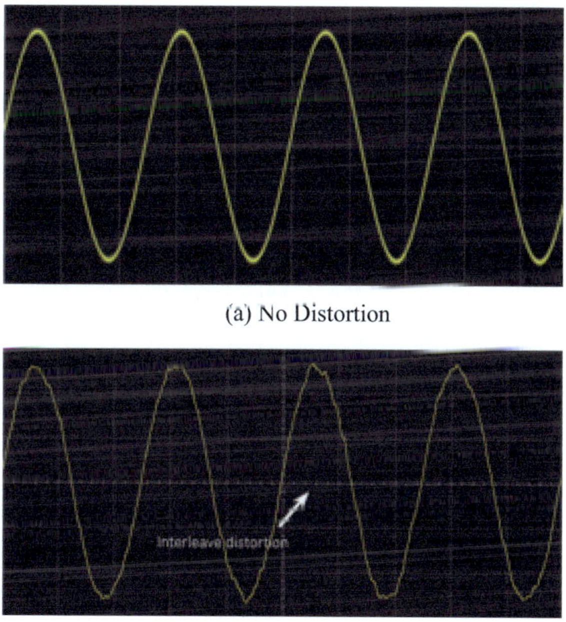

(a) No Distortion

(b) Distortion

Fig. 5.29 Distortion by non-equal interval sampling [19]

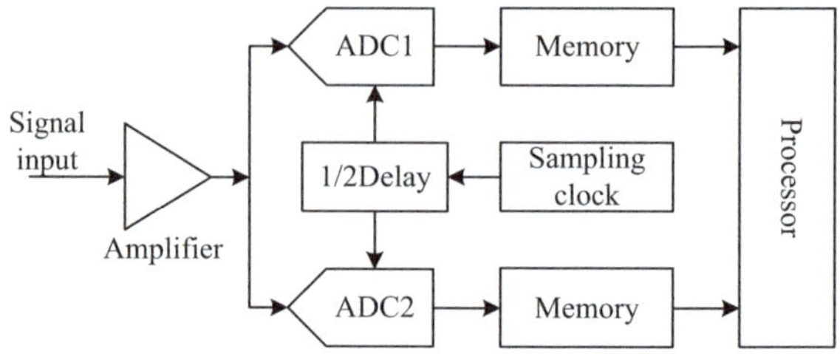

Fig. 5.30 ADC cross sampling principle

by a clock signal with a 180° phase difference. The sampled results from these two ADCs are then combined by the CPU, effectively doubling the sampling rate.

Controlling the sampling time precision of multiple ADCs can be challenging, as even slight errors can lead to cross-distortion. Therefore, when selecting such oscilloscopes, it's important to pay special attention to the occurrence of cross-distortion rather than solely focusing on achieving high sampling rates.

5.3.1.4 Oscilloscope Selection

Based on the discussions above, we've clarified the performance requirements for SiC MOSFET double-pulse test with oscilloscopes, which include high bandwidth, high sampling rate, high resolution, low noise, and low jitter. While bandwidth and sampling rate provide quantifiable requirements, achieving high resolution, low noise, and low jitter poses greater challenges, which may require a comparative analysis to determine the optimal solution.

For SiC MOSFET double-pulse test, it's common to choose mid-range high-resolution oscilloscopes from various manufacturers, typically with bandwidths of 500 MHz or 1 GHz. For example, the Tektronix 5 Series B MSO [20] and 6 Series B MSO [21] mixed-signal oscilloscopes are popular choices, as shown in Fig. 5.31.

The 5 Series B MSO and 6 Series B MSO mixed-signal oscilloscopes feature 12-bit hardware ADCs, offering lower noise, higher resolution, and improved test accuracy and consistency. They support various channel configurations including 4, 6, and 8 channels, allowing simultaneous observation of multiple signals for convenient device characterization and driver circuit analysis. The 5 Series B MSO has

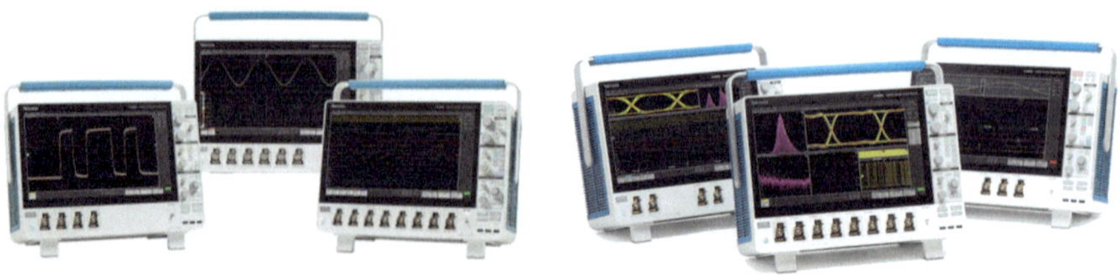

Fig. 5.31 Tektronix 5 Series B MSO and 6 Series B MSO mixed-signal oscilloscopes

a maximum bandwidth of 2 GHz, while the 6 Series B MSO offers a maximum bandwidth of 10 GHz. They can be optionally equipped with built-in arbitrary waveform/function generators. Additionally, they come with a 15.6-inch capacitive touchscreen supporting pinch-zoom-slide gestures and optional Windows 10 operating system. Moreover, the oscilloscope application WBG-DPT [22] provides convenient calculation of switching and reverse recovery characteristics based on test results.

5.3.2 Voltage and Current Measurement Methods

5.3.2.1 Voltage Probe

Voltage probes are one of the most familiar measurement tools for electronic engineers, indispensable in both learning and work environments. With numerous models available from various manufacturers, ranging from tens to even hundreds, the wide variety of voltage probes can be overwhelming [22–25]. These probes, based on different principles and structures, come with varying bandwidths, measurement ranges, and loading effects, catering to different application scenarios and measurement requirements.

Voltage probes can be simply categorized into passive probes and active probes based on whether they require power. Passive probes consist of passive components such as resistors, capacitors, and cables, essentially forming a passive voltage divider network. On the other hand, active probes contain amplifiers internally and require power supply. Furthermore, probes can be classified into more detailed categories based on measurement range, impedance, single-ended or differential measurement, as illustrated in Fig. 5.32.

(1) **Ordinary Passive Probe**

The general passive probe is shown in Fig. 5.33. The most common is a probe with a 10:1 attenuation ratio, which is commonly known as a ten-fold passive probe, which

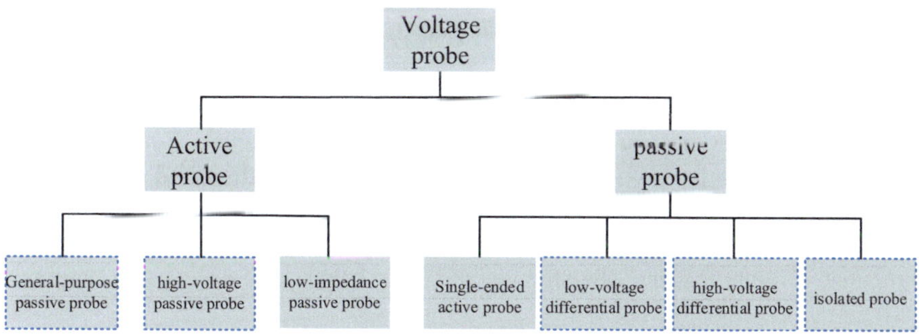

Fig. 5.32 Classification of voltage probe

Fig. 5.33 Ordinary passive probe

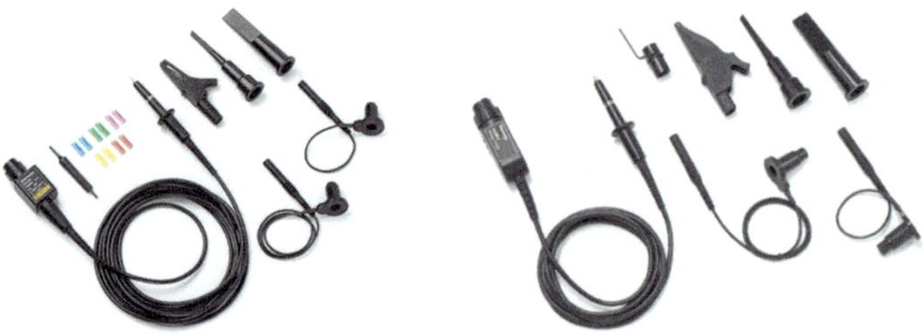

Fig. 5.34 High-voltage passive probe

is the standard voltage probe of the oscilloscope. Its bandwidth can be as high as 500 MHz to 1 GHz, and the measurement range is generally below 500 V. In addition, there is a probe of 1: 1, and its bandwidth is below 50 MHz. Because of its cheap price and easy use, it has been widely used in various fields.

(2) **High-Voltage Passive Probe**

The high-voltage passive probe, as shown in Fig. 5.34, has higher input impedance and attenuation ratio compared to general-purpose passive probes. It can achieve a bandwidth of up to 800 MHz and a maximum measurement range of several kilovolts. It is mainly used in power supply design, power semiconductor test, and similar applications.

(3) **Low-Impedance Passive Probe**

The low-impedance passive probe depicted in Fig. 5.35 features a relatively low input impedance, typically ranging from hundreds to a few thousand ohms, and is also known as a transmission line probe. With a bandwidth of several gigahertz, it is a cost-effective and highly reliable high-bandwidth probe. It is primarily used in computer, communication, data storage, and other high-speed design applications.

Fig. 5.35 Low-impedance passive probe

High-impedance passive probes have high input impedance but limited bandwidth, while low-impedance passive probes have high bandwidth but low input impedance and significant loading effects. Amplifiers, on the other hand, feature high input impedance, high bandwidth, and sufficient drive capability to drive 50 Ω transmission lines. Leveraging these characteristics of amplifiers, active probes overcome the weaknesses of passive probes and can simultaneously meet high input impedance and high bandwidth requirements.

(4) **Single-Ended Active Probe**

Single-ended active probes, as shown in Fig. 5.36, have a high-speed amplifier at the front end, with a single-ended input. These probes have a high bandwidth, reaching several gigahertz. However, the dynamic range of single-ended active probes is small, generally within a few volts. They are primarily used for debugging high-speed designs, signal integrity analysis, jitter analysis, and timing analysis.

(5) Low Voltage Differential Active Probe

Unlike single-ended active probes, the front end of a differential active probe is a differential amplifier, with a differential input. Therefore, it has common-mode rejection capabilities, as shown in Fig. 5.37. Low-voltage differential probes are designed specifically for high-speed signal test, with bandwidths reaching several tens of gigahertz. The dynamic range is generally within a few volts, although some probes with lower bandwidths may have dynamic ranges extending to several tens of volts.

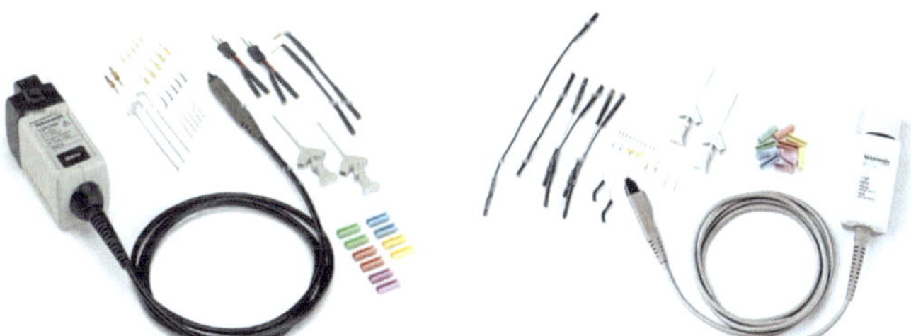

Fig. 5.36 Single-ended active probe

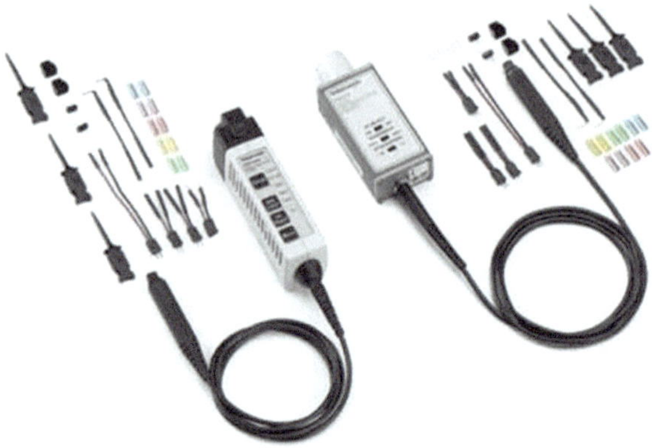

Fig. 5.37 Low voltage differential active probe

(6) **High-Voltage Differential Probe**

The high-voltage differential probe is commonly used by power engineers and is often mistaken for an isolation probe. Its most distinctive feature is its boxy appearance, as shown in Fig. 5.38. It typically has a bandwidth below 200 MHz but offers a wide measurement range, extending to several kilovolts. It is primarily used for power supply and low-speed differential bus measurements.

(7) **Isolation probe**

Isolation probes utilize isolation amplifiers or opto-isolation technology to achieve isolated sampling, providing high isolation voltage to protect personnel and equipment. They boast advantages such as high common-mode rejection ratio and strong anti-interference capabilities.

Based on the above discussion, we find that 10:1 passive probes, high-voltage passive probes and isolation probes may be suitable for SiC MOSFET double-pulse test.

Fig. 5.38 High-voltage differential probe

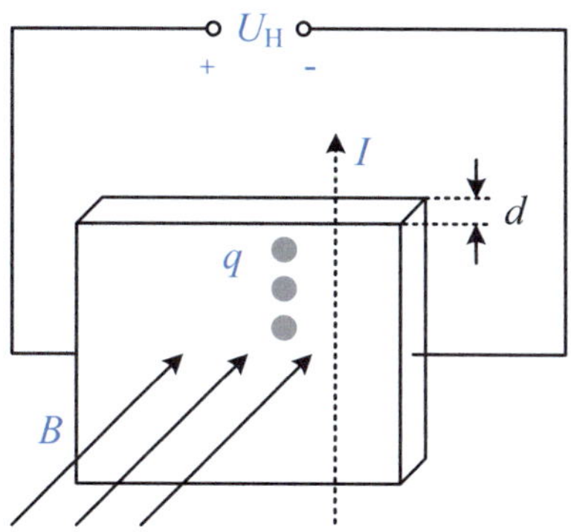

Fig. 5.39 Principles of Hall effect

5.3.2.2 Current Sensor

When measuring current, various types of current sensors are used to convert the current into a proportional voltage signal before feeding it into the oscilloscope. These sensors are manufactured based on different principles, with Hall sensors, current transformers, Rogowski coils, and shunt resistors being the most common among them.

1. **Hall Current Probe**

When a conductive material with current flowing through it is placed in a magnetic field, the charge carriers experience a force known as the Lorentz force, causing them to accumulate at the sides of the material. This accumulation results in the generation of a Hall voltage U_H, perpendicular to both the current I and the magnetic field B. This phenomenon, known as the Hall effect, was discovered by American physicist E.H. Hall in 1897, as depicted in Fig. 5.39.

The voltage U_H is called the Hall voltage, and its direction can be determined using the left-hand rule, as given by (5.22).

$$U_H = \frac{I \times B}{n \cdot q \cdot d}, \tag{5.22}$$

where n is the carrier density, q is the charge of the carrier, and d is the thickness of the conducting material.

Hall devices utilizing the Hall effect convert the motion of objects into voltage signals by detecting changes in magnetic fields, ultimately achieving sensing or switching functions. Hall devices have found wide-ranging applications in displacement, pressure, mass, speed, angle detection, and more.

Current measurement is one of the most familiar applications of the Hall effect for power engineers. Hall current sensing chips and Hall current sensors are widely used in power converters, and Hall current probes are the most commonly used current measurement tools during converter debugging. The most common type of Hall current probe is the clamp-type structure, where the measured wire is clamped

Fig. 5.40 Hall current probe

inside, as shown in Fig. 5.40. Due to this structure, special space needs to be allocated to accommodate the current-carrying circuit into the probe. For double-pulse test, this disrupts the principle of overlapping circuit traces, leading to a significant increase in the main power switch current loop inductance.

Hall current probes can perform isolated measurements of both DC and AC currents, but they have limited bandwidth and the bandwidth decreases continuously as the range increases. Moreover, there is a risk of bias magnetization due to overcurrent.

2. **Current Transformer**

The current transformer is composed of a closed magnetic core, a primary and secondary coils, and a detection resistor R_{sense}, as shown in Fig. 5.41. The primary winding N_1 is directly connected in series to the detected circuit, and the output terminal of the secondary winding N_2 is connected to R_{sense}. The induced magnetic flux generated by the changing current i_p on the primary side will generate a current i_s on the secondary side, and generate a voltage U_{CT} on R_{sense}.

$$U_{\text{CT}} = R_{\text{sense}} \cdot i_p \cdot \frac{N_1}{N_2} \tag{5.23}$$

Generally R_{sense} is very small, so the current transformer can be regarded as a transformer operating in a short circuit.

Current transformers are widely used in many fields such as beam instruments, surge current testing, particle accelerators, EMI testing, plasma research, capacitor discharge, and medical applications.

Fig. 5.41 Principle of current transformer

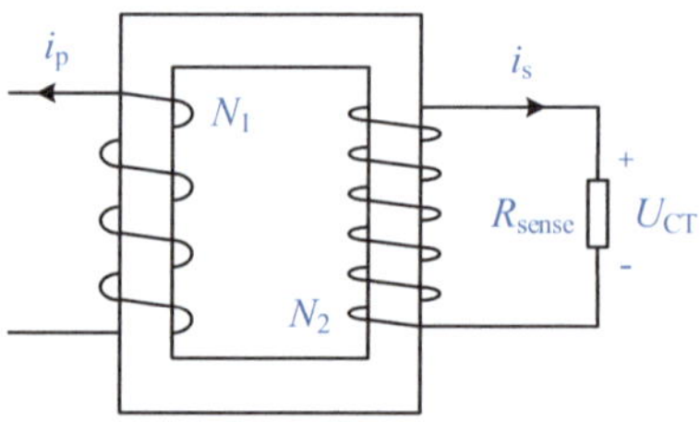

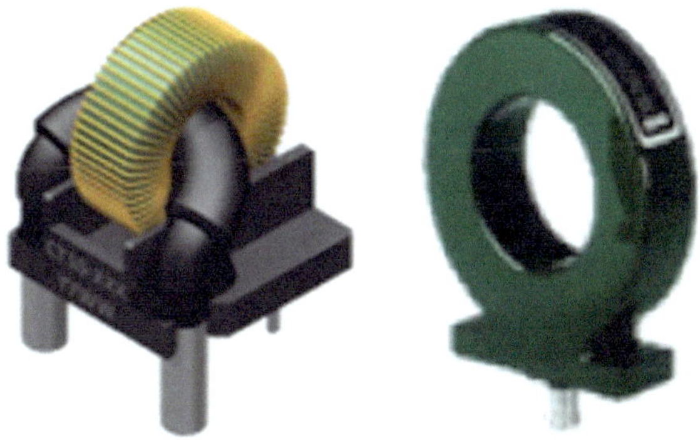

Fig. 5.42 Current transformer

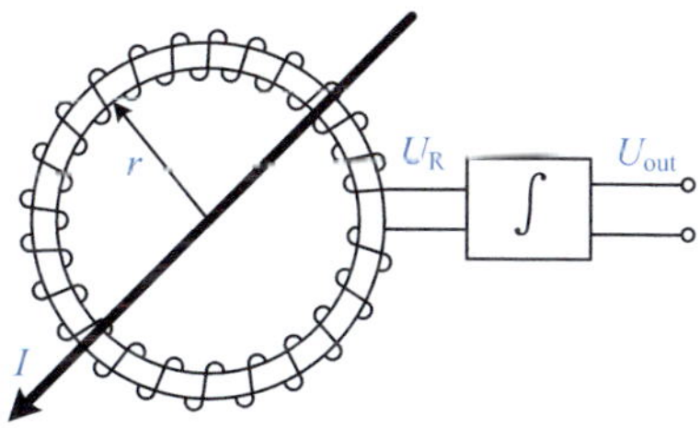

Fig. 5.43 Principle of Rogowski coil

The most common current transformer is a closed loop, in which the measured wire is inserted when in use, as shown in Fig. 5.42. Like the Hall current probe, the current transformer will destroy the principle of overlapping the line routing, greatly increasing the inductance of the main power commutation circuit.

Like Rogowski coils, current transformers are limited to measuring AC currents and cannot measure DC currents. Additionally, their bandwidth is constrained due to the high number of turns in the coil, and they suffer from biasing and magnetic saturation issues.

3. Rogowski Coil

The Rogowski coil, named after the German scientist Walter Rogowski, is a hollow coil, as shown in Fig. 5.43, generally made by winding a wire around a non-magnetic material. Assuming the coil radius is r, the cross-sectional radius is much smaller than r, the cross-sectional area is A, the number of coil turns is N, and a conductor with current is passed through the Rogowski Coil. According to Faraday's law, the changing current will generate an induced voltage U_R at both ends of the coil, given by (5.24), where μ_0 is the vacuum permeability.

$$U_R = -\mathrm{d}i/\mathrm{d}t \cdot \frac{\mu_0 \cdot N \cdot A}{2\pi \cdot r} \tag{5.24}$$

Fig. 5.44 Rogowski coil probe

This shows that the Rogowski coil can sense the change of the measured current. Further, by integrating U_R with an integrator, the real-time current value can be obtained, which is given by (5.25). Where k is the gain of the integrator (Fig. Fig. 5.43).

$$U_{out} = -k \cdot \frac{\mu_0 \cdot N \cdot A}{2\pi \cdot r} \cdot \int \mathrm{d}i/\mathrm{d}t + U_{out}(0) \tag{5.25}$$

Since Rogowski coils do not have hysteresis and magnetic saturation problems, the range is not limited, and they are particularly suitable for experimental measurement of high-frequency pulse currents. They have been widely used in current measurement in power distribution systems, short-circuit test systems, electromagnetic transmitters, slip ring induction motors, lightning strike tests, power semiconductor research and development, bearing current measurement of large machinery, power measurement, and other fields.

The Rogowski coil probe consists of a fine ring, which encloses the current being measured, as shown in Fig. 5.44. Compared to the Hall current probe, the Rogowski coil probe is smaller in size, making it more convenient and flexible. It can easily measure the current of power devices, and even accommodate the main power conversion circuit's key lines without increasing the circuit's inductance.

The Rogowski coil can only isolate and measure AC currents; it cannot measure DC currents. Its measurement accuracy is influenced by the relative position between the measured current and the coil. Additionally, the high number of coil turns results in a lower bandwidth, typically not exceeding 30 MHz.

4. **Sampling resistor**

The sampling resistor is based on the most basic Ohm's law. When the sampling resistor R_{sense} is connected in series to the main power circuit, the current i will produce a voltage drop U_{sense} across R_{sense}.

$$U_{sense} = i \cdot R_{sense} \tag{5.26}$$

Sampling resistors are available in surface-mount and coaxial types, as shown in Fig. 5.45. Surface-mount sampling resistors are commonly used as current sensing elements in converters but are not suitable for current measurement in double-pulse

(a) Patch sampling resistor

(b) Coaxial sampling resistor

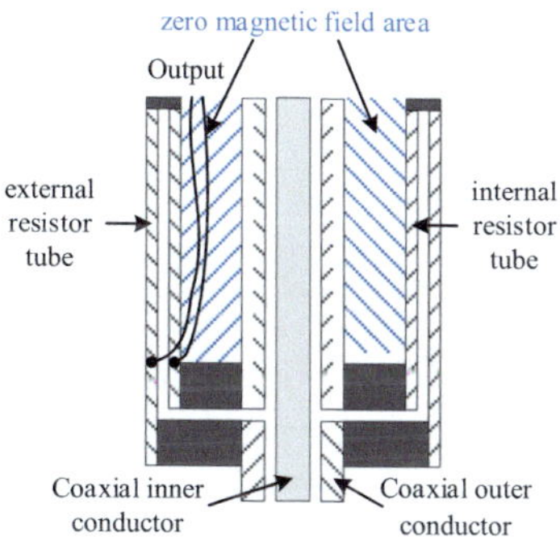

(c) Structure of coaxial sampling resistor [26]

(d) Coaxial sampling resistor installation method

Fig. 5.45 Sampling resistor

testing. This is because surface-mount resistors have higher equivalent series inductance, making them unsuitable for measuring the transient processes of high-speed pulse currents. The magnetic fields generated by high-speed currents during device switching can easily couple into the measurement lines, causing interference.

Coaxial sampling resistors, with their special structural design, significantly reduce their equivalent series inductance and are unaffected by the current being measured. They offer high bandwidth and can be directly mounted on PCB boards, allowing PCB routing to follow overlapping principles, which minimizes additional loop inductance.

Coaxial sampling resistors can perform non-isolated measurements for both DC and AC currents and provide high bandwidth. However, they may be damaged by overcurrent, and non-isolated measurements could pose safety issues in high-voltage applications. T&M offers a wide range of coaxial sampling resistor products.

In summary, due to their physical characteristics, Hall current probes and current transformers are not suitable for SiC MOSFET double-pulse test. However, Rogowski coils and coaxial resistors may be applicable for this purpose.

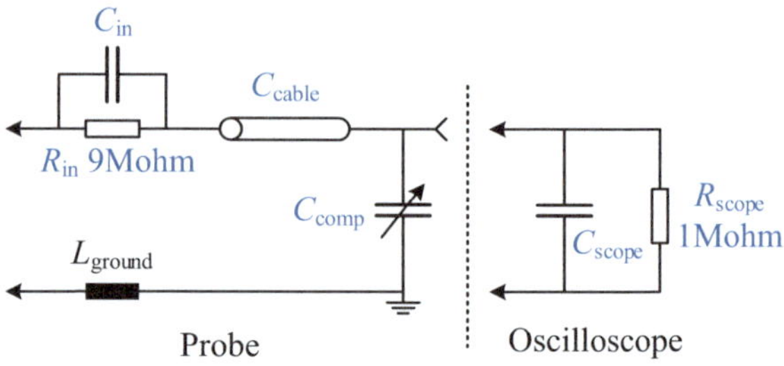

Fig. 5.46 Equivalent circuit of 10:1 passive probe

5.3.3 V_{GS} Measuring Gate-Source Voltage V_{GS}

5.3.3.1 10:1 Passive Probe

The 10:1 passive probe has a bandwidth ranging from 500 MHz to 1 GHz and a measurement range typically below 500 V, suitable for measuring the gate-source voltage VGS of SiC MOSFETs.

It is the most common type of voltage probe, with an equivalent circuit depicted in Fig. 5.46. At the probe's front end, there is an input impedance consisting of input resistor R_{in} and input capacitor C_{in}, along with a ground lead inductance L_{ground}. The front end is connected to the probe tip via a cable, which includes a cable capacitance C_{cable}, and at the tip, there is a compensating capacitor C_{comp}. The probe tip is connected to the oscilloscope input, which comprises the parallel combination of input resistor R_{scope} and input capacitor C_{scope}. With R_{in} at 9 MΩ and R_{scope} at 1 MΩ, a 10:1 attenuation ratio is achieved.

When using a 10:1 passive probe, it's crucial to match its impedance with that of the oscilloscope to achieve a flat gain within the bandwidth. Failure to do so can severely affect its high-frequency characteristics. The impedance matching condition is $R_{in} \cdot C_{in} = R_{scope} \cdot C_{scope}$. Since the value of C_{scope} may slightly differ for each oscilloscope, adjustment of C_{comp} for compensation is necessary. One method involves using the probe to measure the oscilloscope's built-in low-voltage square wave signal. By rotating the compensation adjustment knob on the probe's tip using a screwdriver, the square wave signal should exhibit no overshoot or underdamped response when compensation is correct. Additionally, some oscilloscopes nowadays can recognize the probe type and perform automatic compensation.

The impact of impedance matching on the measurement of the gate-source voltage (V_{GS}) of the SiC MOSFET when using + 15 V/− 4 V for drive is depicted in Fig. 5.47. When the probe is under-compensated, the rise time of V_{GS} is slow, requiring more time to reach + 15 V. Conversely, when the probe is over-compensated, significant overshoot is observed in V_{GS}, even reaching up to + 20 V. Both under-compensation and over-compensation lead to erroneous measurement results, potentially causing misleading conclusions.

As described in Sect. 5.2.3, crosstalk is an accompanying dynamic process in the switching of half-bridge circuits and poses a significant challenge when using SiC

Fig. 5.47 The impact of the impedance matching of the 10:1 passive probe on the switch V_{GS} measurement

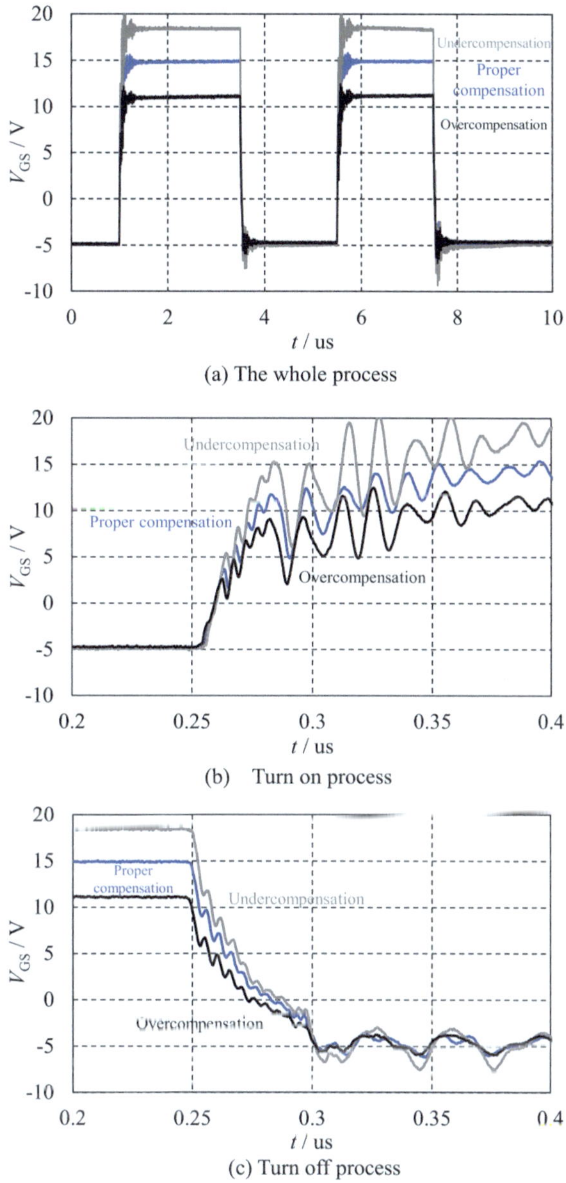

MOSFETs. Therefore, accurately measuring the crosstalk V_{GS} waveform is essential for analyzing and resolving crosstalk issues. Although crosstalk will be specifically introduced in Chap. 8, its test also employs the double-pulse test. Hence, the discussion on crosstalk test is presented here in advance.

The effect of impedance matching on the crosstalk V_{GS} measurement when using a 10:1 passive probe is shown in Fig. 5.48. When the probe is under-compensated,

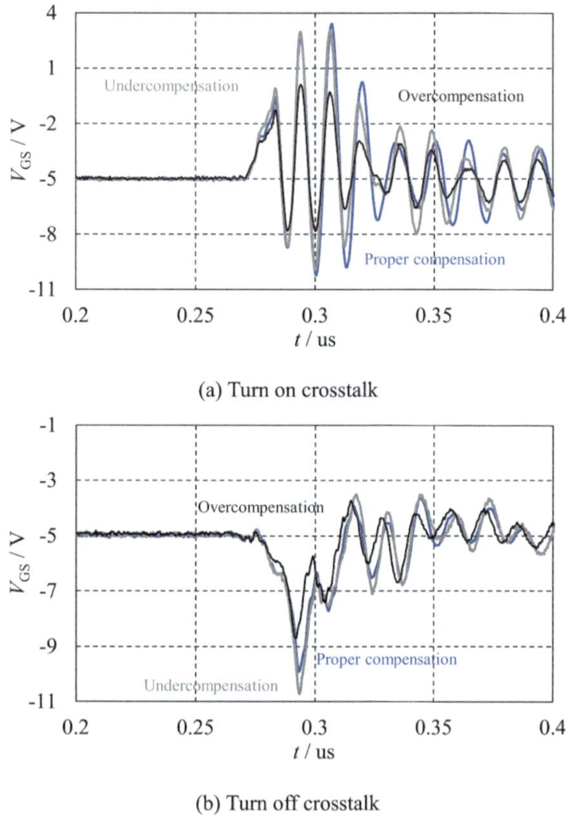

Fig. 5.48 The impact of V_{GS} measurement of the matching of impedance

the measured crosstalk waveform peak is lower and the degree of crosstalk is underestimated. When the probe is over-compensated, the measured crosstalk waveform peak is higher and the degree of crosstalk is overestimated.

In addition, the probe's ground lead inductance L_{ground} affects the probe's high-frequency characteristics. First, L_{ground} causes ringing, making the waveform oscillations appear more severe than they actually are. Moreover, L_{ground} acts like an antenna, picking up surrounding electromagnetic radiation, which is particularly problematic in high di/dt environments such as SiC MOSFET double-pulse test.

To minimize the impact of L_{ground} on measurements, it's important to shorten the ground lead length and reduce the loop area of the probe test connections as much as possible. Common methods include using manufacturer-provided PCB adapters and short spring ground leads. Additionally, you can create your own short ground leads, which are more cost-effective and offer greater flexibility in choosing measurement points, as shown in Fig. 5.49.

The V_{GS} is measured by using long -connected ground wires, short spring ground lines, and winding short -connect lines. V_{GS} waveform is also increasingly clean, as shown in Fig. 5.50. The V_{GS} waveform shock measured when using the self-winding short -connected ground wire is the most milder. Based on this result, it is judged that

Fig. 5.49 The ground wire of 10:1 passive probe

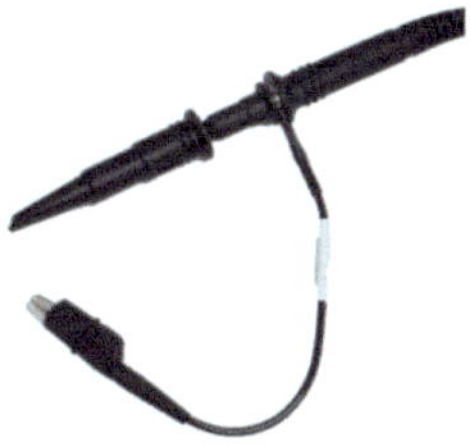

(a) Standard ground wire

(b) Short spring ground wire

(c) PCB adapter

(d) Winding short ground wire

Fig. 5.50 The Influence of the ground wire inductance of the 10:1 passive probe L_{ground} on the switch V_{GS} measurement

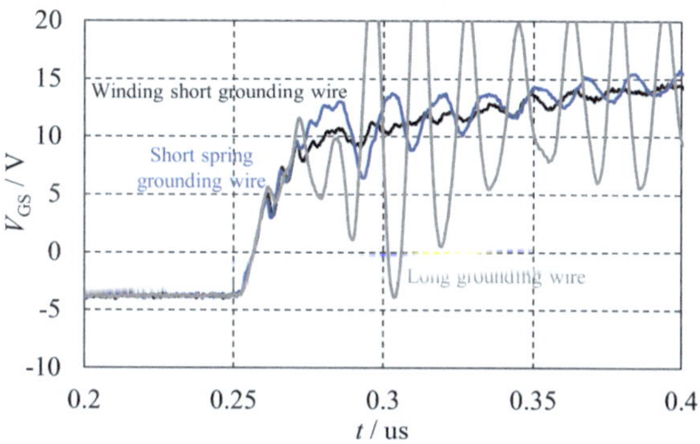

(a) Turn on process

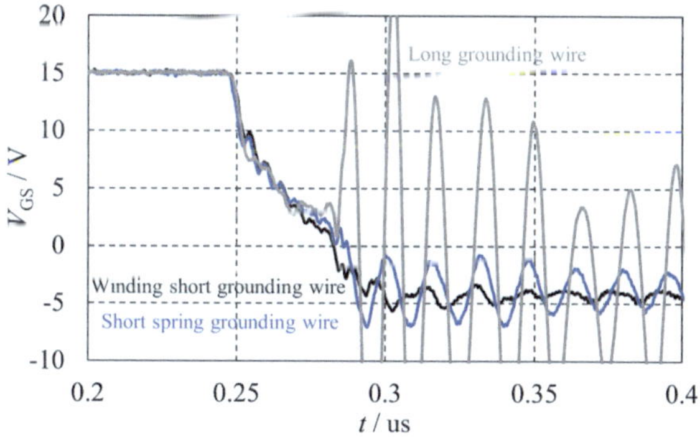

(b) Turn on process

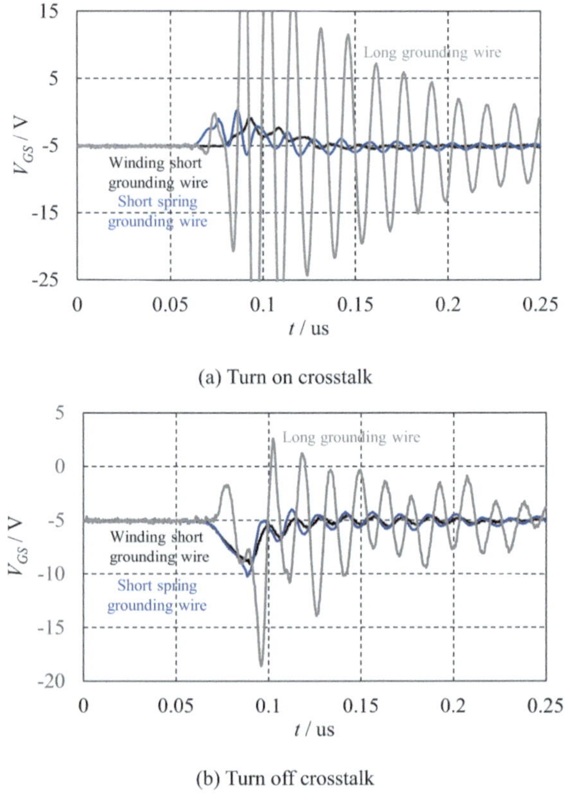

Fig. 5.51 Influence of the ground wire inductance of the passive probe L_{ground} on the string V_{GS} measurement

V_{GS} is within the acceptable range in the application. The V_{GS} waveforms measured during the long ground line were severe, and the peak of the shock exceeded the device gate to resistance. Based on this result, V_{GS} was unacceptable in the application at this time. It can be seen that L_{ground} can cause serious deviations in the measurement results, affecting analysis and judgment.

As shown in Fig. 5.51, the effect of the ground wire inductance L_{ground} on the string V_{GS} measurement.

5.3.3.2 High-Voltage Differential Probe

High-voltage differential probes are designed for differential input with high input impedance, making them ideal for measuring V_{GS} in power converters during development. These probes are typically used for high-voltage signal measurement and feature two 20-cm-long leads at the input end to meet safety requirements and facilitate connection. Similar to 10:1 passive probes, the long leads result in higher inductance in the measurement loop, causing oscillations and susceptibility to external interference, which can affect the accuracy of measurements. Due to the fixed length of the high-voltage differential probe leads, reducing the length of the leads, as done

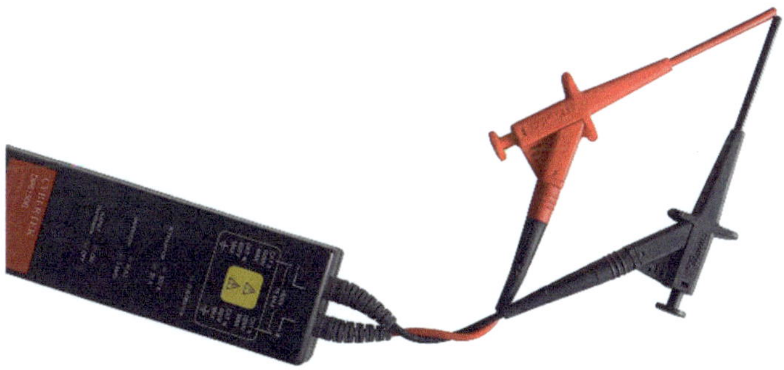

Fig. 5.52 High-voltage differential probe connecting line twisted

Fig. 5.53 Effect of twisted pair of high-voltage differential probe on switch V_{GS} measurement

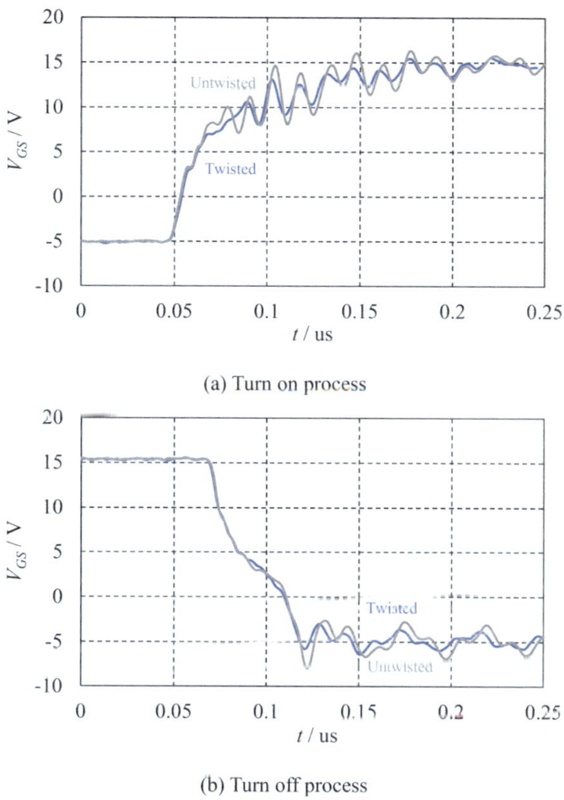

with custom short leads for 10:1 passive probes, is not feasible. Instead, as shown in Fig. 5.52, the measurement accuracy can be improved by twisting the probe leads to minimize loop inductance and reduce interference.

The V_{GS} waveforms measured with the high-voltage differential probe leads untwisted and twisted are shown in Fig. 5.53. Compared with the case of twisted pair, when the twisted pair is not used, there is a serious oscillation on the V_{GS} waveform, which will mislead the analysis and judgment of the waveform. However, after twisting the leads, the amplitude of the oscillations in the V_{GS} waveform is significantly reduced. This comparison highlights that not twisting the leads can lead to significant measurement errors, particularly during the turn-off process where spikes in the waveform might mislead engineers into believing there is a serious risk of false triggering.

The crosstalk V_{GS} waveforms measured when the high-voltage differential probe front-end connection line is not twisted and twisted are shown in Fig. 5.54. By comparison, it can be seen that compared with the use of twisted pairs, the crosstalk V_{GS} waveform when not twisted has serious oscillations, especially the turn-on crosstalk V_{GS}, which will mislead the analysis and judgment of the crosstalk.

High-voltage differential probes typically have two attenuation settings, with higher attenuation providing a larger measurement range. Using the Tektronix TMDP0200 [27] high-voltage differential probe, tests were conducted at 25 × and 250 × attenuation settings. These results were compared to measurements taken with

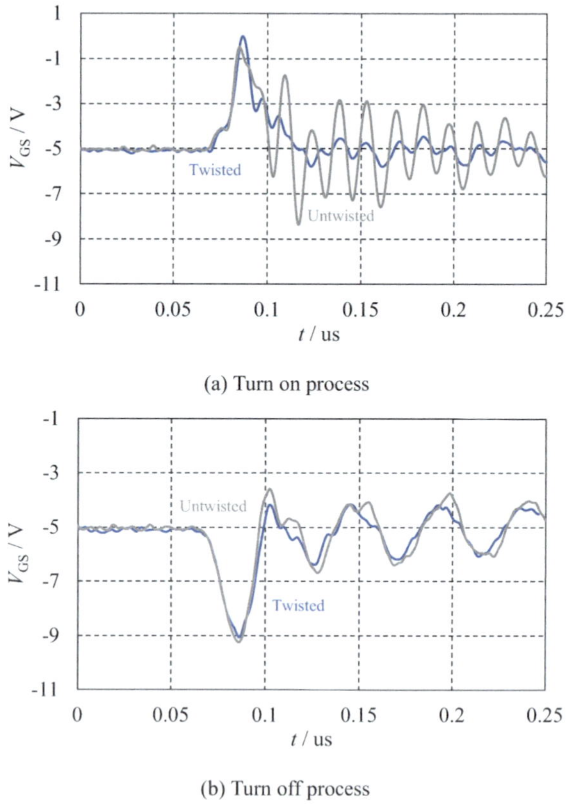

Fig. 5.54 The effects of high-voltage differential probes connected to the string V_{GS} measurement

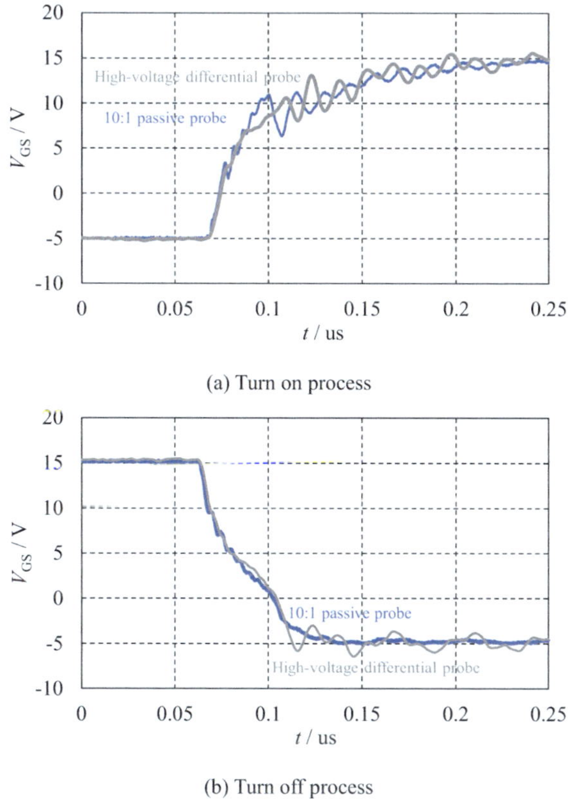

Fig. 5.55 High-voltage differential probes and 10:1 passive probes measured switch V_{GS}

the Tektronix 10:1 passive probe TPP1000 [28], as shown in Fig. 5.55. Compared to the TPP1000, the TMDP0200 measured waveforms appear noticeably thicker and fuzzier, with the 250 × attenuation setting showing more pronounced issues than the 25 × setting. Higher attenuation increases the probe's range but significantly reduces resolution, and the oscilloscope amplifies noise when reconstructing the signal. Therefore, it is advisable to select a high-voltage differential probe with a lower attenuation setting or use the lower attenuation range of the probe. Additionally, the TMDP0200 exhibits more pronounced oscillations than the TPP1000, indicating that even with twisted leads, the performance is inferior to that of the custom-made ground leads.

Using the Tektronix TMDP0200 high-voltage differential probe, tests were conducted at 25 × and 250 × attenuation settings. These results were compared to measurements taken with the Tektronix 10:1 passive probe TPP1000, as shown in Fig. 5.56. The crosstalk waveform characteristics displayed were consistent with the switching waveform.

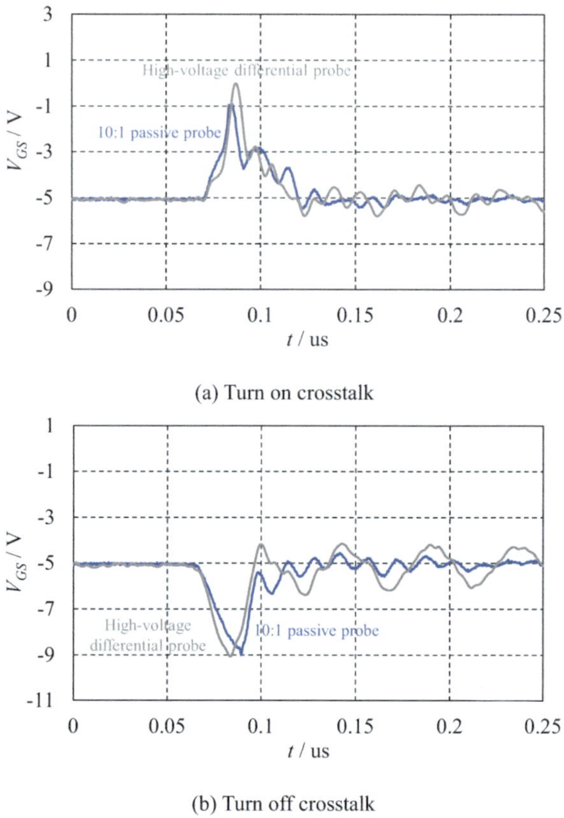

Fig. 5.56 High-voltage differential probes and 10:1 passive probes measure V_{GS}

The source potential of the upper bridge device Q_H is a square wave with an amplitude of V_{Bus}, and the edge transition rate of the square wave is the switching d V_{DS}/dt of the transistor, which constitutes the common-mode voltage for the V_{GS} of Q_H. Measuring the V_{GS} of Q_H is a floating measurement. Since a 10:1 passive probe is grounded along with the oscilloscope, it is not suitable for measuring the V_{GS} of the upper bridge device. Engineers typically use high-voltage differential probes in power converter designs to avoid personal injury and equipment damage.

The results of measuring the switch V_{GS} and the crosstalk V_{GS} of the upper bridge arm device Q_H using the Tektronix high-voltage differential probe TMDP0200 are shown in Figs. 5.57 and 5.58. Comparing with Figs. 5.55 and 5.56, it can be seen that the oscillation in the V_{GS} waveform of the upper bridge arm device is more pronounced than that of the lower bridge arm. However, in the case of a circuit designed to be completely symmetrical, the difference between the V_{GS} waveforms of the upper and lower bridge arm devices should not be so significant. This raises doubts about the measurement results of the high-voltage differential probe, suggesting the need for a probe with better performance for measuring the V_{GS} of the upper bridge arm device.

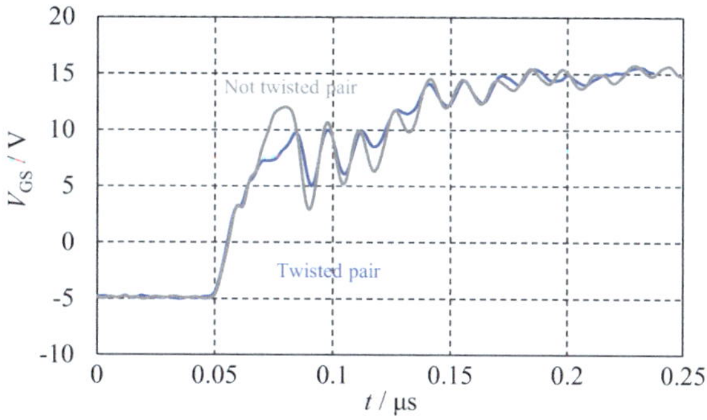

(a) Turn on process

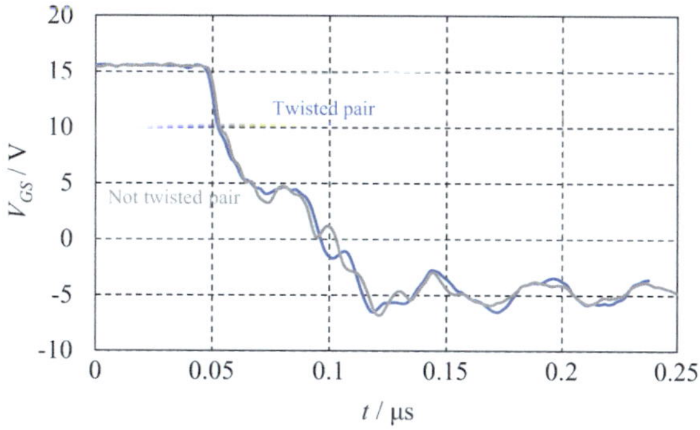

(b) Turn off process

Fig. 5.57 V_{GS} of the high-side device measured by the high-voltage differential probe

5.3.3.3 Light Isolation Probe

The Common-Mode Rejection Ratio (CMRR) of a high-voltage differential probe is an important parameter that reflects its ability to suppress common-mode signals. A higher absolute value of CMRR indicates less influence of common-mode voltage on the measurement results. High-voltage differential probes typically exhibit high CMRR at low frequencies, but it rapidly decreases as the frequency increases. For example, for a typical high voltage differential probe, the CMRR at DC, 100 kHz, 3.2 MHz, and 100 MHz is approximately – 80 dB, – 60 dB, – 30 dB, and – 26 dB, respectively. A low CMRR at high frequencies can lead to erroneous measurement results. To address this issue, voltage probes with high CMRR even at high frequencies are required.

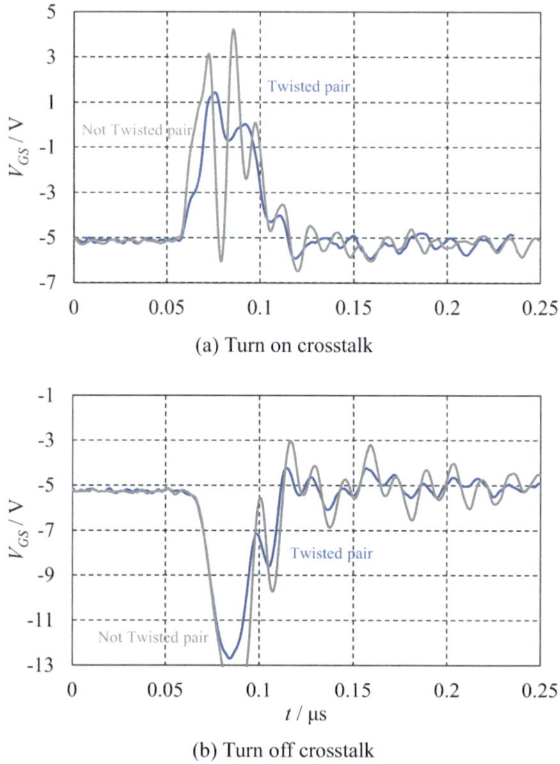

Fig. 5.58 Crosstalk V_{GS} the upper bridge arm device measured by the high-voltage differential probe

Tektronix introduced the second-generation optical isolation probe TIVP in 2020, as depicted in Fig. 5.59. It boasts excellent Common-Mode Rejection Ratio (CMRR) characteristics, as documented in references [29, 30]. The CMRR for different front-end configurations, including SMA Input, TIVPMX50X, and TIVPWS500X, is illustrated in Fig. 5.60.

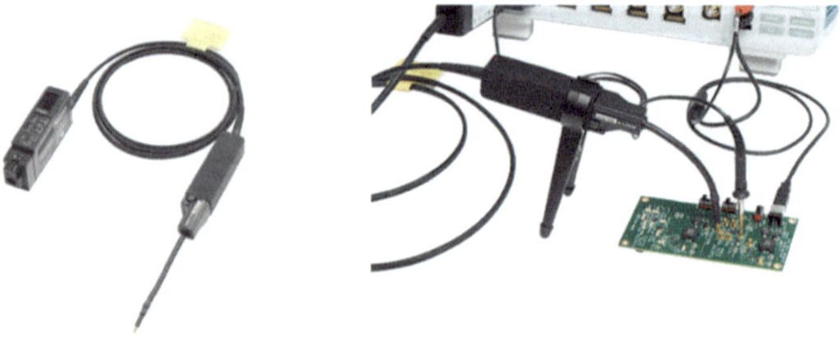

Fig. 5.59 Tektronix light isolation probe TIVP

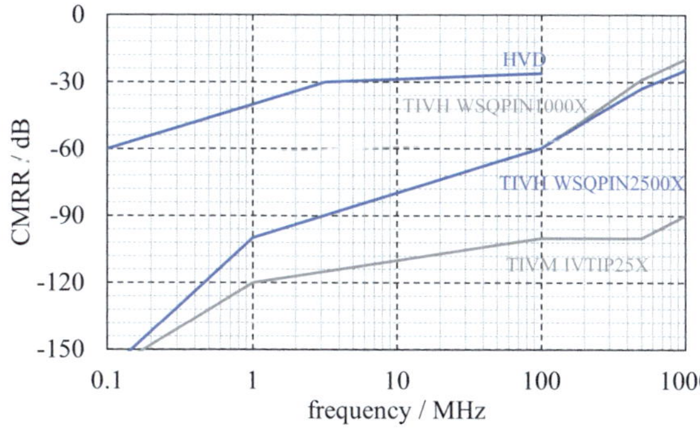

Fig. 5.60 Tektronix light isolation probe TIVP CMRR

The waveform of the gate-source voltage (V_{GS}) of the upper bridge arm device Q_H, measured using Tektronix's optical isolation probe TIVP1 and two high-voltage differential probes, is depicted in Fig. 5.61. The high-voltage differential probes HVD-A and HVD-B exhibit overdamping and underdamping characteristics, respectively, resulting in significant measurement errors. In contrast, the waveform obtained with TIVP1 is a standard square wave, with less pronounced oscillations compared to those measured with the high-voltage differential probes. This improvement is primarily attributed to the outstanding Common-Mode Rejection Ratio (CMRR) of TIVP1, coupled with its MMCX interface at the input terminal, which minimizes loop test and mitigates external interference while measuring circuit inductance.

The crosstalk V_{GS} waveform of Q_H measured by Tektronix's optical isolation probe TIVP1 is shown in Fig. 5.62. The oscillation amplitude of the crosstalk V_{GS} obtained by using the optical isolation probe TIVP1 is much lower than that obtained by using the high-voltage differential probe HVD. At this point, it can be further seen that even if a twisted pair is used, the high-voltage differential probe cannot obtain the correct crosstalk V_{GS} waveform, and will seriously mislead the analysis and judgment of the crosstalk. As shown in Fig. 5.62.

In addition to the measurement of the upper bridge arm device V_{GS}, when the converter is developed, especially when the converter is floating, you can also use the light isolation probe to the lower bridge arm device V_{GS} measures. The light isolation probe is safer than the passive probe 10:1, and the accuracy of the measurement result is far more than the high -voltage differential probe.

In summary, for the measurement of the lower bridge arm device V_{GS}, you can choose the best test results for 10:1 passive probe and light isolation probe; Only when the light isolation probe is selected can the correct waveform be obtained; when the converter development or the requirements for the measurement results are not high, you can choose a high-voltage differential probe.

Fig. 5.61 Light isolation probe measure the upper bridge arm device switch V_{GS}

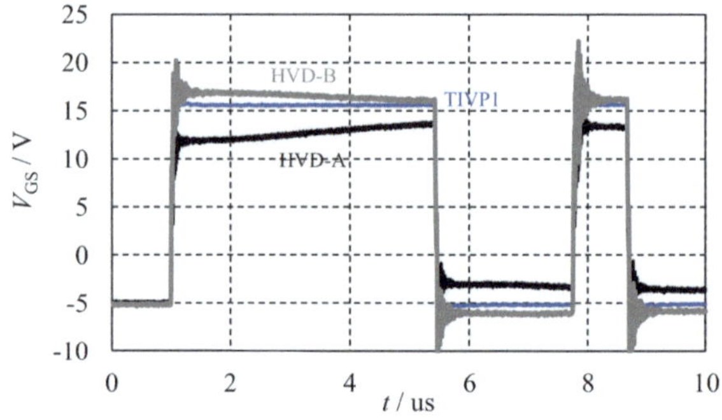

(a) Entire Double-Pulse Process

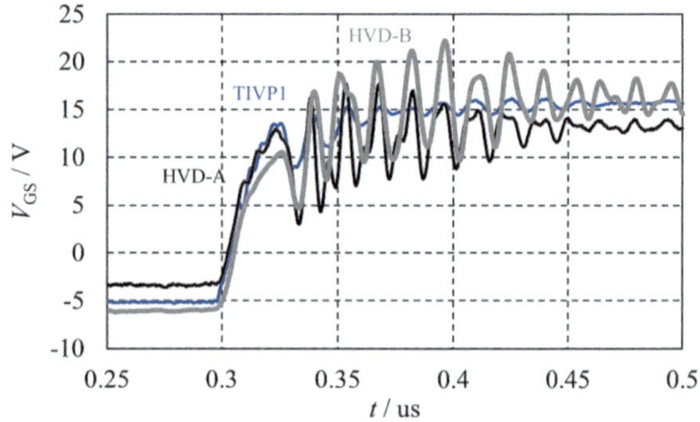

(b) Turn on process

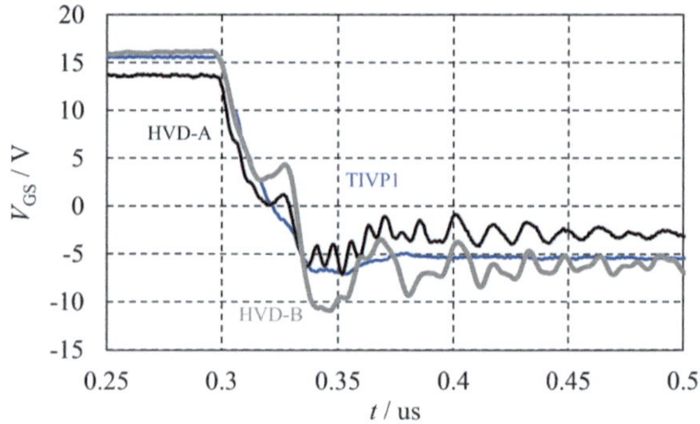

(c) Turn off process

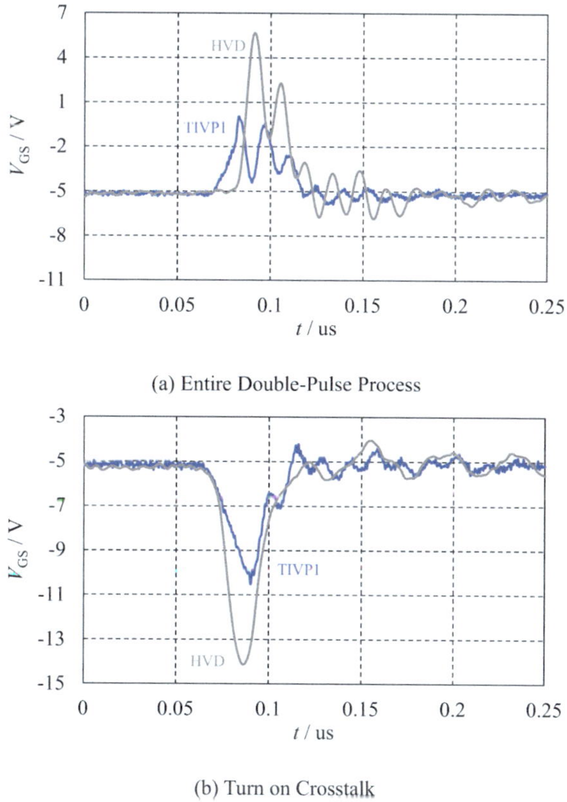

Fig. 5.62 Light isolation probe measures the crosstalk of the upper bridge arm device V_{GS}

5.3.4 *Measure the Drain-Source Voltage V_{GS}*

5.3.4.1 High-Voltage Passive Probe

Increasing the input resistance of a passive probe also increases its attenuation ratio, making it suitable for measuring higher voltages. These probes, known as high-voltage passive probes, typically have input resistances ranging from tens of megohms to hundred of megohms, with attenuation ratios of 50:1, 100:1, or 1000:1. They can achieve bandwidths of up to 800 MHz.

When using the Tektronix high-voltage passive probe TPP0850 (Fig. 5.63) [31] to measure V_{DS}, similar to the 10:1 passive probe, there are issues with impedance matching and ground lead inductance that need attention, as shown in Figs. 5.64 and 5.65. When the probe is under-compensated, the on-state voltage drop of the

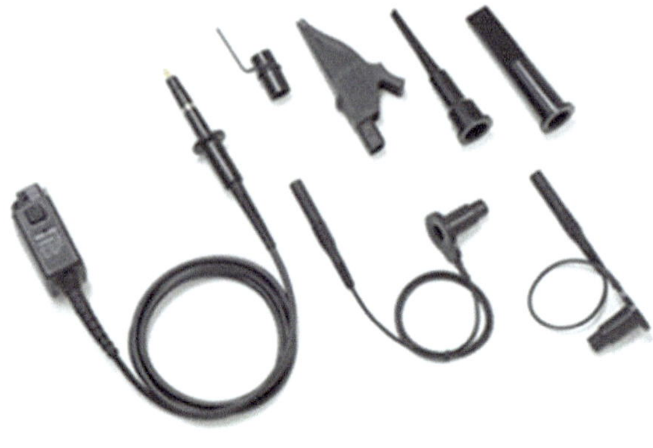

Fig. 5.63 Tektronix high-voltage passive probe TPP0850

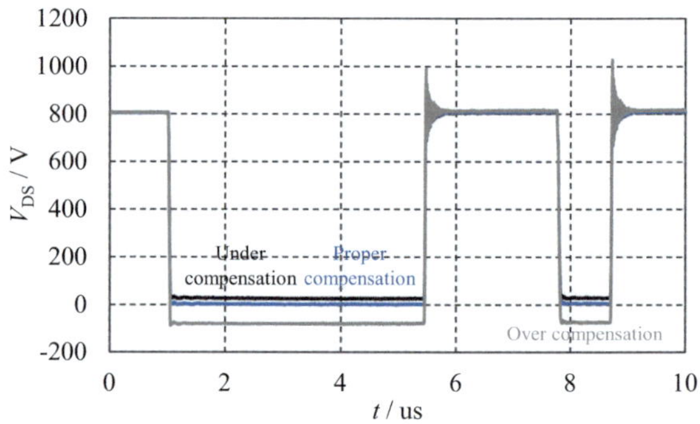

Fig. 5.64 The impact of the impedance matching of high-voltage passive probes on V_{DS} measurement

SiC MOSFET is significantly higher, deviating noticeably from theoretical expectations. Conversely, when the probe is over-compensated, the peak voltage during SiC MOSFET turn-off is significantly higher, and the on-state voltage drop becomes negative, again deviating from theory. When using a long ground lead, after the SiC MOSFET is fully turned on, V_{DS} exhibits severe oscillations and becomes negative. After complete turn-off, V_{DS} does not exhibit a single-frequency oscillation decay, all of which are inconsistent with theoretical expectations.

5.3.4.2 High-Voltage Differential Probe

The voltage measurement range of the high-voltage differential probe meets the requirements for measuring V_{DS}, with certain models having bandwidths that meet these requirements. However, due to the long input connection lines, even with the use of twisted pairs, the measurement accuracy still significantly deviates from that of the high-voltage passive probe TPP0850, as shown in Fig. 5.66. During the decreasing

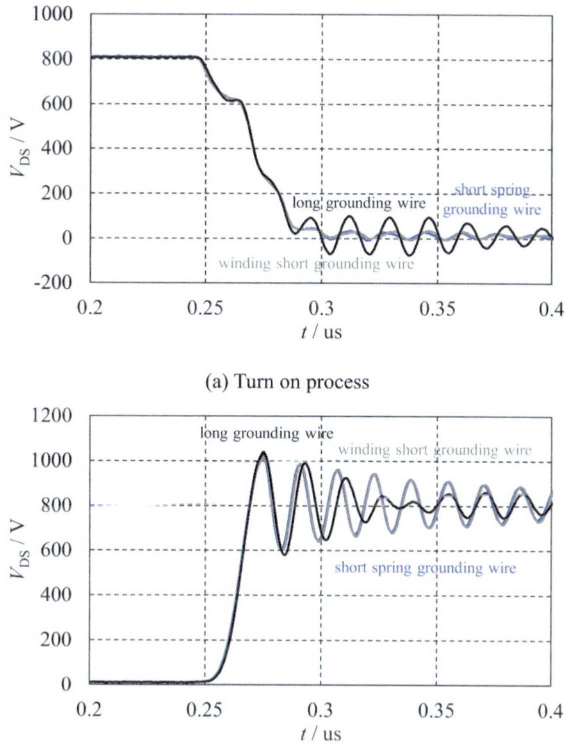

(b) Turn off process

Fig. 5.65 The effect of the ground wire inductance of the high-voltage passive probe L_{ground} on the V_{ds} measurement

phase of V_{DS} in the turn-on process, the high-voltage differential probes HVD-A and HVD-B exhibit noticeable deviations from TPP0850. After the SiC MOSFET is fully turned on, V_{DS} measured by HVD-A shows negative values within the oscillation, while for HVD-B it is entirely negative, which is clearly inconsistent with theory. After complete turn-off of the SiC MOSFET, the oscillation amplitude of V_{DS} measured by HVD-A and HVD-B is significantly higher than that of TPP0850.

Thanks to its differential input characteristic, the high-voltage differential probe can be used for measuring V_{DS} of the upper bridge arm device, as shown in Fig. 5.67. The test results of the high-voltage differential probe HVD-A are good. However, due to the insufficient common-mode rejection ratio and the impact of connection lines, the waveforms of V_{DS} measured by the high-voltage differential probes HVD-B and HVD-C still have issues. The most prominent problem occurs after the SiC MOSFET is fully turned on, where V_{DS} exhibits negative values.

In summary, for measuring V_{DS} of the lower bridge arm device, choosing a high-voltage passive probe would yield the best test results. High-voltage differential

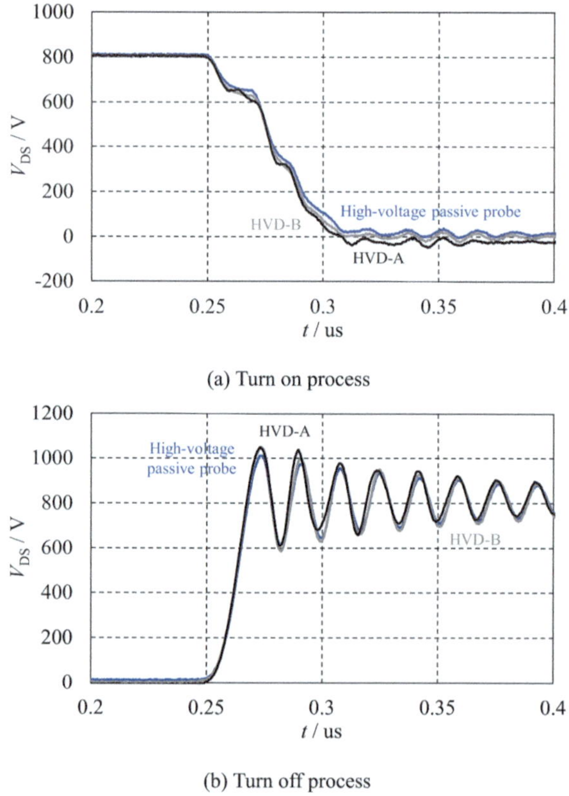

Fig. 5.66 Results of high-voltage differential probes and high-voltage passive probes measure V_{DS}

probes can be an alternative, especially in scenarios where precision measurement is not critical or during power converter development. However, for measuring V_{DS} of the upper bridge arm device, it's advisable to opt for high-voltage differential probes with higher common-mode rejection ratios to ensure accurate measurements.

5.3.5 Drain-Source Current I_{DS} Measurement

When performing Si IGBT module double-pulse test and converter development, the most commonly used current sensor for measuring the collector-emitter leakage current I_{DS} is the Rogowski coil, thanks to its ease of integration into circuits. However, due to its low bandwidth, it is not suitable for precise measurements with SiC MOSFETs. As discussed in Sect. 5.3.2.2 regarding current sensors, coaxial resistors have high bandwidth and low inductance characteristics. Additionally, their unique installation structure can effectively control the degree of increase in the main power loop inductance, making them suitable for double-pulse test of SiC devices.

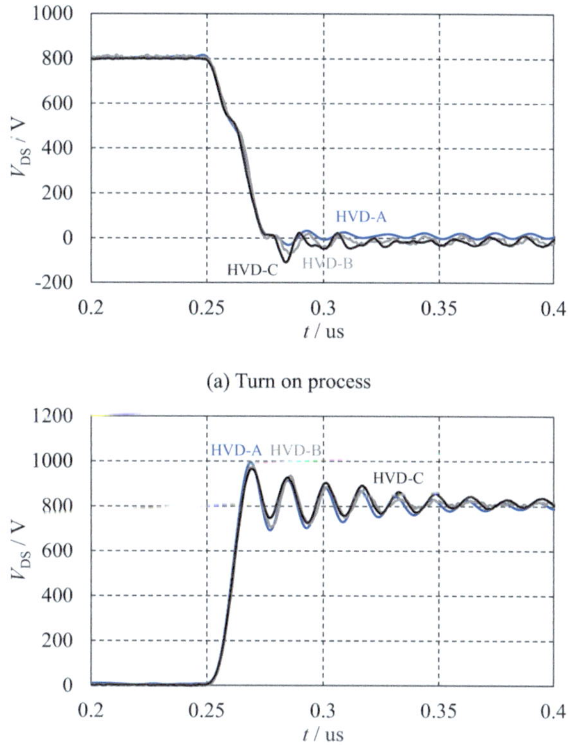

(a) Turn on process

(b) Turn off process

Fig. 5.67 High-voltage differential probe measure the upper bridge arm device V_{DS}

The waveforms of I_{DS} measured using coaxial resistor and Rogowski coil are depicted in Fig. 5.68. It can be observed that during the switching process, the rise and fall rates of I_{DS} measured with the Rogowski coil are slower than those measured with the coaxial resistor. Additionally, the amplitude of I_{DS} during the turn-on process is smaller with the Rogowski coil compared to the coaxial resistor. This discrepancy could lead to significant deviations in switch characteristic analysis and loss calculations. Meanwhile, the oscillation frequency of the I_{DS} waveform measured by the Rogowski coil is also significantly different from that of the coaxial resistor.

When measuring I_{DS} of upper-bridge devices, similar to the issue with the 10:1 passive probe, there are grounding issues that prevent direct use of coaxial resistors. Typically, Rogowski coils with inherent insulation capabilities are employed, but they may still yield suboptimal results due to bandwidth and placement concerns. In such cases, an alternative solution involves using a coaxial resistor paired with an optically isolated probe. This is achieved by using an adapter to convert the coaxial resistor's BNC connector to an MMCX interface, which is then connected to Tektronix's RF isolation current probe TICP10, as illustrated in Fig. 5.69.

Fig. 5.68 I_{DS} measurement of coaxial resistance and Roche coil

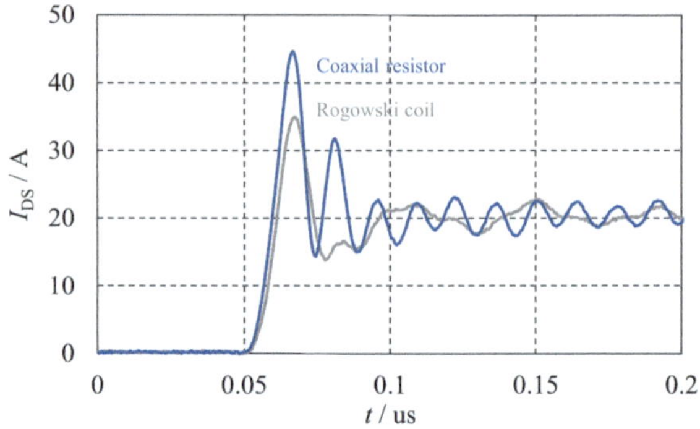

(a) Turn on process

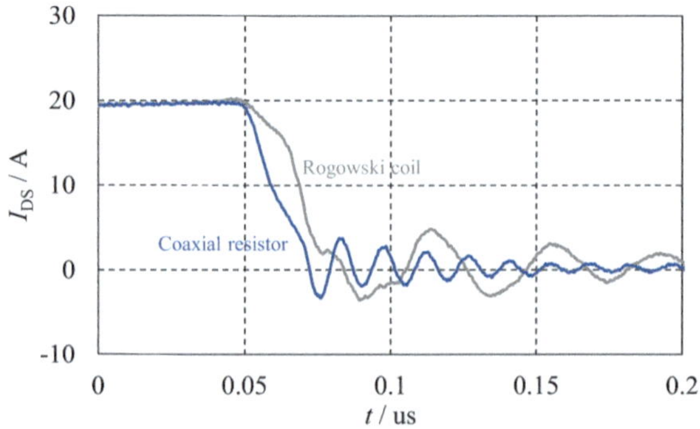

(b) Turn off process

Fig. 5.69 I_{DS} measurement with sampling resistance and RF isolation probe

The waveform of the upper-bridge device's I_{DS} measured using coaxial resistor paired with an optical isolated probe and Rogowski coil is depicted in Fig. 5.70. The difference exhibited in the waveform are similar to those observed when measuring the I_{DS} of the lower-bridge device.

In summary, for measuring the I_{DS} of the lower-bridge and upper-bridge devices, respectively, the optimal choices are coaxial resistor and coaxial resistor paired with optical isolated probe. When developing converters or when precision in measurements is not crucial, a Rogowski coil can be chosen.

Fig. 5.70 I_{DS} measurement with coaxial resistance with RF isolation probe and Rogowski coil on the upper bridge arm device

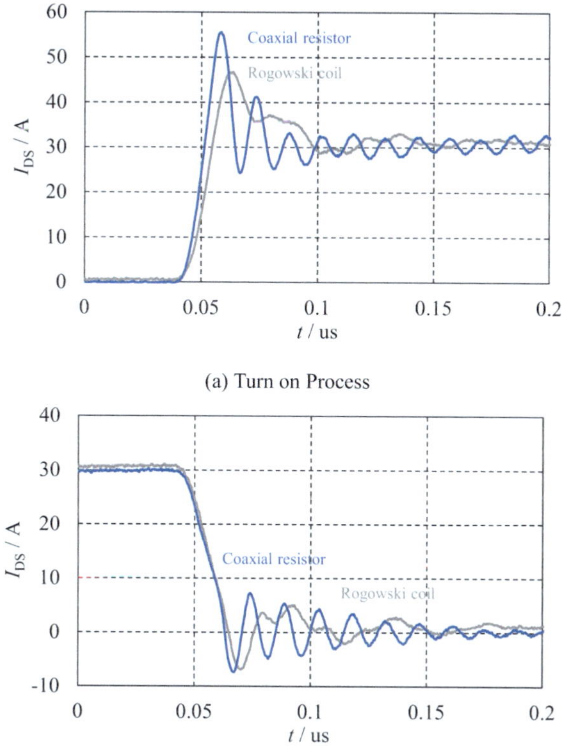

(a) Turn on Process

(b) Turn off Process

5.3.6 *Measurement of Gate Current IG*

In double-pulse test, the gate current I_G is generally not measured directly, but it is crucial for analyzing switching processes, driver circuits, crosstalk, and parasitic parameter compensation methods, which will be introduced later.

The Roche coil, Hall current probe, current sensor and coaxial resistor are not suitable for measuring I_G due to issues related to measurement range, bandwidth and installation method. The gate current IG flowing through the external gate resistor $R_{G(ext)}$ generates a voltage drop. If the voltage across $R_{G(ext)}$, denoted as $V_{R_{G(ext)}}$, can be measured, IG can be calculated. Based on previous discussions, the voltage probes that can be used to measure $V_{R_{G(ext)}}$ include high-voltage differential probes and optical isolated probes. Another method to measure I_G is by using a Rogowski coil to encircle the drive circuit and directly measure the current. Additionally, in 2024, Tektronix launched the IsoVuTMTICP series of RF isolation current probes,

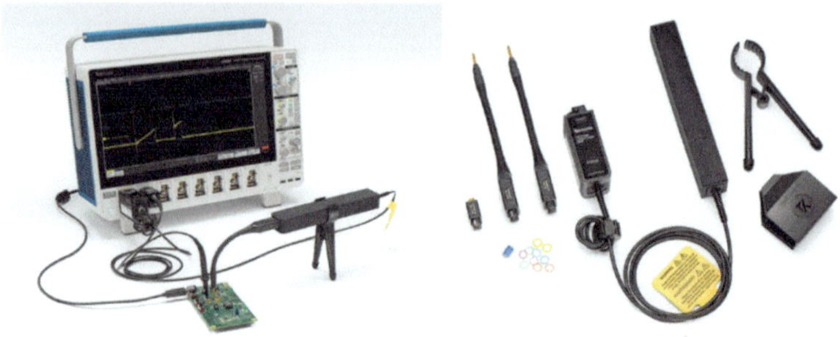

Fig. 5.71 Tektronix TICP Series RF Isolation Current Probe

which are specifically optimized for current measurement using shunt resistors. Even when the shunt resistor is not grounded and is affected by significant common-mode voltage, it can accurately measure the current, as shown in Fig. 5.71. The TICP series RF isolation current probes have the characteristics of high bandwidth, high common-mode rejection ratio, low noise, and the ability to accurately measure very small currents.

Using Tektronix's high-voltage differential probes TMDP0200 and TDP0500 [33], RF isolated probes TIVP, and Rogowski coils to measure the gate current I_G during switching and crosstalk processes, the results are shown in Figs. 5.72 and 5.73. Similar to the measurement of V_{GS}, the TMDP0200, due to its longer connection cable and higher attenuation ratio, shows significant discrepancies in measurement results compared to the TICP100. The Rogowski coil has a low bandwidth and its placement will affect the measurement results. In addition, its I_G is small and its range is large. The waveform measured by it has serious oscillations. The measurement results are also far different from those measured by TICP100. Although the TDP0500 is also a high-voltage differential probe, its lower 5 × attenuation setting and the 2.54 mm pitch pin interface at the probe front end help mitigate the weaknesses of the TMDP0200. However, even with these improvements, the TDP0500's I_G waveform exhibits more pronounced oscillations compared to the TICP100, and abnormal waveforms can still occur. This highlights the importance of selecting the appropriate probe for accurate gate current measurement, particularly for high-speed switching applications involving SiC MOSFETs. The optical isolated probes, with their superior common-mode rejection ratio (CMRR) and minimal loop inductance due to their MMCX interface, provide the most reliable measurements in such scenarios.

When measuring the gate current I_G of the upper bridge device, the TDP0500, with its common-mode voltage range limited to 35 V, is no longer suitable. Using Tektronix's high-voltage differential probe THDP0200, the optical isolated probe TICP100, and Rogowski coils to measure the switching and crosstalk processes of

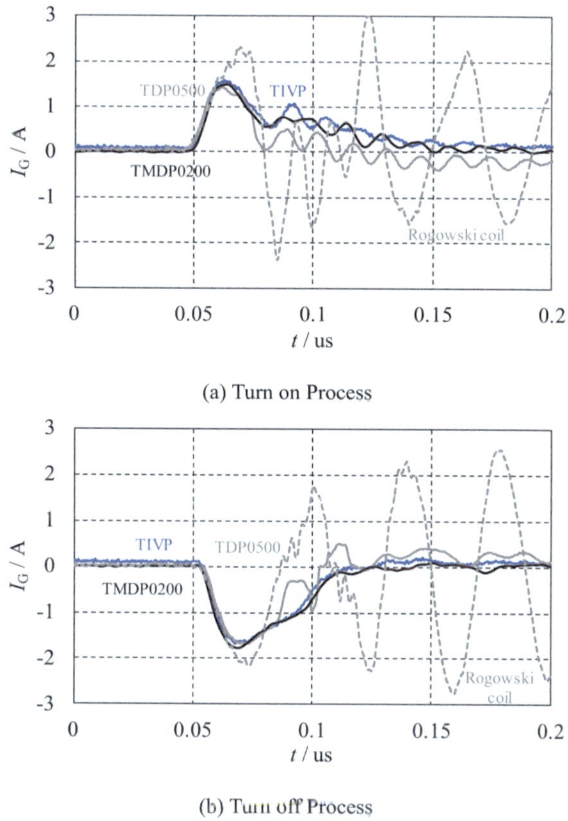

Fig. 5.72 I_G Measurement with high-voltage differential probe, RF isolation probe, and Rogowski coil

I_G, the results are shown in Figs. 5.74 and 5.75. It is evident that only the RF isolated probe TICP100 provides accurate results, while the TMDP0200 and Rogowski coils exhibit significant discrepancies due to their inherent limitations. This observation underscores the critical importance of choosing the right probe for precise gate current measurements in high speed switching applications involving SiC MOSFETs. The TIVP's superior common-mode rejection ratio (CMRR) and minimal loop inductance, achieved through its MMCX interface, ensure the most reliable measurements, particularly for high-voltage and high-frequency applications where accurate signal capture is essential.

In summary, only the optical isolated probes can provide accurate waveforms when measuring IG in high-speed switching applications involving SiC MOSFETs.

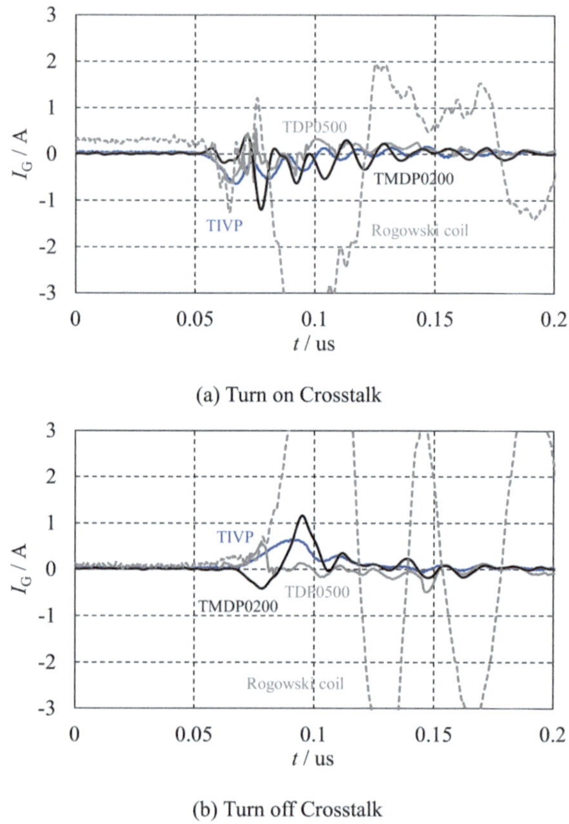

Fig. 5.73 I_G Crosstalk measurement with high-voltage differential probe, optical isolation probe and Rogowski coil

5.3.7 *Time Offset*

5.3.7.1 Effect of Time Offset

When performing double-pulse test, three different probes are typically connected to three oscilloscope channels to simultaneously measure the V_{GS}, V_{DS}, and I_{DS} of the device under test. While the time offset between the oscilloscope channels is minimal (below a few picoseconds) and can be neglected, the time offset between the three probes can be significant, ranging from a few nanoseconds to over ten nanoseconds. This discrepancy causes the measurement results to be out of sync on the time axis, leading to serious inaccuracies in loss calculations and causing unnecessary confusion in waveform analysis.

Taking the switching on and off of a SiC MOSFET at 800 V/30 A as an example, a comparison of the waveforms before and after time offset correction is shown in Fig. 5.76. Before time offset correction, I_{DS} leads V_{DS}, resulting in calculated turn-on and turn-off energies of 866 μJ and 53 μJ, respectively. After time offset correction,

Fig. 5.74 Switch I_G measurement with high-voltage differential probe, optical isolation probe and Rogowski coil on upper bridge arm

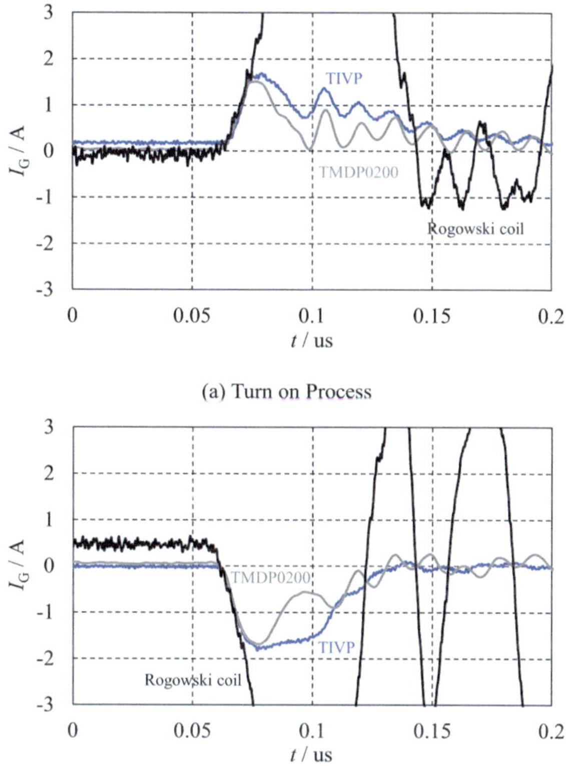

(a) Turn on Process

(b) Turn off Process

the calculated turn-on and turn-off energies are 700 μJ and 180 μJ, respectively. The difference in turn-on and turn-off energies before and after correction is substantial, particularly the turn-off energy, which differs by more than a factor of three. This demonstrates the necessity of performing time offset correction.

5.3.7.2 Time Offset Correction

1. **Time Offset Correction Fixture**

The principle of performing time offset correction is straightforward: it involves measuring the same signal simultaneously with the probes that need to be calibrated. For different voltage probes, they should simultaneously measure the same voltage rising edge. By setting the channel delay to overlap the waveforms, the offset correction for voltage probes is achieved. The set channel delay at this point represents the time offset between the voltage probes. Similarly, when calibrating the offset between voltage probes and current probes, they should simultaneously measure

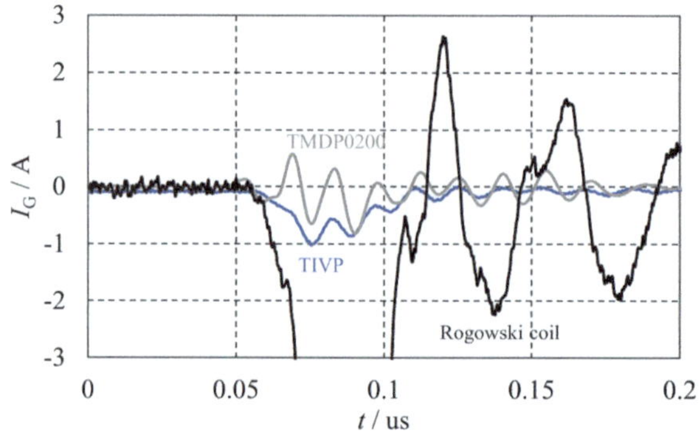

(a) Turn on Crosstalk

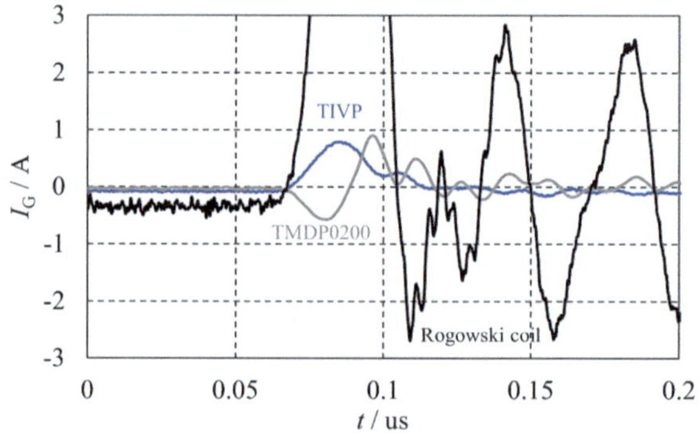

(b) Turn off Crosstalk

Fig. 5.75 I_G crosstalk measurement with high-voltage differential probe, optical isolation probe and Rogowski coil to measure of the upper bridge arm device

synchronous voltage and current rising edges. The channel delay is then adjusted until the waveforms overlap, completing the offset correction.

Based on this principle, various manufacturers provide specialized tools that offer synchronous voltage and current rising edges for offset calibration. For instance, Tektronix provides the Differential Pulse Generator (TEK-DPG) and the Power Measurement Correction Fixture (067-1686-03) [34, 35], as shown in Fig. 5.77. However, these fixtures are designed for use with Hall effect current probes and are not suitable for coaxial shunt resistors.

2. **Ohm's Law of Resistance**

The key to performing offset calibration for voltage probes and coaxial resistors is generating synchronous voltage and current rising edge signals. This can be achieved using a resistive load circuit, as shown in Fig. 5.78, which includes a switching transistor Q, a power resistor R_L, a coaxial sampling resistor R_{sense}, and a bus capacitor C_{Bus}. When Q is turned on, according to Ohm's law, the voltage across the power

Fig. 5.76 Waveforms before and after time offset correction

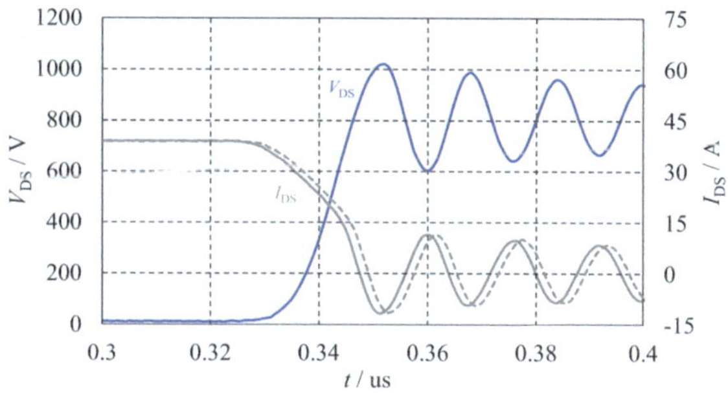

(a) Turn on Process

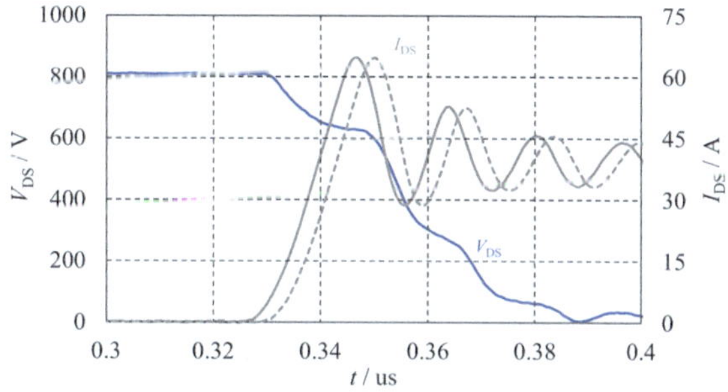

(b) Turn off Process

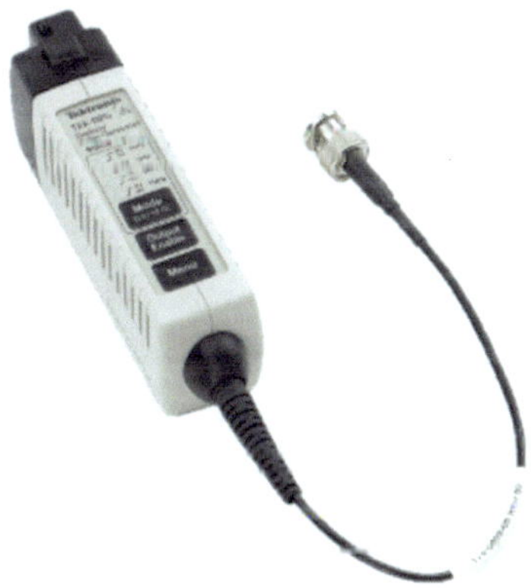

(a) TEK-DPG

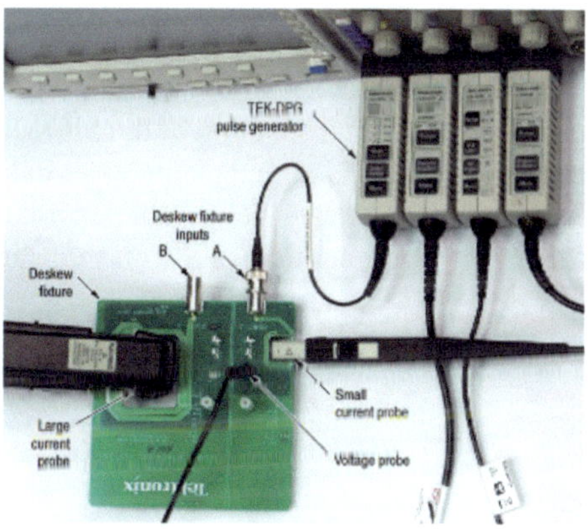

(b) 067-1686-03

Fig. 5.77 Tektronix TEK-DPG and 067-1686-03 time offset correction fixture

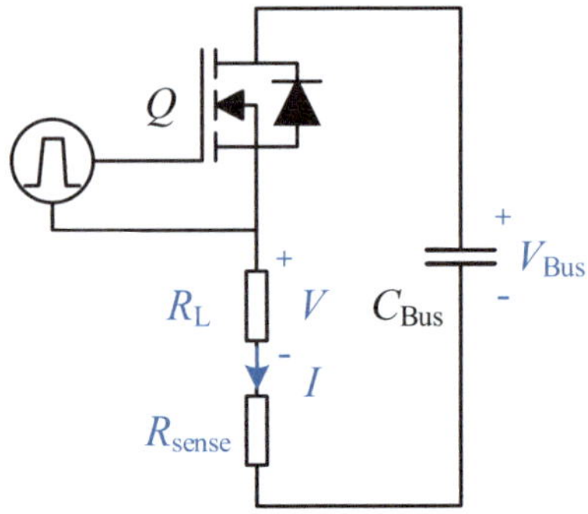

Fig. 5.78 Resistive load circuit

resistor and the current through it rise simultaneously, thus providing synchronous voltage and current rising edge signals. By using a voltage probe to measure the voltage across the power resistor and a coaxial resistor to measure the current flowing through the power resistor, offset calibration can be accomplished.

It's important to note that this calibration method is based on a purely resistive circuit, and the presence of inductance between voltage measurement points can disrupt this premise. Therefore, it's essential to use inductance-free resistors, and the voltage measurement points should be selected at the base of the resistor pins to minimize the influence of inductance. Common examples of inductance-free resistors include Caddock's MP900 series resistors [36] and Vishay's RTO 50 series resistors [37], as shown in Fig. 5.79.

3. **Waveform Characteristics of the Turn on Process**

According to the introduction of the SiC MOSFET turn-on process in Chap. 4, when the SiC MOSFET is turned on under an inductive load, the current I_{DS} starts to climb upward from zero, and the current generates a voltage drop on the inductance of the main power commutation circuit and appears as a voltage drop on the V_{DS} waveform.

(a) Caddock MP900

(b) Vishay RTO 50

Fig. 5.79 Non-inductive power resistor

The current rise and voltage drop should occur at the same time, so we can adjust the channel delay to make the I_{DS} waveform and the V_{DS} waveform meet this feature.

It should be noted that since the starting point of the current rise and voltage drop is sometimes not particularly obvious, human judgment introduces a large error. Therefore, it is recommended to perform multiple tests at a faster switching speed under different working conditions and take the average value as the time offset correction value.

4. **Voltage Reference Waveform of the Turn on Process**

According to the introduction of SiC MOSFET in Chap. 4, when SiC MOSFET is turned on in an inductive load circuit, the circuit formed by SiC MOSFET, coaxial resistor R_{sense} and freewheeling diode terminal satisfies Kirchhoff's voltage law (5.27).

$$V_{DS} = V_{Bus} - V_D - I_{DS} \cdot R_{sense} - L_{Loop} \cdot dI_{DS}/dt \tag{5.27}$$

V_{DS} and I_{DS} are the terminal voltage and current of the SiC MOSFET to be turned on, V_D is the terminal voltage of the freewheeling diode, R_{sense} is the coaxial resistance for measuring the current of the MOSFET, and L_{Loop} is the main power loop inductance. In, the current I_D of the freewheeling diode continues to decrease, which is the difference between the load current I_L and I_{DS}. However, as long as the freewheeling diode is in the forward conduction state, its terminal voltage is its forward conduction voltage drop, which is determined by the temperature and conduction current, and can be easily obtained by testing its forward conduction characteristics. Then (5.27) can be rewritten as,

$$V_{DS}(t) = V_{Bus} - V_D(I_L - I_{DS}(t), T_J) - I_{DS} \cdot R_{sense} - L_{Loop} \cdot dI_{DS}(t)/dt \tag{5.28}$$

Then, when V_{Bus}, $I_{DS}(t)$, $V_D(I_L - I_{DS}(t), T_J)$ and L_{Loop} are known, $V_{DS(cal)}$ can be calculated according to (5.28). Then the calculated $V_{DS(cal)}$ is completely synchronized with I_{DS}, that is, there is no time offset between the two. At this time, $V_{DS(cal)}$ is used as the voltage reference waveform to compare with the measured "V_{DS}" waveform, and the time offset between V_{DS} and I_{DS} can be determined. Tektronix's double-pulse test software WBG-DPT performs time offset calibration based on this principle [38].

Analyzing Eq. (5.28) reveals that obtaining accurate values for $V_D(I_L - I_{DS}(t), T_J)$ and L_{Loop} is crucial for obtaining the correct voltage reference waveform $V_{DS(cal)}$ and involves some difficulty. Since V_D is smaller compared to V_{Bus}, selecting a higher current-rated Si diode as the freewheeling diode allows V_D to be treated as a fixed value, thus improving the accuracy of V_D and simplifying the calculation process. To obtain an accurate value for L_{Loop}, one can conduct a single pulse test first and calculate L_{Loop} based on the peak turn-off voltage and the corresponding $dI_{DS}(t)/dt$ at that moment. It's important to note that the calculated L_{Loop} based on the test waveform excludes the parasitic inductance in the main power loop from the

V_{DS} measurement points, so it's crucial to ensure consistency between the test for obtaining L_{Loop} and the delay calibration test for the V_{DS} measurement points.

5.4 Parasitic Parameters Between Voltage Measurement Points

5.4.1 Basic Principles of Parasitic Parameters Introducing Measurement Deviations

With the increasing demand for higher performance and reliability of power converters, the development of power devices and converters is inevitably moving towards more refined approaches. This necessitates the acquisition of more accurate dynamic waveforms of power devices, especially for SiC devices with higher switching speeds. To address this, many researchers and engineers have been actively involved in related work, focusing primarily on instrument selection, measurement setup, and modeling analysis, as discussed in Sect. 5.3.

Regarding parasitic parameters, what is usually discussed is the impact of SiC MOSFET chip parasitic parameters, package parasitic parameters and test circuit parasitic parameters on the dynamic characteristics of SiC MOSFET, but little attention is paid to whether these parasitic parameters will affect the measurement of dynamic process waveforms.

Taking the TO-247-4 package as an example, as shown in Fig. 5.80, its structure consists of a thin cubic SiC MOSFET chip. The backside serves as the drain (D) terminal, while the top surface contains the gate (G) and source (S) terminals. The chip is soldered onto a metal substrate, with the drain terminal directly connected to the substrate via a pin. Multiple bonding wires connect the source terminal of the chip to the source (PS) pin to meet the current-carrying requirements. Due to the low drive current, only one thin bonding wire each is used to connect the source (S) and gate (G) terminals of the chip to the source (KS) and gate (G) pins, respectively. These pins, connected to the chip through bonding wires, exhibit inductance characteristics at high frequencies. Therefore, parasitic inductances of each terminal need to be considered in the practical dynamic equivalent circuit.

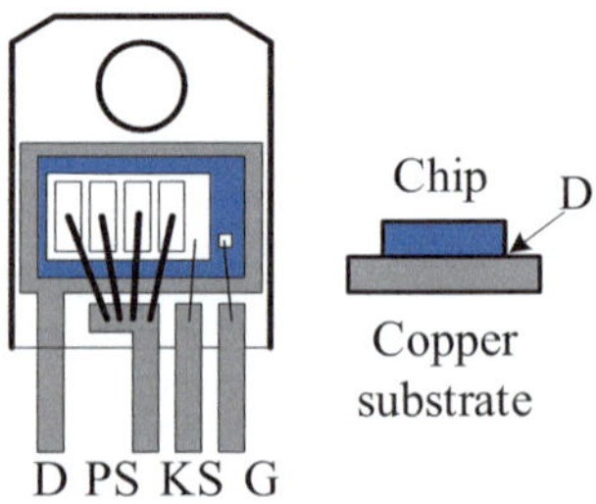

Fig. 5.80 Structure of TO-247-4 package device

The focus of the research analysis is on the terminal voltages of C_{GS} and C_{DS}, representing the gate-source voltage V_{GS} controlling the channel on the SiC MOSFET chip and the drain-source voltage V_{DS}, respectively. However, during the measurement of V_{GS} and V_{DS}, voltage probes cannot directly contact the chip but are clamped onto the pins of the device package, as shown in Fig. 5.81a. The corresponding equivalent circuit is illustrated in Fig. 5.81b. The measured voltages are $V_{GS(M)}$ and $V_{DS(M)}$ between the measurement points, not V_{GS} and V_{DS}. These measurement points include not only C_{GS} and C_{DS} but also bonding wire inductance and a portion of pin inductance. Specifically, $V_{GS(M)}$ measurement point includes $L_{G(pkg\text{ - }M)}$ and $L_{KS(pkg\text{ - }M)}$, collectively referred to as $L_{GS(M)}$; $V_{DS(M)}$ measurement point includes $L_{D(pkg\text{ - }M)}$ and $L_{KS(pkg\text{ - }M)}$, collectively referred to as $L_{DS(M)}$. When there is a varying current passing through $L_{GS(M)}$ and $L_{DS(M)}$, voltage drops will occur, which will also be measured by the voltage probe. Additionally, the SiC MOSFET chip has an internal gate resistance $R_{G(int)}$, and I_G will also generate voltage drops across it, which are also measured. Therefore, the actual measured $V_{GS(M)}$ and $V_{DS(M)}$ deviate from the actual V_{GS} and V_{DS} of the chip, which can be calculated using (5.29) and (5.30).

$$V_{GS(M)} = V_{GS} + I_G \cdot R_{G(int)} + \left(L_{G(pkg\text{ - }M)} + L_{KS(pkg\text{ - }M)}\right) \cdot \mathrm{d}I_G/\mathrm{d}t \tag{5.29}$$

$$V_{DS(M)} = V_{DS} + L_{D(pkg\text{ - }M)} \cdot \mathrm{d}I_{DS}/\mathrm{d}t + L_{KS(pkg\text{ - }M)} \cdot \mathrm{d}I_G/\mathrm{d}t \tag{5.30}$$

The equivalent circuit shown in Fig. 5.81b is designed for the scenario where the measurement points are on the device pins. However, often, for the sake of probe connection convenience, measurement points may be placed on the PCB. In such cases, $L_{GS(M)}$ and $L_{DS(M)}$ in Eqs. (5.29) and (5.30) not only include all the inductance of the looped pins but also may encompass a portion of the PCB trace inductance. This would lead to even more noticeable measurement deviations. We won't delve further into discussing this scenario.

The analysis above is based on the TO-247-4 package device, and it similarly applies to other Kelvin-source (emitter) package devices. However, due to different package structures, the introduction of measurement deviations from parasitic parameters between voltage measurement points on TO-247-3 package devices cannot be expressed using Eqs. (5.29) and (5.30). This specific issue will be discussed in Chap. 10.

It's important to note that the measurement deviations introduced by parasitic parameters between voltage measurement points are not unique to SiC devices; they apply to Si and GaN power devices as well.

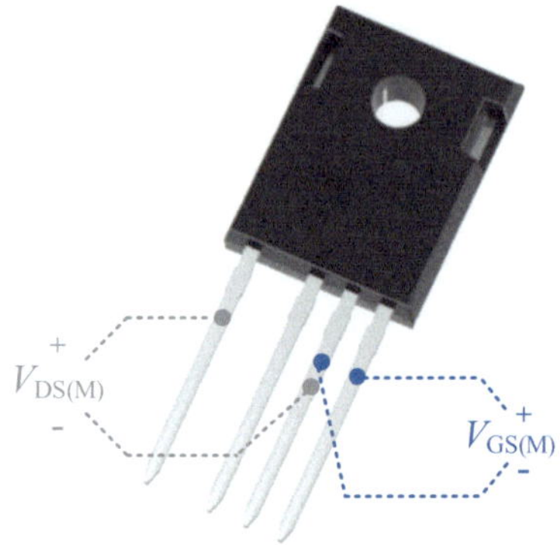

(a) Voltage Measurement Point

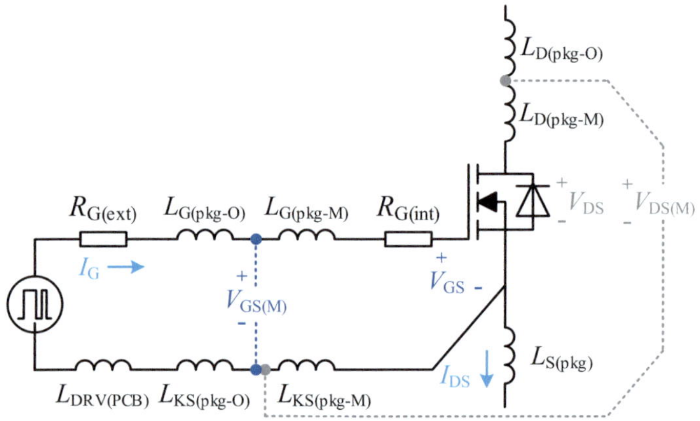

(b) Equivalent Circuit

Fig. 5.81 Voltage measurement points and equivalent circuits

5.4.2 Measurement Errors Introduced by Parasitic Parameters

Based on the discussion in Sect. 5.2.4, the dynamic processes of SiC devices include switching, reverse recovery, and crosstalk. Switching measurement involves V_{GS} and V_{DS}, reverse recovery measurement involves V_{DS}, and crosstalk measurement involves V_{GS}. Below, we'll specifically analyze the measurement deviations introduced by parasitic parameters between voltage measurement points during the switching and reverse recovery processes. The principles and measurement deviations of crosstalk will be addressed separately in Chap. 8.

5.4.2.1 Measurement Error During the Opening Process

As shown in Fig. 5.82, the simulated waveforms during the switching process reveal a significant deviation between the measured $V_{GS(M)}$ and $V_{DS(M)}$ and the desired V_{GS} and V_{DS}.

1. **V_{GS} Measurement**

(1) $t_0 - t_1$

At time t_0, the drive voltage rapidly transitions from $V_{DRV(off)}$ to $V_{DRV(on)}$. The gate current I_G rises swiftly to its peak value with a fast rate and large magnitude. Due to the combined effects of $L_{GS(M)}$ and $R_{G(int)}$, the rate of rise of $V_{GS(M)}$ is significantly greater than that of V_{GS}, rapidly creating a gap between the two.

(2) $t_1 - t_2$

As I_G gradually decreases, although its descent rate is slow, its magnitude remains significant. $V_{GS(M)}$ still notably exceeds V_{GS}, with the deviation primarily manifesting as a voltage drop across $R_{G(int)}$.

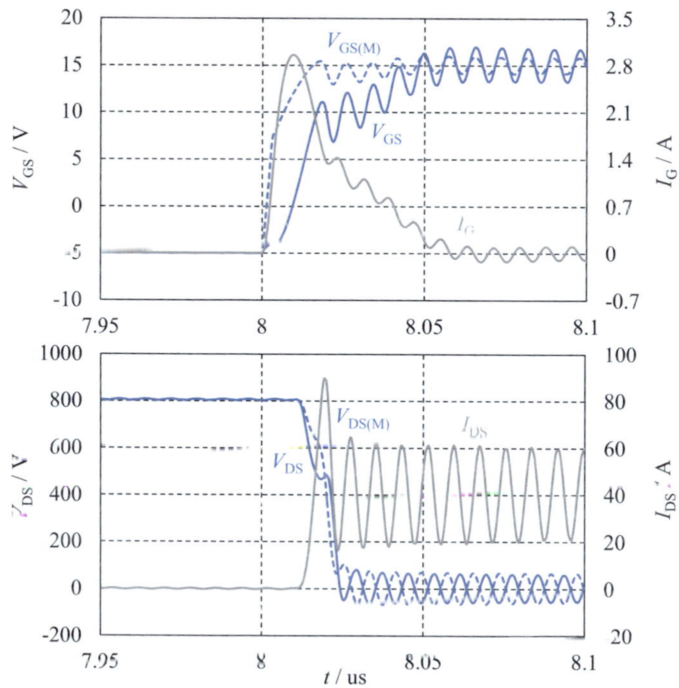

Fig. 5.82 Measurement error of the switching process (simulated waveform)

During this stage, the input capacitance C_{iss} remains essentially constant. V_{GS} and I_G can be represented, respectively, by Eqs. (5.31) and (5.32), where R_G represents the total drive resistance ($R_G = R_{G(ext)} + R_{G(int)}$).

$$V_{GS}(t) = V_{DRV(on)} - (V_{DRV(on)} - V_{DRV(off)}) \cdot e^{-\frac{t-t_0}{R_G \cdot C_{iss}}} \tag{5.31}$$

$$I_G(t) = C_{iss} \cdot \mathrm{d}V_{GS}/\mathrm{d}t = \frac{V_{DRV(on)} - V_{DRV(off)}}{R_G} \cdot e^{-\frac{t-t_0}{R_G \cdot C_{iss}}} \tag{5.32}$$

From the waveform analysis, it's evident that $V_{GS(M)}$ clearly doesn't conform to the RC charging process as depicted in Eq. (5.31). The deviation between $V_{GS(M)}$ and V_{GS} can be expressed using Eq. (5.33).

$$V_{GS(M)} = V_{GS} + \left(\frac{R_{G(int)}}{R_G} - \frac{L_{GS(M)}}{R_G^2 \cdot C_{iss}}\right) \cdot (V_{DRV(on)} - V_{DRV(off)}) \cdot e^{-\frac{t-t_0}{R_G \cdot C_{iss}}} \tag{5.33}$$

(3) $t_2 - t_3$

At time t_2, when I_{DS} begins to rise, corresponding to V_{GS} reaching the threshold voltage $V_{GS(th)}$, $V_{GS(M)}$ is noticeably higher. However, before this point, when $V_{GS(M)}$ reaches $V_{GS(th)}$, I_{DS} hasn't started to rise, indicating that $V_{GS(M)}$ clearly doesn't conform to the turn-on characteristics of a SiC MOSFET. Subsequently, I_G continues to slowly decrease while still maintaining a significant magnitude, with $V_{GS(M)}$ remaining notably higher than V_{GS}, with the deviation primarily manifesting as a voltage drop across $R_{G(int)}$. During this stage, since the input capacitance C_{iss} remains essentially constant, the deviation between $V_{GS(M)}$ and V_{GS} still satisfies Eq. (5.33).

(4) t_3-

At time t_3, when I_{DS} reaches its peak value, V_{DS} rapidly decreases, causing C_{GD} to increase rapidly. This leads to a brief slight increase in I_G before it continues to decrease at a slower rate than before. Consequently, the gap between $V_{GS(M)}$ and V_{GS} slightly increases before gradually decreasing. Meanwhile, due to the effect of $L_{GS(M)}$, the oscillation amplitude of $V_{GS(M)}$ is smaller than that of V_{GS}.

2. **V_{DS} Measurement**

(1) $t_0 - t_2$

Before t_2, V_{GS} hasn't reached $V_{GS(th)}$, I_{DS} remains unchanged, and there is no voltage drop across $L_{D(pkg\text{ - }M)}$. Although earlier analysis suggests there might be a voltage drop across $L_{KS(pkg\text{ - }M)}$, it's negligible compared to the bus voltage V_{Bus} that V_{DS} experiences, making it difficult to observe. Therefore, during this stage, there is no deviation between $V_{DS(M)}$ and V_{DS}.

(2) $t_2 - t_3$

As I_{DS} rapidly increases, it induces a voltage drop across $L_{D(pkg\text{ - }M)}$, resulting in a smaller drop in $V_{DS(M)}$ compared to V_{DS}. This discrepancy can lead to an overestimation of the turn-on energy calculation result compared to reality.

Table 5.3 Effect of measurement deviation on turn-on characteristic calculation

	Using $V_{GS(M)}$ and $V_{DS(M)}$	Using V_{GS} and V_{DS}
$t_{d(on)}$/ns	19.5	17.5
t_r/ns	29.5	32.4
I_{spike}/A	62.5	62.5
E_{on}/ uJ	931.4	904.0

(3) $t_3 - t_4$

As I_{DS} falls from its peak value, it induces a voltage drop across $L_{D(pkg - M)}$, causing the rate of decrease in $V_{DS(M)}$ to be greater than V_{DS}. In terms of magnitude, at time t_3, $V_{DS(M)}$ is higher than V_{DS}, while at time t_4, it becomes lower than V_{DS}.

(4) t_4–

After t_4, V_{DS} decreases to a low value. With $L_{D(pkg - M)}$ dominating, the oscillation amplitude of $V_{DS(M)}$ is greater than that of V_{DS}, leading to an underestimation of the turn-on time and turn-on energy calculation results compared to reality.

Subsequently, the SiC MOSFET is fully on, and V_{DS} is the conduction voltage drop. V_{DS} is positive and slightly oscillates with I_{DS}. However, under the influence of $L_{D(pkg - M)}$, $V_{DS(M)}$ exhibits significant oscillations and is negative, clearly not conforming to the conduction characteristics of a SiC MOSFET.

Using the measured $V_{GS(M)}$ and $V_{DS(M)}$ compared to the desired V_{GS} and V_{DS} to calculate the turn-on characteristics of the SiC MOSFET, as shown in Table 5.3. It can be observed that,

5.4.2.2 Measurement Deviation During Shutdown Process

As shown in Fig. 5.83, it is the simulation waveform of the turn-off process. It can be seen that there is an obvious deviation between the measured $V_{GS(M)}$ and $V_{DS(M)}$ and the expected V_{GS} and V_{DS}.

1. **V_{GS} Measurement**

(1) $t_0 \sim t_1$

At t_0, the driving voltage quickly changes from $V_{DRV(on)}$ to $V_{DRV(off)}$, and I_G quickly rises to the peak value with a fast rate and large amplitude. Under the joint action of $L_{GS(M)}$ and $R_{G(int)}$, the decrease rate of $V_{GS(M)}$ is much greater than that of V_{GS}, and the gap between the two quickly widens.

(2) $t_1 \sim t_2$

I_G gradually decreases. Although the rate of decrease is slow, its amplitude is still large. $V_{GS(M)}$ is still significantly lower than V_{GS}. The deviation is mainly reflected in the voltage drop on $R_{G(int)}$.

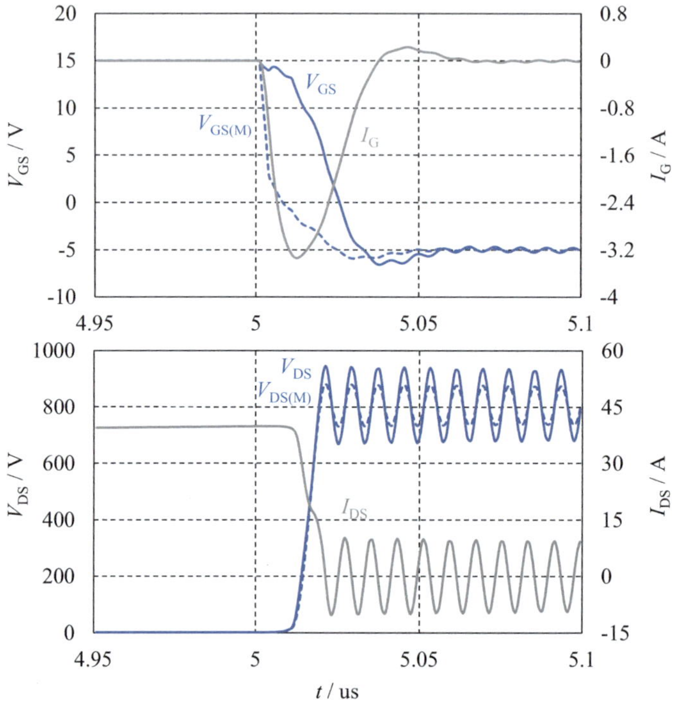

Fig. 5.83 Measurement deviation of the shutdown process (simulated waveform)

In this stage, the input capacitance C_{iss} remains basically unchanged. V_{GS} and I_G can be expressed by (5.34) and (5.35) respectively, where R_G is the total driving resistance ($R_G = R_{G(ext)} + R_{G(int)}$).

$$V_{GS}(t) = V_{DRV(off)} - (V_{DRV(off)} - V_{DRV(on)}) \cdot e^{-\frac{t-t_0}{R_G \cdot C_{iss}}} \tag{5.34}$$

$$I_G(t) = C_{iss} \cdot dV_{GS}/dt = \frac{V_{DRV(off)} - V_{DRV(on)}}{R_G} \cdot e^{-\frac{t-t_0}{R_G \cdot C_{iss}}} \tag{5.35}$$

Combined with the waveform, it can be seen that $V_{GS(M)}$ obviously does not conform to the RC discharge process shown in (5.34). The deviation between $V_{GS(M)}$ and V_{GS} can be expressed by (5.36).

$$V_{GS(M)} = V_{GS} + \left(\frac{R_{G(int)}}{R_G} - \frac{L_{GS(M)}}{R_G^2 \cdot C_{iss}} \right) \cdot (V_{DRV(off)} - V_{DRV(on)}) \cdot e^{-\frac{t-t_0}{R_G \cdot C_{iss}}} \tag{5.36}$$

It should be noted here that the SiC MOSFET is still on at this stage, and the value of C_{iss} in (5.34)–(5.36) is significantly greater than (5.31)–(5.33).

(3) $t_2 \sim t_3$

As I_{DS} gradually decreases, V_{DS} gradually increases, and I_G continues to decrease. Although its descent rate is slow, its magnitude remains significant. $V_{GS(M)}$ still notably lags behind V_{GS}, with the deviation primarily manifesting as a voltage drop across $R_{G(int)}$.

(4) $t_3 \sim$

I_G continues to decrease, the gap between $V_{GS(M)}$ and V_{GS} gradually narrows. Meanwhile, under the influence of $L_{GS(M)}$, the oscillation amplitude of $V_{GS(M)}$ is smaller than that of V_{GS}.

2. **V_{DS} Measurement**

(1) $t_0 \sim t_2$

Before t_2, I_{DS} remains unchanged, and there is no voltage drop across $L_{D(pkg - M)}$. Although earlier analysis suggests there might be a voltage drop across $L_{KS(pkg - M)}$, it's negligible compared to the bus voltage V_{Bus} that V_{DS} experiences, making it difficult to observe. Therefore, during this stage, there is no deviation between $V_{DS(M)}$ and V_{DS}.

(2) $t_2 - t_3$

As I_{DS} rapidly decreases, it induces a voltage drop across $L_{D(pkg - M)}$, causing the rate of increase in $V_{DS(M)}$ to be smaller than that of V_{DS}. This discrepancy can lead to an underestimation of the measurement results of the voltage spike during turn-off and a smaller calculated turn-off energy compared to reality.

(3) $t_3 \sim$

As I_{DS} continues to oscillate, it induces a voltage drop across $L_{D(pkg - M)}$, resulting in the oscillation amplitude of $V_{DS(M)}$ being smaller than that of V_{DS}.

Using the measured $V_{GS(M)}$ and $V_{DS(M)}$ compared to the desired V_{GS} and V_{DS} to calculate the turn-off characteristics of the SiC MOSFET, as shown in Table 5.4. It can be seen that the measured $t_{d(off)}$ is obviously too large, which is because $V_{GS(M)}$ decreases faster than V_{GS}, the measured E_{off} is slightly too small, and the measured V_{spike} is also too small.

Table 5.4 Effect of measurement error on turn off characteristic calculation

	Use $V_{GS(M)}$ and $V_{DS(M)}$	Use V_{GS} and V_{DS}
$t_{d(off)}$/ns	14.4	10.9
t_f/ns	11	11
V_{spike}/V	962.1	1004.9
E_{off}/uJ	307.2	316.8

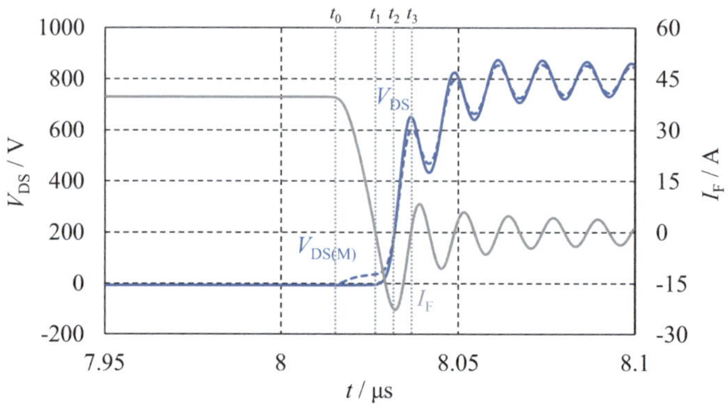

Fig. 5.84 Measurement deviation of reverse recovery process (simulated waveform)

5.4.2.3 Measurement Error in the Reverse Recovery Process

As shown in Fig. 5.84, the simulated waveforms during the reverse recovery process reveal a significant deviation between the measured $V_{DS(M)}$ and the desired V_{DS}.

(1) $t_0 \sim t_1$

As I_F decreases but remains greater than zero, the body diode is in forward conduction state, and V_{DS} is the conduction voltage drop and is negative. Under the influence of $L_{D(pkg\text{ - }M)}$, $V_{DS(M)}$ changes from negative to positive and continues to increase. This clearly does not conform to the conduction characteristics of a body diode.

(2) $t_1 \sim t_2$

As I_F decreases to zero and continues to decrease to its peak value, under the influence of $L_{D(pkg\text{ - }M)}$, $V_{DS(M)}$ remains higher than V_{DS}.

(3) $t_2 \sim t_3$

As I_F reversely recovers to zero, under the influence of $L_{D(pkg\text{ - }M)}$, $V_{DS(M)}$ is lower than V_{DS}.

(4) $t_3 \sim$

After the reverse recovery process concludes, I_F continues to oscillate. Under the influence of $L_{D(pkg\text{ - }M)}$, the oscillation amplitude of $V_{DS(M)}$ is smaller than that of V_{DS}.

5.4.2.4 Effect of Measurement Error

Through the analysis above, it's evident that the parasitic parameters between voltage measurement points lead to significant deviations between the measured waveforms $V_{GS(M)}$ and $V_{DS(M)}$ and the desired waveforms V_{GS} and V_{DS}. The following outlines the negative impacts of measurement deviations in device development, application, and related academic research. This includes a detailed discussion on crosstalk and the case of TO-247–3 package devices, which will be further addressed respectively in Chaps. 8 and 10. This approach aims to provide a more focused and comprehensive reflection of this issue.

1. **Analysis of switching characteristics**

The presence of measurement deviations can lead to situations where the analysis of switching processes based on measured waveforms contradicts theoretical expectations, potentially causing engineers to become stuck in dead ends. For instance, $V_{GS(M)}$ exhibits almost vertical changes at the moment of drive signal reversal, which is not consistent with the theoretical RC charging and discharging process. Even after $V_{GS(M)}$ exceeds $V_{GS(M)}$, there is still no current flow. In the case of TO-247–3 package devices, there are instances of overshoots observed on $V_{GS(M)}$ that should not exist.

2. **Analysis of crosstalk characteristics**

For TO-247-4 package devices, the waveform of $V_{GS(M)}$ due to crosstalk closely resembles V_{GS} but with a lower amplitude. This will result in an underestimation of the severity of crosstalk and also lead to a mismatch between theoretical analysis and test results. For TO-247-3 package devices, the waveform of $V_{GS(M)}$ during turn-on crosstalk is completely different from V_{GS} and has a significantly higher amplitude. Similarly, the waveform of $V_{GS(M)}$ during turn-off crosstalk is similar to VGS but with a significantly higher amplitude. This indicates that measurement deviations will lead to incorrect analysis of crosstalk in TO-247-3 devices.

3. **Device modeling and simulation**

Utilizing device models allows for the rapid study and analysis of device switching characteristics, reverse recovery characteristics, and crosstalk effects, as well as the evaluation of converter efficiency, temperature rise, and electromagnetic interference (EMI). Only when the model is sufficiently accurate can it provide valuable results. Therefore, it is necessary to compare simulation results based on the model with measured values before making adjustments to the model. Many models include parasitic parameters, so it is essential to ensure that the parasitic parameters between voltage measurement points in simulation are consistent with those in actual measurements during model verification; otherwise, the accuracy of the model cannot be effectively verified. Additionally, loss calculations and crosstalk assessments based on simulation results should exclude measurement deviations caused by parasitic parameters to avoid misleading judgments. Unfortunately, current device modeling and simulation efforts have not yet addressed this issue.

4. **Converter Design**

During converter design, dynamic processes of devices can be evaluated through double-pulse test, enabling loss calculations and crosstalk assessments. Measurement deviations can mislead these evaluations. For instance, measurement deviations can lead to inaccuracies in loss calculations, affecting thermal design, with higher switching frequencies exacerbating the impact. A larger internal gate resistance ($R_{G(int)}$) can result in underestimating crosstalk, misleading engineers into believing that the crosstalk risk is lower, increasing the risk of the converter.

5. **Safe Operation Estimate**

During the turn-off process, the peak of $V_{DS(M)}$ is lower than V_{DS}, underestimating the risk of device overvoltage breakdown. For TO-247-4 packaged devices, the crosstalk waveform of $V_{GS(M)}$ closely resembles V_{GS}, but with a lower magnitude, leading to underestimation of the severity of crosstalk. For TO-247-3 packaged devices, it's not possible to effectively assess the risk of bridge arm direct conduction caused by turn-on crosstalk, while also overestimating the severity of turn-off crosstalk.

6. **Academic Research**

In academic research, the highest demands are placed on the accuracy of measurement results since they are crucial for analysis and evaluation. However, in existing research works, the dynamic process waveforms obtained have not taken into account the influence of measurement biases. This is especially pronounced in studies focused on TO-247-3 packaged devices, where the impact of measurement biases is even greater. For instance, the equivalent circuit used for analysis in a study might be based on the TO-247-4 package, while the actual measurements are conducted on TO-247-3 packaged devices. Alternatively, although the equivalent circuit used for analysis in a study may be based on the TO-247-3 package, the effects of measurement biases are not considered, significantly compromising the credibility of the research conclusions. Furthermore, utilize waveform analysis with significant discrepancies from actual values to study the parasitic effects of TO-247-3 packaged devices and propose and validate methods to mitigate these effects.

7. **Professional and Vocational Education**

Indeed, due to the presence of measurement biases, a situation often arises in both university education and engineering vocational training where understanding the switching characteristics of devices becomes a case of hard to learn correctly. Textbooks and professional literature typically explain the switching process based on ideal scenarios without considering the influence of parasitic parameters between measurement points, and they fail to provide explanations or analyses on this matter. Consequently, when students or engineers attempt to analyze test waveforms according to the textbooks, it naturally leads to confusion.

5.4.3 Compensation Method for Measurement Deviation

As described above, the consequences of measurement biases introduced by parasitic parameters between voltage measurement points are significant. Therefore, it is imperative to propose methods to eliminate their influence and obtain correct voltage waveforms. According to Eqs. (5.29) and (5.30), the measured $V_{GS(M)}$ and $V_{DS(M)}$ are composed of two parts: the desired V_{GS} and V_{DS} and the measurement bias, where the measurement bias arises from the voltage drops across $L_{GS(M)}$, $L_{DS(M)}$, and $R_{G(int)}$. Rewriting Eqs. (5.29) and (5.30) as Eqs. (5.37) and (5.38), the square brackets represent the measurement bias. Therefore, by subtracting the measurement bias from the measured $V_{GS(M)}$ and $V_{DS(M)}$, we obtain the desired V_{GS} and V_{DS}.

$$V_{GS} = V_{GS(M)} - [I_G \cdot R_{G(int)} + (L_{G(pkg\text{ - }M)} + L_{KS(pkg\text{ - }M)}) \cdot dI_G/dt] \quad (5.37)$$

$$V_{DS} = V_{DS(M)} - [L_{D(pkg\text{ - }M)} \cdot dI_{DS}/dt + L_{KS(pkg\text{ - }M)} \cdot dI_G/dt] \quad (5.38)$$

The parameters on the right side of Eqs. (5.37) and (5.38) can be categorized into three types. The first type consists of voltage and current waveforms, including $V_{GS(M)}$, $V_{DS(M)}$, I_G, and I_{DS}. The second type comprises the rates of current change, i.e., dI_G/dt and dI_{DS}/dt. The third type includes parasitic parameters between measurement points, such as $R_{G(int)}$, $L_{G(pkg\text{ - }M)}$, $L_{KS(pkg\text{ - }M)}$, and $L_{D(pkg\text{ - }M)}$.

1. **Waveform Measurement**

The precision of measurements is essential to ensure accurate results, and suitable instrumentation can be selected based on the information provided in Sect. 6.3.

(1) **Oscilloscope**

The mid-level high-resolution oscilloscope is employed. For subsequent test, a Tektronix MSO54B oscilloscope with a bandwidth of 1 GHz and 12-bit ADC resolution is utilized.

(2) **$V_{GS(M)}$ Measurement**

A 10 × passive probe is utilized. For subsequent test, a Tektronix TPP1000 probe is employed, featuring a bandwidth of up to 1 GHz, a 10 × attenuation ratio, and built-in capabilities for delay compensation and impedance matching.

(3) **$V_{DS(M)}$ Measurement**

A high-bandwidth high-voltage passive probe is selected. For subsequent test, a Tektronix TPP0850 probe is utilized, featuring a bandwidth of 800 MHz, a 10 × attenuation ratio, and built-in capabilities for delay compensation and impedance matching.

(4) **I_G Measurement**

The voltage across the drive resistor is measured using an optically isolated probe. For subsequent test, a Tektronix TICP100 probe is employed, featuring a bandwidth of 1 GHz, a 10 × attenuation ratio, a range of ± 5 V, and built-in delay compensation functionality.

(5) I_{DS} Measurement

The measurement is performed using a coaxial resistor with high bandwidth and low inductance. For subsequent test, a T&M SDN414 resistor is utilized, featuring a bandwidth of 1 GHz and a resistance value of 50 mΩ.

2. Current Conversion Rate Calculation

The waveform of I_G and I_{DS} obtained from measurements consists of a series of equally spaced sampled data points, thus numerical differentiation can be used to calculate dI_G/dt and dI_{DS}/dt. For equally spaced sampled data, the derivative value corresponding to each sample point can be expressed as a linear combination of the values of adjacent sample points, which is the basic idea of numerical differentiation. To obtain the highest possible accuracy, the five-point method can be chosen for calculation. Taking the numerical differentiation of the k-th point, denoted as DP_k, as an example, it can be represented as a linear combination of the values of the $k + 2$, $k + 1$, $k - 1$, and $k - 2$ points, as shown in (5.39), where H is the sampling interval. It is important to note that the sampling range must exceed the range of the dynamic process, ensuring the requirements on the data in (5.39) are met.

$$\mathrm{DP}_k = \frac{-P_{k+2} + 8P_{k+1} - 8P_{k-1} + P_{k-2}}{12h} \tag{5.39}$$

The truncation error calculated using this method is proportional to h^4, where H is the sampling interval. Taking a sampling frequency of 12.5 GHz as an example, the relative truncation error does not exceed 10^{-40}.

3. Parasitic Parameter Extraction

(1) $L_{G(pkg-M)}$, $L_{KS(pkg-M)}$ and $L_{D(pkg-M)}$

The parasitic inductance values between measurement points can be extracted using finite element analysis simulation software. The specific method involves: creating a geometric model of the device package, defining conductivity, permeability, and other material parameters relevant to the solution for each part of the geometric model, and setting hypothetical current entry and exit surfaces based on the positions of the measurement points. This process allows for the determination of parasitic inductance values between measurement points.

To extract the parasitic inductance values of different parts of the package, hypothetical current entry and exit surfaces need to be set based on the positions of the corresponding measurement points. When extracting $L_{G(pkg - M)}$, the hypothetical current entry surface is the section of the G terminal pin, and the exit surface is the contact surface between the G terminal bonding wire and the chip's G terminal. When extracting $L_{KS(pkg - M)}$, the hypothetical current entry surface is the section of the KS terminal pin, and the exit surface is the contact surface between the KS terminal bonding wire and the chip's S terminal. When extracting $L_{D(pkg - M)}$, the hypothetical current entry surface is the section of the D terminal pin, and the exit surface is the contact surface between the metal substrate and the chip's G terminal.

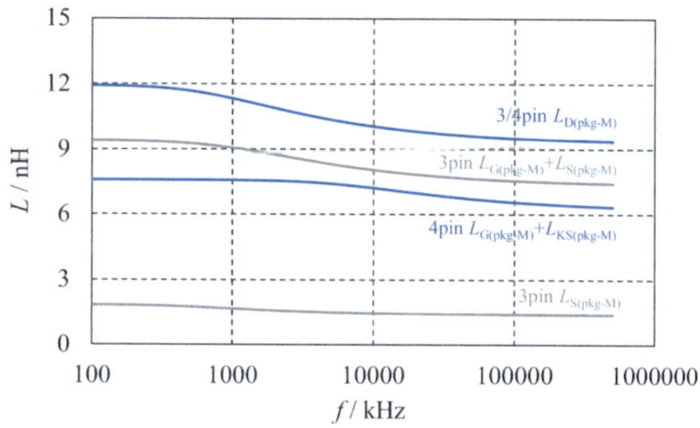

Fig. 5.85 Frequency characteristics of parasitic inductance

The results obtained through the aforementioned method show that the parasitic inductance varies with frequency, as shown in Fig. 5.85. Although frequency-domain analysis can provide the most accurate voltage drop across the parasitic inductance, the method is overly complex. Since the parasitic inductance varies little with frequency, a constant value can be chosen based on the equivalent bandwidth frequency of the switch transient current. This value can then be multiplied by the rate of change of current to obtain the voltage drop across the parasitic inductance. The equivalent bandwidth frequency of the switch transient current (f_{eq}) is calculated using (5.40), where t_r and t_f represent the rise time and fall time, respectively.

$$f_{eq} = \max\left\{\frac{0.35}{t_r}, \frac{0.35}{t_f}\right\} \tag{5.40}$$

(2) $\boldsymbol{R}_{\mathbf{G(int)}}$

The internal gate resistance $R_{G(int)}$ can be obtained by test with a power device analyzer.

5.4.4 *Compensation Effect of Measurement Deviation*

Consistent with the previous sections, this section primarily focuses on compensating for measurement errors during the switching and reverse recovery processes of TO-247 packaged SiC devices. Discussions regarding crosstalk and TO-247-4 packaged devices will be addressed in Chapters Eight and Eleven, respectively.

5.4.4.1 Compensation Effect of Measurement Deviation During Turn on Process

The compensation effect of measurement errors during the turn-on process is illustrated in Fig. 5.86. The test conditions are I_L = 30A, V_{Bus} = 800 V, and $R_{\text{G(ext)}}$ = 5Ω.

Before compensation, $V_{\text{GS(M)}}$ exhibits an almost vertical rise at the beginning of the turn-on process, rather than the first-order inertial rise as theoretically analyzed. At the moment when I_{DS} begins to rise, $V_{\text{GS(M)}}$ is 4.33 V higher than the tested $V_{\text{GS(th)}}$, which significantly deviates from the theoretical expectation. After compensation, the rise of V_{GS} conforms to the theoretical analysis. At the onset of I_{DS} rising, V_{GS} differs from the tested $V_{\text{GS(th)}}$ by only 0.08 V. Additionally, after compensation, the oscillation amplitude of V_{GS} is smaller than that of $V_{\text{GS(M)}}$, with a maximum difference of 3 V, reflecting the actual oscillation of V_{GS}.

During the rapid increase of I_{DS}, the drop in V_{DS} is greater than $V_{\text{DS(M)}}$, with a maximum deviation of 65 V between them. When I_{DS} decreases to a low amplitude, the oscillation amplitude of V_{DS} is smaller than that of the compensated $V_{\text{DS(M)}}$, with a maximum difference of 65 V between them. After the SiC MOSFET is fully turned on, $V_{\text{DS(M)}}$ exhibits significant oscillations and negative values, which does not comply with the conduction characteristics of SiC MOSFETs. After compensation, V_{DS} is positive and oscillates slightly with I_{DS}, reflecting the actual situation. In addition, the $t_{\text{d(on)}}$ and E_{on} calculated based on $V_{\text{GS(M)}}$ and $V_{\text{DS(M)}}$ are 23.8 ns and 507.4 μJ, and the $t_{\text{d(on)}}$ and E_{on} calculated based on V_{GS} and V_{DS} are 22.2 ns and 444.5 μJ.

From (5.37) and (5.38), it can be inferred that measurement errors vary under different operating conditions. When $R_{\text{G(ext)}}$ is smaller, the amplitude and rate of change of IG increase, leading to larger voltage drops on $R_{\text{G(int)}}$ and $L_{\text{GS(M)}}$, resulting in larger measurement errors between $V_{\text{GS(M)}}$ and V_{GS}. Similarly, when $R_{\text{G(ext)}}$ is smaller, faster changes in I_{G} and I_{DS} result in larger voltage drops on $L_{\text{GS(M)}}$, leading to larger measurement errors between $V_{\text{DS(M)}}$ and V_{DS}. Moreover, as I_{L} increases, the rate of change of I_{DS} slows down, resulting in smaller voltage drops on $L_{\text{DS(M)}}$, and consequently smaller measurement errors between $V_{\text{DS(M)}}$ and V_{DS}.

5.4.4.2 Compensation Effect of Measurement Deviation During Turn off Process

The compensation effect of the measurement error during the turn-off process is shown in Fig. 5.87, under the test conditions of I_L = 30 A, V_{Bus} = 800 V, and $R_{\text{G(ext)}}$ = 5 Ω.

After compensation, the rate of decrease of V_{GS} during the turn-off process is significantly smaller than $V_{\text{GS(M)}}$ initially, with a maximum deviation of 8 V. After compensation, the oscillation amplitude of V_{DS} is larger than that of $V_{\text{DS(M)}}$. Without compensation, the measured result of the turn-off voltage spike will be 15 V smaller than the actual value. In addition, the $t_{\text{d(off)}}$ and E_{off} calculated based on $V_{\text{GS(M)}}$ and

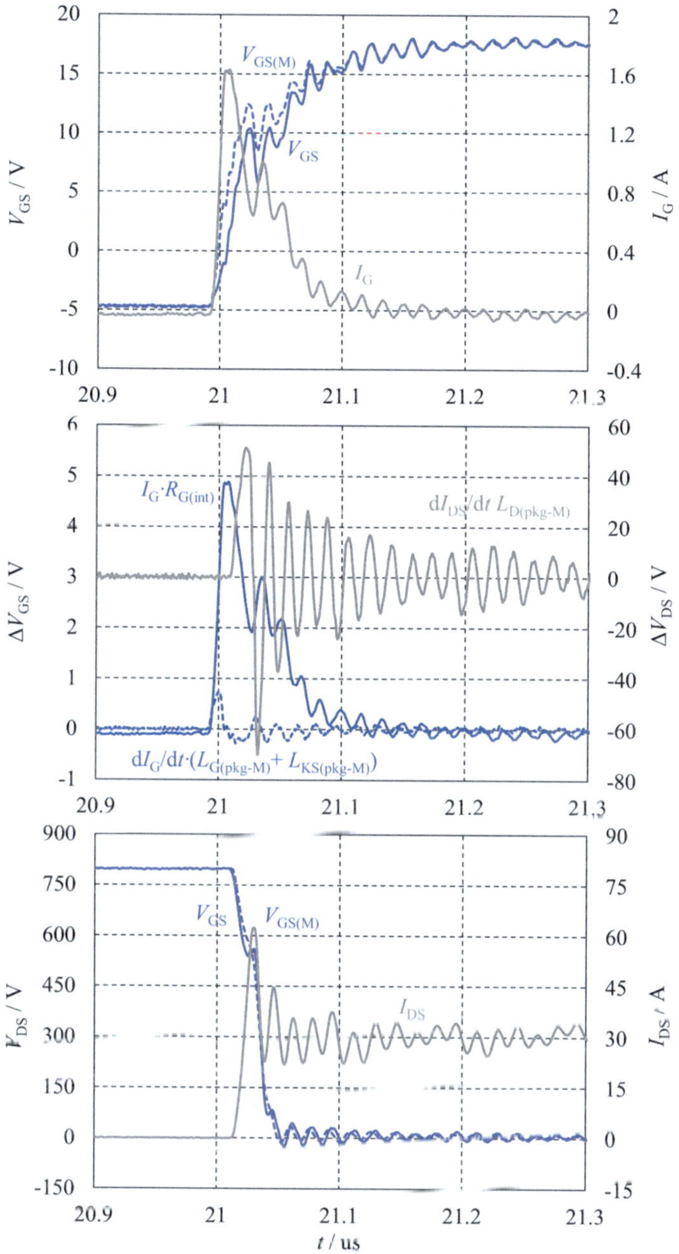

Fig. 5.86 Compensation effect of measurement deviation during opening process

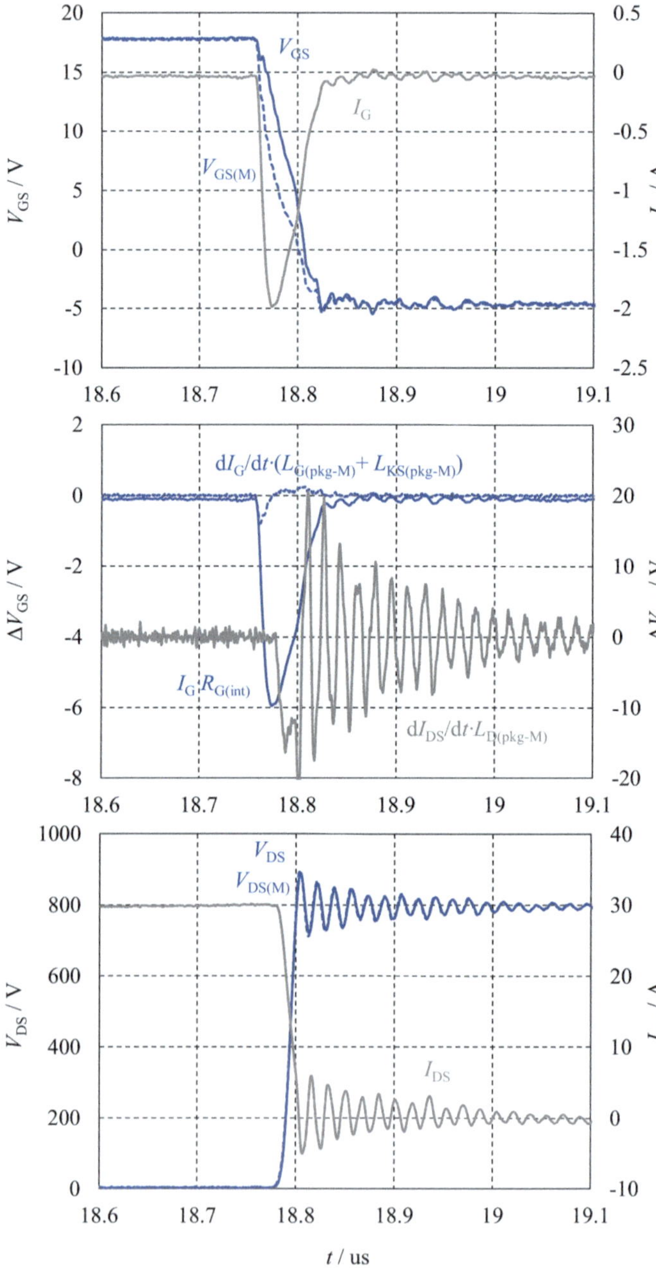

Fig. 5.87 Compensation effect of measurement deviation during turn off process

$V_{DS(M)}$ are 26.9 ns and 53.1 μJ, and the $t_{d(off)}$ and E_{off} calculated based on V_{GS} and V_{DS} are 24.4 ns and 80.3 μJ.

From (5.37) and (5.38), it can be seen that the measurement deviation will be different under different working conditions. When $R_{G(ext)}$ is smaller, the amplitude and rate of change of I_G increase. Consequently, the voltage drop on $R_{G(int)}$ and $L_{GS(M)}$ increases, leading to a larger measurement error between $V_{GS(M)}$ and V_{GS}. Similarly, a smaller $R_{G(ext)}$ results in faster variations in I_G and I_{DS}, causing a larger voltage drop on $L_{DS(M)}$ and hence a larger measurement error between $V_{DS(M)}$ and V_{DS}. Moreover, a higher I_L leads to faster changes in I_{DS}, resulting in a larger voltage drop on $L_{DS(M)}$ and consequently a larger measurement error between $V_{DS(M)}$ and V_{DS}.

5.4.4.3 Compensation Effect of Measurement Deviation During Reverse Recovery Process

The compensation effect of the measurement error during the reverse recovery process is shown in Fig. 5.88, with test conditions of $I_F = 30$ A, $V_{Bus} = 800$ V, and D $dI_F/dt = 1000$ A/ns. During the forward conduction state of the diode, $V_{DS(M)}$ transitions from negative to positive and continues to rise, which does not conform to the conduction characteristics of a body diode. After compensation, V_{DS} reflects the conduction voltage drop and remains negative, which accurately reflects the real scenario, with a maximum deviation of 40 V. Furthermore, during the gradual turn-off process of the body diode, the compensated V_{DS} exhibits a larger oscillation amplitude than $V_{DS(M)}$, with a maximum difference of 60 V, which reflects the actual oscillation behavior of V_{DS}.

As shown in Fig. 5.89, the measurement error and compensation effect during the reverse recovery process vary with different dI_F/dt. When dI_F/dt is larger, the voltage drop across $L_{DS(M)}$ increases, resulting in a larger measurement error between $V_{DS(M)}$ and V_{DS}.

5.5 Evaluation of Dynamic Process Test Results

5.5.1 *Measurement Accuracy and Repeatability*

Before proposing criteria for assessing the results of dynamic process test, let's review two fundamental concepts related to measurement: accuracy and repeatability. Accuracy refers to the degree of agreement between a measured or calculated quantity and its actual (true) value. The closer they are, the higher the accuracy of the measurement. Repeatability (also known as precision) refers to the degree to which repeated

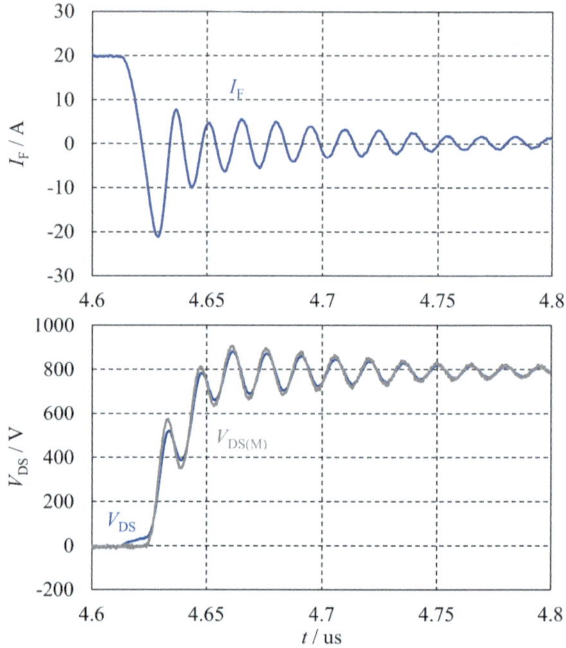

Fig. 5.88 Compensation effect of measurement deviation during reverse recovery process

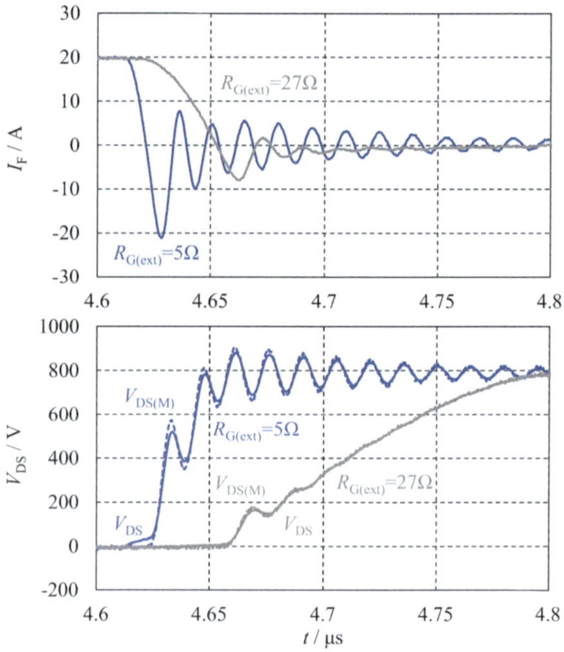

Fig. 5.89 Compensation effect of reverse recovery process under different dI_F/dt

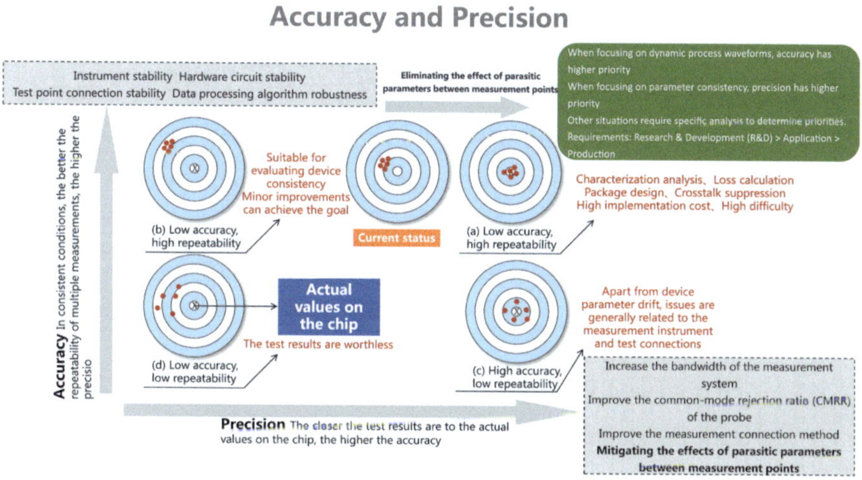

Fig. 5.90 Visualization of measurement accuracy and repeatability

measurements or calculations yield the same or similar results. The closer the results of repeated measurements are, the higher the repeatability of the measurement.

A visual analogy to illustrate accuracy and repeatability is target shooting. In this analogy, the bullseye represents the actual value, and the bullet holes represent the measured values. The closer the bullet holes (measured values) are to the center of the bullseye (actual value), the higher the accuracy. Similarly, the closer the bullet holes (measured values) are to each other in multiple shots, the higher the repeatability. Based on the levels of accuracy and repeatability, test results can be classified into four scenarios, as shown in Fig. 5.90.

(1) High Accuracy and High Repeatability

In Fig. 5.90a, all measurement results are located near the bullseye center, with close distances between them, indicating high accuracy and high reproducibility of the measurements. Such test results are exactly what engineers aim for and can be used for rigorous analysis and research. It's worth noting that achieving high accuracy and high reproducibility in measurements often requires sacrificing test efficiency and incurring higher test costs.

(2) Low Accuracy and High Repeatability

In Fig. 5.90b, the measurement results are overall deviated from the center of the bullseye, but they are still close to each other, indicating low accuracy and high reproducibility of the measurements. At this point, although there is a significant deviation between the measured values and the actual values, this deviation is relatively stable. Therefore, such measurement results can still be used for relative comparison of parameters, analysis of parameter distribution, and detection of parameter changes. Ensuring stability while sacrificing accuracy can significantly improve test efficiency and reduce test costs.

(3) **High Accuracy and Low Repeatability**

In Fig. 5.90c, although the measurement results are all around the center of the bullseye, they are still far apart from each other, indicating high accuracy and low reproducibility of the measurements. At this point, individual test results cannot provide useful information, and obtaining results close to the actual values would require averaging multiple measurements. However, this significantly reduces test efficiency.

(4) **Low Accuracy and Repeatability**

In Fig. 5.90d, the measurement results are far from the center of the bullseye, and they are also far apart from each other, indicating low accuracy and low reproducibility of the measurements. Such measurement results have no value.

Through the above analysis, it is evident that measurement results of practical value must possess high repeatability. For dynamic process test of SiC devices, when the measurement results are utilized for purposes such as device characteristic analysis, loss calculation, packaging design, and interference suppression research, the highest demands are placed on both accuracy and repeatability. Even when the test results are intended for production, there is still a stringent requirement for repeatability, although the demand for accuracy can be relatively reduced.

For dynamic process test of SiC devices, the actual values consist of the gate-source voltage V_{GS} controlling the channel, the drain-source voltage V_{DS} across the chip, and the drain-source current I_{DS}. It is worth noting that parameters such as switching time and switching energy, calculated based on V_{GS}, V_{DS}, and I_{DS}, are indirect numerical values. Therefore, when evaluating the measurement results of dynamic processes, the assessment focuses on the entire waveform of the dynamic process.

Evaluating the repeatability of dynamic process test for SiC devices simply involves overlaying multiple measurement waveforms for analysis. Repeatability primarily relies on ensuring the stability of the test environment, measurement instruments, test connections, and excitation signals. As long as these factors remain stable, measurement consistency can be ensured.

Evaluating the accuracy of dynamic process test for SiC devices is quite challenging because obtaining actual values as standards is not feasible. In such cases, the judgment relies on analyzing whether the test waveforms conform to theoretical characteristics, while ensuring that the device characteristics are normal. Unfortunately, achieving proficiency in waveform analysis and judgment requires a deep understanding of device characteristics and measurement techniques, a skill set that many engineers may not possess. The oscilloscope selection, voltage and current measurement techniques, test connection methods, and time offset correction discussed in Sect. 5.3 all contribute to improving measurement accuracy. Despite the significant efforts made by researchers and test equipment manufacturers in these areas, measurement deviations caused by parasitic parameters between voltage measurement points cannot be entirely avoided. Only by employing the compensation methods discussed in Sect. 5.4 can we elevate the measurement accuracy to the level where we can obtain the actual values for the gate-to-source voltage VGS controlling the channel and the drain-to-source voltage V_{DS} across the chip, further enhance the measurement accuracy to a new level

5.5.2 *Dynamic Process Test Scenarios and Result Evaluation Criteria*

The dynamic process test of SiC devices finds application in various fields, encompassing both industry and academia. In the industry, it spans across device research and development, manufacturing, packaging test, and system applications. In the academic realm, it extends to device research, packaging research, and application studies. This wide range of applications underscores the importance of dynamic process test for SiC devices, as it spans across industries and academic disciplines, covering various stages of the device industry chain and spanning the entire lifecycle of the devices.

The different sectors in both industry and academia can be further subdivided into various test scenarios. Due to the varying test objectives across these scenarios, the required test items and corresponding evaluation criteria naturally differ, and cannot be generalized. SiC device dynamic test scenarios and items are provided in Table 5.5.

1. **Device R&D and Manufacturing**

(1) **Product Research**

During device development, benchmarking against competitors is crucial. Test on the same equipment to collect data helps avoid discrepancies arising from different test conditions and equipment, including in dynamic processes. Based on these test results, one can analyze the strengths and weaknesses of various products and establish parameter design goals. Inaccurate test results can lead to misguided development directions. Therefore, during product research, there is a high demand for accuracy and repeatability in dynamic process measurements, making it necessary to compensate for measurement deviations caused by parasitic parameters between voltage measurement points.

(2) **Application Research**

Based on transformer-based test research, specific application requirements for device characteristics can be extracted, and the performance of different products in transformers can be compared. At this stage, test results serve as the basis for analysis and suggest further optimization directions for devices. Since this involves

Table 5.5 Scenarios and test items for dynamic testing of SiC devices

Stage	Test scenarios	Switch	Reverse recovery	Crosstalk
Chip R&D and manufacturing	Product research	√	√	√
	Application research	√	√	√
	Engineering sample characteristic evaluation	√	√	√
	Specification preparation	√	√	
	Chip manufacturing	√	√	
Packaging test	Packaging design	√	√	√
	Packaging manufacturing	√	√	
System application	Incoming material inspection	√	√	
	Device selection	√	√	√
	Drive design	√		√
	Circuit debugging	√	√	√
Academic research	Chip research	√	√	√
	Packaging research	√	√	√
	Application research	√	√	√

system-level research, the requirements for accuracy and repeatability in dynamic process measurements are slightly lower than those for product research. However, compensation for measurement deviations caused by parasitic parameters between voltage measurement points is necessary, especially for measuring crosstalk.

(3) **Engineering Sample Characterization Evaluation**

During device development, it's essential to evaluate the characteristics of each round of engineering samples, and dynamic processes are indispensable for this purpose. The test results will be analyzed alongside the data collected during product research to assess the outcomes of device development. Therefore, the requirements for accuracy and repeatability in measurements during engineering sample characterization are the same as those during product research.

(4) **Preparation of Specifications**

Specifications typically provide engineers with typical values for dynamic processes as a reference. However, dynamic processes are significantly influenced by the test circuit, and there can be substantial differences between the test circuits used in the specifications and those used in actual converter circuits. Therefore, when creating specifications, the requirements for accuracy and repeatability in dynamic process measurements are slightly lower than those for product research and engineering sample characterization. It is not necessary to compensate for measurement deviations caused by parasitic parameters between voltage measurement points. It's important to note that the test circuit used in creating specifications is generally the same as that used in product research and engineering sample characterization.

(5) **Chip Manufacturing**

The main purpose of test device parameters during the manufacturing process is to ensure product consistency and to eliminate defective products with parameters deviating significantly. Therefore, in chip manufacturing, there is a high requirement for the repeatability of measurements in dynamic processes, while the requirement for accuracy is relatively low, as long as the deviation from the actual value is stable.

2. **Packaging Test**

(1) **Packaging Design**

In the design of packaging, especially in power module design, it is necessary to identify the impact of packaging on the dynamic processes of devices. The test results will be used for packaging design assessment and to provide optimization directions. Therefore, in product research, there is a higher requirement for the accuracy and repeatability of measurements in dynamic processes. In particular, compensation for measurement deviations caused by parasitic parameters between voltage measurement points is necessary when measuring crosstalk.

(2) **Packaging Manufacturing**

Similar to chip manufacturing, test device parameters during the packaging manufacturing process is primarily to ensure product consistency and to eliminate defective products with parameters deviating significantly. Therefore, in packaging manufacturing, there is a high requirement for the repeatability of measurements in dynamic processes, while the requirement for accuracy is relatively low, as long as the deviation from the actual value is stable.

3. **System Application**

(1) **Incoming Inspection**

Power electronics companies typically conduct incoming inspections on components provided by suppliers, usually through sampling, to verify whether the device parameters are within the specifications or if the parameters of devices from different batches are stable. Since dynamic processes are heavily influenced by the test circuit, incoming inspection mainly focuses on the stability of devices from different batches. Therefore, during incoming inspection, the requirement for dynamic process measurements is the same as in chip manufacturing and packaging manufacturing, with greater emphasis on test stability.

(2) **Device Selection**

During converter design, engineers often face the challenge of device selection, choosing the most suitable devices from numerous products of various brands. Measurement of dynamic processes allows for the calculation of switch and reverse recovery losses, enabling efficiency calculation, temperature rise estimation, completion of thermal design, and assessment of crosstalk risks. Therefore, during drive design , there is a high requirement for the accuracy and repeatability of dynamic process measurements, and it is necessary to compensate for measurement deviations caused by parasitic parameters between voltage measurement points.

(3) **Drive Design**

The drive circuit is the circuit that works most closely with the device. Test can validate and compare the impact of the drive circuit on the device's dynamic processes. A well-designed drive circuit ensures the full utilization of device capabilities and guarantees the safe operation of the device. Therefore, during drive design, there is a high requirement for the accuracy and repeatability of dynamic process measurements, and it is necessary to compensate for measurement deviations caused by parasitic parameters between voltage measurement points.

(4) **Circuit Debugging**

During converter debugging, test of device dynamic processes is required to confirm whether the circuit design and devices are operating within a safe range. Since it is mainly used for design validation and auxiliary debugging, the requirement for the accuracy and repeatability of dynamic process measurements is slightly lower than during device selection, but it is still necessary to compensate for measurement deviations caused by parasitic parameters between voltage measurement points.

4. **Academic Research**

In academic research related to chips, packaging, and applications, the measurement results of dynamic processes serve as the basis for selecting research directions and validating research results. Therefore, there is a high requirement for the accuracy and repeatability of dynamic process measurements, and it is necessary to compensate for measurement deviations caused by parasitic parameters between voltage measurement points.

Based on the discussion of double-pulse test technology in this chapter and the focus on dynamic process test in different scenarios, the evaluation criteria for dynamic process test results can be summarized as follows:

(1) The stability of test results forms the foundation of their value.
(2) Measurement results should align with theory, and waveform anomalies should be addressed. Any deviations should be investigated to provide reasonable explanations, and based on the test objectives, determine if the deviation is acceptable.
(3) In the production manufacturing process, emphasis is placed on the stability of measurements, allowing for a reduction in the requirement for precision.
(4) Detailed analysis of device characteristics, such as in engineering sample evaluation or academic research, requires compensation for measurement deviations caused by parasitic parameters between voltage measurement points.

5.6 Key Points for Test Equipment Selection

5.6.1 Self-built Laboratory Test Platform

Referring to the introductions in Sects. 5.2 and 5.3, one can construct a double-pulse test system suitable for laboratory test. Building your own test system allows for the utilization of existing oscilloscopes and probes in the laboratory, and it also enables the design of different test boards to achieve more functionalities and meet diverse requirements, offering higher flexibility and lower costs. However, since it requires the design of test boards and data processing programs, engineers and researchers involved need to have experience in circuit design and software programming. With the application of artificial intelligence, this will not become a bottleneck for building one's own testing system.

For self-built double-pulse test platforms for SiC MOSFETs with voltage ratings up to 1200 V, Table 5.6 provides reference instruments, and Fig. 5.91 depicts an example of a self-built test platform.

5.6.2 Laboratory Test Equipment

The double-pulse measurement equipment used in laboratories for power semiconductor devices is mainly designed for IGBT modules. Due to their high-voltage and current ratings and slow switching speeds, these test devices are primarily designed for safety and ease of operation, which results in their instrument selection and test circuit design not being suitable for SiC devices. Additionally, there are very few dynamic characteristic test devices available for individual power transistors, which also do not meet the test requirements for SiC devices. The high switching speed of SiC devices makes dynamic characteristic test a common requirement, leading to an increasing demand for related equipment. Below, we will use Tektronix's Power Device Dynamic Characterization System DPT1000A as an example to introduce the key points for selecting laboratory test equipment (Fig. 5.92).

1. **Testable Items**

As outlined in Sect. 5.2.3, double-pulse test enables evaluation of the switching, reverse recovery, and crosstalk processes of SiC devices. However, commercial equipment that can only perform these three test items may seem overpriced and lack cost-effectiveness. To minimize the cost of test equipment, manufacturers strive to reuse the instruments and test circuits required for double-pulse test, as described in Sect. 5.2.4, and provide power device dynamic characterization systems with integrated additional test functions to engineers.

Tektronix's Power Device Dynamic Characterization System DPT1000A not only completes switching, reverse recovery, and crosstalk tests but also includes tests for

Table 5.6 Self-built double-pulse test platform instrument example

	Instrument	Description
High-voltage power supply	MAGANA XR3000-0.6	
Auxiliary power supply	Tektronix 2220	Multi-channel, overcurrent protection
Oscilloscope	Tektronix MSO 4B / 5B / 6B	12Bit ADC
Signal generator	Tektronix AFG31000 series	Double channel, with double-pulse test app, Improve test efficiency and safety
Voltage probe for measuring VGS	Tektronix TPP1000A	Passive probe, high precision, Low cost, 1 GHz bandwidth
	Tektronix TMDP0200	High-voltage differential probe
Voltage probe for measuring VGS of upper tube	Tektronix TIVP1	Only optical isolation probe can measure correct upper tube VGS waveform
	Tektronix THDP0200	High-voltage differential probe
Voltage probe for measuring VDS	Tektronix TPP0850	Passive probe, high precision, Low cost, 800 MHz bandwidth
	Tektronix THDP0200	High-voltage differential probe
Voltage probe for measuring VDS of upper tube	Tektronix THDP0200	High-voltage differential probe
Measurement scheme for I_{DS} of upper tube	T&M SDN414	Coaxial resistor, high precision, high bandwidth
Measurement scheme for I_{DS} of upper tube	Tektronix TIVP100	High precision, high bandwidth
Measurement for IG	Tektronix TIVP100	
Measurement for IG of upper tube	Tektronix TIVP100	With sufficiently high common-mode voltage and CMRR

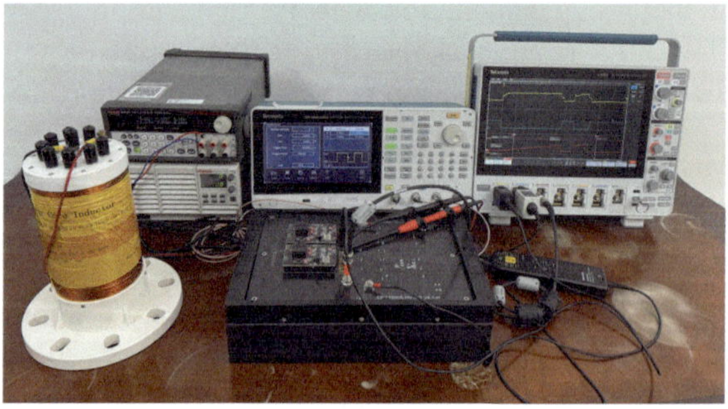

Fig. 5.91 Self-built laboratory test platform

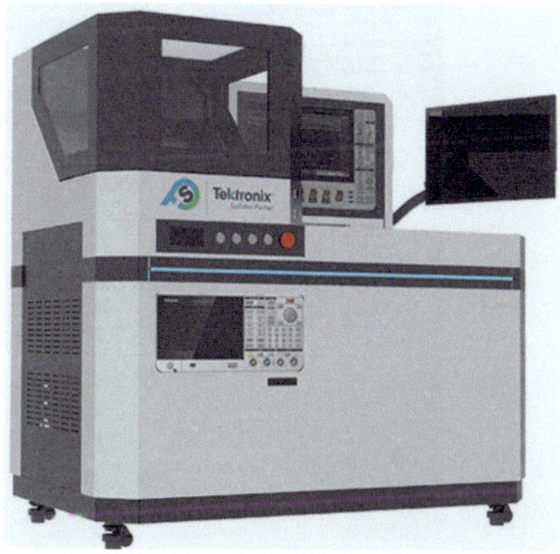

Fig. 5.92 Tektronix power device dynamic characteristics test system DPT1000A

gate charge, avalanche, short circuit, RBSOA, and other test content. The DPT1000A adopts a solution that reuses high-voltage power supplies, auxiliary power supplies, driver boards, load inductors, oscilloscopes, and probes, while providing dedicated test circuit boards for each test item. This approach enables the provision of more test functions while ensuring optimal test results. Additionally, engineers can design their own test boards within the interface rules, enhancing the DPT1000A's scalability and further reducing instrument costs.

2. **Test Scope**

Power device dynamic characterization equipment can cover a range of test aspects including the types of devices under test, the test voltage, and the test current.

(1) **Types of Devices Under Test**

If a test device can only test SiC devices, it often fails to meet the requirements. This is because a single device manufacturer typically produces multiple types of devices, and a single power electronics company may use various types of devices. It is not practical to purchase separate test equipment for each type of device.

Different types of devices have different drive voltages, especially since SiC MOSFETs are still maturing, and the drive voltages of SiC MOSFETs from different manufacturers vary. This requires test equipment to provide adjustable drive voltages that meet the requirements of different types of devices. Generally, $V_{DRV(on)}$ and $V_{DRV(off)}$ need to be adjustable within the ranges of 5 V to 20 V and −10 V to 0 V, respectively. The DPT1000A can provide adjustable ranges for $V_{DRV(on)}$ and $V_{DRV(off)}$ of 4.5 V to 20 V and 0 V to − 20 V, respectively.

Furthermore, Si MOSFETs include both n channel MOSFETs and p-channel MOSFETs, with opposite voltage endurance directions and drive voltage polarities. The DPT1000A specifically provides dedicated double-pulse test boards, avalanche test boards, and driver boards for p-channel MOSFETs. Additionally, the DPT1000A

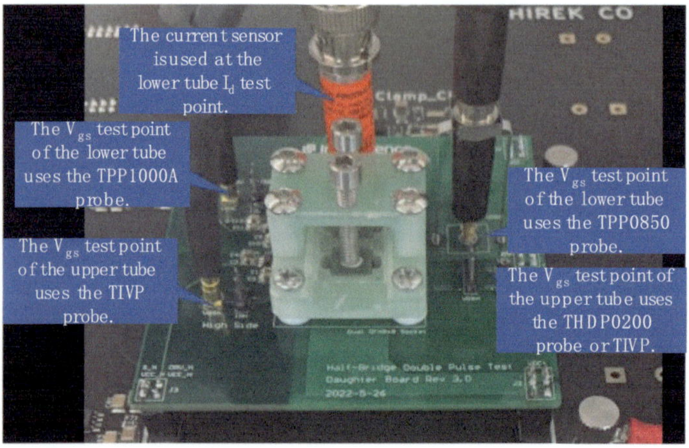

Fig. 5.93 DPT1000A GaN test solution

offers specialized test boards and drive circuits for GaN HEMTs, as shown in Fig. 5.93.

(2) **Test Voltage**

The technical specifications of general test equipment usually only provide the maximum test voltage, which does not necessarily mean it can meet all test requirements below this maximum voltage.

Different types of devices have significant differences in voltage ranges. For example, SGT MOSFETs typically operate below 200 V, while the minimum voltage rating for SiC devices is 400 V. Even within the same type of device, voltage ranges can vary widely. For instance, SiC devices can range from a minimum voltage rating of 400 V to a maximum of 10 kV. Meeting a broader voltage range poses the following challenges for equipment.

As indicated by (5.7), during double-pulse test, when the test current I_{set} and the load inductance L are the same, the first pulse width τ_1 required for low-voltage devices will be much greater than that for high-voltage devices. However, an excessively long τ_1 can cause serious device heating, thereby affecting the device's dynamic characteristics. Choosing a smaller load inductance can effectively shorten the length of τ_1, which requires the equipment to provide a variety of inductances to meet the needs of devices with different voltage levels. The DPT1000A can provide various inductance values according to test requirements and can also be customized as needed. It is important to note that the inductance value should not be reduced blindly, and compliance with the requirements of (5.9) still needs to be observed.

As stated in (5.11), during double-pulse test, when the test current I_{set} and the inductance L are the same, the required bus capacitance C_{Bus} value increases as the test voltage V_{set} decreases. Although choosing a smaller L can reduce the requirement for C_{Bus}, the impact of V_{set} is significantly greater due to its quadratic influence. Furthermore, higher voltage levels are associated with lower capacitance density

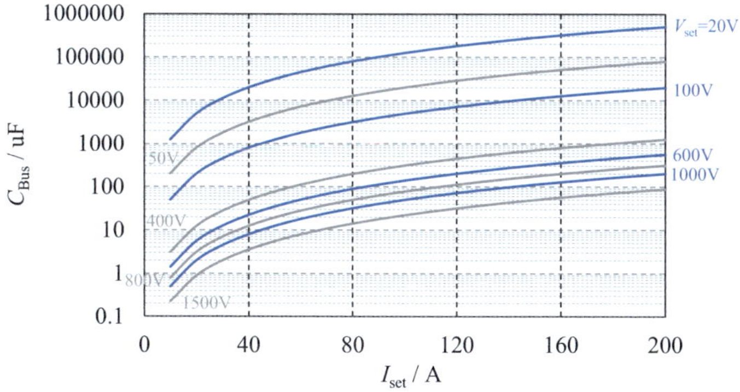

Fig. 5.94 $L = 100\,\mu H$, $K_v = 0.01$, required C_{Bus} capacitance under different V_{set}

and higher unit prices for capacitors. If the capacitance C_{Bus} is chosen based on the highest voltage level device, the required number and cost of capacitors to meet the test requirements for lower voltage levels will be substantial. It's worth noting that this also applies to short-circuit test. According to (5.11), during double-pulse test, when $L = 100\,\mu H$, $K_v = 0.01$, and I_{set} ranges from 10 to 200A, the required C_{Bus} values under different V_{set} conditions are illustrated in Fig. 5.94.

To address this issue, the DPT1000A offers different voltage-level test boards for double-pulse test, including 150 V low-voltage boards, 1000 V medium-voltage boards, and 2000 V high-voltage boards. The capacitance values of C_{Bus} on these test boards are 12mF, 190 μF, and 47.5 μF, respectively. By providing these test boards with different voltage levels, the DPT1000A not only covers a wider range of voltage levels but also minimizes the cost and reduces the size of C_{Bus} to the maximum extent possible.

Power device voltage levels range from tens of volts to several kilovolts. To achieve optimal measurement accuracy, it's essential to choose voltage probes with suitable ranges. When V_{set} is below 200 V, a 10:1 passive probe can be selected. For higher V_{set}, high-voltage passive probes or high-voltage differential probes can be chosen. The DPT1000A can provide appropriate voltage probes according to requirements, helping engineers achieve the best measurement results.

As the maximum voltage of test equipment continues to rise, the costs associated with high-voltage power supplies, bus capacitors, and safety regulations also increase sharply. When selecting test equipment, it's important not to blindly pursue higher test voltage capabilities; it's sufficient to meet the requirements. Additionally, once V_{set} exceeds 1500 V, it's no longer appropriate to integrate test for low-voltage devices within the same equipment setup. Instead, specialized equipment designed for high-voltage devices should be chosen.

(3) Test Current

In general, the test current provided in the technical specifications of test equipment refers to the maximum current, but it does not offer a specific definition, nor does it imply that the maximum current can be reached under all operating conditions.

As indicated by (5.11), during double-pulse test, I_{set} is influenced by V_{set}, L, K_v, and C_{Bus}, while L is also constrained by (5.9). This implies that the lower the Vset, the smaller the maximum I_{set} that can be achieved. Therefore, when evaluating equipment specifications for double-pulse test, it's necessary to calculate the I_{set} range based on the equipment's C_{Bus}, given a clear understanding of V_{set}, L, and K_v. Taking the DPT1000A medium-voltage test board as an example, with $C_{Bus} = 50$ μF, $K_v = 0.01$, L from 20 μH to 1 mH, the I_{set} range corresponding to V_{set} from 400 to 1000 V is shown in Fig. 5.95.

Similarly, relying solely on the test current provided in the technical specifications of test equipment is insufficient for determining the current test range for avalanche and short-circuit tests. Similar calculations as described above are required for clarification.

3. Supported Package Types

Whether discrete devices or power modules, there are a wide variety of package types available. The more package types a dynamic test equipment can support, the greater its value to engineers.

The DPT1000A adopts a shared test board and provides dedicated adapters for devices with different packages, as shown in Fig. 5.96. This approach not only meets the requirements of various package types but also helps save costs as much as possible. Additionally, engineers can design their own adapters.

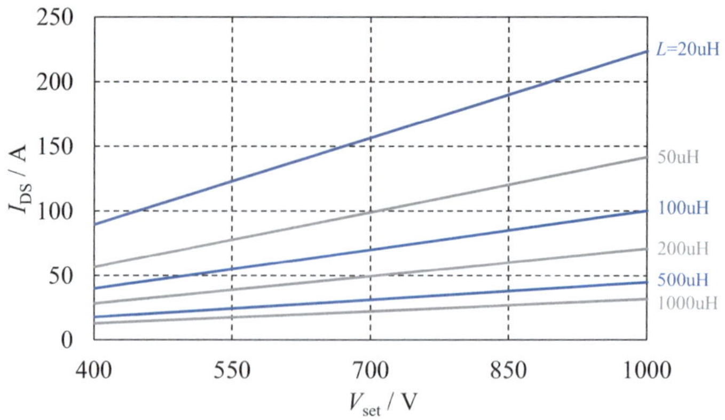

Fig. 5.95 DPT1000A medium voltage test board I_{set} Range ($C_{Bus} = 190$ μF, $L = 100$ μH, $K_v = 0.01$)

Fig. 5.96 Adapter board for different packaged devices

4. **Measuring Instrument Indicators**

Dynamic characteristic test equipment typically mentions measurement accuracy in its technical specifications. Engineers generally assume that this refers to the accuracy of dynamic characteristic test results (such as switching time, switching energy, reverse recovery time, etc.). However, this is a misunderstanding. The measurement accuracy mentioned in the technical specifications refers to the measurement system, namely the accuracy of the probe and oscilloscope, and it does not necessarily equate to the accuracy of dynamic characteristic test results.

According to Sect. 5.3, when examining the measurement capabilities of dynamic characteristics test equipment, the frequency response type, bandwidth, ADC bit number, sampling rate and voltage of the oscilloscope, the type, bandwidth, measurement range, and connection method of the voltage probe should be considered. Tests conducted in the laboratory usually require the highest possible accuracy. For the selection of oscilloscopes and probes, refer to Sect. 5.3. Do not select test equipment that is obviously inappropriate for the configured oscilloscope and probe, for example: the oscilloscope sampling rate and bandwidth are too low, the V_{GS} of the upper bridge arm SiC MOSFET is measured with a high-voltage differential probe, the I_{DS} of the SiC MOSFET is measured with a Rogowski coil, the voltage level of the SGT MOSFET with a voltage probe of 1500 V is measured with a voltage probe of 80 V, and the voltage measurement point is on the test board and includes a section of the main power or drive PCB circuit.

In order to optimize test performance and meet the high-speed voltage and current signal test requirements of SiC devices, the DPT1000A has chosen superior measurement instruments and probe connection methods to ensure accurate waveform acquisition to the greatest extent possible. The gate-source of the lower transistor is connected to the test using a TPP1000A 1 GHz single-ended probe, while the gate-source signal of the upper transistor is tested using a TIVP optical isolation probe, both connected to the test circuit via MMCX interfaces. Current test utilizes a 2 GHz bandwidth current sensor provided by T&M company, while the source-drain voltage is tested using a TPP0850 800 MHz high-voltage single-ended probe, ensuring that the system bandwidth meets test requirements.

5. **Correctness of the Tested Waveform**

Satisfying the requirements of measurement instruments ensures the correct measurement of waveforms, but this does not guarantee obtaining accurate test results. This is because, as described in Sect. 5.2.4.1, the meaningfulness of measurement results relies on the device being tested in the correct state and the dynamic waveform conforming to theory or being explainable by theory. This necessitates that the test board, under the premise of the device's normal characteristics and accurate measurements, should not exhibit abnormal waveforms in the vast majority of test conditions, such as false turn-on, false turn-off, voltage spikes exceeding device limits, and significant discrepancies between switching waveforms and theory.

Typical erroneous test waveforms, as shown in Fig. 5.97, yield results of little value. Therefore, when selecting dynamic characteristic test equipment, one should not only consider dynamic characteristic indicators calculated based on measured waveforms but also carefully analyze the waveform morphology. This applies equally to gate charge test, avalanche test, and short-circuit test. To ensure the correctness of test waveforms, the DPT1000A optimizes circuit parasitic parameters in its test board design.

6. **Convenience of Operation**

Compared to self-built laboratory test platforms, convenience of operation is one of the advantages of the entire test equipment. In order to improve test efficiency and reduce the burden on engineers, the DPT1000A has made significant efforts in both hardware and software aspects.

(1) **Mother-Daughter Board Solution**

The DPT1000A adopts a daughter board scheme for the test circuit and driver board, with the driver board plugged into the test circuit. This allows for quick replacement of the driver board, while also saving on maintenance costs and reducing downtime in case of driver board failure.

(2) **Test Board Replacement**

The DPT1000A provides dedicated test boards for different test items and voltage levels of devices. All test boards have a standardized interface, allowing for quick installation and removal from the test bench.

(3) **Gate Resistance Adjustment**

In laboratory test, it's common to test the characteristics of devices under different drive resistances. The DPT1000A adopts a resistor socket approach, allowing engineers to quickly replace the drive resistors without the hassle of soldering. This method is more flexible and cost-effective compared to relay switching schemes.

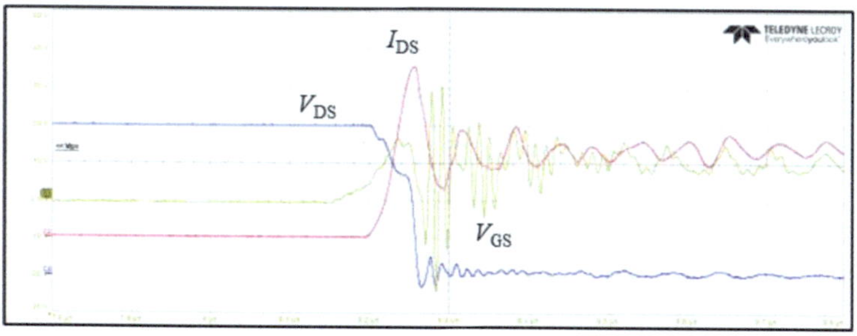

(a) The V_{GS} oscillation during the turn on process is severe

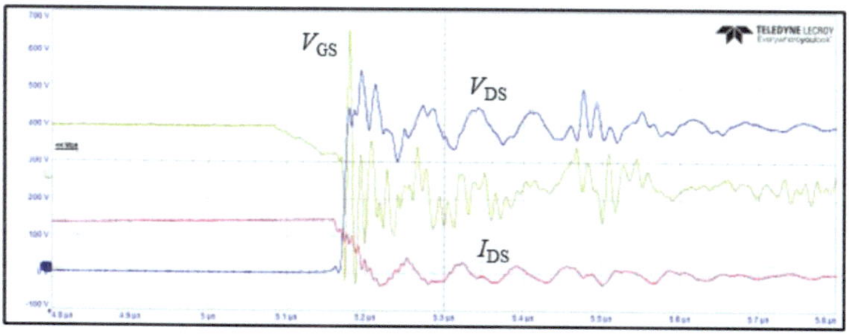

(b) The oscillation waveform of V_{DS} during the shutdown process is abnormal

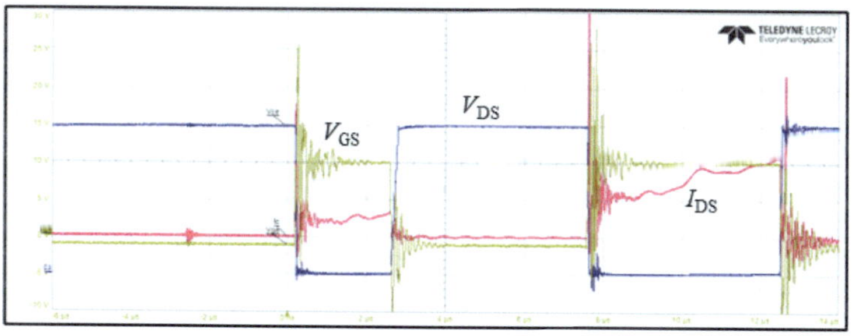

(c) The waveform of I_{DS} during the switching process is abnormal

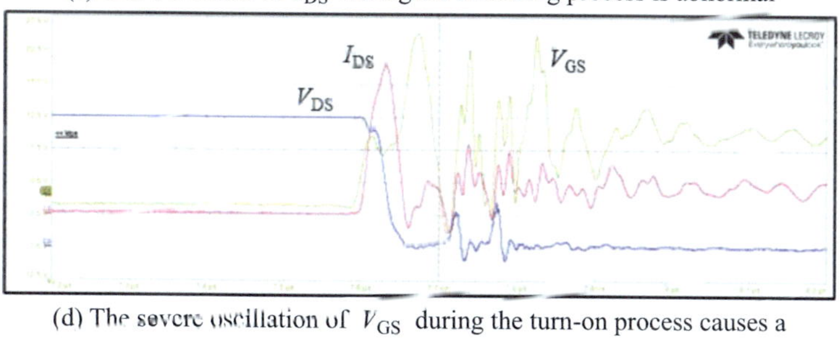

(d) The severe oscillation of V_{GS} during the turn-on process causes a secondary turn-off

Fig. 5.97 Typical wrong test waveform

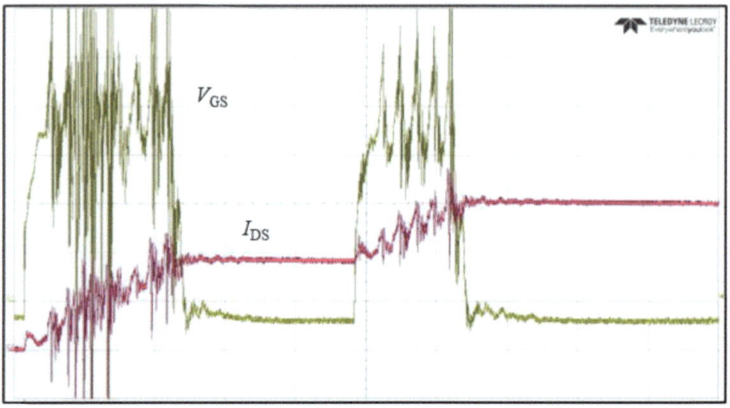

(e) Abnormal oscillation caused by interference of V_{GS} during the switching process

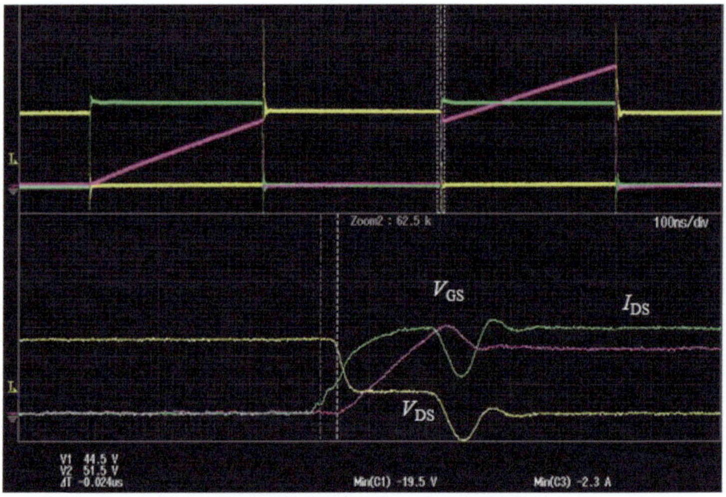

(f) The parasitic parameters between the measurement points cause the waveforms ofV_{GS} and V_{DS} to be abnormal.

Fig. 5.97 (continued)

(4) **Driving Voltage Adjustment**

Different types and brands of devices have varying drive voltages. The DPT1000A communicates with the drive board via a host computer to automatically set the drive voltage.

(5) **Inductor Replacement**

Switching, reverse recovery, and avalanche tests typically require different load inductances. The DPT1000A provides multiple reserved slots on the panel, allowing engineers to easily replace the inductance by simple wiring.

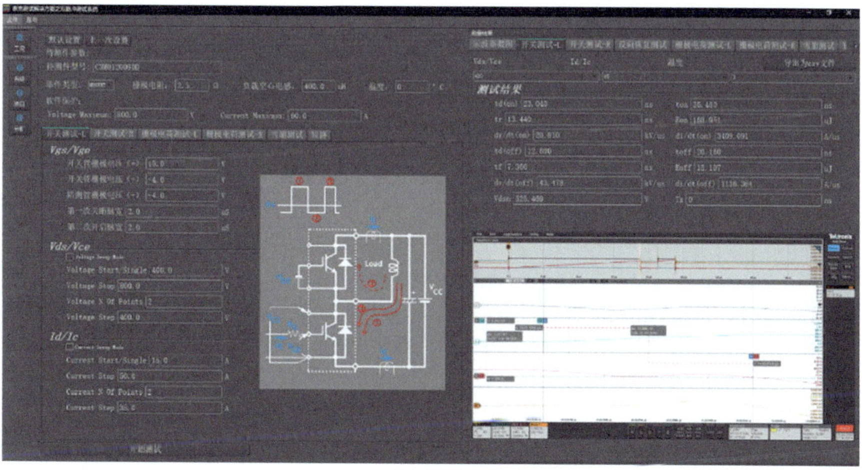

Fig. 5.98 DPT1000A operation interface

(6) Equipment Control

During test, parameters such as test voltage, test current, drive voltage, and test pulse width need to be specified, which can be set through the DPT1000A's user interface. Additionally, raw measurement data, test waveforms, and test result reports are provided to the user (Fig. 5.98).

7. Data Processing Function

Due to the lack of unified test standards for SiC devices, the software of DPT1000A is specifically designed with the capability to set measurement conditions according to different test standards. It can automatically determine test results and process data based on these settings. The software of DPT1000A offers high flexibility, enabling engineers to freely choose between IEC standards, industry standards, company standards, and custom standards. Engineers can set different sampling positions and criteria to meet the requirements of various test users, thus achieving parameter test under customized conditions (Fig. 5.99).

5.6.2.1 Power Module Test Equipment

With the development of SiC power devices in electric vehicles and new energy fields, SiC chips are packaged into various types of modules to meet the demands of high-power, high-power-density application scenarios. Compared to discrete devices, SiC power modules have higher current ratings and larger short-circuit currents, which can reach 8~12 kA. Therefore, SiC module dynamic characterization test equipment, while meeting requirements for test bandwidth and accuracy, also needs to have higher voltage and larger current testing capabilities.

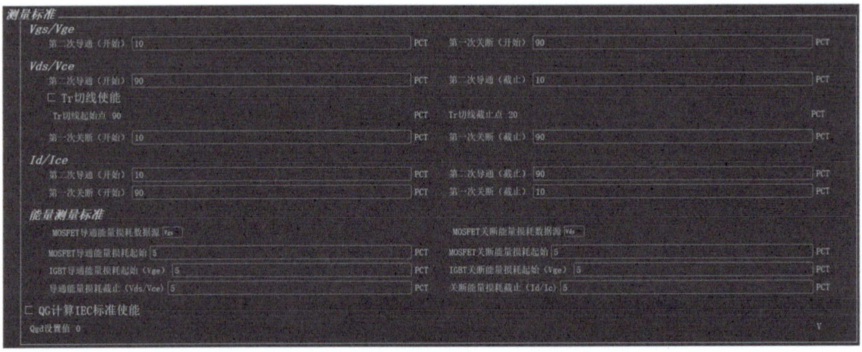

Fig. 5.99 DPT1000A data processing settings interface

Taking Hangzhou Firstack Technology Co., Ltd.'s power device dynamic characterization test system ME400D as an example, the technology and key selection points for laboratory power module dynamic test equipment are introduced below, as shown in Fig. 5.100.

1. **System Structure**

Laboratory power module dynamic test equipment primarily consists of two core systems: the hardware system and the software system. Figure 5.101 shows the system structure of the ME400D. The hardware system provides the hardware environment for testing, mainly composed of a drive unit, bus capacitor unit, charge/discharge unit, snubber capacitor unit, load inductor unit, and data acquisition unit

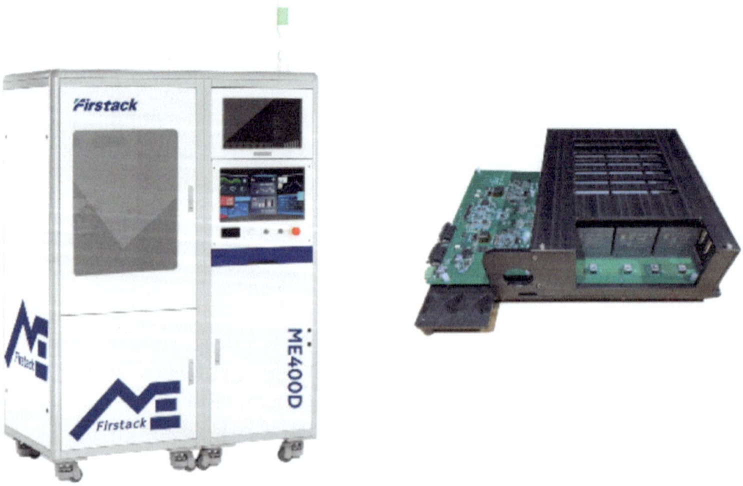

Fig. 5.100 Firstack Technology's power device dynamic characterization test system ME400D

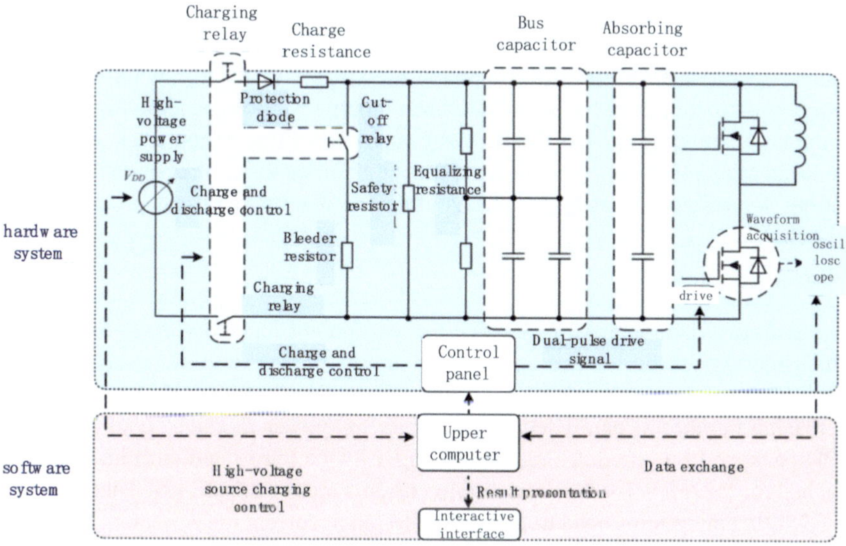

Fig. 5.101 System structure of the ME400D test system

(oscilloscope and voltage/current sensors). The software system mainly consists of the equipment's internal process and logic control unit, data analysis unit, and human-machine interface unit.

Compared to the self-built manual test platform introduced in Sect. 5.6.1, the characteristic of a complete test system is its automated test procedure editing function, capable of automatically completing tests as well as data analysis and processing. This not only saves the human resources required for testing but also reduces the risk of poor test result consistency caused by manual operations.

2. **Test Items**

Laboratory power module dynamic test equipment primarily performs turn-on, turn-off, reverse recovery, and short circuit characteristic tests to evaluate the characteristics of the device under test and verify datasheet specifications. Simultaneously, as laboratory-grade test equipment, it is required to evaluate the power system in which the device operates; therefore, Safe Operating Area characteristics such as RBSOA and SCSOA are also important test items. Benefiting from its intelligent and automated software system, the ME400D test system can automatically extract and process data for multiple operating points of the SOA, efficiently completing SOA testing for both the device and its power system.

3. **Test Circuit Switching**

As SiC power modules gradually penetrate various application fields, their topology types have expanded from simple half-bridge two-level topologies to three-level clamped topologies and flying capacitor topologies. Testing a power module may now involve multiple test circuits and measurement points, requiring applying excitation only to the circuit being tested. To improve test efficiency and reduce the risk of manual operation errors, the test equipment needs to automatically switch test circuits and measurement points based on the power module's topology, applying excitation to the corresponding device. While test circuits for different topologies vary, they are often relatively fixed. Therefore, test equipment generally has built-in topology options for testers to easily select, avoiding the need to edit test programs from scratch.

4. **Test Range**

The main test parameters for dynamic test equipment include voltage and current, which must meet the voltage and current ratings of the SiC power module under test. SiC power module voltage ratings are generally 1200 V and above. With the increasing number of paralleled chips within SiC power modules, commonly used HPD packaged SiC modules require testing at twice their rated current, which has reached the 2 kA level, with short-circuit currents up to 8 kA–12 kA. Therefore, SiC test equipment systems need to possess wide-range current measurement capability. In test equipment, multiple sets of current sensors are configured for single-pulse, double-pulse, and short-circuit conditions to achieve wide-range, high-precision current measurement.

The ME400D test system can provide different voltage and current range configurations as needed to meet the testing requirements of power modules of various specifications. Its maximum voltage capability is 6 kV, and maximum current capability is 12 kA.

5. **High-Temperature and Low-Temperature Testing**

Automotive-grade SiC power modules often require more comprehensive and stringent testing, including not only dynamic characterization tests at room temperature (around 25 °C) but also tests at high temperature (around 200 °C) and low temperature (around −50 °C). Therefore, the test system is equipped with high-temperature and low-temperature platforms to control the device under test SiC power module within the required temperature range.

Figure 5.102 shows the high-low temperature test system ME400D-ATC configured for the ME400D test system. It is a purely mechanical refrigeration device capable of stabilizing the temperature control plate at the required temperature within a closed, thermally isolated environment. Its system includes a main unit, flexible hoses, and a temperature control plate. The ME400D-ATC adopts a contact-based temperature control scheme, achieving test efficiency more than twice that of traditional temperature chamber methods. When the power module reaches the specified temperature, the test ambient temperature is maintained within 0 °C to 85 °C, eliminating the impact of high/low-temperature environments on measurement instruments and probes found in traditional chambers. Simultaneously, by injecting low-dew-point air into the relatively sealed cavity, condensation during low-temperature testing is prevented. The ME400D-ATC's temperature control plate is highly efficient and flexible, allowing customization of plate size to adapt to different power

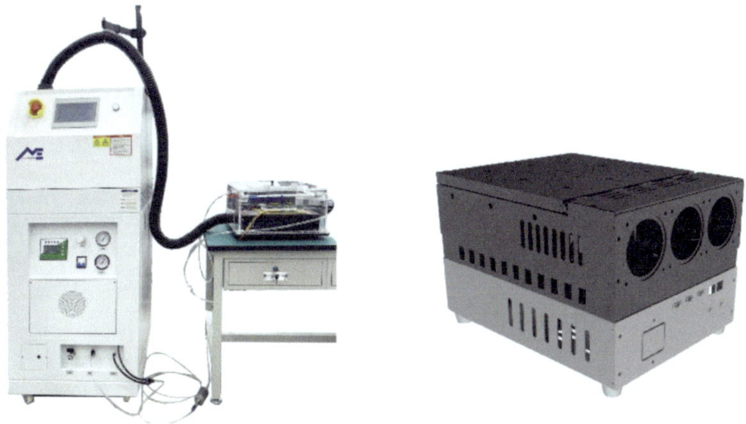

Fig. 5.102 High-low temperature test system ME400D-ATC

module dimensions and interfaces. Furthermore, temperature settings, historical data record viewing, and temperature curves for the control plate and device NTC can be managed via a high-definition touch screen or remote communication interface.

6. **Test Fixture**

The test fixture serves as the bridge connecting the device under test module and the test equipment. Its main functions include: receiving control signals from the host computer, providing drive signals to the device, offering protection during module test failures, and ensuring safe isolation of power and signals.

Figure 5.103 shows the platform-based test fixture configured for the ME400D test system, consisting of a Drive Core and an Adapter Board. The Drive Core receives drive signals from the host computer, provides adjustable drive voltages, and includes protection functions (Miller clamp, short-circuit protection, staged turn-off, etc.). The Adapter Board contains drive resistors and capacitors, protection functions, part of the drive circuit, and provides measurement points for voltage and current probes.

The drive voltage range provided by the test fixture needs to cover all Si and SiC devices: positive voltage 10~30 V, negative voltage -20~-1 V, pulse width 0.2~1000 μs, sufficient for normal double pulse testing and short pulse testing. The Drive Core can be adapted to various module packages; one core drives one switch position. Testing different packages only requires changing the Adapter Board. This provides high versatility, compatibility with different device types and packages, while effectively saving costs. Additionally, the Adapter Board connects directly to the power module, resulting in shorter drive loops and smaller parasitic parameters, leading to more accurate test results. Furthermore, the test fixture is equipped with a wireless module, eliminating signal harnesses, enabling higher voltage isolation, and supporting online configuration of gate voltage and enabling/disabling protection functions.

Fig. 5.103 Platform-bbased test fixture for the ME400D test system

7. **Drive Parameter Matching**

To comprehensively evaluate SiC power module characteristics, testing is often required under different drive resistors and capacitors, necessitating frequent resistor/capacitor changes. Manual replacement is time-consuming, labor-intensive, and prone to errors. To address this, FlySight Technology developed a Drive Parameter Matching Module capable of automatically switching drive resistors and capacitors, as shown in Fig. 5.104.

The Drive Parameter Matching Module contains a series of parallel resistors and capacitors internally. Receiving control signals allows automatic selection, enabling automatic switching of the turn-on resistor, turn-off resistor, and gate capacitor. Concurrently, through a laminated loop design, the module keeps the drive loop

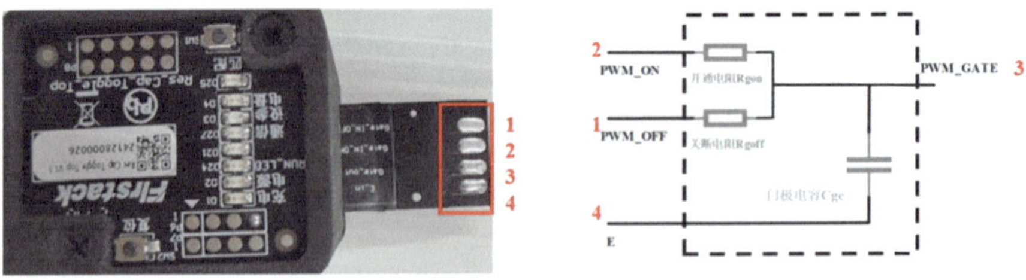

Fig. 5.104 Platform-bbased test fixture for the ME400D test system

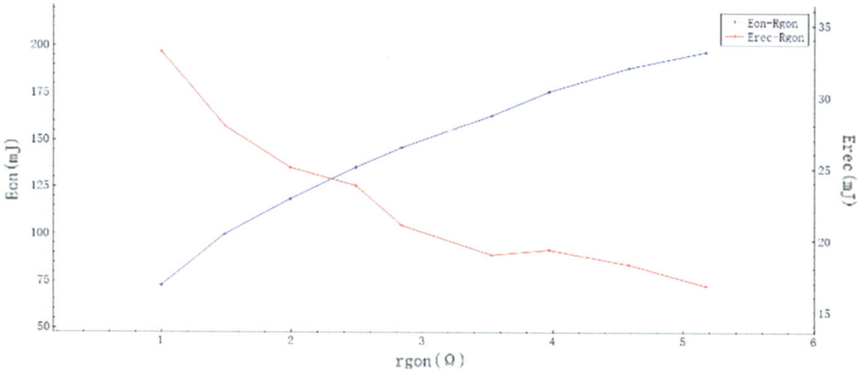

Fig. 5.105 $E_{on}/E_{off} - R_{G\,(out)}$ multi-operating-point curve

inductance below 10 nH. Utilizing gate power harvesting and wireless communication, it eliminates additional wiring harnesses except for the gate connection itself.

8. **Data Analysis and Processing**

After completing dynamic tests, waveform analysis and processing are critical tasks for dynamic test equipment, accelerating analysis speed and avoiding instability from manual readings. The ME400D test system supports functions like zero-drift compensation, waveform overlay, data re-calculation, etc.

High-voltage differential probes and opto-isolated probes can exhibit zero drift, severely impacting test result accuracy. The zero-drift compensation function allows obtaining accurate test results even when probes drift. Secondly, to visualize test results, after a series of automated tests on a power module, the ME400D test system can automatically generate graphs and waveform overlays, facilitating testers in judging result correctness and conducting further analysis. Simultaneously, dynamic characteristics are not only represented by test waveforms but also require calculation of key dynamic parameters from the waveforms according to relevant definitions, including switching times, switching delays, switching energies, etc. The ME400D test system supports custom parameter definitions, enabling rapid parameter calculation. Figures 5.105 and 5.106 show the multi-operating-point curve and the waveform overlay comparison graph automatically generated by the ME400D test system, respectively.

5.6.3 *Production Line Test Equipment*

During the production process of SiC power devices, to ensure parameter consistency and reliability, each product needs to undergo dynamic characterization testing. Test items generally include: single-pulse test, double-pulse test, multi-pulse test,

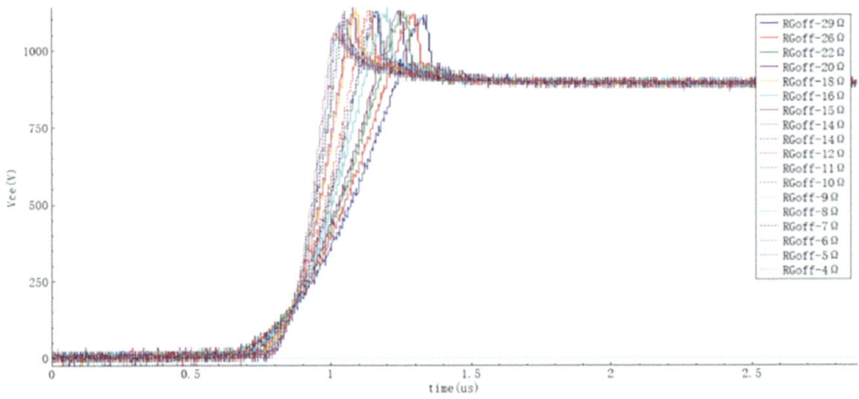

Fig. 5.106 $V_{DS}/R_{Goff\,(out)}$ waveform overlay comparison graph

and short-circuit test. Key dynamic characteristic parameters to monitor include: switching times and energies, reverse recovery time and energy, gate charge, di/dt, dv/dt, etc. Figure 5.107 shows Firstack Technology's production line power device dynamic test system ME100D-AM.

1. **Test Sequence**

In production line dynamic testing, a "pre-test" is typically performed first to ensure device and equipment safety and testing efficiency. This is necessary to verify that the device's contact status, on/off state, leakage current indicators, threshold voltage, etc., are all normal before proceeding with dynamic characterization tests. Furthermore, after completing dynamic characterization tests, static parameter tests are usually performed again to ensure that the device's basic static parameters have not changed or shifted. Therefore, during device production, the test flow usually follows a "static test—dynamic test—static test" sequence.

Fig. 5.107 Firstack technology's power device dynamic characterization test system ME100D-AM

2. **Equipment Reliability**

Production line test equipment bears production tasks. During high-volume production testing, device failures are inevitable. If such failures cause equipment downtime, significant losses can occur. For mass production test equipment, the Mean Time To Repair (MTTR) is typically required to be less than 30 minutes; its Mean Time Between Failure (MTBF) is typically required to be greater than 168 hours.

Dynamic testing of SiC power devices involves high voltage and high current. Coupled with the relative immaturity of SiC MOSFET devices, the possibility of failure during dynamic characterization and short-circuit tests is higher. When a device fails, it often presents as a short circuit, causing the current in the loop to rise rapidly to extremely high values, potentially leading to device explosion, posing threats to personnel and equipment safety. This demands that production line dynamic test equipment has "self-protection capability," enabling it to "brake" promptly to avoid device explosion and prevent the test equipment itself from malfunctioning or being damaged. Therefore, SiC device production line dynamic test equipment is equipped with protection units capable of cutting off the faulty circuit within 2 μs upon device failure.

Equipment reliability can be verified by continuously subjecting dozens, or even hundreds, of devices to short-circuit or overcurrent conditions until they fail. If the handler, test circuit, or drive circuit faults or gets damaged during this process, targeted reinforcement design is necessary.

3. **Test Efficiency**

Compared to laboratory dynamic test equipment, the test efficiency of production line dynamic test equipment directly impacts device production efficiency, i.e., the unit device test cost, making it particularly crucial. Efficiency is typically characterized by UPH (Units Per Hour). Currently, the industry generally requires a UPH of 300–360 for production dynamic testing of half-bridge topology modules, and a UPH of 100–120 for testing three-phase full-bridge topology modules.

4. **Loop Inductance**

Production line testing generally employs automated loading/unloading devices. This can result in potentially long connection distances between the DUT and the test module, leading to excessive main power loop inductance and drive loop inductance. This ultimately causes severe waveform oscillation and high voltage spikes during testing.

Firstack Technology has developed solutions to better connect the device under test, significantly reducing connection inductance through methods like laminated busbars. For discrete device test circuits, the main power loop inductance has been reduced to the range of 50~60 nH (including the device under test circuit). For power module test circuits, the main power loop inductance has been reduced to below 15 nH.

Currently, test equipment manufacturers continue striving to further reduce main power loop inductance, but the necessity of this requires re-evaluation. This is because during device mass production, leakage current is tested by applying a voltage exceeding the device's nominal rating for tens of milliseconds. In contrast, the duration of the turn-off voltage spike during dynamic switching is only tens of nanoseconds. Therefore, controlling the turn-off voltage spike during dynamic testing to a very small range holds little special value.

One possible reason for this focus is that applications using SiC devices require minimizing main power loop inductance to ensure turn-off voltage spikes under all operating conditions are below the device's nominal voltage rating. This requirement is sometimes mistakenly assumed by engineers to also apply to production test equipment. However, the purposes and requirements of application and production testing differ. The goal of production testing is to apply stress to the device to screen out defective units from the production process. Therefore, when selecting production test equipment, the numerical value of the main power loop inductance should not be the sole criterion. Instead, the criterion should be that the turn-off voltage spike during testing does not exceed the voltage applied during leakage current testing, while also monitoring stress parameters like di/dt and dv/dt on the device. This approach helps screen out abnormal devices while reducing the technical difficulty faced by the equipment, thereby lowering equipment costs.

It is important to note that the turn-off voltage spike is influenced by switching speed; faster turn-off speeds result in higher voltage spikes. Simultaneously, many abnormal devices can only be screened out at high switching speeds. Therefore, the judgment of whether the turn-off voltage spike meets requirements needs to be made at high switching speeds. However, when the switching speed is too fast, severe oscillations will occur, which may lead to the failure of the device's gate. At this time, reducing the inductance of the drive loop will effectively alleviate the oscillations.

In summary, current production test equipment should focus primarily on drive loop inductance, as the main power loop inductance is already sufficiently low.

5. **Device Interface**

During production testing, the production line dynamic test equipment needs to connect to automated handlers. This typically requires test boards to carry probe fixtures or test socket components as intermediate interfacing mechanisms. Figure 5.108 shows Firstack Technology's automated interface mechanism for production dynamic test equipment, including static and dynamic test boards and a test socket. Figure 5.109 illustrates Firstack Technology's production dynamic test board + test socket solution. The device under test connects electrically to the tester via the test socket. To meet the requirement of an average repair time of less than half an hour for production line equipment, the modular design of the test board + test socket solution ensures rapid component replacement and equipment maintenance in case of part failure.

6. **Test Data Requirements**

For production line testing, the consistency and statistical patterns of test data are particularly important for production. When the same tester repeats a test 100 times

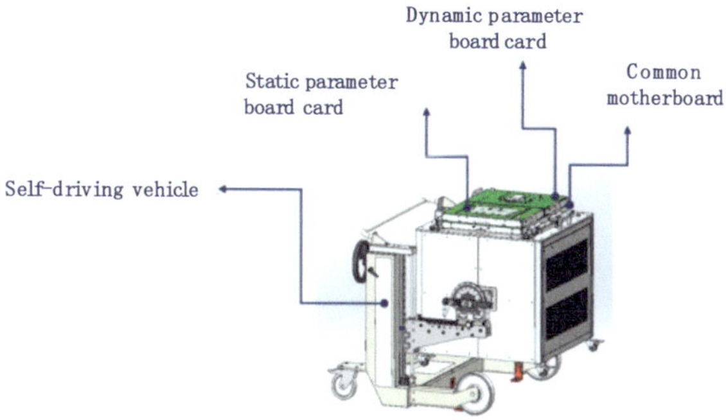

Fig. 5.108 Firstack technology's automated interface mechanism for production dynamic test equipment

Fig. 5.109 Firstack technology's production dynamic test board + test socket solution

under identical conditions, the variation in dynamic parameters must be $\leq 5\%$. When multiple testers of the same model repeat a test 100 times under identical test conditions, the total variation in dynamic parameters must be $\leq 10\%$. Simultaneously, the test equipment needs to support statistical and screening methods such as Part Average Testing (PAT) and Dynamic Part Average Testing (DPAT) to remove devices under test exhibiting abnormal characteristics (outliers).

7. **Correctness of Test Waveforms**

As emphasized in previous sections, the correctness and stability of waveforms are the foundation for obtaining valuable test results. Currently, especially in the production line testing phase, engineers still often do not pay sufficient attention to this issue, a situation more severe than in laboratory testing. This could lead to scenarios on the production line where the act of testing is completed, but whether the test meets requirements and whether the results reflect testing value remains highly questionable.

References

1. Cree, Inc. KIT8020-CRD-5FF0917P-2 (2018) Evaluation board for Cree's SiC MOSFET in a TO-247-4 package. Application Note, CPWR-AN20
2. Cree, Inc. KIT-CRD-CIL12N-XM3 (2019) Wolfspeed's XM3 half-bridge module dynamic evaluation board user guide. CPWR-AN31 Revision C
3. Littelfuse, Inc (2018) Dynamic characterization platform. Application Note
4. Littelfuse, Inc (2018) Gate drive evaluation platform. Application Note
5. Infineon Technologies AG (2017) Evaluation board EVAL-1EDI20H12AH-SIC. AN2017-14, Revision 1.0
6. Infineon Technologies AG (2017) Evaluation board for CoolSiC™ Easy1B half-bridge modules. AN 2017-41, Revision 1.0
7. ROHM Co., Ltd (2019) TO-247-4L Half-bridge evaluation board product specification. User's guide, No. 62UG025C, Revision 001
8. XP Power, Inc (2018) FJ Series 120W regulated high voltage DC power supplies. Datasheet
9. XP Power, Inc (2018) FR series 300W regulated high voltage DC power supplies. Datasheet
10. Magna-Power Electronics, Inc (2019) XR series 2U programmable DC power supply. Datasheet
11. Magna-Power Electronics, Inc (2019) SL series 1U programmable DC power supply. Datasheet
12. AMETEK Inc (2019) XG 850 series 850 W, 1U half rack programmable DC power supplies. Datasheet
13. Heinzinger electronic GmbH (2018) EVO series high-voltage power supplies. Datasheet
14. Keithley Instruments, Inc (2013) 2220, 2220G, 2230, 2230G multi-channel programmable DC power supplies. Data Sheet
15. Tektronix, Inc (2018) AFG31000 series arbitrary function generators. Data Sheet, 75W-61444-2
16. Tektronix, Inc (2019) Double pulse test with the Tektronix AFG31000 arbitrary function generator. Application Note, 75W-61623-0
17. Keysight Technologies, Inc (2019) Infiniium S-series. Datasheet, 5991-3904EN
18. Keysight Technologies, Inc (2019) Evaluating oscilloscope signal integrity. Application Note, 5991-4088EN
19. Keysight Technologies, Inc (2017) Evaluating oscilloscope sample rates versus sampling fidelity. Application Note, 5989-5732EN
20. Tektronix, Inc (2023) 5 series B MSO mixed signal oscilloscope datasheet. Datasheet, 48W-73851-4
21. Tektronix, Inc (2023) 6 series B MSO mixed signal oscilloscope datasheet. Datasheet, 48W-61716-10
22. Tektronix, Inc (2023) Wide bandgap—double pulse test analysis. Application Datasheet, 61W-73888-03
23. Keysight Technologies, Inc (2020) Infiniium oscilloscope probes and accessories. Datasheet, 5968-7141EN
24. Tektronix, Inc (2020) Probe selection guide. 60W1423210
25. Teledyne LeCroy, Inc (2020) Oscilloscope probes and probe accessories. Probe Catalog
26. Johnson CM, Palmer PR (1994) Current measurement using compensated coaxial shunts. IEE Proc Sci Meas Technol 141(6):471–480
27. Tektronix, Inc (2023) High-voltage differential probes TMDP0200-THDP0200-THDP0100-P5200A-P5202A-P5205A-P5210A. Datasheet, 51W-11195-15

28. Tektronix, Inc (2022) Passive voltage probes TPP1000·TPP0500B·TPP0502·TPP0250 Datasheet. Datasheet, 51W-26151-10
29. Tektronix, Inc (2020) Isolated measurement systems TIVP1, TIVP05, TIVP02 datasheet. Datasheet, 51W-61655-0
30. Tektronix, Inc (2019) Complete isolation extreme common mode rejection. Whitepaper, 51W-60485-1
31. Tektronix, Inc (2023) Passive high voltage probes P5100A-TPP0850-P5122-P5150-P6015A datasheet. Datasheet, 56W-10262-15
32. Tektronix, Inc (2017) 1 GHz and 500 MHz high voltage differential probes TDP1000, TDP0500, P6251 datasheet. Datasheet, 51W-19997-10
33. Tektronix, Inc (2008) Power measurement deoffset & calibration fixture instructions. Primary User, 071187202
34. Tektronix, Inc (2008) TEK-DPG Deoffset pulse generator instructions. Primary User, 071234100
35. Cree, Inc (2015) SiC MOSFET double pulse fixture. Application note, CPWR-AN09
36. Caddock Electronics Inc (2004) MP900 and MP9000 series Kool-Pak® power film resistors TO-126, TO-220 and TO-247 style. Datasheet, 28IL102.1004
37. Vishay Intertechnology, Inc (2018) 50 W power resistor, thick film technology, TO-220. Datasheet, 50035
38. Tektronix, Inc. A New Software Deoffset Approach Accelerates Double Pulse Test. https://www.tek.com/en/blog/a-new-software-deoffset-approach-accelerates-double-pulse-test

Further Reading

39. Witcher JB (2003) Methodology for switching characterization of power devices and modules. Virginia Polytechnic Institute and State University
40. Tong CF, Nawawi A, Liu Y et al (2015) Challenges in switching waveforms measurement for a high-speed switching module. IEEE Energy Convers Congr Expos 2015:6175–6179
41. Zhang Z, Guo D, Wang FF et al (2017) Methodology for wide band-gap device dynamic characterization. IEEE Trans Power Electron 32(12):9307–9318
42. Mondal B, Pogulaguntla RT, Arun Karuppaswamy B (2022) Double pulse test set-up: hardware design and measurement guidelines. In: 2022 IEEE international conference on power electronics, drives and energy systems (PEDES), pp 1–6
43. Yang S, Mao S, Wang Z, Lu X, Chen H, Zeng K (2021) Automated SiC MOSFET power module switching characterization test platform. IEEE Worksh Wide Bandgap Power Dev Appl Asia 2021:195–201
44. Li H, Gao Z, Chen R, Wang F (2023) Improved double pulse test for accurate dynamic characterization of medium voltage SiC devices. IEEE Trans Power Electron 38(2):1779–1790
45. Xia Y-K, Li X-Y (2022) Calculation and experiment of stray inductance of PCB double-pulse test circuit based on three-dimensional simulation. IEEE Access 2022:58769–58776
46. Massarini A, Kazimierczuk MK (1997) Self-capacitance of inductors. IEEE Trans Power Electron 12(4):671–676
47. Grandi G, Kazimierczuk MK, Massarini A et al (1999) Stray capacitances of single-layer solenoid air-core inductors. IEEE Trans Ind Appl 35(5):1162–1168
48. Ayachit A, Kazimierczuk MK (2017) Self-capacitance of single-layer inductors with separation between conductor turns. IEEE Trans Electromagn Compat 59(5):1642–1645
49. Tektronix, Inc (2019) XYZs of oscilloscopes. Primer, 03Z-8605-7
50. Keysight Technologies, Inc (2019) Evaluating oscilloscope bandwidths for your application. Application note, 5989-5733EN
51. Keysight Technologies, Inc (2017) Bandwidth and rise time requirements for making accurate oscilloscope measurements. Application note, 5991-0662EN

52. Keysight Technologies, Inc (2017) Understanding oscilloscope frequency response and its effect on rise-time accuracy. Application Note, 5988-8008EN
53. Keysight Technologies, Inc (2019) Understanding ADC bits and ENOB. Application note, 5992-3675EN
54. Keysight Technologies, Inc (2017) Evaluating oscilloscope vertical noise characteristics. Application note, 5989-3020EN
55. Walt K (2008) Understand SINAD, ENOB, SNR, THD, THD + N, and SFDR so you don't get lost in the noise floor. Tutorial, MT-003, Revision A. Analog Devices, Inc
56. Teledyne LeCroy, Inc (2011) Computation of effective number of bits, signal to noise ratio, & signal to noise & distortion ratio using FFT. Application note
57. Pupalaikis PJ (2017) Understanding vertical resolution in oscilloscopes. Teledyne LeCroy, Inc
58. Ken J (2019) Comparing high resolution oscilloscope design approaches. Teledyne LeCroy, Inc
59. Wolfgang M (2006) Digital timing measurements: from scopes and probes to timing and Jitter. Springer, Dordrecht
60. Gabriele M, Dave R (2015) Interleaving ADCs: unraveling the mysteries. Analog Dev 2015(July):1–5 (Analog Dialogue)
61. Tektronix, Inc (2019) Oscilloscope selection guide. 46W-31080-5
62. Teledyne LeCroy, Inc (2020) Oscilloscope feature, options, and accessories catalog (low bandwidth). Catalog, lbw-foa-catalog-26aug19
63. Teledyne LeCroy, Inc (2020) Oscilloscope features, options, and accessories catalog (high bandwidth). Catalog, hbw-foa-catalog01oct19
64. Tektronix, Inc (2019) ABCs of probes. Primer, 60Z-6053-14
65. Ken J (2019) Probing in power electronics what to use and why: part one. On demand webinars. Teledyne LeCroy, Inc., October 2019. https://go.teledynelecroy.com/l/48392/2019-10-24/7sxnsd
66. Ken J (2016) Practical considerations in measuring power and efficiency on PWM and Distorted waveforms during dynamic operating conditions. APEC 2016 industry session. Teledyne LeCroy, Inc
67. Teledyne LeCroy, Inc (2019) Getting the most out of 10x passive probe. Application note
68. Vinci A (2019) Can you trust your power measurements? Webinars, Tektronix, Inc. https://buzz.tek.com/en-tek-academy-webinar-series/power-electronics-measurements-2?aliId=eyJpIjoiNStDY2tZS1dIYzg1VlcxUCIsInQiOiJyZHRFbkNnSFVVcnVjbitZY2M0dU5BPT0ifQ%253D%253D
69. ROHM Co., Ltd (2019) Power devices, switching regulators, and gate drivers method for monitoring switching waveform. Application note, No. 62AN072E, Revision 001
70. ROHM Co., Ltd (2019) Precautions during gate-source voltage measurement. Application note, No. 62AN085E, Revision 001
71. Tektronix, Inc (2019) Fundamentals of floating measurements and isolated input oscilloscopes. Application note, 3AW-19134-2
72. Niklaus PS, Bonetti R, Stäger C, Kolar JW, Bortis D (2022) High-bandwidth isolated voltage measurements with very high common mode rejection ratio for WBG power converters. IEEE Open J Power Electron 3:651–664
73. Ziegler S, Woodward RC, Iu HC et al (2009) Current sensing techniques: a review. IEEE Sens J 9(4):354–376
74. Xin Z, Li H, Liu Q, Loh PC (2022) A review of megahertz current sensors for megahertz power converters. IEEE Trans Power Electron 37(6):6720–6738
75. Zhang W, Zhang Z, Wang F, Brush EV, Forcier N (2021) High-bandwidth low-inductance current shunt for wide-bandgap devices dynamic characterization. IEEE Trans Power Electron 36(4):4522–4531
76. Keysight Technologies, Inc (2008) U1880A deoffset fixture. User's guide, 1st edn. U1880-97000
77. Keysight Technologies, Inc (2017) U1882B measurement application for infiniium oscilloscopes. Date Sheet, 5989-7835EN

78. Teledyne LeCroy, Inc (2017) DCS025 deoffset calibration source. Data Sheet
79. Tektronix, Inc (2020) SiC and GaN power converter analysis kit. Instruction Guide, 48W-61538-0
80. Keysight Technologies, Inc (2019) PD1500A series dynamic power device analyzer/double-pulse tester. Data Sheet, 5992-3942EN
81. Keysight Technologies, Inc (2020) Keysight PD1500A Si/SiC test fixture and modules installation and use guide, 1st edn. PD1500-90004
82. Roy T (2019) New generation power semiconductor dynamic characterization test system. Bodo's Power Syst 2019(August):24–27
83. Roy T, Bernhard H, Michael Z et al (2020) The challenges of obtaining repeatable and reliable double-pulse test results—measurement science. Bodo's Power Syst 2020(June):28–32
84. Roy T, Bernhard H, Mike H (2020) Overcoming challenges characterizing high speed power semiconductors. Bodo's Power Syst 2020(April):36–39

Chapter 6
SiC Device Testing and Failure Analysis Techniques

Testing and failure analysis are crucial methods for detecting, verifying, and analyzing SiC devices during their development, production, and application processes. Based on test data and detection results, engineers can optimize devices and troubleshoot issues.

The testing of SiC devices is divided into single-device testing and system-level testing. Single-device testing includes parameter testing and reliability testing, while system-level testing examines the device's performance within an application system. Additionally, SiC device testing can be categorized into laboratory testing and mass production testing. The purpose of laboratory testing is to obtain precise device parameters and verify device reliability, while mass production testing aims to conduct large-scale device inspection and screening. Furthermore, failure analysis of SiC devices can be classified into three aspects: material, chip, and packaging.

Since SiC devices are not yet fully mature, their testing and failure analysis have garnered increasing attention across various stages of the industry chain to accelerate chip development and improve device yield. Many power electronics companies have established laboratories for device characterization, reliability testing, and failure analysis.

This chapter will introduce the basic principles and related equipment for SiC device testing and failure analysis.

6.1 Principles and Challenges of Parameter Testing

6.1.1 Testing Equipment

In experiments where device parameters are tested, high precision and accuracy in test results are crucial. Therefore, the performance metrics of the testing equipment used must meet stringent requirements.

Y. Gao and Y. Zhang, *Silicon Carbide Power Devices*, CPSS Power Electronics Series,
https://doi.org/10.1007/978-981-96-3480-4_6

To measure the DC parameters of devices, a Source Measure Unit (SMU) is used, as shown in Fig. 6.1, which features a commonly used SMU product from Tektronix. To measure resistance (R), conductance (G), and capacitance (C), an LCR meter is employed, as illustrated in Fig. 6.2, showing a typical LCR meter product from Keysight. For thermal resistance measurement, a thermal resistance tester is used, as depicted in Fig. 6.3, which shows commonly used products from Siemens and Analysis Tech. Dynamic characteristic testing systems, such as the DPT1000A from Tektronix, introduced in Sect. 5.6.2, are used to measure switching, reverse recovery, gate charge, and short-circuit characteristics. Additionally, some manufacturers integrate SMUs and LCR meters into a single system, known as a curve tracer, which is the main equipment for power semiconductor parameter testing. Figures 6.4 and 6.5 show commonly used curve tracer products, such as the B1506A from Keysight and the 2600-PCT from Tektronix.

In mass production testing, the equipment used is typically a production tester, which generally consists of a main unit and modular structure. Each test module completes a specific type of test and can include one or multiple tests. Engineers can select different test modules based on the testing requirements and integrate them into the main unit. For instance, the STS8200 testing platform from Beijing Acco TEST Technology Co., Ltd. is shown in Fig. 6.6. This platform is based on a fully floating source design and utilizes an open software and hardware architecture. The same tester can be configured with different hardware resources to test analog chips, power management chips, power devices, and modules. The system is mature, stable, flexible, easy to use, and offers fast testing speeds. For SiC power devices, different hardware resources can be configured to cover the entire process from SiC wafers and

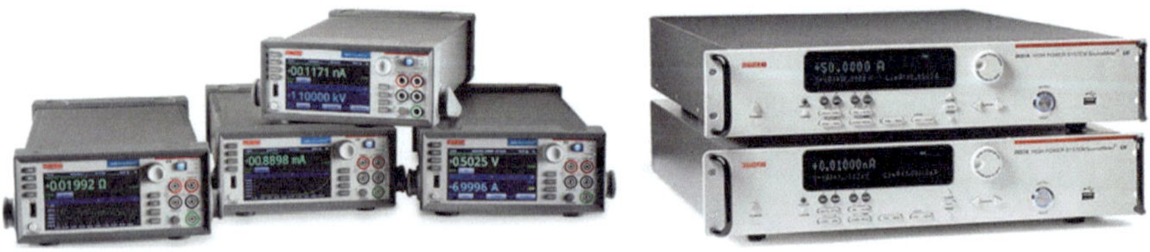

Fig. 6.1 Tektronix source meter unit

Fig. 6.2 Keysight LCR meter

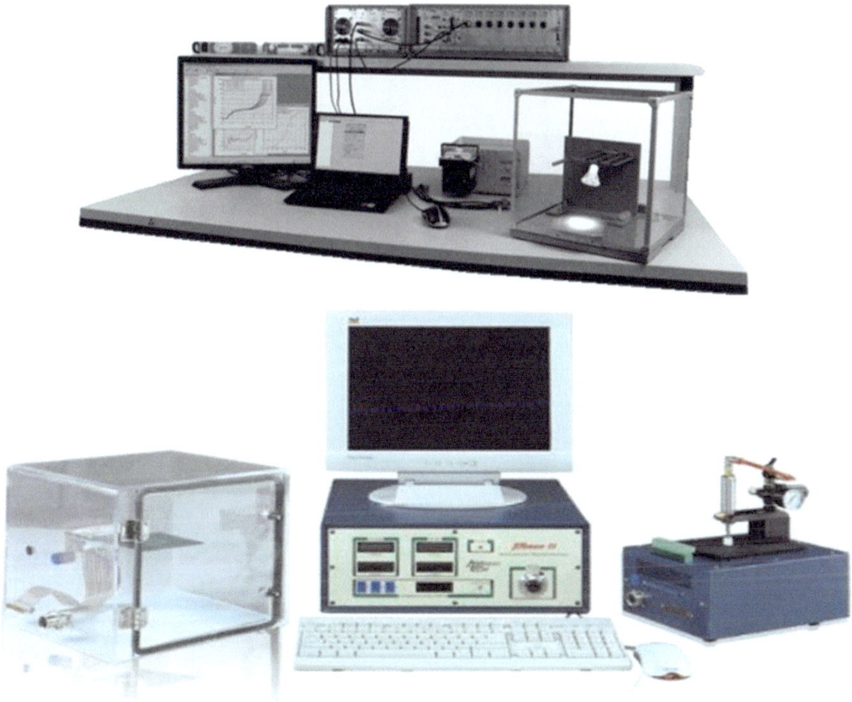

Fig. 6.3 Thermal resistance testers from Siemens and analysis tech

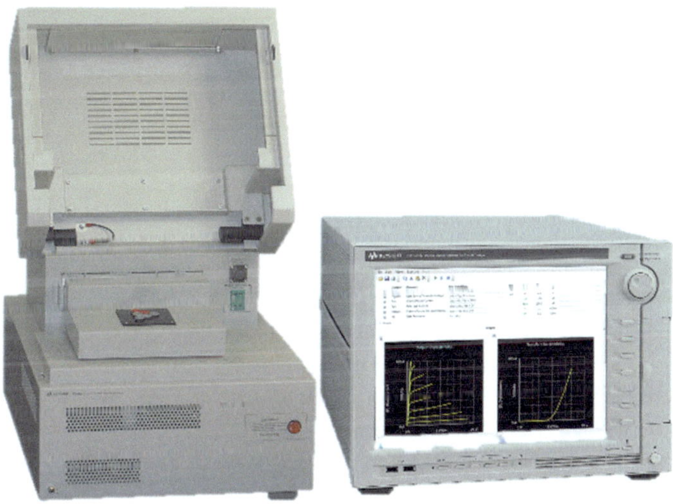

Fig. 6.4 Keysight B1506

Fig. 6.5 Tektronix 2600-PCT

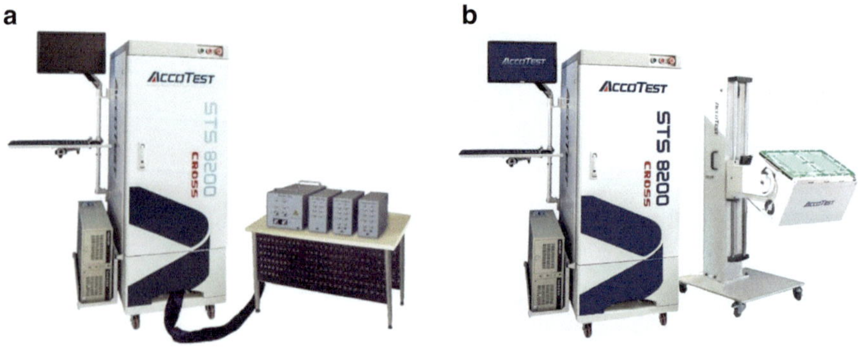

Fig. 6.6 AccoTEST Technology Co., Ltd. STS8200STS8200 test platform

bare die to single transistors and power modules in fully automated mass production testing. This system has been widely adopted by many renowned international and domestic power device (IDM) customers for their mass production needs.

In the STS8200 testing platform, the Source Measure Units (SMUs) SVI40, HVS2K, and HCB200, in conjunction with the DC Test Box, can perform DC parameter testing. The UIS200 module is used for avalanche testing, the DVX900 module for DV/DS testing, the ZMU module for R_G, and Junction Capacitance C testing, and the DSU1200 module for dynamic characteristic testing, as illustrated in Fig. 6.7.

The waveforms for SiC MOSFET parameter tests in the following sections are all based on the STS8200 test machine platform.

(a) SVI40

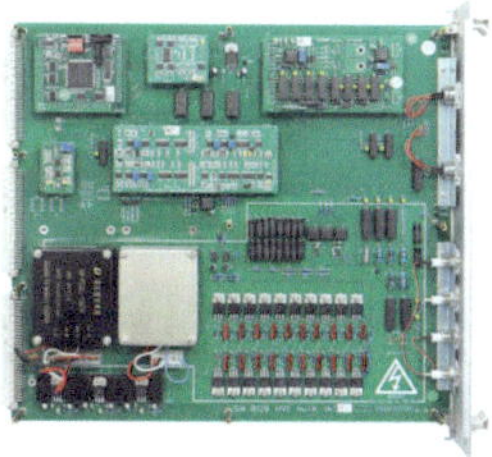

(b) HVS2K

(c) HCB200

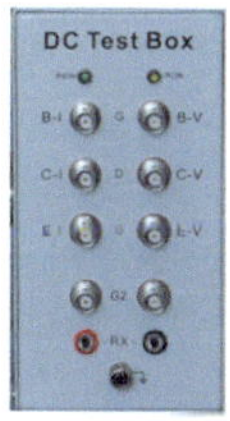

(d) DC Test Box

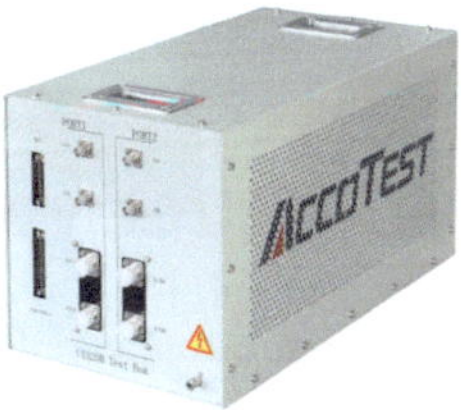

(e) UIS200

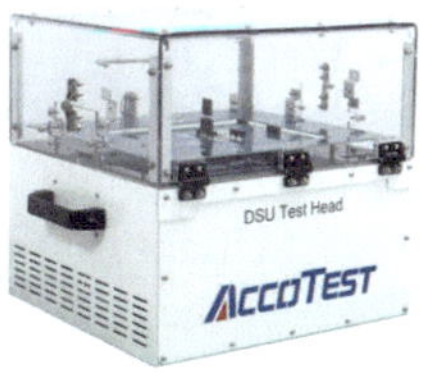

(e) DSU1200

(f) ZMU

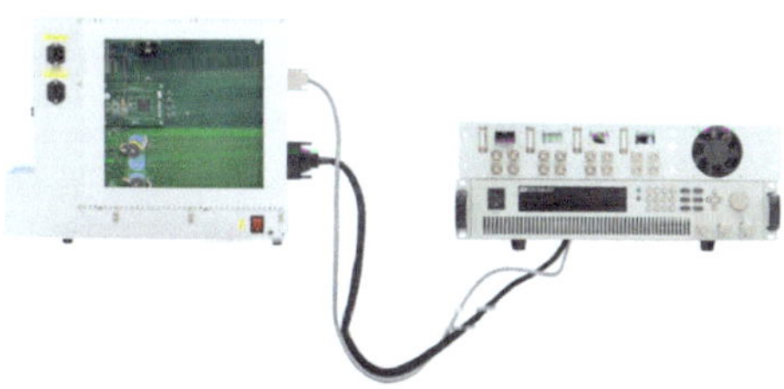

(g) DVX900

Fig. 6.7 The test module of the STS8200 test machine platform of Beijing AccoTEST Technology Co., Ltd

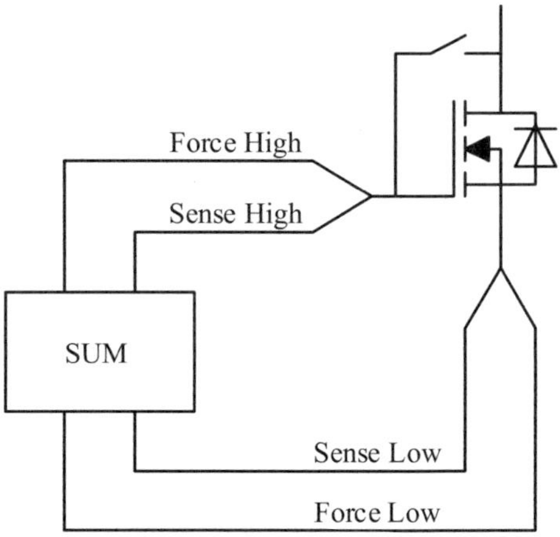

Fig. 6.8 Threshold voltage $V_{GS(th)}$ test circuit

6.1.2 Threshold Voltage $V_{GS(th)}$

6.1.2.1 $V_{GS(th)}$ Test Principle

The definition of the threshold voltage $V_{GS(th)}$ for SiC MOSFETs is described in Sect. 3.2.1. The test circuit for measuring $V_{GS(th)}$ is shown in Fig. 6.8. In this setup, the gate and drain of the SiC MOSFET are shorted and connected to one Force terminal of the Source Measure Unit, with the other Force terminal connected to the source. The two Sense terminals of the SMU are connected to the gate and source. The SMU measures the V_{GS} and I_{DS} of the SiC MOSFET. It is important to note that the SMU has four terminals: two Force terminals and two Sense terminals. The Sense terminals are used for voltage measurement and are connected closer to the device under test than the Force terminals. This method, known as the four-wire measurement or Kelvin measurement, prevents the introduction of voltage drops from the measurement leads, which can occur when measuring voltage directly with the Force terminals.

During the measurement process, the voltage at the Force terminal of the SMU gradually increases, causing the V_{GS} of the SiC MOSFET to increase as well, allowing I_{DS} to flow. When I_{DS} reaches the threshold current set point, V_{GS} stabilizes and I_{DS} remains constant at the threshold current. At this point, V_{GS} is measured to obtain $V_{GS(th)}$. After completing the measurement, the voltage at the Force terminal of the SMU returns to 0 V. The period from when the Force terminal voltage begins to increase until the measurement is called the Force Time, and the period during which the measurement is taken is called the Measuring Time.

The waveforms of $V_{GS(th)}$ for SiC MOSFETs at I_{DS} values of 50 μA, 1 mA, and 10 mA are shown in Fig. 6.9. Since I_{DS} is relatively small, only the V_{GS} waveforms are provided. It can be observed that as I_{DS} increases, the corresponding $V_{GS(th)}$ also increases.

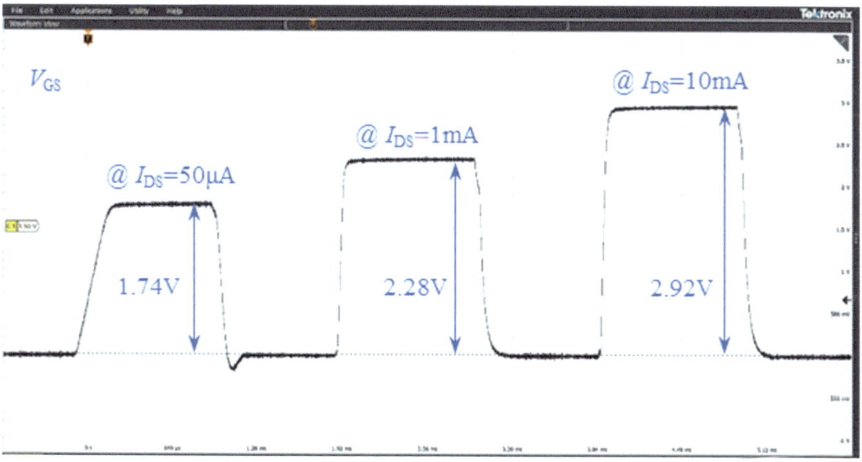

Fig. 6.9 Test waveform of threshold voltage $V_{GS(th)}$

6.1.2.2 $V_{GS(th)}$ Testing Challenges and Notes

(1) In mass production, $V_{GS(th)}$ is typically the first test item used to check whether the device under test has failed and to detect any anomalies in the contact of the manipulator or probe. If the test circuit is open, the measured $V_{GS(th)}$ will be the open-circuit voltage; if the test circuit is shorted, the measured $V_{GS(th)}$ will be close to 0 V. Additionally, $V_{GS(th)}$ can be measured again after other high-power or high-voltage parameter tests to verify whether the device has been damaged after such tests. The test equipment must have a failure stop function, which halts further tests if the device under test fails or if there is a contact anomaly, to prevent further damage.

(2) Since I_{DS} is usually in the μA to mA range, the heat generated in the device is generally negligible.

(3) But the Force Time for I_{DS} should not be too long to avoid excessive temperature rise, with the principle being waveform stability and no impact on the final test time.

(4) For SiC MOSFETs, applying voltage stress to the gate or drain or significantly heating the device during other parameter tests before measuring $V_{GS(th)}$ may cause changes in the subsequent $V_{GS(th)}$ measurement results. This must be considered when adjusting the sequence or conditions of parameter tests.

(5) The gate of SiC MOSFETs is more fragile compared to Si devices, so the V_{GS} waveform during testing must be smooth, with no abnormal voltage overshoot, to avoid gate damage which could lead to device failure or reduced reliability in future applications.

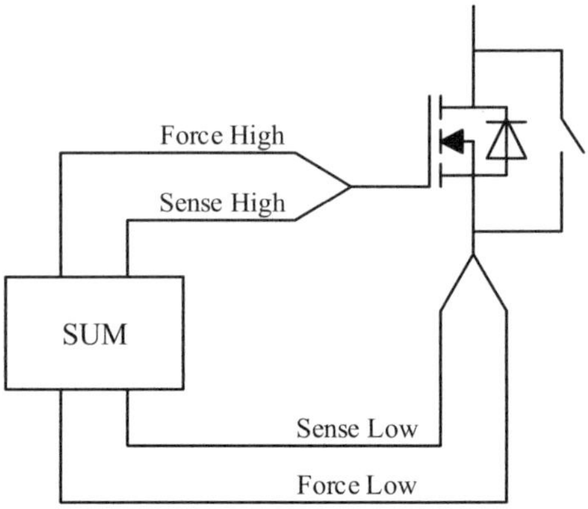

Fig. 6.10 Gate leakage current I_{GSS} test circuit

6.1.3 Gate Leakage Current I_{GSS}

6.1.3.1 Testing Principles of I_{GSS}

The measurement circuit for the gate leakage current I_{GSS} of a SiC MOSFET is shown in Fig. 6.10. In this setup, the drain (D) and source (S) of the SiC MOSFET are shorted. The Force High and Sense High terminals of the Source Measure Unit (SMU) are connected to the gate (G), while the Force Low and Sense Low terminals are connected to the source (S).

During the measurement, the Source Measure Unit Force terminal outputs the specified voltage V_{GS}.. The current flowing into the gate (G) is then measured to obtain I_{GSS}. After the measurement is completed, the voltage at the SMU Force High terminal returns to 0 V. The period from when the SMU Force High terminal voltage begins to increase until the measurement is called the Force Time, and the period during which the measurement is taken is called the Measuring Time.

The waveforms of I_{GSS} for a SiC MOSFET at V_{GS} values of 25, 20, – 15, and – 10 V are shown in Fig. 6.11. Since I_{GSS} is very small, only the V_{GS} waveforms are provided.

6.1.3.2 I_{GSS} Testing Challenges and Considerations

(1) I_{GSS} is typically in the range of μA to nA, and even in the pA level. During actual production, due to the input capacitance C_{ISS} of the SiC MOSFET and the distributed capacitance on the test lines (such as PCB boards, cables, etc.), a sufficient amount of time is needed to charge these capacitances and reach stability during the establishment of I_{GSS}. Otherwise, the charging current may be mistakenly included in I_{GSS}. Accordingly, after the measurement, a corresponding time is also needed to reduce V_{GS} to 0 V to ensure the safety of the device.

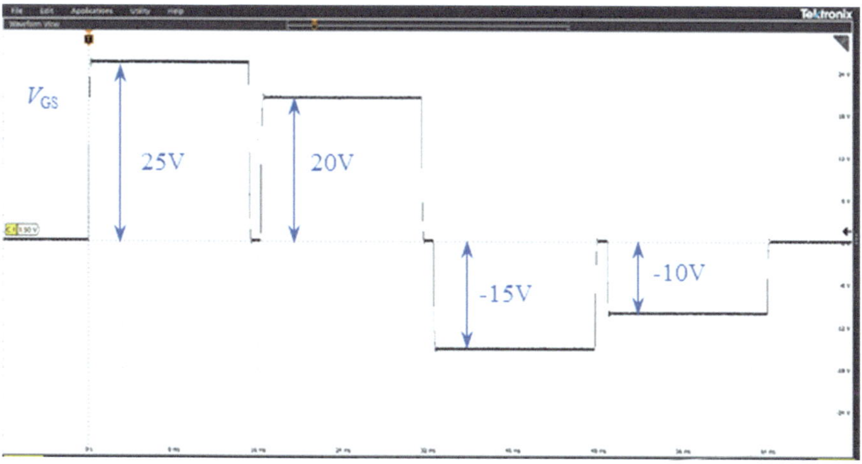

Fig. 6.11 I_{GSS} testing challenges and considerations

(2) I_{GSS} values are very small and often close to the zero point of the current range of the test equipment. When a fault occurs in the measurement circuit of I_{GSS}, such as a cable open circuit or relay open circuit, the measurement result will also be close to the zero point of the current range of the test equipment. However, the reading at this time is a false value, which can lead to measurement errors and quality risks. Therefore, it is usually necessary to perform an online diagnostic of the key lines before measuring I_{GSS} to ensure that the lines are properly connected before conducting the test.

(3) For nA or pA level I_{GSS} testing, during production testing, mechanical arms, probe stations, and other ancillary equipment used in conjunction with the test equipment can generate interference such as power frequency and high frequency, which can affect the measurement of I_{GSS}. Therefore, the corresponding lines should be designed to filter out power frequency and high-frequency interference, and proper hardware shielding and software anti-interference measures should be implemented to obtain accurate and stable data.

(4) Unlike Si devices, SiC MOSFET gates are relatively fragile and have poor negative gate voltage tolerance. Therefore, when measuring negative I_{GSS}, care should be taken to ensure that the voltage does not exceed the maximum negative gate voltage specified in the device manual. Additionally, during the measurement, V_{GS} should have a smooth waveform without abnormal overshoots or glitches.

(5) When using an external oscilloscope to observe V_{GS}, if the current range of the source meter is relatively small and the equivalent input resistance of the oscilloscope voltage probe is also small, it may prevent the establishment of V_{GS}. In this case, you can increase the current range or use the built-in oscilloscope function of the test equipment to observe the voltage and current waveforms.

Fig. 6.12 Breakdown voltage $V_{(BR)DSS}$ test circuit

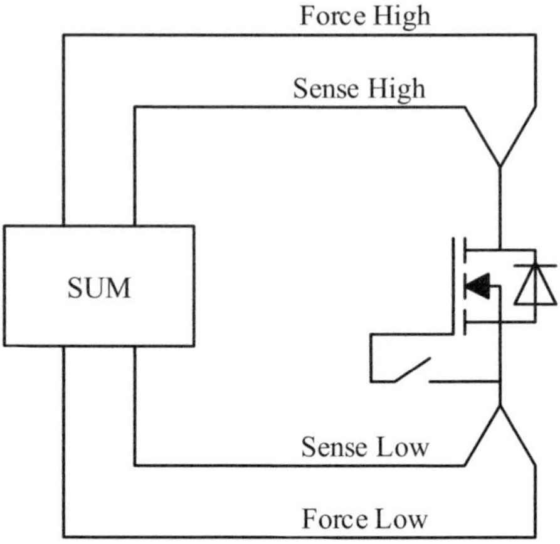

6.1.4 Breakdown Voltage $V_{(BR)DSS}$

6.1.4.1 Testing Principle of $V_{(BR)DSS}$

The definition of the SiC MOSFET breakdown voltage $V_{(BR)DSS}$ is as described in Sect. 3.1.1. The test circuit for $V_{(BR)DSS}$ is shown in Fig. 6.12. The gate and source of the SiC MOSFET are shorted together, with the Force High and Sense High terminals of the source meter connected to the drain, and the Force Low and Sense Low terminals connected to the source.

During the measurement, the voltage at the Force terminal of the source meter gradually increases, causing V_{DS} of the SiC MOSFET to rise and I_{DS} to flow. When I_{DS} reaches the set value, V_{DS} no longer increases. At this point, measuring V_{DS} yields $V_{(BR)DSS}$. After completing the measurement, the Force High terminal voltage of the source meter returns to 0 V. The time during which the voltage at the Force High terminal of the source meter increases before the measurement is called the Force Time, and the time period during which the measurement is conducted is called the Measuring Time.

The waveforms of $V_{(BR)DSS}$ for I_{DS} values of 100 μA, 1 mA, and 10 mA are shown in Fig. 6.13. Since I_{DS} is very small, only the waveform of I_{DS} is provided. It can be seen that as I_{DS} increases, the corresponding $V_{(BR)DSS}$ also slightly increases.

6.1.4.2 $V_{(BR)DSS}$ Testing Challenges and Notes

(1) $V_{(BR)DSS}$ of the SiC MOSFET is a high-voltage parameter, with test results typically ranging from 650 to 2000 V, or even higher. During actual production, ensuring personal safety is crucial. Test cables, PCB boards, connectors, and other peripheral components must comply with safety regulations. Additionally,

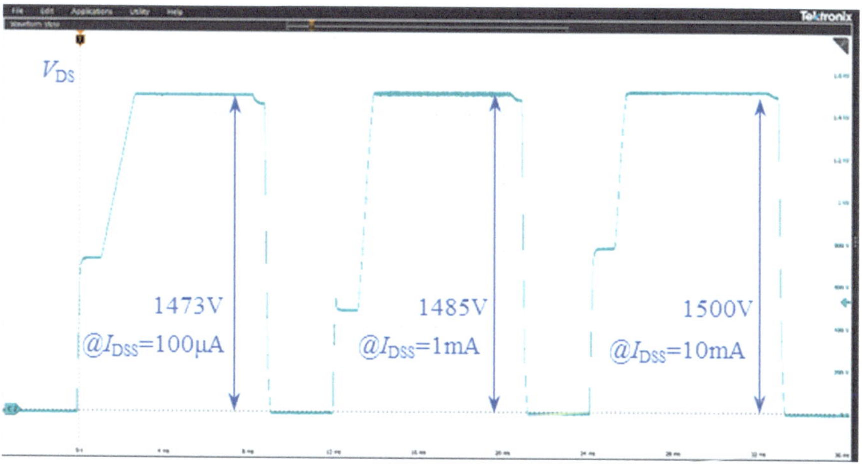

Fig. 6.13 Test waveform of breakdown voltage $V_{(BR)DSS}$

safety protection for the test equipment and related ancillary devices must be properly implemented.

(2) Test circuit boards, test fixtures, gold fingers, and probe spacing must meet high-voltage testing creepage distance and insulation requirements to prevent abnormal breakdowns and arcing.

(3) When the breakdown state is reached, the power dissipation on the device under test is the product of $V_{(BR)DSS}$ and I_{DS}. When testing SiC MOSFETs, this power dissipation can be significant, so it is necessary to adjust the Force Time appropriately to minimize testing deviations caused by heating of the device.

(4) Due to the output capacitance C_{OSS} of the SiC MOSFET and the distributed capacitance between the drain and source on the test lines (such as PCB, cables, etc.), a sufficient amount of time is required to charge these capacitances and stabilize to $V_{(BR)DSS}$ during the $V_{(BR)DSS}$ test. After the test, V_{DS} must be reduced to 0 V to ensure that the charge in these capacitances is discharged, ensuring the safety of the next parameter test.

6.1.5 Drain Leakage Current I_{DSS}

6.1.5.1 Testing Principle of I_{DSS}

The definition of I_{DSS} for SiC MOSFETs is as described in Sect. 3.1.1. The test circuit for I_{DSS} is shown in Fig. 6.14. The gate and source of the SiC MOSFET are shorted together, with the Force High and Sense High terminals of the source meter connected to the drain, and the Force Low and Sense Low terminals connected to the source.

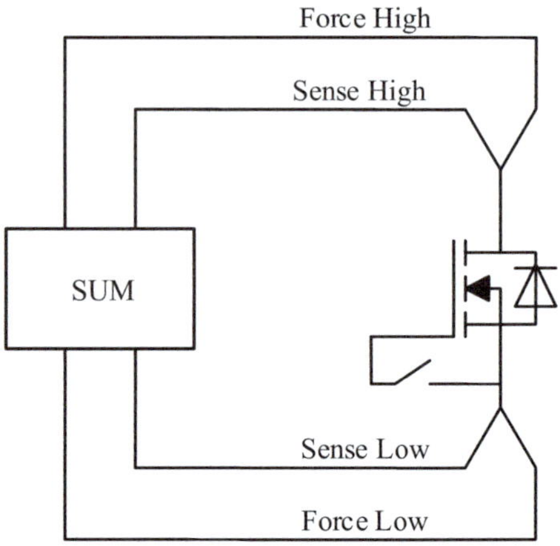

Fig. 6.14 Drain leakage current I_{DSS} test circuit

During the measurement, the voltage at the Force terminal of the source meter gradually increases, causing V_{DS} of the SiC MOSFET to rise, and I_{DS} to flow. When V_{DS} reaches and stabilizes at the set value, measuring I_{DS} yields I_{DSS}. After the measurement, the Force High terminal voltage of the source meter returns to 0 V. The time from when the voltage at the Force High terminal of the source meter starts to increase until the measurement is conducted is called Force Time, and the time period during which the measurement is conducted is called Measuring Time.

The waveforms of I_{DSS} for V_{DS} values of 900, 1200, and 1400 V are shown in Fig. 6.15. Since I_{DSS} is very small at this point, only the waveform of V_{DS} is provided.

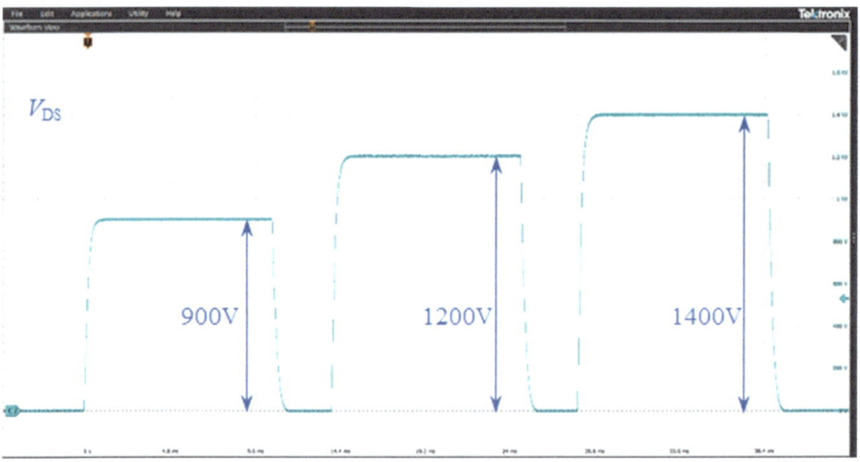

Fig. 6.15 Test waveform of drain leakage current I_{DSS}

6.1.5.2 Challenges and Considerations for Testing I_{DSS}

(1) I_{DSS} is a high-voltage parameter, with V_{DS} typically ranging from 650 to 2000 V, or even higher, during testing. In actual production, personal safety must be ensured first. Test cables, PCB boards, connectors, and other peripheral components must comply with safety regulations, and appropriate safety measures must be implemented for the test equipment and related devices.
(2) The design of the test circuit board, test fixtures, gold fingers, and probe spacing must meet high-voltage testing creepage distance and insulation requirements to prevent abnormal breakdowns and arcing.
(3) At room temperature, I_{DSS} is usually in the nA to μA range. The power dissipation on the device under test is the product of the breakdown voltage V_{DS} and the drain leakage current I_{DSS}. While this power is usually small, in high-temperature testing scenarios, I_{DSS} may reach mA levels. In such cases, this power can be significant, so Force Time should be adjusted appropriately to prevent additional heating of the device that could cause testing deviations.
(4) I_{DSS} typically ranges from μA to nA. Due to the input capacitance (Coss) of the SiC MOSFET and the distributed capacitance between the drain and source on the test lines (such as PCB boards, cables, etc.), a sufficient amount of time is required to charge these equivalent capacitances and stabilize to V_{DS} during the I_{DSS} test. After the test, V_{DS} must be reduced to 0 V to ensure that the charge in these capacitances is discharged, ensuring safety for the next parameter test.

6.1.6 On-Resistance $R_{DS(on)}$

6.1.6.1 Testing Principle of $R_{DS(on)}$

The definition of $R_{DS(on)}$ for SiC MOSFETs is as described in Sect. 3.2.2. The test circuit for $R_{DS(on)}$ is shown in Fig. 6.16. Two source meters are used in the test: the gate-source meter's Force High and Sense High terminals are connected to the gate, and the drain-source meter's Force High and Sense High terminals are connected to the drain. The Force Low and Sense Low terminals of both source meters are connected to the source.

During the measurement, the gate-source meter outputs the set V_{GS} to turn on the device under test at the specified drive voltage. Meanwhile, the drain-source meter outputs the set I_{DS}, and V_{DS} is measured. $R_{DS(on)}$ can be calculated as V_{DS}/I_{DS}. After the measurement, the Force High terminal voltages of both the gate-source meter and drain-source meter return to 0 V. The time from when the voltage at the drain-source meter Force High terminal starts to increase until the measurement is conducted is called Force Time, and the time period during which the measurement is conducted is called Measuring Time.

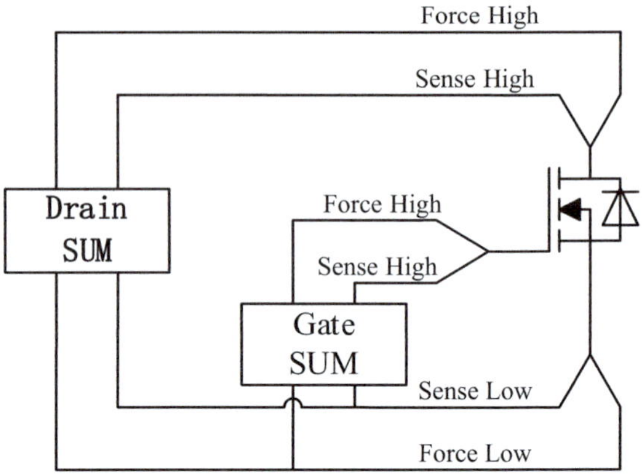

Fig. 6.16 Testing principle of on-resistance $R_{DS(on)}$

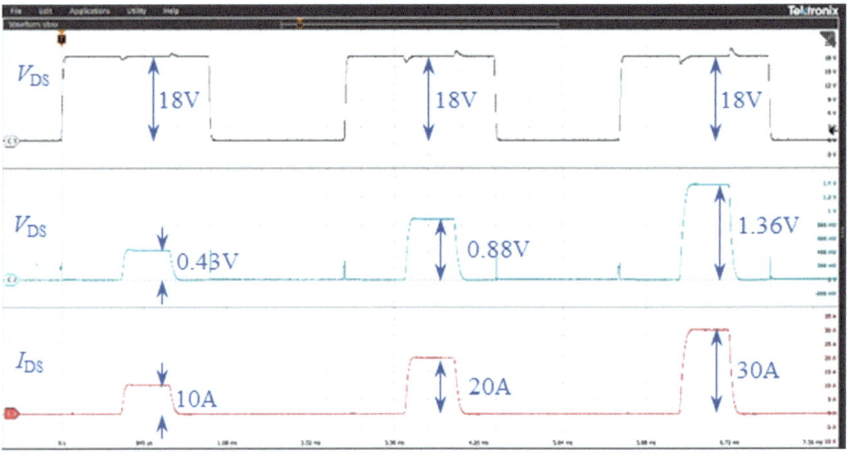

Fig. 6.17 Test waveform of on-resistance $R_{DS(on)}$

The waveforms of $R_{DS(on)}$ for I_{DS} values of 10, 20, and 30 A at V_{GS} of 18 V are shown in Fig. 6.17. It can be seen that as I_{DS} increases, the corresponding V_{DS} also increases.

6.1.6.2 Challenges and Considerations for Testing $R_{DS(on)}$

(1) $R_{DS(on)}$ is a large current parameter, and voltage drops on the test lines can introduce measurement errors. Therefore, a four-wire Kelvin connection must be used. In actual production, before each test of $R_{DS(on)}$,, the Kelvin connection state should be checked to ensure that connections in cables, gold fingers (probe cards), and other intermediate stages are normal.

(2) When testing large current parameters like $R_{DS(on)}$,, if the resistance in the Force line (including contact resistance in gold fingers, cables, PCB traces, etc.) is too high, it can cause significant voltage drops, leading to deviations in the test equipment. Testing devices often use built-in or additional circuits to detect the resistance in the Force line. If the resistance exceeds the set value, the system should generate an error or indicate device failure.

(3) Test equipment usually calculates $R_{DS(on)}$ using the measured V_{DS} and the set I_{DS}. However, due to various non-ideal factors, if the actual I_{DS} is less than the set value, the calculated $R_{DS(on)}$ will be lower than the true value, leading to measurement errors. Therefore, test equipment should simultaneously measure V_{DS} and I_{DS} to calculate $R_{DS(on)}$ using actual measurement data.

(4) $R_{DS(on)}$ is affected by temperature, so pulse testing should be used during the measurement. Pulse current duration is typically between 300 and 500 μs to prevent temperature rise from affecting the test results.

6.1.7 Transconductance G_{FS}

6.1.7.1 Testing Principle of G_{FS}

G_{FS} Of the SiC MOSFET represents the change in drain current I_{DS} resulting from a change in gate voltage V_{GS}. The test circuit for G_{FS} is shown in Fig. 6.18. Two source meters are used in the test: the gate-source meter's Force High and Sense High terminals are connected to the gate, and the drain-source meter's Force High and Sense High terminals are connected to the drain. The Force Low and Sense Low terminals of both source meters are connected to the source.

During the measurement, the drain-source meter outputs the set V_{DS}. Hardware closed-loop feedback is applied to both the gate- and drain-source meters. Adjust the gate-source meter's output V_{GS} so that I_{DS} reaches the set I_{DS1} and measure the corresponding V_{GS1}. Then, adjust the gate-source meter's output V_{GS} again so

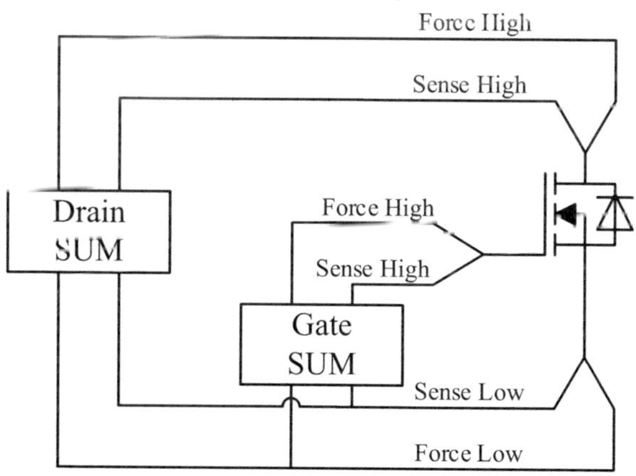

Fig. 6.18 Test circuit for transconductance G_{FS}

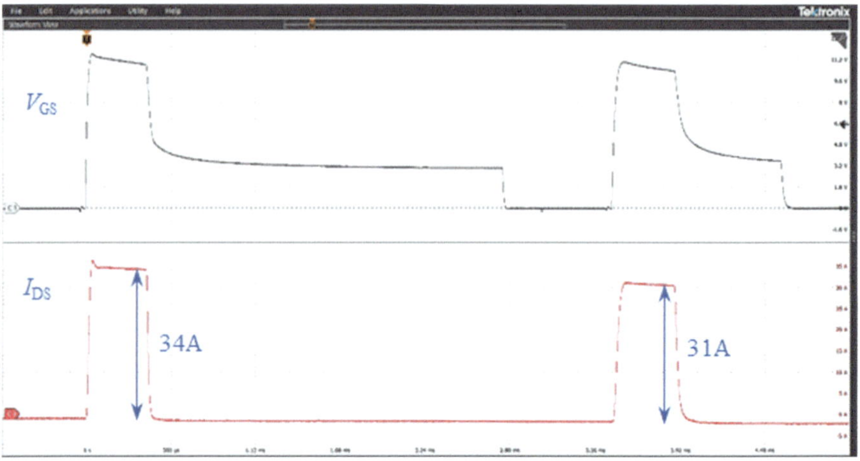

Fig. 6.19 Test waveform of transconductance G_{FS}

that I_{DS} reaches the set I_{DS2} and measure the corresponding V_{GS2}. G_{FS} is calculated as $(I_{DS2} - I_{DS1})/(V_{GS2} - V_{GS1})$. After the measurement, the Force High terminal voltages of the gate- and drain-source meters return to 0 V. The time from when the voltage at the Force High terminal starts to increase until the measurement is conducted is called Force Time, and the time period during which the measurement is conducted is called Measuring Time.

The waveforms of G_{FS} for I_{DS} values of 31 and 34 A are shown in Fig. 6.19. It can be seen that the V_{GS} was tested at these I_{DS} levels.

6.1.7.2 Challenges and Considerations for Testing G_{FS}

(1) G_{FS} is a large current parameter, and voltage drops on the test lines can introduce measurement errors. Therefore, a four-wire Kelvin connection must be used. In actual production, before each G_{FS} test, the Kelvin connection state should be checked to ensure that connections in cables, gold fingers (probe cards), and other intermediate stages are normal.

(2) When testing large current parameters such as G_{FS},, if the resistance in the Force line (including contact resistance in gold fingers, cables, PCB traces, etc.) is too high, it can cause significant voltage drops, leading to deviations in the test equipment. Testing devices should include built-in or additional circuits to detect the resistance in the Force line. If the resistance exceeds the set value, the system should generate an error or indicate device failure.

(3) During the G_{FS} test, the device under test experiences significant power dissipation. This can lead to temperature rise, causing deviations in V_{GS} due to the negative temperature coefficient, resulting in fluctuations in the G_{FS} data. The magnitude and order of I_{DS1} and I_{DS2} can also cause significant deviations in the actual G_{FS} test results.

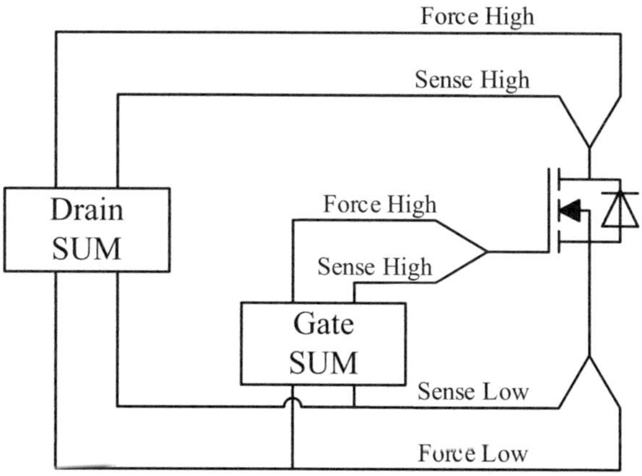

Fig. 6.20 Test circuit for body diode forward voltage drop V_F

6.1.8 Body Diode Forward Voltage Drop V_F

6.1.8.1 Testing Principle of V_F

The definition of the body diode forward voltage drop V_F for SiC MOSFETs is as described in Sect. 3.2.3. The test circuit for V_F is shown in Fig. 6.20. Two source meters are used in the test: the gate-source meter's Force High and Sense High terminals are connected to the gate, and the drain-source meter's Force High and Sense High terminals are connected to the drain. The Force Low and Sense Low terminals of both source meters are connected to the source.

During the measurement, the gate source meter is set to 0 V or a negative V_{GS}. The drain-source meter outputs the set I_{SD}, and the forward voltage drop V_{SD} of the device under test is measured to obtain V_F. After the measurement, the Force High terminal voltages of both the gate-source meter and drain-source meter return to 0 V. The time from when the voltage at the Force High terminal starts to increase until the measurement is conducted is called Force Time, and the time period during which the measurement is conducted is called Measuring Time.

The waveforms of V_{SD} for I_{SD} values of 20A at V_{GS} of 0 and − 5 V are shown in Fig. 6.21. It can be seen that V_{SD} is slightly higher when V_{GS} is − 5 V compared to when it is 0 V.

6.1.8.2 Challenges and Considerations for Testing V_F

(1) V_F is a large current parameter, and voltage drops on the test lines can introduce measurement errors. Therefore, a four-wire Kelvin connection must be used. In actual production, before each V_F test, the Kelvin connection state should be checked to ensure that connections in cables, gold fingers (probe cards), and other intermediate stages are normal.

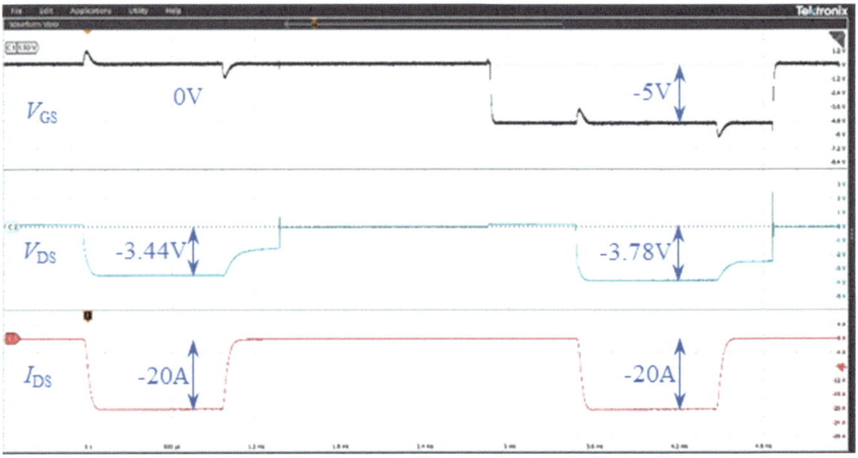

Fig. 6.21 Test waveform of body diode forward voltage drop V_F

(2) When testing large current parameters such as V_F, if the resistance in the Force line (including contact resistance in gold fingers, cables, PCB traces, etc.) is too high, it can cause significant voltage drops, leading to deviations in the test equipment. Testing devices should include built-in or additional circuits to detect the resistance in the Force line. If the resistance exceeds the set value, the system should generate an error or indicate device failure.

(3) Due to various non-ideal factors, the actual I_{SD} may be lower than the set value, which can result in V_F being lower than the true value, leading to measurement errors. Therefore, the test equipment should measure I_{SD} to determine the validity of V_F measurements and prevent errors.

(4) V_F is affected by temperature, so pulse testing should be used during the measurement. The pulse current duration is generally between 300 and 500 μs to prevent temperature rise from affecting the test results.

6.1.9 Avalanche UIS

6.1.9.1 Testing Principle of UIS

The definition of avalanche UIS for SiC MOSFETs is as described in Sect. 3.4.2. The measurement principle of UIS is shown in Fig. 6.22.

During the measurement, the gate drive source outputs the set conduction drive voltage $V_{GS(on)}$, turning on the device under test and charging the inductor L_{load}. When the current through L_{load} reaches the set avalanche current I_{av}, the gate drive source outputs the off-drive voltage $V_{GS(off)}$. Subsequently, the device under test enters avalanche breakdown. The test equipment measures parameters such as breakdown

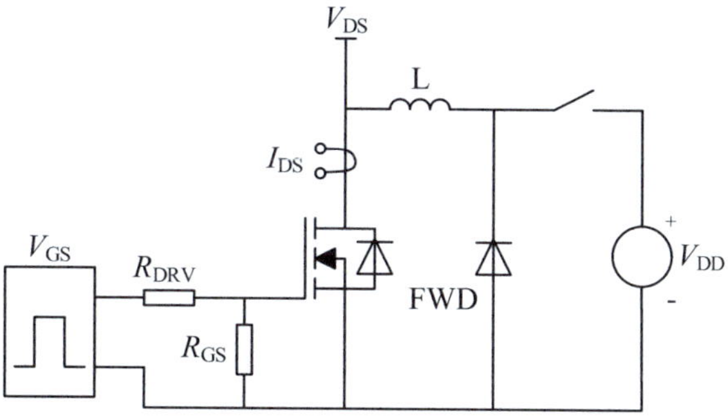

Fig. 6.22 Testing principle of avalanche breakdown UIS

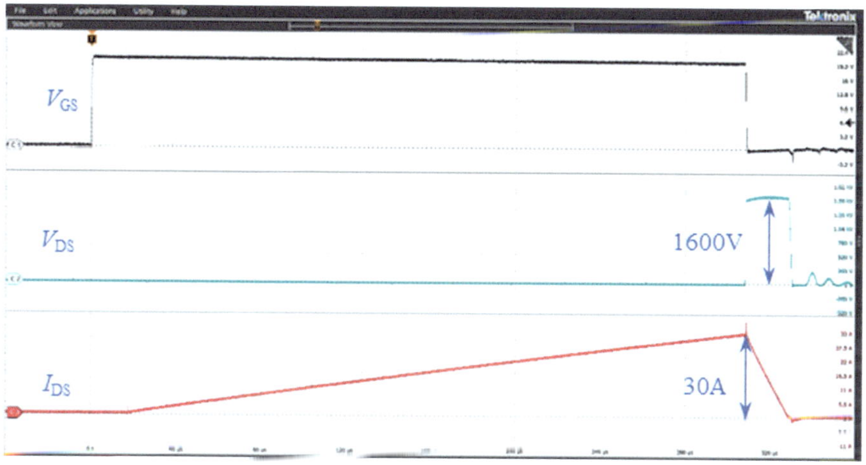

Fig. 6.23 Test waveform of avalanche UIS

voltage, avalanche current, time, and energy during the avalanche breakdown and makes a determination. After the measurement, all source meters are set to zero.

The actual waveforms of UIS measurement are shown in Fig. 6.23. It can be seen that the avalanche current is 30A, and the avalanche voltage is around 1600V.

6.1.9.2 Challenges and Considerations for Testing UIS

(1) UIS is a high-voltage parameter, and the avalanche breakdown voltage is generally at least 1.2 times higher than its nominal voltage rating. In actual production, personal safety must be ensured first. Test cables, PCB, connectors, and other peripheral components must comply with safety regulations, and appropriate safety measures must be implemented for the test equipment and related devices.

(2) UIS is a large current parameter. High resistance in the test lines can slow the charging time, reducing the actual avalanche energy applied to the device under test, affecting the test results. In production, it is essential to control the resistance of cables, gold fingers (probe cards), and other intermediate stages.
(3) UIS involves high power testing, and the device under test is at risk of avalanche failure. Therefore, the test equipment should have fault detection capabilities. During energy application, if abnormal conditions in the device under test or test circuit are detected, energy recovery and protection should be executed quickly to prevent prolonged abnormal energy application that could cause hardware damage.
(4) SiC MOSFETs have fast switching speeds and weak gate voltage tolerance. The gate drive source should have an independent return path to prevent stray inductance in the drive circuit from causing abnormal gate overshoot or negative spikes that could damage the gate of the device.
(5) Before and after UIS testing, the SiC MOSFET should be parameter-tested to ensure the device's condition is normal before and after the UIS test.

6.1.10 Transient Thermal Resistance DVDS

6.1.10.1 Testing Principle of DVDS

There are many parameters that characterize the transient thermal resistance of a SiC MOSFET. In production, DVDS is usually used, which specifically refers to the difference in on-state voltage drop before and after a short-term current flow heating. The test circuit for DVDS is shown in Fig. 6.24. Two source meters are used in the test: the gate-source meter's Force High and Sense High terminals are connected to the gate, and the drain-source meter's Force High and Sense High terminals are connected to the drain. The Force Low and Sense Low terminals of both source meters are connected to the source.

During the measurement process, first measure the forward voltage drop V_{SD1} of the body diode under a small current I_M. Then, apply a transient voltage V_P and current I_P to the device under test using internal closed-loop feedback from the gate-source and drain-source meters. After a period of heating time P_T, stop the heating, and after a delay D_T, measure the forward voltage drop V_{SD2} of the body diode under I_M. Calculate DVDS as $V_{SD1} - V_{SD2}$. After the measurement, set the Force High terminal voltage of the gate-source and drain-source meters back to 0 V.

The actual waveform of the DVDS test is shown in Fig. 6.25. It can be observed that I_P is 20 A, but since I_M is only 10 mA, it is not visible in the waveform.

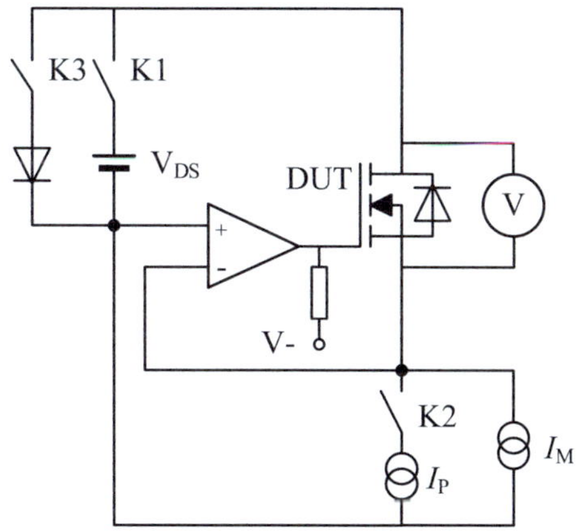

Fig. 6.24 Test circuit for transient thermal resistance DVDS

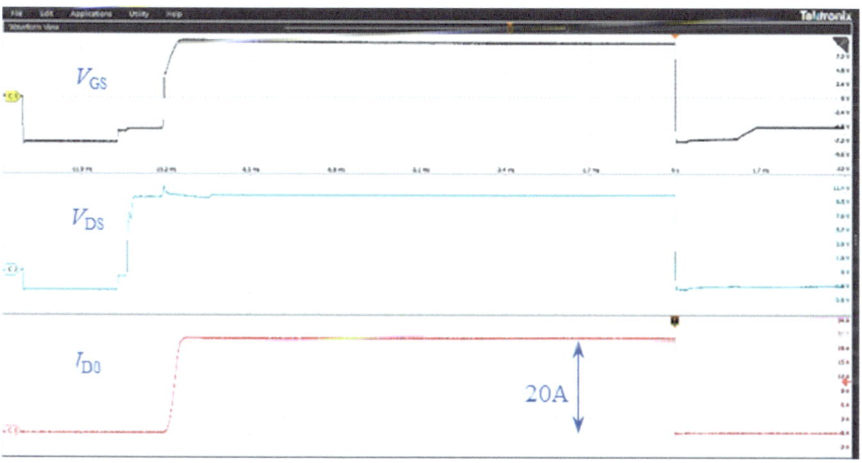

Fig. 6.25 Waveform of transient thermal resistance DVDS test

6.1.10.2 Challenges and Considerations for Testing DVDS

(1) DVDS is a high-current parameter, and the voltage drop on the test circuit can introduce measurement errors. Therefore, a four-wire Kelvin connection is essential. Before each DVDS measurement in production, check the Kelvin connection status to ensure that cables, probe cards, and other intermediate connections are properly connected.

(2) The heating time P_T during DVDS testing typically ranges from tens to hundreds of milliseconds. Even with a proper Kelvin connection, if the resistance in the Force line (including probe contact resistance, cables, PCB traces, etc.) is too high, it can cause significant voltage drops and lead to deviations in the test results. Often, additional circuitry built into or added to the test equipment is

needed to monitor the resistance in the Force line. If the resistance exceeds a set value, the system should report an error or determine that the device is faulty.

(3) DVDS testing will cause the temperature of the device under test to rise. When conducting other parameter tests after DVDS testing, it is necessary to check if the temperature rise will affect the results.

6.1.11 Junction Capacitors C_{iss}, C_{oss}, C_{rss}

6.1.11.1 Testing Principles for Junction Capacitors C_{iss}, C_{oss}, C_{rss}

The definitions of junction capacitors C_{iss}, C_{oss}, and C_{rss} are described in Sect. 3.3.1. The test circuits are shown in Fig. 6.26. For testing C_{iss}, connect the H_{pot} and H_{cur} terminals of the LCR meter to the gate, and the L_{pot} and L_{cur} terminals to the source. Apply a high-voltage bias through an inductor to the drain. For testing C_{oss}, connect the H_{pot} and H_{cur} terminals of the LCR meter to the drain, and the L_{pot} and L_{cur} terminals to the gate and source. Apply a high-voltage bias through an inductor to the drain. For testing C_{rss}, connect the H_{pot} and H_{cur} terminals of the LCR meter to the drain, and the L_{pot} and L_{cur} terminals to the gate and source. Apply a high-voltage bias through an inductor to the drain.

When measuring, the LCR meter is set to output a small AC signal with a frequency of f and an amplitude of V_{AC} from the H terminal and flow back from the L terminal. The phase angle θ is obtained according to the return signal, and then according to the impedance calculation formula $Z = R + jX$, the desired junction capacitance is in the imaginary part, and the junction capacitance is obtained by $C = |Z| \cdot \sin\theta$, $X_c = 1/2\pi f$, |Z| is the modulus of the return signal.

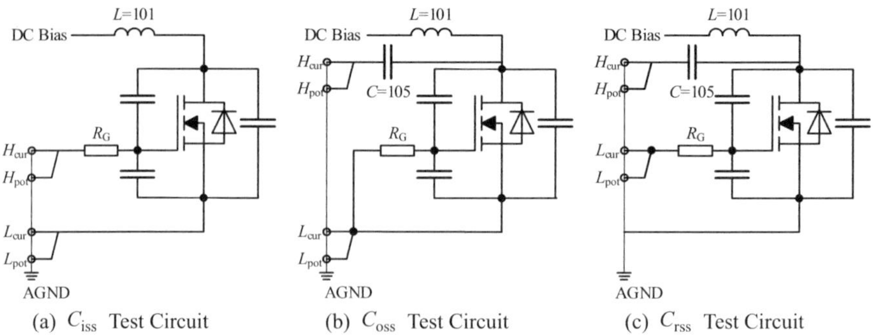

Fig. 6.26 Test circuits for junction capacitors C_{iss}, C_{oss}, C_{rss}

6.1.11.2 Challenges and Considerations for Testing Junction Capacitors C_{iss}, C_{oss}, C_{rss}

(1) When using an LCR meter to measure junction capacitors, high frequency AC signals are used, which are highly susceptible to interference. During testing, if soft cables are used for measurements, secure them to prevent fluctuations that can cause measurement phase shifts and result in unstable measurements. If measurements are made through PCB traces, ensure the measurement traces are free from other signal interference, use dedicated traces, and establish a reference ground or shielding to improve signal immunity.

(2) Before using the LCR meter for testing, compensate the measurement circuit, including open and short-circuit compensation and load compensation, to eliminate parasitic resistances in the measurement circuit, such as traces and test fixtures, to avoid adding these parasitic resistances to the measurement value and causing deviations.

(3) During the measurement of C_{iss}, ensure the applied signal amplitude V_{AC} is less than the device's $V_{GS(th)}$ to avoid incorrect conduction of the device, which can affect the accuracy of the test.

(4) During testing, when applying DC bias to the drain, an inductor should be used in series to block AC signals. For testing C_{oss} and C_{rss}, a large capacitor (much larger than the measurement capacitor) should be connected in series to the drain to prevent high-voltage bias signals from damaging the LCR meter.

6.1.12 Gate Resistor R_G

6.1.12.1 Testing Principles for Gate Resistor R_G

The gate resistor R_G of a SiC MOSFET is the parasitic resistance within the gate terminal of the chip. The test circuit for R_G is shown in Fig. 6.27. Connect the H_{pot} and H_{cur} terminals of the LCR meter to the gate, and the L_{pot} and L_{cur} terminals to the source.

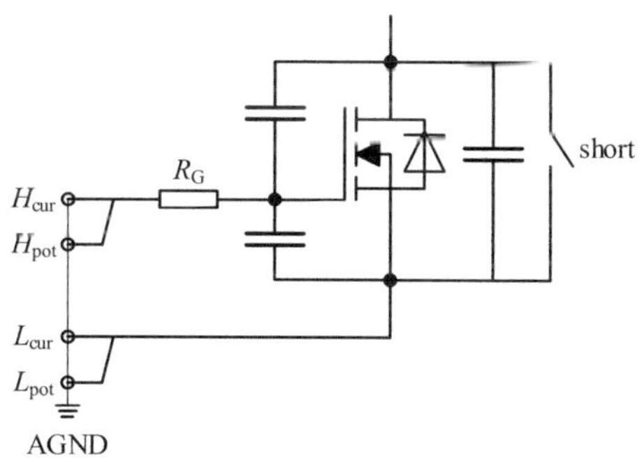

Fig. 6.27 Test circuit for gate resistor R_G

When measuring, the LCR meter is set to output a small AC signal with a frequency of f and an amplitude of V_{AC} from the H terminal and flow back from the L terminal. The phase angle θ is obtained according to the return signal, and then according to the impedance calculation formula $Z = R + jX$, the desired junction capacitance is in the imaginary part, from $R = |Z| \cdot \sin\theta$, $X_c = 1/2\pi f$, |Z| is the modulus of the return signal, thus obtaining |Z| is the modulus of the return signal, thus obtaining |Z| is the modulus of the return signal, thus obtaining R_G.

6.1.12.2 R_G Testing Challenges and Considerations

(1) Measuring junction capacitance with an LCR meter uses high-frequency AC signals, which are highly susceptible to interference. Ensure that measurement cables are fixed to prevent fluctuations that can cause phase shifts and unstable measurements. If measurements are made through PCB traces, eliminate other signal interferences, use dedicated traces, and set up reference grounds or shielding to improve signal immunity.
(2) Before testing with the LCR meter, compensate the measurement circuit for open and short-circuit conditions and load compensation to remove parasitic resistances from the measurement circuit, such as traces and test fixtures, to avoid deviations in measurement results.
(3) During the measurement of R_G, ensure that the applied signal amplitude V_{AC} is less than the device's $V_{GS(th)}$ to avoid unintentional conduction of the device which could affect the test.

6.1.12.3 Switching Characteristics and Gate Charge Q_G

The definitions, testing principles, and technical details for the switching characteristics and gate charge Q_G of SiC MOSFETs have been detailed in Sect. 3.3.3 and Chap. 5. Therefore, they will not be repeated here. In production testing, the following challenges and considerations are important:

(1) Switching characteristics are high-voltage test parameters. Ensure personal safety, and verify that cables, PCB boards, connectors, and other components comply with safety regulations. Also, ensure that test equipment and related devices have adequate safety protection.
(2) Switching characteristics involve high-current and high-speed testing. Control parasitic inductance in the test loop, contact resistance, and parasitic capacitance in the inductive load to minimize their impact on the test results.
(3) For SiC MOSFETs, switching speeds are faster compared to Si IGBTs. Ensure that the gate driver's dynamic response speed is adequate and that the gate drive return path is separate from the main power current return path.
(4) Use high-speed measurement probes for voltage and current, and ensure that the bandwidth and sampling rate of the acquisition card or oscilloscope match the speed of the SiC MOSFET.

(5) Note that $V_{GS(th)}$ drift can cause asymmetry in the gate charge curves for both turn-on and turn-off.

6.2 Production Testing

6.2.1 Overview of Production Testing

Laboratory characterization aims to measure the device's performance as close as possible to the ideal values, while production testing focuses on quickly, stably, and reliably screening large quantities of devices. The testing requirements for production testing differ significantly from laboratory testing. Currently, production tests for SiC devices include CP testing, WLBI testing, KGD testing, DBC testing, PLBI testing, ACBI testing, and FT testing.

6.2.2 CP Testing

6.2.2.1 Basic Principles of CP Testing

CP (Chip Probing) testing, also known as wafer testing, is a crucial process in chip manufacturing, positioned between wafer fabrication and packaging testing. Chips are arranged in a regular pattern on the wafer, and CP testing involves using a prober and tester to evaluate the functional and electrical parameters of the chips on the wafer, marking defective ones.

CP testing holds significant value in chip manufacturing. Firstly, it effectively reduces the cost waste of subsequent packaging tests. As the cost of SiC chips decreases, the proportion of packaging costs in SiC devices, especially power modules, increases. By using CP testing to eliminate defective chips in advance, the packaging cost of these chips can be saved. In other application areas, high-cost advanced packaging methods have been developed to reduce chip packaging area and improve performance, such as MCM packaging for chipsets, memory products, or CPUs. These packaging forms involve more complex processes and higher costs, similarly necessitating CP testing to reduce costs. Secondly, CP testing provides wafer production information that can be fed back to chip designers and wafer manufacturing stages, offering a basis for improving quality, refining processes, optimizing chip designs, and validating development. Additionally, CP testing results can be used to grade chips based on performance and target different markets.

A typical CP testing system includes a prober, tester, and probe card. The prober is responsible for wafer handling, accurately transferring the wafer to the wafer chuck, and aligning the probe card's needles with the pads on the wafer using its built-in vision system. The probe card is fixed on the prober to establish an electrical

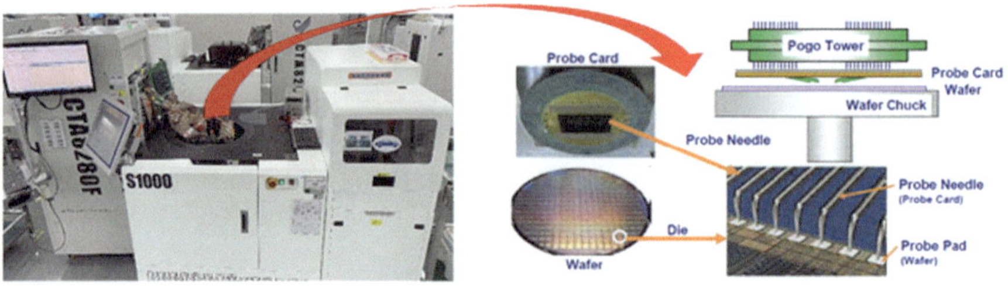

Fig. 6.28 CP testing system

connection between the tester and the pads on the wafer die. The tester performs electrical performance tests on the chips and outputs the results, as shown in Fig. 6.28.

Figure 6.29 shows a typical prober, which consists of a wafer loader and a wafer test stage. The loader handles the accurate transfer of wafers from the cassette to the chuck, including the wafer loader platform, robotic arm, pre-align module for notch positioning and eccentric adjustment, and optical character recognition (OCR) module for wafer ID reading. The stage is responsible for the accurate alignment of the probe card with the wafer pads and the wafer testing, including the chuck, the E1E2 vision system for pad and needle alignment, and the probe card holder.

The prober is the core equipment for CP testing. Electroglas produced the world's first commercial wafer prober in 1963, pioneering the field. In 1982, Electroglas produced the first fully automatic 6-inch ultra-precision wafer testing system, the EG2001, paving the way for automation in testing. The prober has evolved from manual, semi-automatic, and fully automatic to intelligent production, with wafer sizes increasing from 4 and 6 inches to primarily 8 and 12 inches. After decades of development, commercial probers are now highly mature, with fully automatic probers being the market mainstream, with 8/12 inches as the standard sizes. With the development of emerging fields such as artificial intelligence, new energy, and 5G, probers are also advancing towards automation, specialization, and precision.

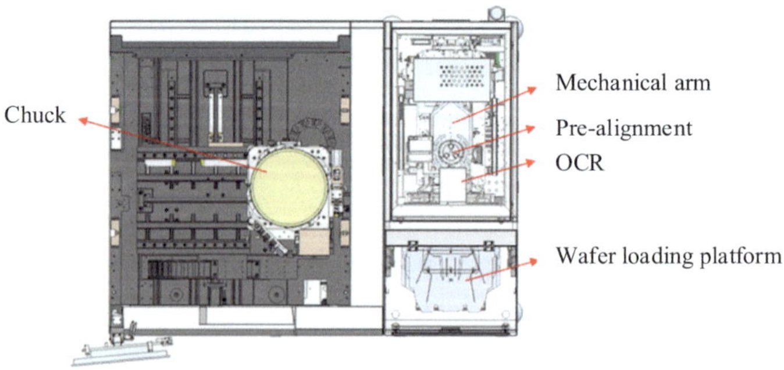

Fig. 6.29 Prober

Due to limitations in SiC raw material sizes and market demand, current SiC wafers are primarily 6 inches, with a small number of 4-inch wafers in production. As the SiC industry transitions to 8-inch wafers, the probe stations used for SiC wafer CP testing will predominantly be for 8-inch wafers in the future.

Currently, SiC wafer thicknesses mainly range from 100 to 700 μm. To further reduce SiC device losses, SiC wafers are moving towards thinner thicknesses. To ensure the safety of thin wafers during testing, probers have developed mechanical structures for handling thin wafers, such as Bernoulli robotic arms for non-contact handling. Additionally, since SiC chips are high-voltage devices, the voltage testing range is from 800 to 3000 V, with some products even requiring 6500–10000 V. Typically, the chuck's leakage current needs to reach the nA or even pA level. This requires probers to handle higher voltages while ensuring low leakage currents to guarantee testing safety and accuracy.

CP testing can be divided into three steps:

(1) The prober transfers the wafer to the chuck, aligns the chip pads with the probes under the vision system's assistance, and connects the probe card to the tester.
(2) The tester applies input signals to the chip and collects output signals to determine whether the chip's functionality and performance meet design requirements under different operating conditions.
(3) The test results are transmitted to the prober via a communication interface, which then marks the chips and creates a mapping of the wafer.

The mapping can show whether devices pass or fail and display the distribution of certain parameters, as shown in Fig. 6.30.

Fig. 6.30 Mapping

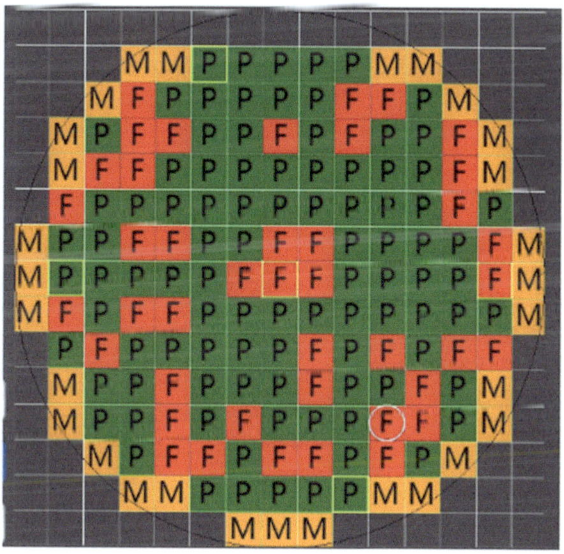

6.2.2.2 Challenges in CP Testing

1. **High-Voltage, High-Current Chuck**

During CP testing, the chuck plays a crucial role in the test results, and the demands for high-voltage and high-current testing present new challenges for chuck design.

Currently, the voltage limit of most probers is below 500 V, and exceeding this can cause chuck breakdown. Additionally, SiC device testing requires very low leakage currents, typically needing to be less than 1nA for voltages above 1 kV. The solution involves improving the insulation design of the chuck and its modules, such as using high-purity ceramic materials and ensuring high insulation resistance, usually greater than 10^{12} Ω, with proper shielding and protection to prevent breakdown under high voltage.

Moreover, SiC device testing has high requirements for parasitic resistance, with the chuck's contact resistance needing to be less than 1 mΩ. This is typically achieved by using materials with uniform conductivity, such as gold-plated or copper alloy-based chucks. Additionally, the flatness and surface roughness of the chuck must be ensured, with flatness less than 10 μm and roughness controlled to Ra 0.04 to reduce wafer contact resistance.

Figure 6.31 shows a dense hole gold-plated suction cup. The wafer carrier on the probe station for SiC wafers also adopts the same design concept. Compared with the traditional groove design, the wafer carrier with dense hole adsorption structure can not only use the suction cup as a drain and achieve full contact with the backside of the SiC wafer, but also avoid the breakage of thinner SiC products due to the suspended back groove design during probe testing.

2. **High-Voltage Probe Cards and Protection Strategies**

During CP testing, the probe contacts the chip pads, transmitting the applied voltage or current signals to the pads. Handling the probe's voltage withstand is a challenge. Low-resistance probes are commonly used, with materials chosen for their excellent conductivity to minimize heating and arcing due to high resistance.

Paschen's Law provides the relationship between vacuum level and breakdown voltage, as shown in Fig. 6.32. Point B represents standard atmospheric pressure, and

Fig. 6.31 Gold-plated chuck with micropores

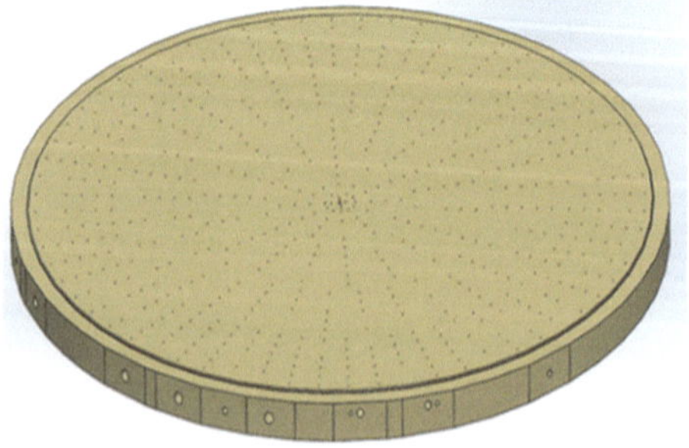

increasing pressure towards B (positive pressure) or decreasing pressure towards A (vacuum) can enhance voltage withstand.

Positive pressure schemes involve introducing dry air, nitrogen, or SF6 at pressures above 5 bar into the testing chamber to avoid discharge or arcing. Alternatively, specialized probe card designs, such as those shown in Fig. 6.33, use channels to deliver compressed dry gases into the sealed probe card chamber to reduce dry gas consumption.

The vacuum scheme uses vacuum insulation and arc extinction technology with high vacuum levels, generally ranging from 10^{-2} to 10^{-5} Pa. This vacuum provides much higher insulation strength than standard atmospheric air and SF6, even exceeding transformer oil insulation strength, allowing for very small electrical gaps.

Another method of preventing arcing is spraying fluorinated oil mist onto the wafer surface using compressed air, creating a dense oil film to isolate the wafer from the air, as shown in Fig. 6.34.

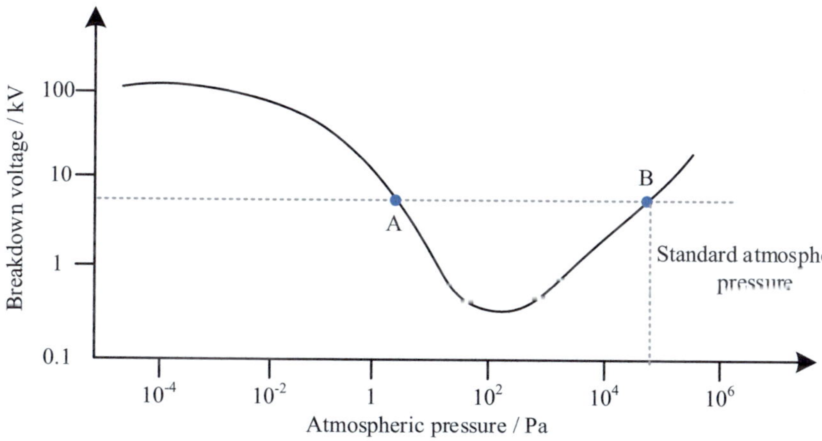

Fig. 6.32 Relationship between vacuum level and breakdown voltage

Fig. 6.33 Lupo probe card structure

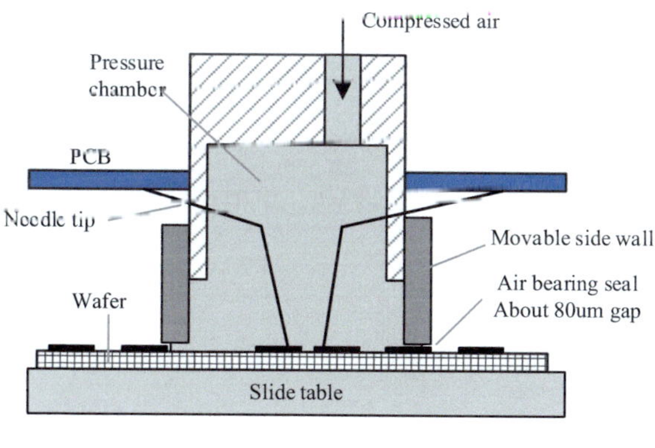

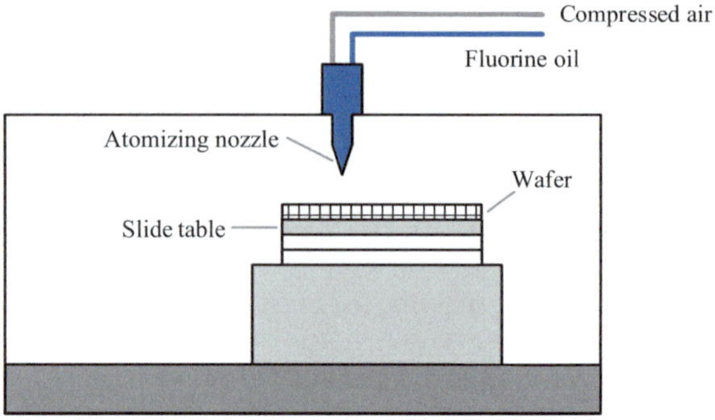

Fig. 6.34 Fluorinated oil arcing prevention scheme

3. **Thin Wafer Handling**

Due to the brittleness of thinned SiC wafers, special designs are required for handling to ensure wafer safety. Bernoulli non-contact adsorption robotic arms and pre-align structures allow for non-contact wafer handling, and the chuck structure should be replaced with a micropore design to reduce local contact stress.

As shown in Fig. 6.35, the Bernoulli robotic arm uses multiple cyclone-type suction cups, with air flowing through the suction cups to create a rotating airflow and negative pressure in the central area. This creates an upward adsorption force on

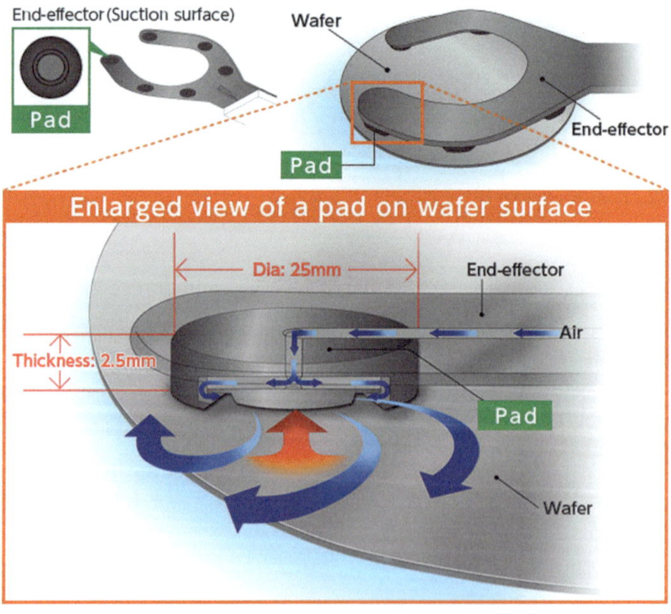

Fig. 6.35 Bernoulli robotic arm principle [1]

the wafer, holding it onto the suction cup. Pressure valves can adjust the air pressure to accommodate various wafer types, including thin, Taiko, and standard wafers.

Thinned SiC wafers are very fragile, and traditional groove-style chuck suction cups can easily cause breakage upon vacuum release or result in ring-shaped failures. For these reasons, thinned wafer testing typically uses gold-plated chuck suction cups with radial micropore designs, which can handle wafer warping and improve safety with vacuum zone release.

4. **High-Power Testing Heating Issues**

High-voltage, high-current testing of SiC wafers presents significant heating issues. If the heat generated during testing is not dissipated in time, it can lead to unstable test data or even chip damage. Currently, using gold-plated micropore chucks can ensure good wafer contact and reduce contact resistance, partially alleviating heating issues.

Additionally, air cooling systems can be used to address heating problems by configuring structures that blow ambient or cooled air to carry away the heat generated during testing, maintaining the required wafer testing temperature. This structure can also adjust airflow or inlet temperature to maintain the target temperature.

Based on current testing requirements and the development status of SiC chips, key parameters for mainstream probe cards for SiC wafers are summarized in Table 6.1

Using the S1000 series probe station independently developed by Hangzhou Changchuan Technology Co., Ltd. as an example, it meets the requirements of SiC CP testing, as shown in Fig. 6.36. This probe station is mainly used for testing power and analog products. It can automatically handle and test Taiko and fully-thinned wafers with thicknesses ranging from 50 to 1250 μm. Under normal temperature

Table 6.1 Key parameters of mainstream probe stations for SiC wafers

No.	Content	Index
1	Cassette	4″, 6″, 8″ Cassette
2	Wafer size	100, 150, 200 mm
3	Wafer thickness	100–700 μm
4	Wafer warpage	± 3 mm
5	Grain size	250–100,000 μm
6	XY test accuracy	± 1.5 μm
7	Z test accuracy (um)	± 2.5 μm
8	Normal high temperature	Normal temperature 150 °C ± 1 °C
9	Suction cup	Gold Plated Chuck with Micropores
10	Flatness	≤ 15 μm
11	Contact resistance	< 1 mΩ
12	Leakage current	< 1 nA
13	Withstand voltage/current conduction capacity	3000 V+/200 A+

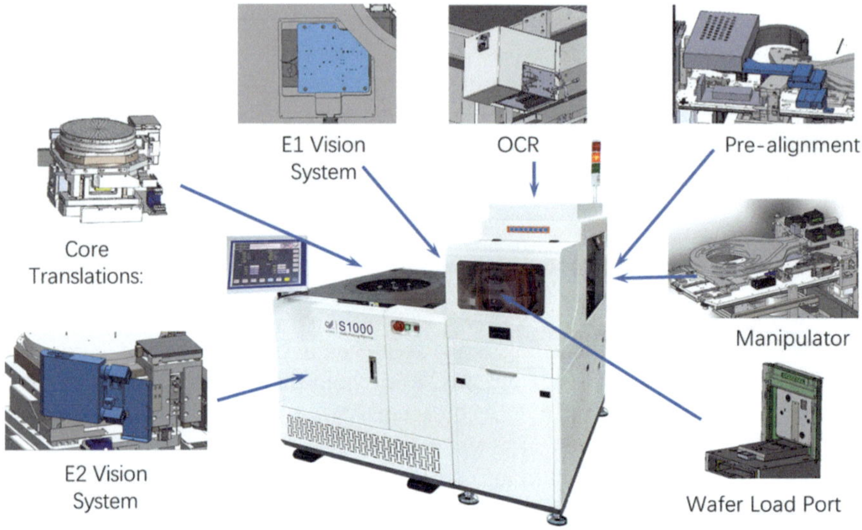

Fig. 6.36 S1000 series probe station by Changchuan Technology

conditions, the whole machine leakage current is in the pA range, and under high-temperature conditions of 150 °C, the leakage current is below 1nA. It meets the requirements for 3000 V/200 A breakdown voltage and current conduction testing, with a contact resistance below 1 mΩ. Additional features such as rapid cooling, anti-arcing, air blow, and manual wafer loading are optional, providing customizable processing capabilities. This machine has been mass-produced in several FAB plants and has received positive feedback from customers. The production yield and test result stability of the machine can benchmark the mainstream machines in the industry. Continuous R&D investments are being made to address future needs such as high voltage/high current, low noise, and low leakage.

6.2.3 WLBI Testing

6.2.3.1 Basic Principles of WLBI Testing

The reliability requirements for chips in main drive inverters and other high-end applications are extremely high. However, the reliability issues of SiC MOSFETs have not been completely resolved, posing major challenges and research directions. Despite this, the demand for SiC MOSFETs in electric vehicles continues to increase at present, creating a contradiction with the immature technology of SiC MOSFETs.

Aging is the most widely used method for early reliability screening. The aging process subjects semiconductor devices to high stress conditions, such as high temperature and high voltage, causing the devices to operate under heavy load, thereby inducing failures in any defects present in the devices. Devices that undergo aging screening will enter the random failure zone early, significantly reducing the device failure rate and achieving failure screening. Therefore, to minimize failures due to reliability issues during the use of SiC MOSFETs, WLBI testing can be used to pre-select defective devices that have a risk of external failure in applications. Additionally, WLBI testing helps to effectively reduce the entry of defective products into subsequent production and packaging processes, reducing overall manufacturing costs and improving production yields.

Currently, the aging projects used in WLBI testing are High-Temperature Reverse Bias (HTRB) and High-Temperature Gate Bias (HTGB), whose specific principles and purposes will be detailed in Sect. 6.3. HTRB test conditions generally select 80% of the nominal breakdown voltage value and the maximum operating junction temperature or 5 °C below the maximum storage temperature. HTGB test conditions generally select 80% of the nominal gate breakdown voltage value and the maximum operating junction temperature or 5 °C below the maximum storage temperature.

The process of WLBI testing for SiC MOSFETs can be selected based on the actual situation of the product, generally including the following three methods:

(1) Conduct static parameter testing before aging, then perform wafer aging, and conduct static parameter testing again after aging. Defective chips are screened out by comparing the static parameter changes before and after aging.
(2) Do not conduct static parameter testing before aging, directly perform wafer aging, and conduct static parameter testing after aging. Defective chips are screened out by judging whether the static parameters are within the qualified range.
(3) Do not conduct static parameter testing before aging, directly perform wafer aging, and use the leakage current real-time detection function and $V_{GS(th)}$ testing function of the WLBI test system during the aging process. Defective chips are screened out by judging the change rates of the leakage current and threshold.

Based on the basic principles and actual needs of WLBI described above, the current WLBI test system mainly consists of several units including wafer fixtures, heating units, automatic loading/unloading units, aging power supply, vacuum/nitrogen pipelines, and user interaction computers. Its functional structure is shown in Figs. 6.37 and 6.38 shows the actual picture of the WLBI aging system by Suzhou Semight Instruments Co., Ltd.

(1) Wafer Fixtures: Contact and power up each bare chip pad on the SiC wafer.
(2) Heating Unit: Heats the SiC wafer, supporting a maximum temperature of at least 175 °C.
(3) Automatic Loading/Unloading Unit: Uses mechanical arms and other mechanical devices to automatically load and unload SiC wafers and fixtures.
(4) Source Meter: Consists of sub-units such as the aging power supply, parameter measurement circuit, and leakage current collection circuit. The aging power supply provides the voltage stress required for HTGB and HTRB aging

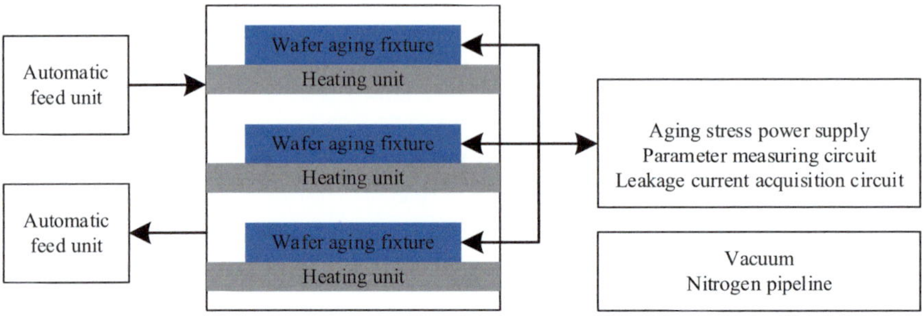

Fig. 6.37 Functional structure block diagram of WLBI test system

Fig. 6.38 WLBI aging system by Suzhou Semight Instruments Co., Ltd

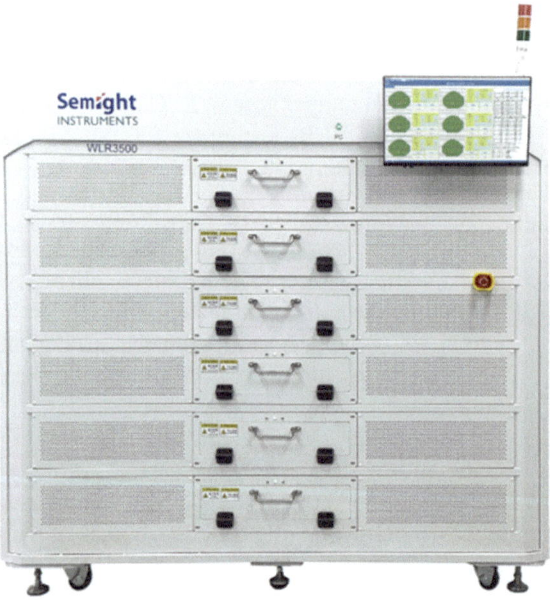

modes. The parameter measurement circuit measures typical parameters of SiC MOSFETs before, during, and after aging, including but not limited to threshold voltage, body diode forward voltage, gate-drain leakage current, and source-drain leakage current. The leakage current collection circuit collects and saves the leakage current in real time during the aging process.

(5) Vacuum/Nitrogen Pipelines: Provide the vacuum and nitrogen required by the aging system.
(6) User Interaction Computer: Provides users with interactive software to complete aging configuration information such as aging time, aging conditions, and aging temperature.

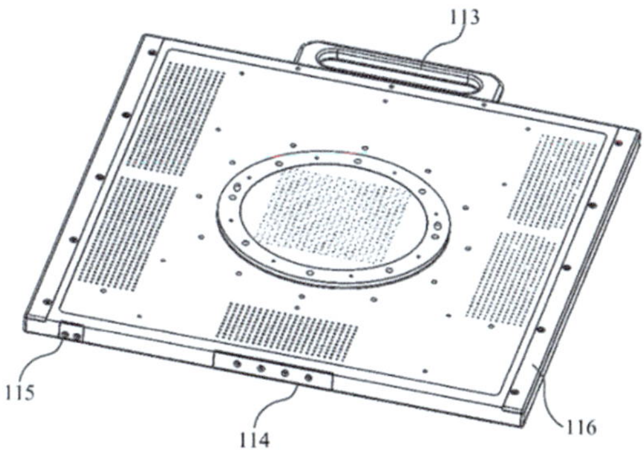

Fig. 6.39 Wafer aging fixture

6.2.3.2 Challenges of WLBI Testing

1. **Wafer Fixtures**

Wafer fixtures are the core units of the system and the most challenging to design. They need to accurately connect the PADs of each chip to the source meter, applying and measuring voltage stress.

(1) **Contacting Large-Area Chips**

SiC MOSFET chips used in main drive inverters typically measure 5 mm × 5 mm, which means a 6-inch wafer contains approximately 500 chips and an 8-inch wafer contains about 900 chips. If the chip size is smaller, the number of chips per wafer increases. Ensuring effective contact with the gate, drain, and source PADs of each SiC chip, and connecting them to the source meter for aging and testing, is a significant challenge in fixture design.

Figure 6.39 shows the design of a wafer aging fixture by Suzhou Semight Instruments Co., Ltd. The center uses spring probes (Fig. 6.40) to press onto the aged SiC wafer. The industry-leading wafer aging probe card can handle up to 3600 channels, supporting the aging and testing of 3600 SiC bare chips, covering 4-inch, 6-inch, and 8-inch SiC wafers. The maximum voltage support is up to 3500 V, and the minimum pitch of the spring probes can reach 0.1 mm (edge-to-edge).

Fig. 6.40 Spring probe

(2) **Precise Control of Spring Probe Marks**

The reliability of gate oxide is a critical issue for SiC MOSFETs. To ensure that no new issues are introduced to the gate PAD during testing, it is essential to strictly control the depth and area of the spring probe marks (Fig. 6.40).

Controlling the marks of each spring probe and ensuring that the probe marks on each large-area chip are within effective control ranges is a challenge for automated measurement systems. One feasible solution is to strictly control the extension path of the spring probes to manage the mark depth (vertical direction) and design the shape and curvature of the probe heads to control the mark area (horizontal direction).

Currently, the industry-leading depth control can be within 0.2 μm, and the horizontal area diameter can be within 40 μm. Of course, the performance of probe marks varies with different materials of SiC wafers, such as nickel-palladium-gold and aluminum. The results discussed here represent the highest levels under ideal conditions.

2. **High-Voltage Arcs**

When conducting HTRB aging on SiC MOSFET chips, the reverse voltage applied is typically the nominal voltage rating of the chip, generally ranging from 650 to 6500V. Under normal conditions, high voltages can cause discharge phenomena, i.e., high-voltage arcs, within the fixture via the spring probes, preventing proper HTRB aging.

Various methods exist to suppress high-voltage arcs, including using vacuum environments, injecting high-pressure inert gases, and injecting special chemical gases. The vacuum environment's suppression ability is limited, challenging support for voltages above 3000 V. Injecting high-pressure inert gases requires gas pressures much higher than atmospheric pressure, demanding higher rigidity from the fixture structure. Injecting special chemical gases can meet arc suppression requirements for voltages above 3000 V but poses safety risks to the operating environment.

3. **Applying Stress**

The WLBI testing system must handle a large number of large-area SiC MOSFET chips, compatible with both HTRB and HTGB modes, while ensuring circuit protection to avoid system power damage due to short-circuiting failed chips. Calculating based on the industry-leading level of 3600 channels, the cost of the source meter for testing a wafer is 3600 times the cost of the circuit for one channel. If the system supports the simultaneous aging of multiple wafers, the system's cost will further increase. Balancing the number of channels, leakage current measurement accuracy, compatibility with HTRB and HTGB, and system cost has become a significant challenge for the WLBI testing system.

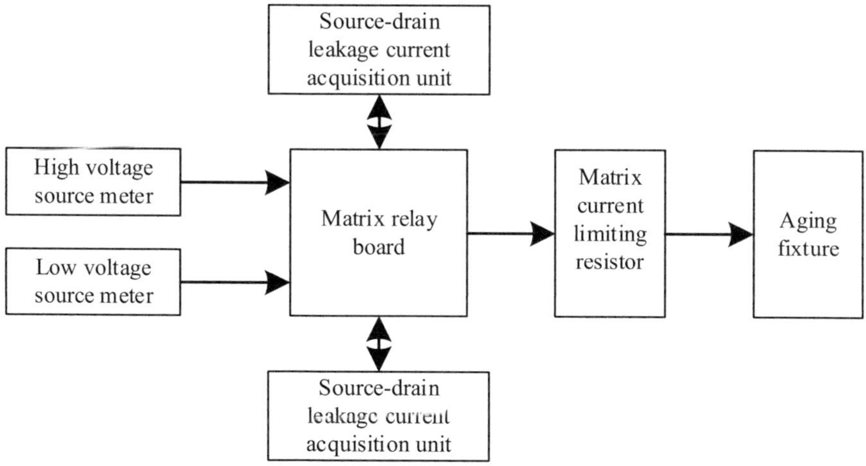

Fig. 6.41 Typical hardware structure for applying stress in WLBI

(1) Multi-Channel High-Voltage Source

Figure 6.41 shows the typical hardware structure for applying stress in WLBI testing system. A high-voltage source meter provides the high voltage needed for SiC MOSFET reverse bias aging. Each high-voltage channel passes through a source-drain leakage current collection unit for real-time monitoring of leakage current during aging. Each high-voltage channel connects a current-limiting resistor to protect against short circuits caused by failed devices during aging.

(2) Switching Between HTRB and HTGB

HTRB requires voltages of thousands of volts, whereas HTGB only needs − 20 to + 50 V. To meet these vastly different voltage requirements, the system uses a dual power supply design, with high-voltage and low-voltage source meters respectively providing voltage stress for HTRB and HTGB. This design ensures both the voltage range and voltage accuracy required for different voltage conditions.

The switching between HTRB, HTGB, and $V_{GS(th)}$ measurements is achieved through a matrix relay board, as shown in Fig. 6.42. During HTRB aging, the high-voltage source meter works with the high-voltage drive board, and the source-drain relay board is integrated into the system to start collecting leakage current. During HTGB aging, the low-voltage source meter works with the low-voltage drive board, and the gate-source relay board is integrated into the system to start collecting leakage current. During $V_{GS(th)}$ measurements, the relay board switches to measurement mode, and a high-precision source meter connects to the system to complete the threshold voltage measurement of the tested SiC MOSFET.

(3) Current Limiting Protection

During the aging process, a failing chip may exhibit a short-circuit condition, necessitating a short-circuit protection function in the system to prevent the failing chip

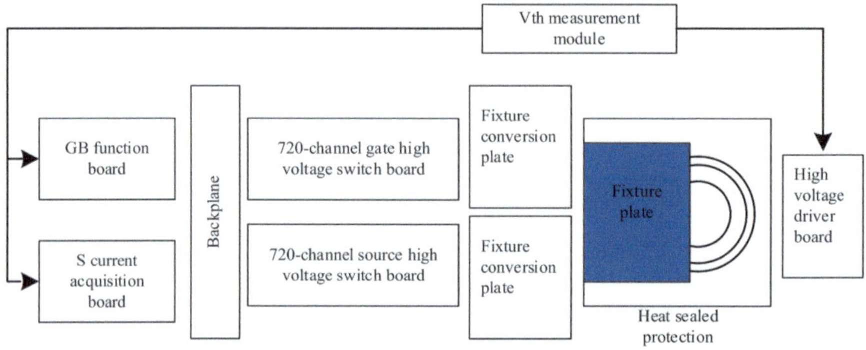

Fig. 6.42 Switching of HTRB, HTGB, and $V_{GS(th)}$ measurement

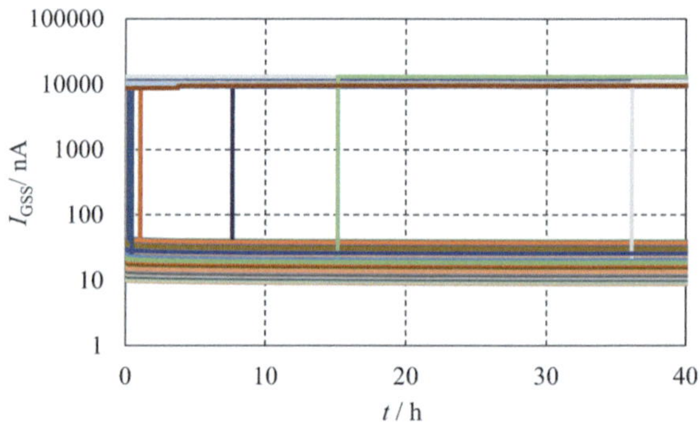

Fig. 6.43 SiC MOSFET HTGB aging failure

from affecting other functional chips and channels. When a SiC MOSFET fails during HTGB aging, it manifests as a short condition, with a sudden increase in leakage current, as shown in Fig. 6.43.

As shown in Table 6.2, the high resistance characteristic of the SiC MOSFET, combined with the extremely low leakage current during normal aging, can be cleverly utilized by using a current-limiting resistor to achieve current limiting during the aging process. During the aging of normal devices, the leakage current is at the microampere level, and the voltage distributed to the current-limiting resistor is limited under different voltages. Calibration and compensation can ensure that the voltage at the load end remains within accuracy and error control ranges under the condition of series current-limiting resistors.

4. **High-Temperature Heating**

To meet the requirements of industry customers and industry standard guidelines, and to quickly and effectively screen out early failures of chips, the WLBI testing system needs to support aging and testing temperatures of at least 175 °C. SiC wafers exhibit warping and thermal expansion, and their hardness is much higher than that of Si, presenting challenges in heating the SiC wafers. To avoid unintended failures

Table 6.2 Suppression capability of nitrogen environment at different pressures on high-voltage arcs

Source meter voltage (V)	Current-limiting resistor (MΩ)	Leakage current (μA)	Resistor power (W)	Resistor voltage drop (V)	Chip voltage (V)	Chip resistance (MΩ)	Chip power (W)
2000	2.5	200	1.0E−01	500	1500	7.5	3.0E−01
1300	2.5	40	4.0E−03	100	1200	30	4.8E−02
1500	2.5	50	6.3E−03	125	1375	27.5	6.9E−02
2000	2.5	300	2.3E−01	750	1250	4.2	3.8E−01
2000	2.5	400	4.0E−01	1000	1000	2.5	4.0E−01
2000	2.5	600	9.0E−01	1500	500	0.8	3.0E−01
1200	2.5	20	1.0E−03	50	1150	57.5	2.3E−02
1200	2.5	400	4.0E−01	1000	200	0.5	8.0E−02
1200	2.5	0.5	6.3E−07	1.25	1198.75	2397.5	6.0E−04
25	2.5	0.1	2.5E−08	0.25	24.75	247.5	2.5E−06
30	2.5	0.2	1.0E−07	0.5	29.5	147.5	5.9E−06
15	2.5	0.005	6.3E−11	0.0125	14.9875	2997.5	7.5E−08

due to non-ideal factors affecting the tested chips, the WLBI testing system's high-temperature heating control system needs to have a fast heating rate, good temperature uniformity, and minimal heating overshoot.

There are two common heating methods: one is oven-style heating, where the tested product or aging fixture is placed in an oven for heating; the other is conduction heating, where a heating rod or other heat source conducts heat to the tested product. In either method, it is necessary to calibrate the chip junction temperature and the ambient temperature to ensure that the chip junction temperature remains within the set aging conditions.

5. **Wafer Size Compatibility**

Currently, SiC wafers are primarily 6 inches in size, with some international brands moving towards 8-inch wafers. Considering the trend in wafer size and the cost of testing equipment, the WLBI testing system and wafer fixtures need to be compatible with both 6-inch and 8-inch wafer sizes. There are typically two solutions: the first is to design test fixtures for 6-inch and 8-inch wafers separately, allowing compatibility with both sizes on one testing system by switching fixtures; the second is to design the wafer fixture for the larger 8-inch wafer size and switch probe cards to accommodate the smaller 6-inch wafers. The second solution is lower in cost but places higher demands on material management at the testing site.

Table 6.3 Effect of chip yield on module yield

Chip yield (%)	Module yield (12 Chips) (%)	Module yield (36 Chips) (%)
99	89	72
98	78	52
90	28	3

6.2.4 KGD Testing

6.2.4.1 Basic Principle of KGD Testing

CP testing uses a probe station to test the entire wafer before dicing, unable to cover tests such as short-circuit, switching, reverse recovery, and gate charge, thus not fully assessing chip parameters. Additionally, dicing after CP testing can cause chip damage, resulting in a yield loss of about 1–2%. Directly packaging chips that have undergone CP testing and dicing into power modules can significantly reduce module yield due to chip abnormalities, as shown in Table 6.3.

If the chip yield is 99%, the yield of a power module containing 12 chips is 99% to the 12th power, which is only 89%, and the yield of a module containing 36 chips is only 72%. If the chip yield is 90%, the module yield drops further to just 3% for a 36-chip power module, which is unacceptable. As previously mentioned, CP testing cannot fully assess chip parameters, and wafer dicing can cause chip damage, making it impossible to achieve 100% chip yield.

To mitigate the impact of these issues, SiC MOSFET chips used in main drive inverter power modules undergo KGD (Known Good Die) testing. KGD testing typically includes static parameter tests at different temperatures, avalanche tests, short-circuit tests, and dynamic tests. Additionally, main drive inverter power modules use multiple chips in parallel. KGD test results allow for parameter matching and BIN classification, selecting multiple chips with consistent parameters for parallel packaging, thereby maximizing the advantages of SiC material.

KGD testing is multi-station testing, with test items, conditions, and sequences generally determined by the chip or module manufacturer based on chip characteristics. Figure 6.44 shows a typical 4-station KGD testing process: the first station is high-temperature DC testing to evaluate DC parameters of SiC MOSFETs after wafer dicing in a high-temperature environment; the second station is high-temperature AC and short-circuit testing, performing dynamic parameter testing and short-circuit capability testing in a high-temperature environment; the third station is room-temperature avalanche testing for avalanche robustness evaluation; and the fourth station is room-temperature DC testing to re-verify chip performance after potentially destructive AC, short-circuit, and avalanche tests. Chips that pass all test stations are classified by parameters and proceed to packaging.

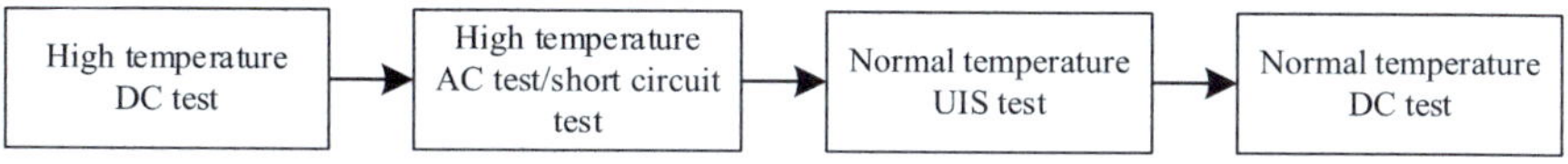

Fig. 6.44 Typical 4-station KGD testing process

There are many suppliers of KGD testing systems, each with different approaches. Figure 6.45 shows a block diagram of a mainstream KGD testing system structure, and Fig. 6.46 shows a physical image of the Suzhou Semight Instruments Co., Ltd KGD testing system, which has certain universality and representativeness.

(1) Loading/Unloading Area: Mechanical handling units complete the transportation of specified material boxes to

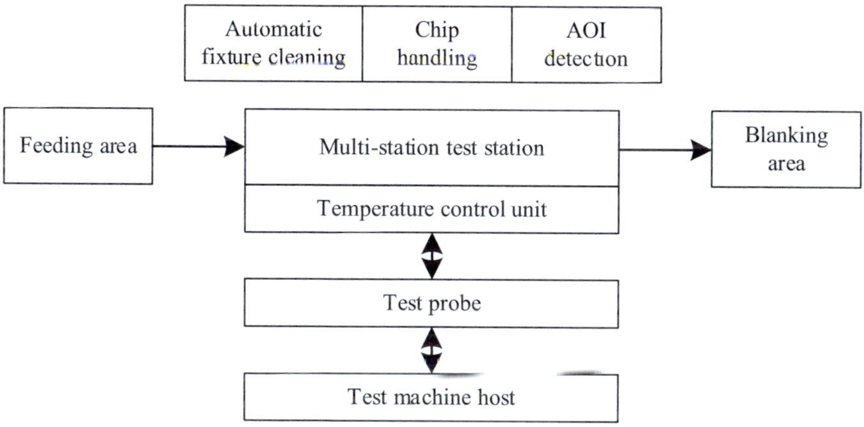

Fig. 6.45 Block diagram of KGD testing system structure

Fig. 6.46 Suzhou Semight Instruments Co., Ltd KGD testing system

(2) Test station: Multiple test stations complete different test contents, including DC test, avalanche test, short-circuit test and dynamic test. They need to have high-temperature control function to meet the performance evaluation of SiC chips at high temperatures.
(3) Chip handling unit: responsible for completing the transportation of chips from the loading station to the testing station and from the testing station to the unloading area.
(4) AOI inspection function: including AOI inspection in the loading area and the unloading area. The AOI in the loading area is responsible for detecting whether the cutting process has caused damage to the chip, and the AOI in the unloading area is responsible for detecting whether the KGD test has caused damage to the chip.
(5) Automatic fixture cleaning unit: responsible for cleaning the fixture to prevent the test fixture from being contaminated by cutting debris, glue, and chip failure and explosion.
(6) Test probe: The closer the test probe of the tester is to the chip under test, the more conducive it is to improving the precision and accuracy of the test, especially for AC test projects that require the stray inductance of the link to be as low as possible.
(7) Test machine host: connected to the test probe, providing test power and high-precision current and voltage acquisition and output.

6.2.4.2 KGD Testing Challenges

1. **Test Fixtures**

The test fixture carries the SiC MOSFET chip under test and connects the chip under test and the test head. Its flatness affects the appearance of the chip, its airtightness affects the test safety, and its closure method affects the test efficiency of the system. It is one of the core units of the KGD test machine, and its characteristics directly affect the performance of the KGD test system.

(1) **High-Voltage Arc Suppression**

During DC testing of SiC MOSFETs, the test voltage can exceed 3000 V. Without any environmental treatment, the resulting high-voltage arcs can disrupt testing and severely impact chip quality.

A low-cost solution is to fill the sealed test fixture with high-pressure nitrogen gas above 2.5 bar. This can effectively prevent high-voltage arcing and protect SiC MOSFET chips from oxidation in high-temperature environments.

(2) **Needle Mark Control**

Controlling needle marks is crucial; deep marks affect product reliability, while shallow marks fail to ensure effective test contact, especially under high current tests.

One feasible solution involves precise control of the spring probe's retractable path to manage needle mark depth (vertical direction) and designing the probe tip's shape and curvature to control needle mark area (horizontal direction). KGD testing requires more spring probes due to high current needs, potentially up to hundreds of probes based on single probe current capacity.

2. **Design of Low Parasitic Inductance Pathways**

SiC MOSFETs have faster switching speeds than Si devices, leading to higher turn-off voltage spikes under the same test conditions. Excessive power loop inductance can cause inaccurate switching and short-circuit tests, and may jeopardize the safety of the tested product and the testing system.

This can be mitigated by designing low parasitic inductance test circuits, using busbar capacitors to reduce loop inductance, shortening the connection between the device under test and the test head, and integrating the test fixture directly onto the AC test head.

3. **AOI Visual Inspection**

Various factors such as chip cutting, adhesive residues on the loading blue film, KGD test probes, avalanche, and short-circuit tests can cause chip surface defects. Common defects include dirt, scratches, burn marks, and foreign objects. To address these, the visual inspection system should be equipped with high-performance deep learning algorithms capable of adapting to common defect detection. While the KGD system's AOI function supplements dedicated AOI equipment, it requires nearly 10 μm resolution for effective defect detection.

Figure 6.47 shows common appearance defects of SiC MOSFET chips, including bright spots on the front surface, surface scratches, edge collapse, and dirt.

4. **Overcurrent Protection**

In the case of the same voltage and current ratings, the short-circuit current of SiC MOSFETs is approximately 10 times their current rating, while the short-circuit current of Si IGBTs is about 4–6 times their current rating. Moreover, the chip area

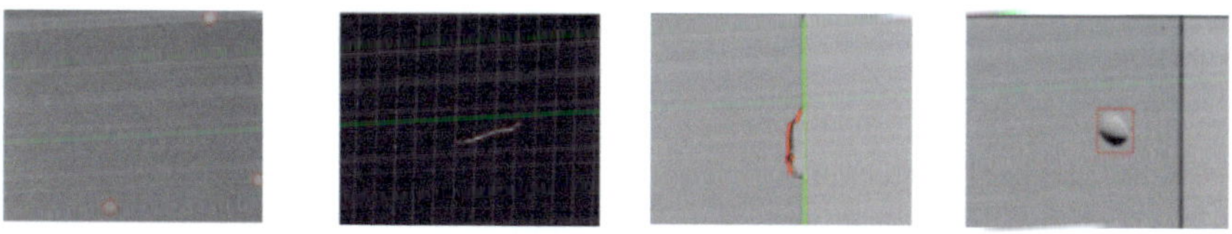

(a) Surface bright spots (b) Surface scratches (c) Edge collapse (d) Contamination

Fig. 6.47 Common SiC MOSFET surface defects

of SiC MOSFETs is roughly 1/3 that of Si IGBTs. Additionally, the short-circuit heating of SiC MOSFETs is concentrated in the epitaxial layer, whereas the short-circuit heating in IGBTs is more evenly distributed. This results in SiC MOSFETs having a short-circuit energy density around 20 times greater than that of Si IGBTs. The short-circuit duration for IGBTs can exceed 10 μs, while for Si MOSFETs, it ranges from 2 to 4 μs, with a higher risk of "blowing the fuse" during a short circuit.

When conducting switching tests, reverse recovery tests, and short-circuit tests, faulty SiC MOSFETs may fail and cause the test circuit to enter a short-circuit state. The current in the circuit increases rapidly, and if the current is not limited, the junction temperatures of both the tested and the replacement devices will rise quickly, potentially leading to failure of the replacement device and "blowing the fuse" in both the tested and replacement devices. These situations can damage probes, test circuits, and test fixtures, increasing maintenance costs and delaying production schedules.

To address these issues, the test equipment should complete detection and current limiting within a short time when overcurrent occurs due to chip anomalies, protecting all components of the test system. Generally, this needs to be completed within 1–2 μs. Additionally, increasing the number of probes and using current sharing techniques can help balance the current flowing through each probe, thereby protecting the system and test fixtures.

5. **High-Temperature Control Design**

To comprehensively test the parameters of SiC MOSFETs and filter out faulty or high-risk chips, high-temperature testing is required for chips in KGD (Known Good Die).

Currently, there are two temperature control solutions in the industry. One solution involves using high-low temperature shock chambers to apply temperature shocks to the test fixture, which allows for rapid temperature changes and supports both high and low-temperature testing. However, this solution has poor temperature uniformity and stability. The other solution integrates heating rods inside or outside the test fixture to control temperature. This method offers high-temperature control precision, good uniformity, and stability, but it only supports high-temperature testing and cannot accommodate low-temperature testing.

6. **Testing Efficiency**

Testing efficiency is a crucial metric for evaluating test equipment, directly affecting production capacity and operational costs. The testing time in a KGD testing system consists of three parts: the first part is the chip handling time, determined by the system architecture and design of moving parts; the second part is the testing time, determined by the product test items and test equipment time; and the third part is the auxiliary function time, including the time for functions such as AOI (Automated Optical Inspection).

Multi-station inline testing can significantly improve testing efficiency, but with the increasing number of test stations, the efficiency does not continuously improve.

On one hand, multi-station inline testing can test different parameters of multiple chips simultaneously, but time savings are limited by the test items that require the longest time. Moreover, more test stations mean more handling operations, as chips need to be moved between different test stations and loading/unloading areas. Additionally, more test stations require more test equipment resources, increasing testing costs. Currently, mainstream KGD testing systems in the industry balance testing efficiency and equipment resources, choosing between 2 and 5 test stations.

6.2.5 *PLBI Testing*

The PLBI (Package Level Burn In) testing for SiC MOSFETs includes two test items: HTRB (High-Temperature Reverse Bias) and HTGB (High-Temperature Gate Bias). The specific purposes and testing principles will be detailed in Sect. 6.3. Currently, there is no industry standard for PLBI testing of SiC MOSFETs; each manufacturer determines the test items (either HTRB, HTGB, or both), test durations, and rotation frequencies of different test items based on the characteristics of the devices and accumulated test data.

SiC MOSFET products include discrete devices and power modules. Depending on the size and characteristics of the packaging, two types of aging test systems have been developed: oven-type and hot-plate-type systems.

6.2.5.1 Oven-Type PLBI Testing System

Taking the BTR-T652 oven-type PLBI testing system from Hangzhou ZoanRel Electronics Co., Ltd as an example, as shown in Fig. 6.48 and Fig. 6.49, it mainly consists of a loading/unloading machine, an AGV (Automated Guided Vehicle) insertion machine, aging test equipment, and an MES (Manufacturing Execution System). It is used for the PLBI testing of discrete SiC MOSFETs.

The loading/unloading machine communicates with the aging test equipment via MODBUS to control the automatic insertion and extraction of aging boards. The

Fig. 6.48 Hangzhou ZoanRel Electronics Co., Ltd' BTR-T652 oven-type PLBI testing system

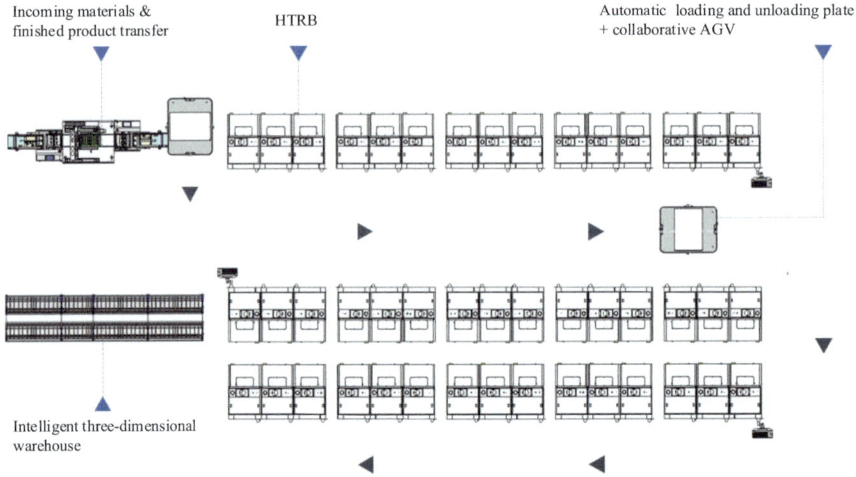

Fig. 6.49 Hangzhou ZoanRel Electronics Co., Ltd' BTR-T652 oven-type PLBI testing system structure

AGV places the aging boards onto the "L"-shaped partition of the aging rack through a side insertion and lifting mechanism, then pushes the boards to the designated position using an extension mechanism. The AGV places the aging boards into the aging cabinet from top to bottom and notifies the upper system, which then automatically closes the cabinet doors.

The maintenance interface of the aging test equipment is on the back of the machine for easy manual maintenance. The operation interface is on the front, equipped with a host computer and a display, which records and saves test data and allows browsing and exporting at any time. Each test channel has its own independent control and sampling system. Real-time test data is transmitted to the host computer control system via RS485 communication for storage, output, and processing, and aging reports and curves can be generated.

The working process of the BTR-T652 oven-type PLBI testing system of Hangzhou ZoanRel Electronics Co., Ltd:

(1) After scanning and calling the MES interface for product inbound, process/batch/product code verification, and tray guiding by the loading/unloading machine, the devices to be tested, transferred from the previous process, are placed onto the aging board. The aging board number and product information (e.g., batch number, product number) are recorded.
(2) The loading/unloading machine assigns the AGV insertion machine to transport and insert the aging board with the devices into the designated channel of the aging test equipment and sends the aging board number and product information to the aging test equipment.
(3) The loading/unloading machine sends the recipe obtained from MES to the suitable channel and starts the aging test.

(4) After the aging test is completed, the loading/unloading machine schedules the AGV insertion machine to remove the tested products and aging boards from the aging test equipment and completes functions such as product screening based on the test results reported by the aging test equipment, result reporting to MES, and outstation (Fig. 6.49).

6.2.5.2 Hot-Plate-Type PLBI Testing System

Taking the T668A hot-plate-type PLBI testing system from Hangzhou ZoanRel Electronics Co., Ltd as an example (As shown in Fig. 6.50), it mainly consists of a robotic arm integrated machine, aging test equipment, and MES. It is used for the PLBI testing of SiC MOSFET power modules.

The pressure testing system is used for pressure testing incoming modules. The incoming pallets are transferred to the pressure testing system via a conveyor line, and an automatic loading and unloading method is used for pressure testing. Qualified products are placed on pallets and transferred to the next process, while unqualified products are placed on a conveyor belt for manual removal.

The automatic loading and unloading system is used for pallet transportation and stacking, including automatic unloading for unqualified products and automatic ID binding functions. It manages the flow of pallets within the equipment system and the unbinding and re-binding of module codes with pallet codes. It includes product conveyor lines, pallet cache stations, pallet positioning stations, and pressure and blocking test NG (Not Good) discharge positions.

The robotic arm's main function is to pick and place modules. After pressure testing and qualification, the modules are picked from the pallet positioning station and placed directly into the blocking test fixture (product scanning is required before placement). After the blocking test, the modules are picked by the robotic arm and placed into empty pallets at the pallet positioning station (product scanning is required before placement).

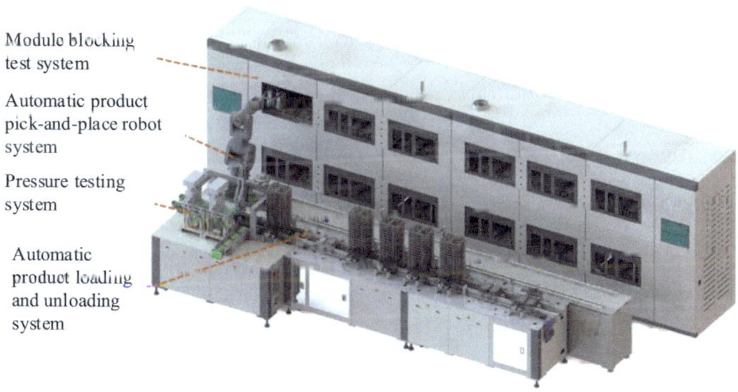

Fig. 6.50 Hangzhou ZoanRel Electronics Co., Ltd' T668A hot-plate-type PLBI testing system

The aging test is conducted at high temperatures. Pressure testing is completed first, followed by high-temperature blocking tests. The test pressure ranges from 0 to 10 kN, and the bottom of the blocking test station is heated electrically to preheat the device. Testing is automatically completed according to the program once the temperature is reached, with a maximum test voltage of 2000 V and a leakage current sample value of 100 mA. This system detects parameters such as terminal voltage, leakage current, case temperature, and junction temperature of SiC MOSFET modules.

Its workflow is the same as that of the oven-type PLBI testing system for single devices.

6.2.6 *ACBI Testing*

6.2.6.1 Basic Principles of ACBI Testing

Although the rapid development of new energy vehicles in China has created a huge demand for power semiconductors, the current stability of batch production for SiC power modules in China is insufficient. This, combined with a lack of experience in high-end applications and a gap in end-use validation compared to international advanced levels, is partly due to the immature technology for validation, reliability testing, and application assessment of SiC power modules by power module manufacturers.

During the development phase of power modules, manufacturers replace power module samples with those used in application endpoints such as inverters, motor controllers, and other power converters to simulate and validate their performance in real applications. This helps to assess the module characteristics, understand its performance in practical applications, identify and eliminate potential risks, and improve product quality and reliability. However, in the mass production phase of power modules, testing typically focuses on individual tests, including static parameters, dynamic parameters, and insulation capabilities. Solely relying on individual tests cannot ensure the stability of batch supply, and failures may still occur during mass production even if the power module has passed client application development validation tests. This leads to poor customer experience and hinders the application of power modules in high-end fields. This issue is more pronounced for SiC power modules, where zero-mileage failure rate is one of the key indicators for automotive manufacturers. If a vehicle experiences a failure, the manufacturer will immediately report the situation to the component supplier and expect a prompt response.

To address these issues in mass production, the industry has introduced ACBI (Advanced Conditioning Burn-In) testing to simulate the actual operating conditions of power modules in motor drive applications. In ACBI testing, the power module under test is connected to a three-phase reactor as the load. Space Vector Pulse Width Modulation (SVPWM) is used to generate six PWM signals, creating a three-phase AC current with a 120° phase difference in the three-phase reactor. The principles of

ACBI testing and SVPWM are illustrated in Figs. 6.51 and 6.52, respectively. Unlike motors, reactors do not have a rotating shaft and thus do not produce rotational mechanical power, making them reactive. During the test, different DC voltages, fundamental frequencies, carrier frequencies, and output AC currents are applied to the power module. The electrical stress assessment incorporates various factors such as conduction loss, switching loss, and temperature. ACBI testing can be conducted for both short-term and long-term evaluations to assess reliability, identify early failures, and predict lifespan.

The following settings and tests can be conducted during ACBI testing:

(1) **Operating Condition Settings**

Parameters are set to simulate different operating conditions and application scenarios, including electrical frequency, carrier frequency, bus voltage, three-phase current, water temperature, water flow, drive voltage, drive resistance, etc.

(2) **Operating State Simulation**

Similar to a driving simulator, control programs can simulate various driving environments and habits. Relevant testing standards can be followed, such as NEDC

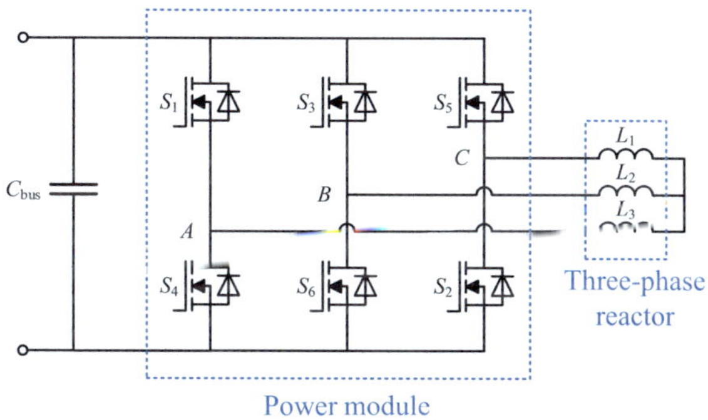

Fig. 6.51 ACBI testing principles

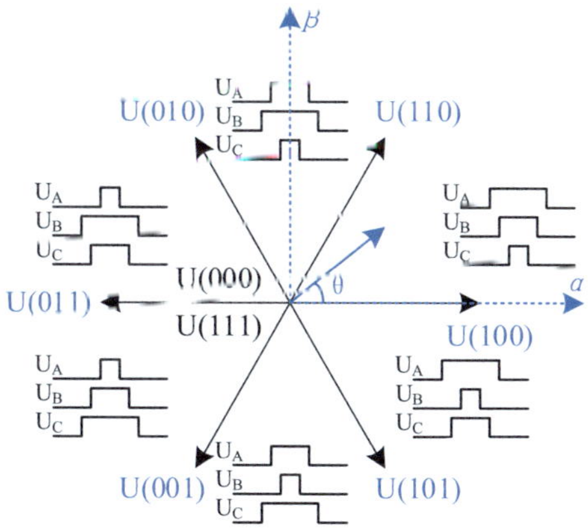

Fig. 6.52 SVPWM principles

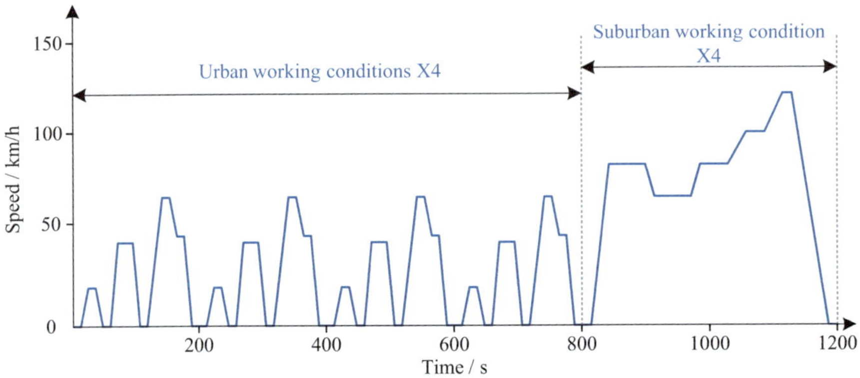

Fig. 6.53 NEDC cycle conditions

(New European Driving Cycle), WLTP (World Light Vehicle Test Procedure), EPA (U.S. Environmental Protection Agency), and CLTC (China Light-Duty Vehicle Test Cycle-Passenger Car). Figure 6.53 shows the NEDC cycle condition.

(3) **Waveform Testing**

During the testing process, waveforms are tested and captured, switching stress parameters of the power module are monitored, and waveform data upon failure is saved. This serves as critical evidence for product reliability verification, ensuring stable operation under various extreme conditions (As shown in Fig. 6.54).

(4) **Current Capability Verification**

By adjusting different parameters to simulate harsh operating conditions, the power module undergoes long-duration high-current operation or withstands pulse current impacts to verify its output capability. For example, at the same junction temperature, change the carrier frequency; according to different vehicle models, change the water

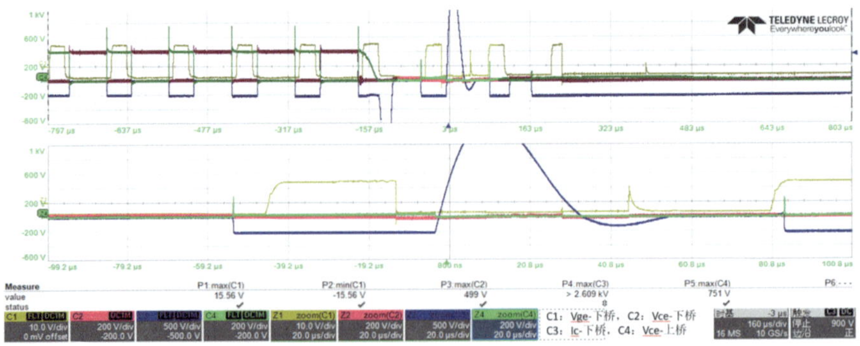

Fig. 6.54 Power module failure waveforms

pump flow rate; based on the battery state (depleted or fully charged), change the bus voltage; according to the motor speed, change the current frequency; water temperature, drive resistance, drive voltage, etc. Conduct long-duration high-current operation under actual operating conditions.

(5) **Locked Rotor Testing**

By simulating motor locked rotor conditions with the control program, waveforms and temperatures under locked rotor conditions are measured. Figure 6.55 shows the data from locked rotor testing, which assesses the performance of the power module under these conditions.

(6) **Junction Temperature Testing**

The power module under test is not filled with silicone or completed with encapsulation. Instead, its surface is sprayed with black paint to facilitate verification of the junction temperature of each chip under different conditions (As shown in Fig. 6.56).

To meet the needs of ACBI testing, the industry has introduced related testing equipment. For example, Yuanlichuang Innovation Technology Co., Ltd has launched the first ACBI testing system in China, suitable for power module manufacturers' production line testing, factory performance inspection, and power module application R&D experiments. As shown in Fig. 6.57.This system, based on SVPWM motor control principles, forms a fully automatic reactive aging testing system for semiconductor power modules, thoroughly evaluating the application performance of power modules. As shown in Fig. 6.58, the ACBI testing system sets the required operating conditions for the power module using an industrial computer and performs tests

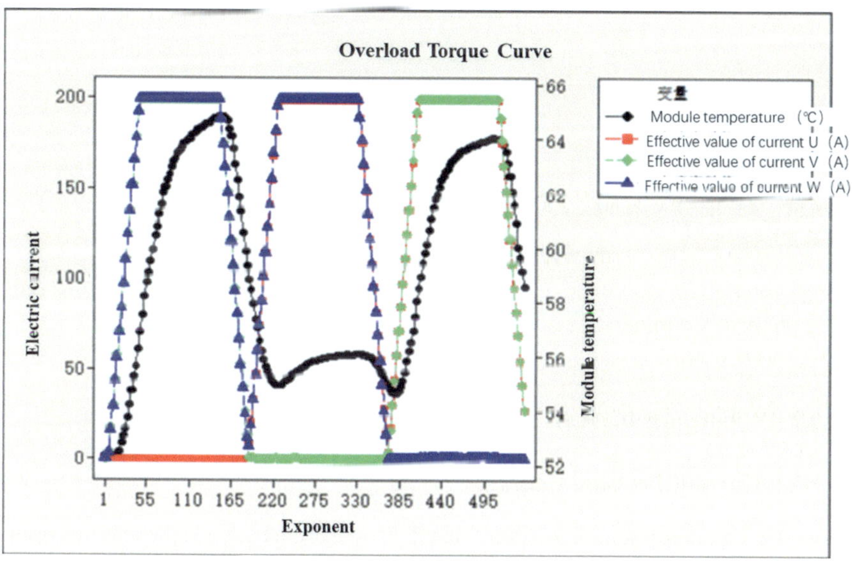

Fig. 6.55 Locked rotor test data

Fig. 6.56 Power module sprayed with black paint

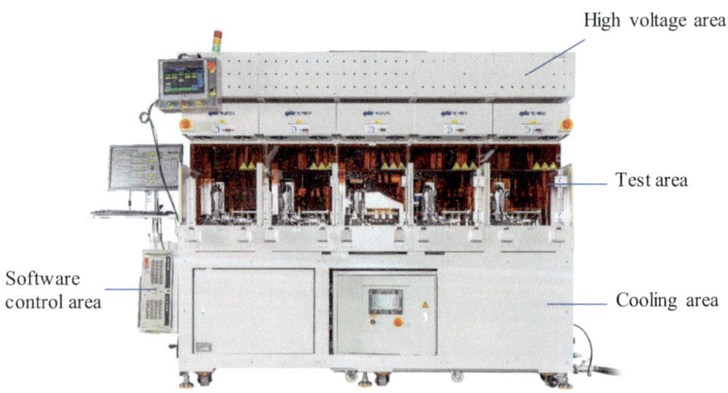

Fig. 6.57 Yuanlichuang company ACBI testing system

such as rated load, peak load, and locked rotor based on these conditions. Test data is automatically compared, evaluated, and archived.

6.2.6.2 Challenges in ACBI Testing

1. **High Current Pressing Contact**

During testing, it is essential to ensure that the connection between the testing equipment and the power module can withstand the high currents involved. Generally, a damage-free pressing solution is used, applying pressure with compression springs to

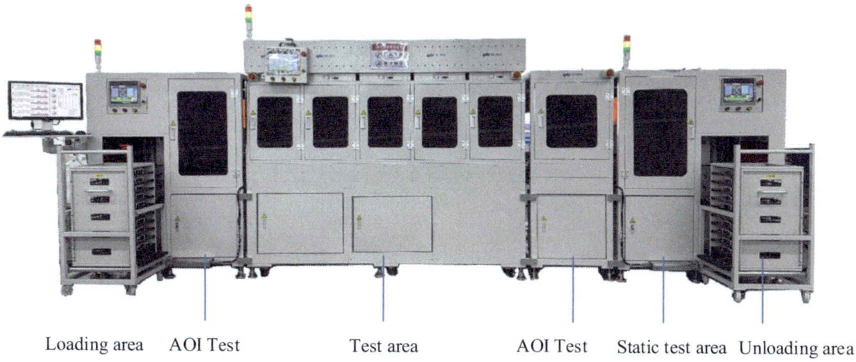

Fig. 6.58 Yuanlichuang company automated ACBI testing system

ensure that the copper busbars and power terminals are tightly connected, achieving quick pressing and good circuit connectivity while maintaining stable temperature at the contact surface under high current (Fig. 6.59).

2. Contact Detection

During testing, poor contact by probes or high-voltage terminals can damage the power module or testing equipment. If the temperature probe does not make good contact, it may lead to inaccurate NTC temperature sampling, failing to protect the power semiconductor in time. Poor contact at the drive terminals can result in an unbalanced output, potentially causing uncontrollable damage to the power module. Poor contact at the high-voltage terminals may cause arcing, affecting the module's normal operation. To address these issues, the ACBI testing system must have contact detection capabilities to quickly identify and stop testing in case of poor contact.

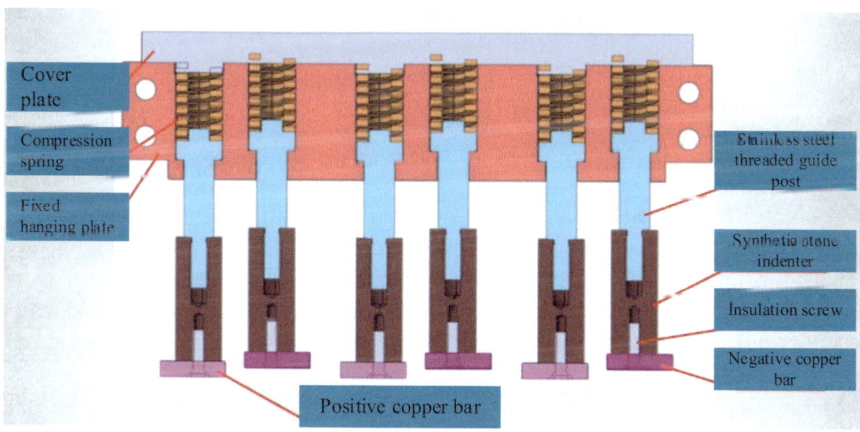

Fig. 6.59 Pressing structure

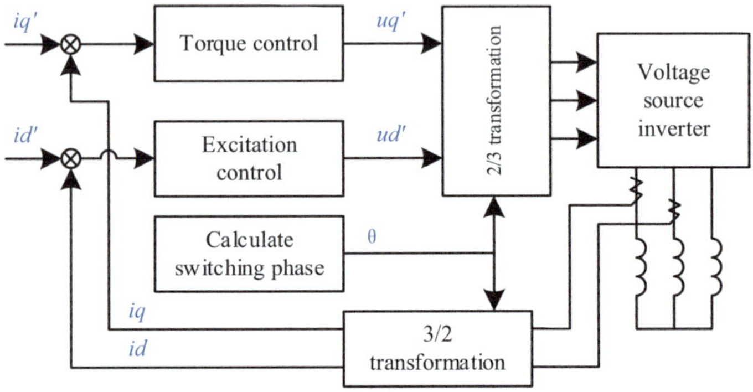

Fig. 6.60 ACBI testing system control principles

3. **Motor Control Simulation**

To simulate the actual operating conditions of power modules in vehicles, a mature electric drive power module drive system is essential. This generally includes a DSP-based power module control unit, a drive unit based on drive chips, and a complete automotive-grade protection and limit hardware protection mechanism. The ACBI testing system uses a three-phase reactor as the load. The testing principle involves controlling a three-phase full-bridge power module to generate three-phase AC current with a 120° phase difference across the load reactor. The control algorithm is similar to that used in motor control, employing vector control technology and space vector pulse width modulation technology, as shown in Fig. 6.60.

4. **Locked Rotor Simulation**

When an electric vehicle is on a slope, the motor is in a locked rotor state, and the power module outputs DC current, operating under the most severe conditions. Specialized simulation of locked rotor conditions is necessary to verify the reliability of the power module. Since ACBI uses a reactor instead of a motor, the control program must simulate locked rotor conditions. Figure 6.61 shows the PWM signal control timing for a locked rotor angle of 0°.

5. **Temperature Control**

To simulate the actual use environment and heat dissipation of the power module, the ACBI testing system must have a temperature control system. This system ensures stable temperature and flow when testing multiple modules simultaneously and independently controls internal and external water circulation. It is crucial to prevent discoloration of the module's appearance due to water contact at high temperatures, often achieved by adding a filtration system.

During the testing process, the temperature of the DUT power module and the busbar temperature at the crimping points must be monitored in real-time to ensure normal system operation. Due to high-voltage interference in the power circuit, an

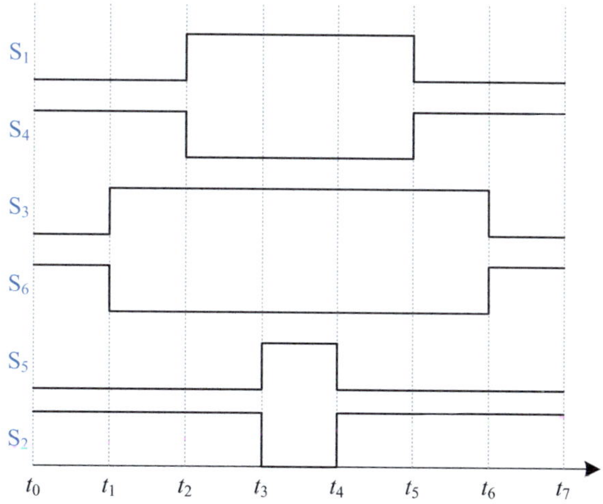

Fig. 6.61 PWM signal control timing at locked rotor angle 0°

infrared pyrometer can be selected for non-contact temperature measurement. This monolithically integrated infrared pyrometer combines the sensor, optical system, and electronic circuitry within a stainless-steel housing. Its standard threaded mount enables rapid connection to installation points.

6. **Module De-Watering**

During testing, power modules require water cooling for heat dissipation. To ensure high-voltage safety and timely product shipment, water residues must be cleaned after testing while meeting production cycle times. Blowing methods can be used for module de-watering. The fixture mechanism locates and seals the module, adjusting the airflow direction to find the optimal blowing angle.

7. **Data Storage and Analysis**

The duration of reactive aging tests generally lasts a few minutes, with data recorded every 0.1 s. This results in thousands of data points over a few minutes, making manual recording impractical and prone to analysis errors. The system's upper-level computer automatically records aging data and analyzes results based on standards, improving test efficiency and providing more intuitive test results.

8. **Visual Inspection**

During testing, improper placement of the product on the testing platform can damage it or prevent proper testing. Therefore, it is necessary to first verify the correct placement of the product before testing. Intelligent visual recognition technology can be employed to automatically inspect the placement of the product, eliminating the need for manual checks, greatly enhancing testing efficiency and reducing the risk of damaging the product under test.

9. **System Electromagnetic Compatibility**

While enabling high-frequency and high-efficiency operation, EMC SiC devices also introduce significant Electromagnetic Compatibility (EMC) challenges. These challenges create systemic issues for test equipment and present difficulties in end applications. For instance, while Product A may function normally, Product B frequently reports errors and fails to operate. An application test system is required to proactively identify the electromagnetic compatibility issues inherent in the power module itself. The extremely high switching speeds (nanosecond-level) of SiC devices result in very high rates of change in current and voltage (high di/dt and dv/dt). This couples through parasitic capacitances to the secondary circuits, generating common-mode noise that interferes with control signal stability. To avoid strong Electromagnetic Interference (EMI) generated during the high-speed switching of SiC devices, effective electromagnetic shielding and filtering measures must be implemented. The fast switching speed of SiC devices necessitates specially designed driving circuits capable of providing appropriate drive signals. These driving circuits require characteristics such as high speed, high power, and low latency to fully exploit the performance of SiC devices. Additionally, electrical isolation and compatibility between the driving circuit and the SiC device must be considered.

6.2.7 *FT Testing*

6.2.7.1 Basic Principle of FT Testing

FT (Final Test) testing, also known as ATE (Automatic Test Equipment) testing in a broader sense, is conducted after the device has been packaged. FT testing helps to remove defective or failed devices caused during the packaging process, ensuring that each shipped device meets relevant standards and avoids negative impacts on users. Generally, devices passing FT testing are ready for shipment. However, for high-requirement products, SLT (System Level Test) or bench testing may also be conducted after FT testing.

Figure 6.62 shows the structure of a typical FT testing system, which mainly consists of a tester and a sorter. The FT testing system's DC, RgCg, and UIL test modules are similar to those used in CP testing, with the addition of AC test modules for switch testing, reverse recovery testing, gate charge testing, and short-circuit testing. The Handler connects the device's pins to the tester's functional modules and automates batch testing.

As shown in the figure, in the FT testing system, the sorting machine mainly consists of contact pushor, the contact blade, the DUT, and the socket. The test system mainly consists of the tester, the test head, and the Loadboard.

Fig. 6.62 FT testing system

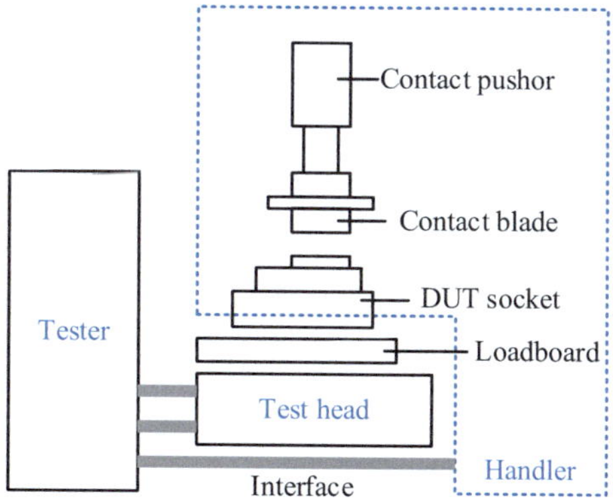

(1) The contact pushor is used to drive the downward movement of the test clamping mechanism.
(2) The contact blade is used to grasp the device under test (DUT) and make contact with the socket under the drive of the downforce mechanism.
(3) The socket is used to place the device. Different devices with different packages require different sockets; one end of the socket contacts the pins of the DUT, while the other end contacts the Loadboard. Through the socket, the DUT is connected to the Loadboard.
(4) The Loadboard is fixed on the test head of the test machine. The test head is adjusted by the test bracket to position and connect the Loadboard with the Handler.
(5) The handler is connected to the tester via an interface board. When the pins of the device come into contact with the pins of the socket, an SOT (Start of Test) signal is sent to the tester via the interface. After testing is completed, the tester sends back the bin value and the EOT (end of test) signal, and the sorting machine performs sorting and classification actions.

According to the transmission method, sorting machines are divided into three types: linear, turret, and gravity. Linear sorting machines use horizontal mechanical arms and vacuum methods to move devices to the test station, which is widely applicable and particularly advantageous for devices with longer test times or advanced packaging, as shown in Fig. 6.63. Turret sorting machines use a main turntable to rotate devices to the test station, suitable for small, lightweight devices with short test times, as shown in Fig. 6.64. Gravity sorting machines use the device's weight to slide to the test station, featuring a simple structure and low investment, as shown in Fig. 6.65.

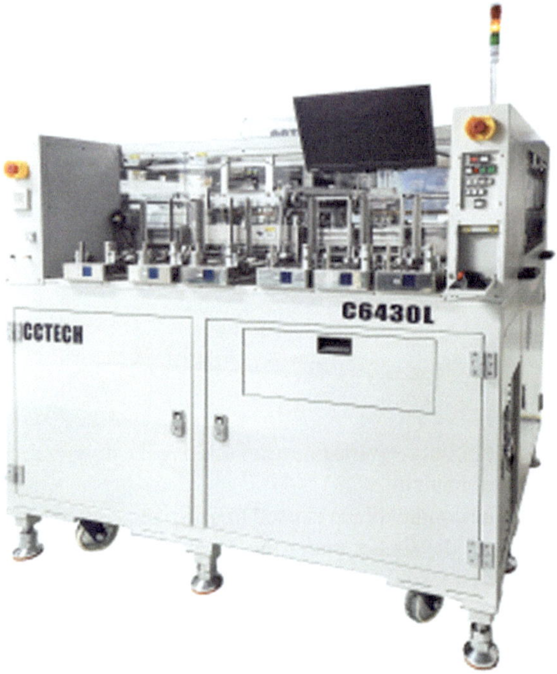

Fig. 6.63 Linear sorting machine from Changchuan Technology

Fig. 6.64 Turret sorting machine from Changchuan Technology

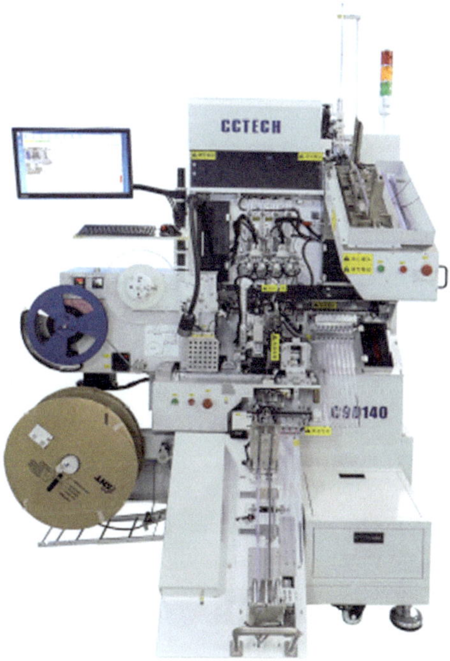

Fig. 6.65 Gravity sorting machine from Changchuan Technology

6.2.7.2 Handler Cases

Below, we introduce Handlers using examples of serial testing for discrete devices and power modules

1. **Gravity Sorting Machine**

SiC diodes and SiC MOSFETs have a variety of package forms, with the TO series (including TO-247, TO-220, TO-252, TO-263) being the most common. Gravity sorting machines can be chosen for FT testing of these packages.

Below is an introduction to the specific structure and functions of Changchuan Technology's C8HT series gravity sorting machine, representing the gravity sorting machine for single-device packages. The C8HT series gravity sorting machine is a fully automatic sorting machine designed for testing power device single package products. It features a direct back-type structure with 2–5 test stations for string testing and 12 automatic tube loading stations. Optional functions include marking and vision inspection. The machine uses a Heater-type heating method, with a maximum heating temperature of 200 °C and a temperature accuracy of ± 3 °C, meeting the testing needs of most single power devices.

The machine mainly consists of five modules: loading module, heating module, testing module, optional module, and unloading module, as shown in Fig. 6.66.

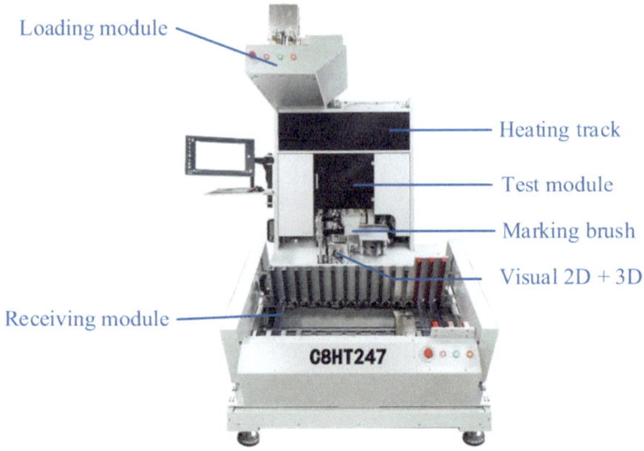

Fig. 6.66 C8HT series gravity sorting machine from Changchuan Technology

(1) Loading Module

The loading module (Fig. 6.67) provides the source of devices for the machine. It uses a manual tube loading method to transport products from the tube to the next process.

(2) Heating Module

The heating module (Fig. 6.68) consists of a control unit, heating elements, and a preheating track. The control unit adjusts the heating elements in real-time using a PID algorithm. The heating elements are heating rods that provide the heat source to create a high-temperature environment. The preheating track acts as the heating carrier, providing a relatively enclosed high-temperature space for heating the products.

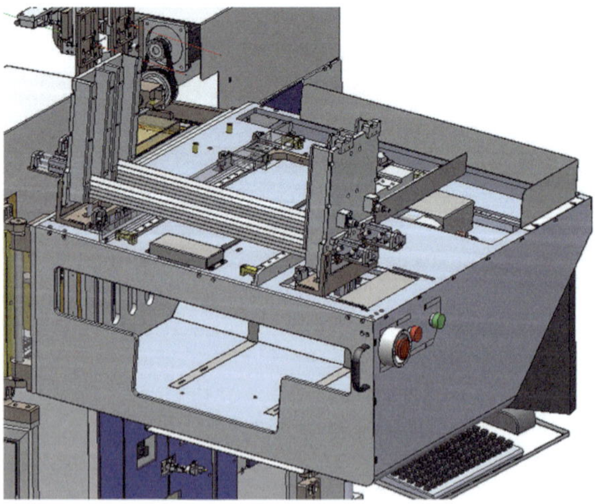

Fig. 6.67 Loading module

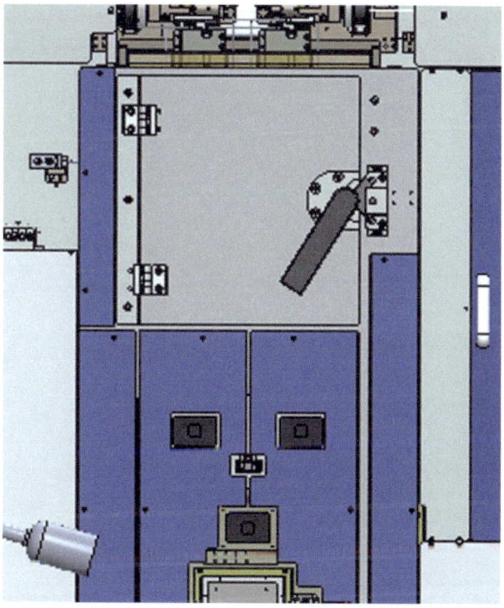

Fig. 6.68 Heating module

Note that at a heating temperature of 150 °C and a preheating time of 2 min, the UPH (units per hour) for different test times is shown in Fig. 6.69. When the test time is less than 0.5 s, the UPH bottleneck is the preheating time, with a maximum of 3000 pcs/h; when the test time is greater than 0.5 s, UPH gradually decreases.

(3) **Testing Module**

As shown in Fig. 6.70, the testing module provides the environment and position for testing the product. The product connects with the test machine via a socket. The C8HT series supports 2–5 test stations and various testing projects, including dynamic testing, static testing, and electrical performance testing.

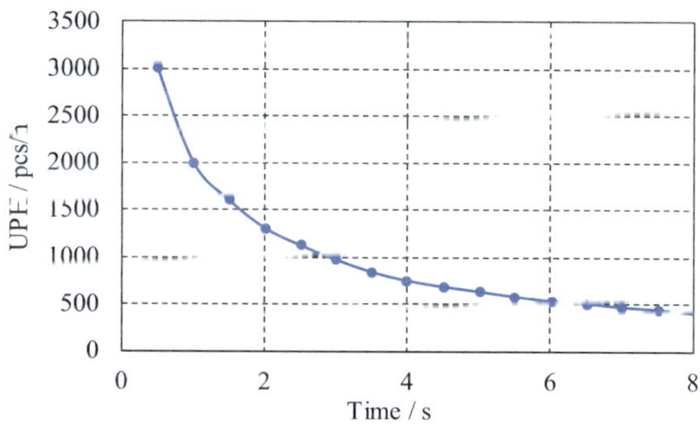

Fig. 6.69 UPH for different test times

Fig. 6.70 Testing module

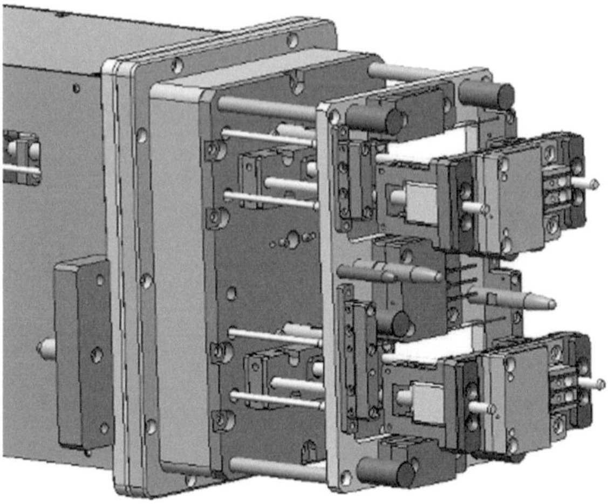

(4) **Optional Module**

The optional module includes 2DID detection, thermal sensing detection, 2D + 3D vision inspection, and laser marking functions, which can be selected based on testing needs. The 2DID detection function is used for QR code scanning and can be bound to test data for easy data analysis and traceability. The thermal sensing function detects the surface temperature of the product to ensure temperature stability. The 2D + 3D vision inspection function detects characters and pins of the product, including character, pin length, pin width, pin spacing, and stand height. The laser marking function marks the product's Mark surface with characters, indicating the batch or template.

(5) **Unloading Module**

As shown in Fig. 6.71, the unloading module classifies tested products according to the test results and places them into corresponding tubes. When the number of tubes reaches the required amount, automatic tube replacement occurs.

In addition, the C8HT series sorting machine supports Hard Docking, effectively reducing dynamic test loop parasitic inductance, as shown in Fig. 6.72.

The main process flow of the C8HT series sorting machine is shown in Fig. 6.73.

2. **Power Module Sorting Machine**

Power modules are another important packaging form for SiC devices. The following describes the specific structure and functions of the PIM series string testing sorting machine from Changchuan Technology, representing the sorting machine for power module FT testing.

The PIM series string testing sorting machine is a fully automatic sorting machine designed for testing power module series package products. The machine mainly

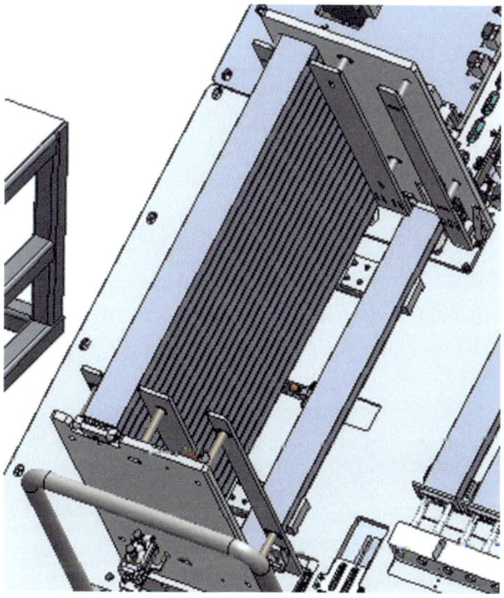

Fig. 6.71 Unloading module

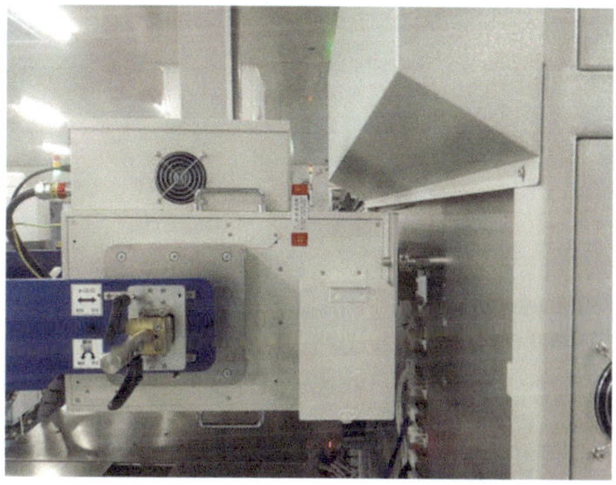

Fig. 6.72 Hard docking

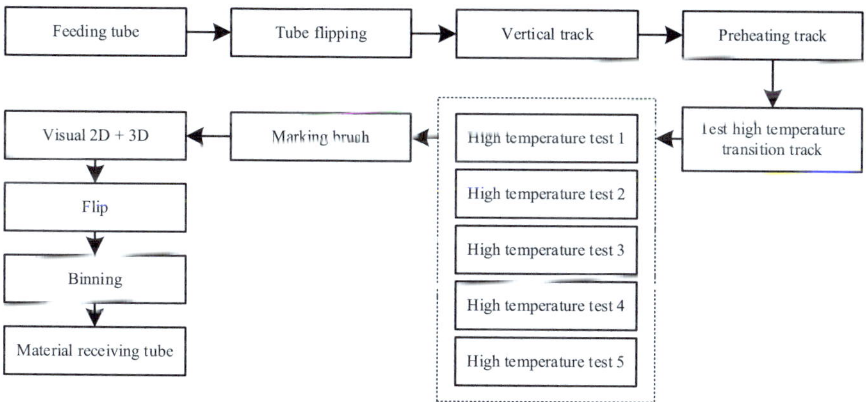

Fig. 6.73 Main process flow of the C8HT series sorting machine

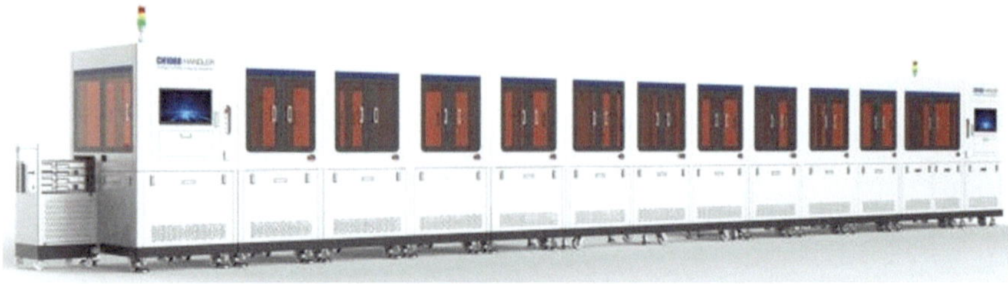

Fig. 6.74 IM series string testing sorting machine from Changchuan Technology

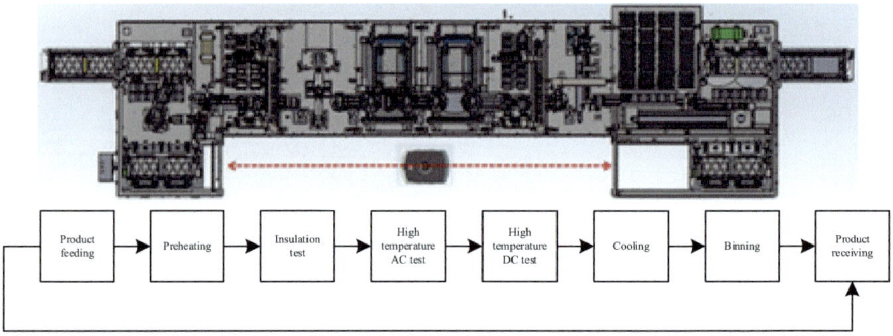

Fig. 6.75 Layout of the PIM series string testing sorting machine from Changchuan Technology

consists of the following modules: loading module, preheating module, testing module, cooling module, visual and curvature inspection module, defective product binning module, and unloading module. Its appearance and layout are shown in Figs. 6.74 and 6.75, respectively.

(1) **Loading Module**

Figure 6.76 depicts the loading module, Power modules are larger and heavier compared to other packaging forms. To avoid damage during handling, power modules are typically loaded and unloaded using trays and carts. The power modules are placed on trays, and the cart carrying the trays docks with the equipment. The robotic arm on the equipment retrieves the trays from the cart and picks up the devices to transfer them to the next process (Fig. 6.76).

(2) **Preheating Module**

As shown in Fig. 6.77, the preheating module heats the power modules to the specified temperature using a preheating plate before transferring them to the test station. Heating rods heat the preheating plate, and temperature sensors installed inside the preheating plate monitor and control the temperature. The control system adjusts the heating rods to maintain the set temperature, with a temperature stability of $\pm$ 1 °C. Temperature fuses prevent overheating and provide over-temperature protection.

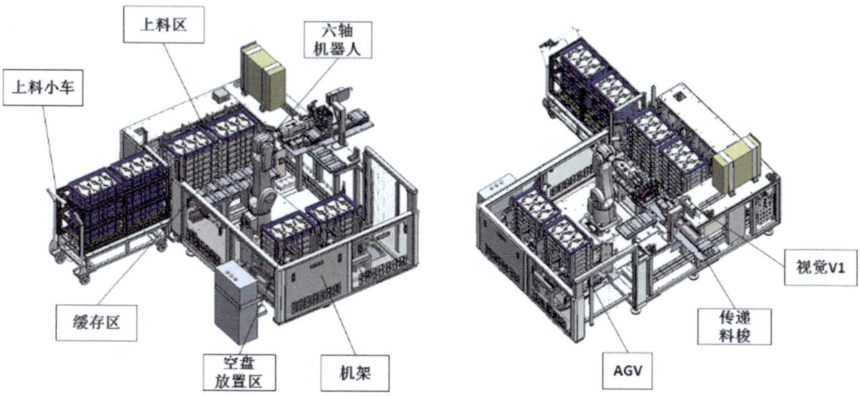

Fig. 6.76 Loading module

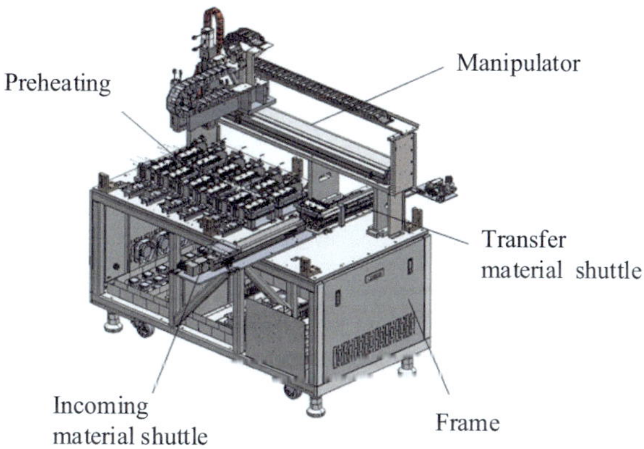

Fig. 6.77 Preheating module

(3) Insulation Testing Module

The insulation testing module uses an ISO tester and test box to perform insulation tests on power modules. It supports Kelvin testing to effectively prevent arcing and provides safety protection for human–machine safety. The testing principle is shown in Fig. 6.78.

The PIM series sorting machine provides insulation testing capabilities of AC 5000 V/100 mA and DC 6000 V/20 mA. It also supports adjustable pressure from 0 to 300 kg, with an accuracy of ± 0.2 kN and a pressure difference between columns within 100N, simulating the device's installation stress.

(4) Cooling Module

After high-temperature testing, power modules are cooled to room temperature in the cooling module (As shown in Fig. 6.79). Typically, water cooling is used for cooling the power modules, and the number of cooling stations is configured based on the device cooling time and overall machine UPH requirements.

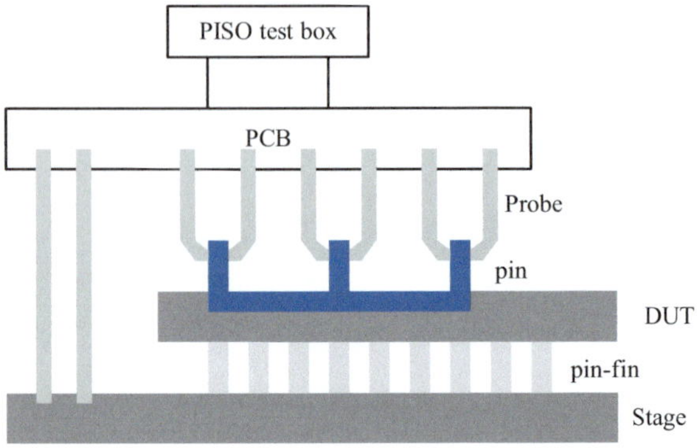

Fig. 6.78 Insulation testing principle

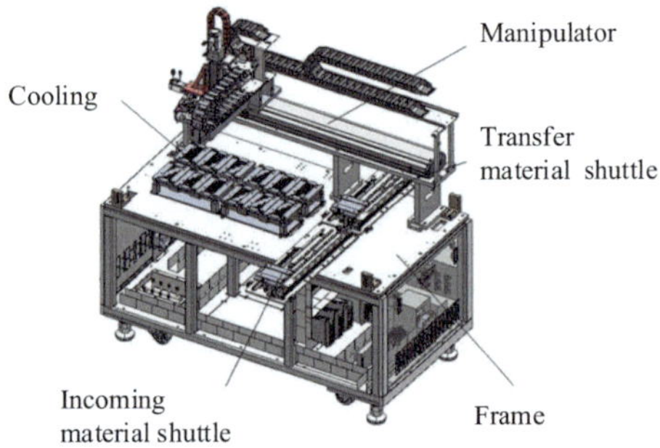

Fig. 6.79 Cooling module

(5) **Visual Inspection and Curvature Inspection Module**

Figure 6.80 depicts the Visual Inspection and Curvature Inspection Module. After testing, the devices undergo appearance and curvature inspections to filter out defective products. Typical inspection items include:

a. Product Mark Surface: Defects such as dirt, scratches, foreign objects, cracks, etc.
b. Pin Terminals: Detection of pin length, pin width, pin spacing, fall-offs, damage, surface scratches, and pin misalignment; defects such as exposed copper, scratches, discoloration, and damage.
c. Marking and Printing: Checks for misprinted codes, missing characters, blurriness, reversed printing, missing characters, and broken characters.
d. Other Appearance Defects: Damage, discoloration of the casing, scratches, dents, discoloration, contamination, and soldering on the product's DBC.
e. Curvature Detection: Identifying deformation in the product's DBC/Pin–Fin, with standard blocks provided for calibrating curvature (Fig. 6.80).

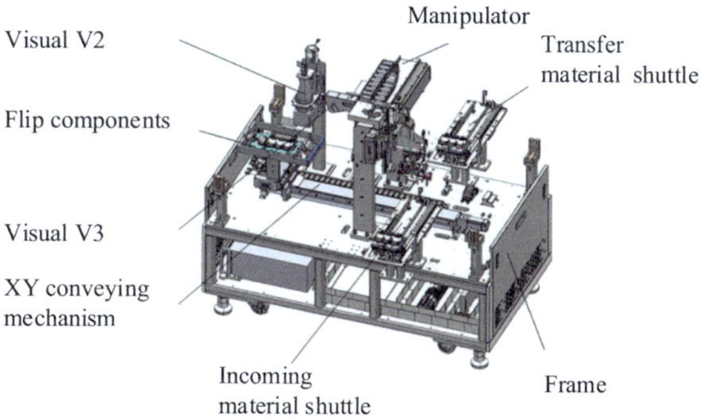

Fig. 6.80 Visual inspection and curvature inspection module

(6) **Defective Product Binning Module**

Figure 6.81 depicts the Defective Product Binning Module. This module collects defective products and is equipped with a multi-layer reject storage system that can classify and collect defective products into different bins.

(7) **Unloading Module**

The unloading module uses a cart with empty trays to dock with the equipment. The equipment places the tested and qualified products into the empty trays for unloading.

The above functional modules are combined to form an automated sorting machine production line system tailored to specific products' manufacturing requirements. The system features scalability and adjustability, accommodating various products or different production processes, achieving flexible manufacturing goals.

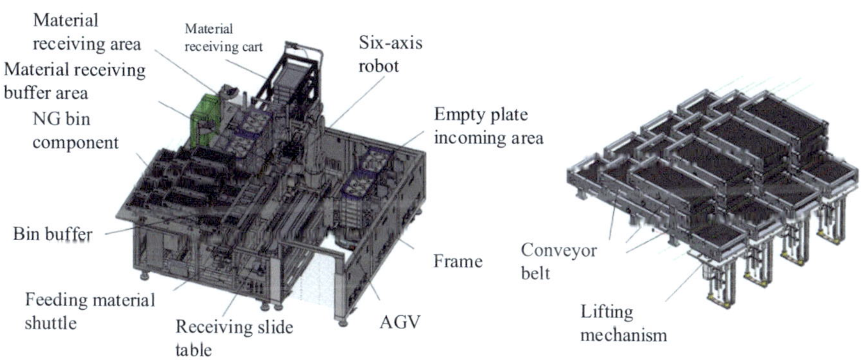

Fig. 6.81 Defective product binning module

6.2.8 *Testing Efficiency and Cost Assessment*

Production testing is highly concerned with testing efficiency and cost. Testing efficiency is typically represented by UPH (Units Per Hour), i.e., the number of devices tested per hour, with higher values generally being better. Testing cost is usually represented by the Hourly Rate, i.e., the cost of testing per hour, with lower values generally being better. Further, using Hourly Rate and UPH, the cost of testing each device (COT, Cost of Test) can be calculated.

When evaluating testing costs, engineers may fall into the trap of only considering the cost of testing equipment, which is an incomplete assessment. Testing cost (COT) is composed of several factors and can be assessed using the following cost model:

(1) **CAPEX**

Capital Expenditure per hour, typically referring to the average fixed capital expenditure for purchasing testing equipment, amortized per hour.

(2) **OPEX**

Operating Expenses, typically referring to the average operating costs per hour.

(3) **Tester Cost**

The cost of the testing machine, usually referring to the price of the testing equipment.

(4) **Test Time**

The time required for the testing machine to complete one test, reflecting the machine's inherent capability. In principle, for the same configuration and cost, shorter test times are preferable.

(5) **Site Count**

The number of parallel testing stations, also known as the number of concurrent tests, which refers to the maximum number of devices the machine can test simultaneously in one test cycle. This parameter reflects the machine's inherent capability. In principle, for the same configuration and cost, a higher number of concurrent tests is preferable.

(6) **Hander Cost**

The cost of the robotic arm (sorting machine), or for CP testing, the cost of the probe station.

(7) **Index Time**

For the robotic arm, this is the time for one material handling cycle. For the probe station, it includes the time for one wafer step plus one probe card's rise and compression time. It also includes communication time with the robotic arm or probe station. This parameter reflects the device's inherent capability. In principle, for the same configuration and cost, shorter index times are preferable.

(8) **Depreciation**

The depreciation period for fixed assets, measured in years. Different companies may have varying depreciation periods, commonly ranging from 5 to 6 years.

(9) **Utilization**

The utilization rate of the equipment, representing the proportion of effective running time within a 24-h day, expressed as a percentage. The theoretical maximum is 100%, but 80% is often used to calculate equipment utilization. This parameter reflects the long-term stability and reliability of the equipment.

(10) **Operator Cost**

The labor cost of operators, measured in "currency/hour". The higher the level of automation in testing equipment and robotic arms (probe stations), the easier it is to use, and the lower the demand for operators, leading to reduced labor costs.

(11) **Electricity**

The electricity cost, usually including the power consumption cost of the testing machine and robotic arm (probe station) per hour, as well as the shared electricity cost for maintaining other factory equipment, measured in "currency/hour".

(12) **Floor Space Cost**

The cost of floor space, representing the comprehensive cost allocated per square meter of factory area per hour, measured in "currency/m^2/hour". This indicator takes into account the volume of the testing equipment and robotic arm (probe station). In principle, for the same capability, equipment with a smaller footprint will incur lower costs in this regard.

(13) **Floor Space**

Refers to the footprint of the equipment.

(14) **Others Cost**

Other costs, including expenses for consumables, software licenses, maintenance services, and other miscellaneous costs.

Based on the above, the following calculations can be made:

$$\text{UPH} = \text{Utilization} \cdot \frac{3600}{(\text{Test Time} + \text{Index Time})} \cdot \text{Site Count} \tag{6.1}$$

$$\text{CAPEX} = \frac{\text{Tester Cost} + \text{Handler Cost}}{\text{Depreciation} * 365.25 * 24} \tag{6.2}$$

$$\begin{aligned}\text{OPEX} = {} & \text{Floor Space Cost} * \text{Floor Space} \\ & + \text{Operator Cost} + \text{Electricity} + \text{Others Cost}\end{aligned} \tag{6.3}$$

$$\text{Hourly Rate} = \text{CAPEX} + \text{OPEX} \tag{6.4}$$

$$\text{COT} = \frac{\text{Hourly Rate}}{\text{UPH}} \tag{6.5}$$

6.3 Reliability Assessment

The reliability of power semiconductor devices refers to their ability to operate stably over an extended period under specific working conditions without failure. Reliability requirements for power semiconductor devices are typically more stringent than those for other electronic devices because they often operate in harsh environments with high power, high temperature, and high voltage, such as in automotive electronics, industrial control, and energy conversion fields.

To improve the reliability of power semiconductor devices, manufacturers typically take measures such as optimizing material selection, enhancing manufacturing processes, and strengthening testing and quality control. Additionally, selecting appropriate designs and application environments for power devices is crucial to ensure stable and reliable operation under intended conditions.

6.3.1 Reliability Standards

Reliability testing of power semiconductor devices is a crucial method for evaluating their reliability. Industry associations and organizations, including device manufacturers, research institutions, and power electronics manufacturers, have developed reliability assessment standards for power semiconductor devices, which are widely followed in the industry. Current main standards for power semiconductor device reliability include JEDEC (Joint Electron Device Engineering Council), AEC (Automotive Electronics Council), ECPE (European Center for Power Electronics), and military standards (MIL-STD).

Currently, reliability assessment standards for SiC devices often use those developed for Si devices. However, these standards may not address the specific characteristics of SiC devices, leading to inadequate and imprecise evaluations. SiC technology is still developing, and related reliability assessment standards are also in the process of being improved. Recently, IEC's IEC/TC47 Semiconductor Devices Technical Committee and JEDEC's JC-70 Wide Bandgap Power Electronics Committee have proposed new methods and standards for SiC device reliability assessment.

The primary application areas for SiC devices are in new energy vehicles, including main drive inverters, OBCs, and DC-DC converters. The former uses module-scale packaging, while the latter two use discrete devices. New energy vehicle applications have high reliability requirements for SiC devices, which must meet automotive standards, specifically AEC-Q101 for discrete devices and AQG-324 for power modules.

6.3.1.1 AEC-Q101

The AEC-Q101 standard is based on failure mechanism testing for automotive discrete devices and was developed by AEC. It serves as a comprehensive reliability testing and certification standard for automotive discrete devices, outlining necessary test items for discrete devices used in automotive applications. It represents a basic threshold for discrete devices in the automotive field.

The latest version of the AEC-Q101 standard is Version E, released in 2021. It specifies five major test groups, including Accelerated Environmental Stress Testing, Accelerated Life Testing, Package Integrity Testing, Chip Manufacturing Reliability Testing, and Electrical Performance Verification Testing, with a total of 57 test items, as shown in Fig. 6.82.

Group A includes Accelerated Environmental Stress Testing, Group B includes Accelerated Life Simulation Testing, Group C includes Package Integrity Testing, Group D includes Chip Manufacturing Reliability Testing, and Group E includes Electrical Performance Verification Testing. The specific test items and standards for each group are summarized in Table 6.4.

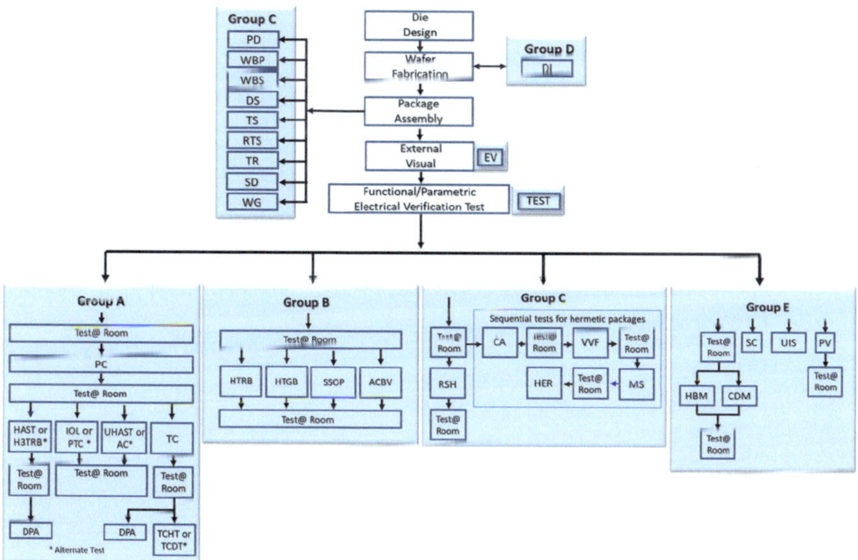

Fig. 6.82 AEC-Q101 standard test flow [2]

Table 6.4 AEC-Q101 standard testing methods [2]

Number	Test item	Abbreviation	Test standard
A1	Preconditioning	PC	JEDEC/IPCJ-STD-020 JESD22-A-113
A2	Highly Accelerated Stress test	HAST	JEDEC JESD22-A-110
A2 alt	High humidity High temp Reverse bias	H3TRB	JEDEC JESD22-A-101
A3	Unbiased HAST	UHAST	JEDEC JESD22-A-118,or A101
A3 alt	Autoclave	AC	JEDEC JESD22-A-102
A4	Temperature Cycling	TC	JEDEC JESD22-A-104 Appendix 6
A4a	Temperature Cycling hot Test	TCHT	JEDEC JESD22-A-104 Appendix 6
A4a alt	Temperature Cycling hot Test	TCDT	JEDEC JESD22-A-104 Appendix 6 J-STD-035
A5	Temperature Cycling hot Test	IOL	MIL-STD-750 Method 1037
A5 alt	Power temperature cycling	PTC	JEDEC JESD22-A-105
B1	High Temperature Reverse bias	HTRB	MIL-STD-750-1 M1038 condition A
B1a	AC blocking Voltage	ACBV	MIL-STD-750-1 M1040 Condition A
B1b	Steady state Operational	SSOP	MIL-STD-750-1 M1038 Condition B
B2	High Temperature Gate bias	HTGB	JEDEC JESD22-A-108
C1	Destructive Physical Analysis	DPA	AEC-Q101-004 Section 6.4
C2	Physical Dimension	PD	JEDEC JESD22-B-100
C3	Wire bond pull Strength	WBP	MIL-STD-750-2 Method 2037 for Au and Al wire AEC Q006 for Cu Wire

(continued)

Table 6.4 (continued)

Number	Test item	Abbreviation	Test standard
C4	Wire bond Shear strength	WBS	AEC-Q101-003 JESD22 B116
C5	Die shear	DS	MIL-STD-750-2 Method 2017
C6	Terminal Strength	TS	MIL-STD-750-2 Method 2036
C7	Resistance to solvents	RTS	JEDEC JESD22-B-107
C8	Resistance to solder heat	RSH	JEDEC JESD22-A-111(SMD) B-106(PTH)
C9	Thermal resistance	TR	JEDEC JESD24-3, 24-4, 24-6
C10	Solderability	SD	JEDEC J-STD-002
C11	Whisker growth Evaluation	WG	AEC-Q005
C12	Constant Acceleration	CA	MIL-STD-750-2 Method 2006
C13	Vibration Variable Frequency	VVF	JEDEC JESD2-B-103
C14	Mechanical shock	MS	JEDEC JESD22 B-104
C15	Hermeticity	HER	JEDEC JESD22-A-109
D1	Dielectric integrity	DI	AEC-Q101-004 Section 6.3
E0	External visual	EV	JEDEC JESD22-B101
E1	Pre- and post-stress electrical test	TEST	Data sheet
E2	Parametric verification	PV	Data sheet
E3	ESD HBM characterization	ESDH	AEC-Q101-001
E4	ESD CDM characterization	ESDC	AEC-Q101-005
E5	Unclamped Inductive Switching	UIS	AEC-Q101-004 Sect. 6.2
E6	Short-circuit characterization	SCR	AEC-Q101 005

6.3.1.2 AQG-324

The AQG-324 guideline, developed by the European Center for Power Electronics (ECPE) working group based on the original German LV-324 standard, offers certification guidelines for automotive power electronic conversion units (PCUs). This guideline defines general procedures for module testing and environmental and lifetime testing for automotive power electronic modules.

The standard specifies the test items, requirements, and conditions necessary for verifying power modules, including those based on discrete devices and special designs. The version released in April 2018 focuses on Si power semiconductor modules, while the latest version published in March 2021 includes additional certification content for SiC power semiconductor modules.

The AQG-324 test framework comprises four parts: QM Module Testing, QC Module Characterization Testing, QE Environmental Testing, and QL Lifetime Testing, as shown in Table 6.5.

QM Module Testing is the foundational test for power module certification and is generally required before and after QC, QE, and QL testing. It includes visual inspection for defects to ensure only defect-free modules proceed to other testing stages.

QC Module Characterization Testing is used to validate the basic electrical performance and mechanical parameters of the power module. The testing process is illustrated in Fig. 6.83. These tests can also detect and assess weaknesses unrelated to functional degradation in the design, including component layout, assembly, interconnection technology, and semiconductor quality. Module characterization testing is a prerequisite for subsequent environmental and lifetime tests. QC Module Characterization Testing requires 18 power modules, where QC-01 parasitic stray inductance and QC-02 thermal resistance tests are non-destructive. After these tests, samples can be reassigned for QC-03 short-circuit capacity and QC-04 insulation tests. Both before and after QC-01 to QC-04 tests, QM Module Testing is required, focusing on the interconnection layer (solder layer, sintering layer) quality and potential degradation from voids, delaminations, or gaps.

QE Environmental Testing assesses the suitability of power modules for automotive environments, including physical analysis, electrical and mechanical parameter verification, and insulation testing. The testing process is illustrated in Fig. 6.84. Environmental testing includes QE-01 temperature shock, QE-03 mechanical vibration, and QE-04 mechanical shock, with QE-02 connectivity not detailed further. Mechanical vibration and mechanical shock tests require QM Module Testing before and after, with data offset values recorded. Temperature shock testing requires QM Module Testing before, during (500 cycles optional), and after the test, with data offset values recorded.

The goal of the QL lifespan test is to trigger the typical degradation of power modules, which primarily involves two failure mechanisms: fatigue near the chip interconnects (chip-end) and fatigue away from the chip interconnects (far-end). The specific testing process is shown in Fig. 6.85. The second-level power cycling (PCsec)

Table 6.5 AQG-324 standard testing items [3]

Test content	Si power module testing items	SiC power module testing items
QM module testing	Gate parameters	Gate parameters
	Rated current and leakage current	Rated current and leakage current
	Saturation voltage	Saturation voltage
	X-ray, SAT/SAM	X-ray, SAT/SAM
	Visual inspection IPI/VI, Optical microscopy OMA	Visual inspection IPI/VI, Optical microscopy OMA
QC module characterization testing	Parasitic stray inductance	Parasitic stray inductance
	Thermal resistance value	Thermal resistance value
	Short-circuit capacity	Short-circuit capacity
	Insulation testing	Insulation testing
	Mechanical parameter testing	Mechanical parameter testing
QE environmental testing	Temperature shock	Temperature shock
	Connectivity	Connectivity
	Mechanical vibration	Mechanical vibration
	Mechanical shock	Mechanical shock
QL lifetime testing	Power cycling (PCsec)	Power cycling (PCsec)
	Power cycling (PCmin)	Power cycling (PCmin)
	High-temperature storage	High-temperature storage
	Low-temperature storage	Low-temperature storage
	High-temperature reverse bias	High-temperature reverse bias
	–	Dynamic reverse bias
	High-temperature gate bias	High-temperature gate bias
	–	Dynamic gate bias
	High-temperature High-humidity reverse bias	High-temperature high humidity reverse bias
	–	Dynamic high-temperature high-humidity reverse bias
	–	High-temperature forward bias
	–	Dynamic high-temperature forward bsias

and minute-level power cycling (PCmin) require samples from at least 6 topological units of 3 different power modules. In addition to the basic tests before and after module testing, it is necessary to record EOL (End-of-Life) lifespan and corresponding lifespan models. After high-temperature storage, low-temperature storage, high-temperature gate bias, high-temperature reverse bias, and high-temperature high-humidity reverse bias tests, OM module testing should be completed, and data

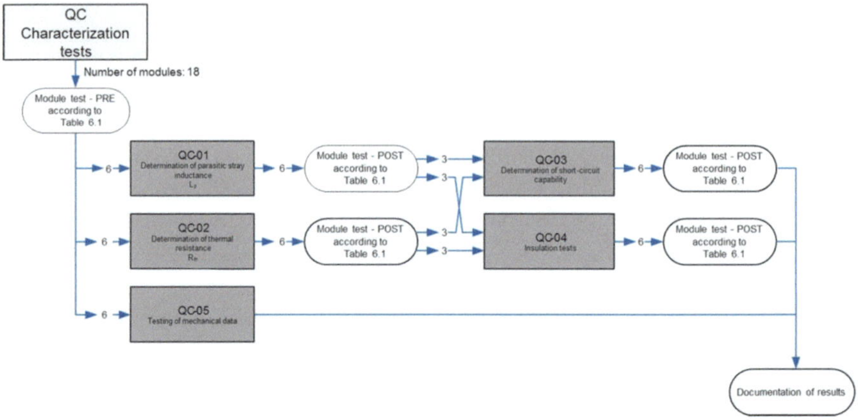

Fig. 6.83 AQG-324 standard QC module characterization testing process [3]

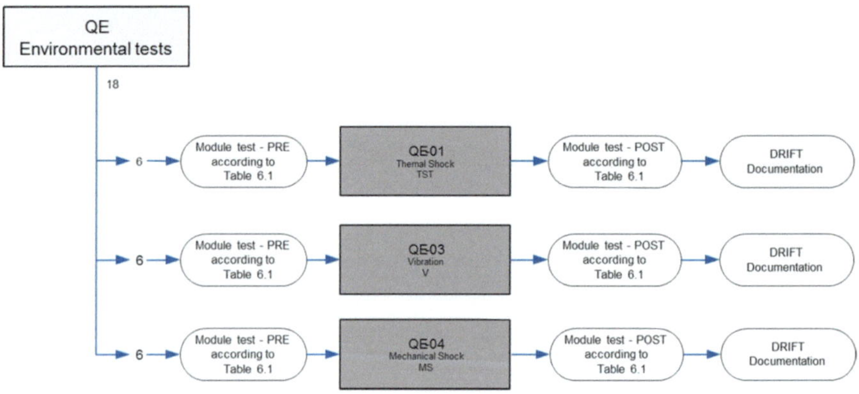

Fig. 6.84 AQG-324 standard QE environmental testing process [3]

shift values recorded. High-temperature storage and low-temperature storage tests also require SAT ultrasonic scanning to monitor interconnect layer degradation.

From the above descriptions, it is evident that although both standards are for automotive-grade power devices, they differ in testing items and requirements. This leads to variations in the rigor of the assessments, and the product range covered by the standards is not clearly defined. This can cause confusion for device manufacturers. For instance, while AEC-Q101 covers discrete devices, AQG-324 covers power electronic modules. Products with medium packages like SOT-227 fall into a gray area between discrete devices and modules, making it difficult to classify them clearly. Common practice is to use AEC-Q101 standards for single-device components and AQG-324 for modules like half-bridge or full-bridge configurations. It is hoped that in future, a unified or combined assessment standard can be achieved.

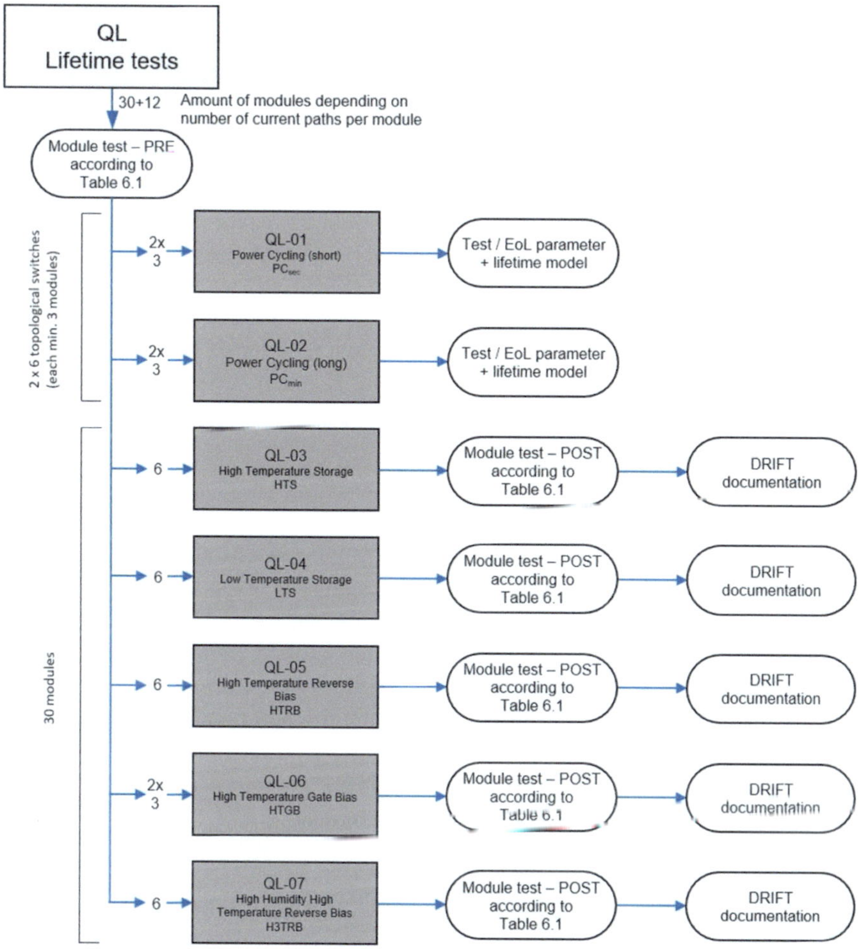

Fig. 6.85 AQG-324 standard ql lifespan testing process [3]

As the gold standard for primary drive module reliability assessment, the AQG-324 standard has gained widespread industry recognition. Its assessment results have been corroborated by numerous products in terms of matching the actual operational lifespan of the products. For example, AccoPower has a full suite of AQG-324 reliability testing capabilities. Its reliability laboratory is equipped with machines for QM, QC, QE, and QL parameter testing, as well as reliability testing equipment and associated facilities, and has a process for interfacing with automotive customers.

6.3.2 Major Reliability Testing Items

Reliability testing is a crucial method to ensure product quality and performance. By testing and evaluating products under various conditions, potential issues and defects can be identified, and corresponding measures can be taken to improve product design and manufacturing processes, thus enhancing product reliability and lifespan. Reliability testing uses specialized equipment to simulate harsher operating environments, including temperature, humidity, and pressure, to accelerate device aging and evaluate overall product reliability lifespan.

Semiconductor device reliability testing involves a series of tests to ensure that devices maintain reliability and stability throughout their entire usage lifespan. The goal is to verify device performance and reliability under various fault conditions and assess long-term stability in actual use. The challenges faced in semiconductor device reliability testing include:

(1) Time and Resources: Reliability testing often requires long durations to simulate the device's actual usage lifespan, necessitating significant time and resources.
(2) Varied Testing Conditions: Device reliability can be affected by various testing conditions, such as temperature, humidity, voltage, and current. These conditions need to be simulated and controlled during testing.
(3) Large-Scale Testing: In R&D and production, a large number of devices need to be tested for reliability. This requires testing equipment capable of handling many devices simultaneously while ensuring the accuracy and consistency of test results.
(4) Data Analysis and Interpretation: Reliability testing generates substantial amounts of data, which requires effective analysis and interpretation. This may involve developing models, statistical analysis, and reliability assessment methods to extract useful information from test data.
(5) Reliability Prediction: Predicting a device's lifespan and reliability based on test results is challenging. Accurate reliability models and algorithms are needed to forecast device lifespan and degradation rates.
(6) New Devices: With the continuous development of new devices, reliability testing must adapt to new challenges. New device structures, materials, and processes may lead to different reliability issues, necessitating continuous improvement and updating of testing methods and processes.
(7) Quality Control: Reliability testing is a key aspect of ensuring semiconductor device quality. Effective quality control processes and metrics must be established to ensure the accuracy and consistency of testing.

The following sections introduce several major reliability testing items, including their testing principles and equipment.

6.3.2.1 HTRB, DRB, and DHTRB

HTRB (High-Temperature Reverse Bias) Test is performed by placing the device under test in a high-temperature environment and applying a reverse voltage across its drain and source terminals while monitoring the leakage current. HTRB testing uses continuous high-temperature bias to simulate the device's operating state and verify the reverse breakdown characteristics of the pn junction at maximum junction temperature. During HTRB testing, the migration of mobile ions or impurities in the passivation layer under high temperature and electric fields increases surface charges, leading to increased leakage current. Eventually, impurities diffuse into the semiconductor, causing device failure. This test can expose degradation effects in the device's edge and passivation areas as well as defects related to the chip edge sealing.

HTRB testing in AEC-Q101 follows MIL-STD-750-1 M1038 (Diode) and MIL-STD-750-1 M1038 (MOSFET) specifications, with a testing duration of 1000 h. The applied voltage is the device's maximum reverse voltage, and the test junction temperature does not exceed the device's maximum operating temperature, typically 175 °C.

In AQG-324, HTRB testing follows IEC 60747-8:2010 (MOSFET) and IEC 60747-2:2016 (Diode) specifications, with a testing duration of at least 1000 h. The applied voltage is no less than 80% of the device's maximum reverse voltage, and the test junction temperature is the device's maximum operating temperature. When a SiC MOSFET can be completely turned off with V_{GS} at 0 V, the gate and source are short-circuited during testing. When a SiC MOSFET cannot be completely turned off with V_{GS} at 0 V, a negative V_{GS} is applied, usually the minimum static V_{GS} allowed by the device. AQG-324 also suggests conducting tests under both 0 V and negative V_{GS} conditions to address potential changes in failure modes. Under negative V_{GS}, the chip experiences stress from the combination of reverse voltage and negative gate voltage, and it remains uncertain which condition is more severe.

Hangzhou ZoanRel Electronics Co., Ltd' BTR-T600pro and BTR-T660pro are high-temperature bias aging test systems for discrete devices and power modules, respectively, as shown in Figs. 6.86 and 6.87. The whole machine includes an oven, an industrial computer, a power supply, a temperature control display module, etc. The configured high-performance lower computer can quickly capture 50–100 data before the device fails.

Besides HTRB testing, AQG-324 requires DRB (Dynamic Reverse Bias) testing for SiC MOSFETs. DRB involves placing the device in a room-temperature environment and applying rapidly changing V_{DS} to accelerate the device's aging through high dv/dt. AQG-324 suggests DRB testing for at least 1000 h at 25 °C, with V_{DS} transition speeds of no less than 50 V/ns, frequencies of no less than 25 kHz, and amplitudes no less than 80% of the device's maximum reverse voltage. Voltage spikes caused by oscillation should not exceed 95% of the device's maximum reverse voltage, which can be managed using clamping circuits. DRB can be performed in passive mode, where external circuits control the drain voltage, or active mode, where the device itself controls the drain voltage. The V_{GS} values should include the device's

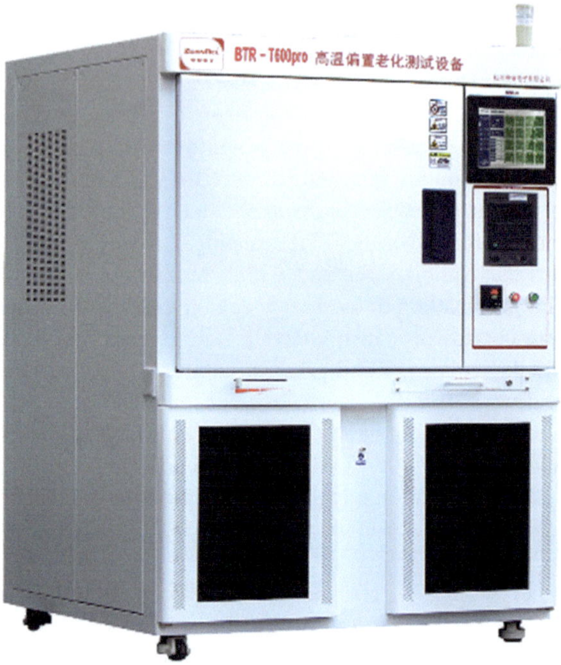

Fig. 6.86 Zhongan Electronics BTR-T600pro high-temperature bias aging test system

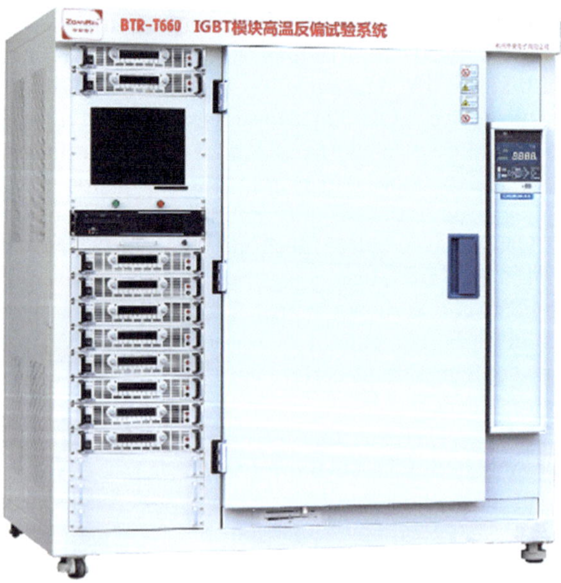

Fig. 6.87 Hangzhou ZoanRel Electronics Co., Ltd BTR-T660pro high-temperature bias aging test system

maximum and minimum gate voltages, and the effects of changing V_{GS} should also be considered.

When the test junction temperature of DRB testing at 25 °C is changed to high temperature, it becomes DHTRB (Dynamic High-Temperature Reverse Bias) testing. Although no standards currently specify this test, it is already applied by some companies.

The BTR-T651 from Hangzhou ZoanRel Electronics Co., Ltd is a dynamic high-temperature reverse bias testing system for SiC MOSFET discrete devices, as shown in Fig. 6.88. The BTR-T651 complies with AQG-324 standards for DRB testing and also supports DHTRB testing. The system features a drawer-type repetitive unit structure that separates the test devices from the parameter control and detection systems, ensuring that high temperatures during testing do not affect system operation and significantly enhancing maintainability. Additionally, the shortest connection methods and high-speed signal impedance matching technologies are used to ensure undistorted dynamic voltage signal loading.

Fig. 6.88 Hangzhou ZoanRel Electronics Co., Ltd BTR-T651 dynamic high-temperature reverse bias testing system

6.3.2.2 HTGB, DRB, and DHTGB

HTGB (High-Temperature Gate Bias) testing involves placing the device under test in a high-temperature environment and applying positive and negative voltages to its gate and source terminals, while monitoring the gate leakage current. The HTGB test applies a DC bias voltage to the gate oxide layer at high temperatures to detect shifts in electrical parameters near the SiC/SiO2 interface caused by charge trapping. This test evaluates the integrity of the gate dielectric, the boundary layer between semiconductor and dielectric, and the effects of mobile ionic contaminants, while observing whether the threshold voltage of the device degrades before and after stress. This test is crucial for assessing the reliability of SiC MOSFETs and is significant for analyzing the quality of the gate oxide layer.

In AEC-Q101, HTGB testing follows JESD22-A108 guidelines, with a test duration of 1000 h. The applied gate-source voltage (V_{GS}) ranges from the maximum to the minimum allowed for the device, with the test junction temperature not exceeding the device's maximum operating temperature, typically 150 °C.

In AQG-324, HTGB testing adheres to IEC 60747-8:2010 (MOSFET) and IEC 60749-23:2011 standards, with a minimum test duration of 1000 h. The test junction temperature is set to the device's maximum operating temperature, and the applied V_{GS} ranges from the maximum to the minimum allowed for the device. 50% of the devices are subjected to positive V_{GS}, while the remaining 50% receive negative V_{GS}.

The BTR-T600pro and BTR-T660pro mentioned earlier not only perform HTGB testing but also support HTGB testing. Engineers can switch between HTGB and HTGB modes using software settings. For gate leakage current measurement, the precision can reach picoamperes, enabling more accurate monitoring of the tested devices.

Besides HTGB testing, AQG-324 requires DGB (Dynamic Gate Bias) testing for SiC MOSFETs. DGB testing involves placing the device under test in a room-temperature environment and applying rapidly varying voltages to its gate and source terminals. Factors like chip paralleling, parasitic parameters, and oscillations can affect the V_{GS} in the switching state, so the DGB results at the module level may differ from those at the chip level. This is why AQG-324 mandates DGB testing at both the chip and module levels. AQG-324 recommends a minimum of 10^{11} cycles for DGB testing, with a test junction temperature of 25 °C, a V_{GS} transition speed of at least 1 V/ns, a frequency of at least 50 kHz, a duty cycle of 20%, and an amplitude ranging from the maximum to the minimum gate voltage allowed for the device. Additionally, V_{DS} should be set to 0 V in RBG only after the chip-level influence of V_{DS} on threshold voltage drift has been ruled out.

DHTGB (Dynamic High-Temperature Gate Bias) testing is essentially the high-temperature version of DGB testing. Although there are no specific standards for this test yet, it is already being used in some companies. The BTR-T652 from Zhong An Electronics is a dynamic high-temperature gate bias testing system for SiC MOSFET discrete devices. It complies with AQG324 standards for dynamic bias testing and allows multiple interrupt points to be set during testing to evaluate the threshold voltage of all tested devices and determine if the devices fail.

Fig. 6.89 Hangzhou ZoanRel Electronics Co., Ltd BTR-T652 dynamic high-temperature gate bias test system

BTR-T652 of Hangzhou ZoanRel Electronics Co., Ltd is a dynamic high-temperature gate bias test system for SiC MOSFET discrete devices, Fig. 6.89 depicts the system appearance. BTR-T652 complies with the AQG324 test standard for dynamic bias. During the test, multiple interruption points can be set to test the threshold voltage of all devices under test to determine whether the current device under test has failed.

6.3.2.3 H3TRB and DH3TRB

H3TRB (High-Humidity, High-Temperature Reverse Bias) testing involves placing the device under test in a high-temperature, high-humidity environment and applying reverse voltage across its drain and source terminals, while monitoring the leakage current. The H3TRB test accelerates the corrosion rate of the device under high temperature and humidity to assess the sealing reliability and electrochemical corrosion resistance of non-hermetic solid-state devices in humid environments over extended periods. This test evaluates weak points in the module structure and verifies the reliability of the chip itself. Most power modules are not hermetically sealed, so moisture can reach the passivation layer over time, affecting the chip's passivation layer structure and edge sealing. Additionally, ionic contaminants introduced during production can migrate with moisture, increasing surface charges and leakage currents under temperature and electric field effects. Corrosive gases introduced during production also affect packaging interconnections and chip reliability. Packaging processes and thermal expansion coefficients (compatibility) can significantly impact passivation integrity and reduce protection against external contaminants, while mechanical stress can exacerbate electrochemical corrosion in power modules.

In AEC-Q101, H3TRB testing follows JESD22-A101 guidelines with a test duration of 1000 h. The test junction temperature is 85 °C, relative humidity is 85%, and the applied voltage is 80% of the device's maximum reverse voltage, but not exceeding 100 V. This limitation is to avoid air discharge under high-humidity conditions. However, since SiC MOSFETs are high-voltage devices, industry practice often involves exceeding the 100 V limit specified in AEC-Q101, applying 80% of the device's maximum reverse voltage, known as HV-H3TRB (High-Voltage—High-Humidity, High-Temperature Reverse Bias) testing.

In AQG-324, H3TRB testing adheres to IEC 60747-8:2010 (MOSFET), IEC 60747-2:2016 (Diode), and IEC 60749-5:2017 standards, with a minimum test duration of 1000 h. The test junction temperature is 85 °C, relative humidity is 85%, and the applied voltage is 80% of the device's maximum reverse voltage without the 100 V limit in AEC-Q101. If the SiC MOSFET can be fully turned off with V_{GS} at 0 V, the gate and source are shorted during testing. If the device cannot be fully turned off with V_{GS} at 0 V, a negative V_{GS} is applied during testing.

The BTR-T670 from Hangzhou ZoanRel Electronics Co., Ltd is a typical H3TRB testing device (Fig. 6.90).

Apart from HTGB testing, AQG-324 requires DH3TRB (Dynamic High-Humidity, High-Temperature Reverse Bias) testing. This involves placing the device under test in a high-temperature, high-humidity environment and applying rapidly varying V_{DS}. AQG-324 recommends a minimum DH3TRB test duration of 1000 h, with a test junction temperature of 85 °C, relative humidity of 85%, and a V_{DS} transition speed of at least 30 V/ns, frequency between 15 and 25 kHz, with an amplitude not less than 50% of the device's maximum reverse voltage. Voltage spikes due to oscillation should be between 80 and 95% of the maximum reverse voltage, and clamping circuits can be used to manage these spikes. For SiC MOSFETs, an active mode can be used to control the drain voltage with the device actively switching, with V_{GS} at the maximum and minimum allowed gate voltages. For SiC diodes, a passive mode using external circuitry to control the drain voltage can be employed.

6.3.2.4 IOL

IOL (Intermittent Operational Life) testing involves using the device's own current to heat up from room temperature to a high temperature and meet the temperature differential requirements, then turning off the current and allowing the device to cool to ambient temperature before heating again. This process is repeated in a

Fig. 6.90 Hangzhou ZoanRel Electronics Co., Ltd BTR-T670 high-temperature and high humidity reverse bias test system

cycle. IOL testing is a power cycling test that applies and removes stress repeatedly to detect defects such as bonding wire and aluminum layer delamination, interface delamination between the chip surface and resin material, and interface delamination between bonding wires and resin material. During IOL testing, the chip, frame, solder paste, and encapsulation materials within the device all experience internal stress from the interactions between different components and the heat generated by the chip under operational conditions. The different thermal expansion coefficients accelerate the device's aging and may lead to failure.

In AEC-Q101, IOL testing follows the guidelines of MIL-STD-750 Method 1037. When the sum of the current heating time t_{on} and natural cooling time t_{off} exceeds 2 min and the temperature differential is greater than 100 °C, 60,000/(t_{on}+t_{off}) cycles are required. If the temperature differential exceeds 125 °C, 30,000/(t_{on}+t_{off}) cycles are required. When the sum of t_{on} and t_{off} is less than 2 min, 15,000 cycles are needed. If the temperature differential exceeds 125 °C, 7,500 cycles are required. The heating current and t_{on} must be calculated and verified based on the device's on-resistance and K factor.

The BTD-T810 from Hangzhou ZoanRel Electronics Co., Ltd is a typical intermittent life testing system (Fig. 6.91). This system can handle up to 80 positions per channel, and its optimized design meets the requirement of testing in identical physical thermal environments. For various packaging types, the system uses a mother-daughter board setup with different sub-boards for different packaging

Fig. 6.91 Hangzhou ZoanRel Electronics Co., Ltd company BTD-T810 intermittent life test system

configurations, enhancing flexibility and reducing costs. During testing, the high-speed real-time junction temperature detection function can complete sampling of the entire board within 500 microseconds, allowing for intuitive monitoring of the ΔT_j of each tested device throughout the test process.

6.3.2.5 PC

PC (Power Cycling) testing follows the same basic principle as IOL testing, using the device's own current to actively cycle the temperature and simulate load changes during actual operation. When the chip heats up, internal stresses occur between the chip and bonding wires, the chip and the AMB (Active Metal Brazing), and even between the AMB and the copper base plate. In AEC-Q101, IOL testing is for discrete devices with smaller heating currents and longer heating times, using natural or forced air cooling. In AQG-234, PC testing is for power modules and is divided into second-level and minute-level PC tests, with larger heating currents, shorter heating times, and water cooling.

In AQG-234, second-level PC testing adheres to IEC 60749-34:2011 guidelines. The current on-time t_{on} must be less than 5 s, with the heating current greater than 85% of the device's rated current. Due to the short t_{on} and t_{off}, the power dissipation of the chip only causes temperature changes in components near the chip, affecting interconnect layers including bonding points and wires above the chip, and interconnect areas below the chip. Consequently, the module case temperature T_C changes more slowly than the chip junction temperature T_j, as shown in Fig. 6.92a. Therefore, second-level PC testing primarily assesses the reliability of chip bond wires and solder layers. During this test, only the temperature rise of T_j needs to be recorded.

In AQG-234, minute-level PC testing follows IEC 60749-34:2011 guidelines. The current on-time t_{on} must be greater than 15 s, with the heating current also greater than 85% of the device's rated current. With sufficient t_{on} and t_{off}, the heat generated by chip power dissipation can reach further, causing T_C to more closely follow T_j, as

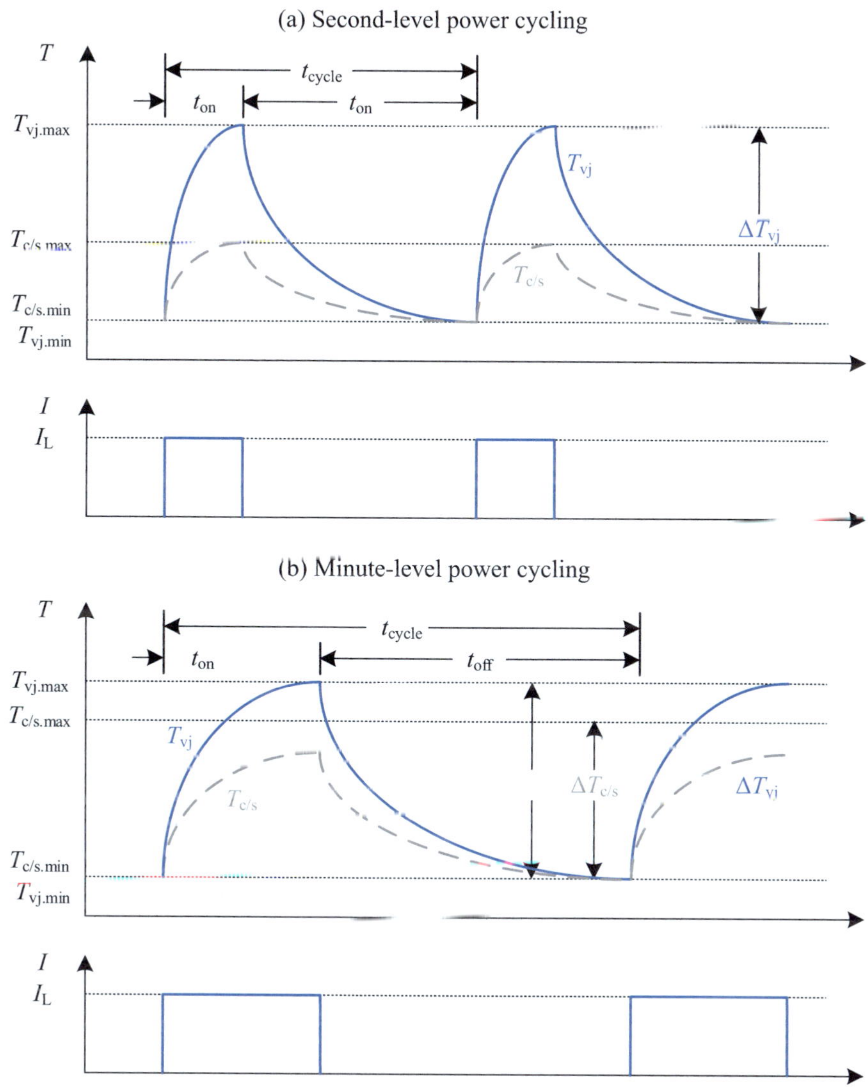

Fig. 6.92 Power cycling test current and temperature curve

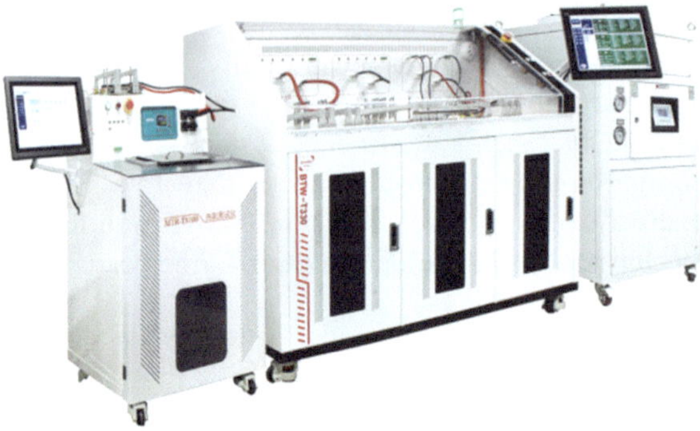

Fig. 6.93 Hangzhou ZoanRel Electronics Co., Ltd BTW-T330 power cycle test system

shown in Fig. 6.92b. Thus, minute-level PC testing mainly assesses the reliability of the substrate and soldering of the heat sink below. During minute-level PC testing, the temperature rise of the substrate or heat sink, and the cooling liquid temperature needs to be recorded.

The BTW-T330 from Hangzhou ZoanRel Electronics Co., Ltd is a typical power cycling test system, suitable for K-factor testing, second-level power cycling, and minute-level PC, thermal resistance testing and structural function analysis of diodes, IGBT modules, and SiC MOSFET modules ranging from 100 to 3600A, as shown in Fig. 6.93. The BTW-T330 primarily consists of a control cabinet and strong current control system, a constant temperature cooling measurement and control system, a high-temperature heating measurement and control system, distributed power supply modules, a core control board, a high-speed data acquisition module, a heating current source, a switch control module, an industrial computer, and centralized control software. This setup provides a comprehensive testing environment for each device under test. With matching water circulation control, the system is compatible with both minute-level and second-level (minimum 0.5 s) PC testing. The single-platform configuration includes a 1200A constant current source, and when multiple power sources are used in parallel, the maximum current on a single platform can reach 3600A.

6.4 Failure Analysis

6.4.1 Overview of Failure Analysis

SiC diodes and SiC MOSFETs have already seen the emergence of relatively mature products. However, there are still many constraints and bottlenecks in various aspects such as the improvement of SiC substrate and epitaxial quality, the optimization of gate oxide process and the enhancement of reliability, as well as the research and testing of high-voltage and large-capacity device structures. The failure problems caused by these deficiencies will all restrict the development of SiC devices. The adoption of reasonable failure analysis techniques can provide reasonable and effective solutions for identifying the causes of device failures and improving reliability.

Firstly, the SiC chip and packaging technology is not yet mature and is still in the development stage. At this point, failure analysis technology needs to be utilized to help identify potential issues that affect the performance and reliability of the product, and then to improve the product design and manufacturing process in a targeted manner. Secondly, during the manufacturing process of the devices, various unexpected situations often lead to product abnormalities. At this time, failure analysis technology can help locate the process steps that cause the abnormalities and reveal the possible problems and defects in the manufacturing process, which is helpful for manufacturers to improve the production process and enhance the quality control level, thereby producing more stable and reliable products. In addition, failure analysis of failed devices in the system application process can assist in determining whether the failure originates from the device's own defects, improper use, or peripheral circuit faults. Thus, targeted optimizations can be adopted to improve the system's reliability. Finally, unlike traditional Si devices, the yield control of SiC devices should start from the defects of the SiC substrate and epitaxial layer. SiC will produce various crystal defects (such as point defects, dislocations, micro tubes, etc.) during the production process, and these defects will expand and extend to the surface during the epitaxial process, forming various surface defects (such as carrot defects, triangular defects, scratches, etc.). These defects have a significant impact on the properties of the fabricated SiC devices, may cause failure problems such as high breakdown field, high reverse voltage, and low leakage current, and will also reduce the production yield and reliability of SiC devices. In large-scale production, surface-type defects can be quickly obtained through chemical etching or optical detection methods for morphological information, but the mechanism by which these defects affect the device performance still needs to be studied and analyzed using electron microscopy products.

The main methods for conducting failure analysis of devices include two categories: non-destructive testing and destructive testing. Common non-destructive tests include appearance inspection, electrical property testing, ultrasonic scanning, and X-ray detection. Common destructive tests include opening, sectioning, layer removal, cratering, lead bonding strength testing, cross-sectional and surface morphology analysis, and failure location. The following provides a brief introduction to the commonly used failure analysis methods.

(1) **Appearance inspection**

The equipment used for the appearance inspection is an optical microscope, which can be employed to observe the surface shape, size, structure, and defects of the device appearance and failure sites.

(2) **Electrical characteristic testing**

The equipment used for electrical characteristic testing is a parameter tester. Based on the test results, the state of the device after failure can be determined, providing information for further analysis and the determination of subsequent failure analysis measures.

(3) **Ultrasonic scanning**

The equipment used for ultrasonic scanning is an ultrasonic scanning microscope. It transmits high-frequency ultrasonic waves into the interior of the inspected sample. Due to the different acoustic impedances of various materials within the device, the absorption and reflection of sound waves vary, resulting in different ultrasonic wave information detected for penetration and reflection. Thus, internal cracks, delamination defects, voids, bubbles and other defects in the inspected sample can be discovered.

(4) **X-ray Inspection**

The equipment used for X-ray testing is an X-ray machine. It generates high-energy electrons through cathode rays and causes them to collide with a metal target, thereby producing X-rays. Due to the different penetrating abilities of different materials within the tested sample to X-rays, defects such as broken wires, connected wires, bubbles, and shattered chips can be detected within the sample.

(5) **Decapsulation**

The purpose of decapsulation is to remove the device's packaging material and expose the die, preparing it for subsequent analysis. Primary methods include mechanical decapsulation, chemical decapsulation, and laser decapsulation.

(6) **De-layering**

De-layering involves the sequential removal of the various layers of the chip, based on an understanding of the chip's structure and materials. Main methods include chemical etching, mechanical polishing, and ion milling.

(7) **Failure Location**

After determining the failure of a device through electrical characteristic measurements, how to locate the failure points on the device is an essential step in tracing the cause of the failure. However, a large portion of electrical failures originate from tiny defects at the nanoscale, which do not show any visible physical damage through general optical detection, and their electrical and thermal signals are extremely weak, making it difficult for conventional infrared thermal imagers to capture them. Secondly, the multi-layer interconnections and complex 3D structures within the chip may bury the defects beneath the surface, and traditional methods are unable to directly observe them. All these pose significant challenges to the location of semiconductor failure defects. Therefore, the failure analysis equipment must have high spatial resolution and multi-angle access capabilities, while also ensuring non-destructiveness as much as possible, so as to accurately locate the failure area without damaging the device.

Although the failure points in semiconductor devices do not have obvious morphological defects, they often exhibit unique physical effects, mainly including infrared thermal radiation, near-infrared micro-light emission, and laser heating-induced resistance changes. Based on these three physical effects, three failure analysis techniques have been developed: Lock-in Thermography (LIT), Emission Microscopy (EMMI), and Optical Beam Induced Resistance Change (OBIRCH). These three techniques are derived from the differences in the physical properties of failure points in terms of heat, light, and resistance change. They are specifically designed for different types of defects and cannot be substituted for each other. Failure analysis engineers need to select appropriate methods based on the specific situation to locate the problem.

(8) **Section and surface morphology analysis**

Section and surface morphology analysis can observe the microstructure and metallographic organization of the failure area, as well as their evolution under stress conditions. The equipment used includes scanning electron microscope, transmission electron microscope and scanning transmission electron microscope.

The failure issues of SiC MOSFET are closely related to the quality of the gate oxide layer. Common failure phenomena such as short circuits often occur when the gate oxide layer is damaged, and the damage often manifests as microscopic abnormalities. At this time, it is necessary to use high-precision positioning equipment, sample preparation capabilities, and analysis equipment such as scanning electron microscopes and transmission electron microscopes to detect them. In the subsequent chapters, specific failure cases will be further introduced. At the same time, the process development of SiC devices and the research on failure mechanisms are still in the exploration stage. The electron microscope system can help us control defects at the microscopic scale and find the microscopic mechanisms behind the failure phenomena, which is an essential tool for improving yield and exploring research in the manufacturing and production of SiC devices. Due to the possibility of conducting more in-depth research on the failure mechanism of the devices, it will be introduced in detail in the following chapters.

6.4.2 *Phase-locked Thermal Imaging*

6.4.2.1 The Basic Principle of Phase-locked Thermal Imaging

Infrared thermal imaging has a long history of being used in non-destructive testing and thermal analysis of electronic devices. LIT, as an active dynamic infrared detection technology, was first developed by manufacturers in the mid-1990s and has the ability to detect material internal defects with high sensitivity. In 2001, the Breitenstein team from the Max Planck Society in Germany first applied micro-scale LIT technology to the failure analysis of integrated circuits, achieving the localization of micro-millimeter defect hotspots within the chips, verifying its feasibility in detecting defects at sub-millimeter and even sub-micron scales in microelectronic devices. In the early 21st century, this technology rapidly became the standard method for failure analysis worldwide. By narrowing the suspected area of faults, it provides precise guidance for precise analyses such as focused ion beam sectioning, significantly improving the analysis efficiency.

LIT is based on the "modulation-locked" measurement principle, which extracts the heat signal of the detected device from the noise, as shown in Fig. 6.94. "Modulation" refers to applying periodic excitation to the tested semiconductor device, such as the alternating current/voltage bias after on-off modulation, causing the defect area to generate a small temperature rise that varies with time. "Locking phase" involves synchronously detecting this fixed-frequency thermal signal, using a high-speed infrared thermal imager to simultaneously capture the subtle changes in the surface temperature of the device, and performing locking processing through specialized software (such as coherent accumulation and Fourier transformation) to obtain amplitude and phase images under the excitation frequency, achieving the amplification and extraction of weak temperature changes. This method is essentially a narrow-band filtering detection, extracting only the temperature signal components with the same frequency as the external excitation, thereby significantly suppressing the noise influence of the environmental background and the steady-state heat generation of the device. Compared with traditional steady-state infrared thermal imaging, LIT significantly improves the signal-to-noise ratio and sensitivity, capable of detecting a tiny temperature rise of 0.1 mK and achieving higher defect resolution capability.

Some failure points can cause semiconductor devices to emit weak thermal radiation due to increased local power consumption, especially in the infrared band. At this time, TIL can be used to locate the leakage path and hidden low-resistance short circuits. Its advantages are non-contact and non-destructive, and it can detect deep defects through backside infrared imaging by penetrating the substrate. Even if the failure points are covered by encapsulation layers or metal layers, the heat flow is likely to penetrate to the surface and generate infrared signals. However, due to the infrared wavelength, the spatial resolution is relatively limited (greater than 2μm), and it is commonly used as a preliminary failure location tool in chip devices.

As shown in Fig. 6.95, the LIT system THERMO 100 of Anhui Luxet Infrared Technology Co., Ltd. is equipped with a high-sensitivity refrigerated mid-wave

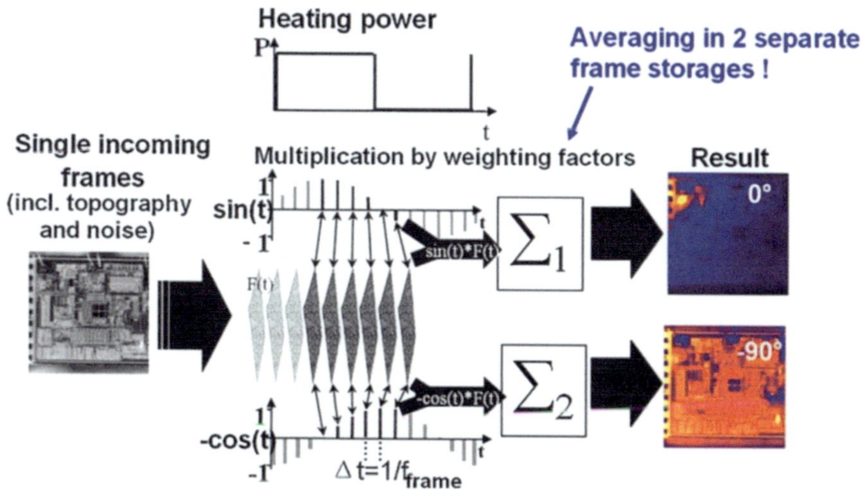

Fig. 6.94 The principle of Phase-locked Thermal Imaging technology [14]

infrared camera, an automatic motion system, a wide-angle lens and a high-magnification microscope lens, a high-voltage source meter, etc. It is suitable for scenarios where high-precision failure point positioning is required, such as chips, advanced packaging, and power devices.

6.4.2.2 Application Cases of Phase-locked Thermal Imaging Technology

The LIT system THERMO 100 was used to locate the failure of the power module that had a short circuit. The module was placed on the test bench, and an alternating voltage was applied for excitation. Within 5 seconds, a clear hotspot location could be obtained. As shown in Fig. 6.96, the imaging results at wide-angle, 1x, and 8x magnification rates are presented. It can be seen that the failure point is in the field ring area. The detection parameters set here are constant voltage mode, lock-in frequency of 5 Hz, and heating power of 18.76 mW.

The LIT system THERMO 100 was used to locate the failure of the faulty digital chip. Hotspots were detected immediately upon the chip being powered on. The results are shown in Fig. 6.97. The detection parameters set here were a test voltage of 456 mV, a heating power of 4.6 mW, and a lock-in frequency of 12.5 Hz.

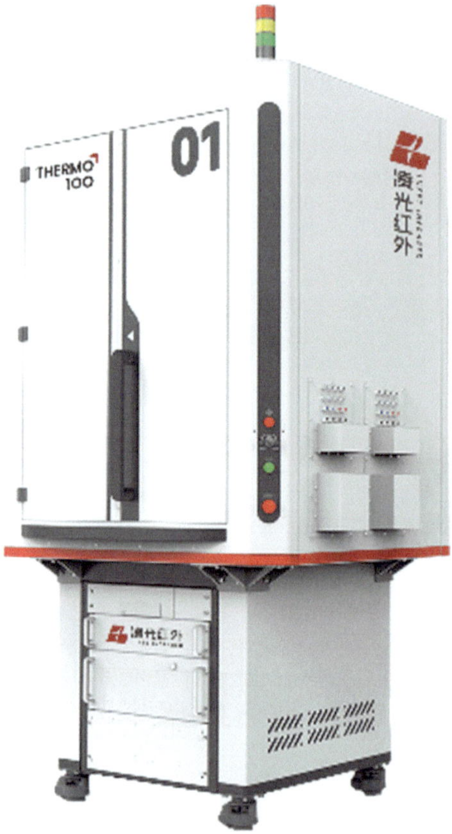

Fig. 6.95 Luxet Infrared Technology 's LIT system THERMO 100

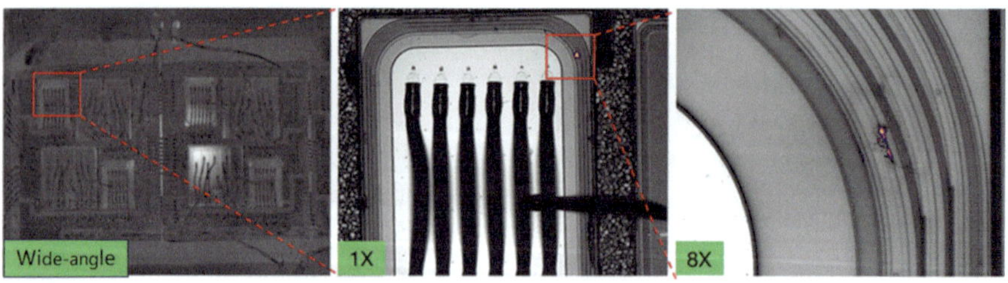

Fig. 6.96 The result of the system's failure location for the power module

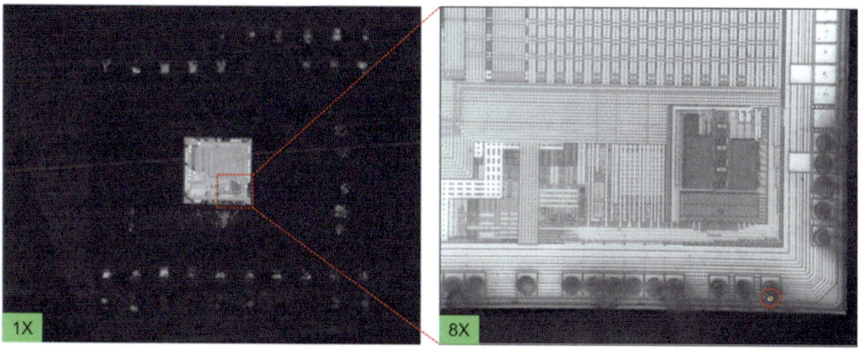

Fig. 6.97 The result of the LIT system in locating the failure of digital chips

6.4.3 *Microscopic Telescope*

6.4.3.1 The Basic Principle of Microscopic Illumination

When semiconductor devices fail or operate abnormally, defect areas will emit weak photons due to high electric fields, carrier recombination, and thermal effects. The wavelength range is generally from 300 to 1700 nm (visible light to near-infrared). The phenomenon of semiconductor device luminescence is usually related to the energy conversion process of carriers (electrons and holes). Especially in the presence of defects, high electric fields, or specific physical mechanisms, the main mechanisms include pn junction luminescence, thermal carrier luminescence, and photon emission from body tube defects.

The pn junction light emission can be divided into reverse bias emission and forward bias emission. The principle and the emission patterns observed using the emission microscopy (EMMI) are shown in Fig. 6.98. When the pn junction undergoes avalanche breakdown or tunneling effect under reverse bias, high-energy carriers collide and ionize to produce secondary electron-hole pairs, which then emit light during recombination. Under forward bias, carriers are injected into the active region and recombination occurs, resulting in light emission. For example, the light emission of photovoltaic devices such as LEDs and laser diodes is designed based on this principle.

When carriers are injected into the high electric field region, such as the breakdown region of MOSFET or the depletion region of pn junction, in the high electric field, the carriers are accelerated and acquire high kinetic energy (thermal carriers), and when colliding with the lattice, they release energy and emit light. The principle and the emission pattern observed using EMMI are shown in Fig. 6.99. The emission caused by thermal carriers has a broad spectral distribution, with the wavelength distribution ranging from visible light to near-infrared light.

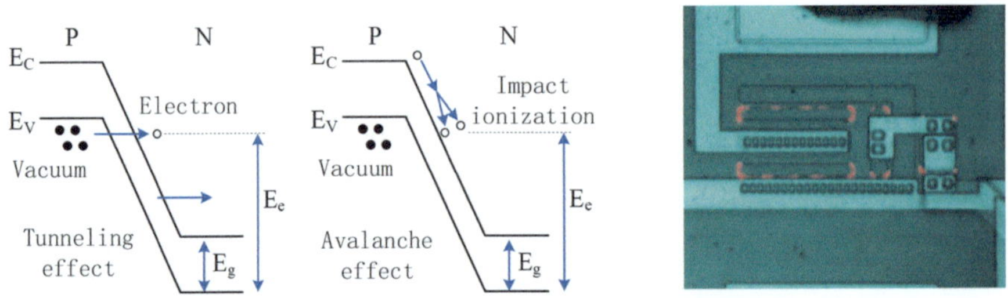

a) PN junction reverse-biased luminescence

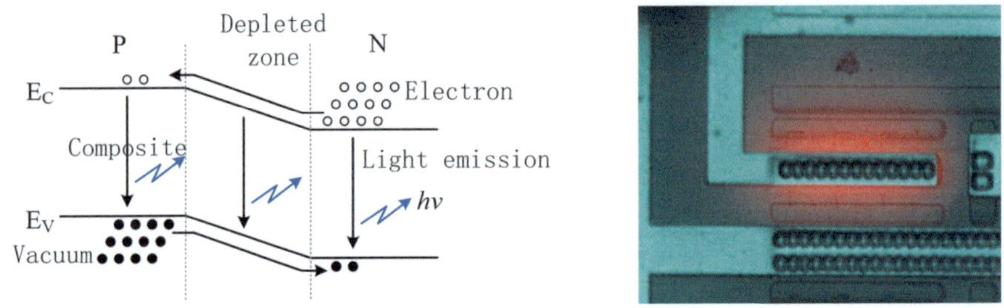

b) PN junction emits light when forward-biased

Fig. 6.98 PN junction light emission

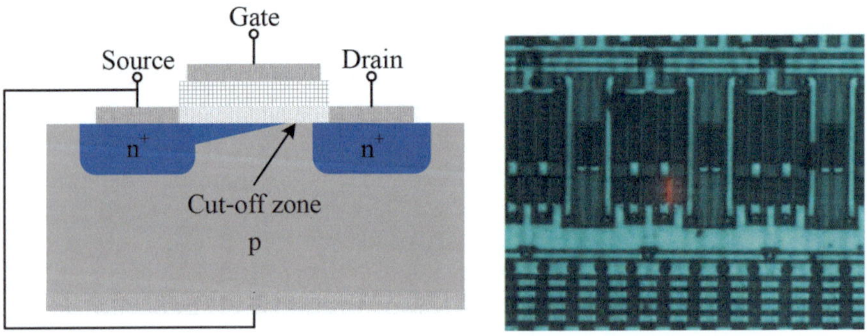

Fig. 6.99 Thermal carrier luminescence

Impurities or lattice defects introduce intermediate energy gap states in the band gap. When carriers pass through the intermediate energy gap, they undergo transitions, promoting electron-hole recombination and generating photons. The principle is the same as that observed by EMMI, and the emission pattern is shown in Fig. 6.100.

Some failed components will emit weak photon signals due to leakage current, breakdown, thermal carrier effect, etc. when powered on. EMMI is extremely sensitive to this phenomenon and can directly display the "bright spots" on the chip. Therefore, EMMI is particularly suitable for locating the failure of front-end active

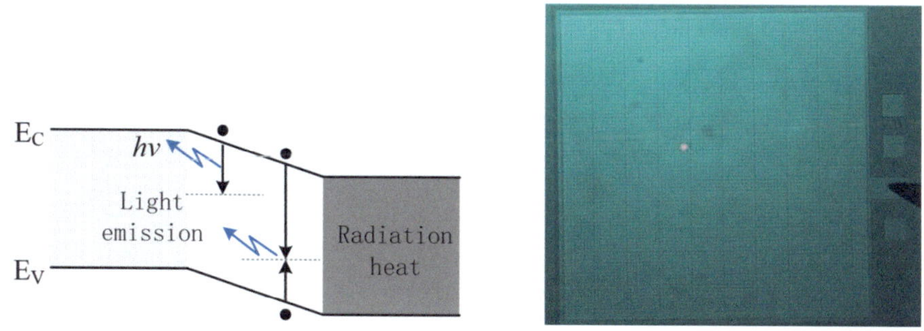

Fig. 6.100 Transistor defect luminescence

components, such as gate oxide breakdown, electrostatic discharge damage, latch-up effect, and PN junction leakage, etc. However, this method can only detect faults accompanied by light radiation. It is powerless against defects that only generate heat but no light (such as pure metal short circuit), and requires the use of other technologies as a supplement.

The origin of EMMI technology can be traced back to the late 1970s and early 1980s. In 1980, at the International Reliability Physics Symposium (IRPS), Richard W. Dutton and R. Subramonian, among others, published the earliest systematic paper describing EMMI technology. In the 1990s, high-sensitivity cooled CCD cameras significantly improved the detection ability of optical signals, capable of capturing even weaker photons. EMMI thus entered the commercialization stage. With the continuous improvement of semiconductor processes and the development of advanced packaging technologies, the EMMI positioning technology on the chip surface has been unable to meet the needs of failure analysis. Therefore, failure location needs to be conducted on the chip 背面. InGaAs cameras detect near-infrared photons in the wavelength range of 900–1700 nm. Near-infrared photons can penetrate the silicon substrate for imaging and hotspot location, thereby enabling backside EMMI testing and gradually replacing CCD cameras.

Figure 6.101 shows the EMMI system InGaAs 100 from Anhui Luxet Infrared Technology Co., Ltd. It is equipped with a fully automatic motion system, a deep-cooling InGaAs camera, different magnification microscopes, and a lock-in measurement mode, which can be applied to the location of failure points in semiconductor devices.

The main components of EMMI include InGaAs cameras, near-infrared objective lenses, probe station, X/Y/Z motion system, air bearing platform, electrical cabinet, industrial control computer and dark box, as shown in Fig. 6.102.

The core components of EMMI include InGaAs cameras, near-infrared lenses, X/Y/Z motion systems, and data acquisition and processing. The optical system composed of the InGaAs camera and the lens determines the resolution of imaging and the sensitivity of photon detection. These two parameters determine the core performance of EMMI and are the most important concerns for users. Taking a system with a maximum magnification of 100 times as an example, its optical resolution is

Fig. 6.101 Luxet Infrared Technology's EMMI system InGaAs 100

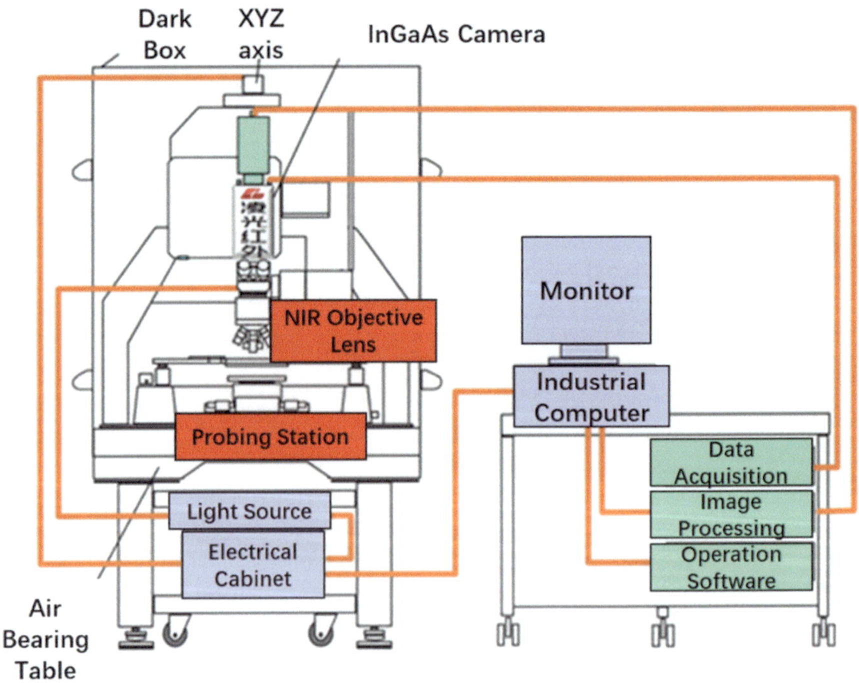

Fig. 6.102 The structure of EMMI

approximately 1 μm. The detection sensitivity is generally constant in the order of leakage current, and generally, EMMI can detect leakage currents of around 1 μA.

Meanwhile, software algorithms for data acquisition and image processing also affect the sensitivity of the system. During the EMMI test, the collected signals will also include system noise, such as the noise from the camera itself and environmental noise. Appropriate software algorithms during the data acquisition and image processing stages can effectively reduce noise and improve the signal-to-noise ratio of the test.

In addition, the movement accuracy and repetition accuracy of the motion system, as well as the rationality of the electrical system design, all will affect the overall performance of the entire system. In summary, the performance of the EMMI system depends on the design and debugging of the entire system, rather than being determined by a single or a few core components.

6.4.3.2 Application Cases of Microscopic Illumination Microscopes

EMMI is widely used in the field of semiconductor failure analysis. Both traditional silicon-based semiconductors and wide-bandgap semiconductors such as SiC/GaN can utilize EMMI for failure location.

A sample with gate oxide integrity (GOI) exhibited a gate leakage of 40uA at a voltage of 3.6V. The I-V re-measurement curve indicated a leakage failure. The sample was tested using the EMMI system InGaAs 100. Luminescent points were detected at the sample edge, which were identified as pinholes in the gate oxide layer. As shown in Fig. 6.103.

One of the components failed during use, and the GND pin of the IO pin leaked electricity and failed. Firstly, the backside of the component was ground and the chip substrate was exposed. Then, the backside EMMI test was conducted using the InGaAs 100 system of the EMMI system. Firstly, an EMMI hot spot was detected

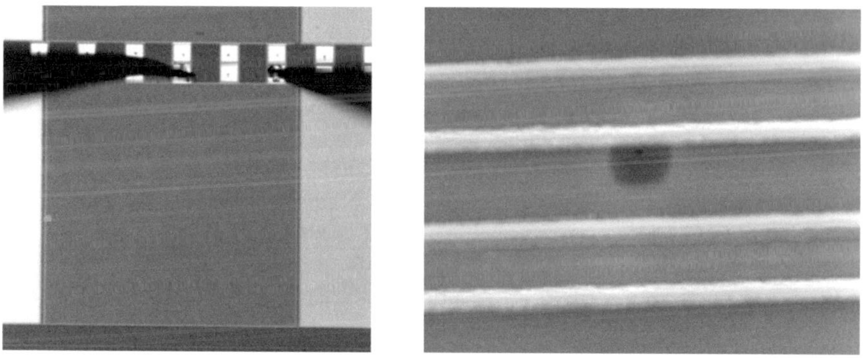

Fig. 6.103 The results of the EMMI system in locating the failure of gate oxide leakage in GOI samples

Fig. 6.104 The result of the EMMI system in locating the failure of ESD

near the failed pin of the component. After acid treatment to remove the metal layer on the chip surface, a scanning electron microscope was used to observe the corresponding position of the hot spot, and it was found that there was a clear breakdown phenomenon in the channel, which is a typical burn-in form of ESD breakdown, as shown in Fig. 6.104.

6.4.4 *Laser-induced hysteresis*

6.4.4.1 The basic principle of laser-induced hysteresis

When the device operates under bias voltage, current flows through the internal traces and nodes. If a focused laser beam is scanned over the device surface at this time, the irradiated local area will absorb the light energy and convert it into heat energy, resulting in a local temperature gradient. According to the temperature-resistance characteristics of the material, the increase in temperature will cause a change in the resistance of this area (for metals, it is a positive temperature coefficient, and the resistance increases; for some semiconductor areas, it may be a negative temperature coefficient, and the resistance decreases). The change in local resistance will redistribute the voltage and current in the circuit. For example, under a constant voltage bias, if the resistance at a certain scanning position slightly increases due to laser heating, the conduction capability of that branch will decrease, and the total current provided by the power supply will be reduced. This slight decrease in current can be detected and recorded by high-sensitivity detection circuits. Similarly, under a constant current bias condition, an increase in local resistance will cause the voltage at that location to rise ($V = IR$ relationship), thereby generating a subtle voltage change at the measurement end.

By synchronously recording the position of the laser beam and the changes in the electrical signal, it is possible to determine at which location the laser caused an extremely large change in resistance. This failure analysis technique is called optical beam induced resistance change (OBIRCH). The correspondence between the position and the signal can be processed to generate an OBIRCH image. The brightness of each pixel in the image represents the degree of resistance change caused by the laser irradiation at that location. Bright spots (high signals) indicate that the laser irradiation at this location caused significant current/voltage changes, while dark backgrounds indicate almost no effect. Generally, areas in the circuit with defects (such as micro short circuits, local high resistance, etc.) often generate more significant signal changes than normal areas due to abnormal current density or resistance when heated by the laser, and these changes are clearly visible in the OBIRCH image.

Due to many electrical failures (such as minor short circuits, open circuits, leakage, etc.) that can cause abnormal local resistance or current distribution, the OBIRCH technology can visually present these abnormalities in the form of image highlights, thereby achieving precise positioning of the defect location. Since the laser beam diameter is only several microns, OBIRCH has extremely high positioning resolution and can detect minor open circuits or short circuits defects, especially suitable for micro bridging and poor contact in metal interconnections in the later stages of failure. It is particularly effective for faults that do not generate light signals but change local resistance, and is a powerful supplement to EMMI. In addition, OBIRCH can also receive infrared laser from the backside of the chip, avoiding the obstruction of the surface metal, and achieve non-destructive positioning of deep defects. When the failure point is covered by a thick encapsulation or metal layer, the resistance change effect caused by laser heating is relatively small, and the sample needs to be opened or masked for processing to enhance the OBIRCH signal.

The OBIRCH system is typically composed of a laser scanning optical system, a precise electrical parameter testing unit, and an imaging processing software. The optical system includes a tunable laser source (typically in the near-infrared range, such as 1.3 μm, to allow for transmission through the backside of the silicon wafer), a microscope objective lens, and a scanning device (such as a high-speed galvanometer), which are used to focus the laser onto the surface of the test chip (either the front or the back) and scan along a pre-set trajectory. The electrical testing unit is responsible for providing bias voltage or current to the chip and monitoring the subtle changes in current/voltage in real time. It often includes low-noise preamplifiers and phase-locked amplifiers to extract the changes corresponding to the laser modulation signal. The imaging software correlates the scanning position with the detection signal, generating grayscale or pseudo-color fault distribution images, and can be superimposed with the optical microscope images to facilitate the location of the physical position of the defect. Modern commercial OBIRCH systems are typically integrated into a large, enclosed cabinet to isolate from environmental light and airflow interference and ensure the safety of laser operation.

Figure 6.105 shows the OBIRCH system LaserSight of Luxet Infrared Technology. This system consists of a laser scanning module (embedded in the protective enclosure, on the left side, the large box) and a control detection module (on the right side, with an operation interface and test instruments). The operation engineer controls the laser to scan the chip through computer software, while monitoring the

Fig. 6.105 Luxet Infrared Technology's OBIRCH system, LaserSight

changes in electrical parameters and generating defect heat maps. To avoid interference from environmental factors, the entire device is placed in a darkroom enclosure and equipped with a shock-proof platform.

When conducting the OBIRCH test, the following precautions need to be followed to ensure the acquisition of reliable results.

(1) Sample Preparation and Contact Methods

Make sure the chip is properly unpacked or opened and the surface to be tested is exposed. For the front OBIRCH, the packaging shell and the surface passivation layer can be removed to reduce laser loss. For the back OBIRCH, the silicon substrate usually needs to be ground to a certain thickness (such as below 100 μm) so that the 1.3 μm laser can penetrate to the active layer.

To provide a stable electrical bias connection, a microprobe or probe card can be used to contact the key nodes of the chip. The contact resistance should be as small and stable as possible to avoid additional noise. At the same time, the probe layout should avoid the laser path and prevent contact failure caused by thermal drift.

(2) Bias conditions and range setting

Based on the type of defect and the nature of the circuit, select the constant voltage or constant current mode and set an appropriate bias amplitude. If the voltage is too low, it will not be able to generate sufficient signals; if it is too high, it is prone to overheating or damaging the device. The monitoring end needs to choose an appropriate current range and filter bandwidth, which can capture the tiny ΔI changes without causing signal saturation.

(3) Scan parameter settings

OBIRCH usually adopts point-by-point array scanning, and the pixel size (such as 512 × 512) and integration time need to be set. The longer the integration time, the higher the signal-to-noise ratio, but it also takes longer. First, use a faster scanning to locate the defect, and then conduct a fine and slow scan in the hot spot area. To reduce the influence of thermal drift, the test can be conducted in a constant temperature environment, or pause the cooling of the sample after scanning a certain area.

(4) System calibration and validation

Before the first use or analysis, the system should be calibrated and its performance verified. By using chips with known defects to detect the hot spot locations and signal intensities, as well as calibrating the laser power, imaging size, and focal plane, the correctness of the detection can be confirmed. After obtaining the OBIRCH hot spot images, the reproducibility should be verified. For example, by changing the laser power or bias amplitude and observing whether the signal changes accordingly, the reproducibility can be tested. If any suspicious artifacts are found, the grounding, electromagnetic shielding, etc. should be checked, and cross-validation with other analysis methods should be conducted.

(5) Restrictions and Risk Management

Under appropriate power, OBIRCH usually does not cause permanent damage to the device. However, it is still necessary to note that if the device operates unstably (such as with oscillations), it will interfere with weak signals; large-area metal traces or shielding layers will reduce the laser heating efficiency or cause rapid heat dissipation, weakening the signal. In such cases, it is possible to remove some of the metal through etching, or use a back OBIRCH to scan directly from the chip's backside, but it is necessary to ensure that there are no metal heat dissipation structures on the backside that could interfere.

6.4.4.2 Application Cases of Laser-Induced Resistance Variation

For subtle short circuits (such as high-impedance bridging between two wires) or incomplete open circuit faults that are difficult to detect through conventional electrical tests, OBIRCH can provide unique positioning capabilities. Weak short circuits often manifest as micro-ampere-level leakage current in the circuit or deviations from the expected resistance value. However, locating their specific positions in complex circuits is like searching for a needle in a haystack. But OBIRCH scans the entire chip and generates local hot spot signals at the locations where short circuit defects exist, turning the hidden "dark leakage current" into bright defect points on the Fig. 6.106.

Modern integrated circuits contain multiple layers of metal interconnections and hundreds or thousands of vertical through-holes (vias). Failures in the interconnection structure (such as metal wire fractures, migration voids, poor via contact, etc.) are common failure modes of the devices. The OBIRCH technology demonstrates

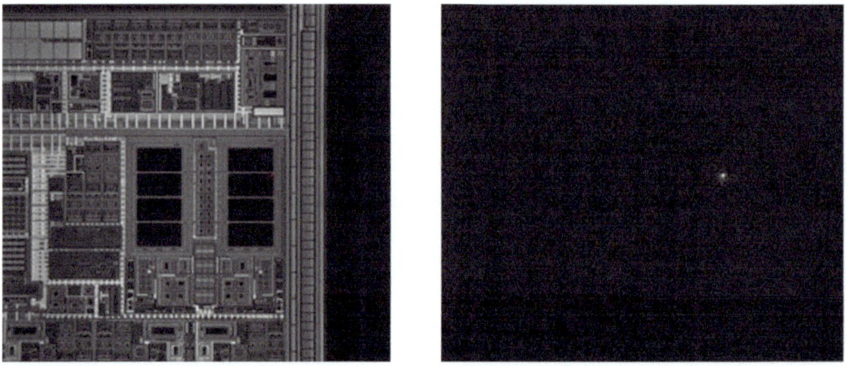

a) Hotspot Overlay Map b) OBIRCH original signal graph

Fig. 6.106 The results of the OBIRCH system in locating the failure of micro-open circuit samples

Fig. 6.107 The result of the OBIRCH system in locating the failure of interconnection faults

unique advantages in locating interconnection layer failures, capable of penetrating multiple layers to discover the hidden failure points. Especially in the chip's backside that cannot be directly observed, OBIRCH scans through the infrared laser passing through the silicon substrate, and the defects of each metal layer may leave "thermal imprints" on the image. In typical cases, interconnection failures lead to an increase in the resistance of a certain trace or a deterioration of the connection of a via, so this area often generates more heat when powered on. During OBIRCH scanning, these heat-generating points are manifested as bright spots detected. Fig. 6.107 shows the result of using the OBIRCH system to locate interconnection failures. Using

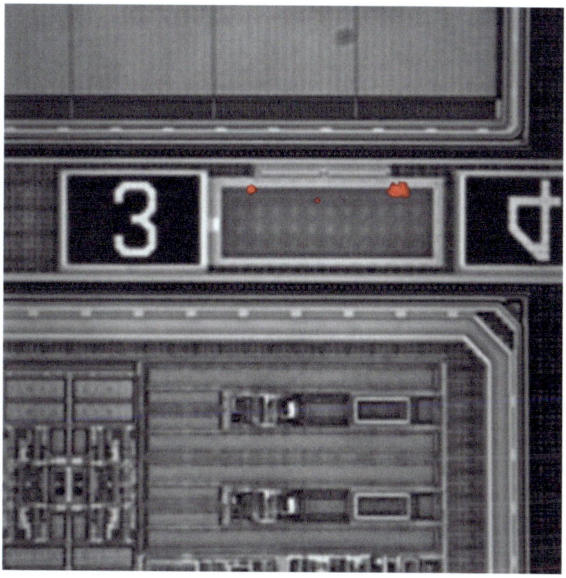

Fig. 6.108 The result of the OBIRCH system in locating the failure caused by electrical migration for defects

the front-side OBIRCH enables the laser to penetrate the silicon substrate, and the heat-generating points (indicated in blue) represent higher heat generation.

Electromigration (EM) refers to the phenomenon where, under high current density conditions, metal atoms are gradually displaced by the momentum of electrons, eventually leading to the formation of voids or protrusions in the metal conductor. This is one of the main causes of metal interconnection failure in modern large-scale integrated circuits. The OBIRCH technology is very useful for the detection and assessment of early defects caused by electromigration. When an electromigration void occurs in a metal wire, the resistance at that location increases, the current is restricted, and local overheating occurs; conversely, if a metal dendrite protrusion appears at a certain location causing a short circuit with adjacent wires, there will be abnormal current distribution when powered on. Both of these situations can be detected through the thermal-induced detection of OBIRCH.

In process and reliability engineering, OBIRCH is commonly used for failure analysis after accelerated aging tests of electrical migration. The specific method is to subject the test samples to high-temperature and high-current aging, and periodically remove the samples to scan the key interconnects with OBIRCH, observing whether any new hotspots appear. Once a bright spot appears in the middle section of a certain metal wire, it indicates that a partial open-circuit electrical migration cavity is likely to have formed here. Figure 6.108 shows the result of using the OBIRCH system to locate the failure of defects caused by electrical migration. Based on this, engineers can assess the reliability lifetime of the interconnection: for example, they can count how many hours after aging the OBIRCH detectable hotspots begin to appear, and thereby calculate the lifetime distribution. In addition, by comparing the OBIRCH hotspots with the cavity sizes in the subsequent FIB sections, the correlation between

OBIRCH signal strength and cavity size can be established, providing a reference for failure criteria.

In conclusion, as a type of laser-induced failure analysis technology, OBIRCH utilizes the resistance changes induced by laser heating to convert invisible electrical defects into visible signal hotspots, bridging the gap from macroscopic electrical measurements to microscopic failure locations. It plays an important role in the micro-defect location of semiconductor devices. For complex integrated circuits such as micro-short circuits, local open circuits, interconnect voids, and electrical migration, OBIRCH provides efficient screening and location methods. In practical engineering applications, OBIRCH is often combined with other analysis tools. For example, LIT is used to lock the approximate area first, and then OBIRCH is used for fine scanning to find the specific defect points; or EMMI distinguishes the luminescence of short circuit hotspots from thermal effects. With the development of semiconductor technology, although circuit density keeps increasing and the structure becomes more complex, OBIRCH still maintains its applicability by continuously improving optical resolution and signal sensitivity (such as developing visible light laser sources, confocal optics, and high-sensitivity phase-locked amplification integration, etc.). For failure analysis engineers, proficiently applying OBIRCH technology and understanding its result implications is one of the essential skills for effectively diagnosing chip faults and improving product reliability. It can be predicted that in the foreseeable future, OBIRCH will remain an indispensable and important technology in failure analysis laboratories, playing a significant role in ensuring the quality of semiconductor devices.

6.4.5 Scanning Electron Microscopy (SEM)

6.4.5.1 Basic Principles of Scanning Electron Microscopy

The invention of the optical microscope opened the door to the microscopic world. However, as semiconductor device processing and production technology become increasingly miniaturized, optical microscopes cannot meet the observation and analysis requirements for micron and nanometer-scale devices due to their resolution limitations. At this point, the ultra-high resolution of scanning electron microscopes (SEM) is required, as shown in Fig. 6.109, which features SEM products from Thermo Fisher Scientific.

The basic components of an SEM are illustrated in Fig. 6.110. An electron beam scans the sample surface in a raster pattern. First, the filament generates free electrons through heating or an external electric field, which are then accelerated into the electron column by a positively charged anode. Both the filament area and the electron column are maintained under high vacuum to prevent contamination, reduce vibrations and noise, and help engineers obtain high-resolution sample images. Without

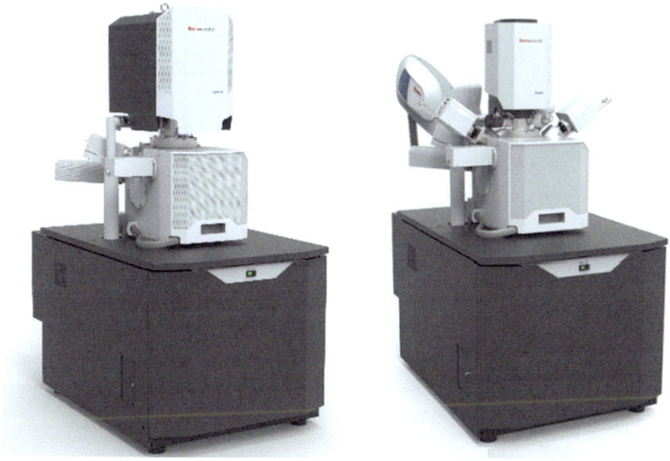

Fig. 6.109 Thermo fisher scientific SEM

Fig. 6.110 Basic components of SEM

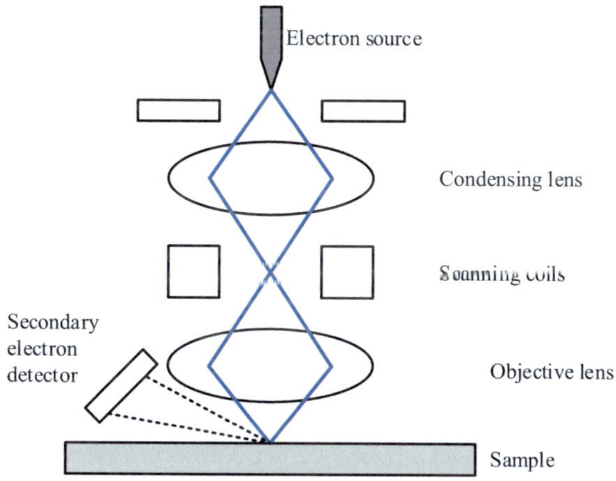

high vacuum protection, volatile substances in the sample would interact with electrons in the electron column, affecting electron beam deflection and reducing detector efficiency, thereby impacting image quality (Fig. 6.110).

Various signals involved in SEM are shown in Fig. 6.111. The primary imaging signals are backscattered electrons (BSE) and secondary electrons (SE). BSE are electrons that are elastically and inelastically scattered after interaction with the sample and reflected back to be collected by the detector. Thus, BSE can reflect the atomic number differences of different materials in the sample, also known as compositional contrast. SE are outer shell electrons from the sample that are excited and ejected from the surface's few-nanometer region, and are collected by the detector. Therefore, SE can reflect the microscopic morphology of the material surface, also known

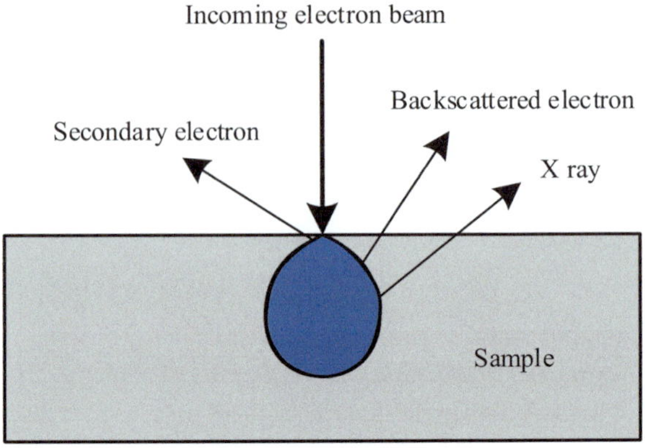

Fig. 6.111 Various signals in SEM

as topographical contrast. Interaction of the incident electrons with the sample also excites characteristic X-rays of the elements themselves. Each substance has X-rays with specific energies, much like each person has a unique fingerprint. By detecting the energy of X-rays in an unknown sample, the elements and their content in the sample can be identified and distinguished.

6.4.5.2 Application Cases of Scanning Electron Microscopy

SiC crystals need to be grown in high-temperature environments and possess high rigidity and chemical stability, which leads to high-density crystal and surface defects in the grown SiC crystals, resulting in poor quality of the substrate and subsequently manufactured epitaxial layers. Various types of defects can cause different degrees of performance degradation in devices, or even complete device failure. To improve yield and performance, fast, high-precision, non-destructive inspection technologies play a crucial role in SiC wafer production lines.

SiC wafer-level SEM defect detection systems are used to detect and analyze defects in SiC wafers, which may include particles, scratches, pits, cracks, and other defects that could affect device performance and reliability. These SEM systems typically include software to analyze captured images and provide quantitative data on the size, shape, and distribution of defects on the wafer. This information is critical for quality control and process optimization in SiC device manufacturing. By using defect detection systems, manufacturers can identify and address potential issues early in the production process, reducing yield loss and improving the overall quality of SiC wafers and their manufacturing equipment (Fig. 6.112).

The primary applications of using SEM for SiC device failure analysis include observing the microscopic surface morphology of chips, locating failure points, and accurately measuring the microscopic dimensions and potential distribution of SiC devices. Additionally, EDS (Energy Dispersive Spectroscopy) can be used to analyze

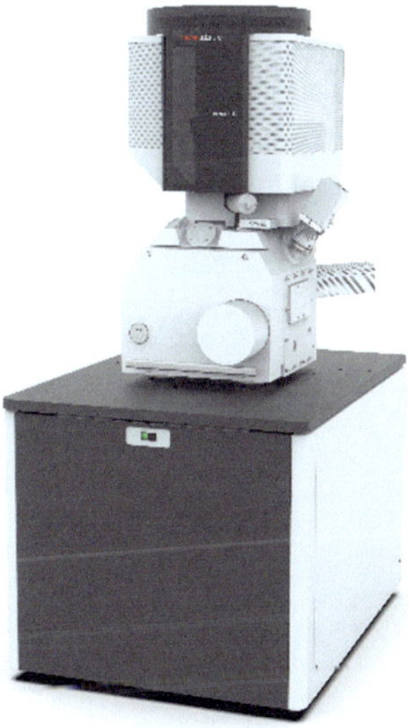

Fig. 6.112 Thermo Fisher Scientific's SiC wafer-level SEM defect inspection system

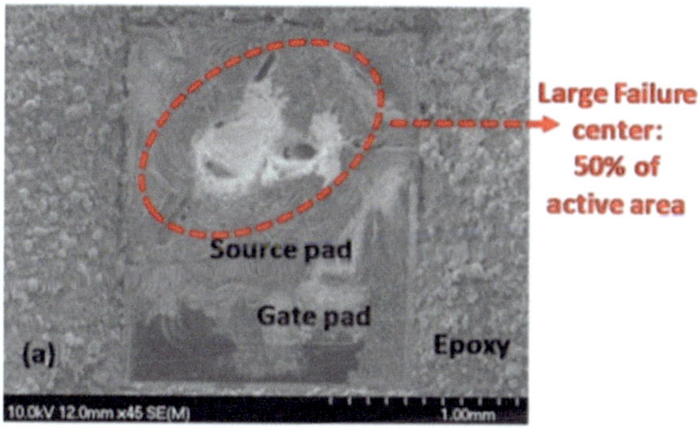

Fig. 6.113 SEM image of a failed SiC MOSFET in low-inductance testing [4]

the elemental composition and content of micro regions, as well as to identify foreign substances or contaminants.

Figure 6.113 shows the SEM image of a failed SiC MOSFET in low-inductance testing

Figure 6.114 shows an SEM photograph of a SiC MOSFET where uneven thermal stress distribution can cause microcracks. The figure shows that many obvious cracks caused by thermal stress are randomly distributed on the surface of the chip.

The SiC MOSFET was subjected to repeated short-circuit tests. Due to the repeated short-circuit, the internal heat of the device under test accumulated, and the normal working time of the gate oxide layer decreased rapidly with the increase of current. Figure 6.115 shows the analysis results of optical microscope and SEM.

The SEM image shown in Fig. 6.116 reveals micro cracks in the gate oxide of a SiC MOSFET that caused device failure.

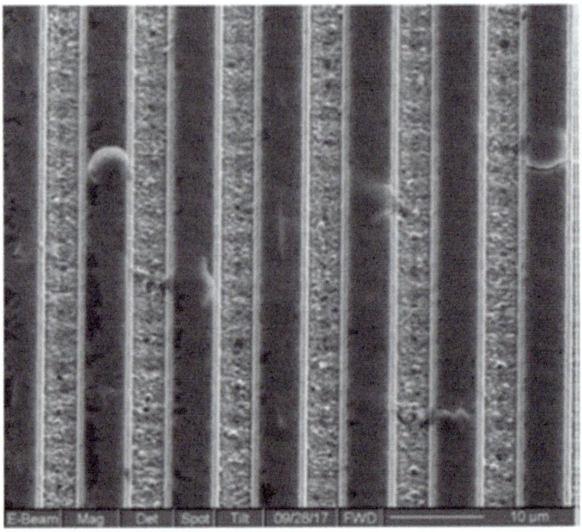

Fig. 6.114 SEM image of a SiC MOSFET with microcracks due to uneven thermal stress [5]

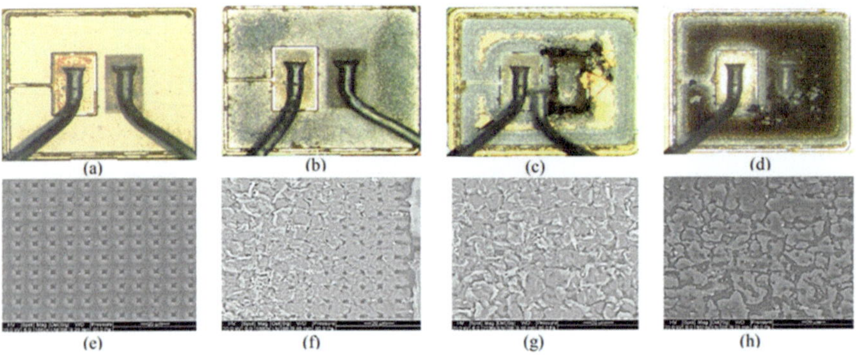

Fig. 6.115 Optical and SEM images of SiC MOSFET under repeated short-circuit testing [6]

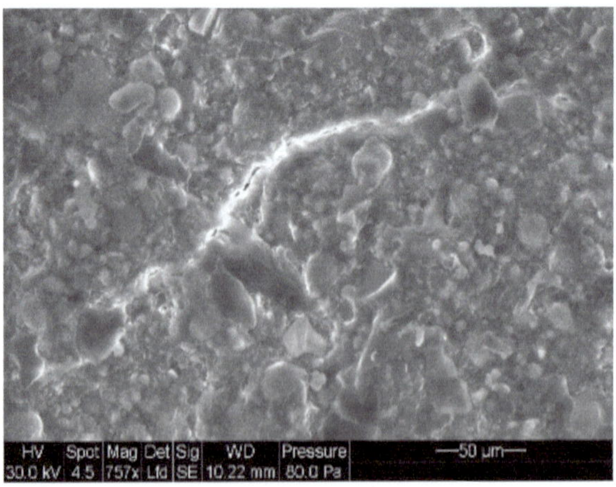

Fig. 6.116 Microcracks in the gate oxide of SiC MOSFET [6]

6.4.6 Dual-Beam Electron Microscope

6.4.6.1 Basic Principle of Dual-Beam Electron Microscope

A Dual-Beam Electron Microscope refers to an instrument equipped with both a Focused Ion Beam (FIB) and a Scanning Electron Microscope (SEM). This device combines the advantages of both beam types, enabling high-precision material processing and imaging, and is widely used in the semiconductor industry, materials science, nanotechnology, and biomedical fields.

As shown in Fig. 6.117, the SEM component of a dual-beam electron microscope is primarily used for observing the surface of samples and cross-sectional morphology after ion beam processing. Its basic operating principles have been detailed in the previous section. Besides observation, the secondary electrons generated on the sample surface can also be used for local metal or non-metal deposition by introducing some gas-phase deposition precursors. This function is usually a supplement to ion beam-induced material deposition, used for surface deposition on materials sensitive to ion beams.

The FIB component of the dual-beam microscope is mainly used for material cutting, embedding, deposition, or etching. FIB systems typically use Liquid Metal Ion Sources (LMIS), such as a gallium source. In a gallium LMIS, liquid gallium is placed on a tungsten needle and heated. The gallium liquid flows to the tip of the needle, where surface tension and the electric field interact to form a structure known as the Taylor Cone. The tip of this cone has a radius of about 2 nm, and an extremely high electric field of over 1×10^8 V/cm is generated at the tip, causing gallium atoms to ionize and field emit. The source ions are usually accelerated to energies between 1 and 50 keV and focused on the sample using electrostatic lenses. LMIS produces ion beams with high current density and very small energy spread. Modern FIB equipment can deliver tens of nanoamperes of current to the sample or modify

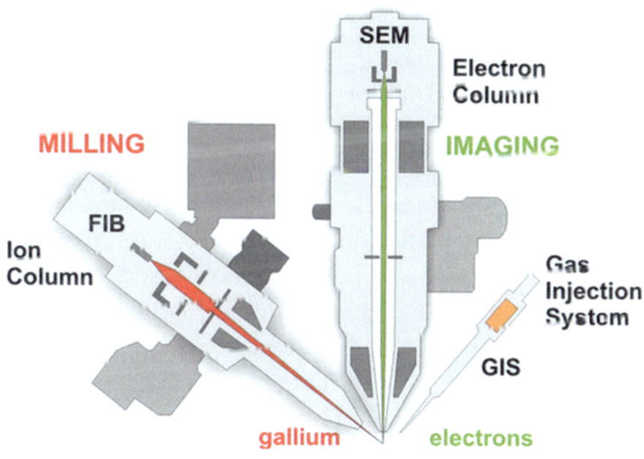

Fig. 6.117 Main functional components of a dual-beam electron microscope

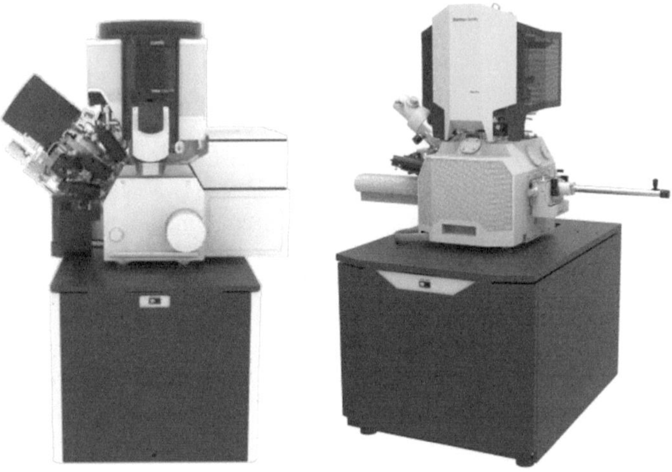

Fig. 6.118 Thermo fisher scientific dual-beam electron microscope

material structures with spot sizes at the nanometer level. In recent years, instruments using plasma beams of inert gas ions, such as xenon, have become more common. FIB can directly etch materials or deposit metal and non-metal materials locally by introducing gas-phase precursors. By varying the beam parameters and precursors, the deposition and etching processes can be precisely controlled. Figure 6.118 shows the dual-beam electron microscope produced by Thermo Fisher Scientific.

6.4.6.2 Application Cases of Dual-Beam Electron Microscope

In the various failure mechanisms of SiC MOSFETs, short circuits are a severe failure mode that can lead to permanent device failure. SiC MOSFETs must withstand high electrical and thermal stress during short circuits without failing before the protection circuit is activated. Studies have found that gate short circuits are a common failure mode following short-circuit transients at low or moderate drain-source voltages. This is related to the aging of the SiO_2 insulation layer in SiC MOSFETs under different temperatures, drain-source voltages, and gate-source biases. By locating the failure point using infrared imaging and then using the dual-beam electron microscope to obtain cross-sections of the failure area and analyze them, images show cracks in the oxide layer. The gate leakage current observed after short-circuit stress is related to the formation of these cracks, which create a conductive path between the polysilicon gate and the source, leading to gate oxide breakdown (Fig. 6.119).

For the failure mechanism of dual-gate trench SiC MOSFETs, the short-circuit location was first detected using Optical Beam Induced Resistance Change (OBIRCH). The failure area was then further located using cross-sectional images

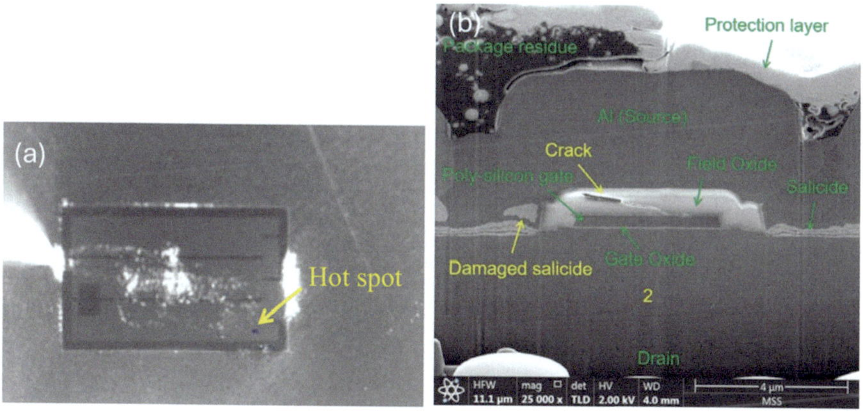

Fig. 6.119 Short-circuit failure in SiC MOSFET

from the dual-beam electron microscope. Figure 6.120a shows the OBIRCH illumination points, while Fig. 6.120b presents a top-view SEM image of the detected failure location, revealing the square cell design of the dual-gate trench SiC MOSFET, where the gate dielectric layers intersect. The damaged area (marked with a red arrow) was identified as the gate-source short-circuit location. Compared to the intact cell on the right side, the damaged area shows surface roughness and protrusions. To further investigate the fault area, tangents (1)–(4) (green dashed lines in Fig. 6.120b) were selected for FIB-SEM cross-sectional imaging. Figure 6.120c shows two separate cells at tangent (1), where a conductive path connects the gate polysilicon and source aluminum in the valley region of the gate dielectric layer. Figure 6.120d shows significant damage to the gate dielectric layer at tangent (2). Figure 6.120e shows a conductive path connecting the gate polysilicon and source aluminum directly on the gate sidewall of the left gate structure at tangent (3). Figure 6.120f shows structural damage to the gate dielectric and gate sidewall at tangent (4), with the conductive path between the gate polysilicon and source aluminum being the reason for the small R_{GS} value and increased V_{GS} after gate closure in the 400 V short-circuit test.

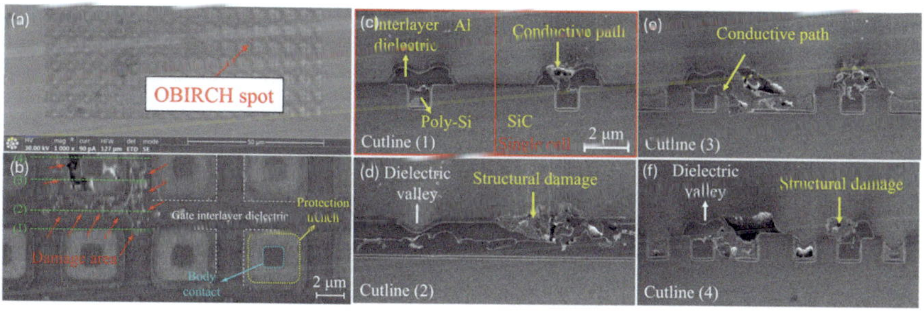

Fig. 6.120 Failure analysis of dual-gate trench SiC MOSFET [7]

Fig. 6.121 Some signals produced after interaction of high-energy electron beam and TEM sample

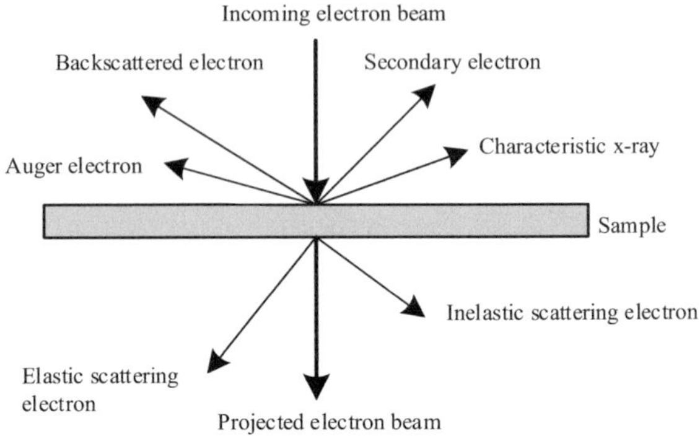

6.4.7 Transmission Electron Microscope

6.4.7.1 Basic Principle of Transmission Electron Microscope

Transmission Electron Microscopes (TEM) are often used for analyzing and characterizing semiconductor material properties and exploring semiconductor device failure mechanisms. TEM typically requires extremely high acceleration voltages (60–300 kV) to project an electron beam onto the sample. As shown in Fig. 6.121, high-energy incident electron beams interact with the sample, producing various secondary signals such as Secondary Electrons, Backscattered Electrons, Elastically Scattered Electrons, Inelastically Scattered Electrons, Auger Electrons, and Characteristic X-Rays. Special signal collectors process these secondary signals to obtain information on the sample's structure, morphology, and chemical composition.

Currently, commercially available field emission TEMs use acceleration voltages up to 200–300 kV. At maximum acceleration voltage, the resolution limit of traditional TEM can reach below 0.5 nm. With continuous iterations in TEM technology, especially with the advent of Scanning Transmission Electron Microscopes (STEM) and spherical aberration correctors, the resolution of TEM has further improved. For example, as shown in Fig. 6.122, the latest Spectra Ultra STEM from Thermo Fisher Scientific achieves a resolution of 50 pm in STEM mode at 300 kV, making it easier to capture atomic images of crystal materials like SiC.

Figure 6.123 shows a high-resolution atomic image of 4H-SiC taken along the <$11\bar{2}0$> crystal direction using a FEI Titan 80–300 spherical aberration-corrected transmission electron microscope at an accelerating voltage of 300 kV. Obviously, the stacking arrangement of atoms can be observed more intuitively using a transmission electron microscope. Further analysis of the stacking of Si atoms (marked with red and yellow triangles in Fig. 6.123) shows that there are two independent stacking faults at the positions marked by red arrows in Fig. 6.123. It can be seen that TEM is very necessary to observe and analyze common crystallographic defects in SiC substrates and epilayers, such as dislocations, stacking faults, etc.

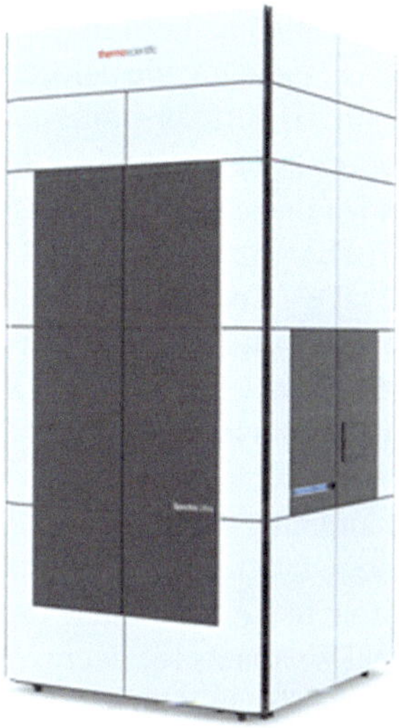

Fig. 6.122 Thermo fisher scientific spectra ultra S/TEM

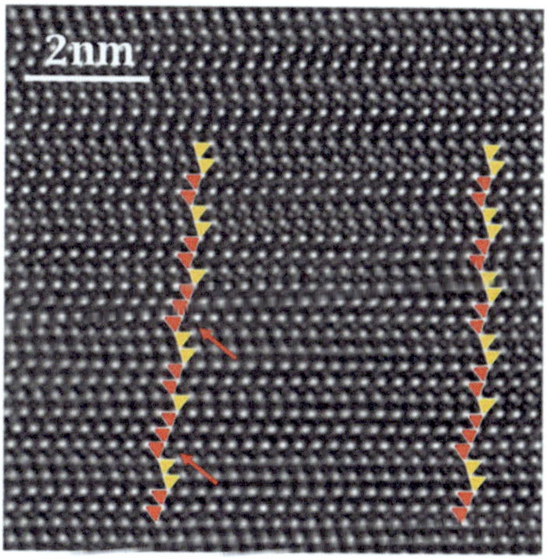

Fig. 6.123 High-resolution atomic image of 4H-SiC taken along the <1120> crystal direction using TEM [8]

6.4.7.2 Application Cases of Transmission Electron Microscope

TEM is primarily used in research on SiC to analyze and characterize defects in SiC substrates and epitaxial layers, and to study the failure mechanisms caused by these defects.

The cost of SiC substrates and epitaxial layers accounts for approximately 70% of the total cost in SiC device production. This is due to the difficulty of producing high-quality, low-defect-density single-crystal SiC. Therefore, analyzing and characterizing defects in SiC substrates and epitaxial layers is crucial. To better illustrate the application of TEM in SiC device failure analysis, the following discussion focuses on crystallographic defects in SiC substrates and epitaxial layers.

In 4H-SiC substrates and epitaxial layers, the primary line defects (i.e., dislocations) are Basal Plane Dislocations (BPDs) and Threading Dislocations (TDs). Threading Dislocations are further classified into Threading Screw Dislocations (TSDs), Threading Edge Dislocations (TEDs), and Threading Mixed Dislocations (TMDs). The Burgers vectors of these dislocations are $\vec{c}$, $a/3\langle 11\bar{2}0\rangle$, and $\vec{c} + \vec{a}$ respectively (where $\vec{a}$ and $\vec{c}$ represent the basis vectors of the 4H-SiC crystal unit cell along ⟨11–20⟩ and ⟨0001⟩ directions). In experiments, Large-Angle Convergent Beam Electron Diffraction (LACBED) was used for systematic studies of TDs in 4H-SiC. LACBED was applied to analyze and characterize all variants of TDs. Combining the interference fringes observed in LACBED images of each TD variant under different crystal orientations, basic information and Burgers vectors of six different TEDs variants (a-f), two different TSDs variants (g-h), and two different TMDs variants (i-t) were obtained. The results show that the types of TDs in 4H-SiC are consistent with and fully match the theoretical defect models for the hexagonal crystal system (Fig. 6.124).

In addition to purely analyzing and characterizing dislocation types using TEM, some studies have also explored the relationship between crystallographic defects and changes in device electrical properties. For instance, one study discussed the connection between TMDs and leakage current in SiC MOSFETs. Figure 6.125a shows leakage current regions detected on a chip with a high concentration of Al ion implantation (implantation concentration of $10^{21}/cm^3$) after KOH etching, with regions *l* and *nl* representing high and low leakage currents, respectively. The leakage current versus reverse voltage curves for each region are shown in Fig. 6.125b. TEM samples were then prepared using a dual-beam electron microscope and characterized using Weak-Beam Dark-Field (WBDF) technique. Dislocation lines were observed under the etch pits in both the nl region (Fig. 6.125c) and the l region (Fig. 6.125d), with the etch pits in the high leakage current region being significantly deeper. Further analysis using LACBED and WBDF techniques revealed that the dislocation type in the high leakage current l region was TMD, with a Burgers vector of ⟨0001⟩ + ⟨11–00⟩. Based on these experimental results, it is suggested that the strain field generated by TMD may cause Al ions to segregate along the TMD, ultimately leading to higher reverse leakage current at the p–n junction.

Another study observed reverse leakage in SiC IGBTs caused by TMDs. Under dual-beam conditions, TEM bright-field images were obtained under the $\vec{g}$ vector conditions of ⟨11–20⟩ and (0004), clearly showing the presence of dislocation lines Fig. 6.126a, b). This indirectly confirmed that the dislocation lines contained both screw and edge dislocation components, i.e., TMD. LACBED confirmed the Burgers vector of this TMD as $\langle \bar{1}\,011\rangle$.

Unlike TDs, which lead to increased reverse leakage current and reduced reliability, BPDs are believed to worsen the forward current in p-i-n bipolar devices. BPDs are actually composed of two Shockley Partial Dislocations (SPDs) and a Shockley Stacking Fault (SSF) formed between two SPDs. Studies have shown that applying

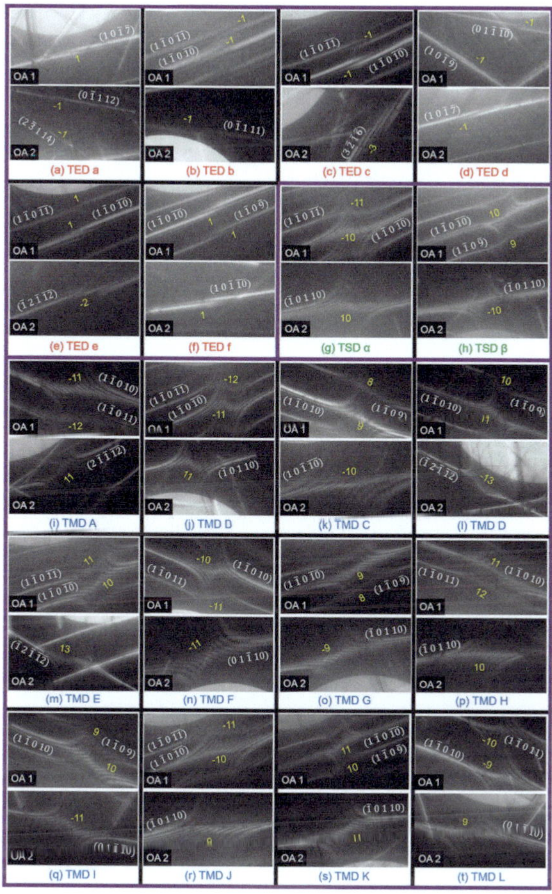

TD		Burgers Vector ($\vec{b}$)		
ID	Type	$\vec{b}_{uvtw}$ (4-index)	$\langle a \rangle$	$\langle c \rangle$
a	TED	$\frac{1}{3}[1\,1\,\bar{2}\,0]$	←	-
b	TED	$\frac{1}{3}[\bar{1}\,2\,\bar{1}\,0]$	↙	-
c	TED	$\frac{1}{3}[\bar{2}\,1\,1\,0]$	↘	-
d	TED	$\frac{1}{3}[\bar{1}\,\bar{1}\,2\,0]$	→	-
e	TED	$\frac{1}{3}[1\,\bar{2}\,1\,0]$	↗	-
f	TED	$\frac{1}{3}[2\,\bar{1}\,\bar{1}\,0]$	↖	-
α	TSD	$[0\,0\,0\,1]$	-	⊙
β	TSD	$[0\,0\,0\,\bar{1}]$	-	⊗
A	TMD	$\frac{1}{3}[\bar{1}\,2\,\bar{1}\,\bar{3}]$	↙	⊗
B	TMD	$\frac{1}{3}[\bar{2}\,1\,1\,3]$	↘	⊙
C	TMD	$\frac{1}{3}[1\,2\,\bar{1}\,3]$	↙	⊙
D	TMD	$\frac{1}{3}[2\,\bar{1}\,\bar{1}\,3]$	↖	⊙
E	TMD	$\frac{1}{3}[1\,1\,\bar{2}\,\bar{3}]$	←	⊗
F	TMD	$\frac{1}{3}[\bar{1}\,\bar{1}\,2\,\bar{3}]$	→	⊗
G	TMD	$\frac{1}{3}[\bar{2}\,1\,1\,\bar{3}]$	↘	⊗
H	TMD	$\frac{1}{3}[1\,\bar{2}\,1\,3]$	↗	⊙
I	TMD	$\frac{1}{3}[\bar{1}\,\bar{1}\,2\,3]$	→	⊙
J	TMD	$\frac{1}{3}[1\,1\,\bar{2}\,3]$	←	⊙
K	TMD	$\frac{1}{3}[2\,\bar{1}\,\bar{1}\,\bar{3}]$	↖	⊗
L	TMD	$\frac{1}{3}[1\,\bar{2}\,1\,\bar{3}]$	↗	⊗

Fig. 6.124 TDs in 4H-SiC obtained using LACBED [9]

forward operating current to 4H-SiC bipolar devices causes the recombination of electron–hole pairs, leading to the sliding of SPDs and the expansion of SSF, which ultimately reduces the stability of the 4H-SiC device. Similar studies have investigated SSF expansion and BPDs by applying forward operating current to 4H-SiC bipolar devices, using X-ray topography (XRT) and TEM. Research has delved into the mechanisms of SSF origin, expansion in the epitaxial layer, transition from BPD to TED [12, 12], and the relationship with forward operating current. Figure 6.127 shows an example of SSF characterization using transmission electron microscopy in a set of related studies.

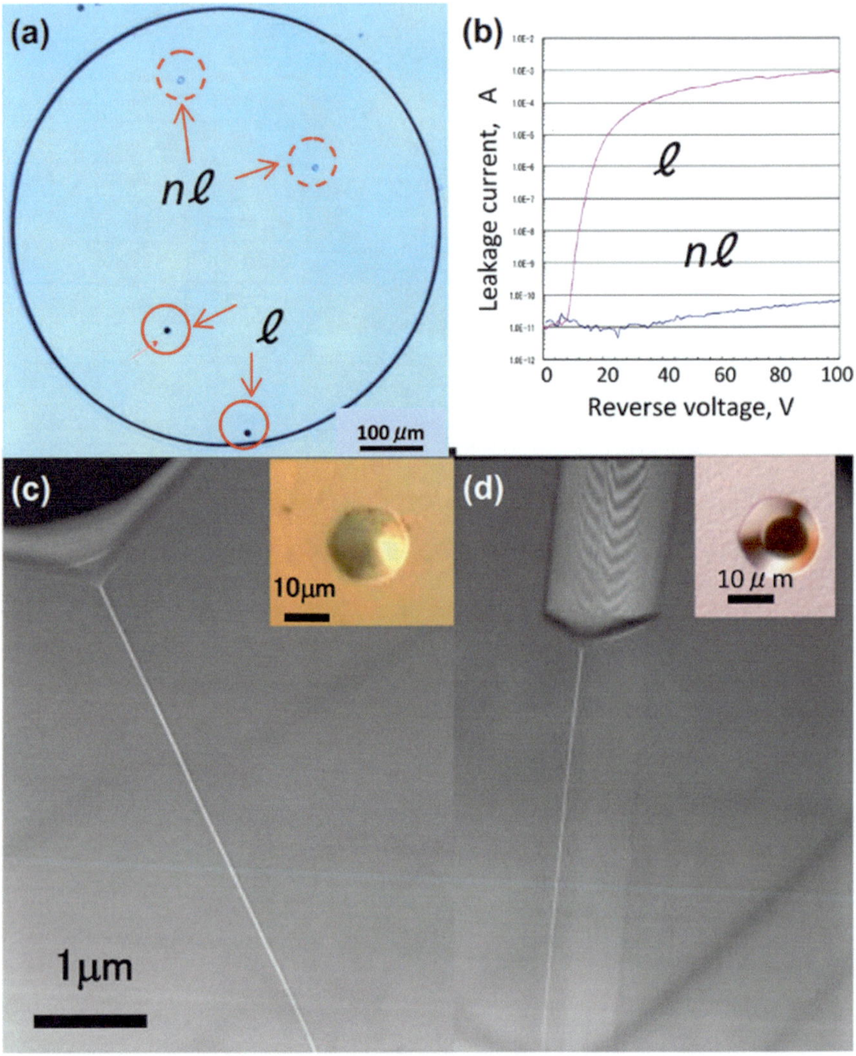

Fig. 6.125 Relationship between TMD and leakage current in SiC MOSFET [10]

Currently, one of the main development directions for 4H-SiC remains the optimization and control of critical defect density in substrates and epitaxial layers. Transmission electron microscopy is essential during the process adjustments. On one hand, it helps to compare defect changes before and after process improvements. On the other hand, studying the nature of defect generation and changes, as well as exploring the mechanisms of defect-induced failures, also aids in guiding the design and development of front-end processes.

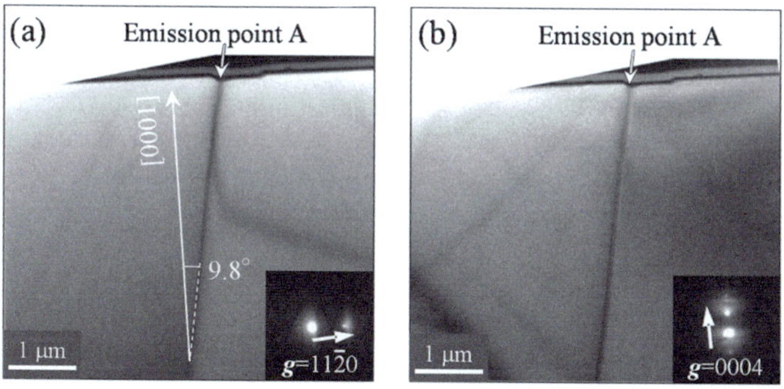

Fig. 6.126 Relationship between TMD and leakage current in SiC IGBT [11]

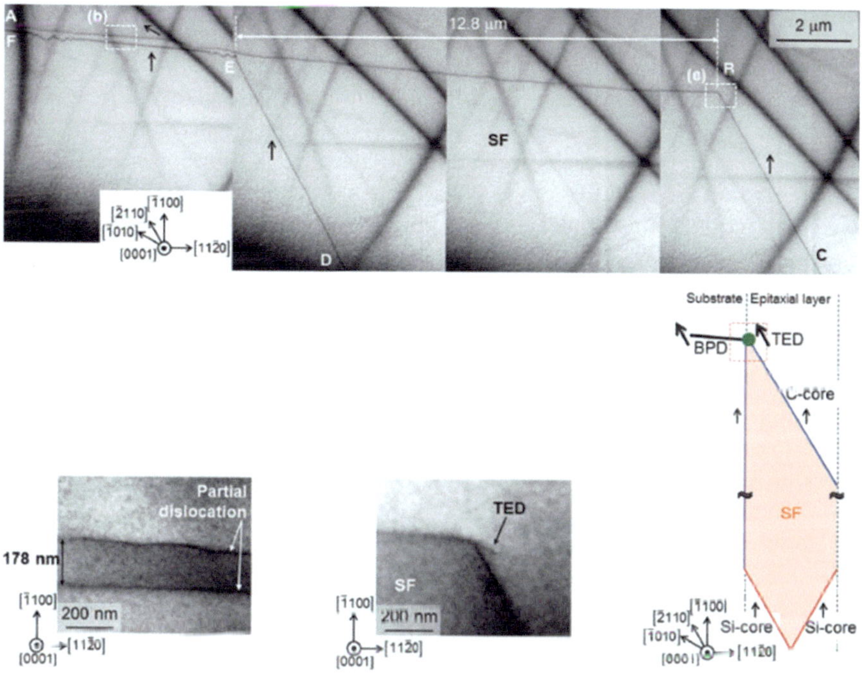

Fig. 6.127 4H-SiC BPDs [13]

6.5 System Application Testing

From traditional analog to high-efficiency switching power converters, there are various types and sizes of products. All companies face complex and dynamic operating environments, with significant changes in equipment load and demand. Even ordinary switch-mode power supplies need to withstand sudden peaks far beyond normal operating conditions.

Ideally, power converters should operate according to the mathematical models used in their design. However, in the real world, imperfect device performance and significant load variations affect power supply linearity and dynamic response. Additionally, environmental changes also alter power supply performance. Furthermore, evolving performance and cost requirements complicate power supply design, requiring power electronics engineers to consider the following issues:

(1) How many watts above the rated capacity can the power supply handle, and for how long?
(2) How much heat does the load generate, what happens when it overheats, and what cooling airflow is needed?
(3) What occurs when the load current significantly increases?
(4) Can the device maintain the rated output voltage?
(5) Can the device maintain the rated output voltage?
(6) What happens when the input voltage changes?

Power electronics engineers are expected to design power converters that are smaller, generate less heat, have lower manufacturing costs, and meet stricter EMI/EMC standards. Achieving these goals requires more precise and rigorous measurements. For engineers accustomed to using high-bandwidth oscilloscopes for measurements (such as DDR, SATA/SAS, PCIe), the lower signal bandwidth in power converters might seem simpler. In fact, power converter testing presents a series of challenges that high-speed circuit designers have never faced. Voltage through switching devices can be very high and "floating", with varying pulse widths, periods, frequencies, and duty cycles, requiring accurate waveform capture and defect analysis.

Choosing the right tools for power converter measurement is crucial and should be set up to perform accurately and repeatedly. Testing a power converter mainly involves power and load testing devices, oscilloscopes, and probes. Power and load testers provide the necessary voltage and current for the converter and measure corresponding functions, while oscilloscopes and probes capture voltage and current waveforms within the converter.

Oscilloscopes used for converter testing must have basic bandwidth and sampling rates to handle SMPS switching frequencies, generally requiring at least two channels—one for voltage and one for current. For three-phase power systems, at least six channels are needed: three for voltage signals and three for current signals, with potentially more channels for testing systems. Using oscilloscopes and probes, the following tests can be performed to locate loss points, validate circuit safety designs, and improve power conversion efficiency and safety:

1. **Input Analysis**

Power converter input sides typically focus on the response to input variations, current and power absorption during operation, and distortion of power frequency currents. Some measurements, such as power consumption, are key indicators of power converters, while others, such as power factor and harmonics, are related to regulations.

(1) **Power Measurement**

Real Power: Measures the actual power of the power converter, in watts (W).

Reactive Power: The imaginary power temporarily stored in inductors or capacitors, in volt-amperes reactive (VAR).

Apparent Power: The absolute value of complex power, in volt-amperes (VA).

Power Factor: The ratio of real power to apparent power.

Phase: The angle between real power and apparent power, in degrees.

(2) **Harmonic Measurement**

Current harmonics occur when nonlinear devices cause current distortion in the circuit. Voltage distortion occurs when this current flows through the distribution system, generating heat that accumulates in cables and transformers. The more power converters connected to the grid, the more severe the harmonic distortion. Several standards have been designed to limit the impact of nonlinear loads on power quality, such as IEC61000-3-2 and MIL-STD-1399.

The IEC61000-3-2 standard limits current harmonics injected into the power grid. It applies to all electrical and electronic devices with input currents up to 16A, connected to public low-voltage distribution systems (230VAC or 415VAC three-phase). This standard further classifies devices into Class A (balanced three-phase equipment), Class B (portable tools), Class C (lighting equipment and dimmers), and Class D (devices with unique current waveform requirements). Measurement equipment can quickly display results graphically and in tabular form, as shown in Fig. 6.128.

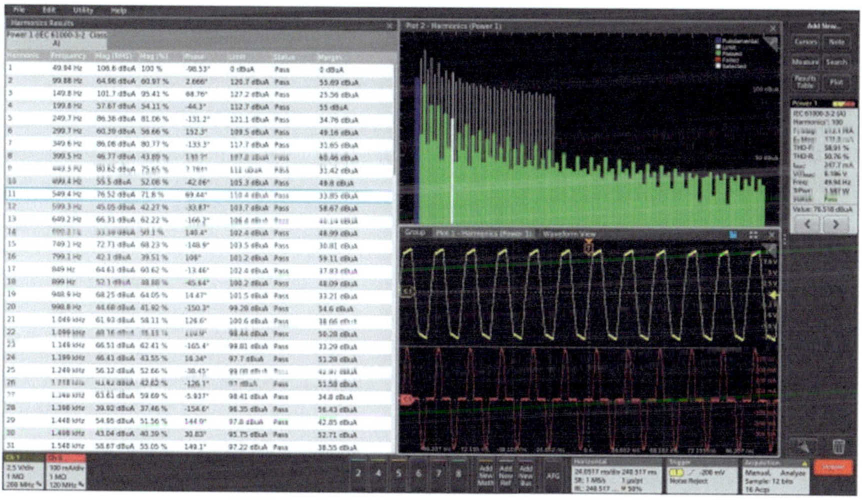

Fig. 6.128 Harmonic test results

(3) **Surge Current**

Surge current refers to the peak current that flows into the power device when it is turned on. This peak current is much higher than the steady-state input current and can affect the safety of the power network and equipment. Power supply designs typically require suppression of excessive surge currents.

2. **Output Analysis**

(1) **Power Frequency Ripple and Switching Ripple**

In short, ripple is the AC voltage superimposed on the DC output of the power supply, expressed as a percentage of the normal output voltage or peak-to-peak voltage. Two types of ripple are displayed on the power supply output: power frequency ripple, which is related to power frequency, and switching ripple, which is detected from the output of the switching power supply according to the switching frequency.

(2) **Efficiency**

The efficiency of the power converter is calculated by dividing the measured output power by the measured input power.

3. **Switching Measurement**

Measuring the switching process of switching devices while the power converter is operating helps confirm whether the converter is working correctly, quantify switching device losses, and ensure the devices operate within normal ranges.

(1) **Switching Losses**

Measure the turn-on loss, turn-off loss, and conduction loss of the switching device under continuous switching conditions. Capture the voltage across the switching device and the current through it, multiply the two to get instantaneous power, then integrate this power over the desired start and end time intervals to calculate the turn-on loss and turn-off loss for each cycle. Oscilloscopes can perform automated measurements and statistical calculations of all results, greatly improving the repeatability and efficiency of measurements.

(2) **Safe Operating Area (SOA)**

The safe operating area (SOA) of switching devices, such as BJTs, MOSFETs, or IGBTs, is typically specified in the product technical documentation and determines the safe current flow at a given voltage. The main challenge in determining the SOA during power converter operation is accurately capturing voltage and current under various load scenarios, temperature variations, and power frequency input voltage changes. Oscilloscopes with automated data capture and analysis functions simplify this task and often provide customizable SOA template testing functions as shown in Fig. 6.129.

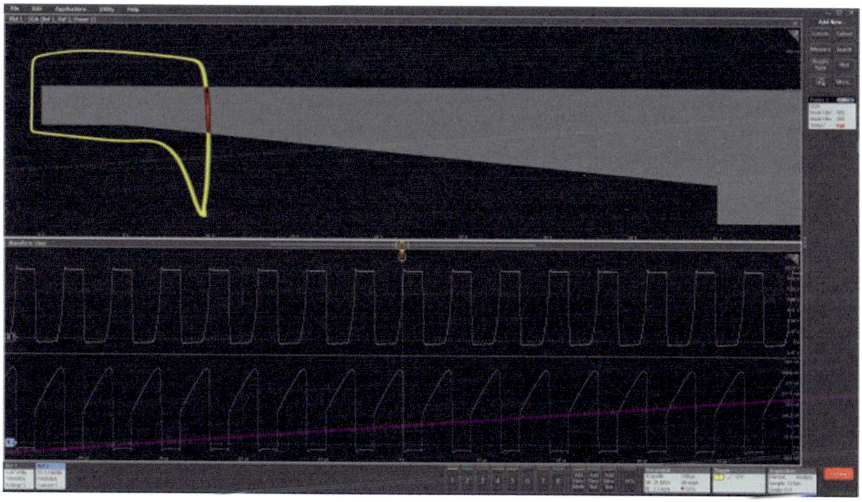

Fig. 6.129 SOA test results

(3) Histogram and Trend Analysis

Analyzing the operating conditions of a power converter based on parameters from each switching cycle is very challenging during continuous high-frequency switching. For example, observing duty cycle variations of PWM waves over several power frequency cycles requires complex data processing of waveforms obtained at long timescales and high sampling rates. Oscilloscope measurement algorithms can calculate based on switching cycles and visually display trends over time, identify abnormal cycles, and present parameter distribution patterns in histogram form, as shown in Fig. 6.130.

4. Magnetic Property Analysis

In power converters, inductors and transformers are used as energy storage components, and inductors are also used for output filtering. Given their critical role in power converters, characterizing these magnetic components under actual operating conditions is crucial for determining the stability and overall efficiency of the power supply.

(1) Inductance Measurement

The impedance of an inductor increases with frequency, with higher frequencies having higher impedance than lower frequencies. To calculate the inductance value, measure the voltage across the magnetic component and the current flowing through it, integrate the voltage over time, and then divide by the change in current.

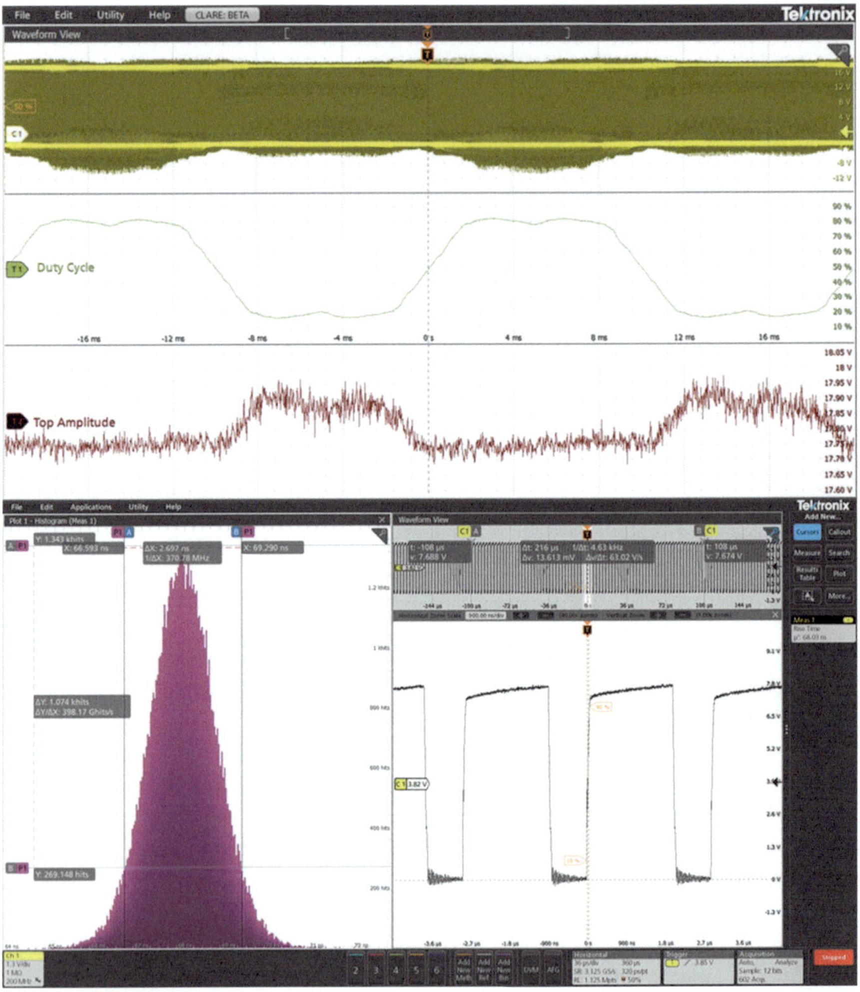

Fig. 6.130 Histogram and trend analysis

(2) **Magnetic Property Measurement and B-H Curve**

Magnetic flux density B refers to the strength of the magnetic field, measured in teslas (T), and determines the force applied to moving charges in the field. Magnetic field strength H refers to the magnetizing force, measured in amperes per meter (A/m). The permeability of a material, measured in henries per meter (H/m), indicates the degree of magnetization induced by an applied magnetic field. Parameters such as magnetic length and the number of coils around the core help determine B and H. The B-H curve is typically used to assess the saturation (or depleteness) of magnetic components in power converters and to measure the energy loss per cycle in the core material. Since both B and H depend on the physical characteristics of the magnetic component, such as magnetic length and coil number, these curves define the performance envelope of the core material. To obtain the B-H curve, measure the voltage across the magnetic component and the current flowing through it.

(3) **Magnetic Loss**

Analyzing the losses in magnetic components is a critical part of the overall loss analysis of power converters. The two main types of magnetic loss are copper loss and core loss. Copper loss arises from the resistance of copper wire coils in the power supply, while core loss is related to eddy current loss and hysteresis loss in the core. Core loss is independent of DC flux but is affected by AC flux swing and operating frequency. Figure 6.131 depicts the test results for magnetic components.

5. **Loop Response Testing**

Loop response testing helps assess the stability of power converters. By using the measured Bode plot to observe the gain and phase of the power supply control loop at different frequencies, this method provides a widely adopted, efficient, and intuitive way to judge loop stability. Specialized loop analyzers, network analyzers, or oscilloscopes with this functionality can be used for testing.

In power converter design, control loop measurements help characterize the converter's response to changes in output load conditions, input voltage, temperature variations, etc. An ideal power converter should have a fast response, maintain a constant output, and avoid excessive ringing or oscillation, typically achieved through rapid switching of devices between the power supply and load. The longer the switching device is open compared to when it is closed, the higher the power

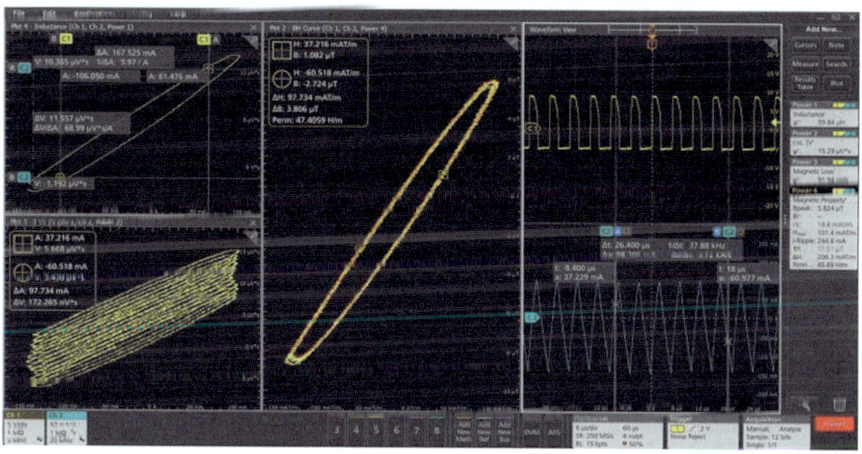

Fig. 6.131 Magnetic component test results

provided to the load. An unstable power supply or regulator may oscillate, leading to large visible ripples in the control loop bandwidth and potentially causing EMI issues.

During control loop response measurements, an excitation source within a certain frequency range needs to be injected into the feedback path of the control loop. For example, using the internal signal source of an oscilloscope and isolating transformer to inject the signal into the loop's feedback. DC-DC converters or LDOs must be equipped with a small (5-10W) injection resistor or termination resistor in their feedback loop to inject interference signals from a function generator. To avoid overloading the control loop, the amplitude of the injected signal must be kept low.

A wide-bandwidth, flat-response injection transformer is connected to the injection resistor, isolating the grounded signal source from the power supply. The choice of injection transformer depends on the frequency of interest. When measuring voltage, low-capacitance, low-attenuation passive probes are recommended for excellent sensitivity. After measurement begins, phase and gain curves will be plotted on the screen. Once the gain curve crosses the 0 dB line, the phase margin will be displayed. When the phase curve crosses the 0° threshold, the gain margin will be shown. Figure 6.132 depicts the loop response measurement results.

6. **Three-Phase Motor Driver Testing**

Measurement and analysis of three-phase power systems are inherently more complex than single-phase systems. The captured voltage and current waveforms require further calculations to derive key power measurements. Power converters

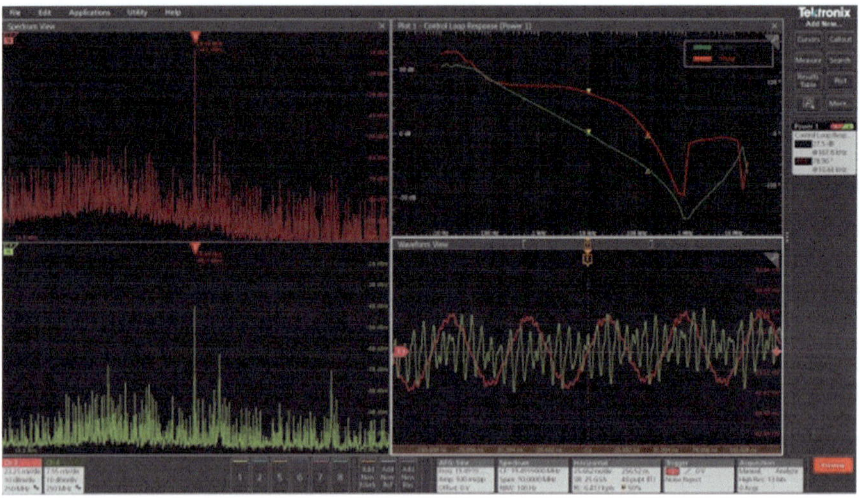

Fig. 6.132 Loop response measurement results

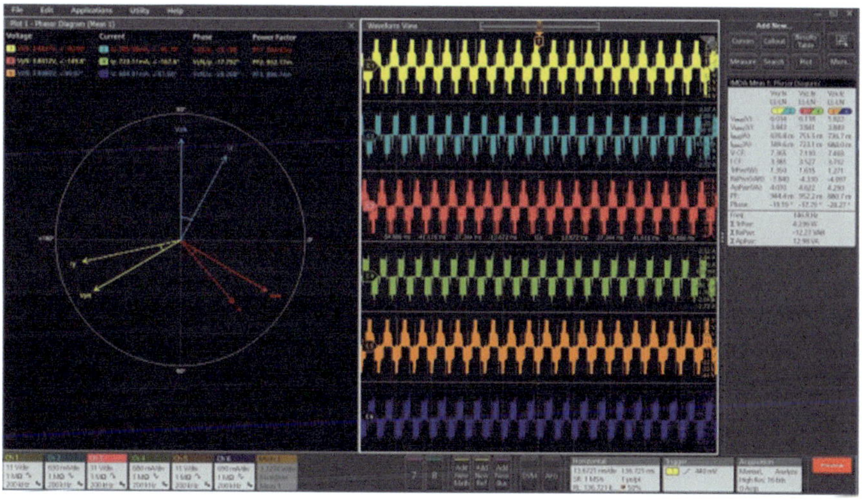

Fig. 6.133 Three-phase motor test results

based on pulse width modulation (PWM), such as frequency converter motor drivers, add complexity to measurements, as accurately extracting the zero-crossing points of PWM signals is crucial. Measurements based on fundamental frequencies or full-spectrum will yield significantly different results. Figure 6.133 depicts the test results for three-phase motors.

(1) **Three-Phase Power Quality**

Key measurements for three-phase power quality include the frequency and RMS amplitude of voltage and current, crest factors of voltage and current, PWM frequency, phase angles of each phase, and active, reactive, and apparent power for each phase and the total three-phase system.

(2) **Three-Phase Harmonics**

The three-phase harmonic plot shows the combined results of all three phases, allowing users to correlate test results between phases. Measurements can be evaluated against IEEE-519 or IEC61000-3-2 standards or custom limits.

(3) **Efficiency**

Efficiency is measured as the ratio of output power to input power (Fig. 6.133).

7. **Direct Quadrature Zero DQ0**

Direct Quadrature Zero (DQ0) parameters are important variables in field-oriented control algorithms. These parameters are often used in control systems but need to be measured in real-time during design optimization and debugging, and related to measured values such as torque output.

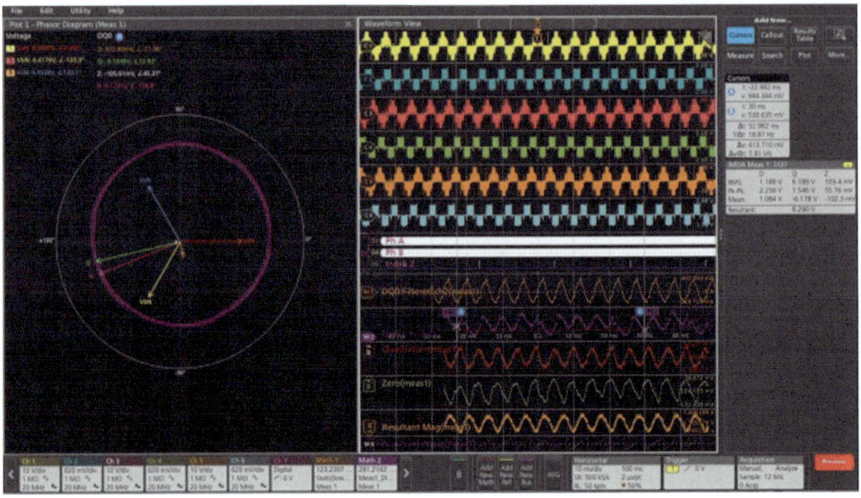

Fig. 6.134 DQ0 test results

Advanced drives for synchronous and AC induction motors typically use vector drive technology. Compared to simple scalar drives, vector drives provide smoother operation, faster acceleration, and superior torque control. Vector drives use field-oriented control (FOC), which, while widely used and efficient, is much more complex than scalar drives. The control system uses Clarke and Park transformations to convert the three-phase voltages applied to the motor into orthogonal D and Q vectors. These simplified vectors can be easily scaled and integrated to maintain the desired speed and torque. Subsequently, an inverse transformation creates pulse-width modulation drive signals within the inverter.

Typically, D and Q values are located within digital signal processing modules (such as FPGAs) and may not be directly measurable. However, new technologies now allow these variables to be displayed on oscilloscopes, making it easier for engineers to view and relate these critical variables to electrical and mechanical parameters. This provides valuable methods for debugging and optimizing drive systems or inverters. Figure 6.134 depicts the DQ0 test results.

References

1. The mechanism of the Bernoulli type end-effector. https://www.jel-robot.com/products/BERNOULLI_CHUCK.html
2. Automotive Electronics Council (2021) AEC-Q101 failure mechanism based stress test qualification for discrete semiconductors in automotive applications. Rev E
3. ECPE European Center for Power Electronics e.V (2021) ECPE guideline AQG 324 qualification of power modules for use in power electronics converter units in motor vehicles review and analysis of SiC MOSFETs' ruggedness and reliability

4. Wang J, Xi J (2020) Review and analysis of SiC MOSFETs' ruggedness and reliability. IET Power Electron
5. Jiang X, Wang J, Jiwu L, Chen J, Yang X, Li Z, Tu C, John Shen Z (2018) Failure modes and mechanism analysis of SiC MOSFET under short-circuit conditions. Microelectron Reliab 88–90:593–597
6. Mulpuri V, Choi S (2017) Degradation of SiC MOSFETs with gate oxide breakdown under short circuit and high temperature operation. IEEE Energy Convers Congr Expos (ECCE) 2017:2527–2532
7. Yao K, Yano H, Iwamuro N (2021) Investigations of short-circuit failure in double trench SiC MOSFETs through three-dimensional electro-thermal-mechanical stress analysis. Microelectron Reliab 122:114163
8. Texier M, De Luca A, Pichaud B, Jublot M, Tromas C, Demenet JL, Rabier J (2013) Evidence of perfect dislocation glide in nanoindented 4H-SiC. J Phys Conf Ser 471:012013
9. Hadorn JP et al (2019) Direct evaluation of threading dislocations in 4H-SiC through large-angle convergent beam electron diffraction. Phil Mag 100(2):194–216
10. Onda S, Watanabe H, Kito Y, Kondo H, Uehigashi H, Hosokawa N, Hisada Y, Shiraishi K, Saka H (2013) Transmission electron microscope study of a threading dislocation with b = 0001 + 1100 and its effect on leakage in a 4H–SiC MOSFET. Philos Mag Lett 93(8):439–47
11. Konishi K, Nakamura Y, Nagae A, Kawabata N, Tanaka T, Tomita N, Watanabe H, Tomohisa S, Miura N (2019) Direct observation and three dimensional structural analysis for threading mixed dislocation inducing current leakage in 4H-SiC IGBT. Jpn J Appl Phys 59(1)
12. Hayashi S, Yamashita T, Senzaki J, Miyazato M, Ryo M, Miyajima M, Kato T, Yonezawa Y, Kojima K, Okumura H (2018) Influence of basal-plane dislocation structures on expansion of single Shockley-type stacking faults in forward-current degradation of 4H-SiC p–i–n diodes. Jpn J Appl Phys 57(4S)
13. Hayashi S, Naijo T, Yamashita T, Miyazato M, Ryo M, Fujisawa H, Miyajima M, Senzaki J, Kato T, Yonezawa Y, Kojima K, Okumura H (2017) Origin analysis of expanded stacking faults by applying forward current to 4H-SiC p–i–n diodes. Appl Phys Exp 10(8)
14. Schmidt C et al (2012) Application of lock-in thermography for failure analysis in integrated circuits using quantitative phase shift analysis. Mater Sci Engg B-Adv Funct Solid-State Mater 177:1261–1267

Chapter 7
Impact and Countermeasures of High di/dt—Turn-Off Voltage Overshoot

Abstract As discussed in Sect. 4.2.4, SiC MOSFETs exhibit significantly faster switching speeds than Si IGBTs, enabling power converters to have lower losses and higher power density. However, this advantage comes with a downside: the higher switching speed results in a faster rate of turn-off current, leading to more severe issues with turn-off voltage overshoot for SiC MOSFETs compared to Si IGBTs. As mentioned in Sect. 3.1.1, SiC MOSFET products from various manufacturers have avalanche voltages significantly higher than the rated breakdown voltage. This is done to address the issue of turn-off voltage overshoot. Although devices already have a certain safety margin, it is essential to avoid situations where the turn-off voltage peak exceeds the device's rated breakdown voltage when designing power converter products. Some companies may even impose stricter derating specifications.

7.1 The Basic Principle of Turn-Off Voltage Overshoot

During the turn-off process of SiC MOSFETs, the current I_{DS} rapidly decreases from the load current I_L. The fast-changing I_{DS} induces a voltage drop across the main power switching loop inductor L_{Loop}, and this voltage drop has a direction opposite to the SiC MOSFET terminal voltage V_{DS}. This results in a noticeable overshoot on V_{DS}. When the overshoot on V_{DS} exceeds the device's breakdown voltage, it can potentially lead to overvoltage failure of the device. Therefore, it is crucial to understand the mechanism of turn-off voltage overshoot and effectively suppress it to ensure the safe operation of the device.

The peak value V_{spike} of the voltage overshoot is influenced by the rate of current decrease $\mathrm{d}I_{DS(off)}/\mathrm{d}t$ at the moment corresponding to L_{Loop} and V_{spike}, following the relationship given in Eq. (7.1). Hence, the larger L_{Loop} or $\mathrm{d}I_{DS(off)}/\mathrm{d}t$, the higher the V_{spike}.

$$V_{spike} = L_{Loop} \cdot \mathrm{d}I_{DS(off)}/\mathrm{d}t. \tag{7.1}$$

Y. Gao and Y. Zhang, *Silicon Carbide Power Devices*, CPSS Power Electronics Series,
https://doi.org/10.1007/978-981-96-3480-4_7

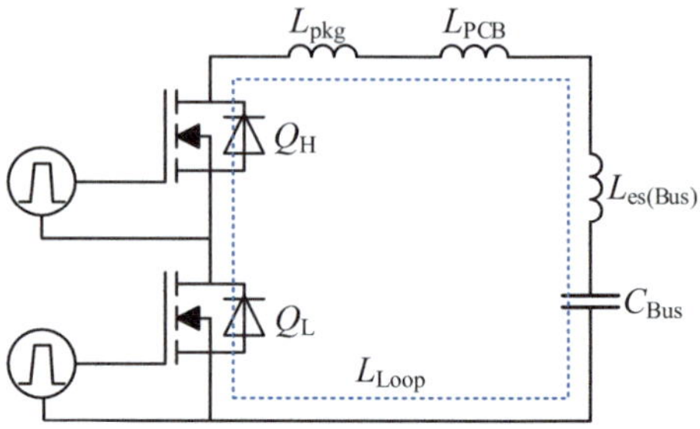

Fig. 7.1 Main power switching circuit model

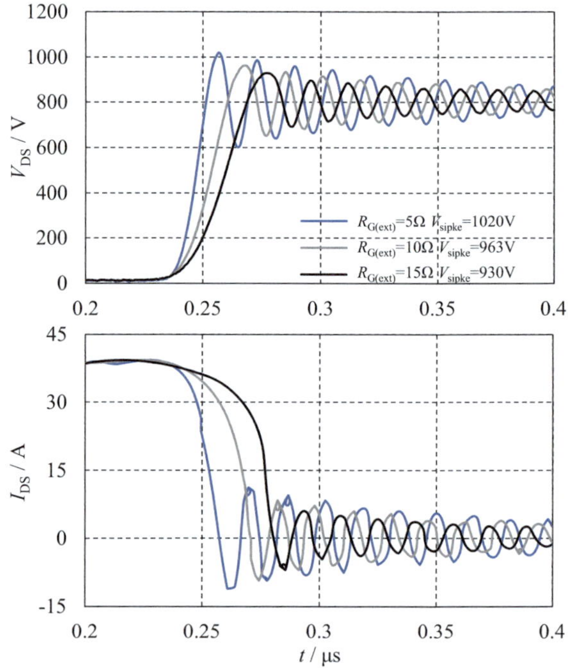

Fig. 7.2 Influence of drive resistance $R_{G(ext)}$

The circuit model of the main power switching circuit is illustrated in Fig. 7.1. The L_{Loop} consists of the device package inductance L_{pkg}, PCB trace inductance L_{PCB}, and the equivalent series inductance (ESL) of the bus capacitor $L_{es(Bus)}$.

When L_{Loop} is 41 nH, the SiC MOSFET's turn-off waveforms under different external drive resistances $R_{G(ext)}$ are shown in Fig. 7.2. It is evident from the waveforms that the turn-off speed of the SiC MOSFET increases as $R''_{G(ext)}$ decreases. The rate of change of the current $dI_{DS(off)}/dt$ increases, V_{spike} continuously rises, and V_{DS} exhibits more pronounced oscillations. Table 7.1 provides corresponding values for $R_{G(ext)}$, $dI_{DS(off)}/dt$ and V_{spike}.

When $R_{G(ext)}$ is 15, 10, and 5 Ω, $dI_{DS(off)}/dt$ is 3.17, 3.97, and 5.36 A/ns, and V_{spike} is 930, 963, and 1020 V, respectively. Thus, under the condition of a fixed

Table 7.1 Effects of drive resistance $R_{G(ext)}$

$R_{G(ext)}$ (Ω)	$dI_{DS(off)}/dt$ (A/ns)	V_{spike} (V)
15	3.17	930
10	3.97	963
5	5.36	1020

drive circuit and power circuit, limiting the turn-off speed can effectively reduce the turn-off voltage spike.

When $R_{G(ext)}$ is 10, the turn-off waveforms of the SiC MOSFET under different L_{Loop} values are shown in Fig. 7.3. It can be observed that the turn-off speed of the SiC MOSFET decreases slightly as L_{Loop} increases. $dI_{DS(off)}/dt$ decreases slightly, V_{spike} continues to rise, and V_{DS} exhibits more pronounced oscillations. Table 7.2 provides corresponding values for L_{Loop}, $dI_{DS(off)}/dt$, and V_{spike}. When L_{Loop} is 41, 52, and 63 nH, $dI_{DS(off)}/dt$ is 3.97, 3.71, and 3.59 A/ns, and V_{spike} is 963, 993, and 1026 V, respectively. Although reducing L_{Loop} accelerates the turn-off speed, leading to an increase in V_{spike}, its dominant effect on reducing V_{spike} is evident. Therefore, when the drive circuit and $R_{G(ext)}$ are fixed, reducing L_{Loop} can effectively limit the turn-off voltage spike.

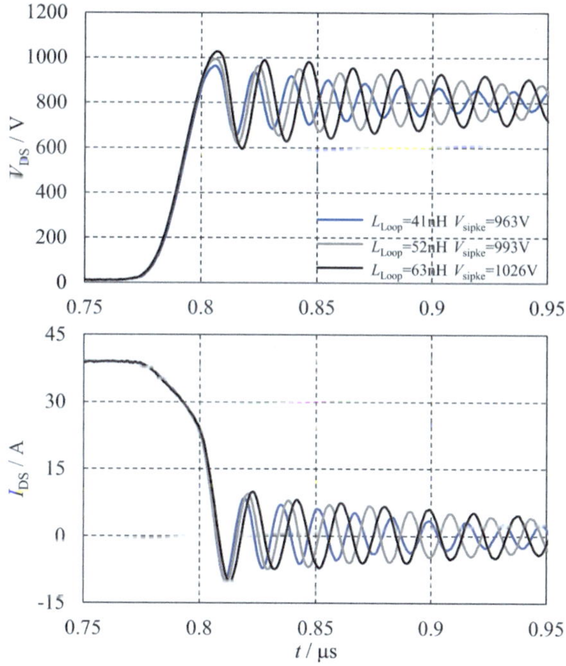

Fig. 7.3 Influence of main power switching circuit inductance L_{Loop}

Table 7.2 Effects of main power switching circuit inductance L_{Loop}

L_{Loop} (nH)	$\mathrm{d}I_{\mathrm{DS(off)}}/\mathrm{d}t$ (A/ns)	V_{spike} (V)
41	3.97	963
52	3.71	993
63	3.59	1026

7.2 Countermeasure 1—Loop Inductance Control

7.2.1 Loop Inductance and Local Inductance

Current generates a magnetic field in space, and the integral of the dot product of the magnetic induction strength through a surface area with the magnetic field passing through it is the magnetic flux. Inductance is the ratio of the magnetic flux through a closed loop surface to the current producing that flux. When the current producing this flux is in the same circuit, it is called self-inductance; when the current producing this flux is in another circuit, it is called mutual inductance. Inductance is the general term for both self-inductance and mutual inductance. Inductance can be calculated from its definition Eq. (7.2), where L″ is inductance, Φ is magnetic flux, I is the current producing the magnetic flux, B is magnetic induction intensity, and $\int_{\mathrm{s}} \mathrm{ds}$ represents the area integral over the closed circuit surface.

$$L = \frac{\Phi}{I} = \frac{\int_s B \cdot \mathrm{d}s}{I}. \tag{7.2}$$

From Eq. (7.2), it is clear that inductance is a property of a closed loop, and the concept of inductance only exists when a closed loop is formed. Therefore, it is commonly referred to as loop inductance, such as L_{Loop} in Fig. 8.1, which is called the main power switching circuit inductance.

Introducing the magnetic vector potential A as in Eq. (7.3),

$$\mathrm{B} = \nabla \times \mathrm{A}. \tag{7.3}$$

The curl of the magnetic vector potential A is the magnetic induction intensity B. Thus, the line integral of A along the closed loop c is equal to the surface integral of B over c. Inductance can also be calculated through Eq. (7.4),

$$L = \frac{\oint_c A \cdot \mathrm{d}l}{I}. \tag{7.4}$$

For the rectangular loop shown in Fig. 7.4, with four sides labeled T, B, L, and R, the inductance calculated from Eq. (7.4) is

$$L = \frac{\oint_{\mathrm{T}} A \cdot \mathrm{d}l}{I} + \frac{\oint_{\mathrm{B}} A \cdot \mathrm{d}l}{I} + \frac{\oint_{\mathrm{L}} A \cdot \mathrm{d}l}{I} + \frac{\oint_{\mathrm{R}} A \cdot \mathrm{d}l}{I}$$

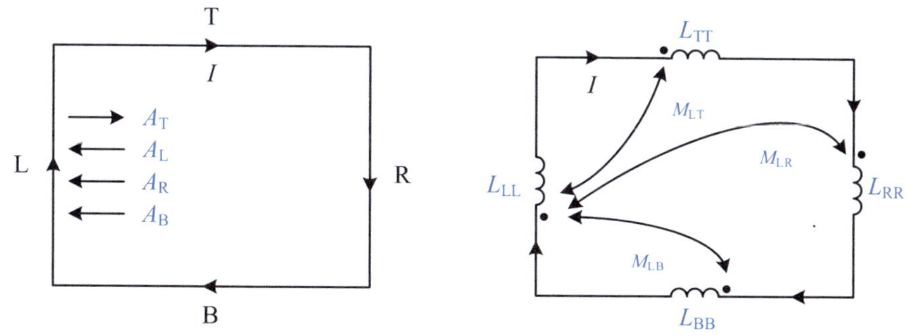

Fig. 7.4 Rectangular line loop inductance

$$= L_T + L_B + L_L + L_R. \tag{7.5}$$

Taking L_L as an example, A_T, A_B, A_L, and A_R are the magnetic vector potentials generated by the respective current segments on the segment L. Thus, L_L is expressed as,

$$\begin{aligned} L_L &= \frac{\oint_L A_T \cdot dl}{I} + \frac{\oint_L A_B \cdot dl}{I} + \frac{\oint_L A_L \cdot dl}{I} + \frac{\oint_L A_R \cdot dl}{I} \\ &= L_{LL} + M_{LB} + M_{LT} + M_{LR}. \end{aligned} \tag{7.6}$$

According to Eqs. (7.5) and (7.6), the loop inductance can be regarded as composed of several segments of inductance. Since these inductances are part of the loop inductance, they are called partial inductances [1]. In Fig. 7.1, L_{pkg}, L_{PCB}, and $L_{es(Bus)}$ represent the concept of partial inductance. At the same time, partial inductance includes both partial self-inductance and partial mutual inductance. For example, in Eq. (7.6), L_{LL} represents partial self-inductance, and M_{LB}, M_{LT}, M_{LR} represent partial mutual inductance. It is important to note that a closed loop is a prerequisite for inductance, and partial inductance needs to be defined in a specific loop, as indicated by the inclusion of partial mutual inductance. Some components such as capacitors and power modules provide parasitic inductance values in datasheets, and as explained earlier, these values essentially represent partial self-inductance, assuming they form a loop with an infinitely distant parallel line.

With the concepts of loop inductance and partial inductance clarified, we have laid the foundation for analyzing line inductance. For convenience of expression, the distinction between loop inductance and partial inductance will not be explicitly made in the following text.

7.2.2 PCB Line Inductance

For a section of PCB line with width w and length l, neglecting its thickness, as shown in Fig. 7.5a, its inductance is calculated by (7.7).The inductance L increases with the increase in l' and decrease in w.

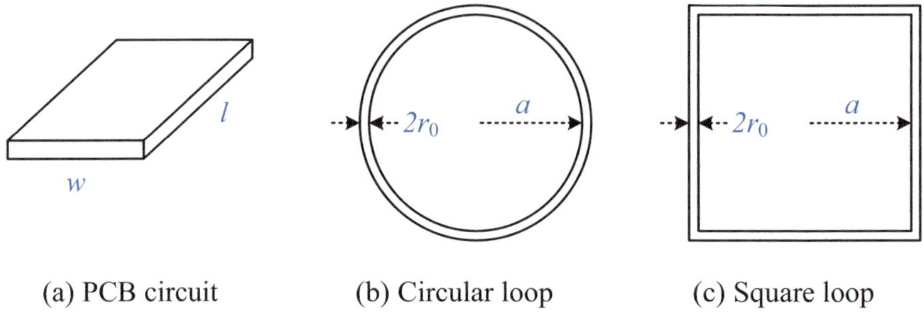

Fig. 7.5 Basic circuits

$$L \approx 2l \cdot \left(\ln \frac{2l}{w} + 0.5 + 0.2235 \frac{w}{l} \right). \tag{7.7}$$

For a circular loop with wire radius r_0 and loop radius a, where $r_0 \ll$ a, as shown in Fig. 7.5b, its inductance is calculated by (7.8), and L increases with the increase in a.

$$L \approx \mu_0 \cdot a \cdot \left(\ln \frac{8a}{r_0} - 2 \right). \tag{7.8}$$

For a square loop with wire radius r_0 and side length a, where $r_0 \ll$ a, as shown in Fig. 7.5c, its inductance is calculated by (7.9), and L increases with the increase in a.

$$L \approx \frac{2\mu_0 a}{\pi} \left(\ln \frac{a}{r_0} - 0.774 \right) \tag{7.9}$$

Figure 7.6a, b depict co-planar line circuits and parallel plate line circuits, respectively, with line width w, line spacing h, and negligible line thickness. The current flowing out of the paper is represented by "·", and the generated magnetic field is shown by solid lines with arrows. The current flowing into the paper is represented by "×", and the generated magnetic field is depicted by dashed lines with arrows. In co-planar and parallel plate line circuits, the magnetic field directions of the two lines are opposite, canceling each other out and reducing the magnetic flux density. According to the definition of inductance, the mutual cancelation of magnetic field lines helps reduce inductance. In parallel plate line circuits, the cancelation of magnetic field lines is more significant, significantly reducing line inductance [2].

The unit length inductance L_{cp} of the co-planar line circuit and the unit length inductance L_{pp} of the parallel plate line circuit can be calculated using Eqs. (7.10) and (7.11), respectively.

$$L_{cp} \approx \frac{\mu_0}{\pi} \cdot \cosh^{-1} \left(\frac{w+h}{w} \right) \tag{7.10}$$

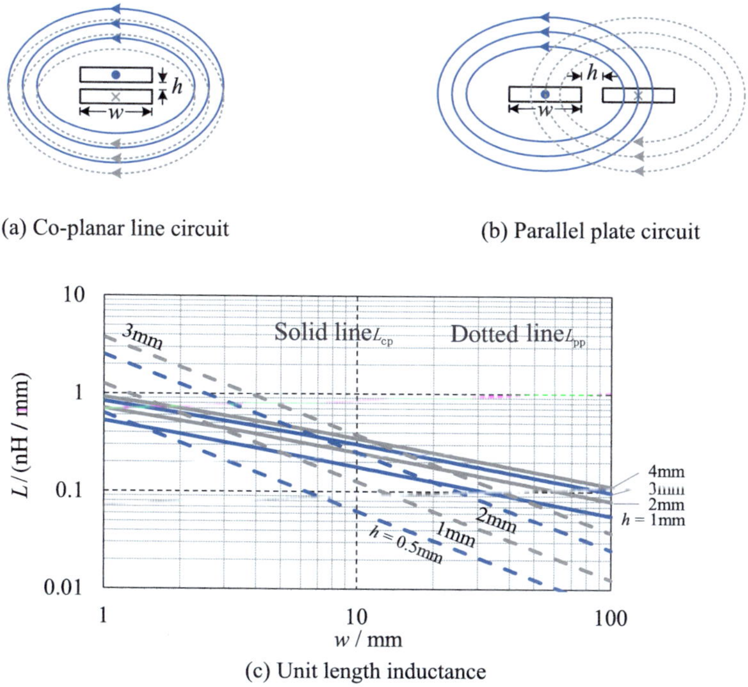

Fig. 7.6 PCB return path inductance pairs

$$L_{pp} \approx \mu_0 \frac{h}{w}. \tag{7.11}$$

Based on Eqs. (7.10) and (7.11), the influence of line width w and line spacing h on L_{cp} and L_{pp} can be determined, as shown in Fig. 7.6c. Both L_{cp} and L_{pp} decrease with increasing w and decreasing h, with L_{pp} exhibiting a more pronounced change. When the main power switching PCB line (busbar copper) adopts a co planar line circuit, it cannot be placed very close due to safety regulations. On the other hand, the parallel plate line circuit utilizes the PCB material (insulating material) as the insulation medium, usually FR-4, with a breakdown field strength greater than 20 kV/mm. Hence, the line spacing can be relatively small. This indicates that designing the main power switching circuit with a parallel plate line circuit can more effectively reduce inductance. Based on the analysis of the above basic circuits, the following wiring methods can be employed to reduce PCB line inductance.

(1) Design the PCB line to be wider and shorter.
(2) Minimize the circuit area of PCB lines.
(3) Arrange the current so that it overlaps on the upper and lower layers of the PCB lines with opposite current flow directions.

7.2.3 Discrete Device Package Inductance

At the current stage, SiC diode discrete devices are available in plug-in packages such as TO-247, TO-247.2, TO-220, TO-220-2, and surface-mount packages including TO-252, TO-252-2, TO-263, TO-263-2, DFN8×8, DFN3×3, DDPAK, etc. SiC MOSFET discrete devices are packaged in plug-in forms like TO-247.3, TO-247.4, and surface-mount packages like TO-263-7.

Plug-in packaged devices are installed by inserting the device pins into PCB mounting holes for soldering. Taking TO-247.3 and TO-247.4 packages as examples, the pin length is approximately 20 mm, and the pin-to-PCB mounting hole distance, denoted as the mounting distance d_{mount}, is determined based on specific requirements during usage. Considering the pin root as the reference point, the relationship between the package inductance L_{DS} introduced into the main power switching circuit and d_{mount} can be simulated, as shown in Fig. 7.7. When $d_{\text{mount}} = 0$, L_{DS} is not zero, indicating that this inductance is caused by the bond wires and pins beneath the encapsulation material. As d_{mount} increases, L_{DS} linearly increases.

Therefore, when using plug-in packaged devices, it is essential to minimize d_{mount} as much as possible to reduce the contribution of the package to L_{Loop}. However, in practical applications, d_{mount} cannot be reduced without limitations due to factors such as mechanical structure, heat dissipation structure, safety regulation distances, and assembly feasibility.

Surface-mount packaged devices are soldered directly onto PCB pads during installation. Represented by TO-263-7, surface-mount packages with exposed pins introduce a fixed L_{DS} when soldered onto the PCB. Since the pin length is significantly shorter than plug-in packages, the introduced L_{DS} is smaller. Surface-mount packages like DFN8×8, which lack exposed pins but have solder pads at the bottom, further reduce L_{DS}.

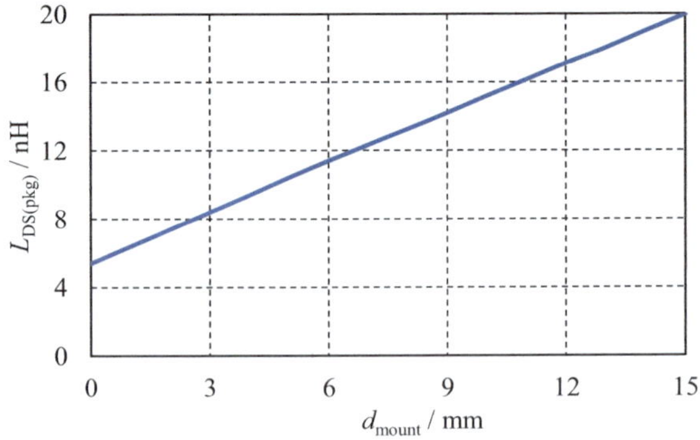

Fig. 7.7 Impact of installation distance d_{mount} on L_{DS}

7.2.4 Power Module Package Inductance

In discrete packaged devices, usually containing only one chip, multiple devices need to be interconnected through the PCB to achieve the desired topology. This approach poses technical challenges in device parallelization, complex topology implementation, electrical insulation, PCB design, thermal management, reliability, and other aspects. In contrast, in power modules, chip interconnection is completed internally, providing a functional module with a certain topology and structure, effectively addressing the aforementioned issues.

In power modules, numerous chips are soldered onto a copper substrate, interconnected through bonding wires to achieve the desired topology. Fig. 7.8 illustrates an example of the internal electrical connections in a power module. It can be observed that the distances between chips, copper substrate, and bonding wires are very close. This is because the module is filled with silicone gel, providing solid insulation. This approach reduces the electrical connection lengths and circuit area needed when using power modules compared to discrete devices, leading to a substantial reduction in L_{Loop}. Additionally, the copper substrate inside the power module, through which current flows, often has a greater width than the pins of discrete devices, contributing to a broader current-carrying path and further reducing L_{Loop}.

Based on the analysis, power modules exhibit shorter electrical connections, wider current paths, and smaller circuit areas, resulting in lower loop inductance. This advantage becomes more pronounced as the topology becomes more complex. To address the issue of high V_{spike} in SiC MOSFETs, various manufacturers have introduced low-inductance power module packages, as discussed in Sect. 1.8.

Fig. 7.8 Internal electrical connections of power modules

Fig. 7.9 Aluminum electrolytic capacitor

7.3 Countermeasure 2—Decoupling Capacitors

7.3.1 Basic Principles of Capacitors

7.3.1.1 Types of Capacitors

The application of capacitors in power converters is very extensive, serving various purposes such as DC connection/energy storage, DC output filtering, AC output filtering, EMC filtering, resonance, and decoupling, among others. When designing converters, the first step is to determine the voltage and capacitance of the capacitors based on the specifications and requirements of the converter. Next, the type of capacitor is selected based on its role in the converter. Finally, various factors, including voltage rating, capacitance value, temperature characteristics, packaging form, size, and cost, are considered to determine the brand and model.

In converters, capacitors are no longer just simple symbols on schematics but rather electronic components with many parameters. To meet the requirements of different applications, a diverse family of capacitors has been developed, with the most common types being aluminum electrolytic capacitors, film capacitors, and multilayer ceramic capacitors (MLCCs) [3].

1. **Aluminum Electrolytic Capacitors**

Aluminum electrolytic capacitors, as shown in Fig. 7.9, use aluminum foil as the anode. After oxidation treatment, a layer of aluminum oxide film is formed on its surface, serving as the dielectric between the anode and cathode plates. The cathode consists of an electrolyte and aluminum foil. They are known for their high-capacitance density, reaching several tens to hundreds of times that of other types of capacitors. Capacitance values can easily range from hundreds to tens of thousands of microfarads, and they are cost-effective. However, drawbacks include polarity sensitivity, high leakage current, and poor thermal stability.

Fig. 7.10 Film capacitor

2. **Film Capacitors**

Film capacitors, illustrated in Fig. 7.10, use plastic film as the dielectric, with common materials including polyethylene terephthalate, polypropylene, polystyrene, and polyethylene naphthalate. They can have either foil electrode which are known as foil-electrode film capacitors or metalized film (Aluminum, zinc, etc.) which are known as metallized film capacitors. The dielectric and electrodes are overlapped and wound into a cylindrical structure. Their characteristics include high insulation resistance, low-temperature coefficient, excellent reliability, and high rated ripple current.

3. **Multilayer Ceramic Capacitors (MLCCs)**

Multilayer ceramic capacitors, as shown in Fig. 7.11, use ceramic film as the dielectric, with materials like titanium oxide and barium titanate. The electrode material is printed onto the ceramic film as electrodes. The ceramic film with printed electrodes is stacked, pressed, fired, and coated to create the capacitor. MLCCs are characterized by their compact size, non-polarity, and high safety. However, they have lower capacitance values and a risk of flexural cracking.

7.3.1.2 Impedance Characteristics of Capacitors

An actual capacitor is not an ideal device; in addition to capacitance C, it also exhibits equivalent series resistance (ESR) R_{es} and equivalent series inductance (ESL) L_{es}, as illustrated in Fig. 7.12. ESR is composed of dielectric losses, electrode, and conductor resistances, leading to energy dissipation. ESL consists of inductance components due to electrode and conductor properties.

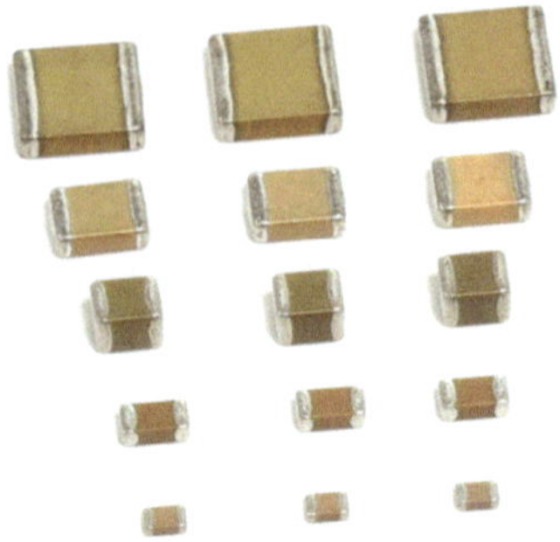

Fig. 7.11 Multilayer ceramic capacitor

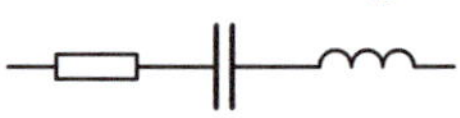

Fig. 7.12 Equivalent circuit of a capacitor

The capacitor C possesses the property of isolating DC while allowing the passage of AC. Its impedance decreases continuously with increasing frequency and tends towards zero. However, due to the presence of ESL and ESR, the capacitor exhibits the characteristics of an LCR series circuit to external observers. The impedance |Z| of the capacitor is given by Eq. (7.12), and the impedance-frequency characteristics are depicted in Fig. 7.13. The impedance-frequency curve takes on a V-shape, and the bottom of the V corresponds to the self-resonance frequency f_o of the LCR series circuit, as given by (7.13). To the left of f_o, in the capacitor region, |Z| is primarily determined by C; to the right of f_o, in the inductor region, |Z| is mainly determined by L_{es}, losing the characteristics of a capacitor. At the self-resonance frequency f_o, the impedance is approximately equal to R_{es}.

$$|Z| = \sqrt{R_{es}^2 + \left(2\pi f_o \cdot L_{es} - \frac{1}{2\pi f_o \cdot C}\right)^2} \tag{7.12}$$

$$f_o = \frac{1}{2\pi\sqrt{L_{es} \cdot C}}. \tag{7.13}$$

Due to differences in types, aluminum electrolytic capacitors, film capacitors, and MLCCs, all with a capacitance value of 10 μF, exhibit distinct impedance-frequency characteristics, as shown in Fig. 7.14. From aluminum electrolytic capacitors to film capacitors and ceramic capacitors, f_o increases, indicating decreasing ESL. The bottom of the V becomes deeper and sharper, indicating decreasing ESR. In the high-frequency range, impedance also decreases.

It is these characteristics that make aluminum electrolytic capacitors suitable for use as bus capacitors, serving the functions of voltage support, energy storage, and power frequency filtering. Film capacitors and aluminum electrolytic capacitors are often used together as bus capacitors, serving the purpose of high-frequency filtering.

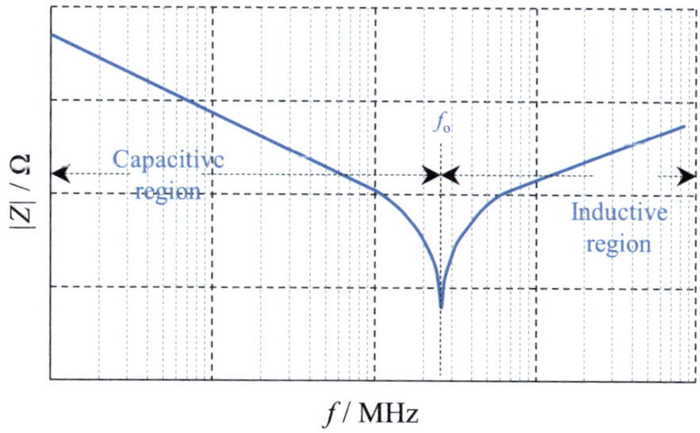

Fig. 7.13 Impedance-frequency characteristics of capacitors

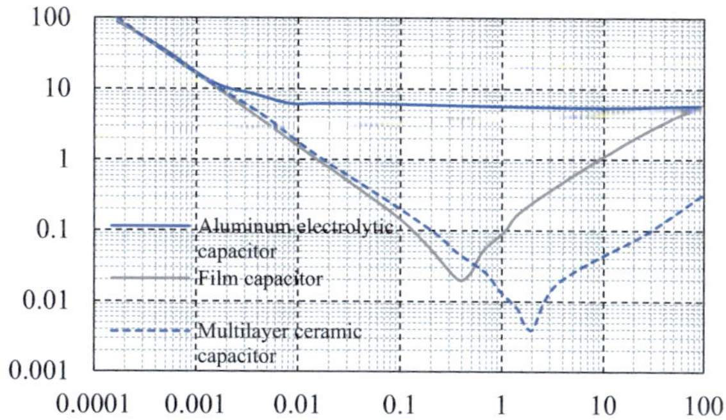

Fig. 7.14 Impedance characteristics of 10 μF capacitors of different types

In cases where the capacitance value requirement is not high, they can also be used as standalone bus capacitors. They are also widely used in resonant converters as resonant capacitors. MLCCs are commonly employed as filtering capacitors and decoupling capacitors in the signal chain. With the increasing demand for high-power density in converters and advancements in capacitor technology, MLCCs are gradually being used as resonant capacitors and bus capacitors.

7.3.2 Decoupling Capacitor Basics

7.3.2.1 Role of Decoupling Capacitors

In converters, the capacitance value of the bus capacitor C_{Bus} is generally high, typically chosen from aluminum electrolytic capacitors and film capacitors. By adding a capacitor C_{Dec} between C_{Bus} and power devices, the original PCB inductance L_{PCB} is split into two parts: PCB inductance between the device and C_{Dec}, $L_{PCB(Dev\text{-}Dec)}$ and PCB inductance between C_{Dec} and C_{Bus}, $L_{PCB(Dec\text{-}Bus)}$. Here, $L_{es(Bus)}$ represents the ESL of C_{Bus}, $L_{es(Dec)}$ is the ESL of C_{Dec}, and L_{pkg} is the device package inductance,

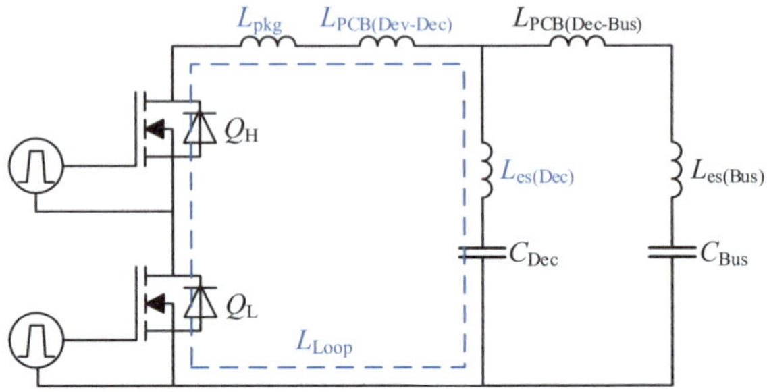

Fig. 7.15 Half-bridge commutation circuit with decoupling capacitor

as shown in Fig. 7.15. When the device is turned off, the commutation current mainly passes through C_{Dec}. At this point, L_{Loop} is reduced compared to before adding C_{Dec} by $L_{PCB(Dec-Bus)}+L_{es(Bus)} - L_{es(Dec)}$, effectively reducing V_{spike}. In this process, C_{Dec} acts as a decoupling capacitor for $L_{PCB(Dec-Bus)}$ and $L_{es(Bus)}$, hence the term decoupling capacitor.

Typically, the capacitance value of C_{Bus} is large, leading to a relatively large volume that often prevents placing C_{Bus} close to power devices, increasing L_{PCB}. In contrast, C_{Dec} has a much smaller capacitance value and a smaller volume. Thus, C_{Dec} can be placed closer to power devices, effectively decoupling $L_{PCB(Dec-Bus)}$ as much as possible. Additionally, when $L_{es(Dec)}$ is less than $L_{es(Bus)}$, L_{Loop} is further reduced.

As shown in Fig. 7.16, when using aluminum electrolytic capacitors as the bus capacitor without a decoupling capacitor, V_{spike} is 1026 V. Adding a film capacitor as a decoupling capacitor reduces V_{spike} to 980 V, a decrease of 46 V. This illustrates the effective reduction of voltage overshoot through the use of decoupling capacitors.

7.3.2.2 Selection of Decoupling Capacitors

From the analysis above, to achieve an effective decoupling effect, the decoupling capacitor needs to have a small ESL. At the same time, since most of the commutation current flows through the decoupling capacitor, it is also required to have a small ESR to prevent excessive temperature rise. According to the impedance-frequency characteristics of various capacitors, it is found that film capacitors and MLCCs are suitable for use as decoupling capacitors. Film capacitors have stable characteristics in all aspects and are not difficult to use. TDK MKP/MFP series metallized polypropylene capacitors can be chosen as film capacitors. However, MLCCs have some unique characteristics, and there are some issues to be aware of in their use.

1. **DC Bias Characteristics**

For high-capacitance MLCCs, the static capacitance will change with different applied voltages. When a direct current (DC) voltage is applied, it is called DC bias characteristics, as shown in Fig. 7.17. This is because high-capacitance MLCCs use a strong dielectric (such as $BaTiO_3$) with spontaneous polarization. When no external

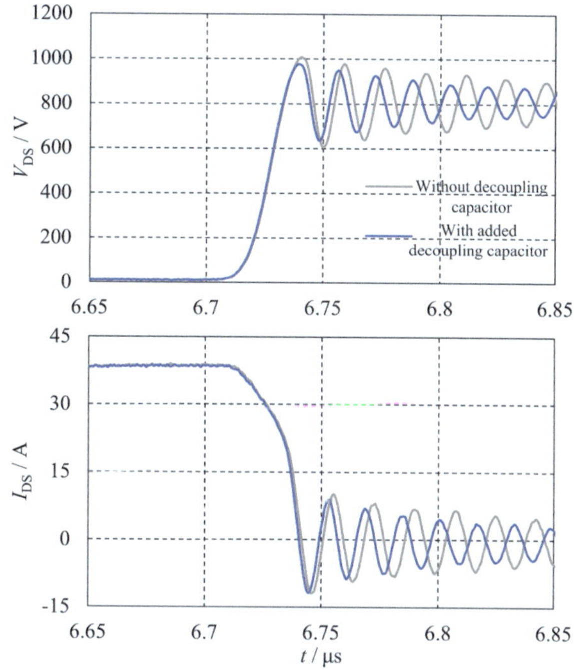

Fig. 7.16 Effect of decoupling capacitor

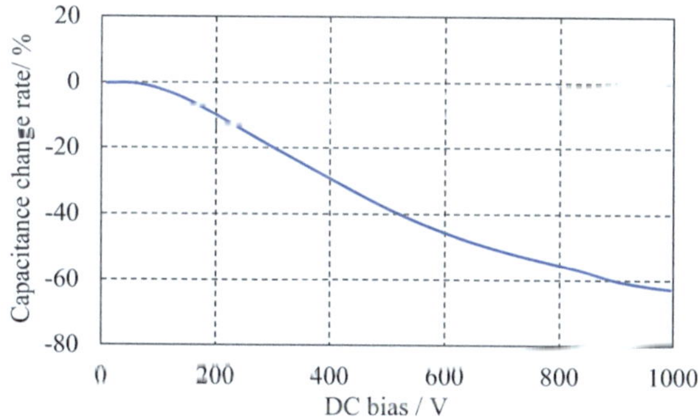

Fig. 7.17 Static capacitance value—DC bias of MLCC

electric field is applied, the spontaneous polarization directions of the strong dielectric are random. When a DC electric field is applied, the spontaneous polarization directions will align with the electric field, increasing the capacitance. When the electric field reaches a certain intensity, saturation occurs, and the dielectric constant decreases, resulting in a decrease in actual static capacitance value.

The DC bias characteristics of MLCCs are influenced by voltage resistance, encapsulation, and materials. The higher the voltage resistance, the smaller the DC bias; the larger the encapsulation, the smaller the DC bias; under the same voltage resistance and encapsulation, X7R has a smaller DC bias than X5R. When using MLCCs as decoupling capacitors, special attention should be paid to the actual capacitance value under the rated voltage. Select capacitors with suitable materials and encapsulation to avoid not meeting the design value and affecting the decoupling effect.

2. **Bending Cracks** [4]

PCBs undergo bending for various reasons throughout the entire life cycle of electronic products, including soldering stress, stress during PCB separation, fixed installation stress during manufacturing, as well as dropping, vibration, and thermal expansion during use.

MLCCs are internally composed of stacked ceramic plates with printed electrodes. Ceramic plates have strong compressive stress resistance but weak tensile stress resistance. Multilayer ceramic capacitors are soldered onto PCBs using solder. When the PCB bends, it is easy to cause bending cracks, as shown in Fig. 7.18. Cracks further lead to a decline in device performance, heating, and consequences such as fires, seriously affecting the safety of equipment and personnel.

To address bending cracks, capacitor manufacturers offer various products:

(1) Generally, adding a conductive resin layer between the Cu and Ni layers at the terminals becomes a soft terminal capacitor. The resin layer can absorb the stress generated by the expansion and contraction of the soldered joint and the bending stress of the substrate, as shown in Fig. 7.19a.
(2) Changing the position of the electrodes on the ceramic plate to create a structure where two capacitors are connected in series inside the capacitor. This significantly reduces the risk of short circuits even if bending cracks occur, as shown in Fig. 7.19b.
(3) Avoiding the overlap of electrodes in areas where cracks may occur, making it an open-circuit mode even in the case of bending cracks, as shown in Fig. 7.19c.
(4) Adding a metal bracket to the capacitor terminals can absorb the stress caused by the bending of the PCB, and this method has the best effect, as shown in Fig. 7.19d.

Considering ESL, DC bias characteristics, resistance to bending cracks, and ripple resistance, Murata's KR3, KC3, and TDK's CeraLink have excellent performance,

Fig. 7.18 MLCC bending crack

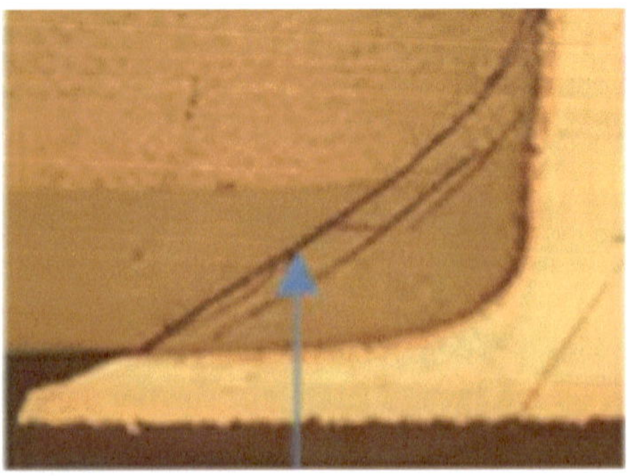

(a) Soft terminal capacitor

(b) Series structure capacitor

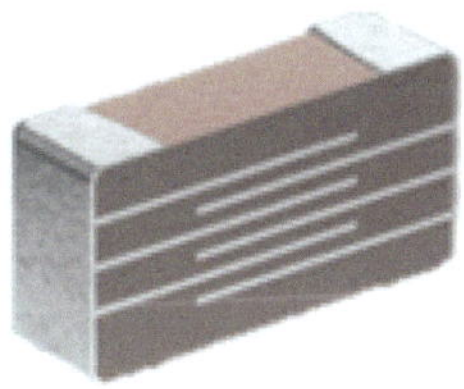

(c) Open circuit capacitor

(d) Metal terminal capacitor

Fig. 7.19 MLCC for bending cracks

(a) Murata KR3、KC3

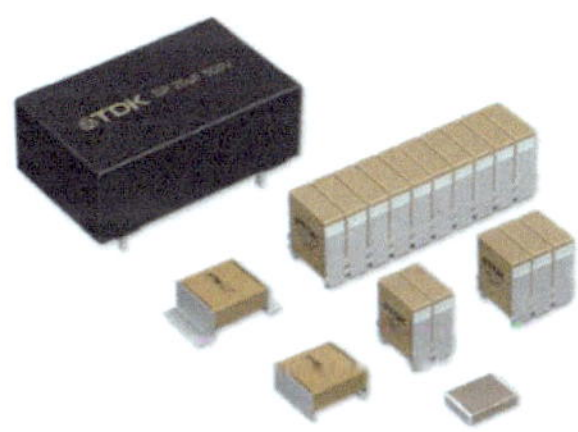

(b) TDK CeraLink

Fig. 7.20 Capacitors for SiC applications

as shown in Fig. 7.20, Murata KR3 [5] has a metal bracket and uses X7T material, improving effective capacitance and ripple resistance compared to traditional materials. TDK CeraLink [6] uses a base on antiferroelectric PLZT ceramic material (lead lanthanum zirconate titanate), with low ESL, low ESR, high-capacitance density, and high reliability. It can provide an extremely compact solution for buffers and DC circuits in high-frequency converters based on SiC and GaN semiconductors.

7.3.3 Small-Signal Model Analysis

7.3.3.1 Without Decoupling Capacitor

In the double-pulse test circuit shown in Fig. 7.21a, during the voltage overshoot and oscillation process after the turn-off of Q_L, the states of various parts of the circuit are fixed. The circuit can be appropriately simplified. After V_{DS} reaches the bus voltage V_{Bus}, the freewheeling current still flows through the channel of Q_L, and Q_L can be equivalent to a current source I_{DS}; the body diode of Q_H conducts for freewheeling, and it can be viewed as a voltage source V_F and a resistor R_F in series; V_{DS} is sufficiently high, and the C_{oss} of Q_L is a constant value; when neglecting the equivalent parallel capacitance of the load inductor, the inductance of the load inductor is very large, and its current can be approximately considered constant throughout the process, making it a constant current source I_L [7]. This gives the equivalent circuit of the turn-off process, as shown in Fig. 7.21b.

The V_{DS} and oscillation occur in the high-frequency range, so I_L can be treated as an open circuit in the equivalent circuit above, and the bus capacitor C_{Bus} and voltage source V_F are treated as short circuits. This results in the small-signal circuit model for the turn-off process, as shown in Fig. 7.22, which can be used for frequency domain analysis.

From the perspective of the current source I_{DS}, i.e., the drain-source terminal of Q_L, it forms an RLC parallel resonant circuit. Taking realistic circuit parameters as C_{oss}= 115 pF, R_F = 40 mΩ, and L_{Loop} = 30 nH, the impedance-frequency characteristics of the RLC parallel circuit are shown in Fig. 7.23. In the low-frequency range, |Z| remains relatively constant, determined by R_F; then, |Z| linearly increases, exhibiting inductance characteristics determined by L_{Loop}; in the high-frequency range, |Z| linearly decreases, showing capacitance characteristics determined by

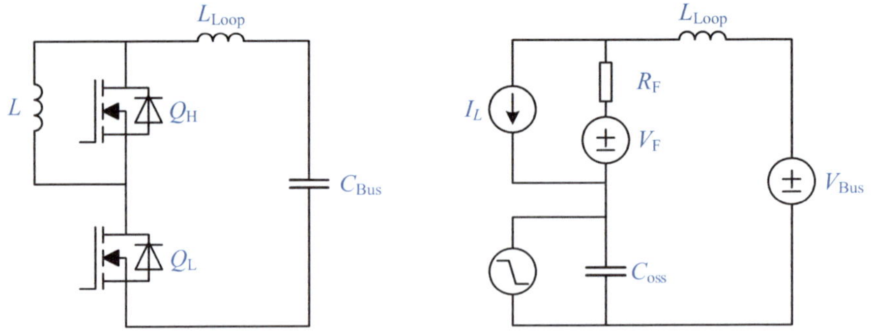

(a) Dual-pulse test circuit

(b) Equivalent circuit during turn-off process

Fig. 7.21 Equivalent circuit during turn-off without decoupling capacitor

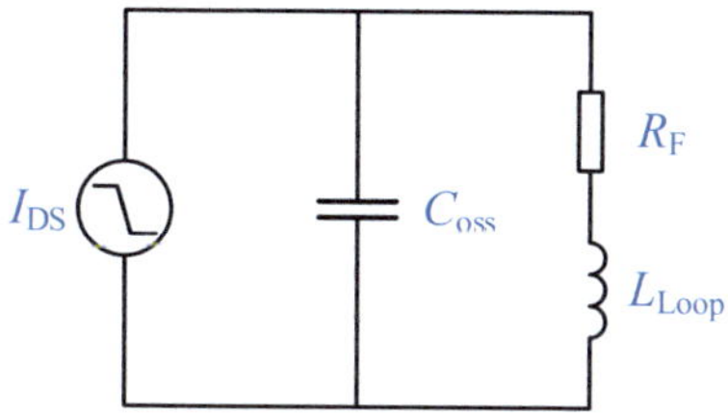

Fig. 7.22 Small-signal circuit model during turn-off without decoupling capacitor

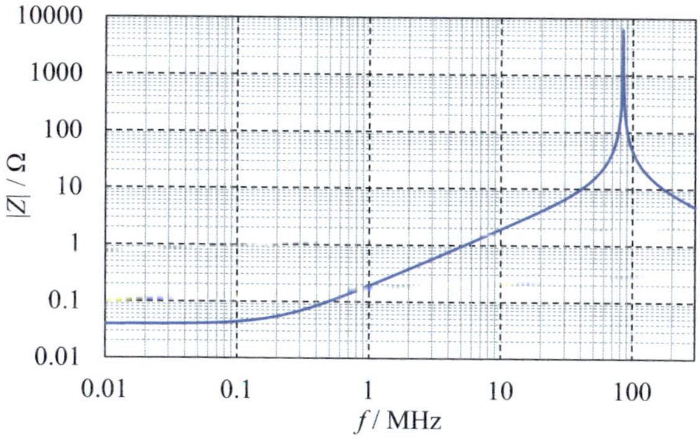

Fig. 7.23 Impendence-frequency characteristics of RLC parallel circuits

C_{oss}; there is a peak between the inductance and capacitance regions, indicating resonance, with the resonant frequency f_o.

The impedance is given by,

$$Z = \frac{s \cdot L_{Loop} + R_F}{s^2 \cdot C_{oss} \cdot L_{Loop} + s \cdot R_F \cdot C_{oss} + 1}. \tag{7.14}$$

The resonant frequency is given by,

$$f_o = \frac{1}{2\pi\sqrt{C_{oss} \cdot L_{Loop}}} \cdot \sqrt{1 - \frac{C_{oss} \cdot R_F^2}{L_{Loop}}}. \tag{7.15}$$

typically, the device and circuit parameters satisfy,

$$\frac{C_{oss} \cdot R_F^2}{L_{Loop}} \ll 1 \tag{7.16}$$

Hence, the resonant frequency can be simplified to

$$f_o = \frac{1}{2\pi\sqrt{C_{oss} \cdot L_{Loop}}}. \tag{7.17}$$

The resonant peak impedance is

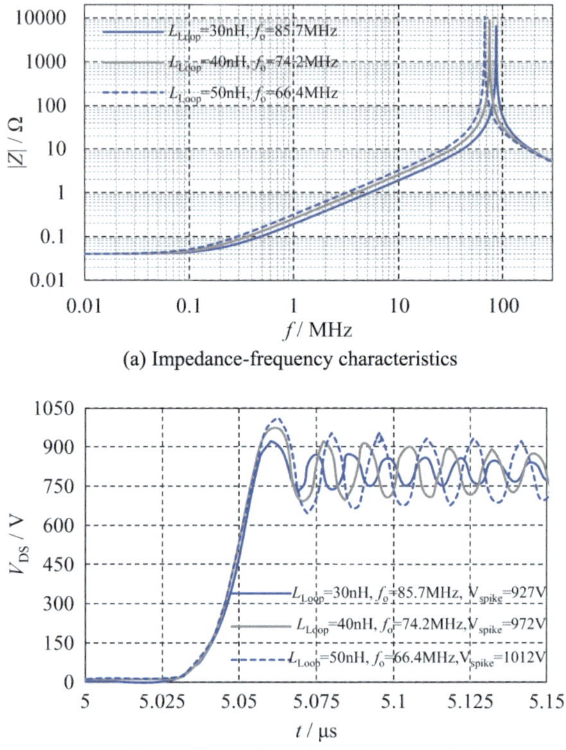

(a) Impedance-frequency characteristics

(b) Turn-off waveform (simulation results)

Fig. 7.24 Influence of L_{Loop} on turn-off voltage overshoot and oscillation without decoupling capacitor

$$Z_o = \frac{L_{\text{Loop}}}{C_{\text{oss}} \cdot R_{\text{F}}}. \tag{7.18}$$

For $C_{\text{oss}} = 115\,\text{pF}$, $R_{\text{F}} = 40\,\text{m}\Omega$, when L_{Loop} is taken as 30, 40, and 50 nH respectively, the impedance-frequency characteristics and the corresponding V_{DS} turn-off waveform are shown in Fig. 7.24. The resonant frequency matches the oscillation frequency of V_{DS}, once again confirming that the oscillation of V_{DS} after turn-off occurs between C_{oss} and L_{Loop}; with the increase of L_{Loop}, the resonant impedance increases, and the peak of the turn-off voltage increases. Therefore, it can be observed that the time-domain waveform of the turn-off V_{DS} has a clear correspondence with the impedance-frequency characteristics, which can be utilized for design and analysis.

7.3.3.2 With Decoupling Capacitor

When employing a decoupling capacitor, the circuit schematic, equivalent circuit, and small-signal circuit model are illustrated in Fig. 7.25.

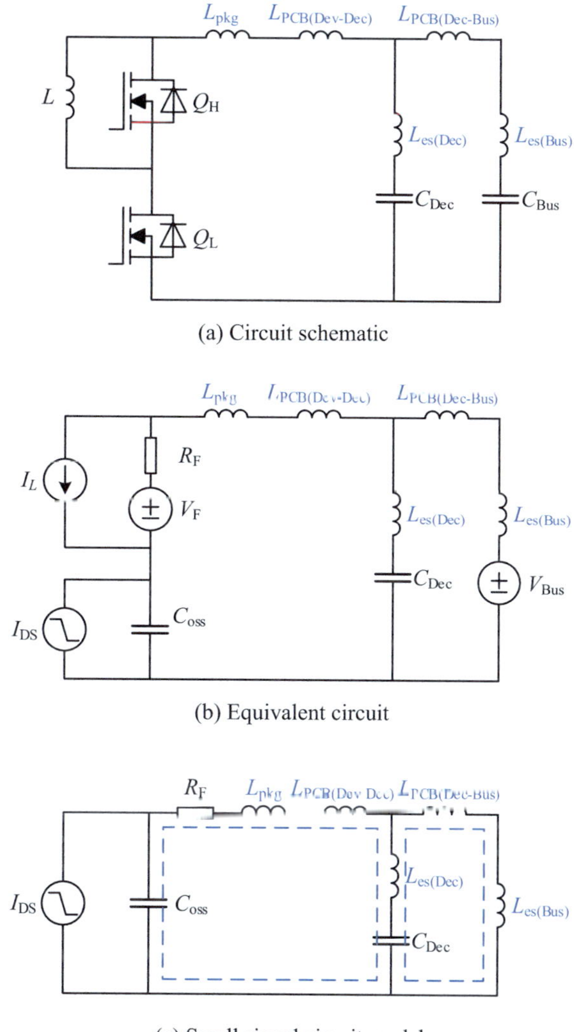

Fig. 7.25 Small-signal circuit model during turn-off with decoupling capacitor

Employing circuit parameters closer to real-world applications, with $C_{oss} = 115\,\text{pF}$, $R_F = 40\,\text{m}\Omega$, $C_{Dec} = 200\,\text{nF}$, $L_{pkg} + \text{L}_{PCB(Dev\text{-}Dec)} = 20\,\text{nH}$, $L_{es(Dec)} = 2\,\text{nH}$, and $L_{PCB(Dec\text{-}Bus)} + \text{L}_{es(Bus)} = 15\,\text{nH}$, the impedance-frequency characteristics are presented in Fig. 7.26. The impedance-frequency curve with a decoupling capacitor exhibits two resonance peaks. Due to $C_{Dec} \gg C_{oss}$, the low-frequency resonance frequency $f_{o(L)}$ and high-frequency resonance frequency $f_{o(H)}$ can be estimated using Eqs. (7.19) and (7.20), respectively, resulting in 100.6 and 2.7 MHz:

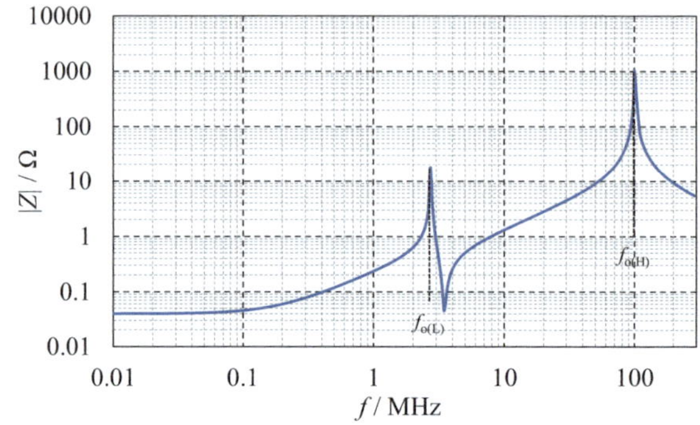

Fig. 7.26 Impedance-frequency characteristics with decoupling capacitors

$$f_{o(L)} = \frac{1}{2\pi\sqrt{C_{Dec} \cdot \left(L_{es(Dec)} + L_{PCB(Dec\text{-}Bus)} + L_{es(Bus)}\right)}} \tag{7.19}$$

$$f_{o(H)} = \frac{1}{2\pi\sqrt{C_{oss} \cdot \left(L_{es(Dec)} + L_{pkg} + L_{PCB(Dev\text{-}Dec)}\right)}}. \tag{7.20}$$

Under these parameters, the simulated turn-off waveform of V_{DS} is depicted in Fig. 7.27. The V_{DS} oscillation includes both high-frequency and low-frequency components, with oscillation frequencies of 102.1 and 2.65 MHz, corresponding to the resonance frequencies in the impedance-frequency characteristics. Combining Eqs. (7.19) and (7.20), it indicates that the high-frequency oscillation of V_{DS} occurs in the loop depicted by the dashed line in Fig. 7.25c, while the low-frequency oscillation of V_{DS} occurs in the loop depicted by the dotted line in Fig. 7.25c.

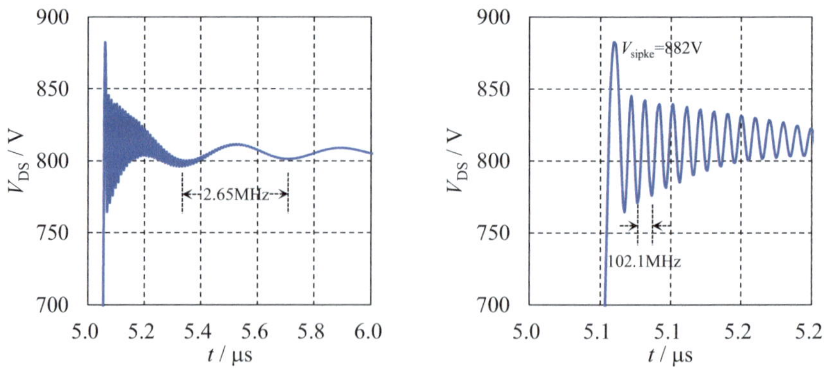

Fig. 7.27 Turn-off waveform with decoupling capacitors (simulation results)

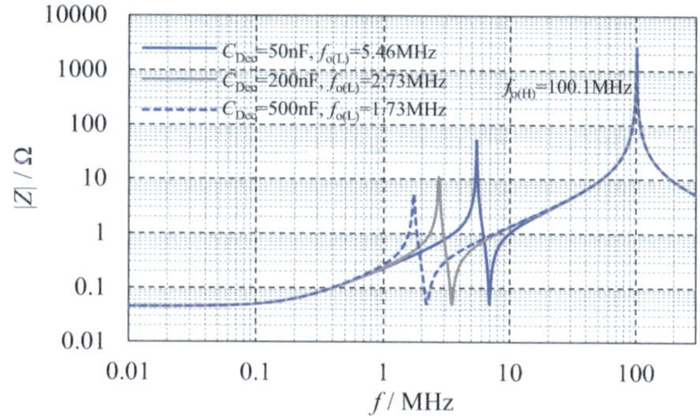

Fig. 7.28 Influence of decoupling capacitor C_{Dec} on impedance-frequency characteristics

When keeping other parameters constant, C_{Dec} is varied at values of 50, 200, and 500 nF, resulting in the impedance-frequency characteristics shown in Fig. 7.28. With an increase in C_{Dec}, both the low-frequency resonance peak impedance and $f_{o(L)}$ continually decrease, reaching 5.46, 2.73, and 1.73 MHz, respectively, while $f_{o(H)}$ and the high-frequency resonance peak impedance remain unchanged.

The simulated turn-off waveform of V_{DS} characteristics in Fig. 7.29 corresponds to the impedance-frequency characteristics shown in Fig. 7.28. As C_{Dec} increases, the low-frequency oscillation amplitude and frequency of V_{DS} continuously decrease, reaching 5.46, 2.73, and 1.73 MHz. The high-frequency oscillation frequency of V_{DS} consistently remains around 102 MHz. When C_{Dec} is 50 nF, the peak of V_{DS} is higher than when C_{Dec} is 200 and 500 nF, which is due to the higher amplitude of low-frequency oscillation.

Keeping other parameters constant, $L_{PCB(Dec\text{-}Bus)} + L_{es(Bus)}$ is varied at values of 15, 35, and 55 nH, resulting in the impedance-frequency characteristics shown in Fig. 7.30. With an increase in $L_{PCB(Dec\text{-}Bus)} + L_{es(Bus)}$, $f_{o(L)}$ continually decreases, reaching 5.46, 1.85, and 1.49 MHz, respectively. The low-frequency resonance peak impedance continuously increases, while $f_{o(H)}$ and the high-frequency resonance peak impedance remain unchanged.

The simulated turn-off waveform of V_{DS} characteristics in Fig. 7.31 corresponds to the impedance-frequency characteristics shown in Fig. 7.30. As $L_{PCB(Dec\text{-}Bus)} + L_{es(Bus)}$ increases, the low-frequency oscillation amplitude of V_{DS} continuously increases, while the frequency continuously decreases, reaching 2.65, 1.78, and 1.43 MHz. The high-frequency oscillation frequency of V_{DS} consistently remains around 102 MHz, and the peak of V_{DS} remains relatively unchanged.

In practical application cases where aluminum electrolytic capacitors are used as bus capacitors, and MLCCs are used as decoupling capacitors with relatively small capacitance, the turn-off V_{DS} waveform for SiC MOSFET is shown in Fig. 7.32. It is noticeable that low-frequency oscillations are prominent, and their peaks are higher than those of high-frequency oscillations. In this scenario, the turn-off voltage peak

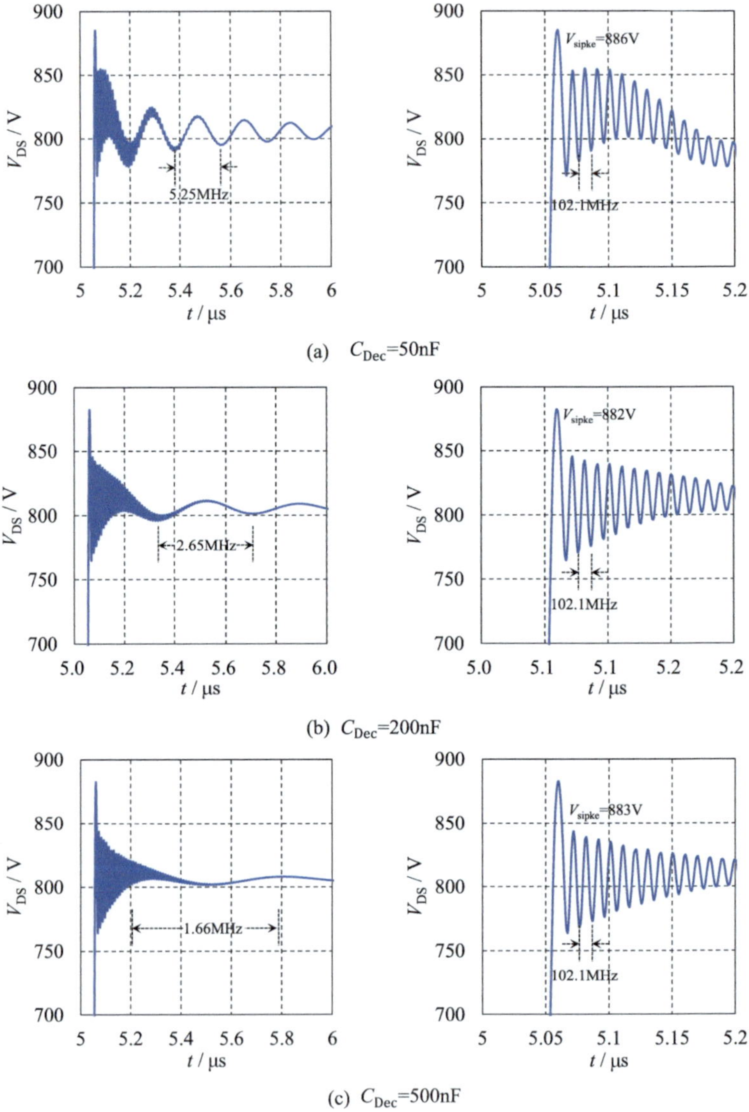

Fig. 7.29 Influence of decoupling capacitor C_{Dec} on turn-off waveform (simulation results)

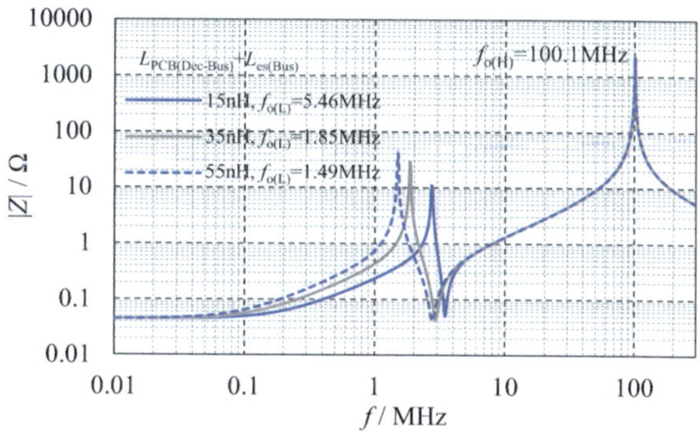

Fig. 7.30 Influence of $L_{PCB(Dec\text{-}Bus)} + L_{es(Bus)}$ on impedance-frequency characteristics

is influenced by low-frequency oscillations. This represents a more extreme situation when using decoupling capacitors, and extra caution is required.

7.4 Countermeasure 3—Reduce Turn-Off Speed

When L_{Loop} is already specified, reducing the SiC MOSFET current turn-off rate $dI_{DS(off)}/dt$ can effectively lower V_{spike}. This can be achieved by increasing the external drive resistance $R_{G(ext)}$ or adding an external gate-source capacitance $C_{GS(ext)}$.

As seen in Fig. 7.33, when $R_{G(ext)}$ is 4.3 Ω, V_{spike} reaches 1031 V, with turn-on energy E_{on} and turn-off energy E_{off} being 588.7 and 165.2 uJ, respectively. As $R_{G(ext)}$ gradually increases, V_{spike} gradually decreases, while both E_{on} and E_{off} gradually increase, indicated by triangular symbols. Maintaining $R_{G(ext)}$ at 4.3 Ω, as $C_{GS(ext)}$ gradually increases, V_{spike} decreases, and both E_{on} and E_{off} gradually increase, with E_{on} showing a more significant increase, indicated by square symbols. When comparing the two methods of increasing $R_{G(ext)}$ and adding $C_{GS(ext)}$ to achieve the same reduction in V_{spike}, both methods have almost the same impact on E_{off}, but adding $C_{GS(ext)}$ leads to a larger increase in E_{on}. It's essential to note that adding $C_{GS(ext)}$ significantly increases driver losses.

The switch waveforms for $R_{G(ext)}$ at 4.3, 10 Ω, and $R_{G(ext)}$ at 4.3 Ω with $C_{GS(ext)}$ at 8.2 nF are shown in Figs. 7.34 and 7.35. It can be observed that larger $R_{G(ext)}$ and added $C_{GS(ext)}$ both result in reduced switch speed and V_{spike}, while E_{on}, E_{off} and switch delay all increase.

The characteristic values of the switch waveforms are presented in Table 7.3. When $R_{G(ext)}$ is 10 Ω, V_{spike} is reduced to 980 V, and E_{off} is 348.7 μJ. When $C_{GS(ext)}$ is 8.2 nF, V_{spike} is reduced to 977 V, and E_{off} is 321.4 μJ, with a small difference between the two. However, during the turn-on process, the influence of $C_{GS(ext)}$ causes both current rise and voltage drop rates to be significantly slower than when $R_{G(ext)}$ is

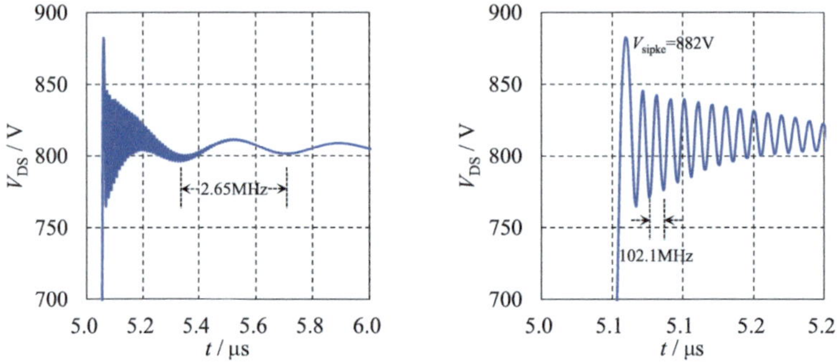

(a) $L_{PCB(Dec\text{-}Bus)}+L_{es(Bus)}$=15nH

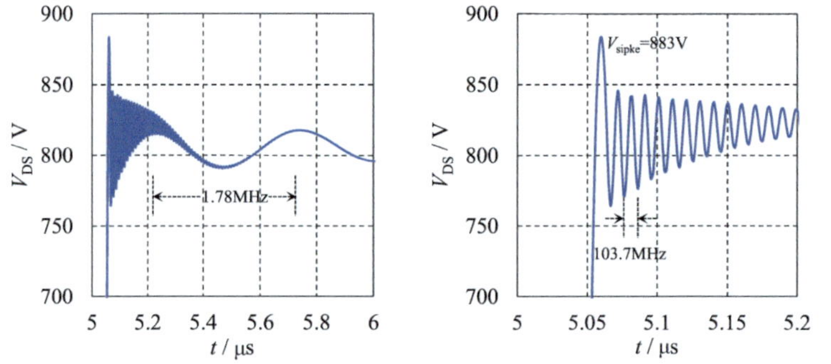

(b) $L_{PCB(Dec\text{-}Bus)}+L_{es(Bus)}$=35nH

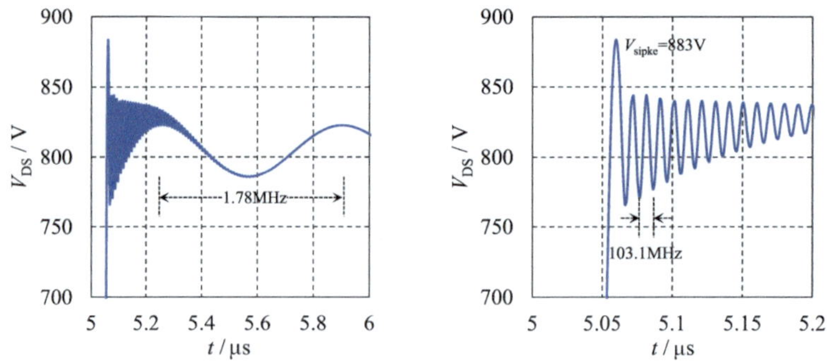

(c) $L_{PCB(Dec\text{-}Bus)}+L_{es(Bus)}$=45nH

Fig. 7.31 Influence of $L_{PCB(Dec\text{-}Bus)} + L_{es(Bus)}$ on turn-off waveform (simulation results)

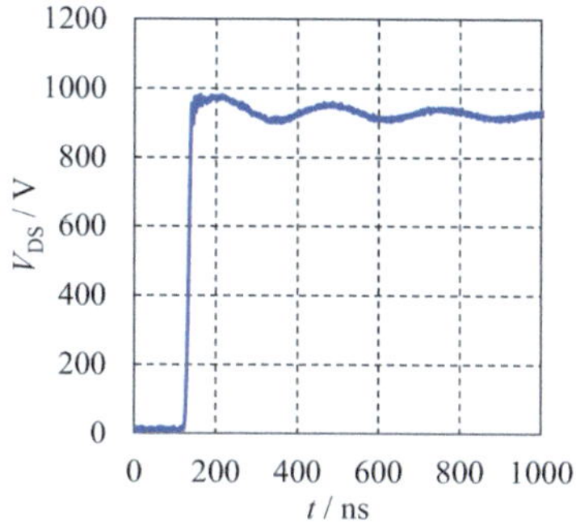

Fig. 7.32 Influence of low-frequency oscillation

Fig. 7.33 Effect of $R_{G(ext)}$ and $C_{GS(ext)}$ on V_{spike} and switching energy

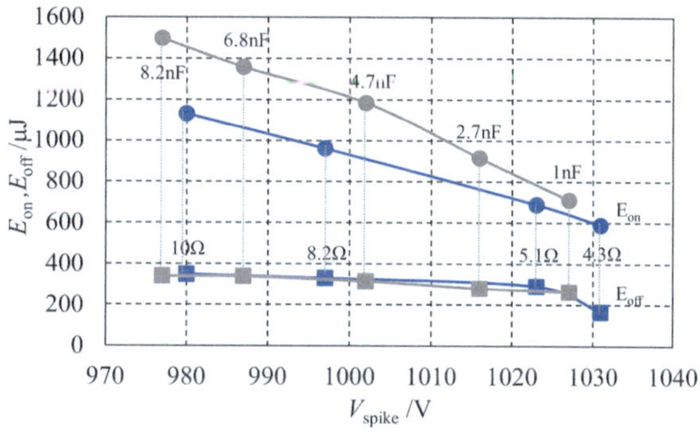

10 Ω. This results in a more substantial increase in E_{on}, with E_{on} being 1131.3 μJ when $R_{G(ext)}$ is 1 Ω, and E_{on} being 1497.4 μJ when $C_{GS(ext)}$ is 8.2 nF.

Based on the above results, it is evident that in practical applications, separating turn-on and turn-off drive resistances, only increasing the turn-off drive resistance $R_{Goff(ext)}$, effectively reduces @@$\mathrm{d}I_{DS(off)}/\mathrm{d}t$, thereby lowering V_{spike}. Meanwhile, keeping the turn-on drive resistance $R_{Gon(ext)}$ unchanged ensures that the turn-on process is not affected, and turn-on energy does not increase. In contrast, adding $C_{GS(ext)}$ simultaneously affects both turn-on and turn-off processes, making it impossible to reduce V_{spike} without increasing turn-on energy.

Fig. 7.34 Reduce switching speed—turn-on process

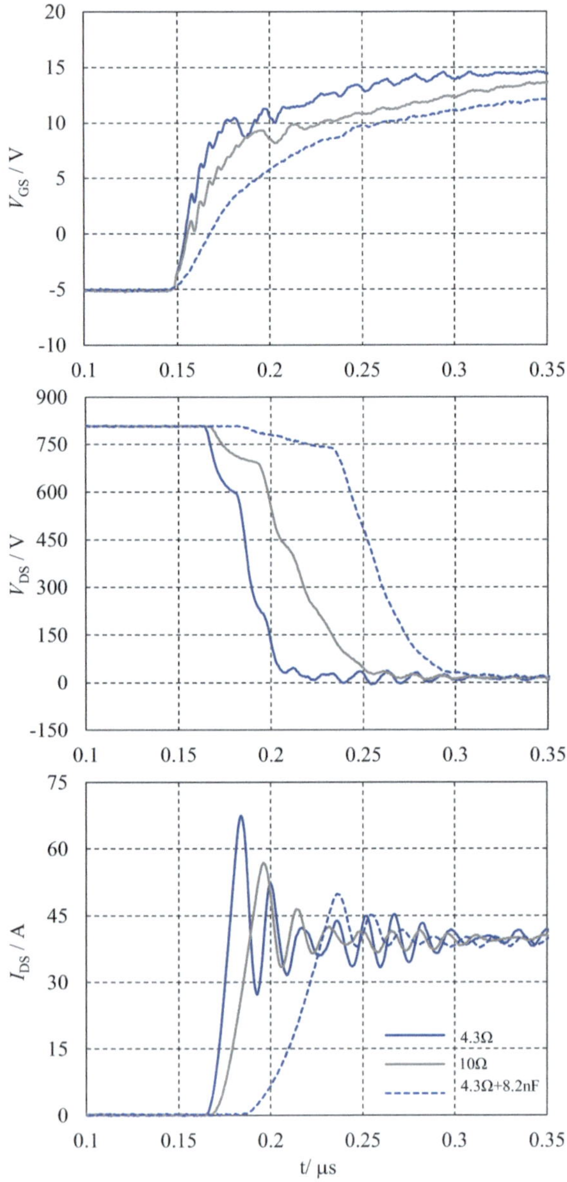

Fig. 7.35 Reduce switching speed—turn-off process

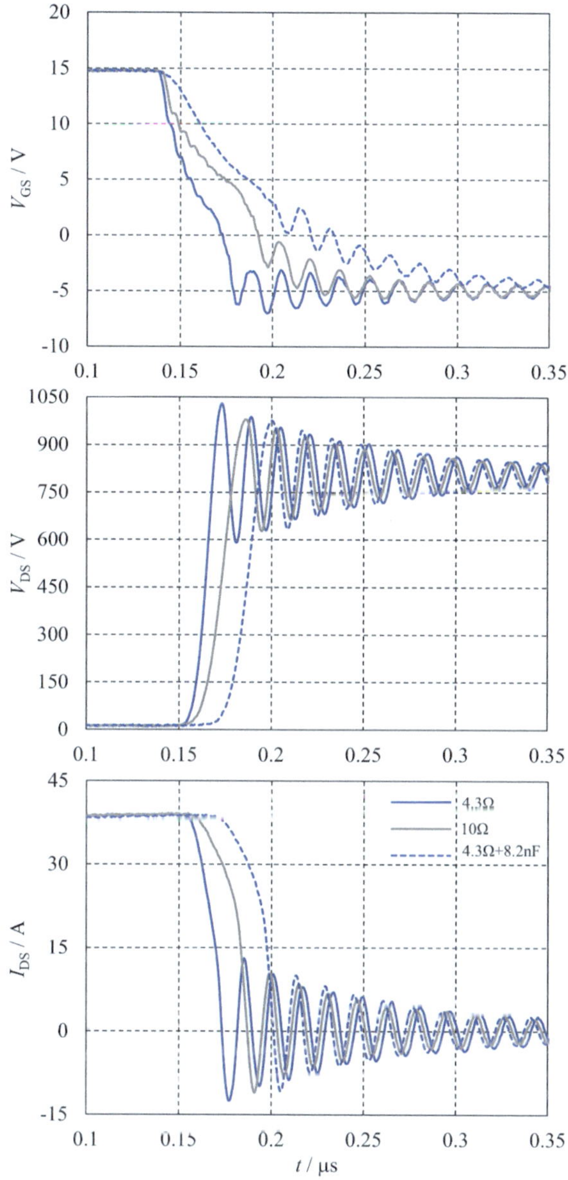

Table 7.3 Switching process characteristics

	$R_{G(ext)} = 4.3\,\Omega$	$R_{G(ext)} = 10\,\Omega$	$R_{G(ext)} = 4.3\,\Omega$ $C_{GS(ext)} = 8.2\,nF$
$t_{d(on)}/$ ns	12.1	23.0	62.8
$t_r/$ ns	33.6	63.4	47.1
$I_{spike}/$ A	67.5	56.9	49.9
$E_{on}/\mu J$	588.7	1131.3	1497.4
$t_{d(off)}/$ ns	15.8	20.6	26.1
$t_f/$ ns	9.8	15.3	15.9
$V_{spike}/$ V	1031.0	980.0	977.0
$E_{on}/\mu J$	165.2	348.7	340.4

References

1. Paul CR (2010) Inductance—loop and partial. Wiley, Hoboken
2. Cree, Inc. (2013) Design considerations for designing with Cree SiC modules Part 2. Techniques for minimizing parasitic inductance. Power application note
3. TDK, Electronics ABC. https://www.tdk.com/en/tech-mag/electronics_primer
4. TDK (2017) Flex crack countermeasures in MLCCs. Solution Guides
5. Murata, KR3 products page. https://psearch.en.murata.com/capacitor/lineup/kr3/
6. TDK, CeraLink® Capacitors products page. https://www.tdk-electronics.tdk.com/en/1195576/products/ceralink-presentation---overview
7. Zheng C (2013) Electrical integration of SiC power devices for high-power-density applications. Virginia Polytechnic Institute and State University

Further Reading

8. Bosshard R (2015) Multi-objective optimization of inductive power transfer systems for EV charging. ETH
9. Montrose MI (2000) Printed circuit board design techniques for EMC compliance: a handbook for designers. Wiley-IEEE Press, New York
10. ON Semiconductor (2023) Impact of stray inductance on EliteSiC power and VE-Trac IGBT module's switching characteristics. Application note, AND90238/D
11. Lamp J. IGBT peak voltage measurement and snubber capacitor specification. Application Note AN-7006, SEMIKRON
12. ROHM Co., Ltd. (2019) SiC MOSFET snubber circuit design methods. Application note, No. 62AN037E, Rev.001
13. Lobsiger Y (2014) Closed-loop IGBT gate drive and current balancing concepts. ETH
14. Nichicon. General descriptions of aluminum electrolytic capacitors. Technical Notes CAT.8101E

Chapter 8
Impact and Countermeasures of High dv/dt—Crosstalk

This Chapter introduces that under high-speed turn-off conditions, the rapid change in off-state current $\mathrm{d}I_{\mathrm{DS(off)}}/\mathrm{d}t$ of SiC MOSFETs leads to more severe turn-off voltage overshoot compared to Si IGBTs, posing challenges in applications. Another effect of high switching speed is the high rate of voltage change $\mathrm{d}V_{\mathrm{DS}}/\mathrm{d}t$, which results in crosstalk issues affecting the gate-source voltage V_{GS}.

Issues reported by power electronics engineers using SiC MOSFETs include: oscillations in the gate-source voltage V_{GS} of non-switching devices, gate breakdown, gate-source short circuits, and bridge arm short circuits. Such problems often consume a significant amount of hardware engineers' efforts, severely impacting product development progress. Frequently, engineers resort to increasing the gate resistance to avoid these issues after attempting various unsuccessful solutions. Crosstalk stands out as a main cause of these issues, posing one of the most challenging problems encountered when using SiC MOSFETs.

8.1 Fundamental Principles of Crosstalk

There are many types of power converter topologies, with a large part of them are bridge circuits such as half-bridge, Totem-Pole PFC, DAB, three-phase full-bridge, and three-level converters, etc. Taking the example of a synchronous buck circuit, the operational waveform is illustrated in Fig. 8.1. It can be seen that there will be glitches on the driving voltage G_2 of the synchronous rectifier tube S_2, and the moment of the glitches coincides with the switching moment of the active switch tube S_1. This shows that in the bridge circuit, the active switching action of the device will affect the driving voltage of the tube, and this phenomenon is crosstalk.

Y. Gao and Y. Zhang, *Silicon Carbide Power Devices*, CPSS Power Electronics Series,
https://doi.org/10.1007/978-981-96-3480-4_8

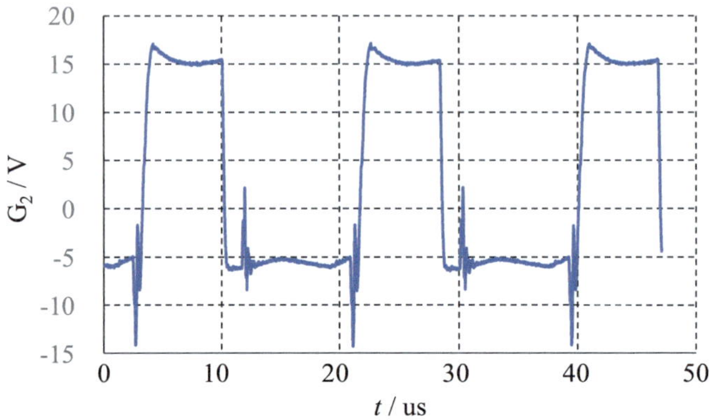

(a) Synchronous buck circuit

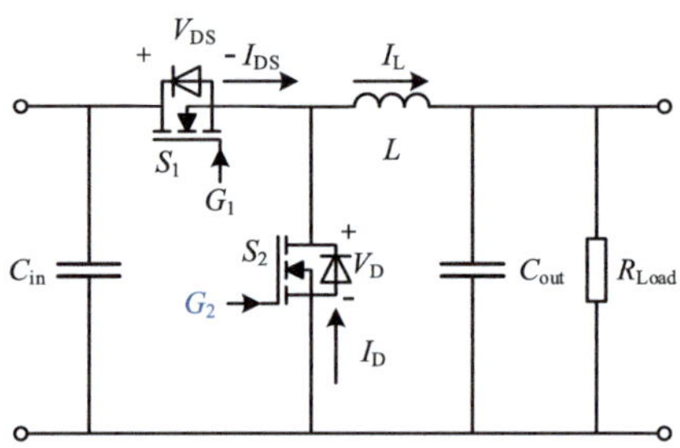

(b) Synchronous buck circuit

Fig. 8.1 Synchronous buck circuit and operating waveforms

8.1.1 Turn-On Crosstalk

In the circuit depicted in Fig. 8.2, Q_H and Q_L represent the SiC MOSFET of the upper and lower bridges, respectively. L stands for the load inductance, $R_{G(int)}$ and $R_{G(ext)}$ represent internal gate resistance and external gate resistance, respectively. C_{GD}, C_{GS}, and C_{DS} are gate-drain capacitance, gate-source capacitance, and drain-source capacitance, respectively. $L_{KS(pkg-O)}$ signifies the source inductance, while $L_{G(pkg-M)}$ and $L_{G(pkg-O)}$, $L_{KS(pkg-M)}$ and $L_{KS(pkg-O)}$, $L_{D(pkg-M)}$ and $L_{D(pkg-O)}$ constitute gate inductance, Kelvin source inductance and drain inductance, respectively. pkg-M is encompassed between the voltage measurement points, while pkg-O denotes external to the voltage measurement points. V_{GS} is the gate-source voltage actually controlling the channel on the chip, $V_{GS(M)}$ is the measured gate-source voltage, V_{DS} is the actual drain-source voltage on the chip, and $V_{DS(M)}$ is the measured drain-source voltage.

Controlled by the drive signal, Q_H transitions from the initial off-state to on-state, while the gate-source voltage V_{GS} of Q_L maintains the off-state voltage $V_{DRV(off)}$

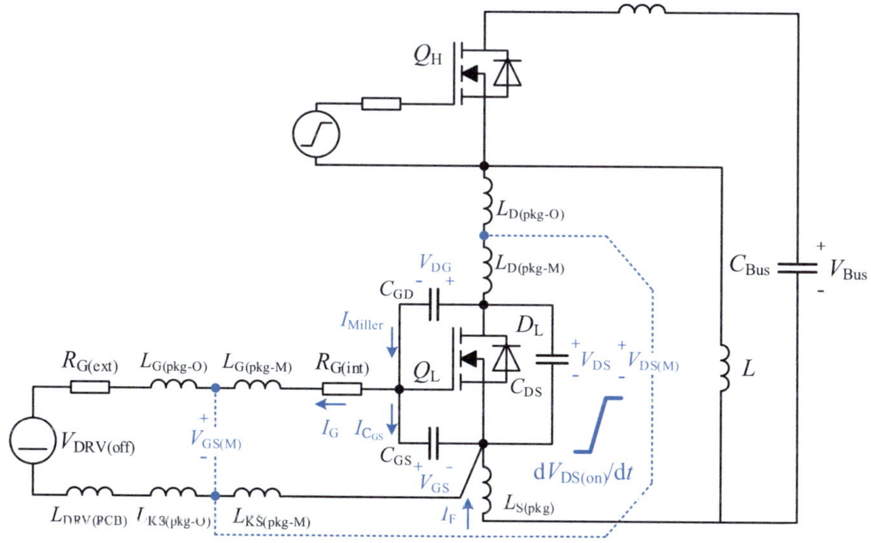

Fig. 8.2 Principle of turn on crosstalk

unchanged. Consequently, the body diode D_L of Q_L transitions from an initial on-state to a reverse recovery state. The drain-source voltage V_{DS} of Q_L rises from the forward voltage drop $-V_F$ of D_L at a rate of $dV_{DS(on)}/dt$ to the bus voltage V_{Bus}. Simultaneously, the terminal voltage V_{DG} of Q_L rises from $-V_F$ to $V_{Bus} - V_{DRV(off)}$ at a rate of $-V_F - V_{DRV(off)}$. It's important to note that $dV_{DS(on)}/dt$ varies with time and is not a constant value. During this process, the changing V_{DG} generates a displacement current I_{Miller} through C_{GD}, the magnitude of which is determined by (8.1).

$$I_{Miller} = C_{GD} \cdot dV_{DS(on)}/dt. \tag{8.1}$$

The I_{Miller} flows into the gate loop through C_{GD}, and I_G induces a voltage drop across the gate resistance R_G ($R_G = R_{G(ext)} + R_{G(int)}$) and the gate loop inductance L_{DRV} ($L_{DRV} = L_{G(pkg\text{-}M)} + L_{G(pkg\text{-}O)} + L_{KS(pkg\text{-}M)} + L_{KS(pkg\text{-}O)} + L_{DRV(PCB)}$). Simultaneously, $I_{C_{GS}}$ charges C_{GS}, causing a variation in V_{GS}.

$$V_{GS} = V_{DRV(off)} + R_G \cdot I_G + L_{DRV} \cdot dI_G/dt. \tag{8.2}$$

Under the combined influence of these two factors, the gate-source voltage V_{GS} is elevated, resulting in an upward spike, as depicted in Fig. 8.3.

(1) t_0–t_1

At t_0, Q_H begins to conduct, and I_F starts decreasing from the load current. V_{DS} is determined by the forward voltage drop V_F across D_L. As I_F decreases, V_{DS} gradually approaches 0V. Despite the small amplitude of V_{DS} variation and slow rate of change

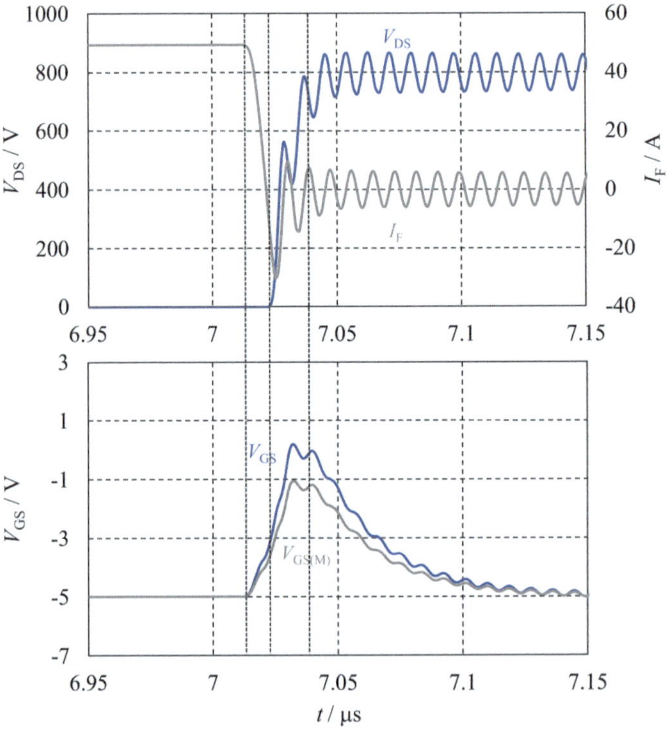

Fig. 8.3 Turn-on crosstalk waveform (simulation waveform)

$\mathrm{d}V_{DS(on)}/\mathrm{d}t$, it's evident from Chap. 4 that C_{GD} is large at this point. Hence, V_{GS} is significantly elevated.

(2) t_1–t_2

At time t_1, D_L begins to withstand the reverse voltage, causing a rapid increase in V_{DS}. Although C_{GD} diminishes with the rise in V_{DS}, the high $\mathrm{d}V_{DS(on)}/\mathrm{d}t$ at this stage swiftly elevates V_{GS}.

(3) t_2–

As $\mathrm{d}V_{DS(on)}/\mathrm{d}t$ gradually decreases, V_{GS} gradually falls back to $V_{DRV(off)}$. Simultaneously, due to the attenuating oscillation once V_{DS} reaches V_{Bus}, V_{GS} oscillates during the decline process.

During this period, Q_L is the disturbed component, with the disturbance originating from the turn-on of Q_H, hence termed turn-on crosstalk. If the peak of V_{GS} exceeds the threshold voltage $V_{GS(th)}$ of the SiC MOSFET, it may lead to false conduction, causing direct conduction of the bridge arm resulting in a short circuit or partial false turn-on leading to additional losses.

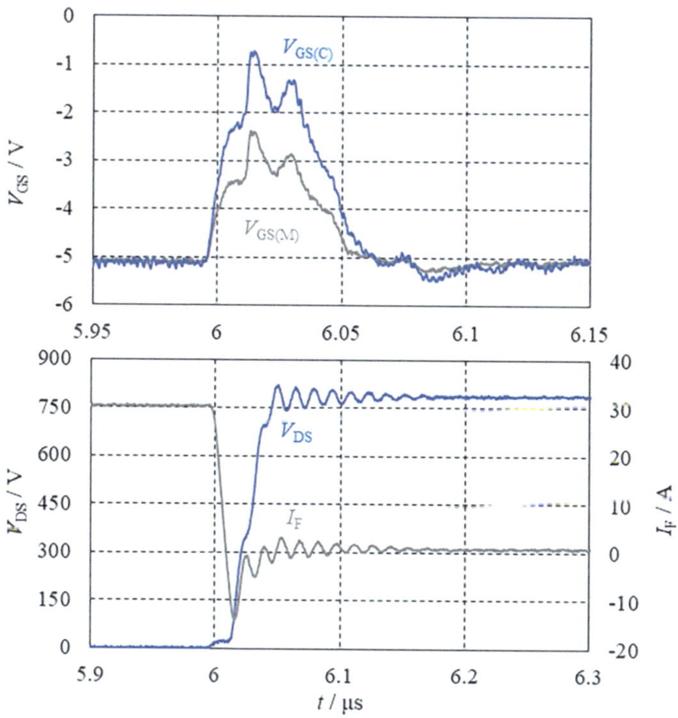

Fig. 8.4 Switching crosstalk waveform

Simultaneously, due to the impact of parasitic parameters between measurement points, the measured $V_{GS(M)}$ is lower than the actual gate-source voltage controlling the channel on the chip, V_{GS}. Consequently, this underestimates the severity of turn-on crosstalk, misleading the circuit design.

Figure 8.4 shows the measured waveform of turn-on crosstalk. $V_{GS(M)}$ is the measured result, while V_{GS} is obtained through parasitic parameter compensation.

At the current stage, the threshold voltage $V_{GS(th)}$ of SiC MOSFETs is concentrated within the range of 2–3 V, and it is the $V_{GS(th)}$ at a low V_{DS} under $T_J = 25\,°C$. As T_J and V_{DS} increase, $V_{GS(th)}$ will significantly decrease. Moreover, due to the high switching speed and high operating voltage of SiC MOSFETs, the risk of false turn-on is higher compared to Si MOSFETs and IGBTs, requiring adequate attention.

8.1.2 Turn-Off Crosstalk

In the circuit shown in Fig. 8.5, under the control of drive signals, Q_H transitions from the initial on-state to the turn-off process, while Q_L's V_{GS} remains at the turn-off voltage $V_{DRV(off)}$ unchanged. Consequently, the body diode D_L of Q_L transitions from

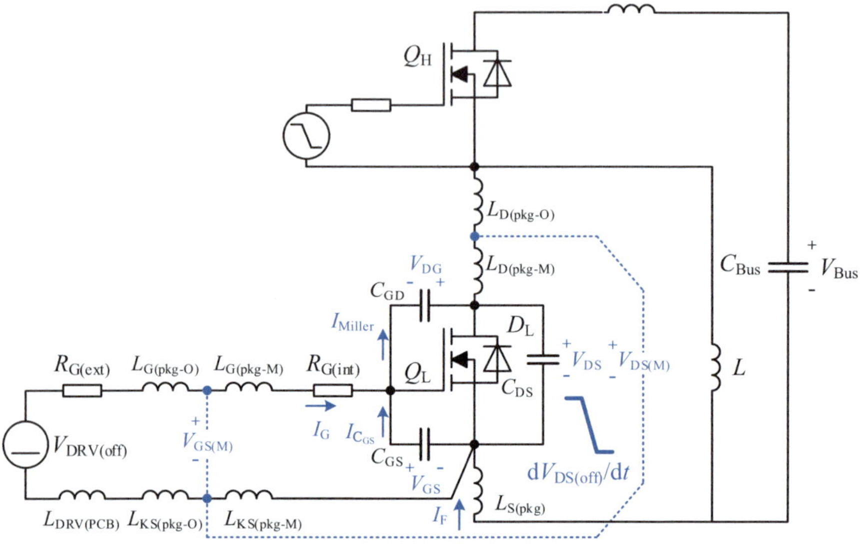

Fig. 8.5 Principle of turn-off crosstalk

an initial off-state to a conducting state. The terminal voltage V_{DS} of Q_L decreases from V_{Bus} at a rate of $\mathrm{d}V_{DS(off)}/\mathrm{d}t$ to the forward voltage drop $-V_F$. Simultaneously, the terminal voltage V_{DG} of C_{GD} of Q_L also decreases from $V_{Bus} - V_{DRV(off)}$ to $-V_F - V_{DRV(off)}$ at a rate of $\mathrm{d}V_{DS(off)}/\mathrm{d}t$. It is important to note that $\mathrm{d}V_{DS(off)}/\mathrm{d}t$ varies with time and is not a constant value. During this, the changing V_{DG} generates a displacement current I_{Miller} through C_{GD}, determined by (8.3).

$$I_{Miller} = C_{GD} \cdot \mathrm{d}V_{DS(off)}/\mathrm{d}t. \tag{8.3}$$

The I_{Miller} flows through C_{GD} into the gate loop. I_G causes a voltage drop across the gate resistance R_G and the drive loop inductance L_{DRV}. Simultaneously, $I_{C_{GS}}$ discharges C_{GS}, resulting in a change in V_{GS}.

$$V_{GS} = V_{DRV(off)} - R_G \cdot I_G - L_{DRV} \cdot \mathrm{d}I_G/\mathrm{d}t. \tag{8.4}$$

Under the combined effect of these two factors, V_{GS} is pulled down, leading to a downward spike as depicted in Fig. 8.6.

(1) t_0–t_1

At t_0, Q_H starts turning off, causing V_{DS} to swiftly drop from V_{Bus} to $-V_F$. Due to the significant value of $\mathrm{d}V_{DS(off)}/\mathrm{d}t$ and the increase in C_{GD} as V_{DS} decreases, V_{GS} is rapidly pulled down.

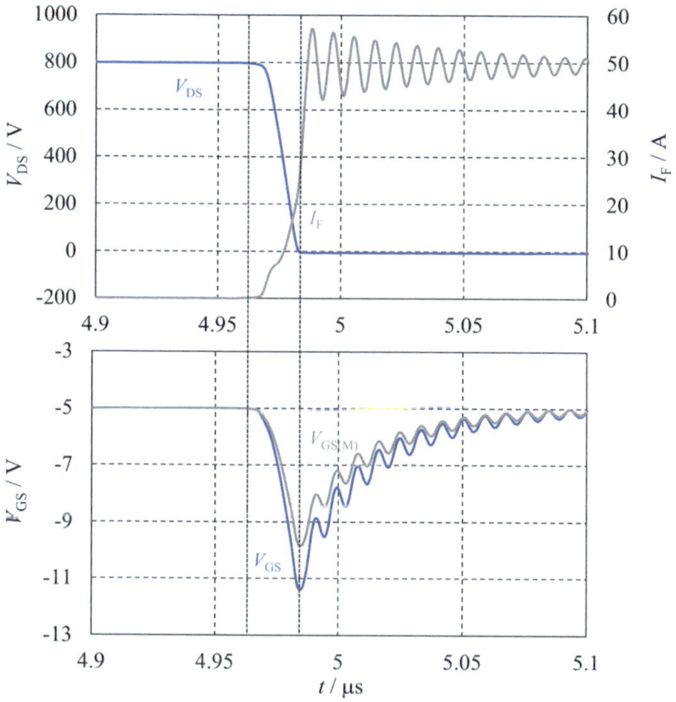

Fig. 8.6 Turn-off crosstalk model (simulation)

(2) t_1–

D_L turns on, causing V_{GS} to gradually rise to $V_{DRV(off)}$. V_{DS} is determined by the forward voltage drop of D_L, and due to the attenuating oscillation of I_F, V_{GS} exhibits oscillations during rise.

At this point, Q_L is the disturbed component, where the disturbance source is the turn-off of Q_H; hence, this process is termed as turn-off crosstalk. If the peak of V_{GS} exceeds the maximum negative voltage tolerance of the SiC MOSFET, it could potentially lead to gate electrode damage and affect the device's lifetime.

Simultaneously, due to the influence of parasitic parameters between measurement points, the measured $V_{GS(M)}$ is larger than the actual voltage V_{GS} of the chip. This underestimates the severity of turn-off crosstalk, misleading the circuit design.

The turn-off crosstalk measured in Fig. 8.7 displays $V_{GS(M)}$ as the obtained measurement result, whereas V_{GS} is the result obtained through compensating for parasitic parameters.

Currently, the maximum gate-source voltage tolerance for SiC MOSFETs is typically limited to − 10 V, and some may be as low as − 7 V or − 3 V, significantly lower than the − 30 V tolerance of Si MOSFETs and Si IGBTs. Due to the robust gate-source voltage tolerance of Si MOSFETs and IGBTs, the impact of turn-off crosstalk is not a major concern during their use. However, when employing SiC

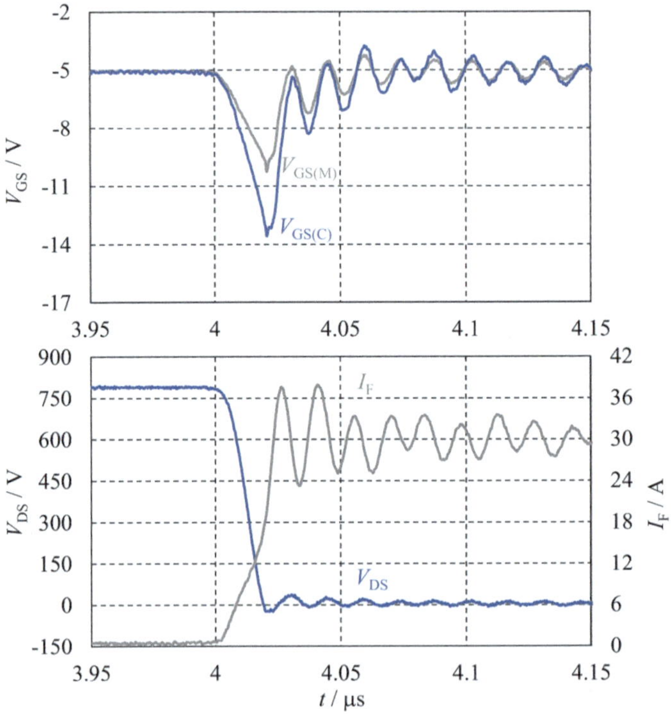

Fig. 8.7 Turn-off crosstalk waveform

MOSFETs, turn-off crosstalk needs to be given equal attention as turn-on crosstalk. Simultaneously, turn-off crosstalk imposes restrictions on the negative value of the turn-off drive voltage $V_{DRV(off)}$. This limitation hinders the utilization of negative voltage turn-off methods to address turn-on crosstalk effectively.

8.2 Main Factors in Crosstalk

8.2.1 Equivalent Circuit Analysis

When neglecting parasitic inductance in the circuit, a simplified equivalent circuit for the disturbed transistor Q_L can be derived, as illustrated in Fig. 8.8 [1]. C_{DS}, C_{GD}, and C_{GS} represent the drain-source, gate-drain, and gate-source capacitance of Q_L respectively. $R_{G(int)}$ denotes the internal gate resistance of Q_L, while $R_{G(ext)}$ represents the external gate resistance, and $V_{DRV(off)}$ stands for the turn-off drive voltage. Influenced by the switching actions of the complementary transistors, the terminal

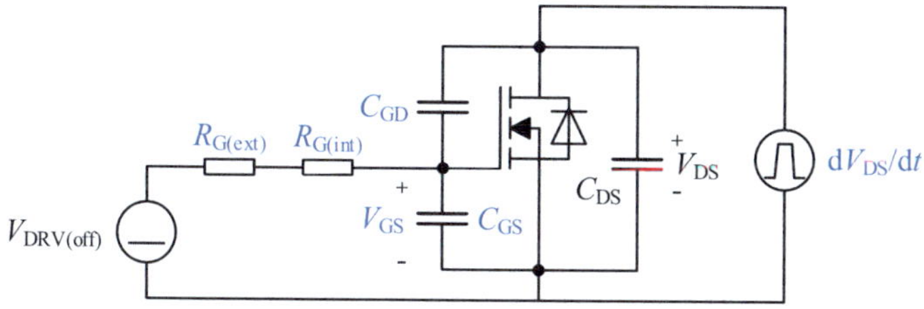

Fig. 8.8 Simplified equivalent circuit of crosstalk

voltage V_{DS} of Q_L rises at a rate of $dV_{DS(on)}/dt$ during turn-on of the complementary transistor and decreases at a rate of $dV_{DS(off)}/dt$ during turn-off of the active transistor.

It is easily obtainable that the forward peak gate voltage $V_{CKon(max)}$ and the reverse peak gate voltage $V_{CKoff(min)}$ in the turn-on and turn-off crosstalk, respectively [1].

$$V_{CK\_on(max)} = V_{DRV(off)} + C_{GD} \cdot dV_{DS(on)}/dt \cdot R_G\left(1 - e^{-\frac{V_{Bus}}{dV_{DS(on)}/dt \cdot (C_{GD}+C_{GS}) \cdot R_G}}\right) \tag{8.5}$$

$$V_{CK\_off(min)} = V_{DRV(off)} - C_{GD} \cdot dV_{DS(off)}/dt \cdot R_G\left(1 - e^{-\frac{V_{Bus}}{dV_{DS(off)}/dt \cdot (C_{GD}+C_{GS}) \cdot R_G}}\right), \tag{8.6}$$

where $R_G = R_{G(int)} + R_{G(ext)}$.

It's important to note that in actual converters, Q_I and Q_{II} are often devices of the same model and use the same R_G. Therefore, the parameters in (8.5) and (8.6) affect each other. Extra attention should be paid when analyzing the impact of each parameter on crosstalk, as detailed below:

(1) C_{GD}

With the same $dV_{DS(on)}/dt$ and $dV_{DS(off)}/dt$, a larger C_{GD} results in increased displacement current, leading to more severe crosstalk. When other conditions remain constant, a larger C_{GD} for the device corresponds to smaller $dV_{DS(on)}/dt$ and $dV_{DS(off)}/dt$, thereby mitigating crosstalk. On one hand, it increases the displacement current, while on the other hand, it reduces both dVDS(on)/dt and dVDS(off)/dt, thereby mitigating crosstalk effects. Concise Engineering Version:

(2) $dV_{DS(on)}/dt$, $dV_{DS(off)}/dt$

With C_{GD} is constant, larger $dV_{DS(on)}/dt$ and $dV_{DS(off)}/dt$ result in increased displacement current, leading to more severe crosstalk.

(3) V_{Bus}

The higher V_{Bus}, the greater energy released by C_{GD} during switch, leading to a more significant impact on the gate.

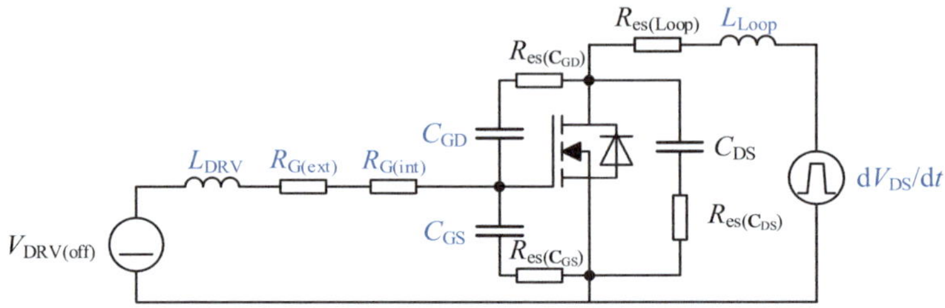

Fig. 8.9 Complete crosstalk equivalent circuit

(4) R_G

Under the same $dV_{DS(on)}/dt$ and $dV_{DS(off)}/dt$, a larger R_G for the disturbed transistor results in a higher voltage drop across it, leading to more severe crosstalk. When the R_G of the switching transistor increases, $dV_{DS(on)}/dt$ and $dV_{DS(off)}/dt$ decrease, mitigating crosstalk. This indicates that it's not accurate to assume that simply increasing or decreasing R_G can mitigate crosstalk.

(5) C_{GS}

C_{GS}: Under same $dV_{DS(on)}/dt$, $dV_{DS(off)}/dt$, and C_{GD}, a larger C_{GS} results in slower charging of the displacement current, contributing to crosstalk. When other conditions is constant, a larger C_{GS} for the device corresponds to smaller $dV_{DS(on)}/dt$ and $dV_{DS(off)}/dt$, similarly mitigating crosstalk.

Considering the inductance of the drive circuit L_{DRV}, the inductance of the main power conversion loop L_{Loop}, along with the equivalent series resistance (ESR) of each diode and line, a complete equivalent circuit is obtained as shown in Fig. 8.9.

Based on this, the expression for the crosstalk becomes too complex, typically requiring numerical methods and simulations for analysis. Some existing literature has conducted relevant research based on a comprehensive circuit model [2]. The specific calculations and analyses are not reiterated here, as the conclusions drawn align with the patterns from simplified equivalent circuit analysis. Next, experimental tests will be used to directly explore the impact of each parameter on crosstalk.

8.2.2 *Experimental Tests Analysis*

Based on the analysis and research findings from relevant literature, considering a given device, specifically when $R_{G(int)}$, C_{GD}, and C_{GS} are fixed, $dV_{DS(on)}/dt$, $dV_{DS(off)}/dt$, I_{DS}, $R_{G(ext)}$, V_{Bus}, and L_{DRV} become the main influencing factors on crosstalk. To avoid mutual influence between these parameters, a controlled variable approach is necessary. Thus, during tests, it's essential that the SiC MOSFET chip, drive circuit, bus capacitor, and decoupling capacitor remain unchanged while facilitating convenient quantitative change of the selected variables.

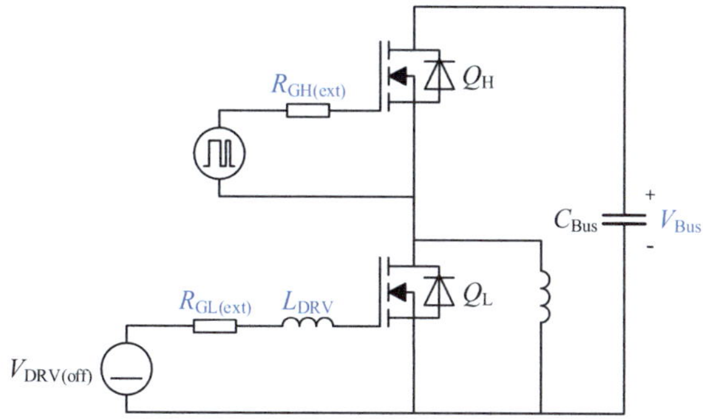

Fig. 8.10 Crosstalk test scheme with control variables

Based on the aforementioned analysis, a control variable crosstalk test scheme is devised, with the circuit shown in Fig. 8.10. In the half-bridge circuit, the upper transistor Q_H is the active device or disturbance source, while the lower transistor Q_L functions as the affected device. Both Q_H and Q_L are SiC MOSFETs in TO-247-4 packages of the same model. Their external gate resistors are $R_{GH(ext)}$ and $R_{GL(ext)}$ respectively. The turn-on and turn-off drive voltages are $V_{DRV(on)}$ and $V_{DRV(off)}$ correspondingly. C_{Bus} is bus capacitance. In the test, $V_{DRV(off)}$ is applied to Q_L to maintain its off-state while a dual-pulse drive signal is applied to Q_H. The first turn-off of Q_H and the second turn-on correspond respectively to the turn-off crosstalk and turn-on crosstalk of Q_L. By measuring V_{DS} and V_{GS} of Q_L, one can obtain $\mathrm{d}V_{DS(on)}/\mathrm{d}t$, $\mathrm{d}V_{DS(off)}/\mathrm{d}t$, and waveforms of crosstalk.

The individual control methods for each parameter are as follows:

(1) $\mathrm{d}V_{DS(on)}/\mathrm{d}t$, $\mathrm{d}V_{DS(off)}/\mathrm{d}t$

These are influenced by $R_{GH(ext)}$ and the load current I_L. You can directly modify $R_{GH(ext)}$ or change the width of the first pulse change $\mathrm{d}V_{DS(on)}/\mathrm{d}t$ and $\mathrm{d}V_{DS(off)}/\mathrm{d}t$.

(2) I_{DS}

Controlled by adjusting the width of the first pulse in dual-pulse signal.

(3) $R_{GL(ext)}$

Simply replace the resistance with a different value.

(4) V_{Bus}

Directly controlled by adjusting the output of the high-voltage power supply.

(5) L_{DRV}

The gate loop is specially designed to precisely adjust L_{DRV}. The gate PCB traces are disconnected and solder pads are set. During testing, just use surface-mount hollow inductors with different values.

When $R_{GL(ext)} = 4.7\ \Omega$, $I_{DS} = 30$ A, and $V_{Bus} = 800$ V, changing the $R_{GH(ext)}$ yields the test results as shown in Fig. 8.11. Where $V_{GS(M)}$ denotes the measured gate-source voltage, and $V_{GS(C)}$ represents the compensated gate-source voltage. It's evident that as $R_{GH(ext)}$ decreases, both $\mathrm{d}V_{DS(on)}/\mathrm{d}t$ and $\mathrm{d}V_{DS(off)}/\mathrm{d}t$ significantly increase, resulting in a more severe crosstalk.

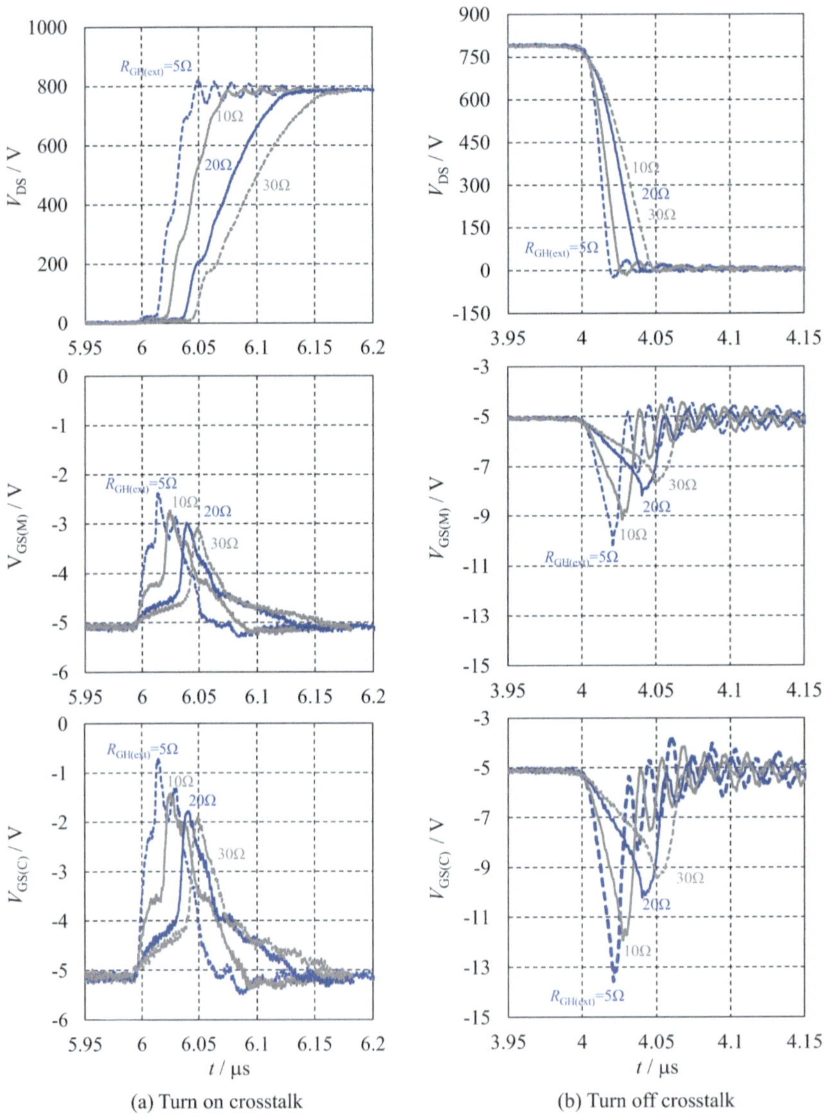

Fig. 8.11 Effect of $\mathrm{d}V_{DS}/\mathrm{d}t$ on crosstalk

Similarly, when $R_{GH(ext)} = 5\ \Omega$, $I_{DS} = 30$ A, and $V_{Bus} = 800$ V, varying the $R_{GL(ext)}$ yields the test results as shown in Fig. 8.12. It's evident that with an increase in $R_{GL(ext)}$, the crosstalk significantly worsens.

Combining results from Figs. 8.11 and 8.12, it's apparent that the impact of the gate resistance $R_{G(ext)}$ on crosstalk is rather complex. Simply adjusting $R_{G(ext)}$ cannot mitigate crosstalk: as $R_{G(ext)}$ increases, dV_{DS}/dt decreases, resulting in smaller crosstalk; however, a larger $R_{G(ext)}$ causes a greater voltage drop across I_{Miller} on

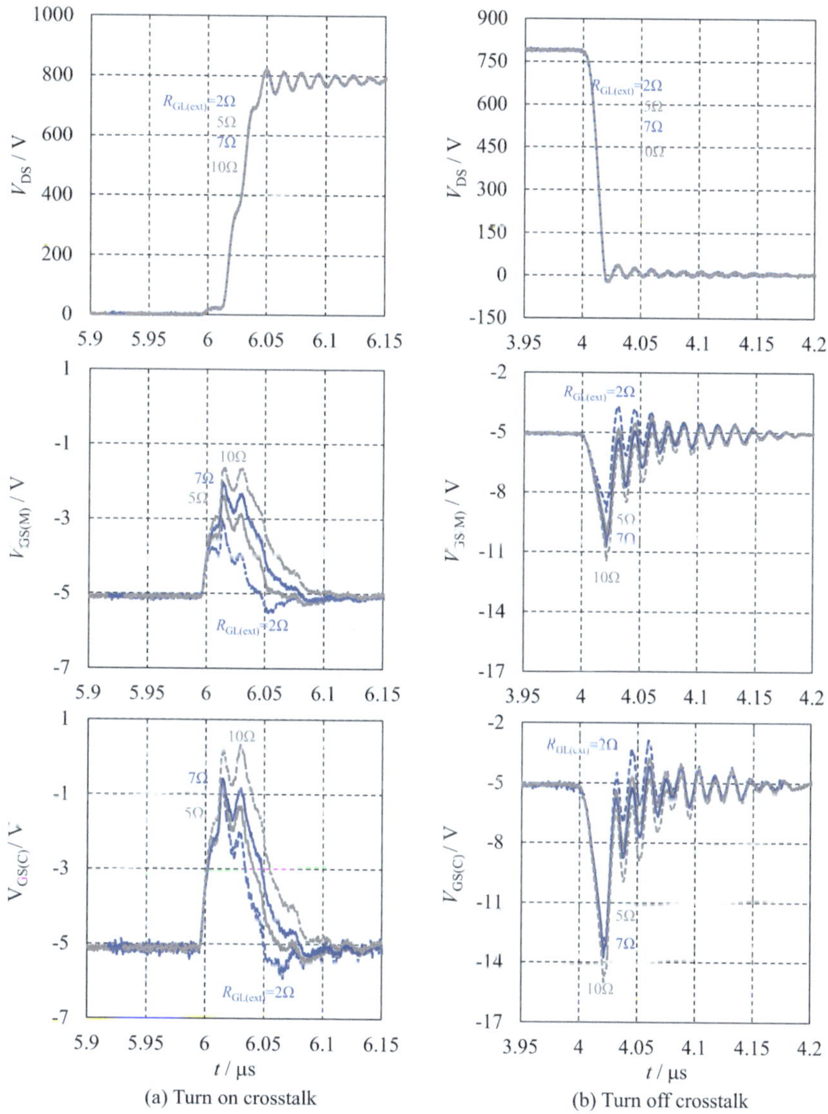

Fig. 8.12 Effect of $R_{GL(ext)}$ on crosstalk

$R_{G(ext)}$, leading to more severe crosstalk. In practical applications, independent turn-on and turn-off gate resistance $R_{Gon(ext)}$ and $R_{Goff(ext)}$ can be used in conjunction with the characteristics of the topology, allowing for flexible adjustment to balance switch speed and crosstalk.

$R_{GL(ext)} = 4.7\ \Omega$, $R_{GH(ext)} = 4.7\ \Omega$, and $V_{Bus} = 800V$, changing the I_{DS} yields the test results as shown in Fig. 8.13. It's noticeable that with an increase in I_{DS}, both $\mathrm{d}V_{DS(on)}/\mathrm{d}t$ and $\mathrm{d}V_{DS(off)}/\mathrm{d}t$ increase, leading to severe crosstalk. Particularly,

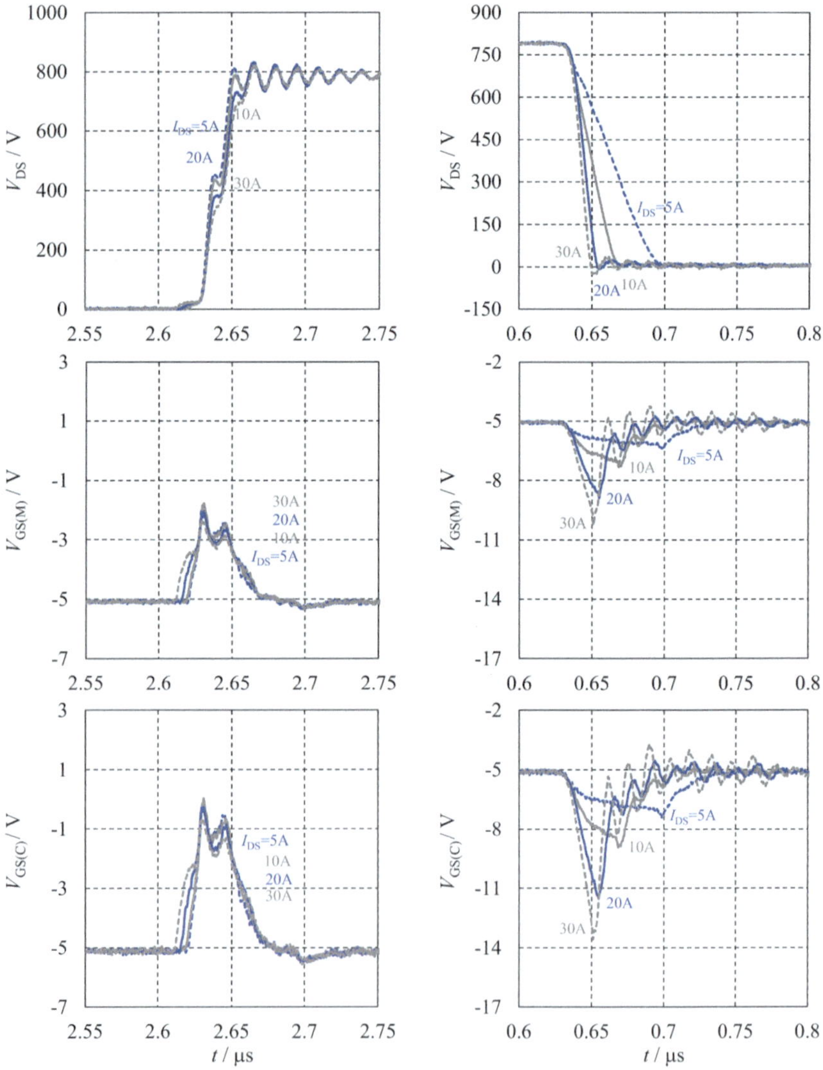

Fig. 8.13 Effect of I_{DS} on crosstalk

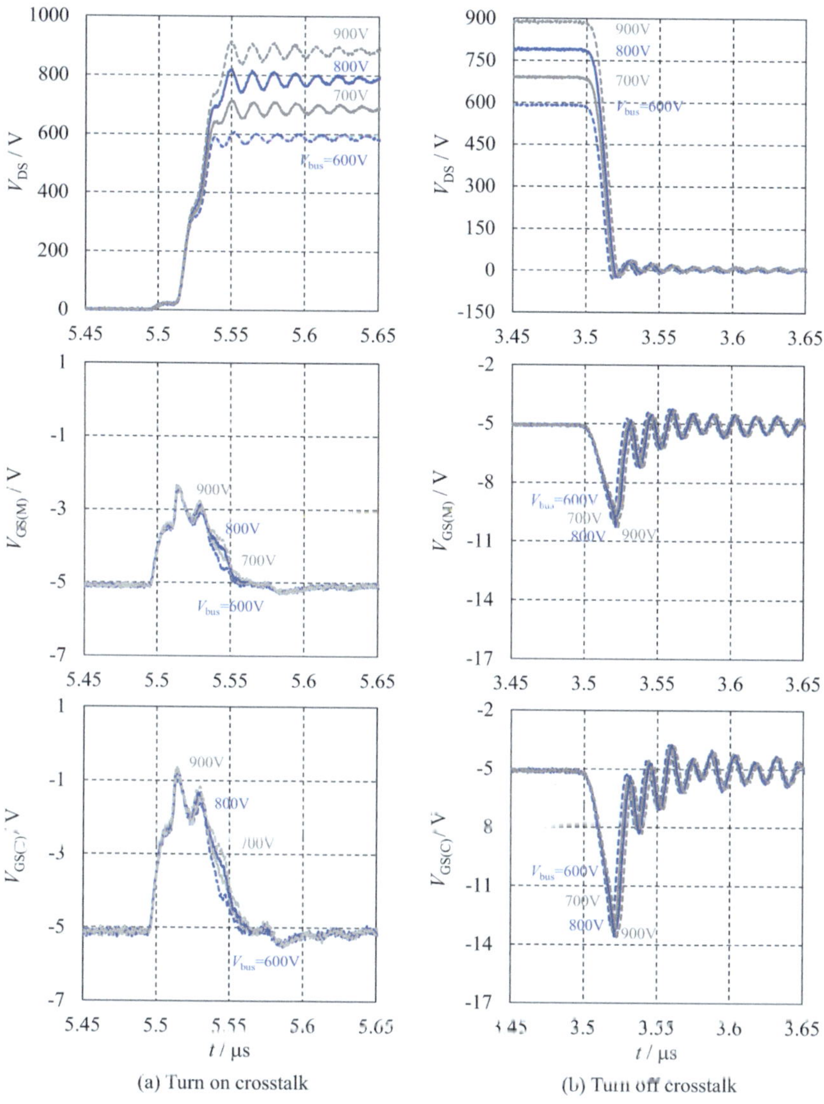

Fig. 8.14 Effect of V_{Bus} on crosstalk

$dV_{DS(off)}/dt$ and its corresponding turn-off crosstalk are significantly affected by I_{DS}, while its impact on the turn-on process is relatively small. This underscores the need for special attention to crosstalk under heavy loads during converter design, as this situation manifests the most severe crosstalk.

$R_{GL(ext)} = 4.7\ \Omega$, $R_{GL(ext)} = 4.7\ \Omega$, and $I_{DS} = 30$A, varying the V_{Bus} yields the test results as shown in Fig. 8.14. It's noticeable that with an increase in V_{Bus}, there's a slight increase in $dV_{DS(on)}/dt$ and $dV_{DS(off)}/dt$, leading to a slightly intensified crosstalk. This emphasizes the need to pay particular attention to the highest level of crosstalk when V_{Bus} is at its peak during converter design, as this situation presents the most severe crosstalk.

$R_{GL(ext)} = 5\ \Omega$, $R_{GH(ext)} = 5\ \Omega$, $I_{DS} = 30A$, and $V_{Bus} = 800V$, the test results for inserting a surface-mount inductor into the drive loop are shown in Fig. 8.15. It can be observed that as the inserted inductance increases, the crosstalk becomes slightly more severe.

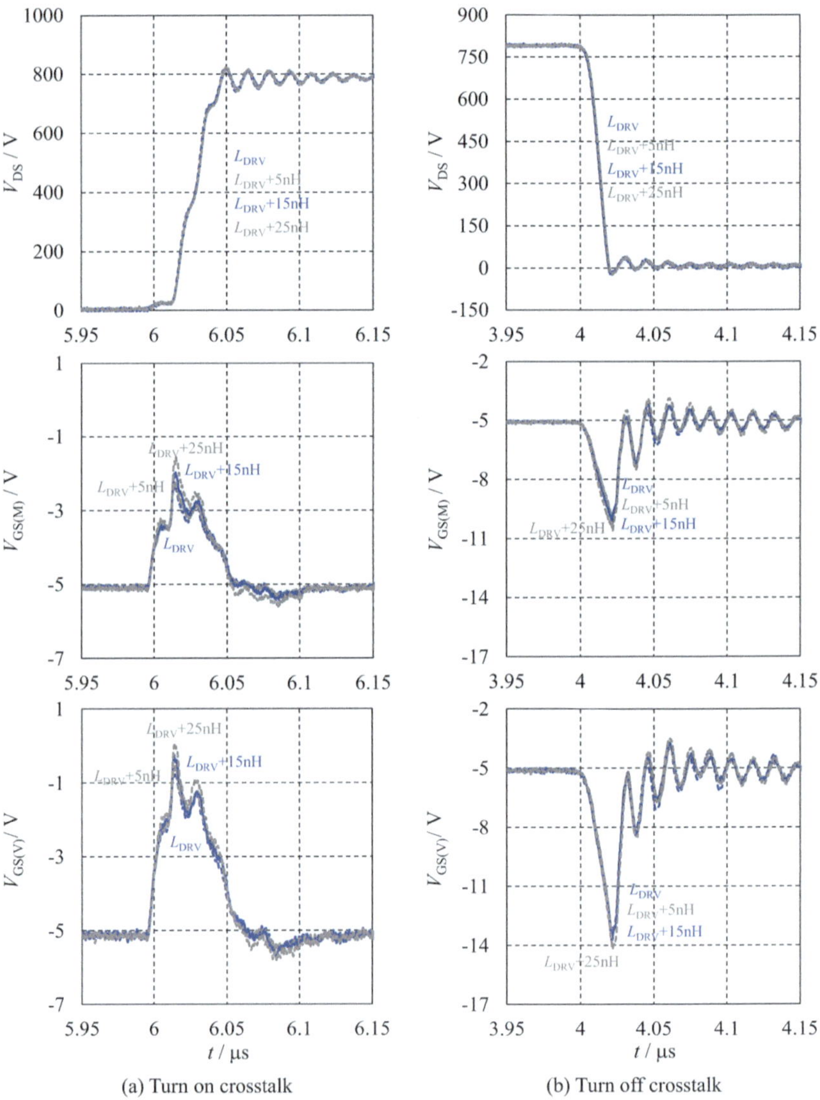

Fig. 8.15 Effect of L_{DRV} on crosstalk

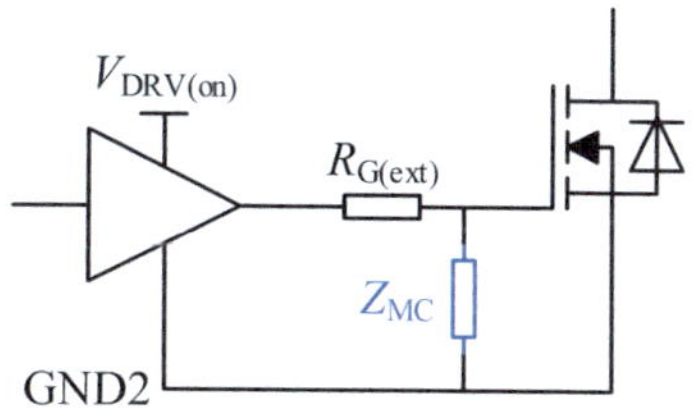

Fig. 8.16 Basic principles of Miller clamping

8.3 Countermeasures 1-Drive Circuit Inductance Control

The essence of crosstalk lies in the current I_{Miller} generated by the displacement of C_{GD}, which induces a voltage drop across R_G and charges/discharges C_{GS}, consequently affecting V_{GS}. Without altering circuit parameters or device switching speeds, if I_{Miller} could be diverted elsewhere, reducing the current flowing through R_G and C_{GS}, it could mitigate crosstalk. Following this approach, a low-impedance path ZMC can be constructed between the device's gate and source to divert most of I_{Miller}, as depicted in Fig. 8.16. As C_{GD} is also known as the Miller capacitance, this method is referred to as Miller Clamping.

8.3.1 Transistor Type Miller Clamping

When the drive circuit is turn off with zero voltage, adding a PNP transistor with its emitter connected to the device's gate, collector connected to the device's source, and base connected to the drive output terminal, forms a PNP transistor Miller Clamping (BJT-MC), abbreviated as shown in Fig. 8.17a. In the event of turn-on noise, I_{Miller} flows into the drive loop. When the component IG across the $R_{G(ext)}$ exceeds a voltage drop of 0.7 V, the emitter voltage of the PNP transistor surpasses the base voltage by 0.7 V, causing the PNP transistor to conduct. Consequently, a portion of I_{Miller} will flow back through the transistor into the power loop, thereby mitigating the turn-on noise. When the drive circuit is turned off with a negative voltage, the connection of the PNP transistor is as depicted in Fig. 8.17b, operating on the same principle as during zero-voltage turn off.

The effectiveness of using BJT-MC to counteract crosstalk is illustrated in Fig. 8.18. Where $V_{GS(M)}$ denotes the measured gate-source voltage and $V_{GS(C)}$ represents the post-compensation value. The compensated waveform using the BJT-MC method is not provided due to unavailability of measured current through the BJT. During the turn-on crosstalk, the forward peak of V_{GS} is somewhat reduced, but there's a noticeable emergence of a negative peak. Originally, in the turn-on crosstalk, the focus was primarily on preventing the V_{GS} forward peak from being excessively

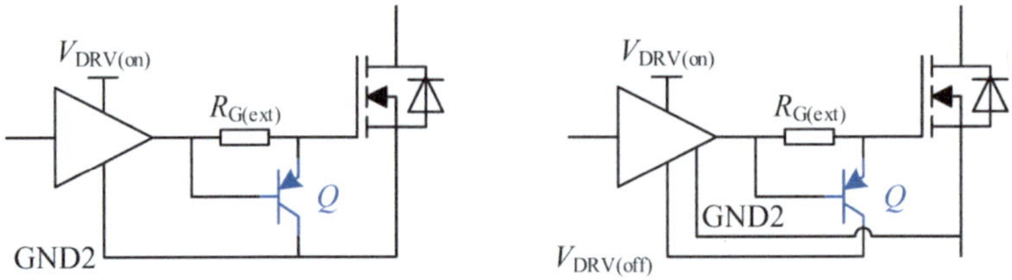

(a) Zero-voltage turn-off (b) Negative-voltage turn-off

Fig. 8.17 PNP transistor type Miller clamp (BJT-MC) circuit

Fig. 8.18 Effect of BJT-MC in turn-on crosstalk suppression

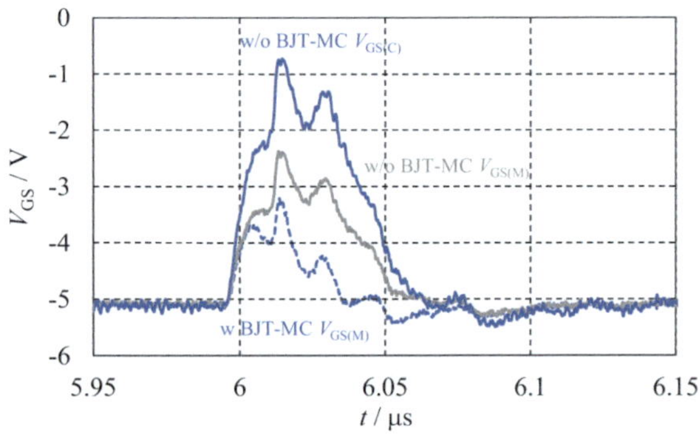

high, leading to a false turn-on. However, with the implementation of BJT-MC, additional attention is required towards the negative peak that was initially not a concern during turn-on crosstalk.

During the turn-off crosstalk, the voltage drop of I_G across $R_{G(ext)}$ causes the emitter voltage of the PNP transistor to be lower than the base voltage, keeping the PNP transistor in the off state. Thus, BJT-MC cannot alleviate turn-off crosstalk, and the turn-off crosstalk waveform remains identical to when BJT-MC is not used, as shown in Fig. 8.19. Moreover, when the active device also employs BJT-MC, upon device turn-off, the drive voltage output transitions from high to low, causing the PNP transistor to conduct. This accelerates the turn-off speed of the SiC MOSFET, exacerbating the turn-off crosstalk. Consequently, it's evident that BJT-MC has significant limitations in mitigating crosstalk.

To mitigate turn-off crosstalk, a diode DMC is reverse-connected in parallel with the PNP transistor's collector–base junction in the BJT-MC circuit. This configuration forms a PNP transistor in parallel with a diode, known as BJT//Diode-MC, illustrated in Fig. 8.20. During the turn-off crosstalk phase, when V_{GS} is lower by 0.7 V than the turn-off drive voltage, DMC conducts, suppressing the turn-off crosstalk.

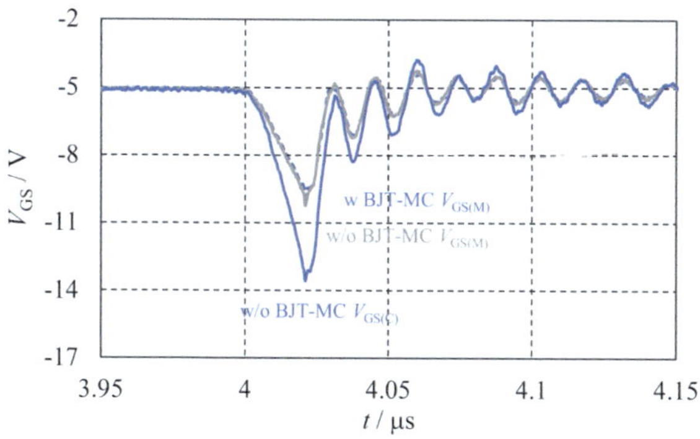

Fig. 8.19 Effect of BJT-MC in suppressing turn-off crosstalk

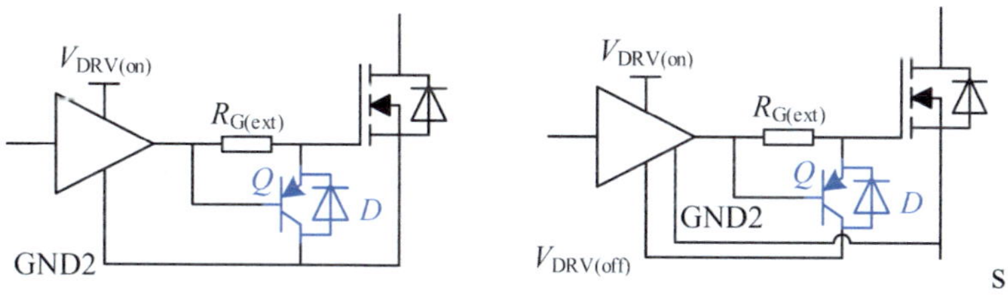

(a) Zero-voltage turn-off (b) Negative-voltage turn-off

Fig. 8.20 PNP transistor parallel diode Miller clamp (BJT//Diode-MC) circuit

The effectiveness of BJT//Diode-MC in addressing crosstalk is depicted in Fig. 8.21. During the turn-off crosstalk, the negative spike in V_{GS} is effectively suppressed. However, a notable positive spike emerges. Previously, the concern during turn-off crosstalk was primarily to prevent excessive negative spikes in V_{GS}. With the use of BJT//Diode-MC, there's an additional focus required on the positive spikes that were not a concern during turn-off crosstalk in the original setup. Yet, in turn-on crosstalk, there's no difference between using BJT-MC and BJT//Diode-MC.

8.3.2 *Active Miller Clamping*

Active Miller Clamping (AMC), widely used to address crosstalk issues, comprises a comparator, MOSFET $S_{MC(int)}$, and logic control circuitry.

Fig. 8.21 Effect of BJT// Diode-MC in crosstalk suppression

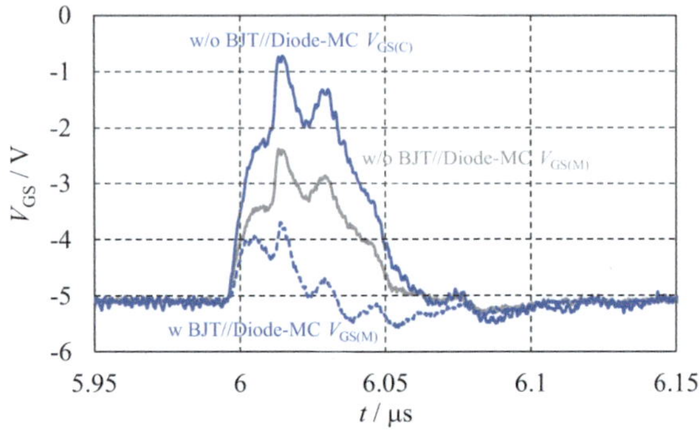

(a) Turn-on crosstalk

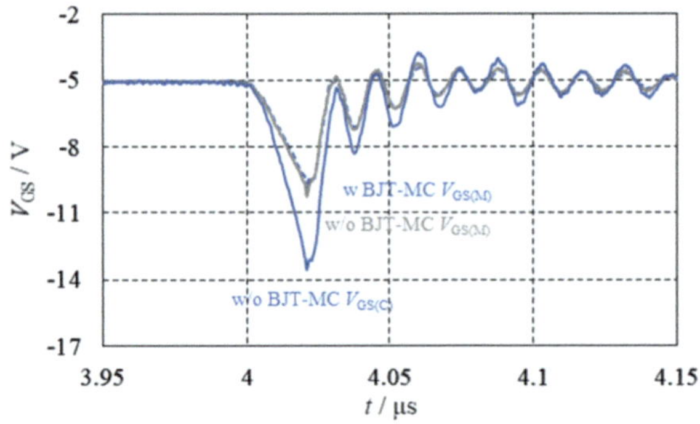

(b) Turn-off crosstalk

During zero-voltage turn-off, the MC is connected to gate of the SiC MOSFET. V_{GS} is fed back to the comparator through the MC, while the GND2is connected to the device's source, as depicted in Fig. 8.22a. When turn-on crosstalk raises V_{GS} to the comparator's threshold voltage $V_{th(MC)}$, the comparator flips, allowing $S_{MC(int)}$ to conduct, thereby suppressing turn-on crosstalk. During turn-off crosstalk, when V_{GS} drops below − 0.7V, the body diode $D_{MC(int)}$ of $S_{MC(int)}$ conducts, suppressing turn-off crosstalk.

During negative voltage turn-off, the circuit configuration shown in Fig. 8.22b is employed. In turn-on crosstalk, $S_{MC(int)}$ switches when V_{GS} exceeds $V_{EE2} + V_{th(MC)}$, while during turn-off crosstalk, $D_{MC(int)}$ switches when V_{GS} drops below $V_{EE2} - 0.7$V.

The efficacy of addressing crosstalk using AMC is illustrated in Fig. 8.23, sharing similar characteristics with BJT//Diode-MC. During turn-on crosstalk, the positive spike in V_{GS} is suppressed more prominently than with BJT//Diode-MC. However, a noticeable negative spike emerges as well. In turn-off crosstalk, it effectively suppresses the negative spike in V_{GS}, yet a prominent positive spike appears.

Fig. 8.22 Active Miller Clamping circuit

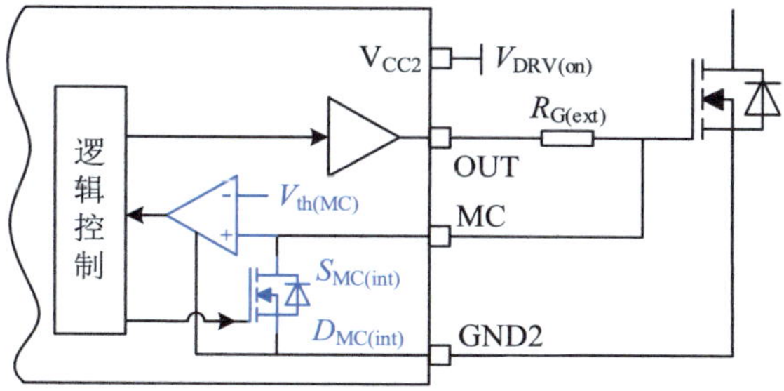

(a) Zero-voltage turn-off

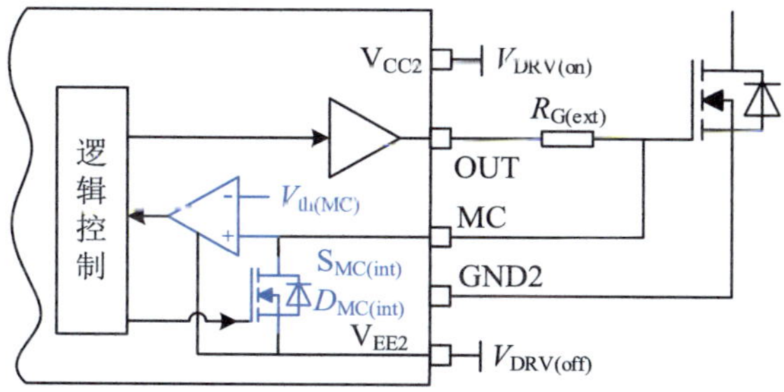

(b) Negative-voltage turn-off

Based on the principles of Miller Clamping, the lower the diversion line impedance, the stronger the crosstalk suppression capability. The current-handling capability of $S_{MC(int)}$ in the AMC circuit is often limited. To enhance crosstalk suppression, an additional PNP transistor $S_{MC(ext)}$ and diode $D_{MC(ext)}$ can be added to it, forming the AMC with additional PNP transistor and diode, abbreviated as ACM-BJT//Diode, as depicted in Fig. 8.24. During turn-on crosstalk, if V_{GS} exceeds the turn-off drive voltage by $V_{th(MC)}$, triggering AMC, $S_{MC(ext)}$ conducts to suppress turn-on crosstalk. During turn-off crosstalk, when V_{GS} drops below the turn-off drive voltage by 0.7 V, $D_{MC(ext)}$ mitigates turn-off crosstalk.

Compared to PNP transistors, MOSFETs exhibit faster switching speeds and lower on-state voltage drops. Therefore, using an external MOSFET will yield better results. However, as depicted in Fig. 8.22, the AMC circuit cannot directly drive an external MOSFET. To address this issue, ROHM has introduced a driver IC, BM6108FV-LB [3], which features an external AMC MOSFET interface, hereinafter referred to as ACM-MOS, as shown in Fig. 8.25. Unlike the previously discussed AMC, when the drive outputs OUT1H and OUT1L of BM6108FV-LB respectively maintain a high impedance and low level, V_{GS} is fed back to the comparator via the

Fig. 8.23 Effect of AMC in crosstalk suppression

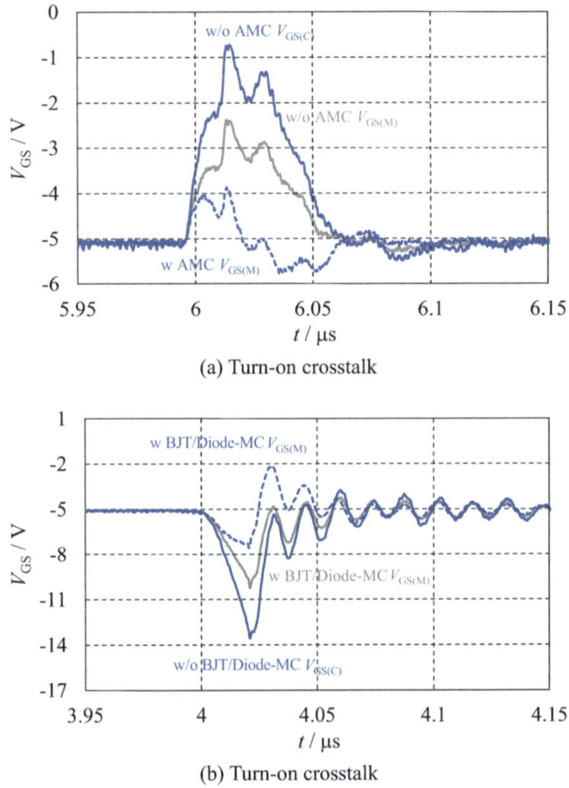

PROOUT. When V_{GS} exceeds $V_{EE2} + V_{th(MC)}$, OUT2 outputs a high level, turning on $S_{MC(ext)}$ directly bypassing $R_{G(ext)}$. Conversely, when the drive outputs OUT1H and OUT1L are respectively high and high impedance, OUT2 immediately outputs a low level, turning off the external MOSFET.

The effects of using ACM-BJT//Diode and ACM-MOS to address crosstalk are illustrated in Fig. 8.26. Due to the lower impedance of the shunt line, ACM-MOS demonstrates better suppression of crosstalk compared to AMC, but its negative effects are more pronounced. However, ACM-BJT//Diode, due to the inferior characteristics of the BJT, actually performs less effectively than AMC.

Fig. 8.24 AMC and PNP transistor parallel diode Miller clamp (ACM-BJT// Diode) circuit

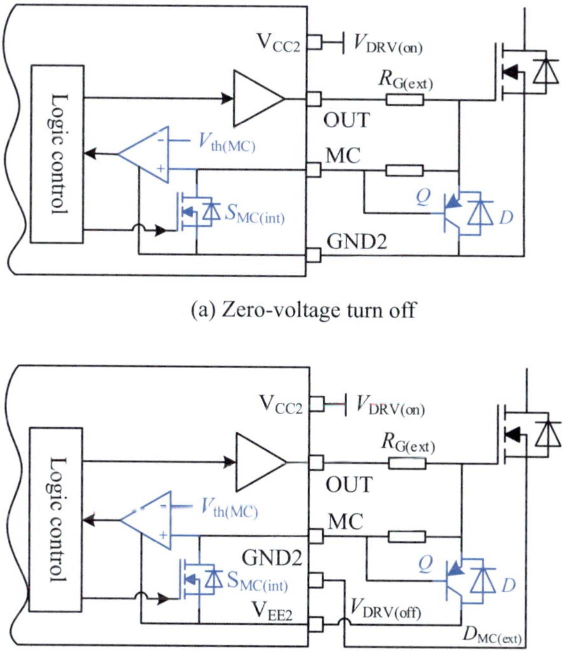

(a) Zero-voltage turn off

(b) Negative-voltage turn off

Fig. 8.25 ROHM BM6108FV-LB ACM-MOS circuit

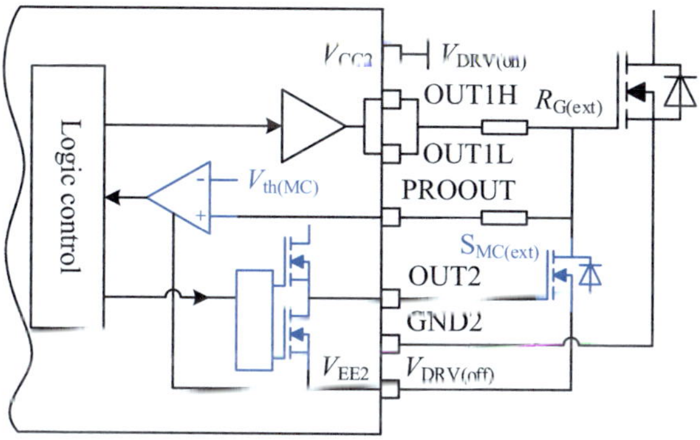

Fig. 8.26 AMC, ACM-BJT/ /diode and ACM-MOS

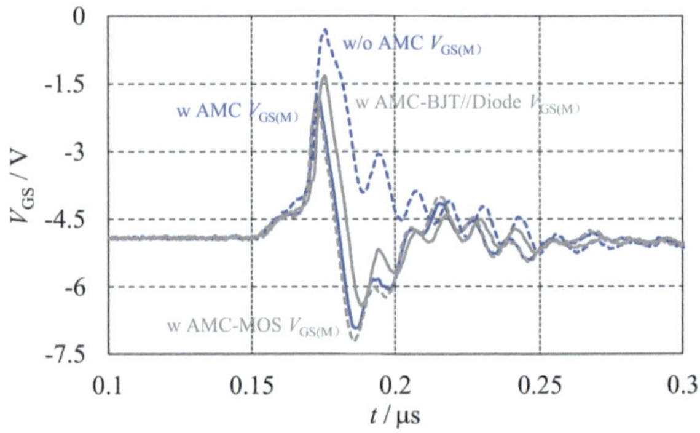

(a) Turn-on Crosstalk

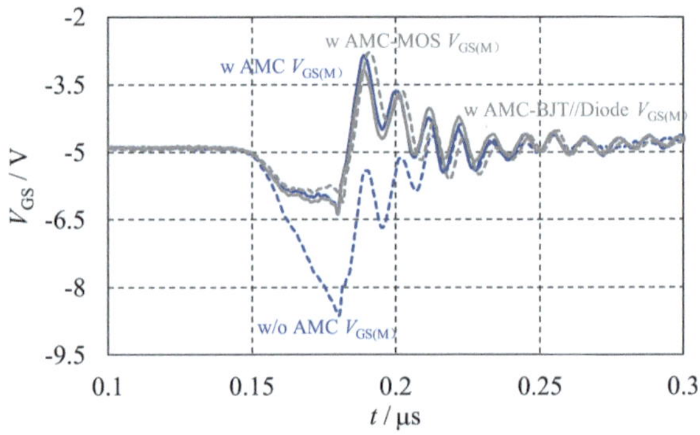

(b) Turn-off Crosstalk

8.4 Countermeasures 2-Drive Circuit Inductance Control

8.4.1 Impact of Drive Circuit Inductance on Miller Clamping

The results in Sect. 8.2.2 indicate that the inductance L_{DRV} in the drive loop affects the crosstalk—higher L_{DRV} leads to more severe crosstalk. The main reason behind this is that with larger L_{DRV}, the voltage drop across I_G increases, causing more significant oscillations in the drive loop. Similarly, when using Miller Clamping, the inductance L_{MC} in the Miller Clamping circuit inevitably impacts its effectiveness. As shown in Fig. 8.27, as L_{MC} increases, the effectiveness of Miller Clamping deteriorates.

Fig. 8.27 Effect of gate Loop Inductance L_{MC} on Active Miller Clamp (AMC) performance

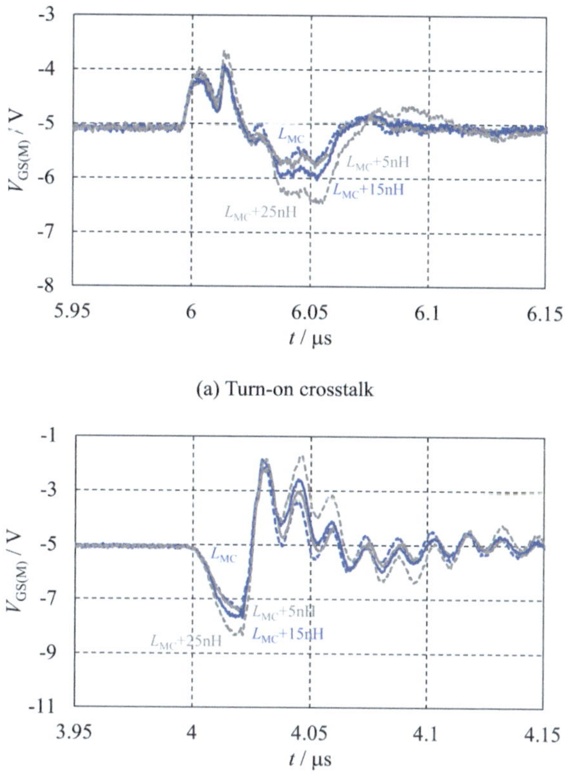

(a) Turn-on crosstalk

(b) Turn-off crosstalk

8.4.2 *Package Integration*

The aforementioned results indicate that L_{DRV} and L_{MC} have negative impacts on crosstalk and Miller Clamping. As L_{DRV} increases, crosstalk worsens, and as L_{MC} increases, the effectiveness of Miller Clamping decreases, leading to more pronounced side effects. Therefore, minimizing L_{DRV} and L_{MC} becomes an inevitable choice when the power devices, gate resistance R_G, and Miller Clamping method have already been determined.

As outlined in Sect. 7.2.2, we can use two methods to reduce gate loop inductance. First, by designing the PCB layout of the drive lines to overlap vertically and second, by shortening the distance between the drive circuit and power devices. However, in scenarios involving discrete devices or standard power modules, limitations arise due to component size and safety regulations, restricting the drive circuit from being placed arbitrarily close to the devices. Furthermore, in some cases, to enhance power density, connectors might be utilized, preventing the implementation of overlapped demagnetization. Simultaneously, as depicted in Fig. 8.28, the pinouts in device packaging, DBC traces, and bonding wires are inevitable factors to consider.

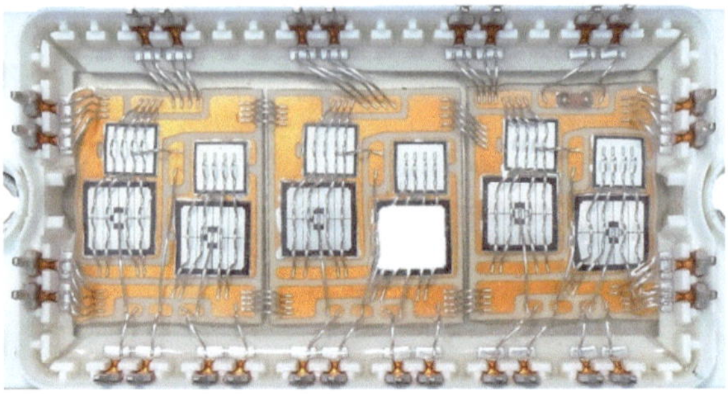

Fig. 8.28 Packaged lead inductance

Through packaging integration techniques, integrating the drive circuit, protection circuit, and sampling circuit into the package resulted in the development of Intelligent Power Module (IPM), significantly reducing the gate loop inductance. IPM commonly exists in two main types: molding and housing. Additionally, while traditional single-device packages typically contain 1–2 power chips, recent advancements by manufacturers have led to the introduction of single-device packages with integrated additional functional circuits for specific applications. These are referred to as Intelligent Power Device (IPD).

8.4.2.1 Molding IPM

The molding IPM is shown in Fig. 8.29. It is a thin black plastic package with an exposed copper substrate on the front for heat dissipation and 10 to 30 pins on both sides. Mitsubishi, Fuji, and Infineon all have such products.

Molding IPM is mainly used in small-capacity variable frequency controller, primarily in household appliances such as air conditioners, washing machines, refrigerators, dishwashers, and exhaust fans. Consequently, these IPMs commonly integrate internal circuits like three-phase full-bridge and PFC circuits. Additionally, they might incorporate other functional circuits as needed to achieve smart and user-friendly functionalities. Taking a three-phase full-bridge IPM as an example, apart from the three-phase full-bridge for successful power conversion, it also includes low-side drive circuits, high-side drive circuits with bootstrap circuits, logic interlock circuits, and various protection functionalities (such as control power undervoltage protection, short-circuit protection, overheat protection, and temperature analog output).

There are now manufacturers introducing SiC MOSFET IPMs. Mitsubishi, for instance, has launched a 600 V/15 A three-phase full-bridge SiC MOSFET IPM, the PSF15S92F6, based on the Super-mini DIPPIM platform. Internally, it integrates low-side drive circuits, high-side drive circuits, and bootstrap circuits, as illustrated in Fig. 8.30.

Fig. 8.29 Molding IPM

(a) Product examples

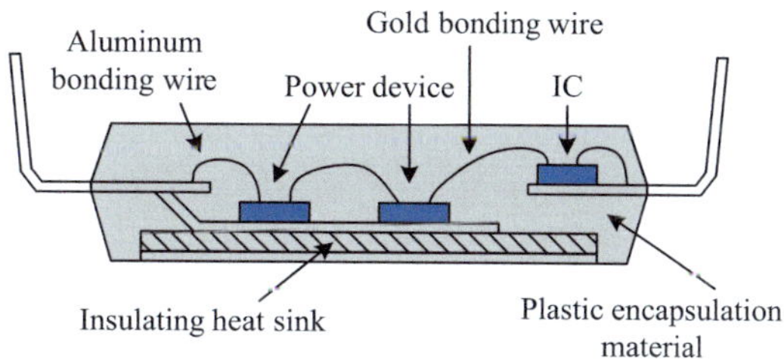

(b) Internal structure

APEX Company has launched the SA110 full-bridge IPM SA110 [6], which is mainly used in DC/AC, AC/DC, motor drive, and other fields. Its internal integrated drive circuit has Active Miller Clamping function.

8.4.2.2 Housing IPM

Unlike molding IPMs, housing IPMs integrate the drive circuit within the housing of the power module. They find primary application in high-power scenarios such as general-purpose inverters, servo controllers, inverters, solar power generation, wind power generation, elevators, and UPS. Both Mitsubishi and Fuji companies offer such products, as depicted in Fig. 8.32. Taking Fuji's IPM as an example, internally, the module comprises two sections: power and drive. The drive circuit is placed on a PCB board, positioned very close to the power devices within the module.

Mitsubishi has launched a 1200 V/75 A three-phase full-bridge SiC MOSFET IPM PMF75CL1A120 based on the IPM L1-series Small-package, which integrates a drive circuit, a short-circuit protection circuit and an over-temperature protection circuit, as shown in Fig. 8.33.

(a) External shape

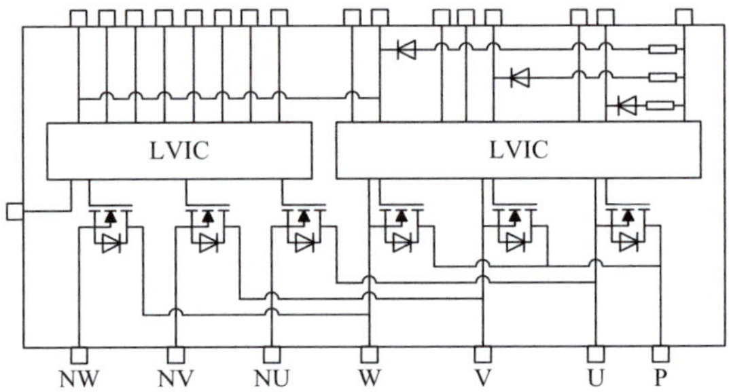

(b) Internal circuit schematic

Fig. 8.30 Mitsubishi PSF15S92F6

(a) External shape

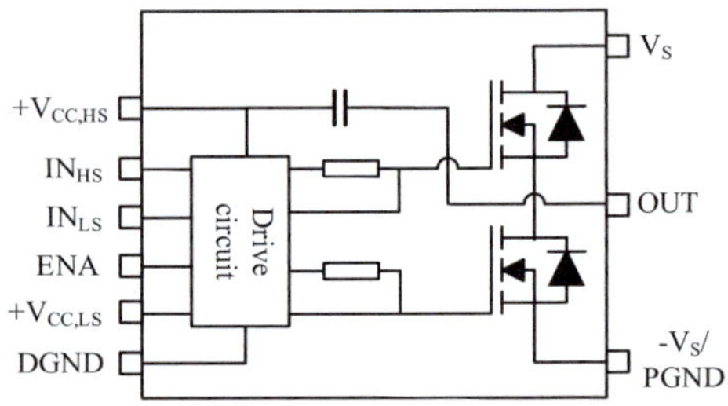

(b) Internal circuit schematic

Fig. 8.31 APEX SA110

8.4.2.3 Intelligent Power Devices

In 2018, Infineon launched the 1300V/20A IGBT single tube product IEWS20R5135IPB for induction cooker applications [9, 10]. It adopts the TO-247-6PIN new package and integrates a driver inside, as shown in Fig. 8.34. It can be seen that in addition to an IGBT and a diode, it also integrates a driver and multiple protection functions, including overvoltage and overcurrent protection, active clamp control circuit, programmable overvoltage threshold, programmable current threshold for

Fig. 8.32 Housing IPM

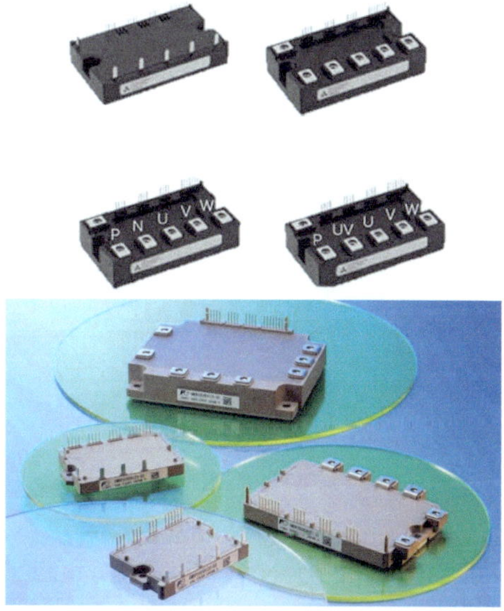

(a) Product examples [7]

(b) FUJI IPM internal structure

(a) External Shape

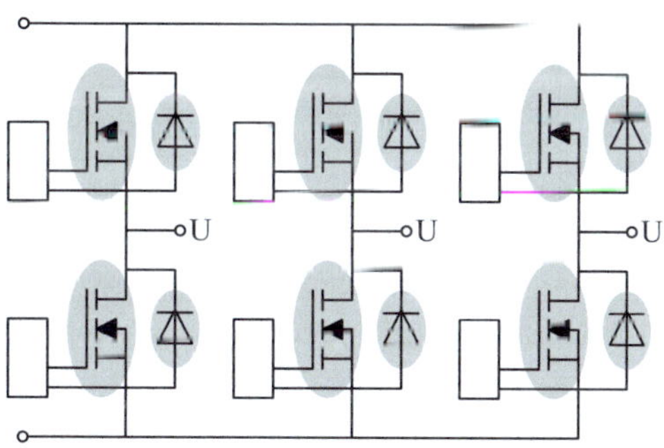

(b) Internal circuit schematic

Fig. 8.33 Mitsubishi PMF75CL1A120

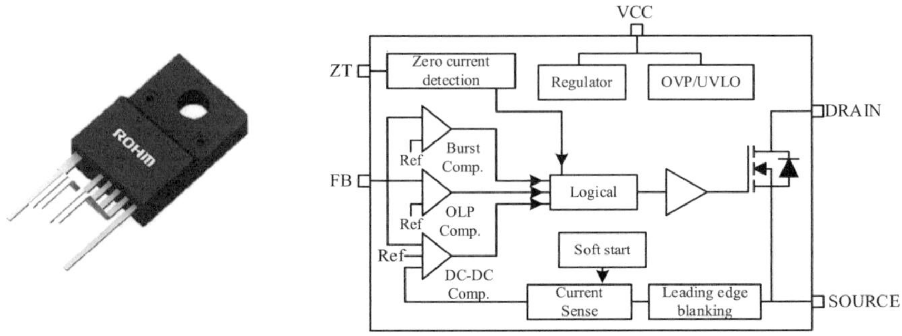

(a) External shape (b) Internal circuit schematic

Fig. 8.34 ROHM BM2SCQ12xT-LBZ

each cycle, temperature warning, overheat protection, VCC UVLO, integrated gate driver, ESD protection and latch immunity on all pins. This enhances overall reliability, reduces replacement/repair costs, lowers circuit board complexity and design investment, simplifies the BOM and lowers the cost of the entire solution.

In 2019, ROHM has launched the AC/DC converter IC BM2SCQ12xT-LBZ [8] with built-in SiC MOSFET, as shown in Fig. 8.34. This product integrates the heatsink and up to 12 components in an integrated package. It has an overwhelming advantage in miniaturization, reduces development cycles and risks, and has built-in protection functions for higher reliability. It is mainly suitable for auxiliary power circuits of various industrial equipment with AC 400 V specifications such as general inverters, AC servos, PLCs, manufacturing equipment, robots, and industrial air conditioners.

Infineon has launched a 1300 V/20 A IGBT single device, the IEWS20R5135IPB [9, 10], designed specifically for induction cooker applications, as illustrated in Fig. 8.35. It employs a new TO-247-6PIN package integrating the driver internally. Besides containing one IGBT and one diode, it integrates a driver and multiple protection features, including overvoltage and overcurrent protection, active clamping control circuits, programmable overvoltage thresholds, programmable current thresholds per cycle, temperature warning, overheat protection, VCC UVLO, integrated gate driver, ESD protection on all pins, and latch-up immunity. These enhancements improve overall reliability, reduce replacement/rework costs, decrease circuit board complexity, minimize design efforts, simplify the BOM, and lower the overall solution cost.

Molding IPMs, housing IPMs, and IPDs all integrate the drive circuit within the package, significantly reducing the drive loop inductance and thereby alleviating crosstalk. It's foreseeable that there will be an increasing number of such SiC-based products in the future.

In 2021, ROHM launched the BM2SC123FP2-LBZ, an AC-DC converter IC incorporating a built-in 1700V-rated SiC MOSFET, as shown in Fig. 8.36. This series utilizes a compact surface-mount package (TO-263), integrating a high-performance SiC MOSFET and control circuitry specifically optimized for industrial equipment auxiliary power supplies. These advantages significantly simplify the development

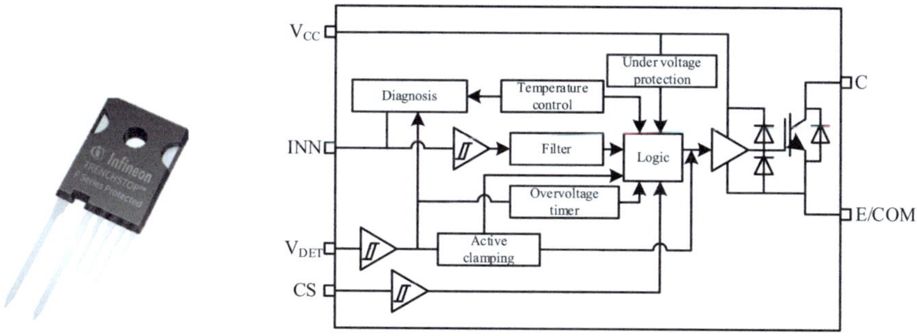

(a) External shape

(b) Internal circuit schematic

Fig. 8.35 Infineon IEWS20R5135IPB

Fig. 8.36 ROHM IC BM2SC123FP2-LBZ

of energy-efficient AC-DC converters. Furthermore, thanks to the surface-mount design, these products can handle outputs up to 48W without requiring a heat sink, thereby helping reduce component count and factory installation costs. The control circuit employs a quasi-resonant topology, which offers lower operating noise and higher efficiency compared to conventional PWM methods, effectively minimizing noise interference with industrial equipment.

References

1. Alan E (2005) Limiting cross-conduction current in synchronous buck converter designs. Application Note, AN-7019, Rev. A., Fairchild Semiconductor Corporation
2. Khanna R, Amrhein A, Stanchina W et al (2013) An analytical model for evaluating the influence of device parasitics on Cdv/dt induced false turn-on in SiC MOSFETs. In: IEEE applied power electronics conference and exposition (APEC), pp 518–525
3. ROHM Co., Ltd. (2015) BM6108FV-LB. Datasheet, Rev.002
4. Mitsubishi Electric Corporation (2019) SiC power devices
5. Wang Y, Watabe K, Sakai S et al (2016) New transfer mold DIPIPM utilizing silicon carbide (SiC) MOSFET. In: PCIM Europe 2016, international exhibition & conference for power electronics, pp 336–341
6. Apex Microtechnology, Inc. SA110 fully integrated half-bridge module. Datasheet, Rev. C
7. Fuji Electric Co., Ltd. (2015) FUJI IGBT V-IPM application manual. In: Application Manuals, REH985b
8. ROHM Co., Ltd. (2019) Quasi-resonant AC/DC converter built-in 1700 V SiC-MOSFET BM2SCQ12xT-LBZ series. Datasheet, Rev. 002
9. Infineon Technologies AG (2018) IEWS20R5135IPB TRENCHSTOP™ feature IGBT protected series. Datasheet, Rev. 2.0
10. Infineon Technologies AG (2018) TRENCHSTOP™ F series protected IGBT: features description and design tips. Application note, AN2018-34, Rev. 1.1

Further Readings

11. Jahdi S, Alatise O, Ortiz Gonzalez JA et al (2016) Temperature and switching rate dependence of crosstalk in Si-IGBT and SiC power modules. IEEE Trans Industr Electron 63(2):849–863
12. ROHM Co.,Ltd. (2024) Quasi-resonant AC/DC converter built-in 1700 V SiC-MOSFET BM2SC12xFP2-LBZ series [Z]. Datasheet, Rev. 002
13. Thomas W (2007) Cdv/dt induced turn-on in synchronous buck regulators. Integrated Rectifier Technologies Inc.
14. Zhao Q, Stojcic G (2007) Characterization of Cdv/dt induced power loss in synchronous buck DC–DC converters. IEEE Trans Power Electron 22(4):1508–1513
15. Miao Z, Mao Y, Wang C et al (2017) Detection of cross-turn-on and selection of off drive voltage for a SiC power module. IEEE Trans Industr Electron 64(11):9064–9071
16. Yuan D, Zhang Y, Wang X, Gao J (2021) A detailed analytical model of SiC MOSFETs for bridge-leg configuration by considering staged critical parameters. IEEE Access 9:24823–24847
17. Guo X et al (2022) Modeling and suppression of the crosstalk issue considering the influence of the parasitic parameters of SiC MOSFETs. IEEE Access 10:114118–114134
18. Zhang Z, Dix J, Wang FF, Blalock BJ, Costinett D, Tolbert LM (2017) Intelligent gate drive for fast switching and crosstalk suppression of SiC devices. IEEE Trans Power Electron 32(12):9319–9332
19. Zhang Z, Wang F, Tolbert LM, Blalock BJ (2014) Active gate driver for crosstalk suppression of SiC devices in a phase-leg configuration. IEEE Trans Power Electron 29(4):1986–1997
20. Zhang B, Wang S (2021) Miller capacitance cancellation to improve SiC MOSFET's performance in a phase-leg configuration. IEEE Trans Power Electron 36(12):14195–14206
21. Hofstoetter N (2015) Limits and hints how to turn off IGBTs with unipolar supply. Application Note AN-1401, Rev 02, SEMIKRON International GmbH
22. Infineon Technologies AG (2005) Driving IGBTs with unipolar gate voltage. Application Note, AN-2006-01

23. ROHM Co., Ltd. (2019) Gate-source voltage surge suppression methods. Application Note, No. 62AN010E, Rev. 01
24. Fairchild Semiconductor Corporation (2012) Active Miller Clamp technology. Application Note, AN-5073, Rev. 1.0.1
25. STMicroelectronics (2019) Mitigation technique of the SiC MOSFET gate voltage glitches with Miller Clamp. Application Note, AN5355, Revision 1
26. Avago Technologies (2010) Active Miller Clamp products with feature: ACPL-331J, ACPL-332J. Application Note 5314
27. STMicroelectronics (2005) TD351 advanced IGBT driver principles of operation and application. Application Note, AN2123, Revision 1

Chapter 9
Impact and Countermeasures of High dv/dt—Common-Mode Current

In addition to interference, high dV_{DS}/dt can also introduce common-mode current, posing another challenge when using SiC MOSFETs. Mentioning common-mode current in power converters generally refers to the common-mode current in the power path, leading to external EMI issues. Due to the higher switching speed of SiC MOSFETs, EMI problems caused by common-mode currents become more severe. Relevant research and mitigation methods are well-documented, and we won't elaborate on them extensively in this book.

When designing power converters based on SiC MOSFETs, issues such as control signal errors and inaccurate sampling measurements may occur, rendering the converter unable to function properly or even causing failure. In such cases, attention needs to be given to common-mode currents. Unlike the common-mode currents mentioned in the power path, the common-mode currents causing the problems above exist in the signal path, and their research and attention are insufficient.

9.1 Common-Mode Current in the Signal Path

9.1.1 Common-Mode Current in Power Converters

On a pair of wires, voltage and current signals with opposite directions constitute a differential-mode signal, and generally, useful signals in circuit operation are differential-mode signals. If the current flowing through all wires in a pair (or multiple pairs) has the same direction, this current is referred to as the common-mode current. Common-mode currents are typically generated due to the displacement current caused by rapidly changing voltages at circuit nodes passing through parasitic capacitances relative to the reference ground.

Shown in Fig. 9.1, a typical boost PFC circuit is depicted, where C_{CM} represents the equivalent distributed capacitance between the switch node and the reference

Y. Gao and Y. Zhang, *Silicon Carbide Power Devices*, CPSS Power Electronics Series,
https://doi.org/10.1007/978-981-96-3480-4_9

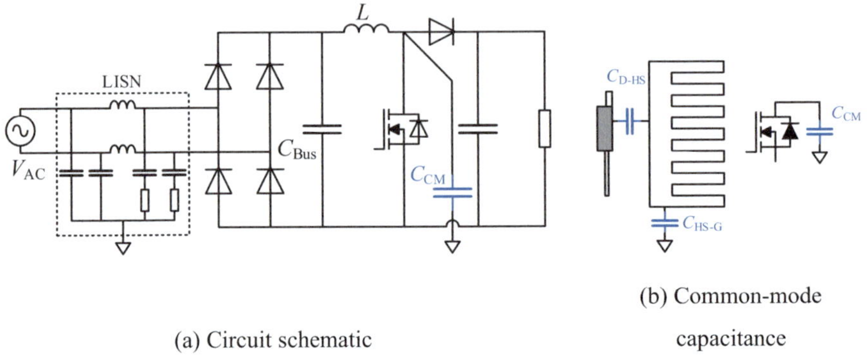

Fig. 9.1 Boost PFC circuit

ground. Assuming the switch transistor is mounted to the heatsink using a thermally insulating pad and a floating metal heatsink, C_{CM} can be considered as the series connection of the distributed capacitances $C_{D\text{-}HS}$ between the switch transistor's drain and the heatsink and $C_{HS\text{-}G}$ between the heatsink and the ground.

In the normal operating mode, the inductor current ripple ΔI_L serves as a source of differential-mode noise, flowing through the inductor L, bus capacitor C_{Bus}, and switching devices. Its circulation path always traverses between a pair of input power lines, as illustrated in Fig. 9.2a. The common-mode noise in this circuit is also caused by the switching action of the transistors, formed by the application of transient voltages on C_{CM}, resulting in displacement currents to the ground. As shown in Fig. 9.2b, the common-mode current is situated between the input power lines and the reference ground, maintaining the same amplitude and direction within each input power line.

Representative of an isolated transformer, the common-mode current path in a flyback circuit is depicted in Fig. 9.3. The flyback circuit has two main common-mode current paths: the first path is formed by the ground displacement current due to the parasitic capacitance $C_{D\text{-}G}$ between the transformer's first-side switch node and the heatsink, indicated by dashed lines with arrows; the second path is formed by the ground displacement current due to the parasitic capacitances $C_{P\text{-}S}$ between the first-side and first, secondary windings of the transformer and the distribution capacitance $C_{L\text{-}G}$ between the load output and ground, represented by Dashed lines with arrows.

With the widespread application of PWM frequency converters in power electronic drives and industrial automation, the issue of common-mode noise caused by high-speed dv/dt and parasitic capacitance has become increasingly prominent. As shown in Fig. 9.4, the switching devices or power modules of the inverter are typically mounted on a heatsink directly connected to the grounded enclosure or frame. This introduces distributed capacitance $C_{D\text{-}G}$ to ground for each switching transient electrical node. The motor serves as the load, and the parasitic capacitance $C_{M\text{-}G}$ between the stator winding of the motor and the grounded motor housing and the

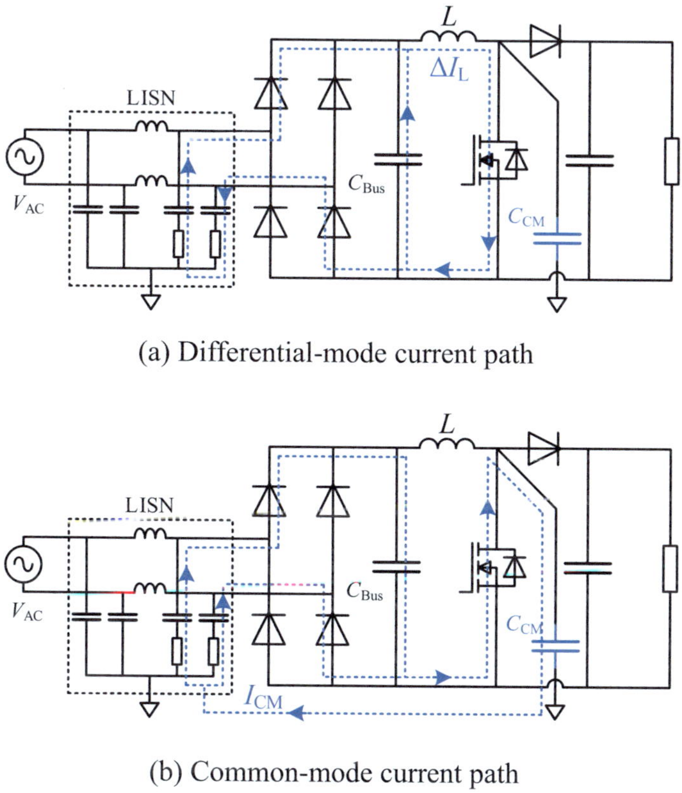

Fig. 9.2 Common-mode and differential mode paths in boost PFC circuit

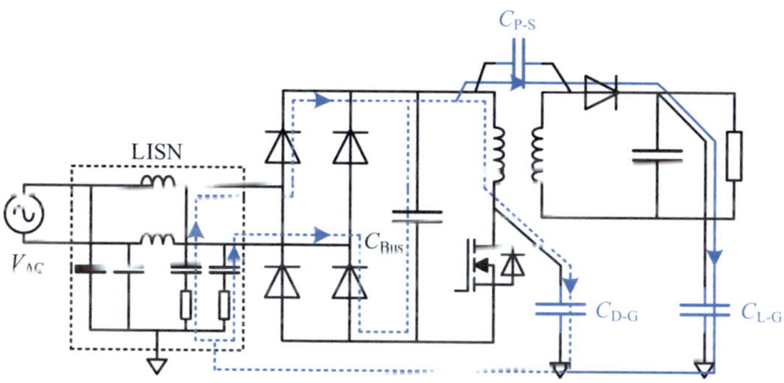

Fig. 9.3 Common-mode current in flyback circuit

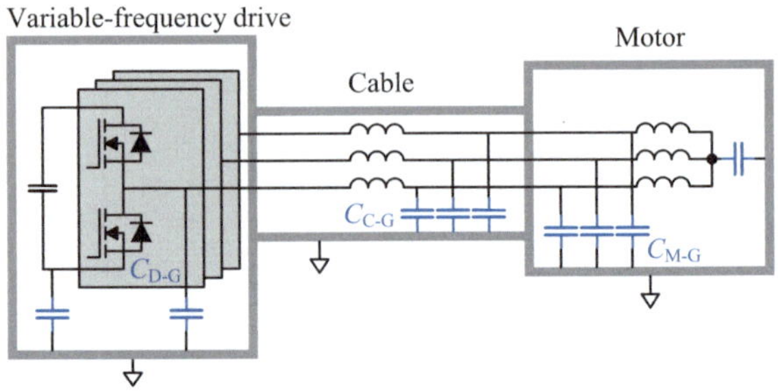

Fig. 9.4 Common-mode current in the PWM frequency converter

rotor bearings, are considerable. Additionally, the cables connecting the frequency converter's inverter bridge arms to the motor windings may introduce significant ground distributed capacitance $C_{C\text{-}G}$ due to their specific cable routing or shielded grounding structure.

The common-mode currents analyzed in the above converters all flow through the power circuit, primarily affecting the external conducted EMC characteristics of the converter. In the converter, there are also common-mode currents flowing through the signal path [1]. Taking the half-bridge circuit as an example, as shown in Fig. 9.5, $C_{D\text{-}G}$ is the midpoint-to-ground distributed capacitance of the bridge arm, $C_{I\text{-}O}$ is the isolation capacitance of the upper bridge arm drive circuit, $C_{C\text{-}G}$ is the ground capacitance of the control circuit; $C_{I\text{-}O(sig)}$ and $C_{I\text{-}O(pwr)}$ are the isolation capacitances of the drive circuit for the isolated drive chip and the isolated power supply, and the voltage V_{CM} across the midpoint of the bridge arm causes changes in the voltage across the isolation capacitance, generating common-mode currents $I_{CM(sig\_H))}$ and $I_{CM(pwr\_H)}$. When the impedance on the common-mode path is asymmetric, common-mode currents will be transformed into differential-mode voltages, leading to differential-mode interference that could result in consequences such as misoperation of the drive chip and logic errors in the control circuit.

9.1.2 *Characteristics of Common-Mode Currents in Signal Paths*

Refining the half-bridge circuit shown in Fig. 9.5 and considering the major parasitic parameters results in an equivalent common-mode circuit as illustrated in Fig. 9.6. The meanings and reasonable values of each parasitic parameter are provided in Table 9.1.

The common-mode currents in each branch of Fig. 9.6 are indicated by arrows. $I_{CM(pwr\text{-}H)}$ and $I_{CM(sig\text{-}H)}$ represent the common-mode currents in the isolated power supply branch and the isolated driver chip branch of the upper bridge arm, respectively. The common-mode current in the upper bridge arm driver circuit $I_{CM(H)}$ =

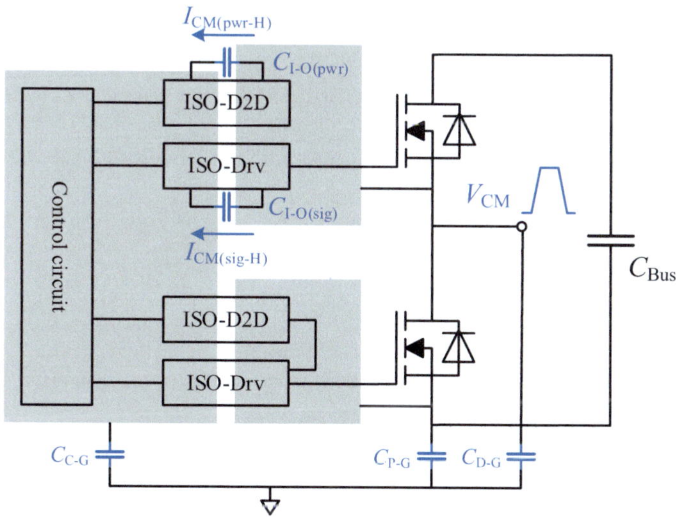

Fig. 9.5 Common-mode current in signal path of half-bridge circuit

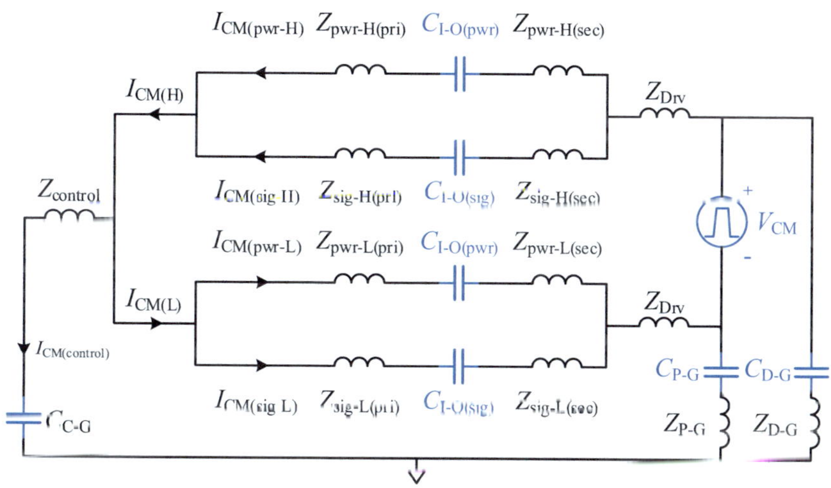

Fig. 9.6 Equivalent common-mode circuit model of half-bridge circuit

$I_{CM(pwr\text{-}H)} + I_{CM(sig\text{-}H)}$. Similarly, $I_{CM(pwr\text{-}L)}$ and $I_{CM(sig\text{-}L)}$ represent the common-mode currents in the isolated power supply branch and the isolated driver chip branch of the lower bridge arm, respectively. The common-mode current in the lower bridge arm driver circuit $I_{CM(L)} = I_{CM(pwr\text{-}L)} + I_{CM(sig\text{-}L)}$. The common-mode current in the control circuit $I_{CM(control)} = I_{CM(H)} + I_{CM(L)}$.

Table 9.1 Parasitic parameters and values

Parasitic parameter	Value	Meaning
$C_{\text{I-O(pwr)}}$	10 pF	Capacitance of isolation capacitor in power circuit of driver circuit
$C_{\text{I-O(sig)}}$	2 pF	Capacitance of isolation capacitor in power circuit of driver circuit
$C_{\text{C-G}}$	100 pF	Parasitic capacitance between control circuit and ground
$C_{\text{D-G}}$	200 pF	Parasitic capacitance between bridge arm midpoint and ground
$Z_{\text{D-G}}$	10 nH + 50 mΩ	Impedance of heatsink and ground
$C_{\text{P-G}}$	500 pF	Parasitic capacitance between main power bus and ground
$Z_{\text{P-G}}$	10 nH + 50 mΩ	Ground impedance
$Z_{\text{pwr-H(pri)}}$, $Z_{\text{pwr-L(pri)}}$	20 nH + 20 mΩ	Impedance of driver circuit primary power line
$Z_{\text{sig-H(pr)}}$, $Z_{\text{sig-L(pri)}}$	20 nH + 20 mΩ	Impedance of driver circuit primary signal line
$Z_{\text{pwr-H(sec)}}$, $Z_{\text{pwr-L(sec)}}$	10 nH + 10 mΩ	Impedance of driver circuit secondary power line
$Z_{\text{sig-H(sec)}}$, $Z_{\text{sig-L(sec)}}$	10 nH + 10 mΩ	Impedance of driver circuit secondary signal line
Z_{Drv}	20 nH + 20 mΩ	Gate drive circuit loop impedance
Z_{control}	10 nH + 50 mΩ	Control circuit line impedance

Utilizing the common-mode equivalent circuit for AC analysis, the admittance representation obtained by the ratio of each branch's common-mode current to V_{CM} indicates that higher admittance results in larger common-mode currents [2]. Assuming a minimum switch time of 20 ns for SiC MOSFETs, the effective bandwidth of the interference source V_{CM} is up to 17.5 MHz. Therefore, the focus is primarily on common-mode currents at this frequency and below.

As shown in Fig. 9.7a, $I_{\text{CM(H)}}$ is significantly greater than $I_{\text{CM(L)}}$, indicating that the majority of $I_{\text{CM(H)}}$ flows into the control circuit, with only a small portion returning to the power circuit through the lower bridge arm driver circuit. As shown in Fig. 9.7b, $I_{\text{CM(pwr-H)}}$ is much larger than $I_{\text{CM(sig-H)}}$, indicating that common-mode currents are mainly generated by $C_{\text{I-O(pwr)}}$, primarily due to the fact that $C_{\text{I-O(pwr)}}$ is greater than $C_{\text{I-O(sig)}}$.

Taking $C_{\text{I-O(pwr)}} = 10$ pF and varying $C_{\text{I-O(sig)}}$ within the range of 0.5–5 pF, with other parameters unchanged, and taking $CC_{\text{I-O(sig)}} = 2$ pF and varying $C_{\text{I-O(pwr)}}$ within the range of 5–15 pF, with other parameters unchanged. The amplitude-frequency characteristics of $I_{\text{CM(sig-H)}}/V_{\text{CM}}$ and $I_{\text{CM(pwr-H)}}/V_{\text{CM}}$ are shown in Fig. 9.8. It can be observed that as $C_{\text{I-O(sig)}}$ and $C_{\text{I-O(pwr)}}$ decrease, $I_{\text{CM(sig-H)}}$ and $I_{\text{CM(pwr-H)}}$ also decrease, with no mutual interference.

Fig. 9.7 Admittance of each common-mode branch

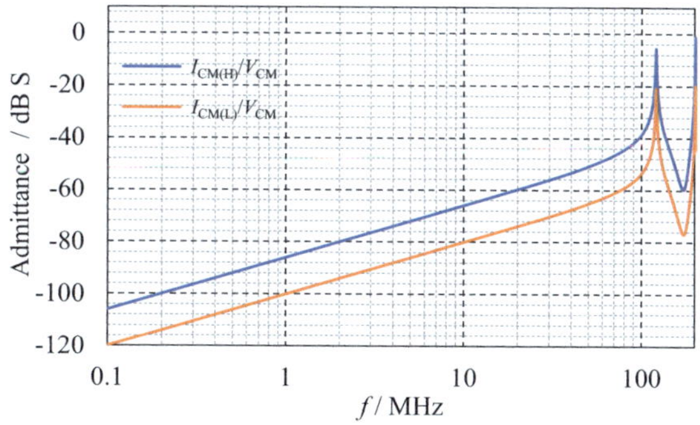

(a) Upper arm branch and lower arm branch

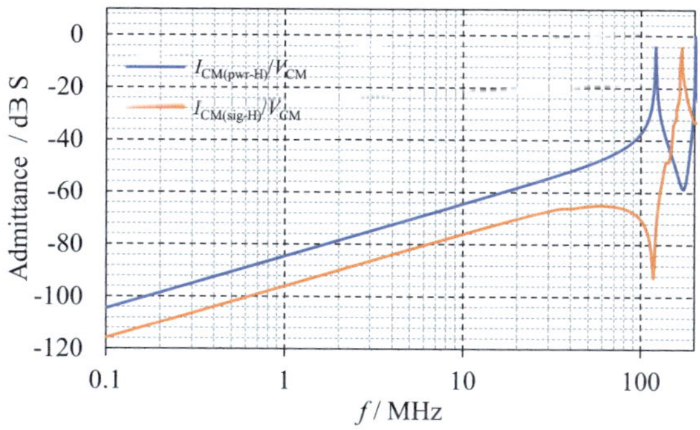

(b) Isolated power supply branch and isolated driver chip branch of the upper bridge arm

9.2 Countermeasure 1—High CMTI Driver Chips

Given that common-mode interference can lead to misoperation in the driver circuit, enhancing the anti-common-mode interference capability of the driver circuit becomes a necessary choice. The ability of a signal isolation transmission circuit to withstand the maximum common-mode voltage transient rate on both sides of the isolation is defined as common-mode transient immunity (CMTI). CMTI is a crucial indicator measuring the anti-common-mode interference capability of a signal isolation transmission circuit, expressed in units of V/ns or kV/us. A higher value indicates stronger resistance to common-mode interference [3].

Figure 9.9 illustrates the testing principle for the CMTI of an isolation driver chip. The input of the isolation driver chip is set to a high or low level, and a voltage pulse with a positive or negative transition is applied between the original and secondary side grounds to cover all possible operating conditions. During testing, voltage pulse signals are measured to determine their amplitude and transition rate. The output of

Fig. 9.8 Impact of isolation capacitors on common-mode currents

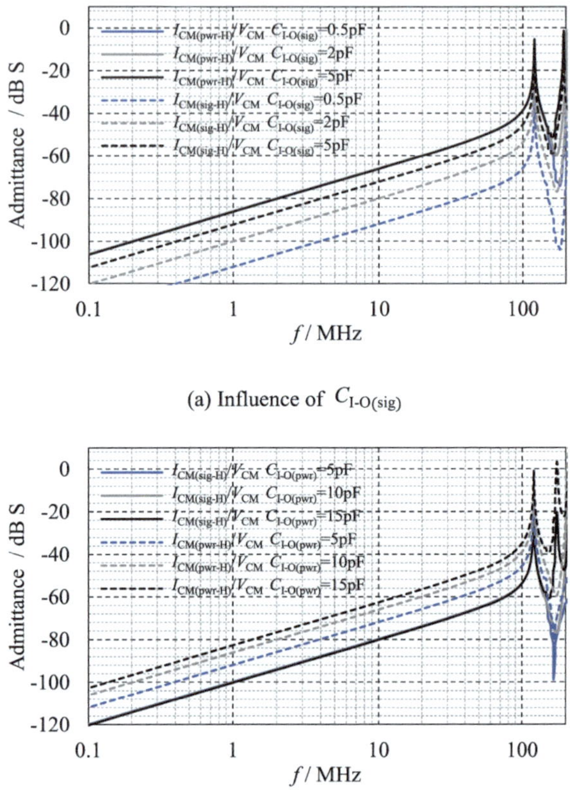

(a) Influence of $C_{\text{I-O(sig)}}$

(b) Influence of $C_{\text{I-O(pwr)}}$

the driver chip is also measured to determine its susceptibility to interference. Non-invasive probes with short grounding wires are generally used to improve testing accuracy. The test is considered successful if the output state of the driver chip remains unchanged under the specified voltage pulse amplitude and transition rate; otherwise, it is considered unsuccessful.

Figure 9.10 provides circuit connection examples for CMTI testing of five common driver chips. To obtain accurate test results, the circuit connections must adhere to the following basic requirements:

(1) Provide power separately to the primary and secondary sides of the chip, with decoupling capacitors configured on the power supply pins;
(2) If the driver output is isolated, short the two output pins for high and low levels;
(3) 若具有Miller Clamping功能，则将对应引脚接驱动输出；If it has Miller Clamping functionality, connect the corresponding pin to the driver output;
(4) If it has DESAT functionality, connect the DESAT pin to the secondary ground;

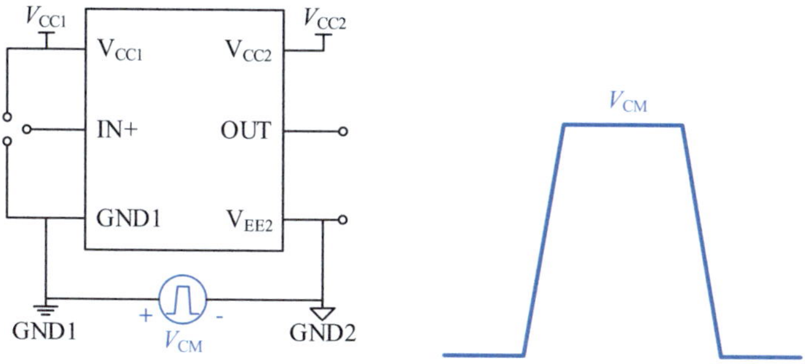

Fig. 9.9 CMTI test principle of isolated driver chip

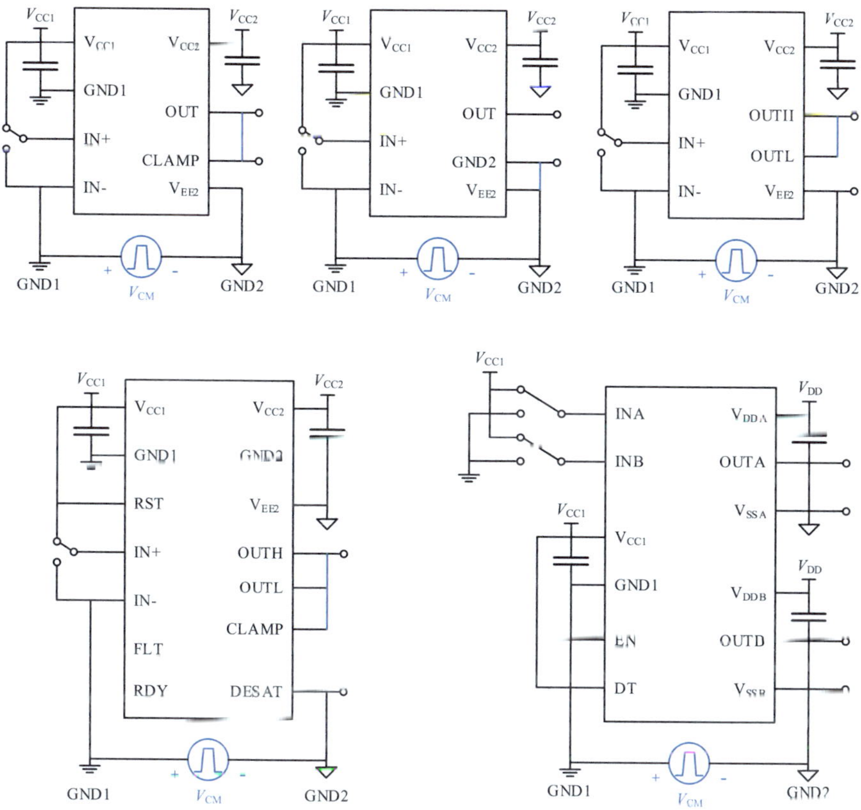

Fig. 9.10 CMTI test circuit connection of isolated driver chip

(5) If the secondary is bipolar powered, connect the negative power supply pin to ground or connect it to a negative power source for closer approximation to the application;
(6) For dual-channel driver chips, short the grounds of the two channels on the secondary side.

During testing, a high-frequency noise simulator can be used to provide voltage pulses for testing. It operates with simplicity, high reliability, and increased safety. Additionally, positive pulses can be generated using a boost circuit, while negative pulses can be generated using a buck-boost circuit [4], as shown in Fig. 9.11.

Taking the example of providing positive pulses with a boost circuit, the testing circuit has a low-side voltage $V_{CMTI(low)}$ and an output voltage $V_{CMTI(high)}$. The voltage transition point connects to the primary side ground GND1, and the high-side output ground connects to the secondary side ground GND2. By turning on switch S for a period t_{on}, a current I_{CMTI} is established in the inductor L.

$$I_{CMTI} = t_{on} \cdot \frac{V_{CMTI(low)}}{L}. \tag{9.1}$$

When switch S is turned off, a pulse voltage V_{CM} ranging from 0 V to $V_{CMTI(high)}$ is generated between GND1 and GND2. The transition rate of V_{CM} depends on the switch S cutoff speed and is influenced by I_{CMTI}, C_{oss}, and C_D. With S and D selected and t_{on} kept constant, adjusting $V_{CMTI(low)}$ controls the transition rate of V_{CM}.

Fig. 9.11 CMIT test pulse generation circuit

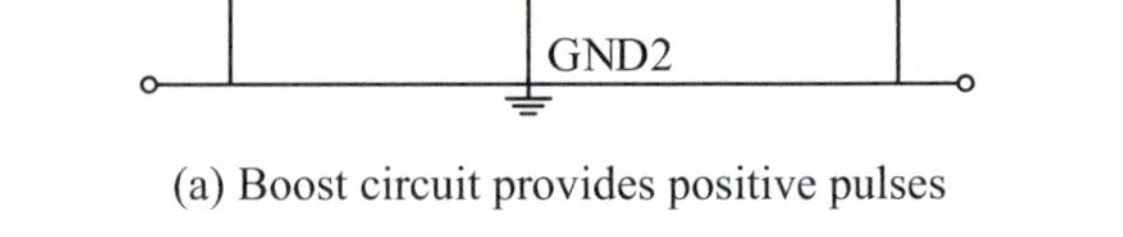

(a) Boost circuit provides positive pulses

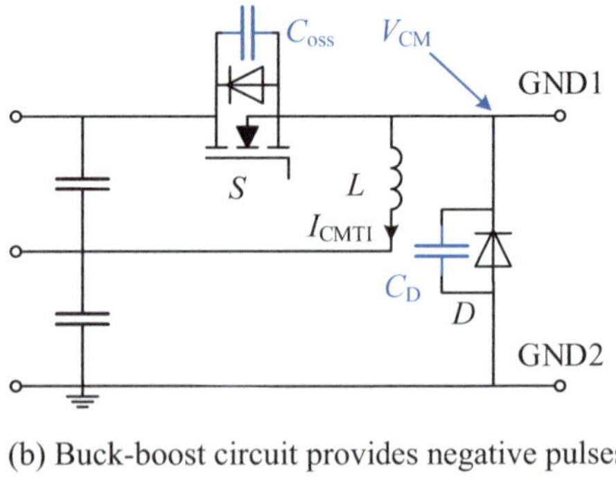

(b) Buck-boost circuit provides negative pulses

$$\frac{dV_{CM}}{dt} = \frac{I_{CMTI}}{C_{oss} + C_D} = \frac{V_{CMTI(low)} \cdot t_{on}}{L(C_{oss} + C_D)}. \quad (9.2)$$

Isolation driver chip CMTI can be obtained from the datasheet, usually provided with the CMTI value and corresponding test conditions. For example, CMTI = 100 V/ns@V_{CM} = 1000 V means that the chip manufacturer guarantees that as long as the dv/dt of V_{CM} is less than 100 V/ns when V_{CM} is 1000 V, no errors will occur. It is important to note that the testing condition V_{CM} is crucial when characterizing the anti-common-mode interference capability of isolation driver chips using CMTI. This is because the essence of interference is energy, and under the same dv/dt, the larger V_{CM}, the greater the energy of the common-mode interference. If two chips have a CMTI of 100 V/ns but have testing conditions V_{CM} of 100 and 1000 V, respectively, the latter's anti-common-mode interference capability is much better than the former. It's worth noting that CMTI is also affected by temperature and decreases with increasing temperature.

Manufacturers continually enhance the CMTI of isolation driver chips through technological innovations, improving from the initial levels of no more than 10 V/ns to the widespread achievement of 100 V/ns, and some even reaching as high as 200 V/ns. This progress has largely met the requirements posed by the high dv/dt in current-stage devices. Specific methods to enhance the CMTI of isolation driver chips include:

1. **Common-mode filtering circuit**

Integration of internal common-mode filtering circuits within the driver chip suppresses common-mode interference, boosting CMTI.

2. **Differential signal transmission**

Most driver chips employ differential modulation, where the detection of the difference between two signals eliminates the common-mode component. This method proves effective in countering common-mode interference.

Additionally, as mentioned earlier, errors are not directly caused by common-mode current but by the conversion of common-mode to differential mode due to impedance asymmetry. Therefore, even with differential transmission, careful attention must be paid to the impedance symmetry of the two differential signal paths. The more symmetrical the impedance, the stronger the resistance to common-mode interference, resulting in higher CMTI.

3. **Reducing isolation capacitance**

Common-mode current is the displacement current generated by the voltage jump at the ends of the isolation capacitor $C_{I\text{-}O}$, and its magnitude is determined by dv/dt and $C_{I\text{-}O}$. Hence, under the same dv/dt, reducing $C_{I\text{-}O}$ decreases the common-mode current, effectively lowering the interference source's size and, consequently, enhancing the CMTI of the isolation driver chip.

Currently, various types of isolation driver chips have managed to control isolation capacitance to very small values, typically around 1 pF. Optocoupler-type driver

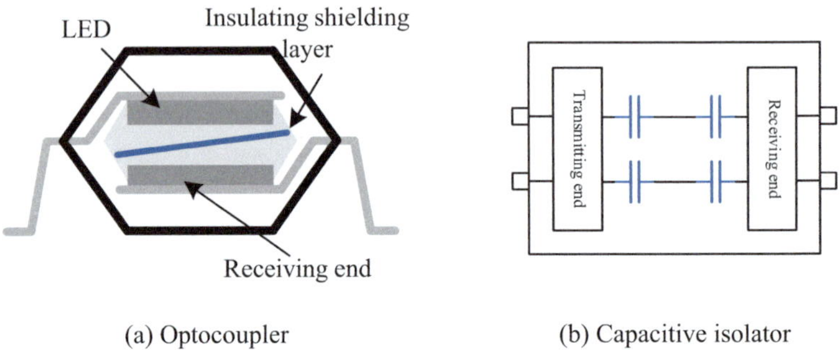

Fig. 9.12 Reducing isolation capacitor

chips use a transparent insulating layer to decrease coupling between the primary and secondary sides, reducing $C_{I\text{-}O}$. Capacitive isolation-type driver chips, aiming to enhance insulation capabilities, adopt a two-stage capacitor series scheme, effectively reducing the overall $C_{I\text{-}O}$, as shown in Fig. 9.12.

CMTI reflects the isolation driver chip's intrinsic capability to resist common-mode voltage transitions, and selecting a high CMTI ensures that it won't experience errors due to high dv/dt in SiC applications. However, this alone doesn't address the impact of common-mode current on the control circuit, requiring the utilization of the methods introduced next to enhance the overall system's resistance to common-mode interference.

9.3 Countermeasure 2—High Common-Mode Impedance

9.3.1 Reducing Isolation Capacitance

As discussed earlier, the smaller $C_{I\text{-}O}$, the less common-mode current is generated, resulting in a smaller impact on the control circuit. The isolation capacitance $C_{I\text{-}O}$ of high CMTI isolation driver chips has already been reduced to as low as 1 pF, and further reduction potential is limited. However, the isolation power supply, responsible for handling significantly more power than the isolation driver chip, has a larger $C_{I\text{-}O}$, playing a dominant role in contributing to common-mode current.

Several power module manufacturers now offer isolation power modules with small isolation capacitance specifically designed for SiC MOSFETs. In Fig. 9.13, products from RECOM, MURATA, and MORNSUN are shown, and it can be observed that the isolation capacitance of 1–2 W module power supplies is 3–5 pF.

Adopting a similar principle of two-stage capacitor series connection seen in capacitive isolation-type driver chips, using a two-stage magnetic core connection in the isolation power transformer significantly reduces isolation capacitance. The driver board of GE's SiC power block [8] employs this approach, as illustrated in

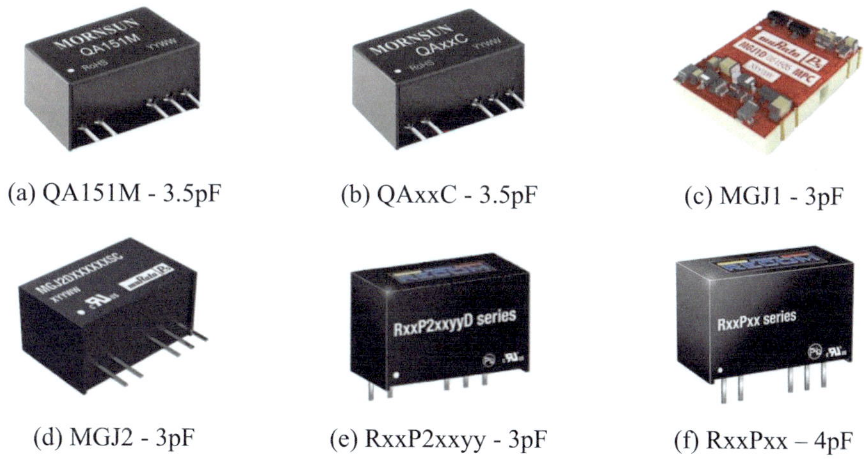

(a) QA151M - 3.5pF (b) QAxxC - 3.5pF (c) MGJ1 - 3pF

(d) MGJ2 - 3pF (e) RxxP2xxyy - 3pF (f) RxxPxx – 4pF

Fig. 9.13 Isolation module power supply with small isolation capacitor [5–7]

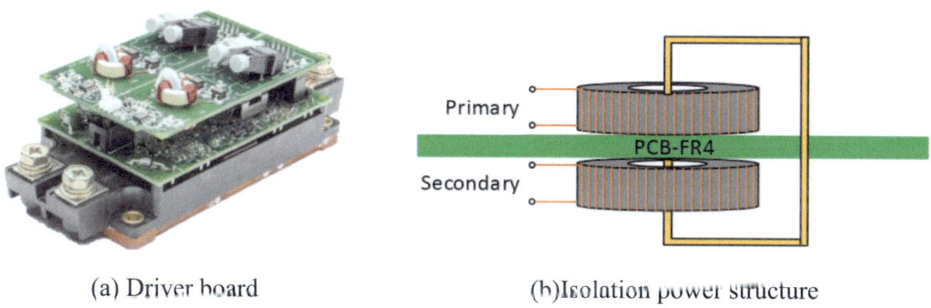

(a) Driver board (b)Isolation power structure

Fig. 9.14 GE SiC power block drive isolation power supply

Fig. 9.14. The transformer consists of two magnetic cores placed on either side of the PCB, dedicated to the primary and secondary windings, coupled by a single-turn coil for primary-secondary coupling. With this structure, the $C_{\text{I-O}}$ is effectively reduced to 0.98 pF.

Beyond the $C_{\text{I-O}}$ of the driver circuit contributing to common-mode current, using a current transformer to measure the bridge midpoint output current also generates common-mode current due to $C_{\text{I-O}}$. This directly enters the sampling analog circuit, impacting the control and stability of the converter. Similarly, employing a two-stage magnetic core connection helps reduce $C_{\text{I-O}}$, mitigating the impact of common-mode current, as depicted in Fig. 9.15.

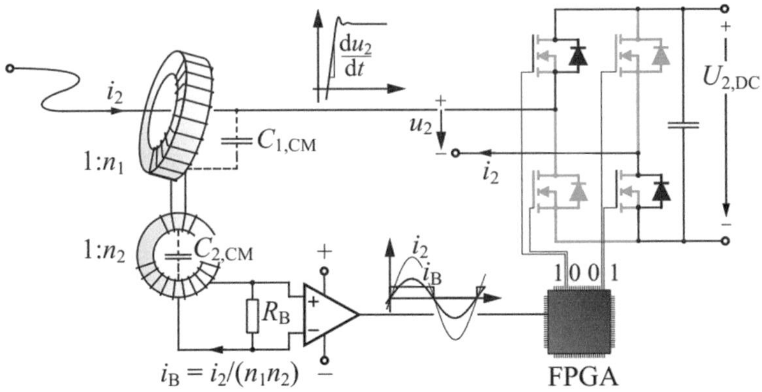

Fig. 9.15 Two-stage current transformer [9]

9.3.2 Common-Mode Inductor

The structure of a common-mode inductor is straightforward, consisting of two sets of coils symmetrically wound on the same magnetic core. When winding the coils, it is crucial to ensure that the dimensions and turns are identical, and the terminals of the same name are the same, as shown in Fig. 9.16. When common-mode current flows through the common-mode inductor, the magnetic flux of the two coils adds up due to the same current direction, resulting in a significant inductance, effectively suppressing common-mode current. When differential-mode current flows through the common-mode inductor, the opposite current directions of the two coils cancel out each other's magnetic flux, leading to minimal inductance, and the differential-mode current remains unaffected.

Due to its "block common-mode, pass differential-mode" characteristics, common-mode inductors find widespread applications in various electronic devices and products, enhancing anti-interference capabilities and system reliability. Common-mode inductors can be designed and fabricated independently or selected directly from standard products offered by component manufacturers such as TDK, Murata, KEMET, and Schaffner. Common-mode inductors for high-speed signal filtering and low-power low-voltage filtering are compact and typically come in chip form. In contrast, common-mode inductors for power converter power filtering, handling high power and voltage, have a larger form factor and come in plug-in form, as shown in Fig. 9.17.

Incorporating common-mode inductors at the input of isolated power supplies, the power supply for isolated driver chips, and digital signals helps suppress common-mode currents in the driver circuit, as illustrated in Fig. 9.18. It's essential to note that after using common-mode inductors, the ground at both ends of the common-mode inductor is separated. Additionally, common-mode inductors can be added to other critical circuits, such as the power supply for controllers and operational amplifiers, to further enhance the suppression of common-mode currents.

(a) Common mode inductor

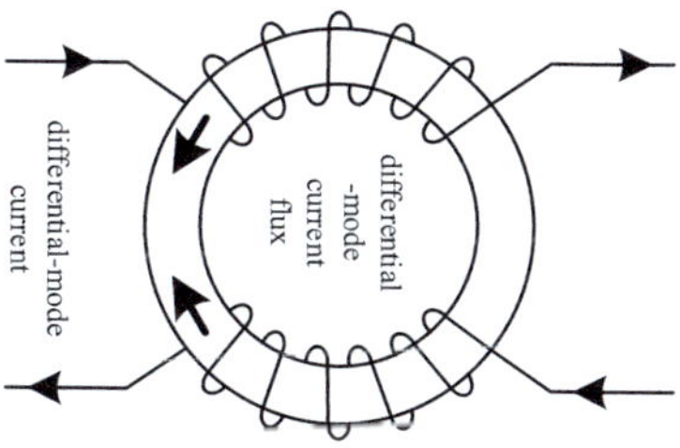

(b) Differential mode current path

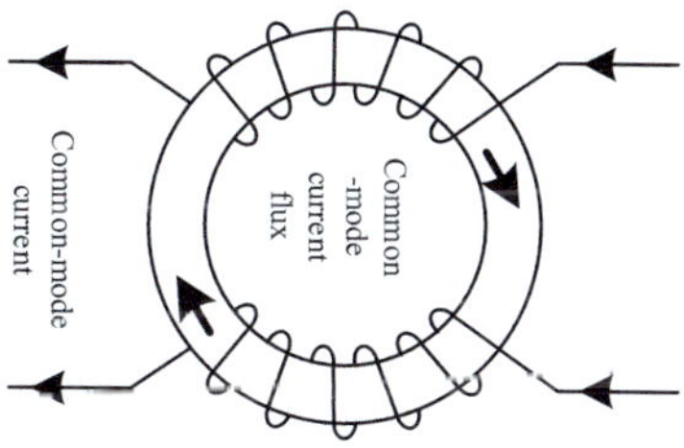

(c) Common mode current path

Fig. 9.16 Structure of common-mode inductor

Fig. 9.17 Common-mode inductor [10]

Fig. 9.18 Using common-mode inductor to suppress common-mode current in drive circuit

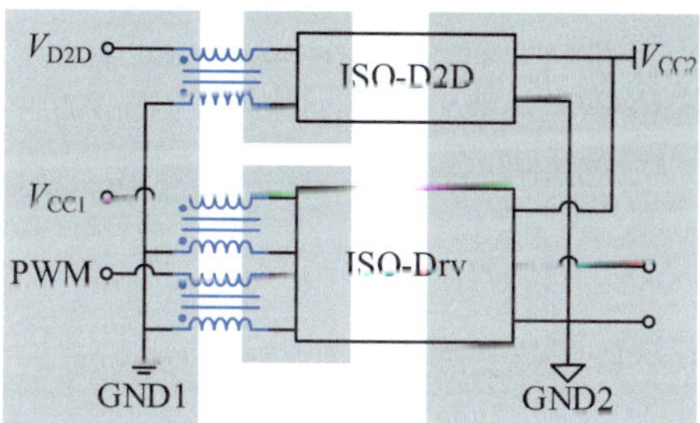

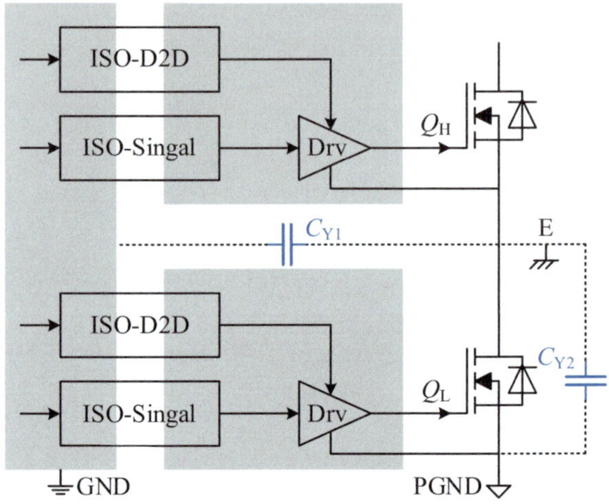

Fig. 9.19 Constructing low-impedance path with Y capacitors

9.4 Countermeasure 3—Common-Mode Current Diversion

9.4.1 Y Capacitor

Utilizing Y capacitors for the diversion of common-mode current is a common technique in handling common-mode currents in converters. The specific approach involves connecting Y capacitor C_{Y1} between the signal ground and the chassis and connecting Y capacitor C_{Y2} between the power ground and the chassis [11], as shown in Fig. 9.19. This configuration creates a low-impedance path with two Y capacitors, which diverts a portion of the common-mode current directly back to the power loop, thereby reducing the common-mode current flowing into the control circuit.

When using a Bootstrap circuit to supply power to the upper bridge arm drive circuit, the diode D_B has a junction capacitance C_{D_B}, and the transition of the bridge midpoint generates common-mode current. To address this, a capacitor C_Y can be connected between the anode of D_B and the power ground to divert the common-mode current to the power loop, reducing the common-mode current flowing into the control circuit, as illustrated in Fig. 9.20.

9.4.2 Parallel Power Supply

In general, the control circuit, sampling circuit, and drive circuit of the converter are powered by the same auxiliary power supply. According to the connection methods of power supply for various functional circuits, there are three ways: series connection, parallel connection, and mixed connection, as shown in Fig. 9.21.

Fig. 9.20 Bootstrap circuit using capacitor to drain common-mode current

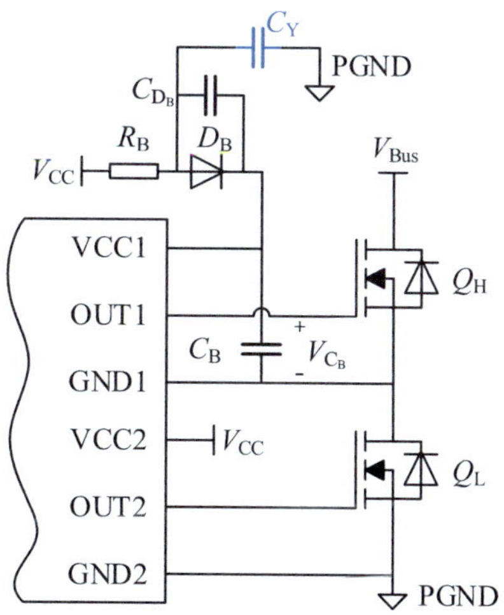

Fig. 9.21 Power supply connection methods

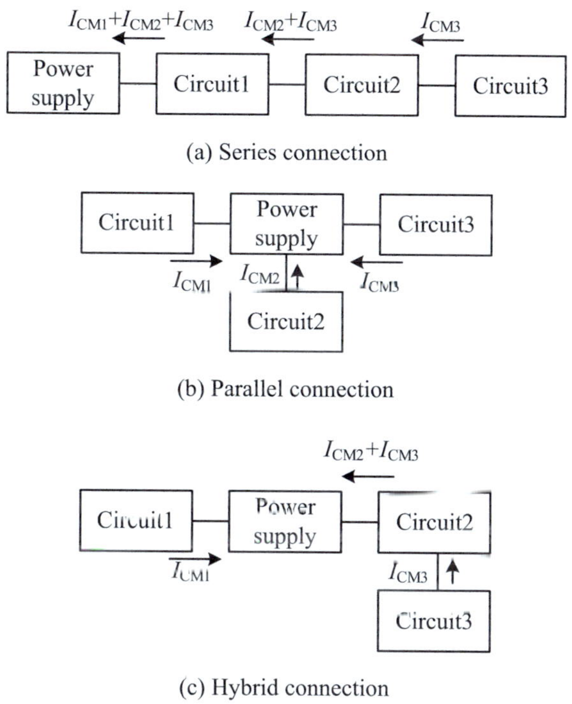

1. **Series Power Supply**

When the auxiliary power supply has a single output, power is sequentially supplied to each section of the circuit. In this arrangement, common-mode current continuously converges from a distance to the closest circuit to the auxiliary power supply, affecting the circuit closest to the auxiliary power supply the most.

2. **Parallel Power Supply**

When the auxiliary power supply has a single output, the ground of each section of the circuit is only connected at the output point of the auxiliary power supply. When the auxiliary power supply has isolated multiple outputs, the grounds of each section of the circuit are completely separated and not directly connected. This approach avoids the situation where common-mode current converges during sequential connection.

3. **Hybrid connection**

Some circuits are connected in series, presenting an overall parallel connection.

Therefore, using parallel or mixed connections can guide the common-mode current generated on the drive circuit to bypass, avoiding it all flowing through the control circuit, which is advantageous for dealing with common-mode currents.

9.4.3 Series-Connected Drive Circuit

To address common-mode interference, researchers have proposed a series-connected drive circuit structure, different from the traditional parallel structure [12], as shown in Fig. 9.22. The power supply and drive signals are split into two paths after the first isolation. One path is used to drive the lower bridge arm, while the other, after isolation or level lifting, is used to drive the upper bridge arm. With this structure, most of the common-mode current generated by the upper bridge arm drive circuit directly flows back to the power circuit, significantly reducing the common-mode current flowing into the control circuit.

The approach introduced in the previous section, using high common-mode impedance, is a "blocking" method. In contrast, the approach discussed in this section directs the common-mode current, reducing its flow to key circuits that are susceptible to interference, using a "diverting" method. When addressing common-mode currents, it is often a combination of "blocking" and "diverting".

Additionally, it is important to note that in topologies with multiple devices in series or multiple modules in series, the switching of lower-level devices can cause potential jumps in all higher-level devices. Common-mode currents have a cumulative effect and require careful handling.

Fig. 9.22 Series-connected drive circuit architecture

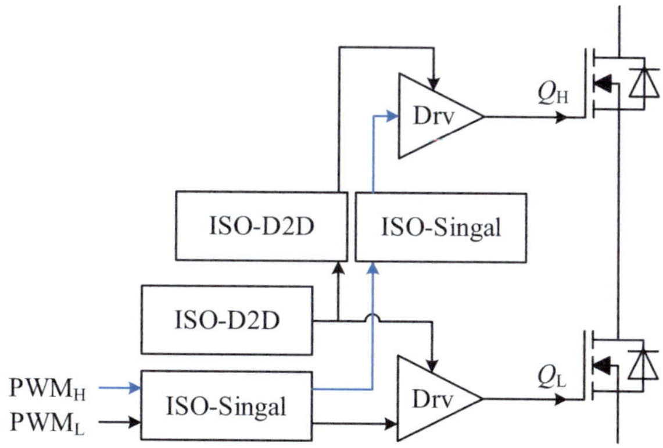

9.5 Differential-Mode Interference Measurement

In most cases, measuring common-mode currents on the signal path is not easy, and sometimes impossible. Consequently, it is not feasible to directly evaluate the control of common-mode currents by measuring their magnitude. On the other hand, the primary harm caused by common-mode currents is the conversion into differential-mode interference, leading to sampling errors, logic level errors, and even the malfunction or damage of the converter. Since common-mode currents are the source and differential-mode interference directly affects the operation of the converter, measuring the differential-mode interference on critical voltage signals during the converter's design is essential. However, accurately measuring interference signals is challenging, and this will be detailed in the following based on practical testing.

9.5.1 *Conventional Voltage Probe Measurement of Differential-Mode Interference*

To investigate the measurement of differential-mode interference, we will conduct tests on a Totem-Pole PFC converter based on SiC MOSFETs. We will measure the controller's power supply $V_{+3.3V}$ and the output voltage sampling conditioning output signal V_{sample}.

Three measurement methods will be employed: a 100:1 high-voltage differential probe, a 10:1 passive probe with a twisted short-circuit ground wire, and a 1:1 passive probe with a twisted short-circuit ground wire. Their bandwidths are 100, 200, and 6 MHz, respectively. The measurement results are shown in Figs. 9.23 and 9.24.

Both $V_{+3.3V}$ and V_{sample} exhibit significant spikes. Analyzing the time intervals of their occurrences suggests that they are caused by the switching actions of SiC MOSFETs. The magnitudes of the spikes differ among the probes, following the

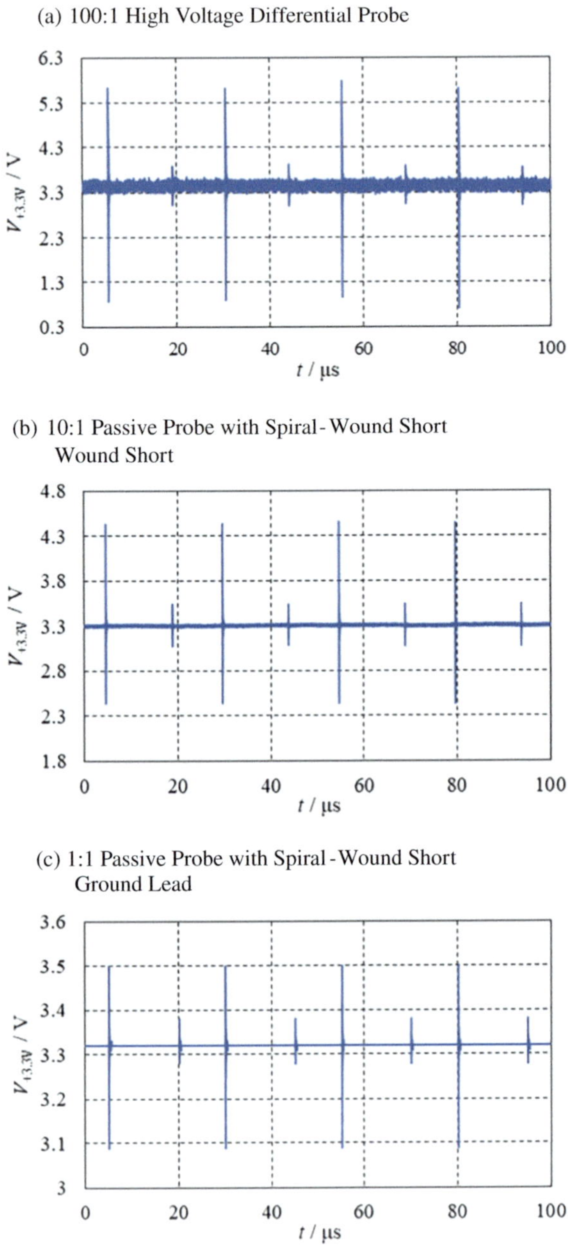

Fig. 9.23 Controller power supply $V_{+3.3\text{V}}$ waveform

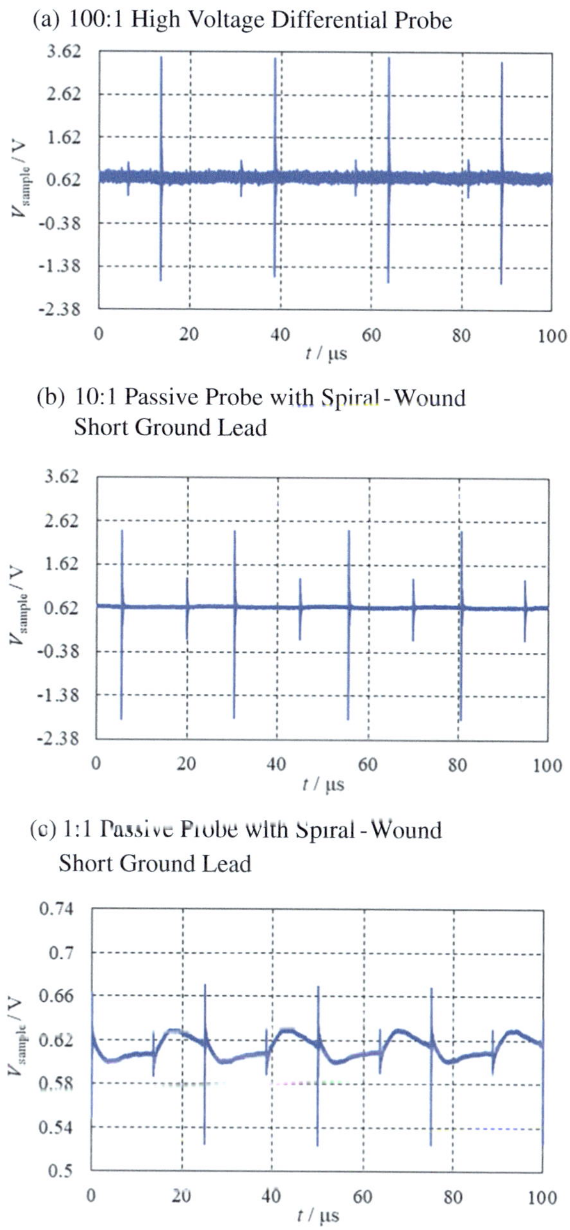

Fig. 9.24 Output voltage sampling conditioning output signal V_{sample} waveform

pattern that larger attenuation ratios result in larger spikes. Additionally, in flat regions without spikes, the waveforms appear coarse due to noticeable noise, with larger attenuation ratios corresponding to higher noise levels. Furthermore, using the 1:1 passive probe, distinct ripples are visible on V_{sample}, corresponding to the PFC output voltage ripple, but this is not evident with the other probes.

Faced with such waveforms, questions arise: how can the converter operate stably under such significant interference in the controller's power supply and the sampling signal? How can sampling accuracy be ensured with such coarse sampling signals? The results from the three measurement methods also differ significantly, so which result is correct?

9.5.2 Power Rail Probe Measurement of Differential-Mode Interference

Various types of voltage probes were introduced in Sect. 5.3, including a special kind known as a power rail probe. Examples include Tektronix's TPR1000 and TPR4000 [13], as shown in Fig. 9.25.

Using the TPR1000 power rail probe to measure $V_{+3.3V}$ and V_{sample}, the measurement results are displayed in Fig. 9.26. The waveforms exhibit only very small glitches, with a peak-to-peak value of 0.04 V for the spikes on $V_{+3.3V}$ and 0.06 V for the spikes on V_{sample}. The ripples on V_{sample} are very clear. Additionally, $V_{+3.3V}$ and V_{sample} have successfully slimmed down, presenting a waveform that is thin and "clean". This is entirely different from the results obtained using conventional voltage probes. However, this aligns with the actual situation of a stable operation of the converter and precise control. Therefore, the measurement results using the power rail probe are the correct ones. When measuring with conventional voltage probes, the observed interference signals are introduced by the measurement and do not exist in the original circuit. The coarse waveform indicates high measurement path noise.

Fig. 9.25 Tektronix TPR1000 and TPR4000 Power rail probes Tektronix TPR1000 and TPR4000

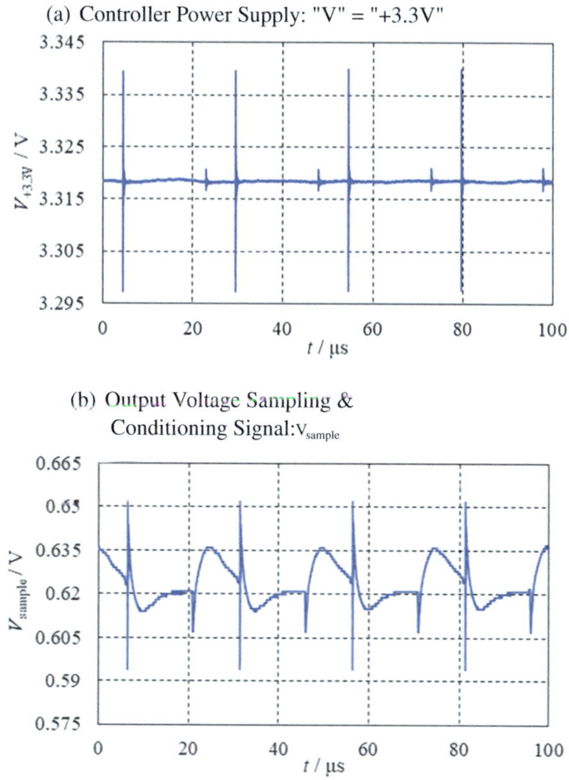

Fig. 9.26 Power rail probe measurement results

The significant differences in measurement results between conventional probes and power rail probes are indeed surprising and once again highlight the importance of choosing the right measurement tools to obtain accurate results.

The unique structure of the power rail probe allows it to deliver correct results, as depicted in Fig. 9.27. Its DC input impedance is 50 kΩ, and high-frequency impedance is 50 Ω, reducing the load effect while leveraging the low-noise characteristics of the 50 Ω path. It features a 1.1:1 attenuation ratio, causing only a 10% increase in the oscilloscope's baseline noise. The ability to set bias voltage allows users to employ the most sensitive oscilloscope settings and center the signal on the screen. With a bandwidth of 1 GHz, it can effortlessly capture rapid transients causing clock and digital data distortion.

Firstly, the power rail probe has the shortest ground wire and the smallest ground loop area, with strong anti-interference capabilities. This prevents the introduction of interference signals that don't exist in reality, avoiding the confusion caused by erroneous test results.

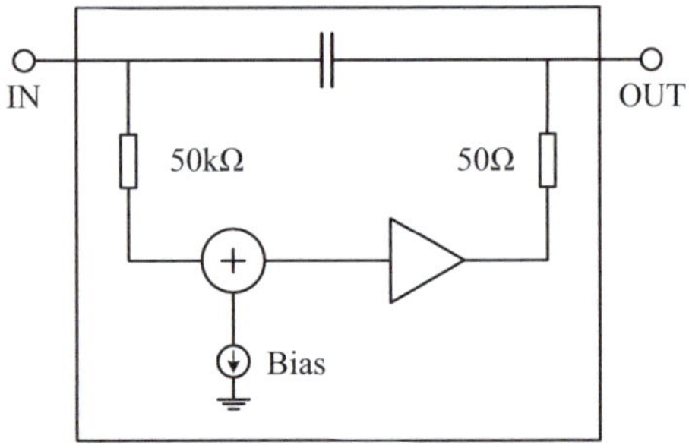

Fig. 9.27 Power rail probe structure

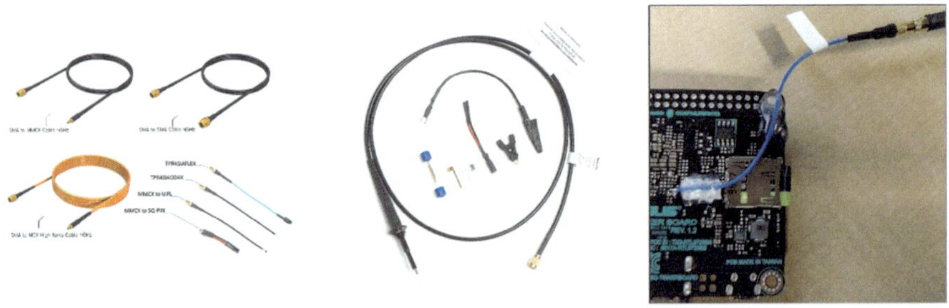

Fig. 9.28 Power rail probe connectors [14, 15]

1. **Shortest ground wire, smallest ground loop area**

When using conventional probes, limited by probe structure and optional accessories, reducing the length of the ground wire and loop area is restricted. Even with a short ground wire, its length is still at least 1 cm, resulting in a large ground loop area.

For different application scenarios, manufacturers provide various probe accessories for power rail probes, enabling engineers to flexibly conduct tests, as shown in Fig. 9.28. Directly soldering the probe end wires to the test point achieves the shortest ground wire and the smallest loop area, nearly reaching the limit. This approach alleviates the ground wire effect and minimizes the coupling of external magnetic fields through the loop area, reducing interference in the test.

2. **Strong anti-interference capability**

Converters, during operation, generate high dv/dt and di/dt, leading to strong high-frequency electromagnetic radiation. It is a significant challenge for testing equipment. While oscilloscopes are specifically designed to resist electromagnetic radiation, the characteristics of probes in this aspect are often overlooked.

Placing a 100:1 high-voltage differential probe, a 10:1 passive probe, and a TPR1000 near a running Totem-Pole PFC converter reveals that the TPR1000 is unaffected, while the other two probes display evident interference signals on their waveforms, as shown in Fig. 9.29. This indicates that the power rail probe has superior shielding performance, mitigating external electromagnetic radiation interference.

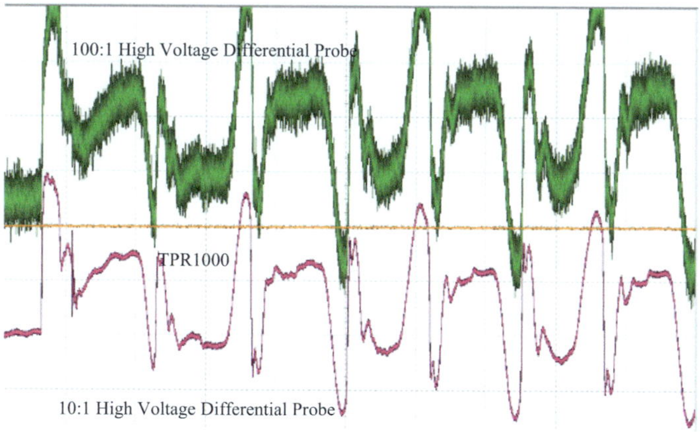

Fig. 9.29 Probe anti-interference capability test

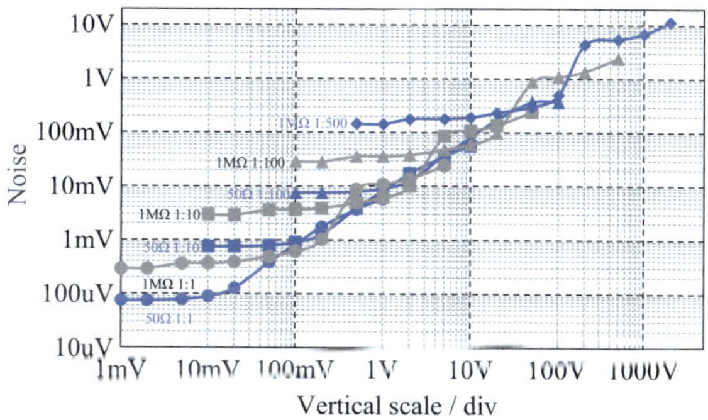

Fig. 9.30 Noise under different input impedances and vertical scales

Secondly, the noise characteristics of the oscilloscope, as shown in Fig. 9.30, reveal that the power rail probe, with a 50 Ω output impedance, 1:1 attenuation ratio, and the ability to set bias voltage, can minimize the noise in the measurement path as much as possible.

1. **50 Ω output impedance**

As discussed in Chap. 3, an oscilloscope has lower noise with a 50 Ω input impedance than with a 1 MΩ input impedance. Conventional probes connect to a 1 MΩ input impedance, while power rail probes connect to a 50 Ω input impedance. This allows the use of an oscilloscope with lower noise to measure the circuit, reducing measurement noise.

2. **1.1:1 Attenuation ratio**

Voltage probes condition the measured signal to fit within the oscilloscope's allowed input voltage range according to a certain attenuation ratio. The oscilloscope samples and converts the signal through analog front end and ADC, then amplifies the result by the reciprocal of the voltage probe's attenuation ratio to restore the actual amplitude of the measured signal.

Both the probe and the oscilloscope's analog front end have noise. When restoring the amplitude of the measured signal, they are amplified together. Therefore, probes with higher attenuation ratios yield waveforms with more significant noise, meaning lower signal-to-noise ratios and coarser waveforms. The power rail probe's attenuation ratio is 1.1:1, effectively controlling noise.

3. **Bias voltage**

The measured signals $V_{+3.3\mathrm{V}}$ and V_{sample} consist of a certain amplitude of DC voltage and interference signals. The focused interference signals are dynamic signals with a small amplitude. For such measurements, a smaller vertical scale on the oscilloscope is more advantageous. Firstly, oscilloscope noise varies with the vertical scale; higher vertical scales result in lower noise. Additionally, higher vertical scales imply smaller ranges, leading to higher measurement resolution and more accurate measurement results.

When measuring $V_{+3.3\mathrm{V}}$ with a conventional probe and setting the oscilloscope to DC coupling, a vertical scale of 500 mV/div is needed, resulting in significant noise. Since the focus is on the interference signal, setting the oscilloscope to AC coupling can remove the DC component, allowing the use of a higher vertical scale to effectively reduce noise. However, AC coupling has limitations; when the signal has a DC offset, using AC coupling prevents observation, leading to the omission of important information.

To address this issue, setting the bias voltage of the power rail probe allows maximizing the use of a smaller vertical scale to reduce noise, improve resolution, and measure the DC component of the signal.

Lastly, the power rail probe's characteristics of high bandwidth and low load effect are crucial for interference signal testing.

1. **High bandwidth**

The power rail probe has a high bandwidth, accurately capturing high-frequency interference without missing any details. However, it's essential to note that higher bandwidth also comes with more significant noise. Instead of blindly using the maximum bandwidth, it's more appropriate to flexibly and reasonably use the oscilloscope's bandwidth limiting function. This is analogous to using a wide-aperture camera lens; a wide aperture provides a shallow depth of field and strong background blur, allowing the use of lower ISO to reduce noise. However, wide apertures make focusing challenging, resulting in suboptimal image quality and more pronounced chromatic aberration compared to smaller apertures. Therefore, photographers don't always use the widest aperture but choose the appropriate aperture based on the actual scene.

2. **Low load effect**

During measurements, lower load effects of testing equipment on the original system are preferred. Especially when measuring sample conditioning signals, excessive load effects can cause sampling errors, leading to the improper operation of the converter. According to the structure of the power rail probe shown in Fig. 9.27, its DC input impedance is 50 kΩ, providing a light load effect, and its high-frequency input impedance is 50 Ω, giving the probe a higher bandwidth.

References

1. Kazanbas M, Schittler A, Araújo S et al (2015) High-side driving under high-switching speed: technical challenges and testing methods. In: PCIM Europe 2015, international exhibition & conference for power electronics, pp 1385–1392
2. Wang J, Shen Z, Dimarino C et al (2016) Gate driver design for 1.7 kV SiC MOSFET module with Rogowski current sensor for shortcircuit protection. In: 2016 IEEE applied power electronics conference and exposition (APEC), pp 516–523
3. Coughlin Chris. Common mode transient immunity. Technical Articles, Analog Devices Inc.
4. Wei Z, Mateo B (2018) Common mode transient immunity (CMTI) for UCC2122x isolated gate drivers. In: Application report, SLUA909, Texas Instruments Inc.
5. Murata Manufacturing Co., Ltd. Website. https://www.murata.com/
6. RECOM Power, Inc. Website. https://recom-power.com/
7. Mornsun Power Website. https://www.mornsun-power.com/
8. She X, Datta R, Todorovic MH et al (2017) High performance silicon carbide power block for industry applications. IEEE Trans Ind Appl 53(4):3738–3747
9. Bosshard R, Kolar JW (2017) All-SiC 9.5 kW/dm^3 on-board power electronics for 50 kW/85 kHz automotive IPT system. IEEE J Emerg Select Top Power Electron 5(1):419–431
10. TDK EMC Components. https://product.tdk.com/info/en/products/emc/index.html
11. Xue L, Boroyevich D, Mattavelli P (2015) Driving and sensing design of an enhancement-mode-GaN Phaseleg as a building block. In: 2015 IEEE 3rd Workshop on Wide Bandgap Power Devices and Applications (WiPDA), pp 34–40
12. Nguyen V-S, Kerachev L, Lefranc P, Crebier J-C et al (2017) Characterization and analysis of an innovative gate driver and power supplies architecture for HF power devices with high dv/dt. IEEE Trans Power Electron 32(8):6079–6090
13. Tektronix, Inc. (2019) TPR1000 and TPR4000 Active Power Rail Probes. Datasheet, 51W-61491-2
14. Tektronix, Inc. (2019) TPR1000 and TPR4000 active power rail probes compliance and safety instructions. User, 071363701
15. Tektronix, Inc. (2019) Getting started with power rail measurements. Application Note, 51W-61562-0

Further Reading

16. Paul CR (2011) Introduction to electromagnetic compatibility, 2nd edn. Wiley, Hoboken
17. Maniktala S (2012) Switching power supplies A–Z, 2nd edn. Newnes, New York
18. Jaroslaw L (2018) High frequency conducted emission in AC motor drives fed by frequency converters: sources and propagation paths, 1st edn. Wiley-IEEE Press, Hoboken

19. Avago Technologies Inc. (2012) Common-mode noise sources and solutions. Application Note 1043, AV02-3698EN
20. Silicon Laboratories, Inc. CMOS digital isolators supersede optocouplers in industrial applications. In: White paper, Rev 0.3
21. DIN VDE V 0884-11 (2017) Semiconductor devices Part 11: magnetic and capacitive coupler for basic and reinforced isolation
22. Vishay Intertechnology Inc. (2013) Optocoupler common mode transient immunity (CMTI)—theory and practical solutions. Application Note 83:83702, Rev. 1.5
23. STMicroelectronics Inc. (2016) Common mode filters. Application note, AN4511, Rev 2, DocID026455
24. Gill L, Ikari T (2019) Analysis and mitigation of common mode current in SiC MOSFET gate driver power supply. In: 2018 IEEE international conference on electrical systems for aircraft, railway, ship propulsion and road vehicles & international transportation electrification conference (ESARS-ITEC). IEEE, pp 1–6
25. Nguyen V, Lefranc P, Crebier J (2018) Gate driver supply architectures for common mode conducted EMI reduction in series connection of multiple power devices. IEEE Trans Power Electron 33(12):10265–10276
26. Wang J, Mocevic S, Burgos R et al (2020) High-scalability enhanced gate drivers for SiC MOSFET modules with transient immunity beyond 100 V/ns. IEEE Trans Power Electron 35(10):10180–10199
27. Huber JE, Kolar JW (2014) Common-mode currents in multi-cell solid-state transformers. In: 2014 international power electronics conference (IPEC-Hiroshima 2014—ECCE ASIA). IEEE, pp 766–773
28. Kasper M, Bortis D, Kolar JW (2014) Scaling and balancing of multi-cell converters. In: 2014 international power electronics conference (IPEC-Hiroshima 2014—ECCE ASIA), pp 2079–2086
29. Karthik R (2019) Digital power management and power rail measurements using high definition oscilloscopes. On Demand Webinars, Teledyne LeCroy, Inc. https://go.teledynelecroy.com/l/48392/2019-08-22/7rvpyy

Chapter 10
Impact and Countermeasures of Common Source Inductance

When SiC MOSFET was first introduced to the market, they used the TO-247-3. Less than a decade later, the TO 247-4 and TO-263-7 have become dominant. The TO-247-3 is no longer recommended for use in new converter.

SiC MOSFET manufacturers increasingly recommend TO-247-4 devices, claiming they exhibit faster switching speeds and lower losses. Power electronics engineers also tend to prefer TO-247-4. In comparison with using TO-247-3, issues such as drive waveform oscillation, crosstalk, and instances of converter failure have decreased. All parties have benefited from the adoption of TO-247-4, but the reasons behind its performance are not thoroughly researched.

This chapter will delve into the structural differences between TO-247-3 and TO-247-4. Through methods such as equivalent model analysis, circuit simulation, and experimental verification, it aims to elucidate the fundamental reasons behind the outstanding characteristics of TO-247-4.

10.1 Common Source Inductance

10.1.1 Basic Principles of Common Source Inductance

TO-247-3 is one of the most familiar single-package forms of power devices for engineers. Figure 10.1 shows physical samples of TO-247 packages from different manufacturers.

Tesla Model X's drive system extensively utilizes IGBT single devices in TO-247 3, marking one of the most prominent applications of TO-247-3 package. In the high-power rear motor control unit, 84 IGBT single devices rated at 600 V/240A are paralleled in sets of 14. In the low-power front electric control unit, 36 IGBT single devices rated at 600 V/160A are paralleled in sets of 6.

Y. Gao and Y. Zhang, *Silicon Carbide Power Devices*, CPSS Power Electronics Series,
https://doi.org/10.1007/978-981-96-3480-4_10

Fig. 10.1 TO-247-3 packaged devices

Amid the maturation of power electronics technology, market competition intensifies, and cost pressures continue to rise. For an equivalent capacity, the cost of power modules is 2–4 times that of using a solution with paralleled discrete devices. Hence, solar inverter, UPS, and variable-frequency drive manufacturers are now transitioning towards using discrete devices to replace power modules. It's anticipated that in these fields, we will witness more cases similar to the widespread use of paralleled discrete devices, as seen in the Model X.

From Fig. 10.1, it's evident that TO-247-3 packaged devices provided by different manufacturers exhibit some visual differences. Moreover, within the offerings of a single manufacturer, multiple variations of TO-247-3 packages might be available. The primary distinctions among these TO-247-3 packages lie in pin lengths, mounting holes, and the outer casing. However, all these variants feature precisely 3 pins spaced at distances ranging from 5.44 to 5.45 mm, corresponding to the gate (G), drain (D), and source (S) terminals of the SiC MOSFET.

TO-247-3 packaged SiC MOSFET devices' internal structure is illustrated in Fig. 10.2. The SiC MOSFET chip is a thin cube, with its backside serving as the drain (D), while the top surface features the gate (G) and source (S). The chip is soldered onto a metal substrate, with the chip's drain (D) connected directly to a pin linked to the substrate. Multiple bonding wires are used to connect the chip's source (S) to the pin to meet current requirements. Due to the smaller drive current, only one finer bonding wire is used to connect the gate (G).

When analyzing the switching process of power devices, considering their high-frequency nature, PCB traces, package pins, and bonding wires are typically treated as inductances. Based on this consideration, an equivalent circuit for the TO-247-3 device can be established. Figure 10.3 illustrates the scenario simultaneously accounting for both the device and the drive circuit. Within the dashed lines, the

Fig. 10.2 Internal structure of TO-247-3 packaged device

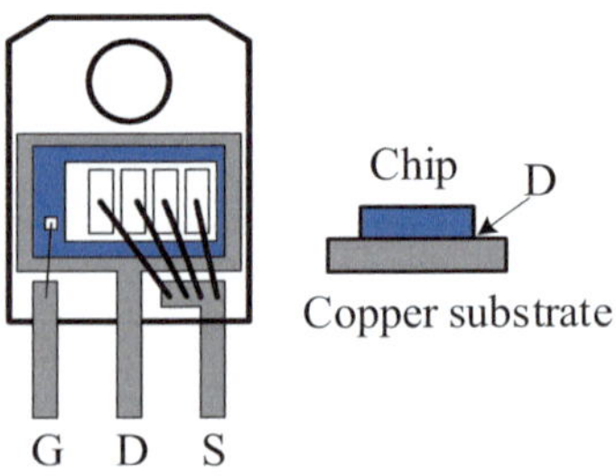

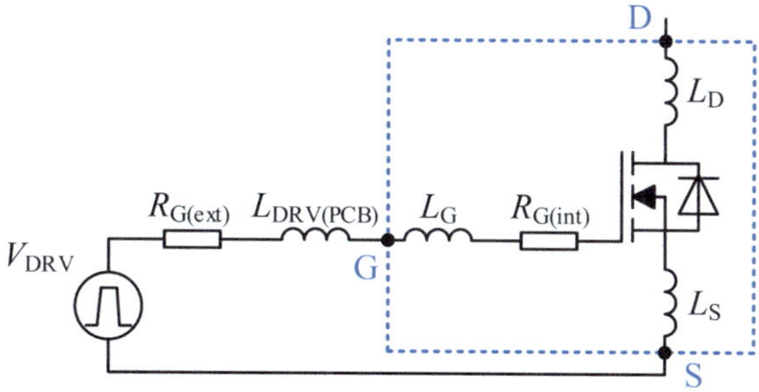

Fig. 10.3 Equivalent circuit of TO-247-3 packaged device

equivalent circuit of the TO-247-3 device is depicted, where L_G, L_D, and L_S respectively represent the parasitic inductances caused by bonding wires and package pins, corresponding to the gate, drain, and source. Outside the dashed lines represents the drive circuit section: $L_{DRV(PCB)}$ is the inductance of the drive circuit PCB traces, $R_{G(int)}$ is the internal gate resistance within the chip, $R_{G(ext)}$ is the external gate resistance, and V_{DRV} represents the drive voltage.

During the on–off transitions of power devices, the drive circuit charges and discharges the input capacitance of the device through the gate loop, while high-frequency currents complete the power switching through the main power conduction path. Hence, these two loops are pivotal in our design and analysis. The gate loop inductance, L_{DRV}, comprises $L_{DRV(PCB)}$, L_G, and L_S, whereas L_D and L_S constitute part of the main power conduction loop's inductance. These loops share a common element, L_S, termed as the common source inductance. During the switching process, rapidly changing drain-source current I_{DS} induces voltage drops across L_S.

The Tektronix isolation probe TIVP1 is utilized to measure the voltage V_{L_S} across the L_S using the principle and probe connection method as illustrated in Fig. 10.4. It's important to note that the actual measurement points are at the ends of the device pins. Therefore, the measured voltage V_{L_S} across the common source inductance does not encompass the voltage across the bonding wires within L_S. Consequently, the measured result of V_{L_S} tends to be slightly lower than the actual voltage across L_S.

The measurement results are shown in Fig. 10.5, including I_{DS} and V_{L_S}. It can be seen that during the switching process, V_{L_S} changes with the change of I_{DS}. The larger, $\mathrm{d}I_{DS/\mathrm{d}}t$, the larger V_{L_S}.

Then during the switching process, V_{L_S} can be regarded as a current-controlled voltage source, following (10.1),

$$V_{L_S} = L_S \cdot (\mathrm{d}I_{DS}/\mathrm{d}t). \tag{10.1}$$

The equivalent circuit can be obtained as shown in Fig. 10.6. V_{DRV} is the drive voltage, V_{GS} is the gate voltage, which will be affected byV_{L_S}. Since SiC MOSFET has fast switching speed and high gate resistance of

$\mathrm{d}I_{DS/\mathrm{d}}t$, under the same L_S, the amplitude of V_{L_S} is also larger. In the above experiment, the drive voltage was − 5/+ 15 V, and the peak value of V_{L_S} reached 15 V,

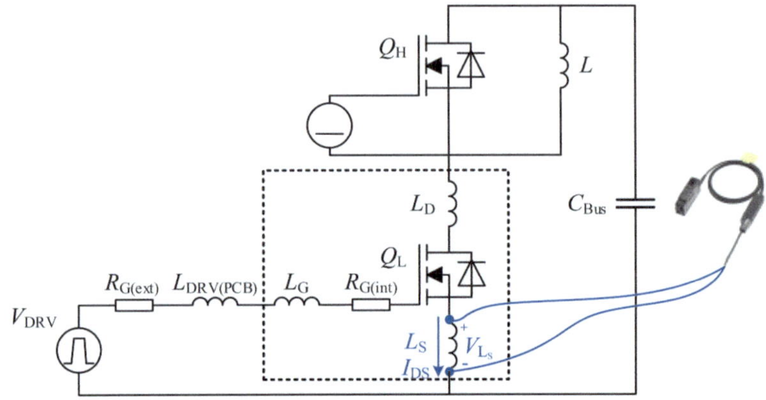

Fig. 10.4 Measurement of common source inductance L_S terminal voltage V_{L_S}

Fig. 10.5 Drain-source current I_{DS} and common source inductance terminal voltage V_{L_S} during the switching process

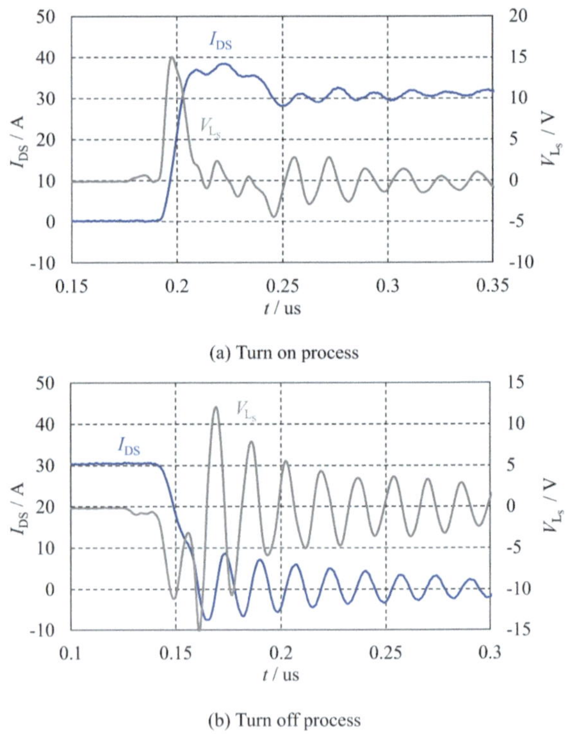

which shows that effect of V_{L_S} on V_{GS} is very obvious. This will have a negative effect on the V_{GS} measurement and switching process, which will be discussed in the next few sections.

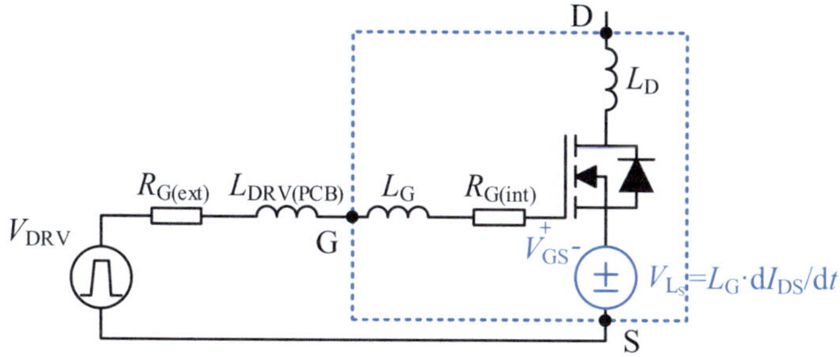

Fig. 10.6 Equivalent circuit diagram of effect of V_{L_S} on V_{GS}

10.1.2 Kelvin Source Package

As technology advances, the switching speed of Si power devices is increasing, highlighting the issues caused by the common source inductance L_S. Various manufacturers have made improvements on the TO-247-3 package and introduced the TO-247-4 package, as shown in Fig. 10.7. The TO-247-4 package is mainly used in high-speed SJ-MOSFETs, high-speed IGBTs, and SiC MOSFETs. A comparison with Fig. 10.1 reveals that, in contrast to the TO-247-3 package devices, the TO-247-4 package devices have two source pins: Kelvin Source (KS) and Power Source (PS). Moreover, there are changes in the sequence and spacing of the pins.

The internal structure of the TO-247-4 packaged SiC MOSFET is shown in Fig. 10.8. It is found that KS and the source of the chip are connected by a thin bonding wire.

Fig. 10.7 TO-247-4 package devices

Fig. 10.8 Internal structure of TO-247-4 packaged device

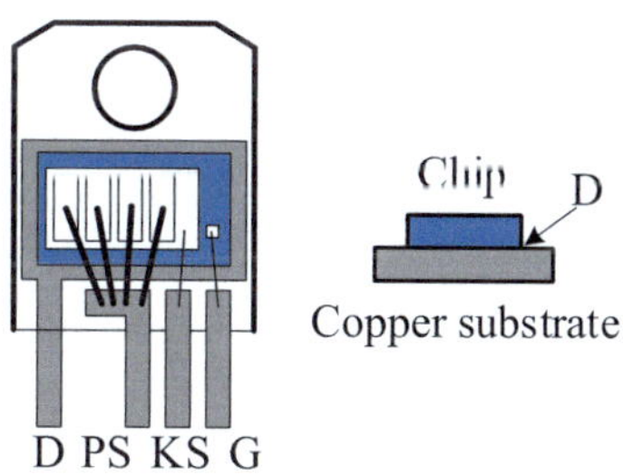

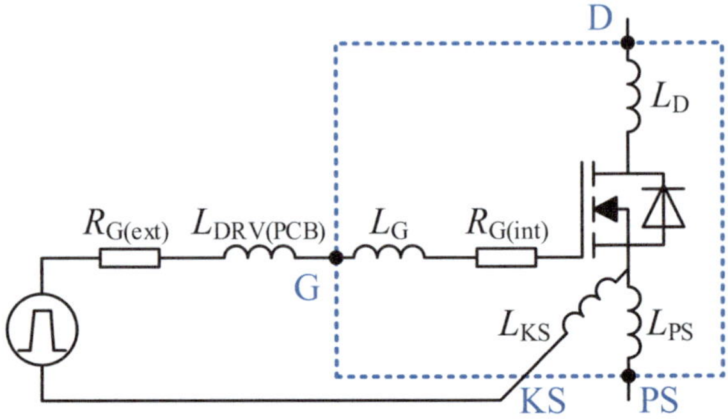

Fig. 10.9 Equivalent circuit of TO-247-4 packaged device

When using TO-247-4 packaged devices, the drive circuit is connected to the G and KS, and the main circuit is connected to the D and PS. In this way, the TO-247-4 package avoids the common lines between the drive circuit and the main power commutation circuit and realizes the decoupling of the two circuits, as shown in Fig. 10.9. It should be noted that although there is no direct electrical connection, there is mutual inductance between the drive circuit and the main power commutation circuit, and there will be an equivalent common source inductance. However, the equivalent common source inductance is small and has limited effect, so it is ignored in the subsequent analysis.

The wiring method described is known as Kelvin connections, widely used in measurement circuits. It's a simple technique used to eliminate effect of voltage drops across wires in a circuit on measurements. It's commonly employed in the four-terminal resistance used for current sampling, showcasing a typical application instance.

When the TO-247-4 package is used for MOSFETs, the Kelvin bond wire connection is made on the source, termed as the Kelvin source package. For IGBTs, when applied in the TO-247-4 package, the Kelvin bond wire connection is made on the emitter, known as the Kelvin emitter package. Apart from the TO-247-4 package, various manufacturers have introduced multiple forms of Kelvin-connected packages based on voltage, current ratings, and specific applications. Examples include TO-263-7, ThinPAK 8 × 8 [3, 4], TOLL [5, 6], DDPAK [7], as depicted in Fig. 10.10a~d, respectively. The selection of Kelvin-connected package forms has become an inevitable choice for high-speed power devices seeking optimal performance.

In the following sections, we will use the TO-247-3 and TO-247-4 packages as examples to conduct a detailed analysis of effects caused by common source inductance L_S. Additionally, we'll verify the efficacy of the Kelvin source package in mitigating these effects.

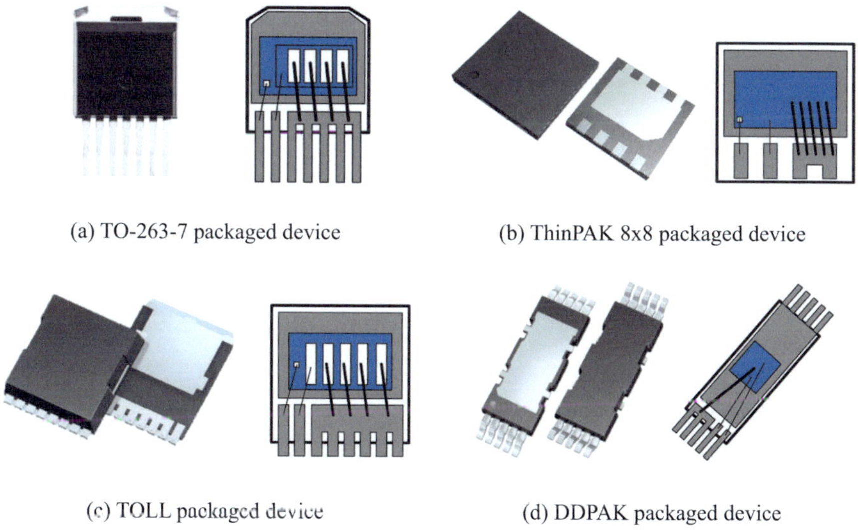

(a) TO-263-7 packaged device

(b) ThinPAK 8x8 packaged device

(c) TOLL packaged device

(d) DDPAK packaged device

Fig. 10.10 Kelvin source (emitter) package

10.2 The Comparative Test

10.2.1 The Traditional Comparative Test

Through experiments to study the negative effect of common source inductance L_S and the performance improvement of Kelvin source package, the easiest way to think of is to use the same model SiC MOSFET chips, respectively TO-247-4 package and TO-247-3 package devices, and then test them, respectively, as shown in Fig. 10.11. Manufacturers usually provide both TO-247-4 and TO-247-3 package products for SiC MOSFETs of the same specifications, making this comparison method very easy to implement.

According to Sect. 5.4, due to parasitic parameters between test points, for TO-247-4 packaged devices, the measured $V_{GS(4\text{-}M)}$, $V_{DS(4\text{-}M)}$ are different forms of $V_{GS(4PIN)}$ used to control the channel on the chip and actual $V_{DS(4PIN)}$, which is obtained from (10.2) and (10.3),

$$V_{GS(4\text{-}M)} = V_{GS(4PIN)} + I_G \cdot R_{G(int)} + (L_{G(pkg\text{-}M)} + L_{KS(pkg\text{-}M)}) \cdot dI_G/dt \tag{10.2}$$

$$V_{DS(4\text{-}M)} = V_{DS(4PIN)} + L_{D(pkg\text{-}M)} \cdot dI_{DS}/dt + L_{KS(pkg\text{-}M)} \cdot dI_G/dt. \tag{10.3}$$

For TO-247-3 packaged devices, the relationship between the measured $V_{GS(3\text{-}M)}$, $V_{DS(3\text{-}M)}$, and the $V_{GS(3PIN)}$ used to control the channel on the chip, actual $V_{DS(3PIN)}$ are calculated by (10.4) and (10.5),

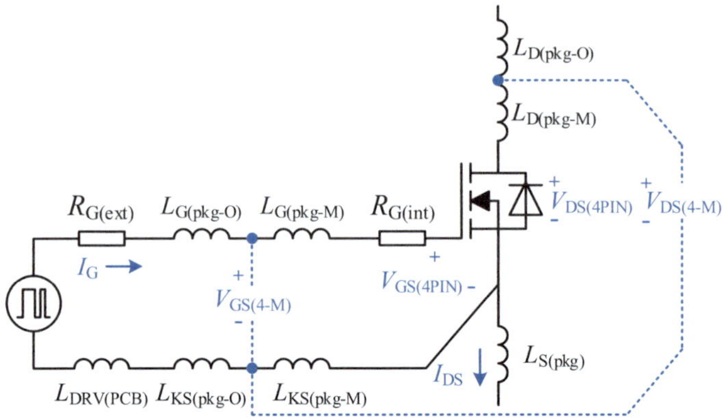

(a) Test schematic of TO-247-4 packaged device

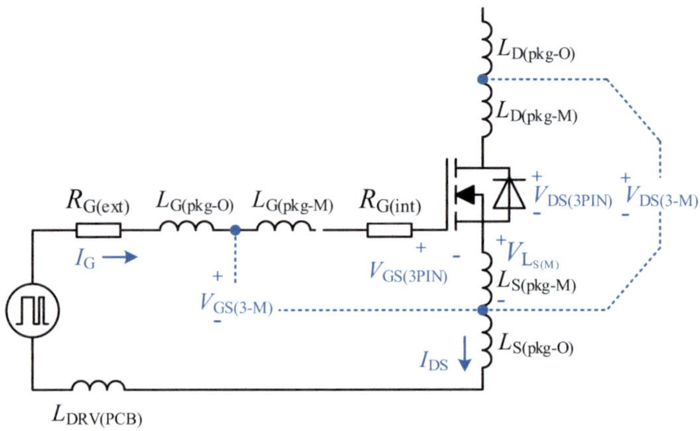

(b) Test schematic of TO-247-3 packaged device

Fig. 10.11 Traditional comparative test

$$V_{\text{GS(3-M)}} = V_{\text{GS(3PIN)}} + I_{\text{G}} \cdot R_{\text{G(int)}} + L_{\text{G(pkg-M)}} \cdot \text{d}I_{\text{G}}/\text{d}t + L_{\text{S(pkg-M)}} \cdot \text{d}I_{DS}/\text{d}t \quad (10.4)$$

$$V_{\text{DS(3-M)}} = V_{\text{DS(3PIN)}} + (L_{\text{D(pkg-M)}} + L_{\text{S(pkg-M)}}) \cdot \text{d}I_{\text{DS}}/\text{d}t. \quad (10.5)$$

For TO-247-3 packaged devices, $V_{\text{GS(3-M)}}$ not only includes the voltage drop of I_{G} at $R_{\text{G(int)}}$ and $L_{\text{G(pkg-M)}}$, but also includes the voltage drop of I_{DS} at $L_{\text{S(pkg-M)}}$. It can be seen from the test results in the previous section that if $\text{d}I_{\text{DS}}/\text{d}t$ is higher, $V_{L_{\text{s}}(\text{M})}$ can significantly affect $V_{\text{GS(3PIN-M)}}$, causing the measurement results to be misleading for analysis.

It should be noted that when using this method of comparative tests using devices of the same type and different packages, since the SiC MOSFET chips being tested are not the same, the differences in SiC MOSFET chip parameters will affect the test results, resulting in inaccurate comparisons rigorous.

10.2.2 The Comparative Test Between 4in3 and 3in4

To address the two issues mentioned above, a test that compares effect of TO-247-4 package and TO-247-3 package on the electrical characteristics of SiC MOSFETs, utilizing solely the TO-247-4 packaged device. This solution comprises two parts: the 4in4 test and the 4in3 test.

1. **4in4 Test**

Regarding circuit settings, jumper resistor R_1 is disconnected while jumper resistor R_2 is short-circuited. The main power loop is connected to both the D and PS, and the drive circuit is connected between the D and KS.

As for measurements, the measurement points for V_{GS} are the G and KS, while for V_{DS}, they are the D and KS. The circuit's wiring method and test points mirror those used with TO-247-4 packaged devices, as depicted in Fig. 10.12. For convenience, this method is referred to as the "4in4 test".

$V_{GS(4PIN)}$ and $V_{DS(4PIN)}$ are the gate-source voltage and the actual drain-source voltage of the control channel on the chip during the 4in4 test, which are the same as when the TO-247-4 package device is used normally, while $V_{GS(4\text{-}M)}$ and $V_{DS(4\text{-}M)}$ are the corresponding measurement results. It also follows Eqs. (10.2) and (10.3).

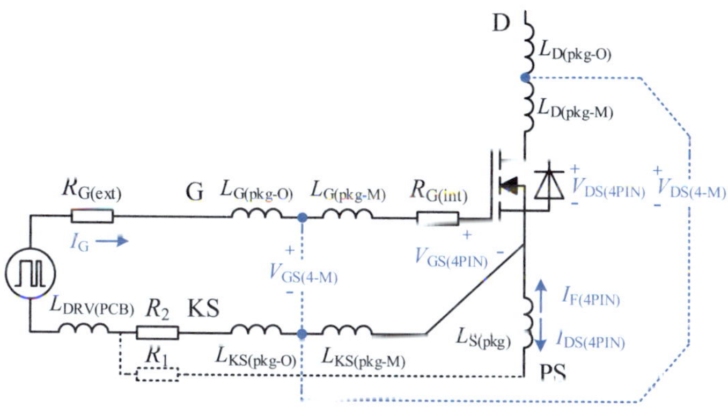

Fig. 10.12 4in4 test circuit schematic diagram

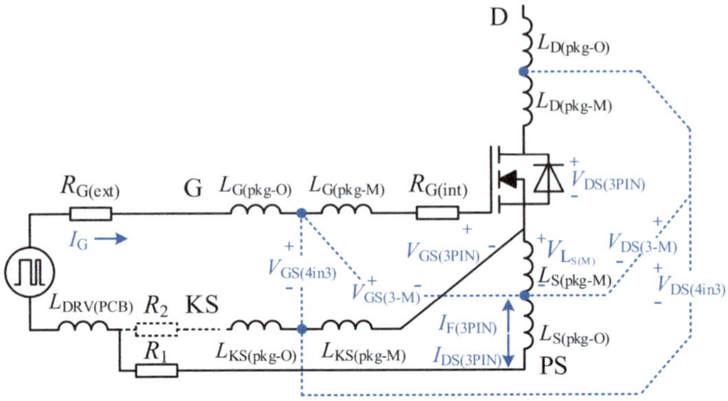

Fig. 10.13 4in3 test circuit schematic diagram

2. **The 4in3 Test**

In terms of circuit settings, the jumper resistor R_1 is short-circuited, R_2 is disconnected, the main power loop is connected to the D and PS. The driver circuit is connected with the G and PS, and the KS pin is left floating (Fig. 10.13).

In terms of measurement, when the measurement points of V_{GS} are G and PS, the measured $V_{GS(3\text{-}M)}$ is the same as the measured results when using TO-247-3 packaged devices. When the measurement points of V_{GS} are G and KS, the measured $V_{GS(4in3\text{-}M)}$ can exclude effect of $V_{L_s(M)}$ to obtain more accurate of V_{GS}. When the measurement points of V_{DS} are the D and PS, the measured $V_{DS(3\text{-}M)}$ is the same as when using TO-247-3 package device. When V_{DS} the measurement points are D and KS, $V_{L_s(M)}$ is eliminated to obtain a more accurate $V_{DS(4in3\text{-}M)}$. This circuit wiring method is to connect the TO-247-4 package device into a TO-247-3 package. This method is called 4in3 test.

$V_{GS(3PIN)}$ and $V_{DS(3PIN)}$ are the gate-source voltage of the control channel on the chip in the 4in3 test and the actual drain-source voltage. Same as when using TO-247-3 packaged devices normally, $V_{GS(3\text{-}M)}$ and $V_{DS(3\text{-}M)}$ follow (10.4) and (10.5), $V_{GS(4in3\text{-}M)}$ and $V_{DS(4in3\text{-}M)}$ follow Eqs. (10.6) and (10.7),

$$V_{GS(4in3\text{-}M)} = V_{GS(3PIN)} + I_G \cdot R_{G(int)} + L_{G(pkg\text{-}M)} \cdot dI_G/dt \tag{10.6}$$

$$V_{DS(4in3\text{-}M)} = V_{DS(3PIN)} + L_{D(pkg\text{-}M)} \cdot dI_{DS(3PIN)}/dt. \tag{10.7}$$

The characteristics of TO-247-4 packaged devices and TO-247-3 packaged devices can be derived using the 4in4 test and 4in3 test, respectively. This allows for experimentation using the same chip, while maintaining consistent main power loop inductance and drive loop inductance. Studying effect of L_S aims to mitigate effects of V_{GS}, V_{DS}, and $V_{L_s(M)}$. Additionally, the parasitic parameter compensation method outlined in Chap. 5 can be utilized to acquire gate-source voltage and actual drain-source voltage closer to the control channel on the chip.

10.3 Effect on the Switching Process

10.3.1 Turn on Process

Through simulation, effect of L_S on the turn on waveform under the same $R_{G(ext)}$ can be obtained. The simulation waveform is shown in Fig. 10.14. In order to simplify the analysis of the turn on process, the inductance on the gate loop is ignored, but

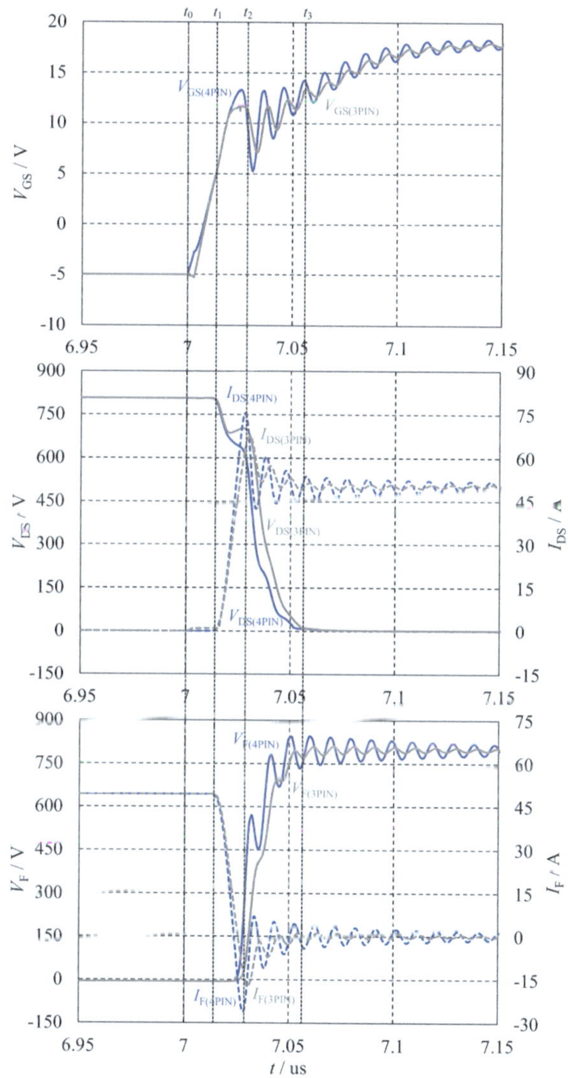

Fig. 10.14 Effect of L_S on the turn on process (simulation results)

in order to obtain a more meaningful waveform, L_{DRV} is taken into account during simulation.

1. $t_1 \sim t_2$

The SiC MOSFET begins to turn on, I_{DS} rises rapidly, producing an upper positive and lower negative V_{L_S} on L_S, then $V_{GS(3PIN)}$ is calculated according to Eq. (10.8)

$$V_{GS(3PIN)} = V_{DRV(on)} - R_{G(ext)} \cdot I_{G(on)} - L_S \cdot dI_{DS(3PIN)}/dt. \tag{10.8}$$

It can be seen from Eq. (10.8) that V_{L_S} weakens the drive ability of $V_{DRV(on)}$ and plays a negative role. Effect of feedback makes $V_{GS(3PIN)}$ smaller than $V_{GS(4PIN)}$. Since the SiC MOSFET is in saturation at this stage, the rise speed of $I_{DS(3PIN)}$ is slower than that of $I_{DS(4PIN)}$, and thus this makes the voltage drop of $V_{DS(3PIN)}$ smaller than $V_{DS(4PIN)}$, and at the same time, the reverse recovery current of the tube body diode is also smaller.

2. $t_2 \sim t_3$

Since the rise speed of $I_{DS(3PIN)}$ in the previous stage is slow, and the reverse recovery current of the body diode is smaller, in this stage, $I_{DS(3PIN)}$'s oscillation amplitude is smaller. And the voltage drop it produces on L_{Loop} is also smaller, which makes $V_{DS(3PIN)}$ and $V_{GS(3PIN)}$ are reduced. At the same time, V_{L_S} also plays a role in weakening $V_{GS(3PIN)}$ oscillation.

The rising speed of $V_{GS(3PIN)}$ is very little affected by V_{L_S}, so $V_{DS(3PIN)}$ has a comparable falling speed to $V_{DS(4PIN)}$.

3. $t_3 \sim$

The SiC MOSFET is fully turned on, $I_{DS(3PIN)}$ shows a waveform of attenuated oscillation, and its oscillation amplitude is smaller than $I_{DS(4PIN)}$. Then the oscillation of $V_{DS(3PIN)}$ is smaller than $V_{DS(4PIN)}$, so the amplitude of $V_{GS(3PIN)}$ is smaller than $V_{GS(4PIN)}$.

The waveforms obtained by using 4in4 and 4in3 tests are shown in Fig. 10.15; $V_{GS(3\text{-}C)}$ and $V_{DS(3\text{-}C)}$ are the result after parasitic parameter compensation. Effect on the turn on characteristics is shown in Table 10.1, which is consistent with the above analysis and simulation results. It can be seen that L_S significantly reduces the turn on speed of SiC MOSFET, the rise speed of I_{DS} and the reverse recovery speed of the body diode.

At the same time, due to effect of parasitic parameters between measurement points, there is a clear difference between the measured $V_{GS(3\text{-}M)}$, $V_{GS(4in3\text{-}M)}$ and the actual $V_{GS(3PIN)}$ and $V_{GS(3\text{-}C)}$ used to control the channel on the chip, as shown in Fig. 10.16.

It can be seen that when using TO-247-3 packaged devices, affected by $L_{S(pkg\text{-}M)}$, the measured oscillations in $V_{GS(3\text{-}M)}$ is significantly higher than that of $V_{GS(3PIN)}$ and $V_{GS(3\text{-}C)}$. A very high peak appears on $V_{GS(3PIN\text{-}M)}$ during the rapid rising stage in

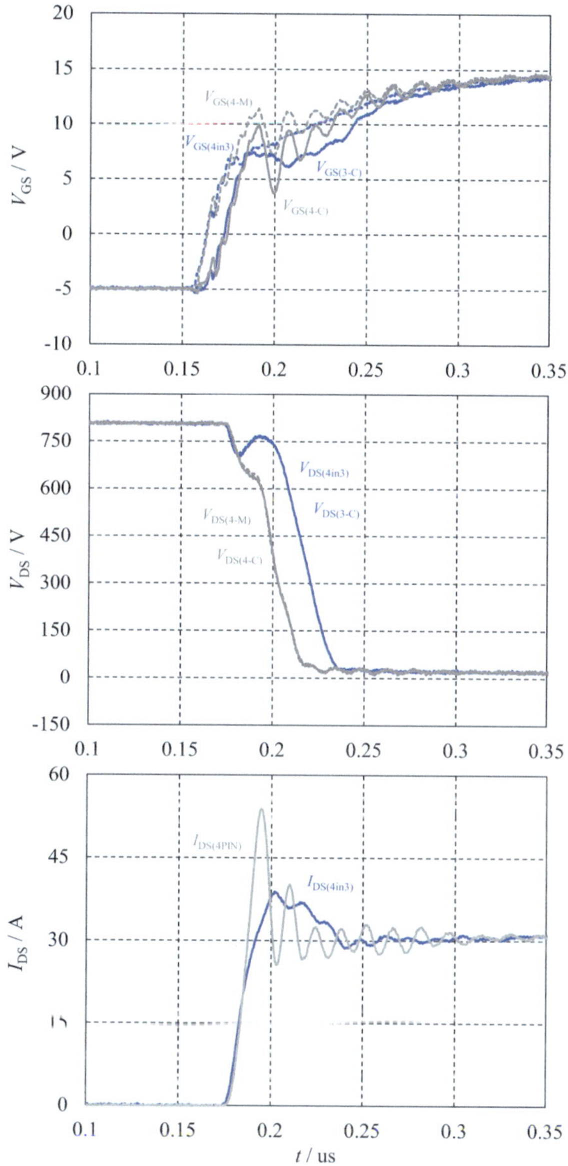

Fig. 10.15 Effect of L_S on the turn on process

Table 10.1 Effect of L_S on turn on characteristics

	4in4 measurement	4in4 compensation	4in3 measurement	4in3 compensation
$t_{d(on)}$/ns	15.2	4.1	14.9	4.4
t_R/ns	33.4	34.3	51.5	52.3
I_{spike}/A	53.9	53.9	38.9	38.9
E_{on}/μJ	524.1	523.7	937.8	922.4

Fig. 10.16 Effect of L_S on the measurement of turn on process V_{GS}

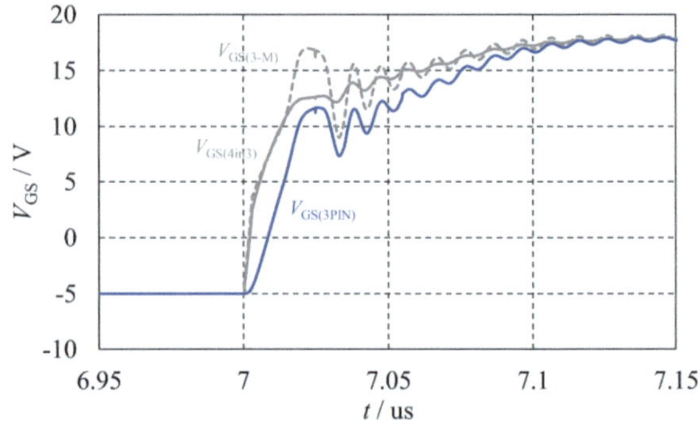

(a) Simulation results

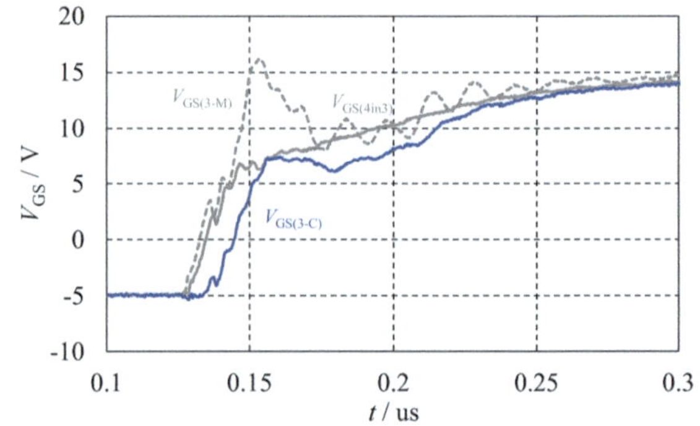

(b) Experimental results

$I_{DS(3PIN)}$, which can easily cause misleading waveform analysis. It can be seen from (10.9) that the influence of $L_{S(pkg\text{-}M)}$ on $V_{GS(4in3PIN\text{-}M)}$ can be avoided by using 4in3 test, However, the effects of $R_{G(int)}$ cannot be eliminated.

$$V_{GS(3\text{-}M)} = V_{GS(4in3\text{-}M)} + L_{S(pkg\text{-}M)} \cdot \mathrm{d}I_{DS(3PIN)}/\mathrm{d}t \tag{10.9}$$

Combining Eqs. (10.4), (10.6), and (10.9), it can be seen that the smaller the driving resistance, the greater the turn-on current, the greater the difference among $V_{GS(3PIN\text{-}M)}$, $V_{GS(4in3PIN\text{-}M)}$, and $V_{GS(3PIN\text{-}C)}$, as shown in Fig. 10.17.

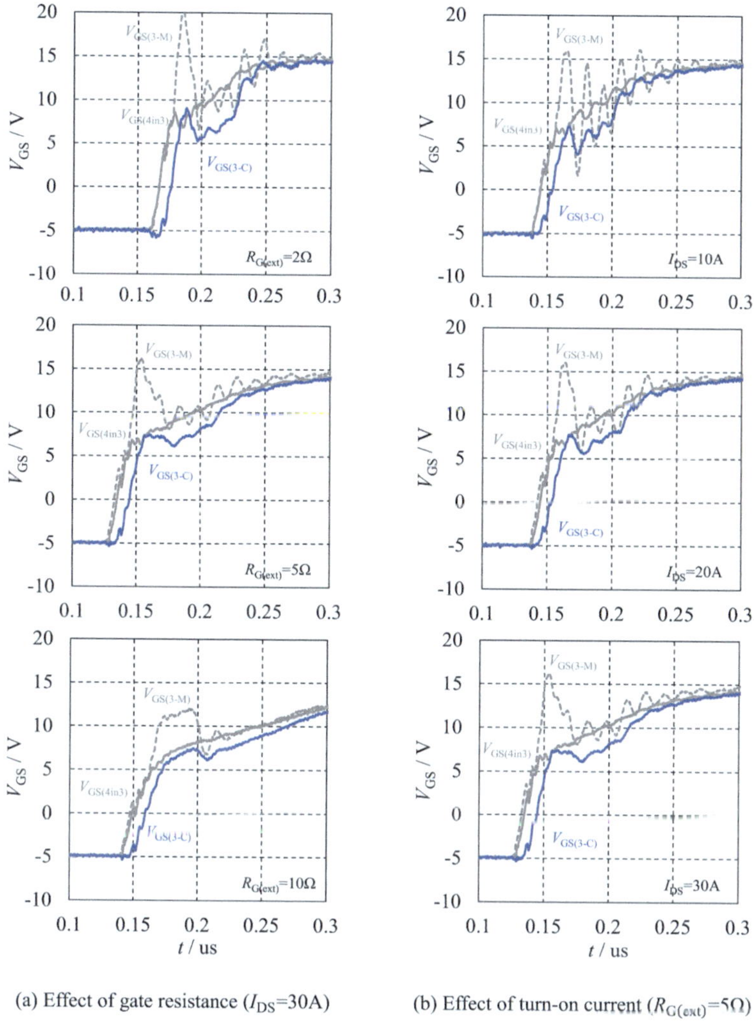

Fig. 10.17 Influence of L_S on the measurement of the turn-on process V_{GS} under different driving resistance and turn-on current

10.3.2 Turn off Process

Through simulation, effect of L_S on the turn off waveform under the same $R_{G(ext)}$ can be obtained. The simulation waveform is shown in Fig. 10.18. In order to simplify the analysis of the turn off process, L_{DRV} is ignored, but in order to obtain a more meaningful waveform, L_{DRV} is in consideration.

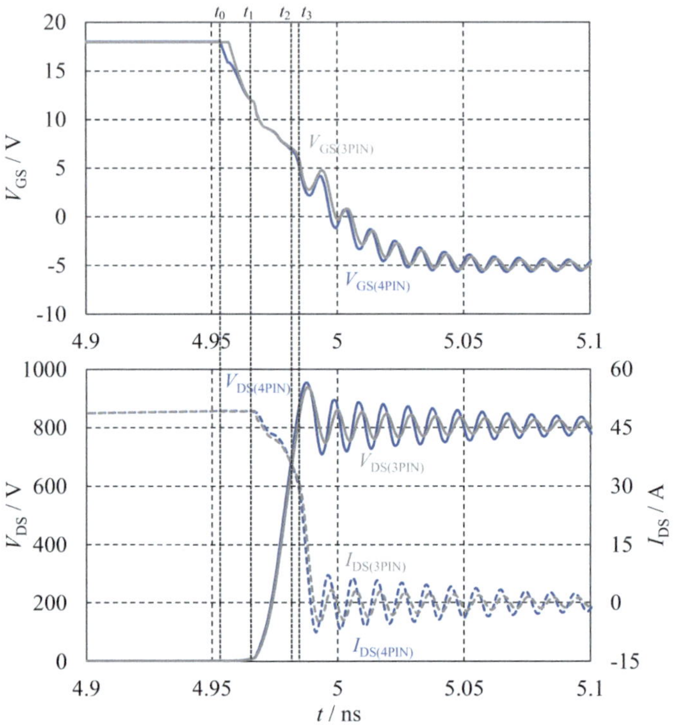

Fig. 10.18 Effect of L_S on the turn off process (simulation results)

1. t_1 ~t_2

I_{DS} begins to decrease, producing an upper negative and lower positive V_{L_S} on L_S, weakening the drive ability of $V_{DRV(off)}$ and acting as a negative feedback. Because at this time $dI_{DS(3PIN)}/dt$ is smaller, making $V_{GS(3PIN)}$ little higher than $V_{GS(4PIN)}$, so the falling speed of $I_{DS(3PIN)}$ is slightly lower than $I_{DS(4PIN)}$.

2. t_2 ~t_3

At t_2, $V_{DS(3PIN)}$ reaches V_{Bus}, and the opposite diode begins to conduct. I_L quickly commutates to the opposite diode. Compared with the previous stage, $I_{DS(3PIN)}$ decreases faster, so V_{L_S} is more Large, $V_{GS(3PIN)}$ is significantly higher than $V_{GS(4PIN)}$, resulting in the falling speed of $I_{DS(3PIN)}$ is lower than that of $I_{DS(4PIN)}$. The rising speed and voltage peak value of $V_{DS(3PIN)}$ are below $V_{DS(4PIN)}$.

3. t_3 ~

The SiC MOSFET is completely off. Due to effect of the previous stage, the oscillation amplitude of $I_{DS(3PIN)}$ is smaller than $I_{DS(4PIN)}$. The oscillation of $V_{DS(3PIN)}$ is smaller than $V_{DS(4PIN)}$, and oscillation of $V_{GS(3PIN)}$ is smaller than $V_{GS(4PIN)}$.

The test waveforms obtained from 4in4 and 4in3 tests are shown in Fig. 10.19. $V_{GS(3\text{-}C)}$ and $V_{DS(3\text{-}C)}$ are the result after parasitic parameter compensation. Effect on the turn on characteristics is shown in Table 10.2, which is consistent with the above analysis and simulation results. It can be seen that L_S significantly reduces the turn off speed of SiC MOSFET and slows down the decrease speed of I_{DS}.

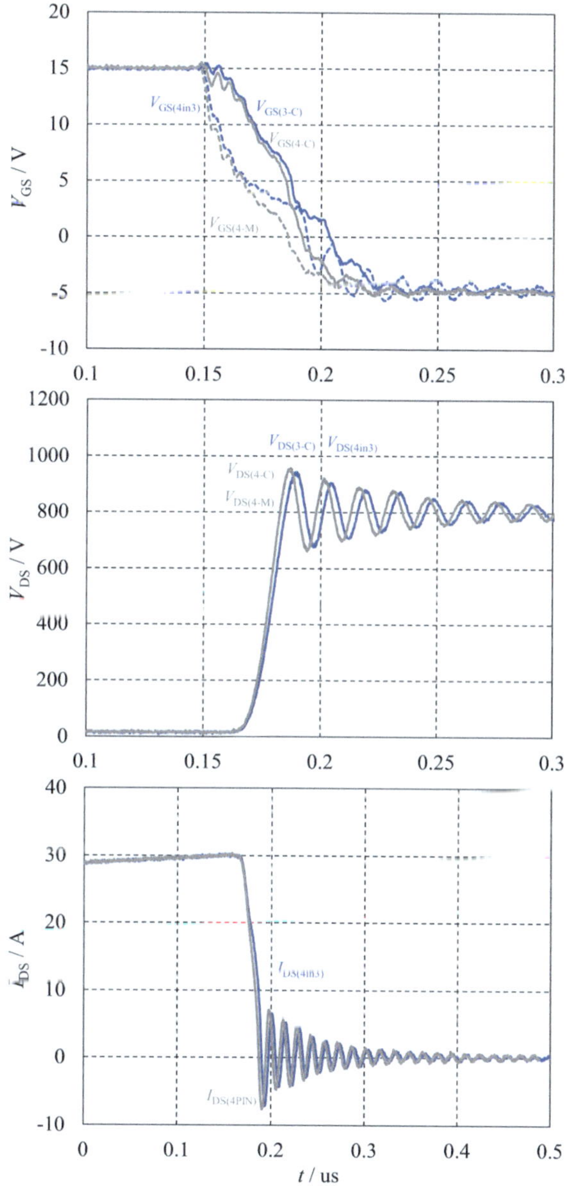

Fig. 10.19 Influence of L_S on the turn off process

Table 10.2 Effect of L_S on turn off characteristics

	4in4 measurement	4in4 compensation	4in3 measurement	4in3 compensation
$t_{d(off)}$/ns	180.8	180.5	169.4	169.4
t_f/ns	11.5	11.8	12.9	12.9
I_{spike}/A	944.7	956.4	931.6	942.8
E_{off}/μJ	202.3	203.5	220.1	221.3

Due to effect of parasitic parameters between measurement points, the measured $V_{GS(3\text{-}M)}$ and $V_{GS(3in4)}$ are different from the $V_{GS(3PIN)}$ used to control the channel on the chip, as shown in Fig. 10.20.

Since the influence of $L_{S(pkg\text{-}M)}$, when use TO-247-3, the measured oscillation of $V_{GS(3\text{-}M)}$ is significantly higher than $V_{GS(3PIN)}$ and $V_{GS(3\text{-}C)}$. Especially in the rapid decrease stage $I_{DS(3PIN)}$, there is an obvious decrease in $V_{GS(3\text{-}M)}$, and it is easy to mislead the waveform analysis. From Eq. (10.9), it can be seen that the influence of

Fig. 10.20 Effect of L_S on the measurement of the V_{GS} of turn off process

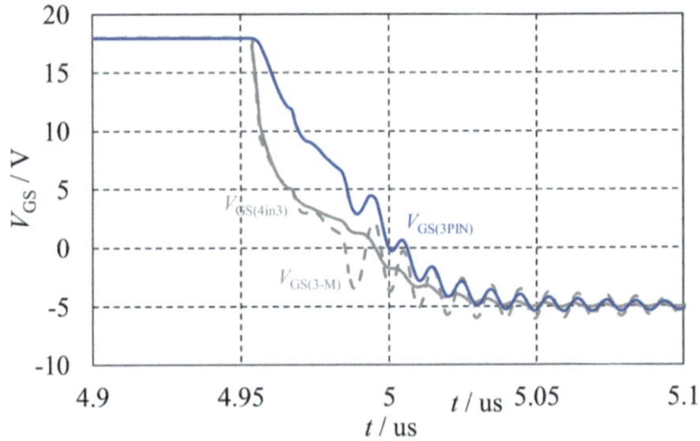

(a) Simulation results

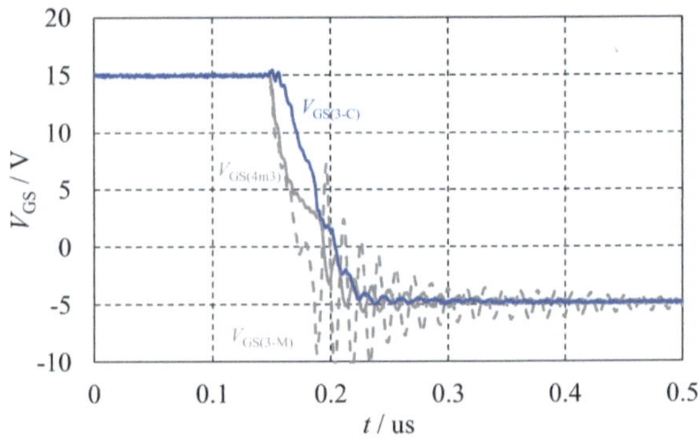

(b) Experimental results

$L_{S(pkg\text{-}M)}$ on $V_{GS(4in3\text{-}M)}$ can be avoided by using 4in3 test. However, the influence of $R_{G(int)}$ cannot be eliminated.

Combining Eqs. (10.4), (10.6), and (10.9), It follows that the smaller the gate resistor, the larger the turn-off and turn-on currents, the greater the difference between $V_{GS(3\text{-}M)}$, $V_{GS(4in3\text{-}M)}$ and $V_{GS(3\text{-}C)}$, as shown in Fig. 10.21.

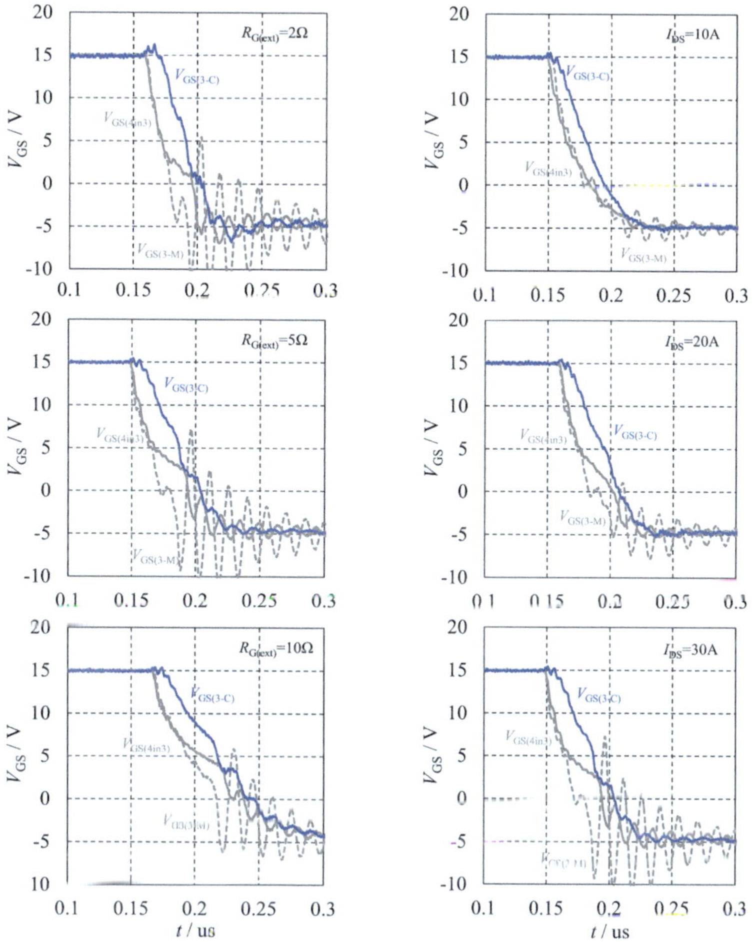

(a) Effect of gate resistance (I_{DS}=30A) (b) Effect of turn-on current ($R_{G(ext)}$=5Ω)

Fig. 10.21 Influence of L_S on the measurement of the turn-off process V_{GS} under different drive resistances and turn-on currents

10.3.3 Switching Energy and dV/dt

The load current scan test is performed under the same $R_{G(ext)}$, and the switching loss changes are shown in Fig. 10.22. The larger the load current, the more L_S increases the switching losses. Therefore, under the same $R_{G(ext)}$, the switching speed of TO-247-4 devices is faster and the switching loss is lower than that of TO-247-3 devices. It seems that it can be seen as an advantage of TO-247-4.

The switching speed of SiC MOSFEST is greatly affected by $R_{G(ext)}$. It is easy to get that if the $R_{G(ext)}$ of the TO-247-3 package device is reduced to smaller than $R_{G(ext)}$ of the TO-247-4, the switching speed of the TO-247-3 will reach even exceeds TO-247-4 devices. However, $R_{G(ext)}$ can only be reduced to 0 Ω at its minimum. Therefore, compared with TO-247-3 packaged devices, TO-247-4 packaged devices can enable the device to achieve the switching speed limit faster. Unfortunately, higher switching

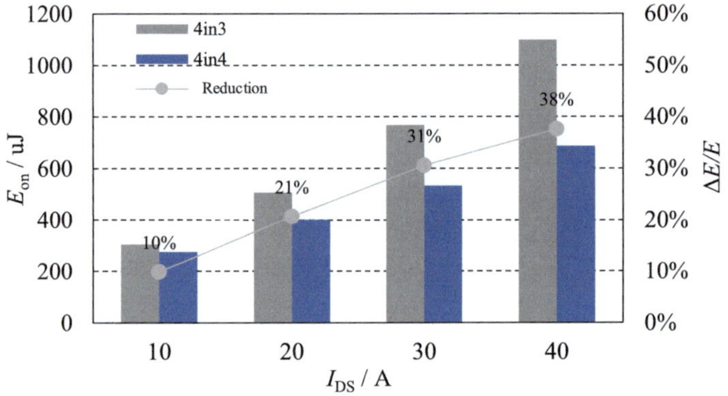

(a) Turn on loss

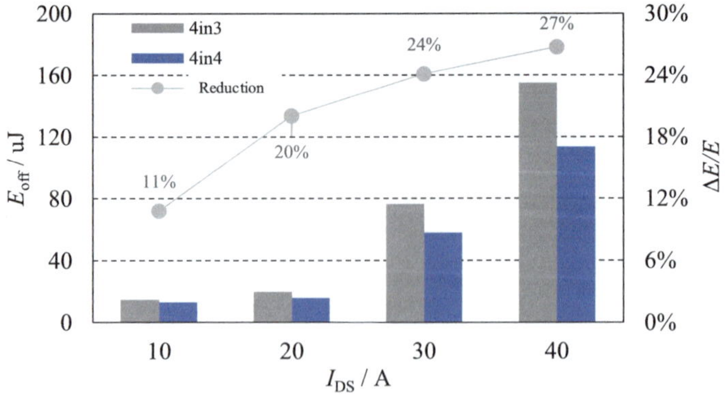

(b) Turn off loss

Fig. 10.22 Effect of L_S on switching loss

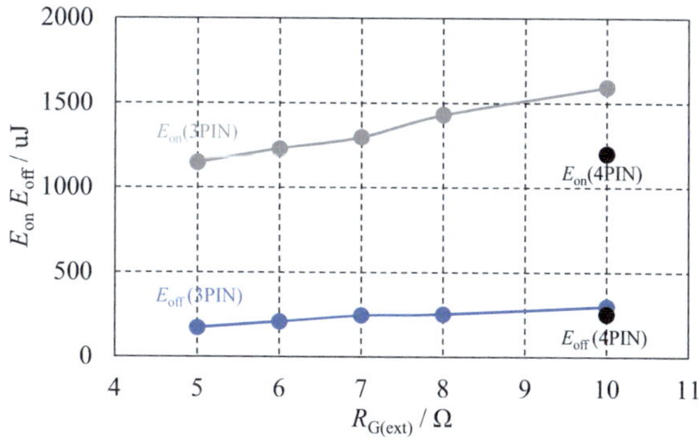

Fig. 10.23 Effect of $R_{G(ext)}$ on switching energy

speed limits are unattractive on in practical applications. Effect of EMI needs to be considered when designing the power supply, which will not make the device work at an excessively fast switching speed, and sometimes even sacrifice switching losses to ensure passes EMI standards. EMI is usually evaluated using the voltage change rate when the device switches dV_{DS}/dt. At the same time, crosstalk, which is particularly important for SiC MOSFETs, is also affected by effect of dV_{DS}/dt.

Continuously reducing the $R_{G(ext)}$ in the 4in3 test, the switching energy also decreases, as shown in Fig. 10.23. When $R_{G(ext)}$ is 10 Ω, E_{on} and E_{off} in 4in4 test are 1198 μJ and 250 μJ, while the E_{on} and E_{off} in the 4in3 test are 1594 μJ and 297 J, respectively. When $R_{G(ext)}$ is 5 Ω, E_{on} and E_{off} in the 4in3 test are 1145 μJ and 171 J, respectively.

The switching waveforms of the 4in4 test when $R_{G(ext)}$ is 10 Ω and the 4in3 test when $R_{G(ext)}$ is 5 Ω are shown in Fig. 10.24 and shown in Fig. 10.25. It can be seen that during the switching process, the dV_{DS}/dt tested in 4in3 is higher when $R_{G(ext)}$ is 5 Ω. This shows that switching losses cannot be reduced simply by reducing $R_{G(ext)}$ without any negative effect.

Under the same $R_{G(ext)}$, L_S versus dV_{DS}/dt is shown in Fig. 10.26, and the data comes from the test in Fig. 10.22. For the turn on process, in the 4in3 test, the starting point for $dV_{DS(on)}/dt$ is turn on V_{DS} platform end time and 10% V_{Bus}. In 4in4 test, $dV_{DS(on)}/dt$'s starting points are 90% V_{Bus} and 10% V_{Bus}, respectively. For the turn off process, the starting points for $dV_{DS(on)}/dt$ are respectively 10% V_{Bus} and 90% V_{Bus}. It can be seen that the dV_{DS}/dt of 4in3 test and 4in4 test are very similar.

Manufacturers that provide TO-247-4 SiC MOSFET devices emphasize that TO-247-4 can enable devices to achieve faster speeds and thereby reduce losses. Combining the results in Fig. 10.22, a more complete conclusion should be: TO-247-4 package makes the device switch faster and the switching loss is lower, and it will not significantly increase dV_{DS}/dt, achieving the best of both.

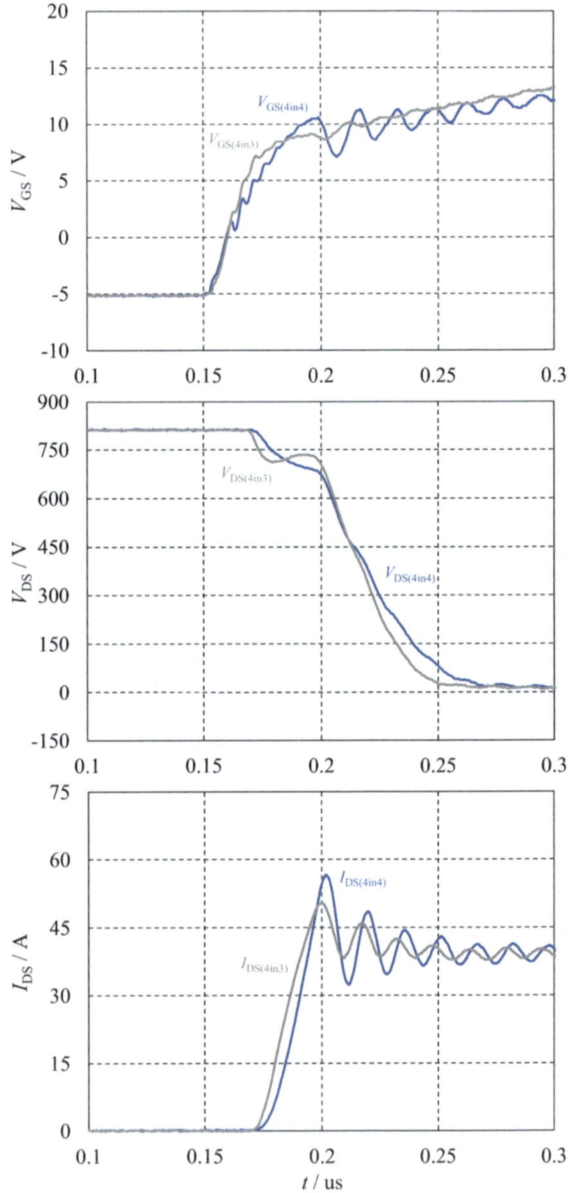

Fig. 10.24 Turn on process

Fig. 10.25 Turn off process

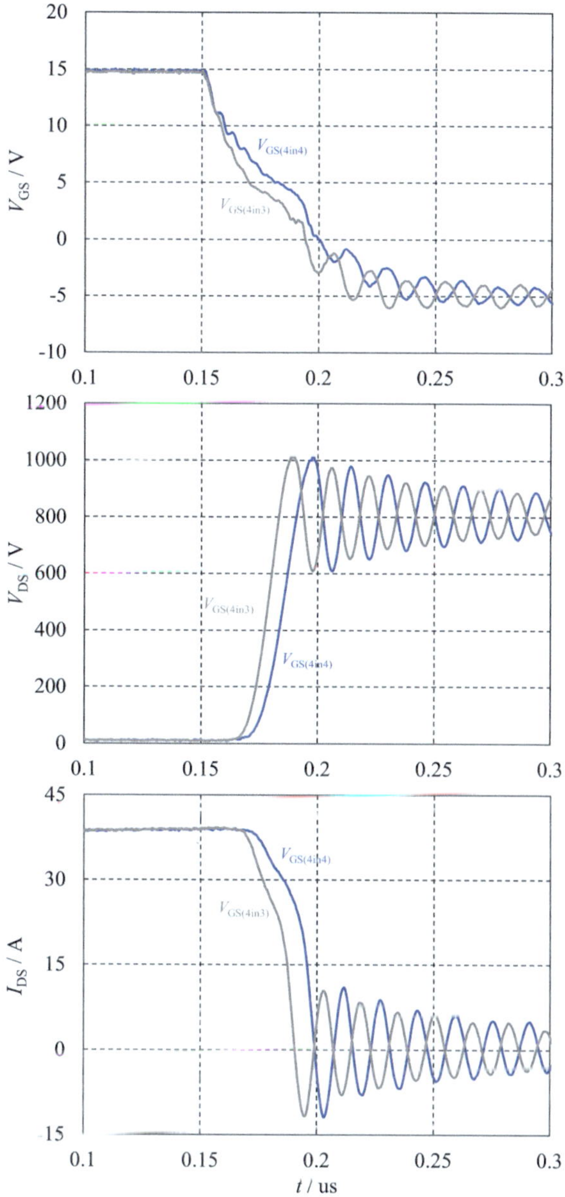

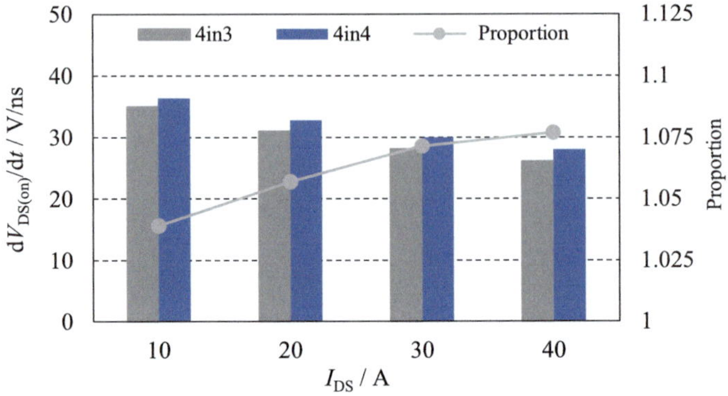

(a) Turn on process $dV_{DS(on)}/dt$

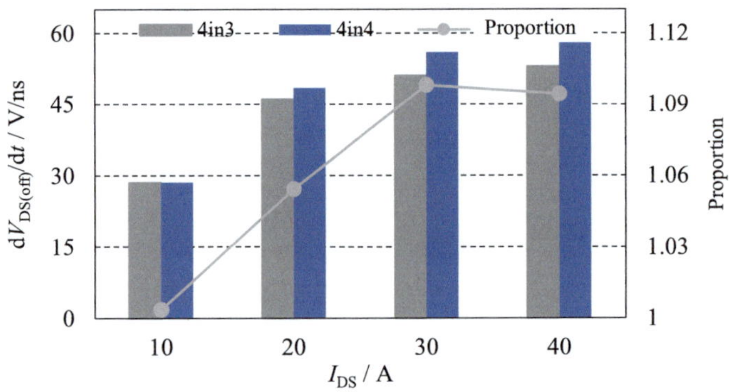

(b) Turn off process $dV_{DS(off)}/dt$

Fig. 10.26 Effect of L_S on dV_{DS}/dt

10.4 Effect on Crosstalk

The above chapters have discussed effect of V_{L_S} on the active switch. During the switching process, there is also a high dI_{DS}/dt on the opposite switch. The generated V_{L_S} will also affect V_{GS}, the measurement results are shown in Fig. 10.27. Combined with the content of Sect. 8.1, it can be seen that the process of V_{L_S} participating in crosstalk and has a significant effect on it.

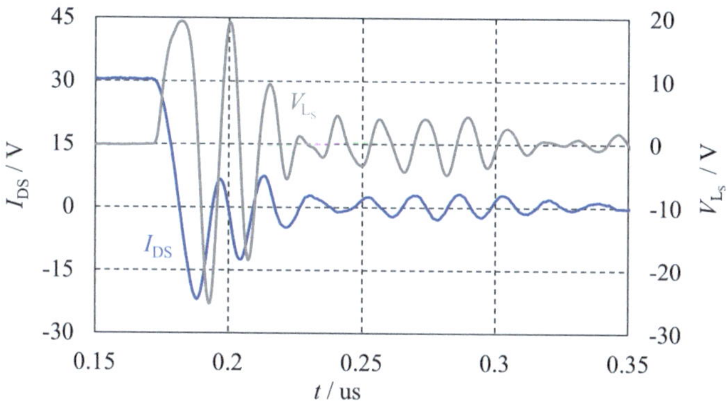

(a) Turn on process of the opposite switch

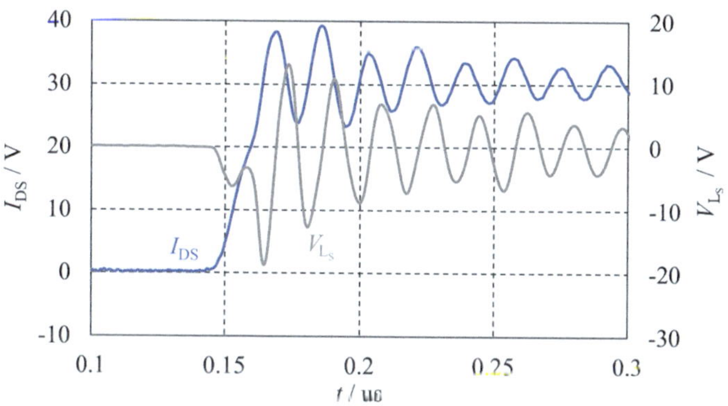

(b) Turn off process of the opposite switch

Fig. 10.27 Drain-source current I_{DS} and common source inductor voltage V_{L_S} during the switching process of opposite switch

10.4.1 Turn on Crosstalk

The research on crosstalk in Chap. 8 was based on Kelvin source package, and I_F did not participate in the crosstalk process. Figure 10.28 shows the circuit schematic diagram of the 4in3 test to turn on crosstalk. Different from the TO-247-4 device, when using the TO-247, the voltage drop V_{L_S} generated by $dI_{F(3PIN)}/dt$ on L_S also participates in crosstalk process. Then V_{L_S} and $V_{GS(3PIN)}$ are

$$V_{L_S} = (dI_{C_{GS}}/dt + dI_{F(3PIN)}/dt) \cdot L_S \tag{10.10}$$

$$V_{GS(3PIN)} = V_{DRV(off)} + R_G \cdot I_G + L_{DRV} \cdot dI_G/dt - V_{L_S} \tag{10.11}$$

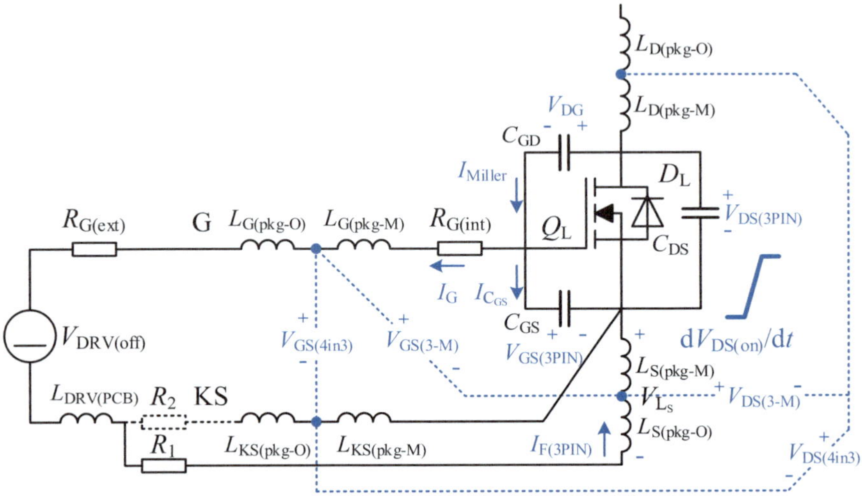

Fig. 10.28 4in3 test turn on crosstalk circuit schematic diagram

$$R_G = R_{G(ext)} + R_{G(int)} \tag{10.12}$$

$$L_{DRV} = L_{G(pkg\text{-}M)} + L_{G(pkg\text{-}O)} + L_{S(pkg\text{-}M)} + L_{S(pkg\text{-}O)} + L_{DRV(PCB)} \tag{10.13}$$

Through simulation, we can see effect of L_S on turn on crosstalk. The simulation waveform is shown in Fig. 10.29. It can be seen that L_S makes the turn on crosstalk waveform different in shape from the TO-247-4 device, and the upward spike is smaller.

1. $t_0 \sim t_1$

At t_0, Q_H starts to turn on, and $I_{F(3PIN)}$ starts to decrease rapidly from the load current. While $V_{DS(3PIN)}$ increases as $I_{F(3PIN)}$ decreases. It can be seen from Eq. (10.11) that the rise of $V_{DS(3PIN)}$ has a lifting effect on $V_{GS(3PIN)}$, the decrease of $I_{F(3PIN)}$ has a pull-down effect on $V_{GS(3PIN)}$, and the two effects are in opposite directions. In the early stage, $V_{DS(3PIN)}$ rises to play a dominant role, and $V_{GS(3PIN)}$ is elevated above $V_{DRV(off)}$. In the later stage, the decrease of $I_{F(3PIN)}$ plays a leading role, $V_{GS(3PIN)}$ is pulled down below $V_{DRV(off)}$.

2. $t_1 \sim t_2$

At t_1, D_L begins to withstand the reverse voltage, $V_{DS(3PIN)}$ rises rapidly, $I_{F(3PIN)}$ is rebounded by the reverse recovery current. And $V_{GS(3PIN)}$ is rapidly raised under the combined effect of the two.

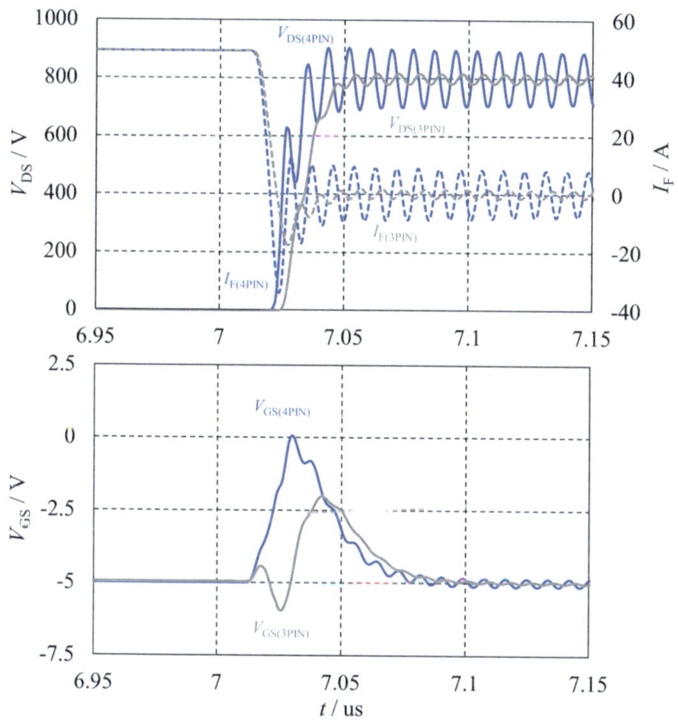

Fig. 10.29 Effect of L_S on turn on crosstalk (simulation results)

3 t_2 ~

$V_{GS(3PIN)}$ gradually falls back to $V_{DRV(off)}$, due to $V_{DS(3PIN)}$ is attenuated and oscillated after reaching V_{Bus}. So $V_{GS(3PIN)}$ is accompanied by oscillation during the fall back process.

Due to effect of parasitic parameters between measurement points, there are obvious differences between the measured $V_{GS(3-M)}$, $V_{GS(4in3-M)}$ and $V_{GS(3-C)}$ used to control the channel on the chip. In addition, the test waveforms obtained by using 4in4 and 4in3 tests are shown in Fig. 10.30. It can be seen that the $V_{GS(3-M)}$ measured when using TO-247-3 packaged devices will seriously misjudge the turn off crosstalk, and even if the 4in3 test is used, it cannot faithfully reflect turn-off crosstalk.

Combining the results of Figs. 10.29 and 10.30, it can be seen that L_S alleviates the turn on crosstalk, but under the combined effect of the drive circuit inductance L_{DRV}, device gate resistors R_G and L_S, the correct shape of the crosstalk waveform cannot be obtained when using TO-247-3 packaged devices, and the severity of turn on crosstalk is grossly overestimated.

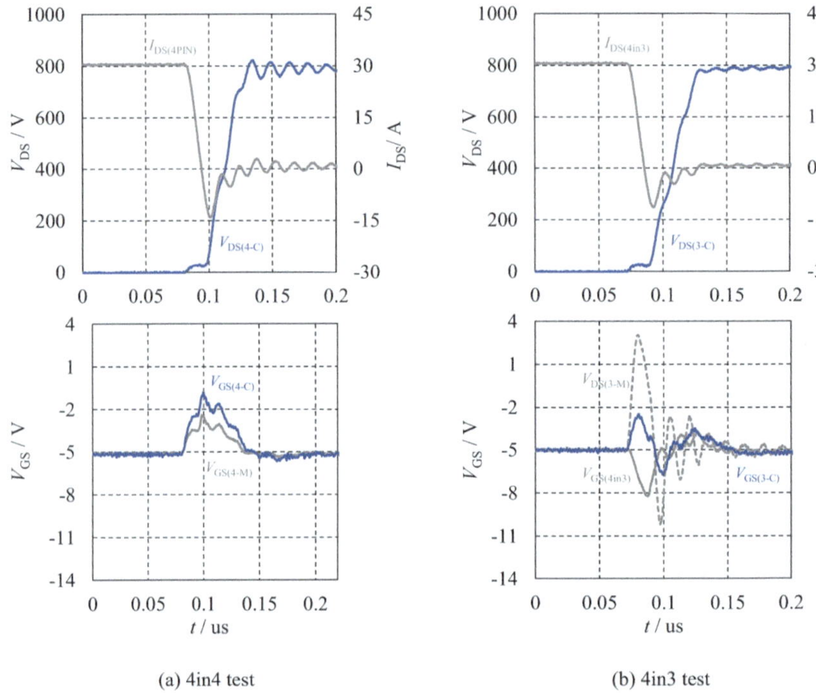

Fig. 10.30 L_S effect of L_S on turn on crosstalk

It can be seen from Sect. 8.3.2 that the premise for AMC to function as designed is to correctly detect V_{GS} of SiC MOSFET. From the above analysis and waveforms, when using TO-247-3 packaged devices, there is a huge deviation between the $V_{GS(3\text{-}M)}$ detected by AMC and $V_{GS(3PIN)}$ used to control the channel on the chip, which will cause the AMC to fail to work as expected, nor can the effects of using AMC be properly evaluated as shown in Fig. 10.31. For the results before and after applying the Active Miller Clamp (AMC), the corresponding $V_{GS(3\text{-}C)}$ waveform data cannot be provided due to unmeasured current through the AMC circuit.

10.4.2 Turn off Crosstalk

Figure 10.32 shows the schematic diagram of the 4in3 test turn off crosstalk circuit. Different from the TO-247-4 device, when using the TO-247 device, the voltage drop V_{L_S} generated by $dI_{F(3PIN)}$ on L_S also participates in the crosstalk process, then V_{L_S} and $V_{GS(3PIN)}$ are respectively

$$V_{L_S} = (dI_{DS(3PIN)}/dt - dI_{C_{GS}}/dt) \cdot L_S \tag{10.14}$$

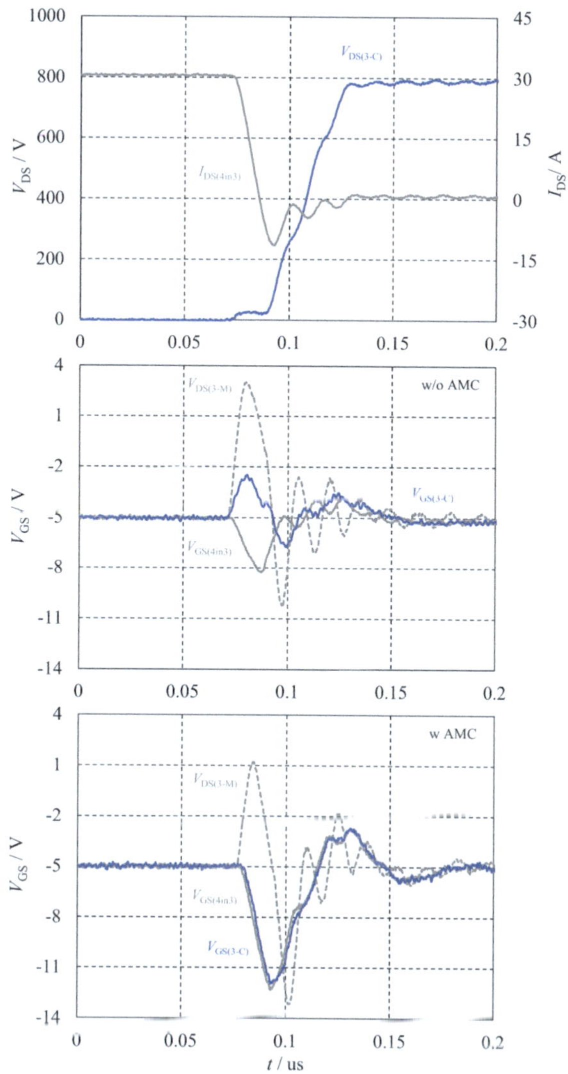

Fig. 10.31 Effect of L_S on AMC of turned on crosstalk

$$V_{GS(3PIN)} = V_{DRV(off)} - R_G \cdot I_G - L_{DRV} \cdot dI_G/dt - V_{L_3} \tag{10.15}$$

Through simulation, we can see effect of L_S on turn off crosstalk. The simulation waveform is shown in Fig. 10.33. It can be seen that L_S makes the turn off crosstalk smaller than the downward spike of the TO-247-4 device, and the difference in shape is small.

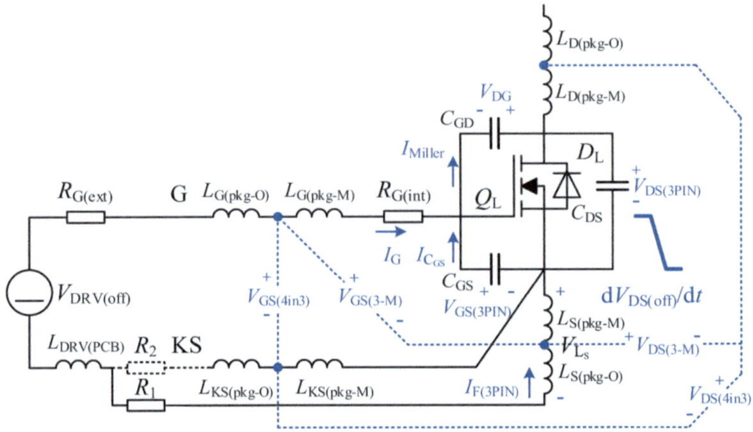

Fig. 10.32 Test turn off crosstalk circuit schematic diagram

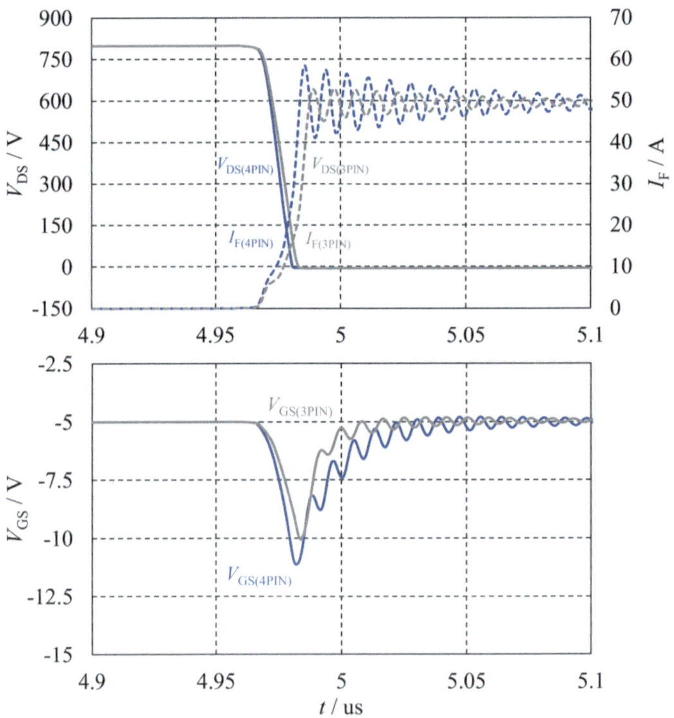

Fig. 10.33 Effect of L_S on turn off crosstalk (simulation waveform)

(1) $t_0 \sim t_1$

At t_0, Q_H begins to turn off, $V_{DS(3PIN)}$ drops quickly from V_{Bus} to V_F while $I_{F(3PIN)}$ rises. From Eq. (10.14), the decrease of $V_{DS(3PIN)}$ has a pull-down effect on V_{GS}, and the rise of $I_{F(3PIN)}$ has a lifting effect on $V_{GS(3PIN)}$, and the two effects are in opposite directions. The decrease of $V_{DS(3PIN)}$ plays a dominant role, and $V_{GS(3PIN)}$ is pulled down below $V_{DRV(off)}$.

(2) $t_1 \sim$

$V_{GS(3PIN)}$ gradually fell back to $V_{DRV(off)}$. Because $V_{DS(3PIN)}$ is attenuated and oscillated after reaching V_{Bus}, so $V_{GS(3PIN)}$ is accompanied by oscillation during the fall back process.

The test waveforms obtained by using 4in4 and 4in3 tests are shown in Fig. 10.34, where $V_{GS(3\text{-}C)}$ is the result after compensation. Due to effect of parasitic parameters between measurement points, there is an obvious difference between the measured $V_{GS(3\text{-}M)}$, $V_{GS(4in3\text{-}M)}$ and $V_{GS(3\text{-}C)}$ used to control the channel on the chip. It can be seen that the $V_{GS(3\text{-}M)}$ measured when using TO-247-3 packaged devices will seriously mislead the turn off crosstalk, and even if the 4in3 test is used, it cannot reflect turn off crosstalk.

Combining the results of Figs. 10.33 and 10.34, it can be seen that L_S alleviates the turn off crosstalk, but under the combined action of the drive circuit inductance

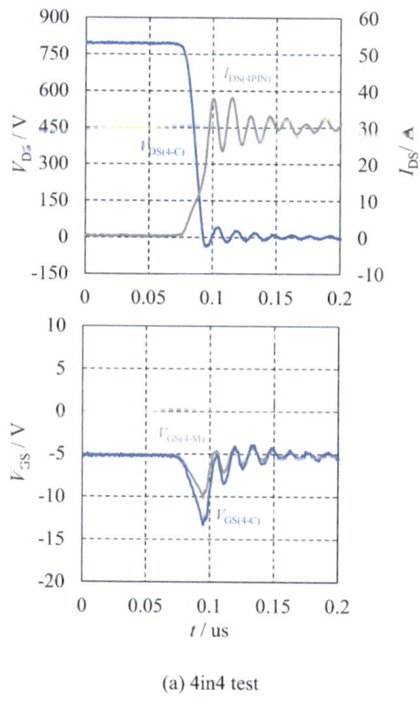

(a) 4in4 test

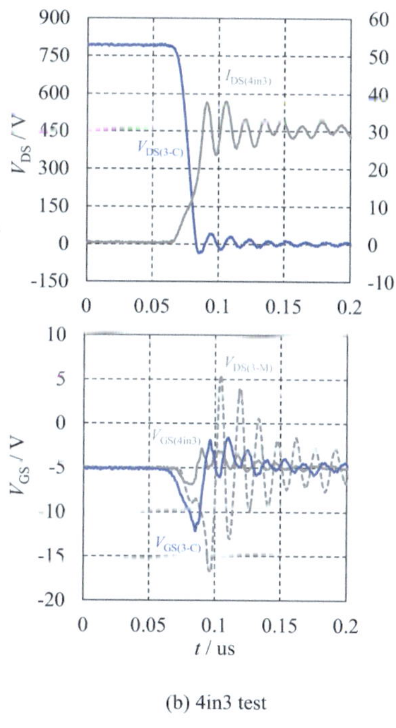

(b) 4in3 test

Fig. 10.34 Effect of L_S on turn off crosstalk

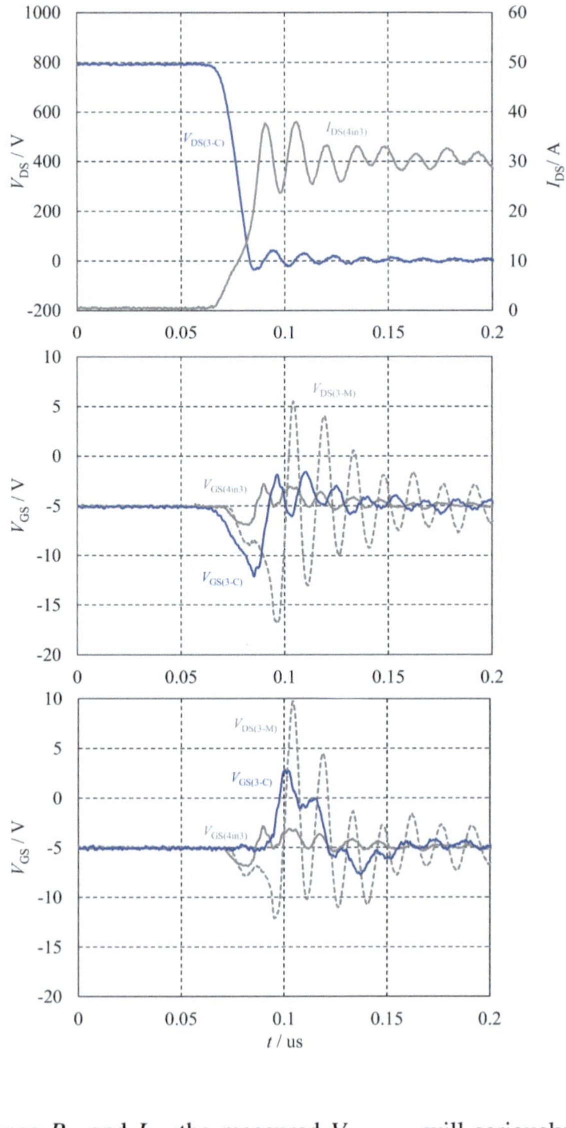

Fig. 10.35 Effect of L_S on AMC of turn off crosstalk

L_{DRV}, the device gate resistance R_G and L_S, the measured $V_{GS(3\text{-}M)}$ will seriously overestimate the severity of turn off crosstalk when using TO-247-3 packaged device.

As for turn off crosstalk, the AMC does not work as intended, nor can the effects of using AMC be properly evaluated as shown in Fig. 10.35. For the results before and after applying the Active Miller Clamp (AMC), the compensated $V_{GS(3\text{-}C)}$ waveform cannot be provided due to unmeasured current in the AMC circuit.

References

1. Infineon Technologies AG (2017) TO-247PLUS description of the packages and assembly guidelines. Application note. AN2017-01, Rev. 2.0
2. Siu K, Franz S (2015) Practical design and evaluation of an 800 W PFC boost converter using TO-247 4pin MOSFET. Application Note, AN_201409_PL52_012, Rev. 1.0, Infineon Technologies AG
3. Infineon Technologies AG (2019) CoolMOS™ in ThinPAK 8x8 the new leadless SMD package for CoolMOS™. Product Brief
4. Infineon Technologies AG (2014) ThinPAK 8x8 new high voltage SMD-package. Additional Product Information, Rev. 1.0
5. Infineon Technologies AG (2017) 600 V CoolMOS™ C7 Gold (G7) a perfect partnership for power application. Application Note, AN_201703_PL52_018, Rev. 1.0
6. Infineon Technologies AG (2016) C7 Gold CoolMOS™ C7 Gold + TOLL = a perfect combination. Application Note, AN_201605_PL52_019, Rev. 1.1
7. Preimel Stefan (2018) 600 V CoolMOS™ G7 and 650 V CoolSiC™ G6 come in a new top-side cooling package—the DDPAK. Application Note, Rev.1.0, Infineon Technologies AG

Further Reading

8. Li Y, Zhang Y, Gao Y, Du S, Liu J (2021) Switching characteristic analysis and application assessment of SiC MOSFET with common source inductance and Kelvin source connection. IEEE Trans Power Electron
9. Infineon Technologies AG (2014) TRENCHSTOP™ 5 IGBT in A Kelvin emitter configuration performance comparison and design guidelines. Application Note, Rev. 1.0
10. Vladimir S, Klaus S (2014) TRENCHSTOP™ 5 in TO-247 4pin evaluation board. Application Note, Rev. 1.0, Infineon Technologies AG
11. Zojer B (2018) A new gate drive technique for superjunction MOSFETs to compensate the effects of common source inductance. IEEE Appl Power Electron Conf Expos 2018:2763–2768
12. Stella CG, Laudani M, Gaito A et al (2014) Advantage of the use of an added driver source lead in discrete Power MOSFETs. IEEE Appl Power Electron Conf Expos 2014:2574–2581
13. Crisafulli V (2015) A new package with Kelvin source connection for increasing power density in power electronics design. In: 2015 17th European conference on power electronics and applications (EPE'15 ECCE-Europe), pp 1–8
14. Zhang W, Zhang Z, Wang F et al (2017) Common source inductance introduced self-turn on in MOSFET turn-off transient. IEEE Appl Power Electron Conf Expos 2017:837–842
15. ROHM Co., Ltd (2018) Gate-source voltage behaviour in a bridge configuration. Application Note, No. 60AN135E, Rev.001
16. ROHM Co., Ltd (2019) Improvement of switching loss by driver source. Application Note, No. 62AN040E, Rev.001

Chapter 11
Drive Circuit

When designing a drive circuit, it is necessary to apply a comprehensive understanding of device static and dynamic characteristics, electromagnetic interference (EMI), circuit analysis, thermal management, and other factors. As a result, the ability to design drive circuits with outstanding performance is one of a key criterion for assessing the capabilities of power electronics hardware engineers.

At the same time, many engineers believe that the design of SiC MOSFET drive circuits is challenging, mainly due to a limited understanding of its characteristics. Although SiC MOSFET has many differences in dynamic and static characteristics compared to Si IGBT and MOSFET, the impact on the drive circuit is not revolutionary, and it imposes heightened requirements on specific performance indicators.

With this consideration, the drive circuit is discussed in this chapter, following the introduction of SiC MOSFET characteristics and potential application challenges in the preceding chapters. The drive voltage is determined by the SiC MOSFET's output characteristics and gate voltage, details covered in Chap. 3. The drive circuit in the half-bridge configuration need to address crosstalk issues, as discussed in Chap. 8. The isolation drive circuit requires dealing with the common-mode current problem, outlined in Chap. 9. The impact of common source inductance on the drive process, along with crosstalk and its measurement, has been explored in Chap. 10. Although the preceding chapters may not explicitly delve into drive circuit design, they form the essential foundation for designing a good drive circuit.

In addition to the drive circuit content covered in the previous chapters, this chapter offers a detailed introduction to drive circuit modules, drive characteristics, model isolation, and short-circuit protection.

Y. Gao and Y. Zhang, *Silicon Carbide Power Devices*, CPSS Power Electronics Series,
https://doi.org/10.1007/978-981-96-3480-4_11

11.1 Fundamentals of Drive Circuit

11.1.1 Structure and Development of Drive Circuit

Similar to Si MOSFET and Si IGBT, SiC MOSFET operates as a voltage-controlled device, and its switching state is dictated by the gate-source voltage V_{GS}. As explained in previous chapters, to control the switching device, it's necessary to charge or discharge its input capacitance C_{iss}. In a power converter, the controller generates a turn-on or turn-off signal, namely a PWM signal, through a control algorithm. Since the controller has a limited current output capability and cannot directly drive the switching device, a circuit with specific functions must be added between the controller and the switching device. This circuit receives the PWM signal, amplifies its power, and thereby accomplishes the drive of the switching device—this is the drive circuit.

The core and basic function of the drive circuit is to receive PWM signals and perform power amplification, that is, the fundamental drive function. This function can be achieved by simply using discrete components in a push-pull circuit, representing the original structure of the drive circuit. Over the past decades, significant changes in power devices, application scenarios, and converter specifications have arisen, imposing increasingly stringent requirements on drive circuits. Through continuous research and practical applications, drive circuits have progressed substantially, acquiring diverse functions, and continuous enhancements in characteristic indicators.

When designing a converter, the choice of drive solutions depends on a comprehensive consideration of topology, application, and cost. Low-power consumer products, like mobile phone chargers, often face significant cost pressures and typically use non-isolation solutions without various protection functions. On the other hand, high-power industrial products, such as high-speed rail traction drives, demand exceptional performance and reliability, necessitating an isolation solution along with the use of multiple protection and detection functions. This parallels the process of buying a car where customers select the brand, series, and configuration based on their needs and budget, considering the car functions available.

Based on isolation characteristics, the drive circuit is categorized into three types: non-isolated, Level Shift, and isolated. Taking the isolated drive circuit as an example, its circuit framework is illustrated in Fig. 11.1.

(1) **Digital interface**

It receives PWM signals and feedback on the drive status. It has logic processing functions and anti-interference capabilities.

(2) **Signal isolation transmission**

This function achieves electrical isolation between control and power stages to ensure the safety of personnel and equipment.

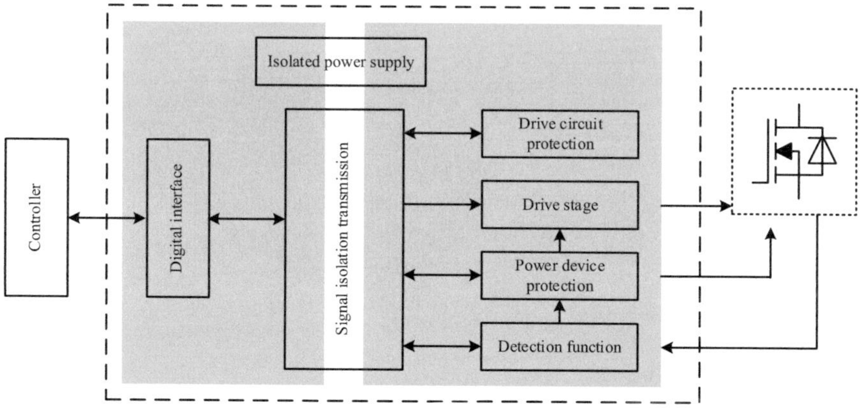

Fig. 11.1 Architecture of the isolated driver circuit

(3) **Drive stage**

This stage provides power output to drive the switching device.

(4) **Isolated power supply**

It supplies power to the power circuit of the isolated drive.

(5) **Detection**

This function includes monitoring of power devices, such as current detection and temperature detection.

(6) **Drive circuit protection**

It detects and protects against abnormal conditions in the drive circuit, such as undervoltage lockout (UVLO) and gate clamping.

(7) **Power device protection**

This function involves detecting and safeguarding against abnormal conditions in power devices, including overcurrent protection, overtemperature protection, Miller Clamping, active clamping, soft turn-off, and two-level turn-off.

The primary characteristic of the isolated drive is the combination of isolated signal transmission and isolated power supply, facilitating the efficient drive of the upper-side device in the half-bridge topology. Additionally, isolation drive methods involve signal isolation transmission combined with a Bootstrap circuit. Another common approach for drive high-side devices is using a Level Shift circuit in conjunction with a Bootstrap circuit. This method is typically implemented through a dual-channel drive IC and is commonly referred to as a half-bridge drive.

Fig. 11.2 Drive circuit based on isolated drive chip

The architecture of the isolated drive circuit reveals that each functional module operates relatively independently. These modules are assembled based on requirements to form a drive circuit. In the early stages of Si MOSFET and IGBT, drive circuit design was exploratory, and functions were gradually proposed, often using discrete devices. With the maturation of drive circuit and the finalization of functions, coupled with the rapid progress in semiconductor, drive chips have now become the mainstream. These chips integrate digital interfaces, signal isolation transmission, drive stages, detection, and protection functions into a single chip. Engineers can now design the drive circuit by adding simple peripheral passive components.

Figure 11.2 depicts a drive circuit based on an isolated drive chip. Its simplicity is evident, thanks to the integration of multiple functions. In comparison to a solution with discrete devices, this design minimizes the complexity of the design, reduces the PCB board area for the drive circuit, and enhances overall reliability.

The drive chip integrates most of functional modules, and the manufacturers also offer engineers guidance through peripheral circuit design methods, reference circuits, and evaluation boards. Selecting the drive chip has become a key task of drive circuit design. Once engineers are well-versed in the principles and specifications of each drive function, they can effectively complete the design and debugging of the drive circuit. This significantly reduces the technical threshold associated with drive circuit design. This trend aligns with the general pattern seen in the development of most technologies: as a technology matures, cost-effective, overall integrated solutions tend to become mainstream.

11.1.2 Functional Modules of Drive Circuit

As mentioned in the previous section, the selection of drive chip has become the primary task of drive circuit design. Therefore, a thorough understanding of the principles and specifications of each functional module is essential to excel in the seemingly simple selection process. The functional modules of the drive circuit are

Table 11.1 Single-ended, differential interface input–output relationship

IN+	IN−	OUT
Low	X	Low
X	High	Low
High	Low	High

well-established and have been extensively covered in various books, facilitating a quick learning curve for engineers.

While SiC MOSFET has not revolutionized the drive circuit, it has set higher requirements for some functional modules in the original drive scheme, which will be elaborated on in subsequent chapters. To provide a comprehensive knowledge framework for everyone, the remaining functional modules will be briefly outlined in the following sections.

11.1.2.1 Digital Interface

The conventional PWM signal input terminals typically comprise a single-ended interface with two pins, IN and GND, and offers two forms of positive and negative logic output. However, in recent years, there has been a growing adoption of differential interfaces in many drive chips, characterized by two pins, IN+ and IN−. The output status is determined by the states of these two pins, as outlined in Table 11.1. The main advantage of the differential interface is its robust resistance to interference, proving particularly advantageous when dealing with high-voltage and high-speed devices like SiC MOSFETs. Since controllers usually cannot directly output differential signals, the use of a single-ended to differential digital interface chip becomes necessary between the controller and the drive chip. However, this addition incurs extra costs.

In addition to the differential method, the input–output relationships presented in Table 11.1 can be flexibly utilized to achieve functions such as single-ended, differential, high enable, high turn-off, and interlocking [1], as illustrated in Fig. 11.3.

The high-level threshold voltage at the input terminal typically ranges from 55 to 70% of the supply voltage, while the low-level threshold voltage is in the range of 30% to 45% of the supply voltage. While traditional digital interfaces are often powered by +3.3 and +5 V, newly released drive chips are designed with digital power supply compatibility ranging from +3 to +18 V. When utilizing a high-voltage power supply, the anti-interference performance is further enhanced.

11.1.2.2 Isolated Power Supply

The isolated power supply supplies power to the secondary (power side) circuit of the isolated drive chip, delivering the necessary voltage and power. Typically, the voltage it provides corresponds to the drive voltage, and the specific voltage value is

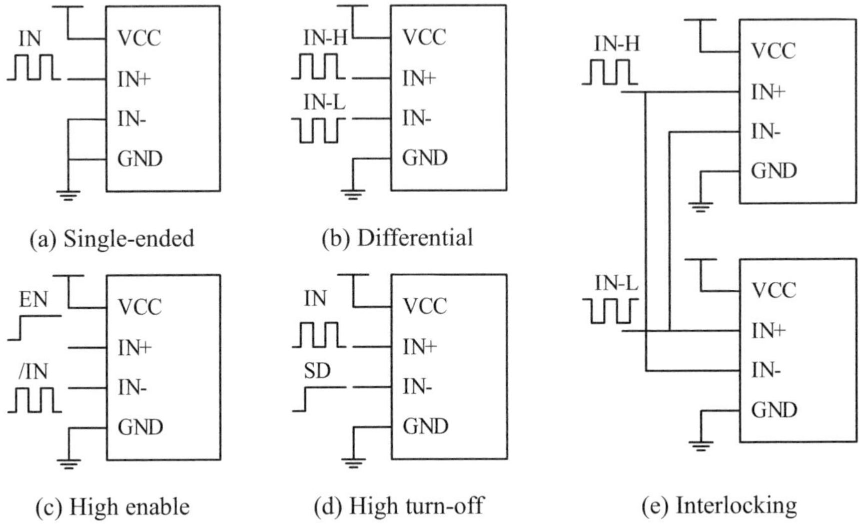

Fig. 11.3 Use of differential interface

determined by the type of power device being used. For Si MOSFET, the commonly used drive voltage is +10 V/0 V, and for Si IGBT, it is +15 V/−9 V. The output power of the isolated power supply must satisfy the power requirements outlined in Eq. (11.1).

$$P_{\mathrm{DRV}} = P_{\mathrm{DRV(bias)}} + P_{\mathrm{DRV(sw)}} = I_{\mathrm{CC2(max)}} \cdot (V_{\mathrm{CC2}} - V_{\mathrm{EE}}) + \Delta V_{\mathrm{GS}} \cdot Q_G \cdot f_s, \tag{11.1}$$

where $P_{\mathrm{DRV(bias)}}$ is the power required for steady-state operation of the secondary circuit of the drive chip, $P_{\mathrm{DRV(sw)}}$ is the power required to drive the switching device, $I_{\mathrm{CC2(max)}}$ is the drive chip The maximum steady-state operating current on the power side, V_{CC2} and V_{EE} are the secondary supply voltages of the drive chip, which respectively correspond to the turn-on drive voltage $V_{\mathrm{DRV(on)}}$ and turn-off drive voltage $V_{\mathrm{DRV(off)}}$, $\Delta V_{\mathrm{GS}} = V_{\mathrm{CC2}} - V_{\mathrm{EE}}$, Q_G is the gate charge of the switching device, and f_s is the switching frequency.

Isolated power supplies commonly employ the simplest isolated DC-DC topologies. For lower power, a flyback circuit or a forward circuit is typically employed. In cases where higher power is needed, a half-bridge circuit or a push-pull circuit is utilized, as depicted in Fig. 11.4.

Considering cost, isolated power supplies typically employ open-loop methods. In instances where high drive voltage stability is crucial, an additional voltage-stabilizing circuit may be incorporated. A common approach involves using a Zener diode or low drop-out (LDO) regulator, as illustrated in Fig. 11.5.

Apart from open-loop methods, closed-loop isolated power supplies offer more stable and reliable drive voltages, particularly for high-end applications. ROHM's

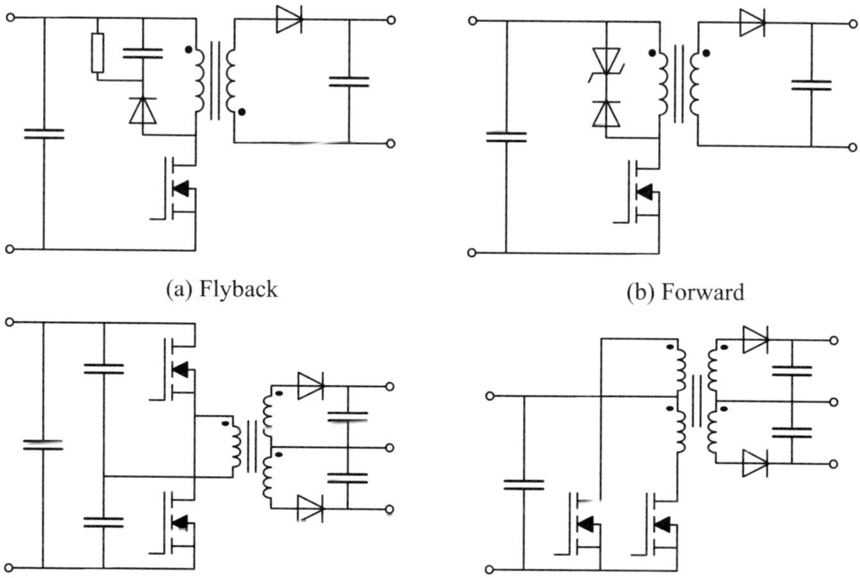

(a) Flyback (b) Forward

(c) Half-Bridge (d) Push-pull

Fig. 11.4 Common topologies for drive isolated power supplies

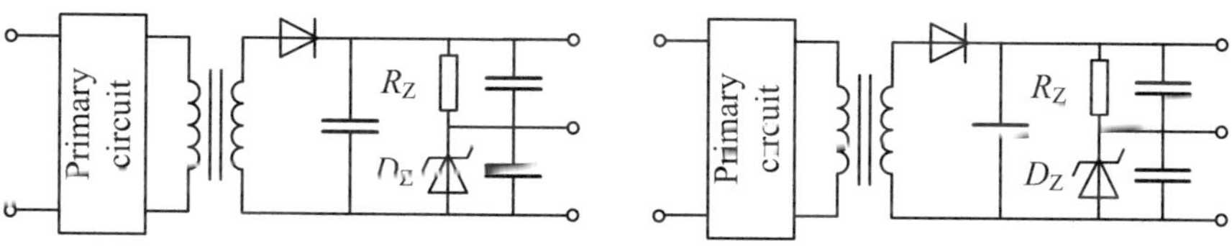

(a) Zener diode (b) LDO

Fig. 11.5 Voltage stabilization method of drive isolated power supply

BD7F100, for instance, is a primary-side feedback flyback controller. Both ROHM and SiliconLab have introduced drive chips featuring flyback controllers, as depicted in Fig. 11.6.

11.1.2.3 Level Shift

Figure 11.7 displays the internal circuit of a Level Shift half-bridge drive chip [4]. The green-shaded portion represents the low-voltage circuit, the red-shaded part signifies the high-voltage circuit, and the dotted line box encloses the Level Shift circuit. This Level Shift circuit is employed to transform the pulse signal referenced to GND into

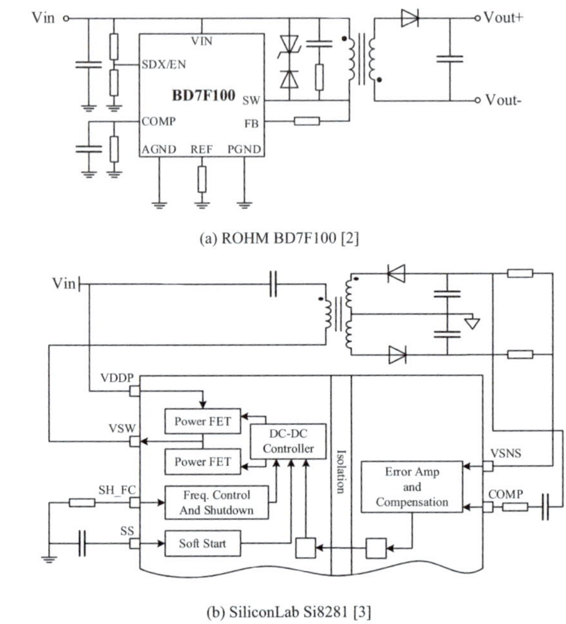

Fig. 11.6 Drive isolated power supply closed-loop mode [2, 3]

(a) ROHM BD7F100 [2]

(b) SiliconLab Si8281 [3]

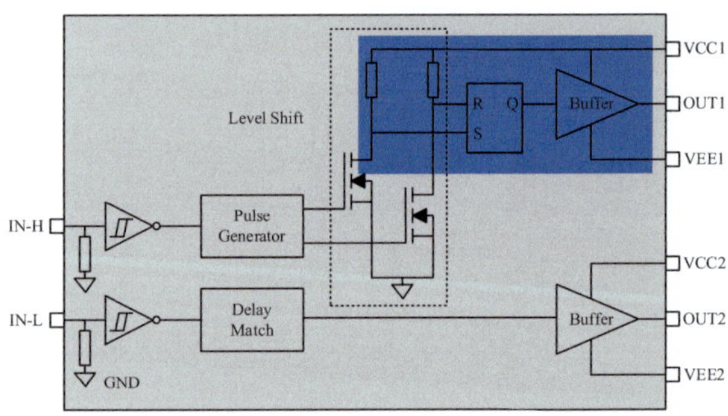

Fig. 11.7 Level Shift half-bridge drive chip internal circuit

a pulse signal referenced to VEE1 for controlling the high-voltage side circuit. Level Shift drive chips are implemented using two semiconductor processes: P-N Junction Isolated (P-N JI) and silicon on insulator (SOI).

(1) P-N JI

P-N JI involves isolating electronic components, such as transistors, with a reverse-biased p-n junction on an integrated circuit. This method employs a semiconductor material with the opposite doping type to the substrate to encircle the transistor, resistor, capacitor, or other components on the IC. The surrounding material is connected to a voltage to reverse bias the P-N junction, creating an electrically isolated well around the component, as illustrated in Fig. 11.8.

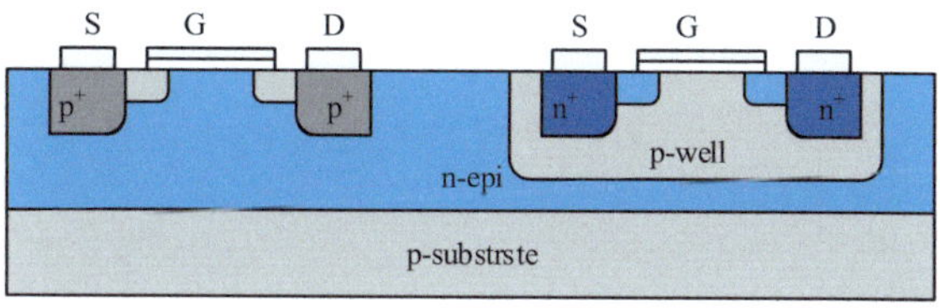

Fig. 11.8 P-N JI

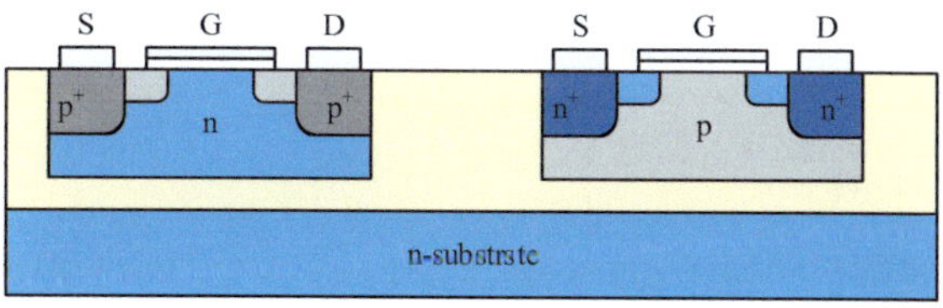

Fig. 11.9 SOI

(2) **SOI**

SOI entails covering a substrate with a layer of silicon dioxide and then fabricating components on top of it. The silicon dioxide layer serves as an insulating barrier between the active layer and the substrate, preventing leakage between adjacent components. Figure 11.9 provides a visual representation of this process.

11.1.2.4 Bootstrap Circuit

The Bootstrap circuit (Bootstrap) is used to power the drive circuit of the upper switch in the half-bridge circuit [5, 6] including resistor R_B and capacitor C_B and diode D_B three passive components, as shown in Fig. 11.10. It has the advantages of simplicity and low cost and is widely used. Its working principle is shown in Fig. 11.11.

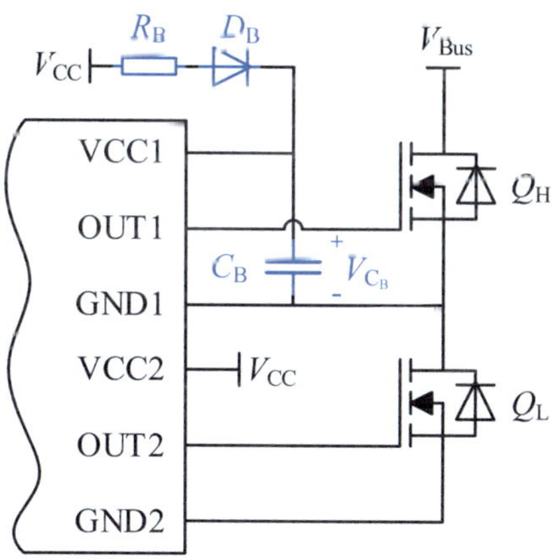

Fig. 11.10 Bootstrap circuit

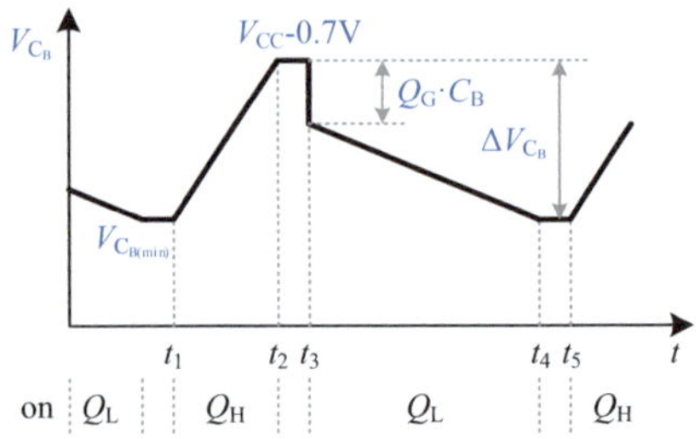

Fig. 11.11 Working principle of bootstrap

(1) $t_1 - t_2$

Lower switch Q_L turns on, upper switch Q_H turns off. Since the forward voltage drop of Q_L is very low, D_B is turned on, and V_{CC} passes R_B to charge C_B, and the voltage across it V_{C_B} continues to rise, reaching $V_{CC} - 0.7\text{V}$;

(2) t_2

Turns off, the bridge arm midpoint voltage quickly reaches the bus voltage, D_B turns off, C_B starts to supply power to the upper device drive circuit;

(3) $t_2 - t_3$

Dead time;

(4) t_3

Q_H is turn on, C_B provides the required drive charge Q_G, the voltage across it drops rapidly Q_G/C_B;

(5) $t_3 - t_4$

Q_H is on, C_B provides the energy for the drive circuit to operate normally and maintain the gate voltage. The voltage at both ends slowly drops to $V_{C_B(\min)}$.

(6) t_4

Q_H turns off;

(7) $t_4 - t_5$

Dead time;

(8) t_5

Q_L turns on, and so on.

In order to avoid excessive conduction loss due to too low drive voltage, it is required that the minimum value of C_B voltage shall not be lower than $V_{C_B(\min)}$, then the voltage fluctuation ΔV_{C_B} of C_B is,

$$\Delta V_{C_B} = V_{CC} - 0.7 - V_{C_B(\min)} < 0.05 \times (V_{CC} - 0.7). \quad (11.2)$$

Under normal circumstances, the dead time is much smaller than the turn-off or turn-on time of the device, then the total discharge charge Q_{C_B} of C_B is:

$$Q_{C_B} = Q_G + I_{G(\text{bias})} \cdot t_{\text{on}}. \tag{11.3}$$

Among them, Q_G is the gate charge of the driven power device, $I_{G(\text{bias})}$ is the drive circuit maintenance The quiescent operating current of the gate voltage is turned on, t_{on} is the time length of $t_3 - t_4$. Combining (11.2) and (11.3), the capacity of C_B needs to satisfy,

$$C_B > Q_{C_B}/\Delta V_{C_B}. \tag{11.4}$$

In order to ensure that under all working conditions, V_{C_B} can reach $V_{\text{CC}} - 0.7\text{V}$ before t_2, the value of R_B must satisfy,

$$R_B < t_{\text{off(min)}}/(5 \times C_B), \tag{11.5}$$

where $t_{\text{off(min)}}$ is the minimum off time of Q_H, that is, $t_1 - t_2$.

It should be noted that the reverse recovery process of the diode D_B in the bootstrap circuit will affect the safety of the drive circuit and generate a large amount of heat. In order to solve this problem, SiC diodes are increasingly being chosen for use in bootstrap circuits.

11.1.2.5 Active Clamping

When the switching device is turned off in an overcurrent or short-circuit state, a higher voltage spike will be generated than when it is turned off under normal operating current. The method introduced in Chap. 8 can no longer effectively suppress the voltage spike. In this case, it is necessary to use active clamping circuit.

The active-clamp circuit consists of a TVS device and a fast recovery diode connected in reverse series [7], as depicted in Fig. 11.12. The fundamental principle is as follows: when the turn-off voltage peak of the switching device surpasses the breakdown voltage of the TVS, the TVS breaks down, allowing the breakdown current i_R to flow into the gate circuit. Consequently, the decreasing gate voltage is elevated, slowing down the turn-off speed of the device and effectively suppressing voltage spikes.

To enhance the performance of active clamping, various iterations of active clamping circuits have emerged. One advancement involves feeding back the breakdown current of the TVS device to the signal input of the drive output stage, forming an enhanced active clamp, as illustrated in Fig. 11.13. This modification can notably enhance dynamic response speed and allow for the use of lower-cost, low-current level TVS devices. This circuit is extensively employed in PI Company's SCALE1 series drives [7].

Fig. 11.12 Active-clamp circuit

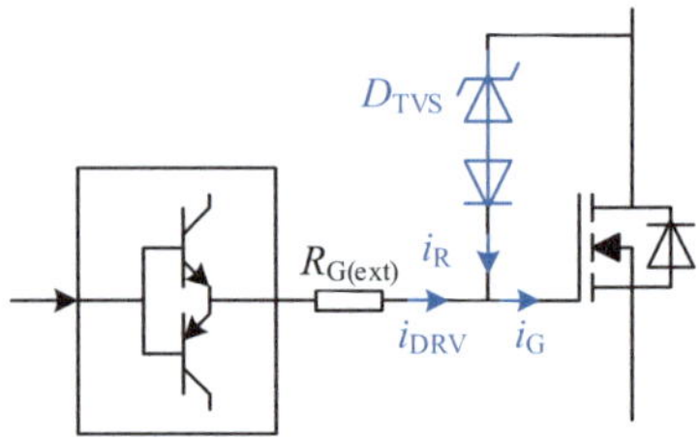

Fig. 11.13 Improved active clamping

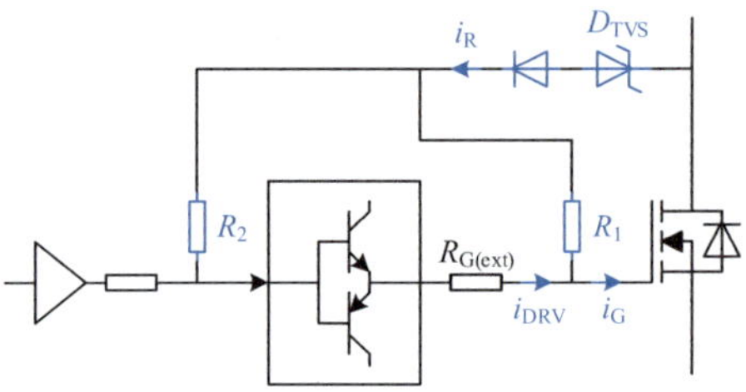

Detect i_{AAC} current, and when the current value reaches the set value, control the drive circuit to slowly turn off the switch. This is advanced active clamping, such as shown in Fig. 11.14. The biggest advantage of this circuit is that the load of the TVS device is very small and works at the rated operating point, so the clamping voltage is very stable. This circuit is widely used in PI's SCALE2 series drive and SIC1182K drive chips [7].

Add a parallel circuit composed of TVS devices and switching devices to form dynamic advanced active clamping, as shown in Fig. 11.15. When the switch is turned on and for a period of time after it is turned off, TD is turned on, and the threshold voltage of the active clamp is low at this time; when the switch is turned off, TD is turned on, and the threshold voltage of the active clamp is increased.. This can effectively solve the problem that under a single clamping threshold voltage, the upward floating of the bus voltage may cause the switching device to always enter the linear region. This circuit is widely used in PI Company's SCALE2 series drives [8].

Fig. 11.14 Advanced active clamping

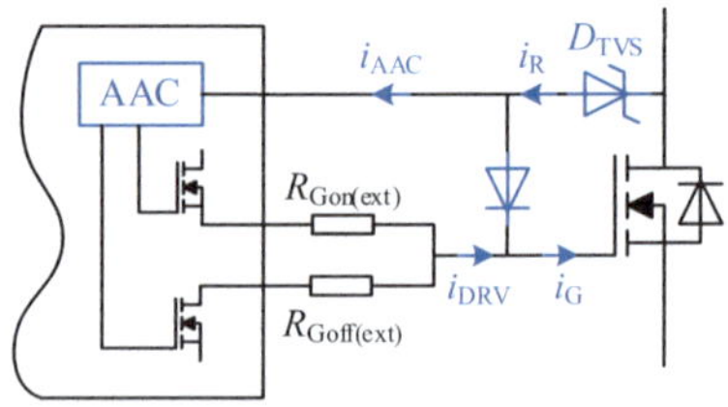

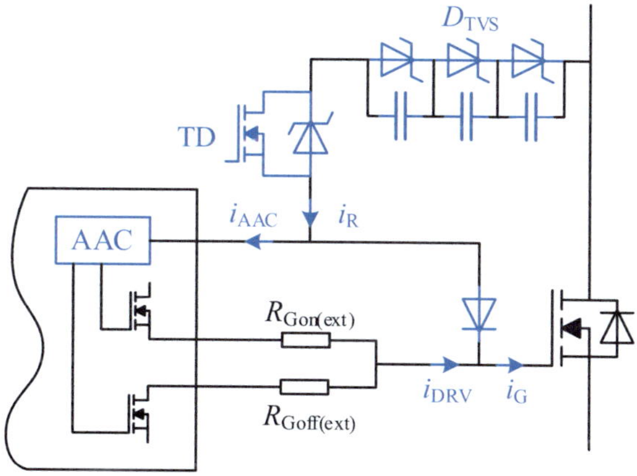

Fig. 11.15 Dynamic advanced active clamping

11.2 Value of Gate Resistor

When designing a drive circuit, a common problem is how much the gate resistor should be. Before answering this question, let's first take a look at the effects of changing the resistance of the gate resistor.

Figure 11.16 shows a typical drive circuit model, $R_{Gon(ext)}$ and $R_{Goff(ext)}$ are the external turn-on gate resistor and the external turn-off gate resistor respectively, $R_{Gon(DRV)}$ and $R_{Goff(DRV)}$ respectively turn on the drive internal resistance and turn off the drive internal resistance of the output stage of the drive circuit, $R_{G(int)}$ is the internal gate resistance of the device, $V_{DRV(on)}$ and $V_{DRV(off)}$ are the turn-on drive voltage and the turn-off drive voltage respectively, C_{GD} is the gate-drain capacitance, C_{GS} is the gate-source capacitance, L_{DRV} is the drive loop inductance.

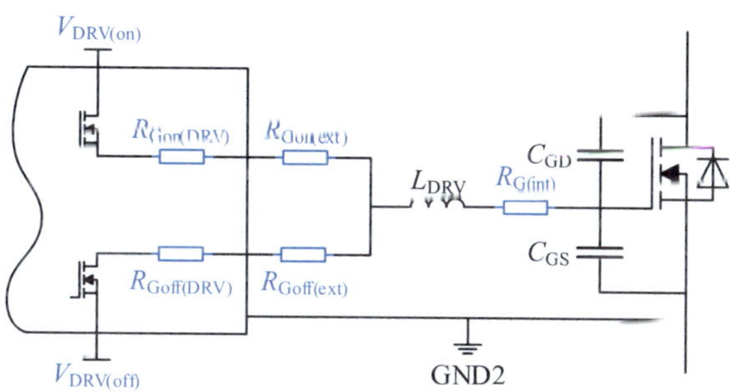

Fig. 11.16 Drive circuit model

11.2.1 Impact on Drive Circuit

The essence of the drive circuit drive the power device is to charge or discharge the C_{GD} and C_{GS} of the power device, which requires drive. The circuit has corresponding current output capabilities. Typical drive waveforms are shown in Fig. 11.17. During the turn-on or turn-off process, the drive current quickly reaches the peak value and then slowly decreases; when the device is fully turned on or turned off, the drive voltage reaches the given turn-on and turn-off drive voltages; the drive current stabilizes at a very small value to maintain the drive voltage.

Turn on drive peak current $I_{Gon(peak)}$ and turn off peak drive current $I_{Goff(peak)}$ can be estimated by (11.6) and (11.7), which is determined by the drive voltage and gate resistance. The smaller the gate resistance, the greater the required peak drive current. When designing the drive circuit, the peak source current and peak sink

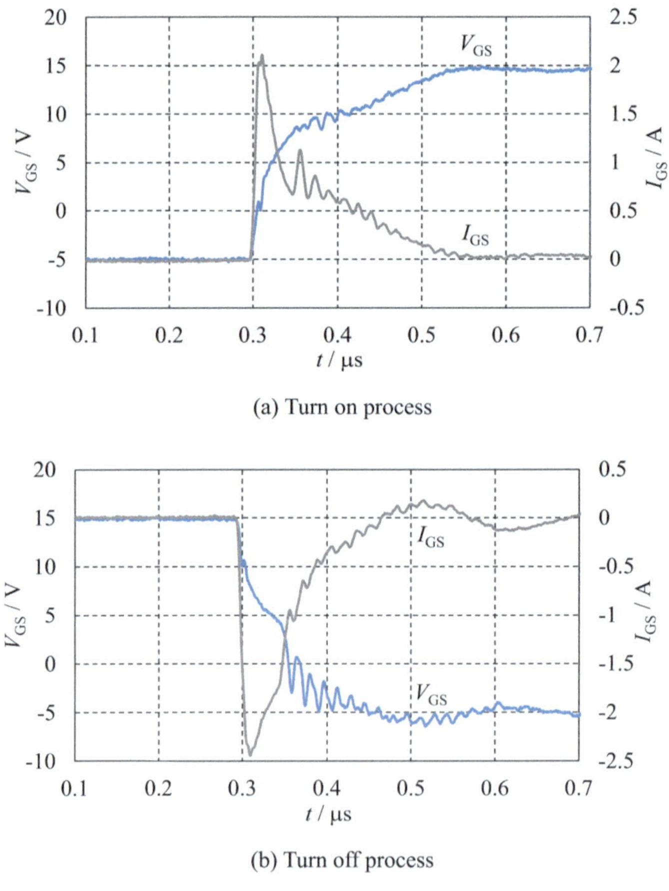

Fig. 11.17 Typical gate drive waveform

current of the selected drive chip need to meet the requirements of the drive peak current. Otherwise, the switching speed will be reduced and the life of the drive stage of the drive chip will be affected.

$$I_{\text{Gon(peak)}} \approx \frac{V_{\text{DRV(on)}} - V_{\text{DRV(off)}}}{R_{G\text{(on)}}} \approx \frac{V_{\text{DRV(on)}} - V_{\text{DRV(off)}}}{R_{\text{Gon(DRV)}} + R_{\text{Gon(ext)}} + R_{G\text{(int)}}} \tag{11.6}$$

$$I_{\text{Goff(peak)}} \approx \frac{V_{\text{DRV(off)}} - V_{\text{DRV(on)}}}{R_{G\text{(off)}}} \approx \frac{V_{\text{DRV(off)}} - V_{\text{DRV(on)}}}{R_{\text{Goff(DRV)}} + R_{\text{Goff(ext)}} + R_{G\text{(int)}}} \tag{11.7}$$

The power dissipation on the internal resistance of the drive stage of the drive circuit and the external gate resistor are respectively

$$P_{R_{\text{Gon(DRV)}}} = \frac{1}{2} Q_G \cdot \left(V_{\text{DRV(on)}} - V_{\text{DRV(off)}}\right) \cdot f_s \cdot \frac{R_{\text{Gon(DRV)}}}{R_{\text{Gon(DRV)}} + R_{\text{Gon(ext)}} + R_{G\text{(int)}}} \tag{11.8}$$

$$P_{R_{\text{Goff(DRV)}}} = \frac{1}{2} Q_G \cdot \left(V_{\text{DRV(on)}} - V_{\text{DRV(off)}}\right) \cdot f_s \cdot \frac{R_{\text{Goff(DRV)}}}{R_{\text{Gon(DRV)}} + R_{\text{Gon(ext)}} + R_{G\text{(int)}}} \tag{11.9}$$

$$P_{R_{\text{Gon(ext)}}} = \frac{1}{2} Q_G \cdot \left(V_{\text{DRV(on)}} - V_{\text{DRV(off)}}\right) \cdot f_s \cdot \frac{R_{\text{Gon(ext)}}}{R_{\text{Gon(DRV)}} + R_{\text{Gon(ext)}} + R_{G\text{(int)}}} \tag{11.10}$$

$$P_{R_{\text{Goff(ext)}}} = \frac{1}{2} Q_G \cdot \left(V_{\text{DRV(on)}} - V_{\text{DRV(off)}}\right) \cdot f_s \cdot \frac{R_{\text{Goff(ext)}}}{R_{\text{Gon(DRV)}} + R_{\text{Gon(ext)}} + R_{G\text{(int)}}} \tag{11.11}$$

The impact of the gate resistor's resistance on the distribution of drive losses is evident. A smaller external gate resistance results in increased losses absorbed by the drive stage of the drive chip. In drive design, it is crucial to account for the influence of drive power dissipation on the operating temperature of the drive chip. Simultaneously, selecting a package that aligns with the power dissipation requirements for the external gate resistor is essential.

11.2.2 Impact on Power Devices

The drive loop is a second-order circuit, which is composed of the drive loop inductance L_{DRV}, the device gate-source capacitance C_{GS} and the gate resistance R_G. During the drive process, the smaller the gate resistance, the faster the switching device is turned on and off, and the more serious the gate voltage oscillation will be, as shown in Table 11.2 and Figs. 11.18 and 11.19

Table 11.2 Effect of gate resistance on the turn-off process

$R_{G(\text{ext})}/\Omega$	2.2	5.1	10	15
$t_{d(\text{on})}$/ns	10.6	13.8	23.2	40.8
t_r/ns	19.9	35.0	55.7	62.3
E_{on}/uJ	77.5	66.0	57.5	53.9
V_{spike}/V	378.1	632.2	1034.0	1412.0
$t_{d(\text{off})}$/ns	15.9	18.2	23.1	31.2
t_f/ns	8.5	11.1	16.6	21.8
E_{off}/uJ	1030.0	1020.0	968.0	940.0
I_{spike}/A	131.2	205.2	346.0	485.9

(1) According to Sect. 7.1, the smaller the gate resistance, the larger the $\mathrm{d}I_{\mathrm{DS}}/\mathrm{d}t$ when the device is turned off, and the greater the voltage spike high.
(2) According to Sect. 8.2, the smaller the gate resistance, the larger the $\mathrm{d}V_{\mathrm{DS}}/\mathrm{d}t$ when the device is turned off, and the more serious the crosstalk will be.

11.2.2.1 Impact on Converter

(1) According to Sect. 4.3.2, the smaller the gate resistance, the faster the switching speed of the device, the smaller the switching loss, and the higher the efficiency of the converter.
(2) The smaller the gate resistance and the larger $\mathrm{d}V_{\mathrm{DS}}/\mathrm{d}t$, the more serious the EMI problem faced by the converter.

The analysis above indicates that selecting the value of the gate resistor involves balancing multiple factors. To minimize switching losses, enhance efficiency, and increase power density, opting for a smaller gate resistor is essential. However, excessively small gate resistance can lead to issues such as excessive turn-off spikes, severe crosstalk, aggravated EMI, and pronounced gate voltage oscillations.

In converter design, the maximum resistor value must align with the converter's efficiency requirements, while the drive circuit should meet peak current demands. Simultaneously, it is crucial to optimize the design to address issues arising from switching speed, including controlling the main power commutation circuit inductance, managing gate drive circuit inductance, and implementing measures like Miller clamp. In instances where certain problems prove challenging to resolve, compromises may be necessary, potentially involving an increase in gate resistance.

Fig. 11.18 Effect of gate resistor on turn-on process

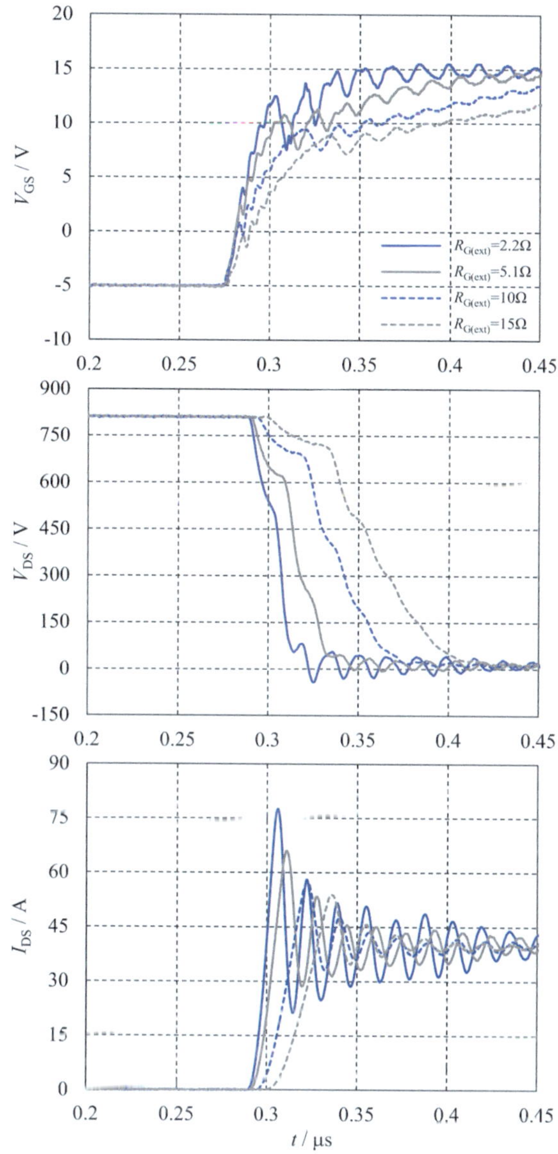

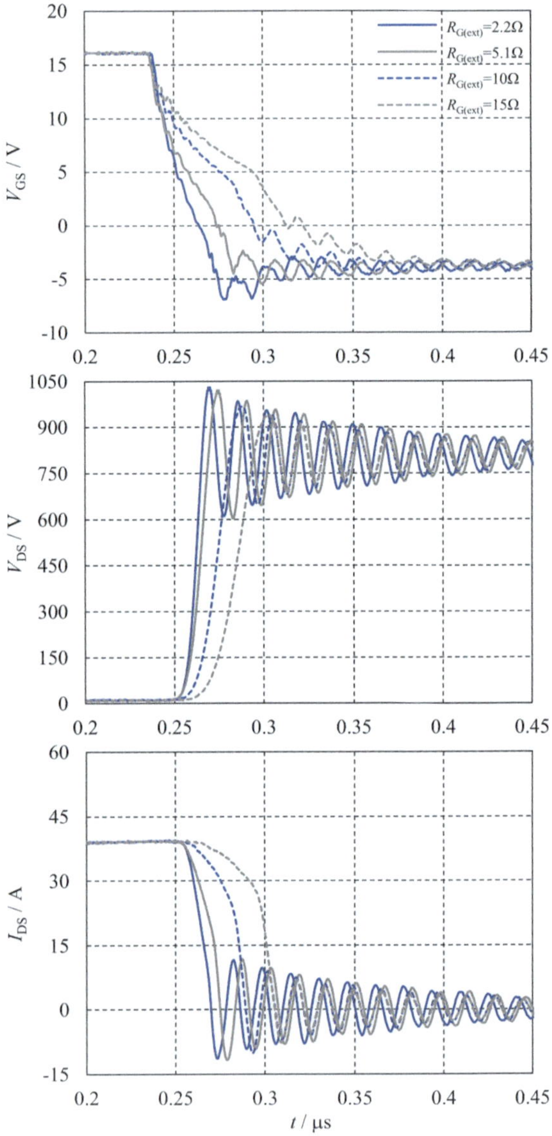

Fig. 11.19 Effect of gate resistor on turn-off process

11.3 Drive Voltage

11.3.1 SiC MOSFET Drive Voltage Requirements

The selection of the power device drive voltage is influenced by factors such as the device gate withstand voltage, output characteristics, and other switching features. With the maturity of Si power devices, drive voltages across various manufacturers are generally standardized. Si MOSFETs commonly use +10 V/0 V or +12 V/0 V, while Si IGBTs typically operate with +15 V/−9 V. However, the development of SiC MOSFETs is ongoing, and recommended drive voltages vary among manufacturers.

It is crucial to ensure that the chosen drive voltage stays within the gate withstand voltage limit to avoid the risk of gate breakdown. Given the potential overshooting of gate voltage during the drive process, a safety margin must be maintained. Table 11.3 provides the gate withstand voltage limits for SiC MOSFETs from different manufacturers.

Secondly, the output characteristics of a SiC MOSFET are shown in Fig. 11.20. When the turn-on drive voltage $V_{DRV(on)}$ is higher than 16 V, the device is fully turned on, $R_{DS(on)}$ changes with $V_{DRV(on)}$. The increase changes very little. If the value of $V_{DRV(on)}$ is too small, it will cause $R_{DS(on)}$ to be too large, resulting in excessive conduction loss.. If $V_{DRV(on)}$ continues to be too small, when $V_{DRV(on)}$ is less than 13 V, SiC MOSEET will work at $R_{DS(on)}$ is the area of negative temperature coefficient area, which makes the devices unable to be used in parallel. It is for these reasons that the drive circuit also needs an undervoltage protection (UVLO) function to detect the drop of the drive voltage in time to ensure the safe operation of the converter.

In addition, according to the introduction in Chap. 10, SiC MOSFET has fast switching speed and low $V_{GS(th)}$, so its crosstalk problem is very serious. At this time, choosing the turn-off drive voltage $V_{DRV(off)}$ as a negative voltage can reduce the risk of bridge arm shoot-through caused by turn-on crosstalk. But on the other hand, the current gate negative voltage withstand voltage of SiC MOSFET is far less

Table 11.3 SiC MOSFET gate withstand voltage limits of each manufacturer

Device	Maximum value/V	Minimum value/V
ROHM 2nd generation	22.0	– 6.0
ROHM 3rd generation	22.0	– 4.0
Wolfspeed 2nd generation	25.0	– 10.0
Wolfspeed 3rd generation	19.0	– 8.0
Infineon	23.0	– 7.0
ST	25.0	– 10.0
OnSemi	25.0	– 15.0
Littelfuse	22.0	– 6.0
Microsemi	25.0	– 10.0

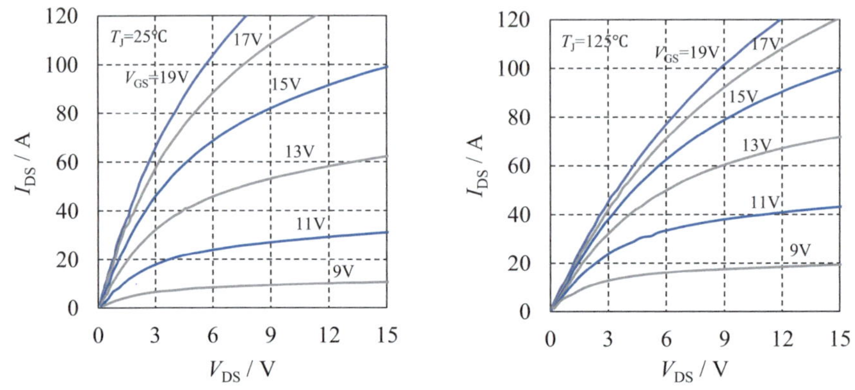

Fig. 11.20 Output characteristics of SiC MOSFET

than the − 30 V of Si devices. Turning off crosstalk can easily cause the gate voltage to exceed the limit, so $V_{DRV(off)}$ value cannot be too low.

Finally, SiC MOSFET also has gate oxide reliability issues, which puts forward new requirements for the selection of $V_{DRV(off)}$. When negative voltage is used for turn-off, as the number of switching times accumulates, the $V_{th(GS)}$ of the SiC MOSFET will gradually increase, causing $R_{DS(on)}$ rises, potentially causing the device to overheat. The lower the turn off negative pressure, the more serious the drift will be. Therefore, the negative pressure cannot be too low during use. In response to this problem, some SiC device manufacturers have given recommended values for negative voltage selection to ensure that $V_{th(GS)}$ The increase in $R_{DS(on)}$ caused by drift is within the acceptable range [9].

Therefore, when designing the drive circuit, it is necessary to comprehensively balance the above four points to reasonably determine the drive voltage value.

11.3.2 Turn Off Negative Voltage Supply

1. **Bipolar Drive Chip**

The secondary side of the bipolar drive chip has three power supply pins: positive power supply, power ground, and negative power supply. They are generally named V_{CC2}, GND2, and V_{EE2}, respectively, which requires bipolar power supply. A bipolar power supply can be used to directly power the drive chip, or a Zener diode can be used to provide bipolar power supply on the basis of a unipolar power supply, as shown in Fig. 11.21. When the output is high, V_{OUT} − GND2 outputs V_{CC2}, which is $V_{DRV(on)}$; When the output is low, V_{OUT} − GND2 outputs V_{EE2}, which is $V_{DRV(off)}$.

Fig. 11.21 Negative voltage power supply circuit of bipolar drive chip

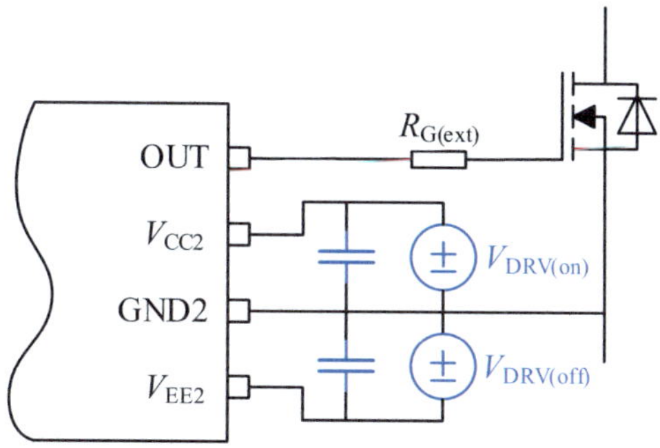

(a) Bipolar power supply

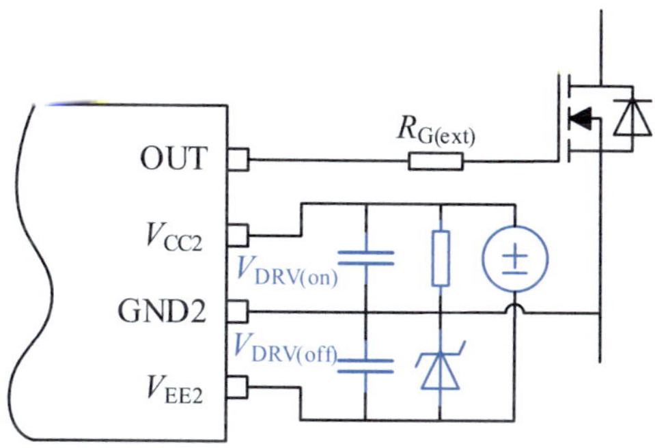

(b) Unipolar power source and zener diode power supply

2. Unipolar Drive Chip

Unipolar drive chips have two power supply pins, positive power supply and power ground. In single-channel drive chips, they are generally named V_{CC2} and GND2 respectively. If it is powered by a single power supply, $V_{DRV(on)}$ is V_{CC2}, $V_{DRV(off)}$ is 0 V, and negative voltage turn off cannot be achieved.

There are three solutions to achieve negative voltage turn off of unipolar drive chips, as shown in Fig. 11.22.

(1) Bipolar Power Supply

The bipolar power supply output is V_P and V_N connected in series. The sum of the two is used to power the drive chip, and the common point of the two is connected. The S pole of SiC MOSFET. When the output is high, V_{OUT} − GND2 outputs $V_P + V_N$, $V_{DRV(on)}$ is V_P; when the output is low, V_{OUT} − GND2 output 0 V, $V_{DRV(off)}$ is $-V_N$, thus achieving negative voltage turn off.

Fig. 11.22 Negative voltage power supply circuit of unipolar drive chip

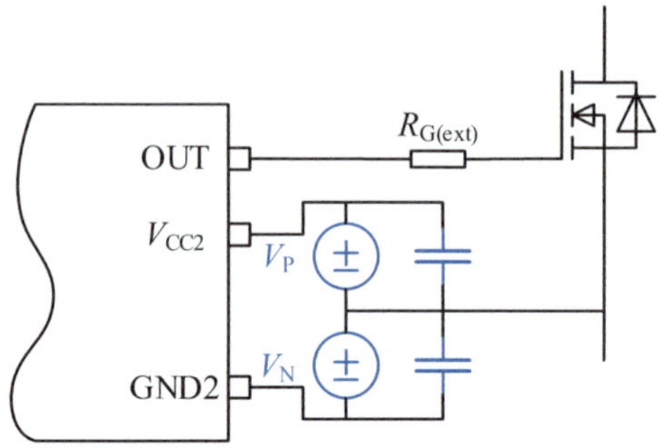

(a) Bipolar power supply

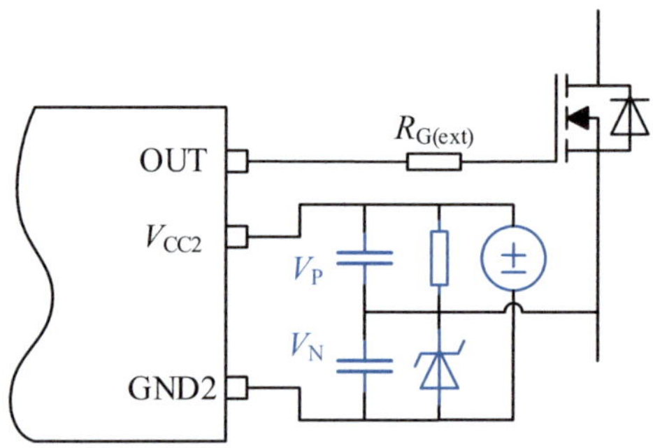

(b) Unipolar power supply parallel zener diode

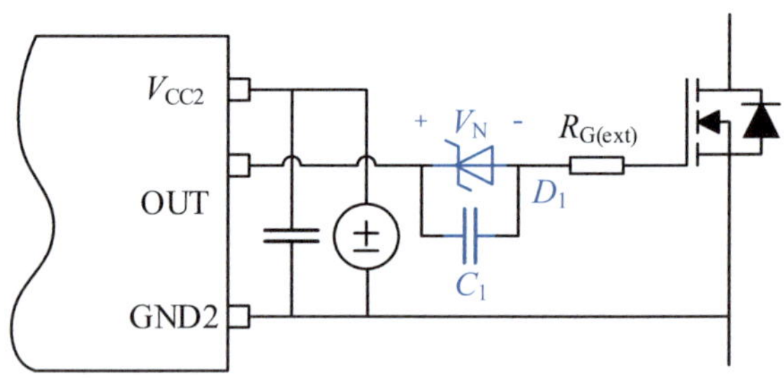

(c) Unipolar power supply series zener diode

(2) **Unipolar Power Source Parallel Zener Diode**

Use Zener diodes to split the unipolar power supply output into series of V_P and V_N, V_N The size is determined by the Zener diode. The sum of the two is used to power the driver chip. At the same time, the common point of the two is connected to the S pole of the SiC MOSFET. Similar to using a bipolar power supply directly, $V_{DRV(on)}$ is V_P, $V_{DRV(off)}$ is $-V_N$, thus achieving negative pressure shutdown.

(3) **Unipolar Power Supply Series Zener Diode**

Use a unipolar power supply V_{CC2} to power the drive circuit, and add a Zener diode D_1 and a parallel capacitor C_1 in series to the drive circuit. When the output is high, V_{OUT} − GND2 outputs V_{CC2}, The voltage across D_1 is V_N, $V_{DRV(on)}$ is $V_{CC2} - V_N$; when the output is low level, V_{OUT} − GND2 outputs 0 V, and the voltage at D_1 is V_N, $V_{DRV(off)}$ is $-V_N$, realizing negative pressure off break.

11.4 Influence of Drive Stage Characteristics

11.4.1 Peak Output Current

The drive current peak value can be estimated by (11.6) and (11.7), but it is not an exact value. The drive loop is a simple LCR series second-order circuit, L_{DRV} is the drive loop inductance, C_{iss} is the input capacitance of the device, R_G is the gate resistance, which is the sum of the internal resistance of the drive stage, the external gate resistance, and the internal resistance of the device gate. The drive signal is a step signal with amplitude V_{DRV}, and the drive current is $i_G(t)$, expressions with $i_G(t)$ as unknowns can be listed,

$$L_{DRV} \cdot \frac{di_G(t)}{dt} + R_G \cdot i_G(t) + \frac{1}{C_{iss}} \int i_G(t) \cdot dt = V_{DRV}. \tag{11.12}$$

Then the second-order constant coefcient linear differential equation,

$$L_{DRV} \cdot \frac{d^2 i_G(t)}{d^2 t} + R_G \cdot \frac{di_G(t)}{dt} + \frac{1}{C_{iss}} \cdot i_G(t) = 0. \tag{11.13}$$

Its damping ratio is,

$$\xi = \frac{R_G}{2}\sqrt{\frac{C_{iss}}{L_{DRV}}}. \tag{11.14}$$

When $\xi > 1$, it is over-damped; when $\xi = 1$, it is critical damping; when $\xi < 1$, it is under-damped.

When $C_{iss} = 1900$ pF, $V_{DRV} = 20$ V, drive peak current $I_{(peak)}$ is affected by R_G and L_{DRV} as shown in Fig. 11.23. When L_{DRV} is the same, the larger R_G is, the smaller $I_{(peak)}$ is; when R_G remains unchanged, the larger L_{DRV} is, the smaller $I_{(peak)}$ is, and the smaller R_G is, the greater the impact of L_{DRV} on $I_{(peak)}$.

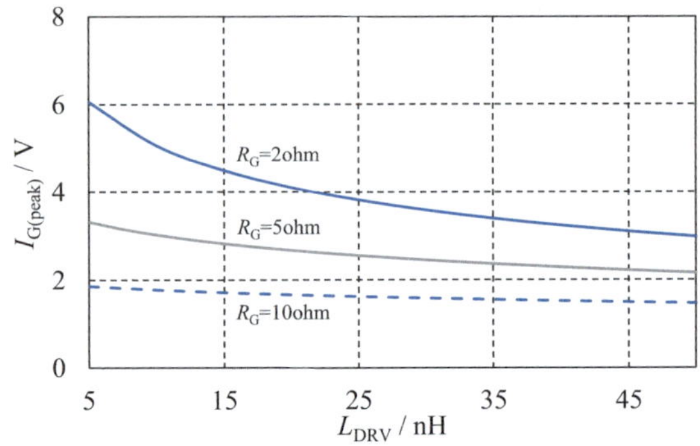

Fig. 11.23 Peak drive current

11.4.2 BJT and MOSFET Current Boost

The drive stages of common drive circuits include BJT emitter follower circuit, P-N MOS push-pull circuit and N-N MOS push-pull circuit, as shown in Fig. 11.24.

(1) **BJT Emitter Follower Circuit**

The BJT emitter follower circuit is composed of an NPN BJT upper device T_H and a PNP lower device T_L, which is the well-known Class B power amplifier. When the drive signal is high, T_H is turned on and T_L is turned off to charge the gate of the device and turn on the device; when the drive signal is low, Normally, T_H is turned off and T_L is turned on to discharge the gate of the device and turn off the device.

(2) **P-N MOS Push-Pull Circuit**

The P-N MOSFET push-pull circuit is composed of P-MOS upper device T_H and N-MOS lower device T_L. The worki principle of the BJT emitter follower circuit is similar. T_H is responsible for turning on the device, and T_L is responsible for turning off the device.

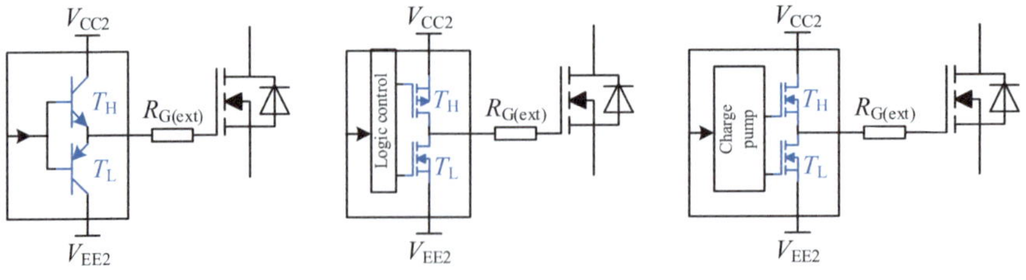

(a) BJT emitter follower circuit
(b) P-N MOS push-pull circuit
(c) N-N MOS push-pull circuit

Fig. 11.24 Common drive stage circuits

(3) N-N MOS Push-Pull Circuit

The N-N MOSFET push-pull circuit is composed of N-MOS upper device T_H and N-MOS lower device T_L, in which also T_H is responsible for turning on the device, and T_L is responsible for turning off the device. Since both the upper and lower devices are N-MOS, their control logic is reversed. In addition, when T_H is turned on, its source voltage will be close to V_{CC2}. In order to ensure that T_H can continue to conduct, a charge pump needs to be used to provide a sufficiently high drive voltage for T_H.

Compared to MOSFET drive stages, BJT drive stages have many disadvantages. Since the saturation voltage drop of the BJT during conduction is unavoidable, the actual drive voltage obtained will be lower than the supply voltage and is not rail-to-rail drive. If this is ignored when designing the drive circuit, low $V_{DRV(on)}$ will lead to increased conduction losses. The conduction voltage drop of MOSFET is very small and it is driven from rail to rail. At the same time, the saturation voltage drop will also cause the loss generated by BJT during the drive process to be much greater than that of MOSFET. The BJT drive stage is not at a disadvantage in terms of switching speed, and even turns on faster, as shown in Figs. 11.25 and 11.26 (Table 11.4).

The driving capability of the driver chip must meet the device's requirements for driving current, specifically corresponding to the driver chip's peak source current and peak sink current. When the driving current requirement exceeds the driver chip's driving capability, especially in high-power applications, it is necessary to add a driver-level circuit with stronger driving capability outside the driver chip. We call this type of circuit current boost, which includes BJT current boost and MOSFET current boost [10]. BJT current boost is a BJT emitter follower circuit that can be directly driven by the driver chip and has been widely used. When a conventional driver chip directly drives a P-N MOS push-pull circuit, it will cause the P-N MOS bridge arm to be directly turned on. When it directly drives an N-N MOS push-pull circuit, it cannot keep the upper tube turned on continuously. Therefore, MOSFET current boost is not as common as BJT current boost.

In order to take advantage of the excellent drive characteristics of the MOSFET drive stage, some manufacturers have launched drive chips with a MOSFET current Boost interface.

AVAGO ACPL-339J [11] has special output stage structure and timing control, as shown in Fig. 11.27. When the control signal changes from high level to low level, V_{OUTP} changes from low level to high level. When it exceeds $V_{CC2} - V_{th(TH)}$, T_H changes from on to off; then V_{OUTN} changes from low level to high level;, when it exceeds $V_{EE2} + V_{th(TL)}$, T_L changes from off to on, turning off the device; there is a delay t_{NLH} between T_L turning on and T_H turning off to avoid P-N MOS push-pull bridge arm straight-through. When the control signal changes from low level to high level, T_L turns off first, and then after a delay t_{NHL}, T_H turns off again and turns on the device.

PI SID1102K [12] has an N-N MOS push-pull output structure, and its N-N MOS push-pull drive signal is derived through the pin to drive the N-N MOS current Boost, as shown in Fig. 11.28.

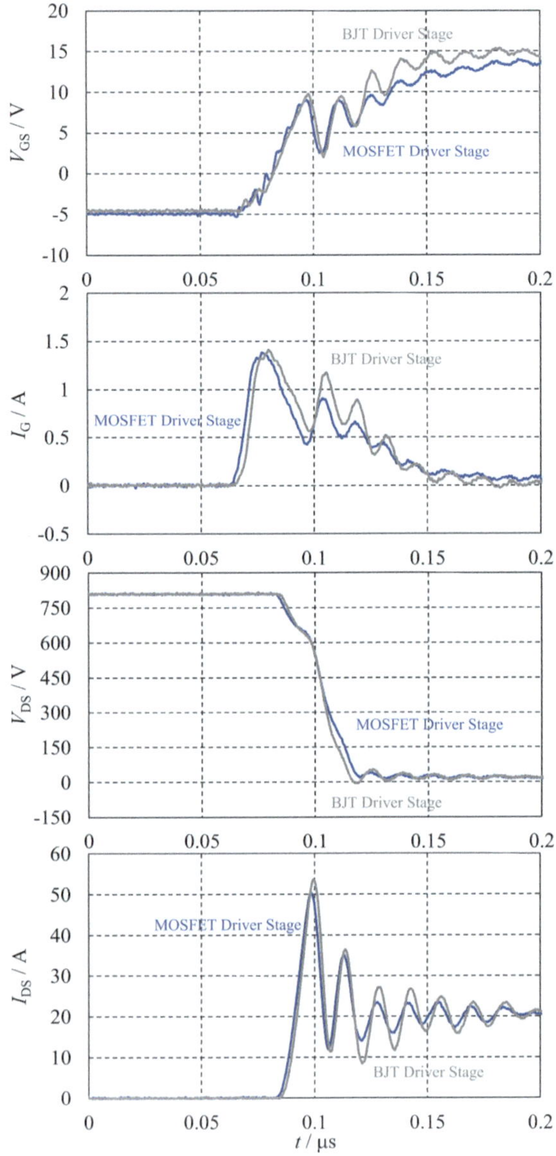

Fig. 11.25 Impact of BJT and MOSFET driver stages on the turn-on process

11.4.3 Drive Capability at Miller Slope

Through the analysis in the previous section, the difference between the BJT drive stage and the MOSFET drive stage is obtained. So in the case of the same MOSFET drive stage, do different drive chips have the same drive effect?

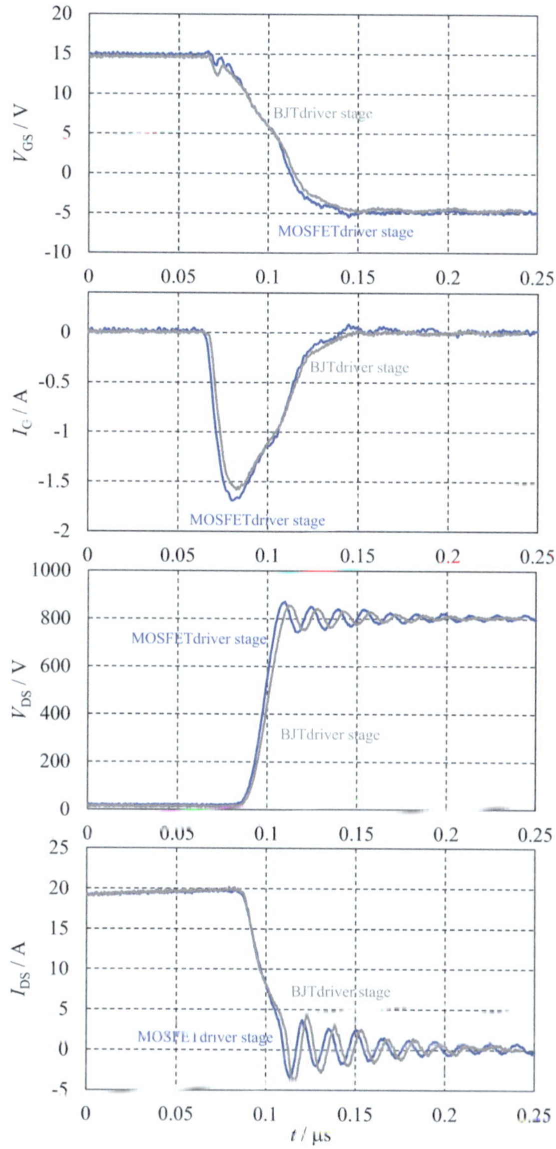

Fig. 11.26 Comparative impact of BJT vs. MOSFET driver stages on turn-off characteristics technical breakdown

The premise for studying this issue is that the drive chip drive stage meets the peak drive current requirements. Only under this condition is the discussion reasonable and effective. The $R_{G(\text{int})}$ of SiC MOSFET is 4Ω, choose $R_{G(\text{ext})}$ to be 5.1Ω, and the drive voltage is +15/−5 V, the peak drive current requirement will not exceed 2A. We selected four drive chips suitable for SiC MOSFET, and their maximum current output capabilities are all greater than 2A, as shown in Table 11.5.

Table 11.4 Impact of BJT drive stage and MOSFET drive stage on switching characteristics

	MOSFET drive stage	BJT drive stage
$t_{d(\text{on})}$/ns	21.2	88.4
t_r/ns	27.7	23.8
E_{on}/uJ I_{spike}/A	50.3	53.6
V_{spike}/V	375.7	338.7
$t_{d(\text{off})}$/ns	14.1	21.4
t_f/ns	14.2	15.8
E_{off}/uJ V_{spike}/V	871.4	856.5
I_{spike}/A	66.5	63.0

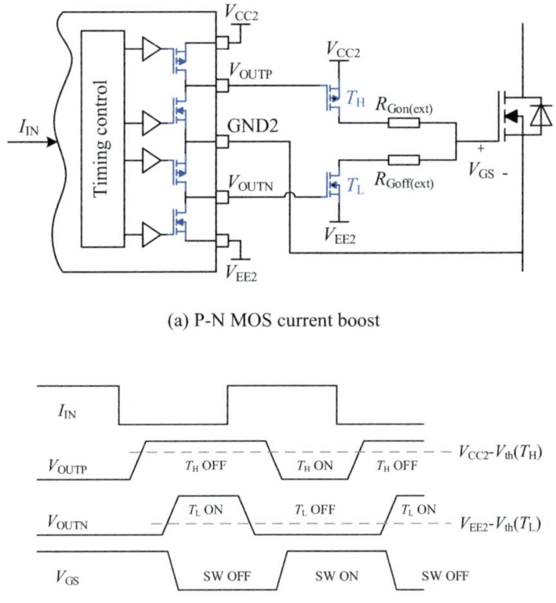

(a) P-N MOS current boost

(b) Timing Control

Fig. 11.27 AVAGO ACPL-339J P-N MOS current boost

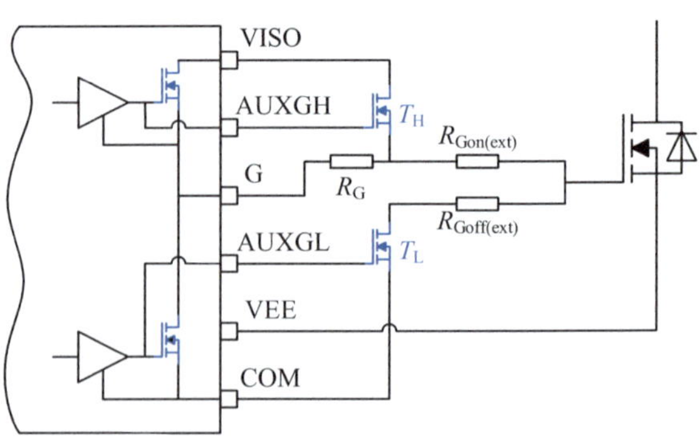

Fig. 11.28 PI SID1102K N-N MOS current boost

Table 11.5 Drive capabilities of drive chip

Type	Peak output current/A		Peak input current/A	
	Minimum value	Typical value	Minimum value	Typical value
A	2.2	3	3.4	5
B	4	9	2.4	3
C		7.1		7.8
D	10	17	10	17

Four driver chips were used to test the same SiC MOSFET, and the corresponding switching waveforms were somewhat different, as shown in Figs. 11.29 and 11.30. Due to the difference in switching waveforms, the switching characteristics of SiC MOSFETs under driver chips that meet the theoretical drive current requirements are different, as shown in Table 11.6. At the same time, combined with Table 11.5, it can be seen that the actual switching loss ranking and size ratio are inconsistent with the drive current peak size ranking given in the datasheet.

The I_G waveforms are shown in Figs. 11.29 and 11.30. It can be seen that there are obvious differences in I_G when using different driver chips, which is the fundamental reason for the different switching characteristics. Therefore, referring only to the driving peak current in the datasheet cannot correctly evaluate the driving capability of the driver chip and its impact on the turn-on characteristics. In the process of driving SiC MOSFET, the driving current that can be provided should be given enough attention, which is affected by the output characteristics of the driver MOSFET.

11.5 Signal Isolated Transmission

11.5.1 Isolation Method

Through isolation, we can avoid electrical connections between different circuit systems, block the current flow between them, and at the same time realize the transmission of power or signals. This can protect operators and low-voltage circuits from high voltage, prevent ground potential differences between systems, and improve anti-interference capabilities.

In power converters, isolation technology is widely used, such as isolated sampling, isolated topology primary and secondary signal transmission, and isolated drive. The isolated drive isolates the low-voltage control circuit from the high-voltage power circuit, which can not only ensure the safety of personnel and equipment, but also facilitate the drive of the upper tube in the half-bridge circuit. The isolation technologies used in the drive circuit are mainly: optical fiber, pulse transformer, optocoupler, magnetic isolation, and capacitive isolation.

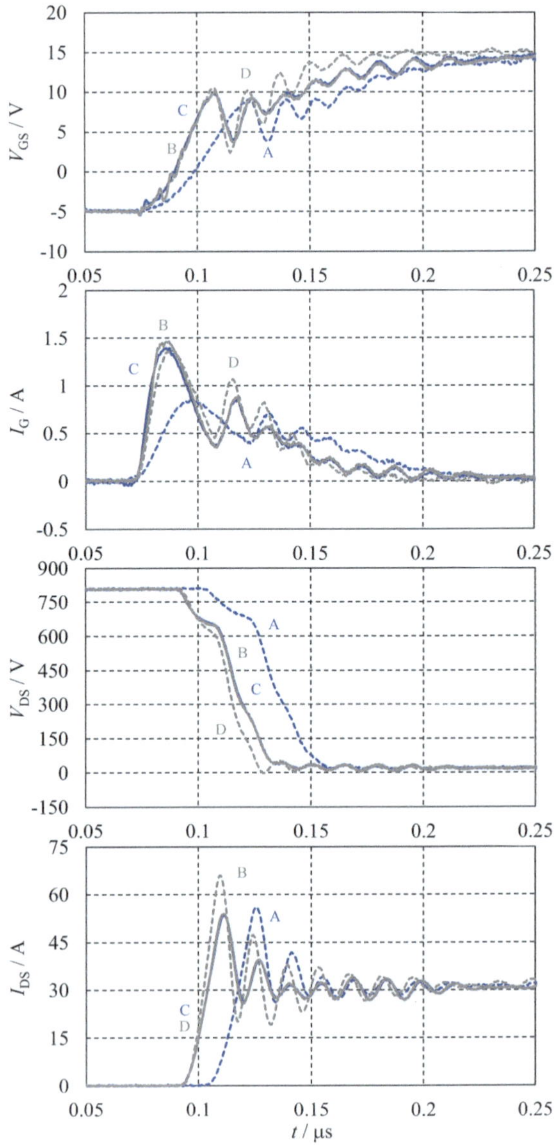

Fig. 11.29 Turn-on process under different driver ICs

1. **Optical fiber**

Optical fiber is a fiber made of glass or plastic. Light is transmitted by total reflection inside it, so it can be used as a light transmission tool. When in use, the transmitting device uses LED to convert electrical signals into optical signals and transmits light pulses to the optical fiber. The receiving device at the other end of the optical fiber

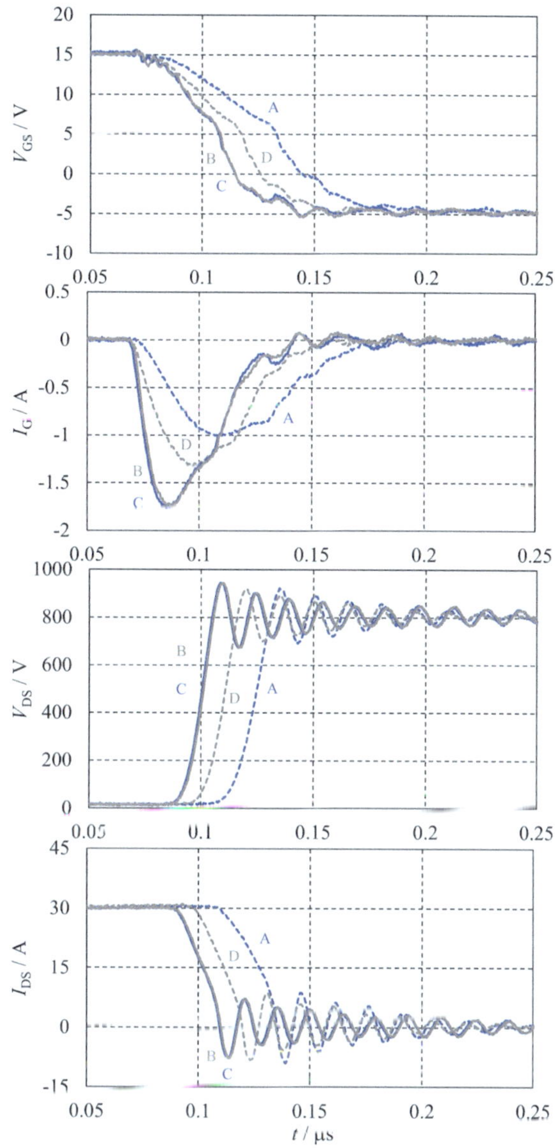

Fig. 11.30 Turn-off process under different driver ICs

uses a photosensitive element to detect the pulses and convert the optical signal into an electrical signal, as shown in Fig. 11.31.

Optical fiber has good insulation ability, high stability and strong anti-interference ability, but it has high cost and large transmission delay. It is often used in power

Table 11.6 Impact of different drive chips on switching characteristics

Drive chip	A	B	C	D
$t_{d(\text{on})}$/ns	34.2	20.9	20.8	21.4
t_r/ns	36.2	33.4	27.2	33.4
E_{on}/uJ	56.0	53.4	65.9	53.9
V_{spike}/V	650.3	510.0	474.9	504.7
$t_{d(\text{off})}$/ns	20.8	8.3	15.4	12.0
t_f/ns	16.2	11.6	13.7	11.7
E_{off}/uJ	922.3	944.9	917.4	943.8
I_{spike}/A	142.8	111.3	105.2	108.8

(a) Fiber Optic Cable (b) Fiber Optic Receiver

Fig. 11.31 Optical fiber

device drives of 3.3 kV and above. The main suppliers include Broadcom, Honeywell, Optek, etc.

2. **Pulse Transformer**

A pulse transformer is a transformer with a magnetic core. This isolation method was commonly used in early drive circuits. It is characterized by low cost and good insulation performance, but it occupies a large PCB area. This problem becomes more prominent when multiple signals need to be transmitted, as shown in Fig. 11.32.

Fig. 11.32 Pulse transformer

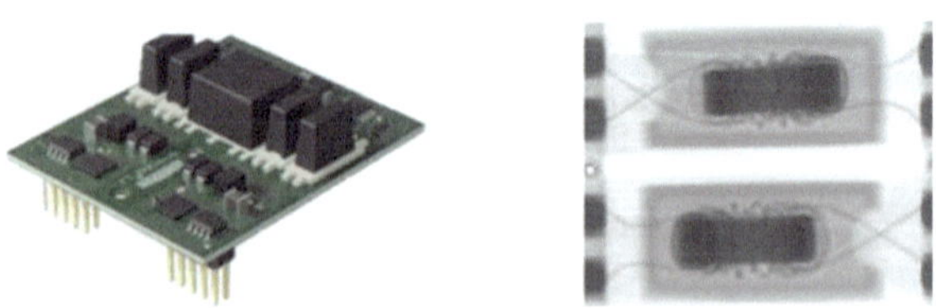

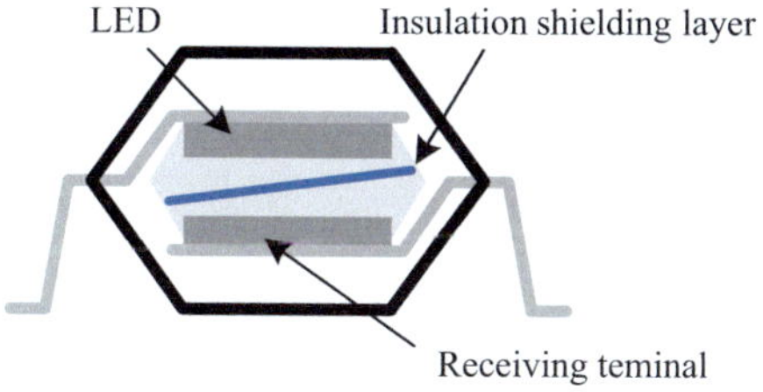

(a) Optocoupler Structure

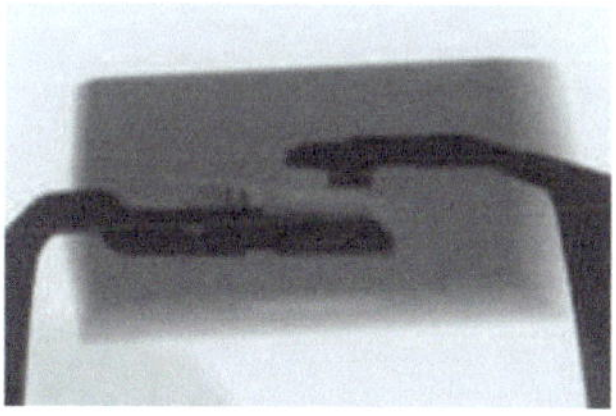

(b) Optocoupler X-Ray Perspective[15]

Fig. 11.33 Optocoupler [15]

3. **Optocoupler**

Like optical fiber, optocouplers also achieve isolation through photoelectric conversion. The difference is that optocouplers integrate the transmitter, optical transmission path, and receiver into a small package. The structure of the optocoupler is shown in Fig. 11.33. Its transmitter and receiver chips are welded on a split lead frame, separated by a distance, and a transparent insulating shielding layer is used to reduce the isolation capacitance.

Compared with optical fiber and pulse transformer, the volume of optocouplers is significantly smaller, but due to the large physical spacing between the primary and secondary sides, its insulation capacity is also good. Because it still needs to perform photoelectric conversion, its transmission delay and power consumption are still high. In addition, optocouplers also have a light attenuation effect. The light intensity decays with the use time, affecting the service life. The main suppliers of optocouplers are Broadcom, TOSHIBA, Vishay, etc.

4. **Coreless Transformer**

The coreless transformer is a transformer without a magnetic core. The biggest difference from the pulse transformer is that its coils are made in a semiconductor structure. The primary and secondary coils are very close to each other by filling silica or organic materials between them. Complete the insulation as shown in Fig. 11.34.

Core-free magnetic isolation can easily isolate multiple signals and occupies a small PCB area, with small transmission delay and low-power consumption. Drive chip suppliers that use non-nuclear magnetic isolation technology include Infineon, ADI, ONSemi, ROHM, Broadcom, ST, power integration, etc.

5. **Capacitor isolation**

Similar to the coreless transformer, capacitor isolation is also realized in the semiconductor structure. The two plates of the capacitor are placed vertically, and silicon dioxide is filled between them as a dielectric, as shown in Fig. 11.35. Capacitive isolation is similar to coreless transformer isolation in all aspects. Driver chip suppliers using capacitive isolation technology include TI, SiliconLab, etc.

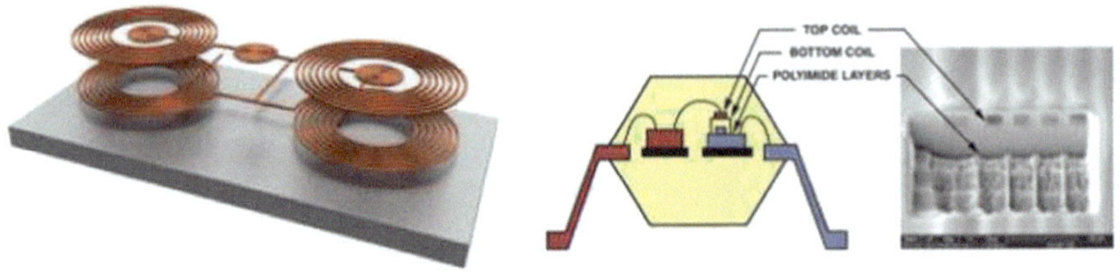

(a) Infineon[16] (b) ADI[17]

Fig. 11.34 Coreless transformer structure [16, 17]

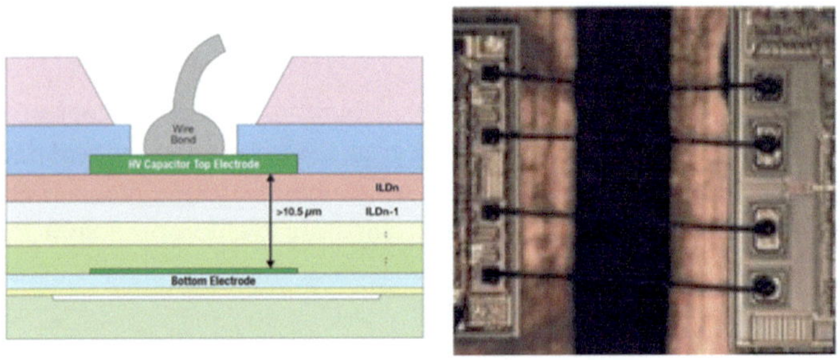

Fig. 11.35 Capacitor isolation structure [18, 19]

11.5.2 Safety and Insulation

In power converters, there are usually high-voltage circuits (such as the main power circuit) and low-voltage circuits (such as drive, control, and human-machine interface HMI classified as SELV safety extra-low voltage). In order to meet safety standards, prevent personal electric shock, and ensure that circuits of different voltage levels can work normally, insulation measures must be used to implement electrical isolation between high and low-voltage circuits (or primary and secondary sides) in accordance with safety standards. As the interface between high and low-voltage circuits, the isolation driver chip needs to meet the insulation requirements of the safety standards corresponding to the actual application scenarios.

11.5.2.1 Isolation Driver Chip Insulation Capability

The insulation capability of the driver chip is divided into internal insulation and external insulation. The internal insulation capability is mainly determined by the internal isolation material and the through-isolation distance (DTI), which is reflected in the insulation parameters of the driver chip. Commonly used insulation materials are polyimide and silicon dioxide. Some manufacturers also provide through-isolation distances in their datasheets. The external insulation capability is determined by the package size and package material, which is reflected in the two safety distances of creepage distance and electrical clearance.

1. **Insulation parameters**

The insulation parameters of the isolated driver chip are defined by relevant standards as shown in Table 11.7.

As shown in Table 11.8, it is the relevant information about the insulation parameters in the datasheet of an isolation drive chip.

The following are the definitions of insulation parameters in Table 11.8. For specific test methods, please refer to relevant standards [20, 21].

(1) **V_{IORM} (maximum repetitive peak isolation voltage)**

V_{IORM} is the maximum repetitive peak isolation voltage, which refers to the repeatable maximum peak voltage that the isolation drive chip can withstand. In applications,

Table 11.7 Isolator related standards

IEC 60747-5-5	Semiconductor devices—discrete devices—part 5-5: optoelectronicdevices—photocouplers
IEC 60747-17	Semiconductor devices—discrete devices—part 17: magnetic and capacitive coupler for basic and reinforced isolation
VDE 0884-5-5	Semiconductor devices—discrete devices—part 5-5: optoelectronic devices—photocouplers
VDE 0884-10	Semiconductor devices—magnetic and capacitive couplers for safe isolation
VDE 0884-11	Semiconductor devices part 11: magnetic and capacitive coupler for basic and reinforced isolation
UL1577	Standard for safety for optical isolators

Table 11.8 Example of insulation parameters from isolation driver chip datasheet

Parameter	Test conditions	Value	Unit
V_{IORM}	AC voltage (bipolar)	2121	V_{PK}
V_{IOWM}	AC voltage (sine wave) time-dependent dielectric breakdown (TDDB) test	1500	V_{RMS}
	DC voltage	2121	V_{DC}
V_{IOTM}	$V_{TEST} = V_{IOTM}$; $t = 60$ s (qualification); $t = 1$ s (100% production)	8000	V_{PK}
V_{IOSM}	Test method per IEC 60065, 1.2/50us waveform, $V_{TEST} = 1.6 \times V_{IOSM} = 12800$ V_{PK} (qualification)	8000	V_{PK}
V_{ISO}	$V_{TEST} = V_{ISO} = 5700$VRMS, $t = 60$ s (qualification); $V_{TEST} = 1.2 \times V_{ISO} = 6840$VRMS, $t = 1$ s (100% production)	5700	V_{RMS}

it is generally required that the maximum withstand voltage spike that the device actually withstands does not exceed V_{IORM}.

(2) $\boldsymbol{V_{IOWM}}$ **(maximum working isolation voltage)**

V_{IOWM} is the maximum working isolation voltage, which refers to the highest effective voltage value that the isolation drive chip can withstand during long-term operation within the promised life. In applications, it generally corresponds to the highest effective voltage withstand of the driven voltage, such as the highest bus voltage in a half-bridge topology.

(3) $\boldsymbol{V_{IOTM}}$ **(maximum transient isolation voltage)**

V_{IOTM} is the maximum transient overvoltage, which refers to the maximum transient overvoltage that the isolation chip can withstand within one minute. In some cases, V_{IOTM} can be used to determine whether the drive chip meets the system overvoltage requirements.

(4) $\boldsymbol{V_{IOSM}}$ **(maximum surge isolation voltage)**

V_{IOSM} is the maximum surge voltage, which represents the ability of the isolation drive chip to withstand extremely high-voltage pulses with specified transient characteristics. In some cases, V_{IOSM} can be used to determine whether the drive chip meets the system's surge resistance requirements.

(5) $\boldsymbol{V_{ISO}}$ **(maximum withstanding isolation voltage)**

V_{ISO} is the maximum withstand isolation voltage, which refers to the highest effective voltage that the isolation drive chip can withstand within one minute.

2. **Safety Distance**

When the voltage difference between the primary and secondary sides of the drive chip is too large, the packaging insulating material will be polarized and become conductive, leading to insulation failure. When the withstand voltage of the primary and secondary edges is constant, in order to reduce the electric field intensity, the distance between the packaging surface of the primary and secondary edges of the isolation drive chip must be large enough. The creepage distance refers to the shortest path between two conductive parts measured along the insulating surface. The creepage distance of the isolation drive chip is shown in Fig. 11.36a. The creepage distance of the drive chip needs to be greater than or equal to the system requirements, which is affected by the comparative tracking index (CTI, comparative tracking index), pollution level and working voltage, and is specifically stipulated by relevant standards [22, 22].

Fig. 11.36 Isolated drive chip safety distance

Table 11.9 Creepage distances and clearances of commonly used packages

	Creepage distance (mm)	Electrical clearance (mm)
SOIC8-D	4	4
SOIC8-DWV	8.5	8.5
SOIC16-DW	8	8
eSOP-R16B	9.5	9.5

When the voltage difference between the primary and secondary sides of the drive chip is too large, air breakdown will occur, leading to insulation failure. When the withstand voltage of the primary and secondary sides is constant, in order to reduce the electric field intensity, the spatial distance between the primary and secondary sides of the isolation drive chip must be large enough. The electrical clearance refers to the shortest spatial distance between two conductive parts. The electrical clearance of the isolation drive chip is shown in Fig. 11.36b. The electrical clearance of the drive chip needs to be greater than or equal to the system requirements, which is affected by the system voltage, overvoltage level, altitude and pollution level, and is specified by relevant standards [22, 23].

Table 11.9 gives the creepage distance and electrical clearance of commonly used packages for isolation drive chips.

11.5.2.2 Insulation Design

The insulation design follows the sequence from system to component. The specific steps are as follows:

1. **Step 1: Determine the insulation type**

Determine the insulation type according to the system circuit type and insulation location, including basic insulation and reinforced insulation. Basic insulation refers to the type of insulation that provides basic protection against electric shock, and

Table 11.10 Motor drive system information

Power module voltage	1200 V
Maximum DC bus voltage	800 V
Altitude	2000 m
Pollution level	2
Pollution level	II

reinforced insulation is the type of single insulation that provides a level of protection against electric shock equivalent to double basic insulation.

2. **Step 2: Determine system information**

The system information that needs to be determined includes: system voltage, overvoltage level, pollution level, altitude, and safety standards that need to be followed. System voltage is related to the power system; overvoltage levels are classified according to how the equipment is connected to the AC power supply and its ability to withstand surges; pollution levels are classified according to the amount of conductive or hygroscopic dust, free gases or salts and relative humidity, and due to hygroscopic or The frequency of events in which condensation causes a decrease in surface dielectric strength or resistivity is graded; the safety standards that need to be complied with are determined based on the system's application field and place of use.

3. **Step 3: Determine safety requirements**

Based on the information determined in the first two steps, determine the creepage distance, electrical clearance and requirements for the isolation parameters of the isolation drive chip in accordance with safety standards.

Taking the motor drive system as an example, it adopts a three-phase full-bridge topology and requires the drive circuit to have basic insulation. It follows the design standard IEC 61800-5-1 [24]. The system information is shown in Table 11.10.

The module voltage is 1200 V, and the calculated system voltage is 849 V (0.707 × 1200 V). From Table 11.11, it is found that the surge voltage of basic insulation under overvoltage level II is 6000 V, and then from Table 11.12 the air gap under pollution level 2 is 5.5 mm. The maximum bus voltage is 800 V, which can be regarded as the working voltage. From Table 11.13, it is found that under the working voltage of 800 V, pollution level 2, and Class I insulation material, the creepage distance is 4.0 mm. At an altitude of 2000 m, there is no need to correct the above safety distance.

Table 11.14 shows the system insulation requirements and alternative drive chip parameters. It can be seen that the alternative drive chip meets the requirements and is suitable for this motor drive system.

Table 11.11 IEC 61800-5-1 Table 7 isolated power supplies for low-voltage circuits

Column 1	2	3	4	5	6
System voltage (4.3.6.2.1) (V)	Impulse voltage (V)				Temporary overvoltage (crest value/r. m. s.)[a] (V)
	Overvoltage category				
	I	II	III	IV	
< 50	330	500	800	1500	1770/1250
100	500	800	1500	2500	1840/1300
150	800	1500	2500	4000	1910/1350
300	1500	2500	4000	6000	2120/1500
600	2500	4000	6000	8000	2550/1800
1000	4000	6000	8000	12,000	3110/2200

Table 11.12 IEC 61800-5-1 Table 9 air gaps

Column 1	2	3	4	5	6
Impulse voltage (Table 7, Table 8, 4.3.6.3) (V)	Temporary overvoltage (crest value) for determining insulation between surroundings and circuits or working voltage (recurring peak) for determining functional insulation (V)	Working voltage (recurring peak) for determining insulation between surroundings and circuits (V)	Minimum clearance mm		
			Pollution degree		
			1	2	3
N/A	≤ 110	≤ 71	0.01	0.2	0.80
N/A	225	141	0.01	0.20	0.80
330	340	212	0.01	0.2	0.80
500	530	330	0.04	0.20	0.80
800	700	440	0.10	0.20	0.80
1500	960	600	0.5	0.50	0.80
2500	1600	1000	1.5		
4000	2600	1600	3		
6000	3700	2300	5.5		
8000	4800	3000	8		
12,000	7400	4600	14		
20,000	12,000	7600	25		

Table 11.13 IEC 61800-5-1 Table 10 creepage distance

Column 1	2	3	4	5	6	7	8	9	10	11	12
Working voltage (r. m. s.) (V)	PWBs"		Other insulators								
	Pollution degree		Pollution degree								
	1	2	1	2				3			
	b	c	b	Insulating material group				Insulating material group			
				I	II	III a	III b	I	II	III a	III b
< 2	0.025	0.04	0.056	0.35	0.35	0.35		0.87	0.87	0.87	
5	0.025	0.04	0.065	0.37	0.37	0.37		0.92	0.92	0.92	
10	0.025	0.04	0.08	0.4	0.4	0.4		1	1	1	
25	0.025	0.04	0.125	0.5	0.5	0.5		1.25	1.25	1.25	
32	0.025	0.04	0.14	0.53	0.53	0.53		1.3	1.3	1.3	
40	0.025	0.04	0.16	0.56	0.8	1.1		1.4	1.6	1.8	
50	0.025	0.04	0.18	0.6	0.85	1.2		1.5	1.7	1.9	
63	0.04	0.063	0.2	0.63	0.9	1.25		1.6	1.8	2.0	
80	0.063	0.10	0.22	0.67	0.95	1.3		1.7	1.9	2.1	
100	0.1	0.16	0.25	0.71	1.0	1.4		1.8	2.0	2.2	
25	0.2	0.3	0.3	0.8	1.1	1.5		1.9	2.1	2.4	
160	0.3	0.4	0.3	0.8	1.1	1.6		2.0	2.2	2.5	
200	0.40	0.6	0.4	1.0	1.4	2.0		2.5	2.8	3.2	
250	0.6	1.0	0.6	1.3	1.8	2.5		3.2	3.6	4.0	
320	0.8	1.6	0.8	1.6	2.2	3.2		4.0	4.5	5.0	
400	1.0	2.0	1.0	2.0	2.8	4.0		5.0	5.6	6.3	
500	1.3	2.5	1.3	2.5	3.6	5.0		6.3	7.1	8.0	
630	1.8	3.2	1.8	3.2	4.5	6.3		8.0	9	10.0	
800	2.4	4.0	2.4	4.0	5.6	8.0		10.0	11	12.5	e
1000	3.2	5.0	3.2	5	7.1	10.0		12.5	14	16	
1250	4.2	6.3	4.2	6.3	9.0	12.5		16	18	20	
1600	1	1	5.6	8.0	11	16		20	22	25	
2000			7.5	10.0	14	20		25	28	32	
2500			10.0	12.5	18	25		32	36	40	

Table 11.14 System insulation requirements, drive chip parameters

System insulation requirements		Drive chip parameters	
Turn off spike maximum	1200 V	V_{IORM}	$2121V_{PK}$
Maximum bus voltage	800 V	V_{IOWM}	$2121V_{DC}$
System surge voltage	6000 V	V_{IOSM}	$8000V_{peak}$
Electrical clearance	5.5 mm	Electrical clearance	8 mm
Creepage distance (Class I CTI)	4.0 mm	Creepage distance (Class I CTI)	8 mm

11.6 Short-Circuit Protection

When using Si IGBT, it is often necessary to provide short-circuit protection, especially in high-power applications. When using Si MOSFETs, short-circuit protection is rarely provided. So, do you need to protect SiC MOSFETs from short circuits when using them?

In fact, whether short-circuit protection is required is not entirely determined by the type of device, but rather by the application and cost. Si IGBTs are often used in high-voltage and high-power applications. The consequences of a short circuit are more serious. In addition, the cost of converters and Si IGBT modules is high. It is very necessary to use short-circuit protection to avoid excessive losses. Si MOSFETs are mostly used in low-power applications, and the loss after short circuit is small. In addition, the cost pressure of the converter is high, so short-circuit protection is often no longer provided.

SiC MOSFET is mainly used in high-voltage and high-power applications, aiming to replace Si IGBT. You can directly refer to the existing Si IGBT solution for short-circuit protection. In addition, although the price of SiC MOSFETs has dropped year by year, it is still much higher than Si devices in the short term. When using a SiC MOSFET power module, if a short-circuit occurs, the entire module will be lost, so it is necessary to provide short-circuit protection. There are many application scenarios for single-device SiC MOSFETs. When the application requires high reliability of the converter, short-circuit protecton is also necessary.

11.6.1 Detection Method of Short-Circuit Protection

To implement short-circuit protection, you need to first detect and identify the short-circuit condition of the device, and then turn off the device. The existing methods of short-circuit detection include the following.

1. **Sampling Resistor**

Connect a sampling resistor R_{shunt} in series in the loop, and measure its terminal voltage V_{shunt} to get the current flowing through the device I_{DS} [25, 26], the corresponding relationship is as shown in Eq. (11.15). When a short circuit occurs, I_{DS} is much greater than the normal operating range, V_{shunt} will be greater than the threshold voltage $V_{\text{th(SC)}}$, triggers the comparator and turns off the device, as shown in Fig. 11.37. The advantages of this method are high bandwidth, low cost, and high measurement accuracy. However, in large current situations, it will cause large losses on R_{shunt}.

$$V_{\text{shunt}} = I_{\text{DS}} \cdot R_{\text{shunt}} \tag{11.15}$$

Fig. 11.37 Sampling resistor

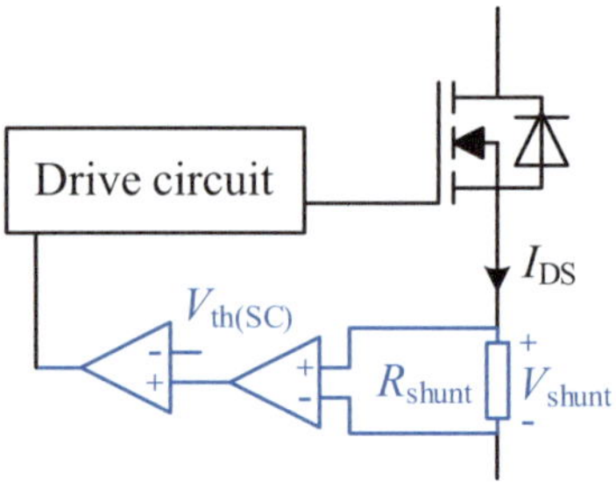

2. **Power Devices with Current Detection**

Mitsubishi has launched a SiC MOSFET chip with current detection [27]. The basic principle is to divide the chip into two parts. The larger T_{power} is the main MOSFET for large flows. Part of the current, the smaller area T_{sense} is the current sensing MOSFET, as shown in Fig. 38a. When using, connect a sampling resistor R_{sense} in series and measure its terminal voltage V_{sense} to calculate the current flowing through the device. The current and corresponding relationship of V_{sense} is affected by the chip and the sampling resistor. When a short circuit occurs, V_{sense} is greater than the threshold voltage $V_{th(SC)}$, the comparator is triggered and turn off the device, as shown in Fig. 11.38b. Both Mitsubishi and Fuji have launched Si IGBT chips with current detection function. However, due to the high cost of this kind of chip, it is mainly used for electric vehicle drive and locomotive traction.

3. **PCB-Based Rogowski Coil**

The basic principles of the Rogowski coil are introduced in Chap. 7. Compared with the first two detection methods, the Rogowski coil has many advantages. Isolated measurement does not require additional isolation processing; no additional loop inductance will be added; based on the law of electromagnetic induction, no excessive loss will occur; since no magnetic core is used, coil saturation will not occur.

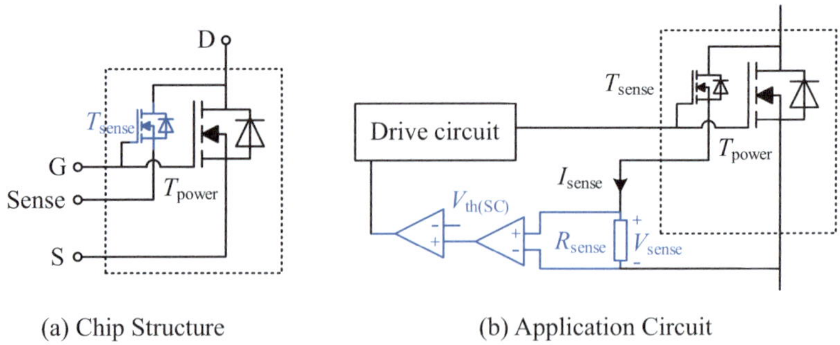

Fig. 11.38 Power device with current sensing

There have been studies using PCB-based Rogowski coils for short-circuit detection, as shown in Fig. 11.39. Since the design of Rogowski coils must consider space placement, safety distance, and integral analog circuits, it has a certain technical threshold and presents many challenges, so it is not widely used in products.

4. **Lead Inductance Voltage Drop**

When a short circuit occurs, the current will increase sharply, and its rate of change dI_{DS}/dt will be much larger than the range during normal operation. The voltage drop caused by dI_{DS}/dt on the loop inductance will also be larger than the normal range. Using this characteristic, the voltage drop V_{sense} across the source inductance of the SiC MOSFET is measured. When a short circuit occurs, V_{sense} is greater than the threshold voltage $V_{th(SC)}$, triggering the comparator and shutting down the device [33], as shown in Fig. 11.40.

Fig. 11.39 Application examples of Rogowski coils based on PCB [29–32]

Fig. 11.40 Lead inductance voltage drop

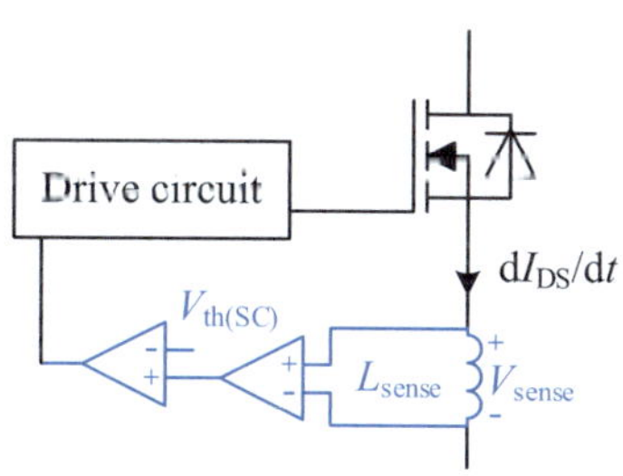

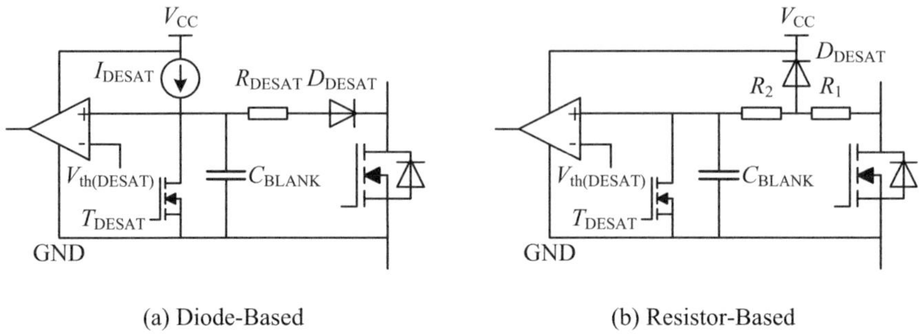

Fig. 11.41 Desaturation (DESAT) detection circuit

5. **Desaturation (DESAT) detection**

Under normal operating conditions, Si IGBT operates in the saturation region when turned on and in the cutoff region when turned off. The turn-on and turn-off processes need to pass through the amplification region. When a Si IGBT is short-circuited, its operating point will move from the saturation region to the amplification region, that is, the region of high voltage and high current, which is called desaturation. By detecting the Si IGBT collector-emitter voltage V_{CE}, the Si IGBT short circuit can be identified. This method is called desaturation detection, and there are generally two implementation methods: diode type and resistor type, as shown in Fig. 11.41.

Under normal operating conditions, the SiC MOSFET operates in the linear region when it is turned on and in the cutoff region when it is turned off. The turn-on and turn-off processes need to pass through the saturation region. When a SiC MOSFET is short-circuited, its operating point will change from the linear region to the saturation region, that is, a region of high voltage and high current. Therefore, by detecting the SiC MOSFET drain-source voltage V_{DS}, the SiC MOSFET short circuit can be identified, and DESAT is still effective for SiC MOSFET [34].

It should be noted that the definitions of the saturation region of Si IGBT and SiC MOSFET are different. Si IGBT saturation refers to the maximum concentration of carriers injected into the base of the Si IGBT transistor. MOSFET saturation refers to the maximum drain current. A short circuit occurs in Si IGBT when the operating point leaves the saturation region. A short circuit occurs in SiC MOSFET when it enters the saturation region, but in essence they both enter the high-voltage and high-current region. DESAT is the name for Si IGBT, while for SiC MOSFET, it should be called "advanced saturation detection". But for convenience of expression, we still use DESAT.

DESAT detection has been widely used in Si IGBT short-circuit protection. Due to its stability and reliability, many drive chips have integrated the DESAT function. If the existing DESAT circuit can be used to complete the short-circuit detection of SiC MOSFET, it will undoubtedly greatly reduce the difficulty of using SiC MOSFET. Since diode-type DESAT is the most widely used, the applicability of DESAT to SiC MOSFET will be discussed based on this.

11.6.2 DESAT Detection

11.6.2.1 Principle of DESAT Detection

The diode-type DESAT detection circuit built based on the driver chip with integrated DESAT detection function is shown in Fig. 11.42. The blue shading part is the circuit integrated into the drive chip, and the yellow shading part is the peripheral circuit that needs to be configured during use. The drive chip contains a logic control unit, current source $I_{DESAT\text{-}charge}$, comparator and discharge control switch $T_{DESAT\text{-}discharge}$ and diode $D_{DESAT\text{-}discharge}$, and the peripheral circuit includes a capacitor C_{BLANK}, diode D_{DESAT} and resistor R_{DESAT}.

1. **No Short Circuit**

When no short circuit occurs, the working principle of the DESAT circuit is shown in Fig. 11.43.

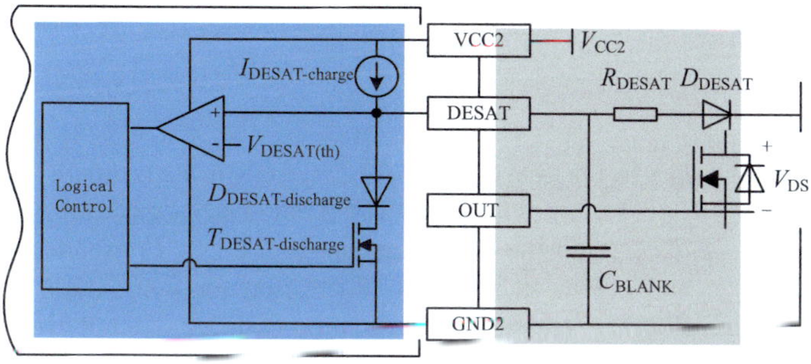

Fig. 11.42 Diode-implemented DESAT detection circuit based on driver IC technical breakdown

Fig. 11.43 Operating principle of DESAT detection circuit under normal (non-fault) conditions

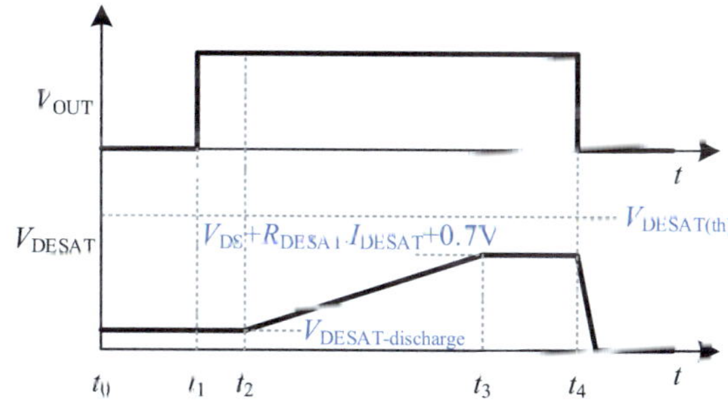

(1) $-t_1$

The input of the drive chip is low level, and the output V_{OUT} is also low level. When the switch device is in the off state, the control signal is high, $T_{DESAT\text{-}discharge}$ is in the on state, and $I_{DESAT\text{-}charge}$ flows to GND2 through $T_{DESAT\text{-}discharge}$ and $D_{DESAT\text{-}discharge}$. Then the terminal voltage V_{DESAT} of C_{BLANK} is the sum of the on-state voltage drops of $I_{DESAT\text{-}charge}$ on $T_{DESAT\text{-}discharge}$ and $D_{DESAT\text{-}discharge}$;

(2) $t_1 - t_2$

The input of the drive chip becomes high level, and the output V_{OUT} also becomes high level. The switch device is in the conducting state, and $T_{DESAT\text{-}discharge}$ remains in the conducting state. Then V_{DESAT} remains unchanged at $V_{DESAT\text{-}discharge}$. This stage is the DESAT delay $t_{d(DESAT)}$.

(3) $t_2 \sim t_3$

At time t_2, the $T_{DESAT\text{-}discharge}$ control signal of goes low, turning off the $T_{DESAT\text{-}discharge}$ circuit. The $I_{DESAT\text{-}charge}$ current source then initiates constant-current charging of the C_{BLANK} capacitor, causing V_{DESAT} to rise linearly from $V_{DESAT\text{-}discharge}$. With the power device in full conduction, its drain-source voltage V_{DS} remains very low, typically below 2.5V.

(4) $t_3 \sim t_4$

At time t_3, V_{DESAT} rises to $V_{DS} + R_{DESAT} I_{DESAT} + V_F$ (where V_F is the forward voltage drop of D_{DESAT}), D_{DESAT} becomes forward-biased. The $I_{DESAT\text{-}charge}$ current then flows through R_{DESAT}, D_{DESAT}, and the power device to GND2. Consequently, V_{DESAT} becomes clamped and ceases to increase. Given that the comparator's reference voltage V_{ref} is typically set between 7V and 9V, this condition will not trigger DESAT protection.

(5) $t_4 \sim$

At time t_4, the input of the driving chip becomes low, and the output V_{OUT} also becomes low, $T_{DESAT\text{-}discharge}$ turns on, C_{BLANK} is discharged to $V_{DESAT\text{-}discharge}$.

2. **Short Circuit**

Assuming that the SiC MOSFET is short-circuited immediately after it is turned on, the working principle of the DESAT circuit is shown in Fig. 11.44.

(1) $-t_1$

The input of the drive chip is low level, and the output V_{OUT} is also low level. When the switch device is in the off state, D_{DESAT} is reversely cutoff. $T_{DESAT\text{-}discharge}$ is in the conductive state and discharges, $I_{DESAT\text{-}charge}$ flows through $T_{DESAT\text{-}discharge}$ and $D_{DESAT\text{-}discharge}$ and to GND2. The terminal voltage V_{DESAT} of C_{BLANK} is the sum $V_{DESAT\text{-}discharge}$ of forward voltage of $I_{DESAT\text{-}charge}$ across $T_{DESAT\text{-}discharge}$ and $D_{DESAT\text{-}discharge}$.

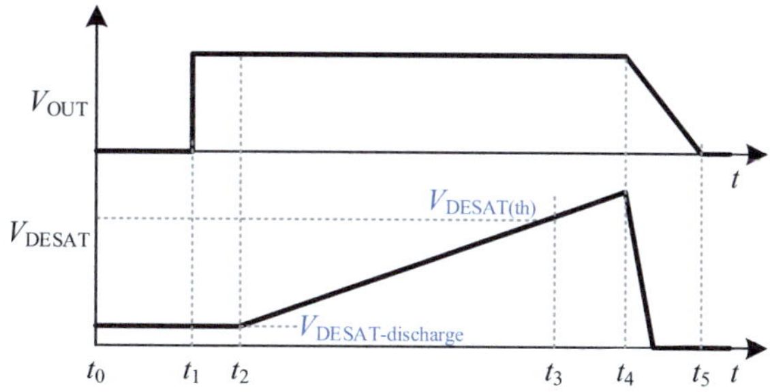

Fig. 11.44 Operating principle of DESAT detection circuit under non-fault conditions

(2) $t_1 \sim t_2$

The input of the drive chip becomes high level, and the output V_{OUT} also becomes high level, the SiC MOSFET is short-circuited. I_{DS} rises rapidly, V_{DS} also maintains at a higher voltage, D_{DESAT} keeps reverse cutoff. $I_{DESAT\text{-}charge}$ remains on, V_{DESAT} remains at $V_{DESAT\text{-}discharge}$. This stage is DESAT delay $t_{d(DESAT)}$.

(3) $t_2 \sim t_3$

At time t_2, the $T_{DESAT\text{-}discharge}$ control signal goes low, turning off the $T_{DESAT\text{-}discharge}$ circuit. The $I_{DESAT\text{-}charge}$ current source then commences constant-current charging of the C_{BLANK} capacitor, causing V_{DESAT} to rise linearly from the $V_{DESAT\text{-}discharge}$ level. Subsequently at time t_3, when V_{DESAT} reaches the threshold voltage $V_{DESAT(th)}$ it triggers the comparator. This entire period constitutes the blanking time t_{BLANK}.

(4) $t_3 \sim t_4$

No action, and this is turn off delay $t_{d(off)}$.

(5) $t_4 \sim t_5$

At time t_4, the $T_{DESAT\text{-}discharge}$ control signal goes low, turning on the $T_{DESAT\text{-}discharge}$ circuit. This rapidly discharges the C_{BLANK} capacitor while simultaneously, the output V_{OUT} starts transitioning from high to low to turn off the power device.

(6) $t_5\sim$

At time t_5, the power device achieves full turn-off, and V_{DESAT} stabilizes at the $V_{DESAT\text{-}discharge}$ level.

The device short circuit may occur at any time after $t_1 - t_3$ and t_3 in Fig. 11.43. Among various situations, the detection and reaction time required for a short circuit to occur just after the device is turned on is the longest. The small value of the SiC MOSFET chip results in a short short circuit withstand time, and protection needs to be completed in a short time. Therefore, when using DESAT, it is necessary to ensure that the SiC MOSFET can be safely turned off even if a short circuit occurs just after it is turned on.

11.6.2.2 Key Points in DESAT Detection Circuit Design

1. **$\mathrm{d}V_{DS(down)}/\mathrm{d}t$**

When the SiC MOSFET is turn on, the waveform of the DESAT circuit is shown in Fig. 11.45.

(1) **$-t_1$**

The SiC MOSFET is in the off state, and $T_{DESAT\text{-}discharge}$ is in the on state. Then V_{DESAT} is $V_{DESAT\text{-}discharge}$, and D_{DESAT} is reverse cutoff, its terminal voltage is $V_{Bus} - V_{DESAT\text{-}discharge}$.

(2) **$t_1 - t_2$**

At t_1, the SiC MOSFET starts to turn on, and its V_{DS} jumps quickly from V_{Bus} to low at change rate $\mathrm{d}V_{DS(down)}/\mathrm{d}t$. At this time, the voltage at D_{DESAT} also jumps from $V_{Bus} - V_{DESAT\text{-}discharge}$ to low at the same rate, and it generates a displacement current $I_{J\text{-}DESAT}$ by $C_{J\text{-}DESAT}$ of D_{DESAT}. Due to the existence of $D_{DESAT\text{-}discharge}$, $I_{J\text{-}DESAT}$ discharges $C_{J\text{-}DESAT}$, and makes V_{DESAT} to appear negative. The negative voltage of V_{DESAT} depends on C_{BLANK}, I_{DESAT} and $I_{J\text{-}DESAT}$, where $I_{J\text{-}DESAT}$ is represented by the product of $\mathrm{d}V_{DS(down)}$ and $C_{J\text{-}DESAT}$.

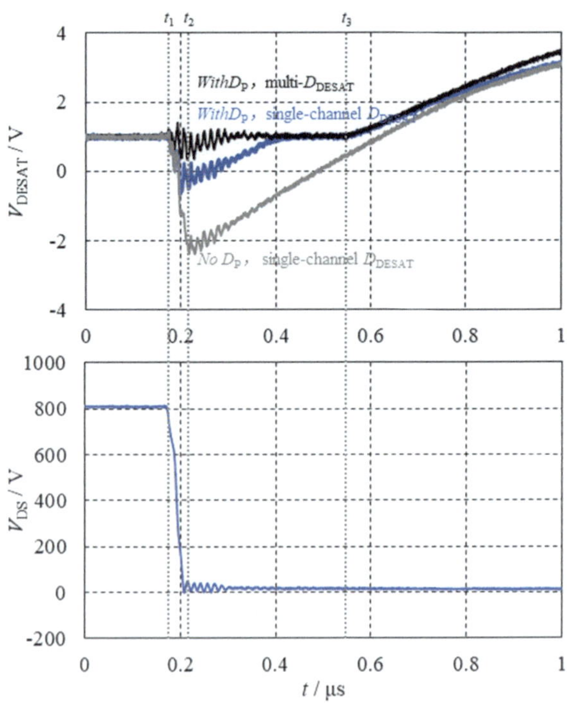

Fig. 11.45 Impact of negative $\mathrm{d}V_{DS(down)}/\mathrm{d}t$ on DESAT performance

At time t_2, the SiC MOSFET is fully turned on, and the voltage transient V_{DS} concludes, stabilizing at the conduction voltage of the SiC MOSFET. Typically, the duration of $t_1 - t_2$ is on the order of tens of nanoseconds, while the DESAT delay $t_{d(DESAT)}$ is generally in the range of hundreds of nanoseconds.

(3) $t_2 - t_3$

As V_{DESAT} is a negative value, $D_{DESAT\text{-}discharge}$ is reverse-biased, causing $I_{DESAT\text{-}charge}$ to charge C_{BLANK}, gradually raising V_{DESAT}. At t_3, the DESAT delay concludes, and the control signal for $I_{DESAT\text{-}charge}$ transitions to a low state.

When $t_2 - t_3$ is sufficiently long, V_{DESAT} will return to and gradually stabilize at $V_{DESAT\text{-}discharge}$. Otherwise, V_{DESAT} will remain below $V_{DESAT\text{-}discharge}$. This behavior is also dependent on $I_{DESAT\text{-}charge}$ and C_{BLANK}.

(4) t_3~

$I_{DESAT\text{-}charge}$ charges C_{BLANK} with a constant current, causing V_{DESAT} to gradually rise.

From the analysis above, it is evident that $dV_{DS(down)}/dt$ has two significant impacts. On the one hand, during the $t_1 - t_2$ interval, when the negative voltage amplitude of V_{DESAT} is excessively high, it may generate a substantial surge current in the circuit, potentially leading to damage to the DESAT pin, which typically has a reverse breakdown voltage of around − 9 V. On the other hand, at the end of the DESAT event, if V_{DESAT} remains below $V_{DESAT\text{-}discharge}$, it implies that V_{DESAT} needs a longer ramp-up time, while simultaneously increasing the blanking time t_{BLANK}, slowing down the response speed of the DESAT short-circuit protection.

To address the aforementioned issues, a diode D_p can be connected in parallel with C_{BLANK} to clamp the negative voltage. For optimal clamping effectiveness, D_p should have a low forward voltage drop. To minimize its impact on turn-off time, D_p should exhibit low leakage current, with the leakage current change overtemperature being as minimal as possible. Simultaneously, a resistor R_{DESAT} with a relatively large value can be chosen to limit the surge current, typically ranging from a few hundred ohms to 1 k ohms, avoiding excessively low DESAT threshold voltage $V_{DESAT(th)}$ due to an excessively large R_{DESAT}. Additionally, selecting a diode D_{DESAT} with a smaller $C_{J\text{-}DESAT}$ effectively reduces the impact of $I_{J\text{-}DESAT}$ generated by $dV_{DS(down)}/dt$, mitigating the influence of $dV_{DS(down)}/dt$ at its source. Furthermore, connecting multiple D_{DESAT} diodes in series can significantly reduce $C_{J\text{-}DESAT}$.

It is important to note that, apart from the turn-on process, other operating conditions that cause a high-to-low transition in V_{DS}, such as a post-commutation device in the event of a hard-switching fault short circuit, will also experience similar influences on V_{DESAT}. This will be discussed in detail in Sect. 11.6.2.4.

2. $\mathbf{dV_{DS(up)}/dt}$

When $T_{DESAT\text{-}discharge}$ is in the off state, and V_{DS} of the SiC MOSFET undergoes an upward transition with $dV_{DS(up)}/dt$, D_{DESAT} experiences reverse recovery, and

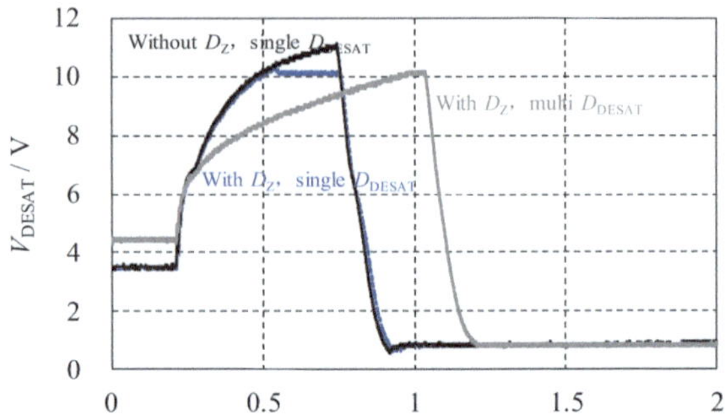

Fig. 11.46 Influence of $dV_{DS(up)}/dt$ on DESAT

its voltage across its terminals also increases at the same rate. At this moment, the current flowing through D_{DESAT}, denoted as $C_{J\text{-}DESAT}$, is the sum of the reverse recovery current and the displacement current. The direction of $I_{J\text{-}DESAT}$ is opposite to that during the turn-on of the SiC MOSFET.

As shown in Fig. 11.46, both $I_{DESAT\text{-}charge}$ and $I_{J\text{-}DESAT}$ simultaneously charge $C_{J\text{-}DESAT}$, causing a continuous elevation of V_{DESAT}. When the amplitude of V_{DESAT} becomes too high, it may lead to damage to the DESAT pin. To protect against this, a diode D_z can be connected in parallel with C_{BLANK}. The voltage rating of D_z should be higher than $V_{DESAT(th)}$ but lower than the reverse breakdown voltage of the DESAT pin.

If, at this point, the SiC MOSFET has not experienced a short circuit or overcurrent, an undesired triggering of the DESAT short-circuit protection may occur if the elevated V_{DESAT} exceeds V_{ref}. The extent to which V_{DESAT} is raised depends on C_{BLANK}, I_{DESAT}, and $I_{J\text{-}DESAT}$, where $I_{J\text{-}DESAT}$ is determined by the product of $dV_{DS(up)}/dt$ and $C_{J\text{-}DESAT}$, as well as the reverse recovery characteristics of D_{DESAT}. Therefore, selecting a D_{DESAT} with a smaller $C_{J\text{-}DESAT}$ and faster reverse recovery characteristics can effectively reduce $I_{J\text{-}DESAT}$ generated by $dV_{DS(up)}/dt$, mitigating the impact of $dV_{DS(up)}/dt$ at its source. Furthermore, connecting multiple D_{DESAT} diodes in series can significantly reduce $C_{J\text{-}DESAT}$.

1. **Response time adjustment**

Based on Sect. 11.6.1, DESAT can be divided into four stages, namely DESAT delay $t_{d(DESAT)}$ $(t_1 - t_2)$, blanking time t_{BLANK} $(t_2 - t_3)$, turn off response delay $t_{d(off)}$ $(t_3 - t_4)$, and off $t_{off}(t_4 - t_5)$, which is required to quickly complete short-circuit detection and turn off. The total duration of these four stages should be shortened as much as possible.

(1) **DESAT delay $t_{d(DESAT)}$ and turn off response delay $t_{d(off)}$**

DESAT delay and turn-off response delay help avoid false triggering. They are determined by the drive chip and are unavoidable. Therefore, it is necessary to choose a drive chip with a smaller sum of the two.

(2) **Blanking time t_{BLANK}**

The length of the blanking time t_{BLANK} represents the speed of charging C_{BLANK}, which is given by (11.16),

$$t_{BLANK} = C_{BLANK} \cdot \frac{V_{DESAT(th)}}{I_{DESAT\text{-}charge}}. \quad (11.16)$$

Then reducing C_{BLANK} can make V_{DESAT} rise faster; but C_{BLANK} is too If it is small, it will be susceptible to interference and cause false triggering. C_{BLANK} generally ranges from 100 to 330pF. Increasing $I_{DESAT\text{-}charge}$ can also shorten the blanking time, so you can choose a drive chip with a relatively larger I_{DESAT}. The drive chip I_{DESAT} is generally around 500uA. When $V_{DESAT(th)}$ is 9 V, when C_{BLANK} is 100pF, the calculated t_{BLANK} is 1800 ns. In order to further shorten t_{BLANK}, you can use a pull up resistor to connect C_{BLANK} to V_{CC2}, greatly increasing the charging current.

(3) **Turn off $t_{(off)}$**

There are two ways to turn off the device: hard turn off and soft turn off. Hard turn off is performed directly through the normal turn off circuit, and the turn off time is short. In order to avoid the turn off voltage spike $V_{DS(spike)}$ exceeding the device withstand voltage during turn off under high current, after triggering the DESAT short-circuit protection, the drive chip performs a high-impedance drive circuit on the device. Slow turn off is called soft turn off.

It should be noted that the turn-off time parameters given in the drive chip datasheet are the result of turning off the output signal under a specific output load, and are different from the actual time it takes to complete the turn-off of the device. Therefore, the turn-off time needs to be determined based on experimental measurements.

(4) **Adjustment of DESAT Threshold Voltage**

The reference voltage for the comparator in the integrated DESAT circuit within the driver chip is denoted as V_{ref}. However, due to the forward voltage drop V_F across D_{DESAT}, the actual DESAT threshold voltage $V_{DESAT(th)}$ is given by (11.17), where n represents the number of D_{DESAT} diodes used.

$$V_{th} = V_{DESAT(th)} - n \cdot V_F - I_{DESAT\text{-}charge} \cdot R_{DESAT}. \quad (11.17)$$

Additionally, it is possible to connect a zener diode D_{zener} in series with D_{DESAT} to further adjust $V_{DESAT(th)}$. This is given by (11.18), where V_z is the voltage drop across D_{zener}.

$$V_{th} = V_{DESAT(th)} - n \cdot V_F - V_z - I_{DESAT\text{-}charge} \cdot R_{DESAT}. \quad (11.18)$$

(5) **Si FRD Forward Recovery**

When using IGBTs, the forward recovery diode (FRD) in reverse parallel with the IGBT experiences a phenomenon known as the forward recovery effect. This occurs when the FRD transitions from the off state to the on state, causing the anode voltage to be much higher than the cathode voltage. The voltage difference can range from a few volts to several hundred volts and can last for several hundred nanoseconds. The forward recovery effect of the Si FRD is influenced by various factors, including higher temperatures, faster turn-on speeds, higher voltage levels, and larger current levels, all of which intensify the forward recovery effect.

The forward recovery effect can cause V_{DESAT} to become negative, potentially leading to damage to the DESAT pin. Therefore, similar to the measures taken to address $dV_{DS(down)}/dt$, a diode D_p can be connected in parallel with C_{BLANK} to clamp the negative voltage. Simultaneously, selecting a resistor R_{DESAT} with a larger value can be used to limit the surge current.

(6) **IGBT Forward Recovery**

The drive voltage for IGBT is high, and in the initial state, there is no current flowing through it. When a sudden current flow occurs thereafter, a significant voltage rise will occur across the terminals of IGBT. The voltage will be much higher than the normal forward voltage drop during IGBT conduction. This phenomenon is referred to as the positive recovery effect of IGBT. In servo systems and resonant circuits, the positive recovery effect manifests when the current crosses zero.

During the occurrence of the positive recovery effect, the excessively high-voltage amplitude causes the reverse cutoff of D_{DESAT} and initiates reverse recovery. Simultaneously, the voltage rise of IGBT is characterized by a high instantaneous rate. At this juncture, akin to the impact of $dV_{DS(up)}/dt$ mentioned earlier, (V_{DESAT}) is elevated. When V_{DESAT} surpasses V_{ref}, inadvertent triggering of the DESAT short-circuit protection may occur. Exceeding the withstand voltage of the DESAT pin by V_{DESAT} could potentially lead to damage to the DESAT pin.

Therefore, the same countermeasures are taken for $dV_{DS(up)}/dt$, choose the one with smaller $C_{J\text{-}DESAT}$ and faster reverse recovery characteristics. D_{DESAT} can also reduce the impact of the forward recovery effect. Furthermore, using multiple D_{DESAT} in series can reduce $C_{J\text{-}DESAT}$ more significantly. At the same time, a Zener diode D_z can be connected in parallel to C_{BLANK}, and its voltage stabilization value needs to be higher than the DESAT threshold voltage $V_{DESAT(th)}$ and lower than DESAT pin voltage resistance value.

Based on the above analysis, it can be seen that in practical applications, the DESAT detection circuit will face voltage jumps or current shocks. At the same time, there are various parasitic parameters in the real circuit, and the parameters of the DESAT circuit need to be adjusted according to the application requirements. This makes the circuit in Fig. 11.42 cannot be used directly in the converter, and it is necessary to use a practical DESAT short-circuit protection circuit as shown in Fig. 11.47. An experimental study on the performance of the DESAT short-circuit protection circuit will be conducted using the circuit shown, and employs the DESAT short-circuit protection circuit depicted in Fig. 11.48.

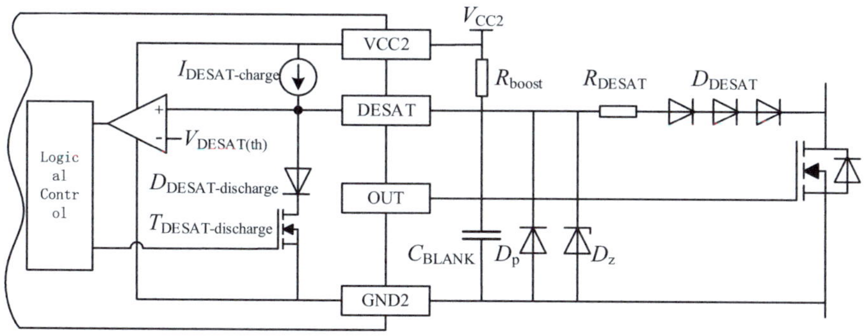

Fig. 11.47 Production-proven DESAT short-circuit protection circuit

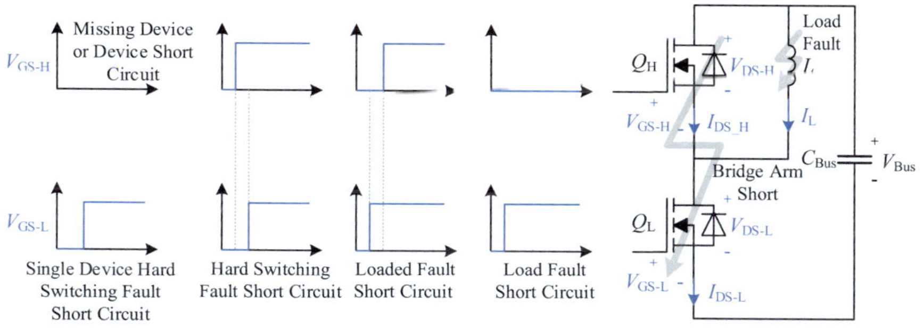

Fig. 11.48 DESAT Test and Protection Circuit

11.6.2.3 Effect of DESAT on Single Hard Switch Fault Short Circuit

The effect of DESAT on single-switch fault short circuit is shown in Fig. 11.49.

(1) $-t_1$

SiC MOSFET is in the off-state, it withstands the bus voltage V_{Bus}. V_{DESAT} is the sum of the conduction voltage drop $V_{DESAT\text{-}discharge}$ of $I_{DESAT\text{-}charge}$ across $T_{DESAT\text{-}discharge}$ and $D_{DESAT\text{-}discharge}$.

(2) $t_1 - t_2$

At time t_1, the SiC MOSFET turns on and enters a short-circuit state. I_{DS} rapidly increases, causing a slight decrease in V_{DS}, which nonetheless remains at a relatively high voltage. This stage represents the DESAT delay, where the $T_{DESAT\text{-}discharge}$ control signal remains high, and $V_{DESAT\text{-}discharge}$ remains constant.

(3) $t_2 - t_3$

At time t_2, the DESAT delay concludes, and the $T_{DESAT\text{-}discharge}$ control signal transitions to a low state. $I_{DESAT\text{-}charge}$. charges C_{BLANK}, causing V_{DESAT} to gradually rise to $V_{DESAT(th)}$.

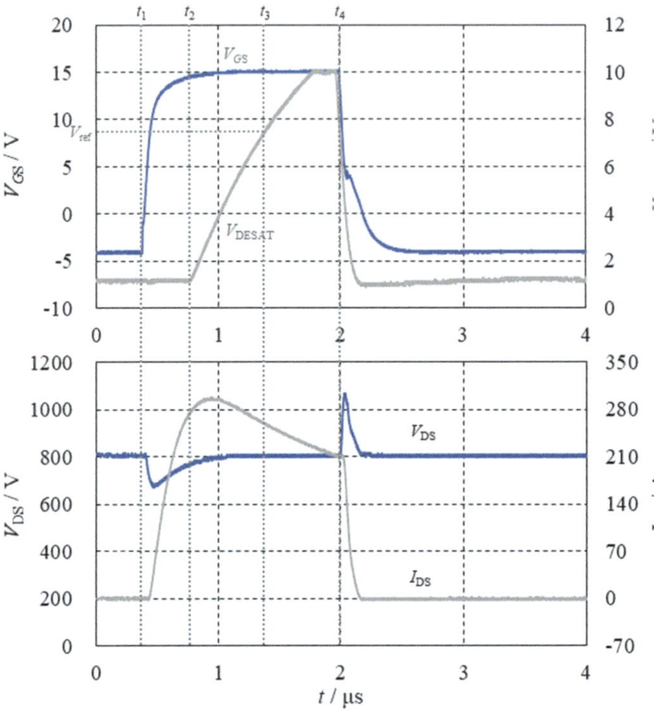

Fig. 11.49 Effectiveness of DESAT protection during single-device hard-switching fault short-circuit

(4) $t_3 - t_4$

V_{DESAT} continues to rise and is clamped by D_z.

(5) t_4-

At t_4, the drive circuit initiates the turn-off of the SiC MOSFET. The $T_{DESAT\text{-}discharge}$ control signal transitions to a higstate. Consequently, I_{DS} gradually decreases to zero, and V_{DESAT}. gradually decreases to $V_{DESAT\text{-}discharge}$. V_{DS} exhibits a peak and oscillations before stabilizing at V_{Bus}.

11.6.2.4 Effect of DESAT on Hard-Switched Fault Short Circuits

The effect of DESAT on switch fault short circuit is shown in Fig. 11.50.

(1) $-t_1$

When Q_H is in the on state, $V_{DESAT\text{-}H}$ stabilizes at the sum of the voltage drops across $I_{DESAT\text{-}charge} \cdot R_{DESAT}$ and D_{DESAT}. When Q_L is in the OFF state, it withstands the

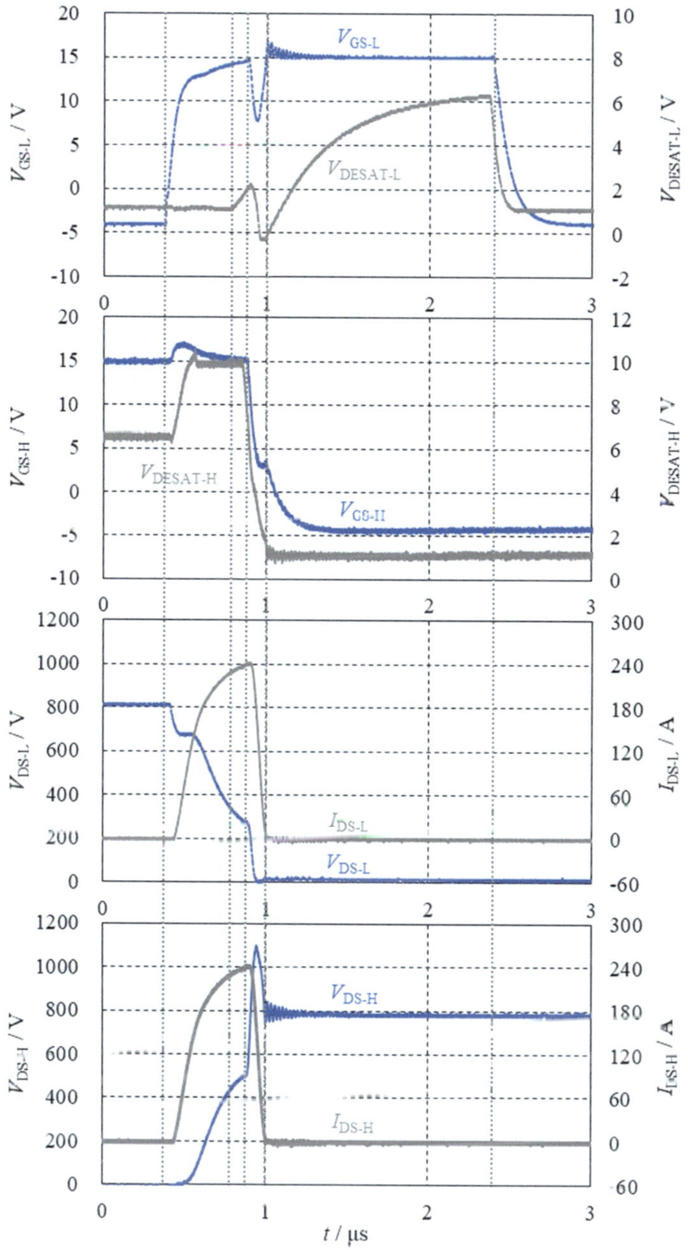

Fig. 11.50 Effectiveness of DESAT protection during hard-switching fault short-circuit events

bus voltage V_{Bus}. $V_{DESAT\text{-}L}$ is the sum of the conduction voltage drops of $I_{DESAT\text{-}charge}$ across $T_{DESAT\text{-}discharge}$ and $D_{DESAT\text{-}discharge}$, resulting in $V_{DESAT\text{-}discharge}$.

(2) t_1–t_2

At time t_1, Q_L turns on, forming a short-circuit loop with Q_H and the bus capacitor. The I_{DS} rises rapidly, $V_{DS\text{-}L}$ decreases, and $V_{DS\text{-}H}$ increases. Under the combined effect of $I_{DESAT\text{-}charge}$ and $dV_{DS\text{-}H(up)}/dt$, $V_{DESAT\text{-}H}$ rises rapidly and exceeds $V_{DESAT(th)}$ eventually being clamped by the Zener diode D_z. Due to the effect of $dV_{DS\text{-}H(up)}/dt$, a spike appears on $V_{GS\text{-}H}$. Since the DESAT of Q_L is in the blanking time, $V_{DESAT\text{-}L}$ remains unchanged.

(3) $t_2 - t_3$

At time t_2, the blanking time of the DESAT of Q_L ends, and $V_{DESAT\text{-}L}$ gradually rises under the effect of $I_{DESAT\text{-}charge}$. It is worth noting that at this moment, the rate of decrease $dV_{DS\text{-}L(down)}/dt$ of $V_{DS\text{-}L}$ is relatively slow, insufficient to cause a decrease in $V_{DESAT\text{-}L}$. Due to the action of D_z, $V_{DESAT\text{-}H}$ remains unchanged.

(4) $t_3 - t_4$

At time t_3, the DESAT protection of Q_H activates and turns it off. As $V_{GS\text{-}H}$ decreases, I_{DS}, $V_{DESAT\text{-}H}$, and $V_{DS\text{-}L}$ decrease, while $V_{DS\text{-}H}$ increases. At this moment, the rate of decrease $dV_{DS\text{-}L(down)}/dt$ of $V_{DS\text{-}L}$ is relatively fast, causing $V_{DESAT\text{-}L}$ to decrease and be clamped by the diode D_p. Simultaneously, $V_{GS\text{-}L}$ experiences a downward drop.

(5) $t_4 - t_5$

At time t_4, I_{DS} decreases to zero, and $V_{DS\text{-}L}$ stabilizes at zero. $V_{DESAT\text{-}L}$ gradually rises under the effect of $V_{DS\text{-}H}$ stabilizes at V_{Bus} after oscillation, and $V_{DESAT\text{-}H}$ stabilizes at $V_{DESAT\text{-}discharge}$.

(6) t_5

At time t_5, the driver circuit initiates the turn-off of Q_L turns of and $V_{DESAT\text{-}L}$ stabilizes at $V_{DESAT\text{-}discharge}$, while other waveforms remain unchanged.

11.6.2.5 Effect of DESAT on Fault Under Load

The effect of DESAT on single-switch fault short circuit is shown in Fig. 11.51.

(1) $-t_1$

Both Q_H and Q_L are in the off state. Q_L withstands the bus voltage V_{Bus}. $V_{DESAT\text{-}H}$ and $V_{DESAT\text{-}L}$ are the sum of the conduction voltage drops of $I_{DESAT\text{-}charge}$ across $T_{DESAT\text{-}discharge}$ and $D_{DESAT\text{-}discharge}$, resulting in $V_{DESAT\text{-}discharge}$.

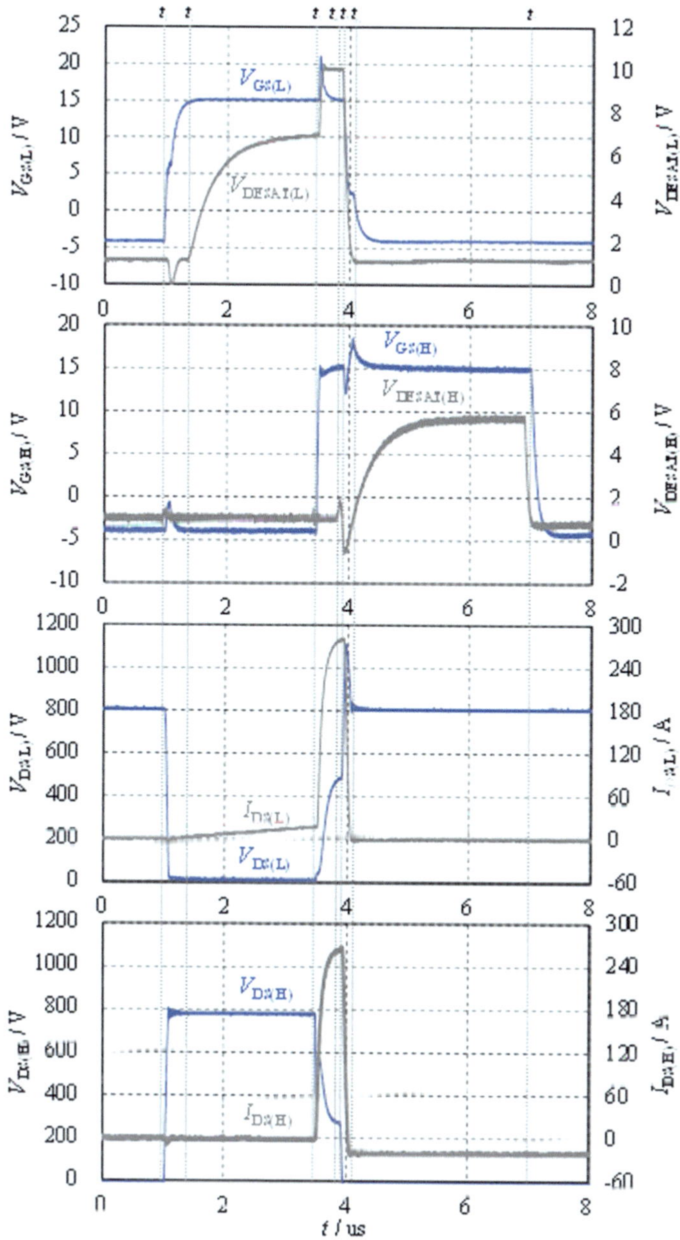

Fig. 11.51 Effectiveness of DESAT protection during loaded short-circuit faults

(2) $t_1 - t_2$

At time t_1, Q_L turns on, $I_{\text{DS-L}}$ linearly rises, and $V_{\text{DS-L}}$ decreases. Under the effect of $\mathrm{d}V_{\text{DS-L(down)}}/\mathrm{d}t$, $V_{\text{DESAT-L}}$. experiences a downwa spike and is clamped by D_p. After $V_{\text{DS-L}}$ stabilizes, $V_{\text{DESAT-L}}$ returns to $V_{\text{DESAT-discharge}}$. $V_{\text{DESAT-H}}$ increases, and under the effect of $\mathrm{d}V_{\text{DS-H(up)}}/\mathrm{d}t$, $V_{\text{GS-H}}$ experiences an upward spike.

(3) $t_2 - t_3$

At time t_2, the blanking time of Q_L's DESAT ends, and $V_{\text{DESAT-L}}$ gradually rises under the effect of $I_{\text{DESAT-charge}}$, stabilizing at the sum of the voltage drops across $I_{\text{DESAT-charge}} \cdot R_{\text{DESAT}}$ and D_{DESAT}. During this stage, $I_{\text{DS-L}}$ continues to linearly rise.

(4) $t_3 - t_4$

At time t_3, Q_H turns on, forming a short-circuit loop with Q_L and the bus capacitor. $I_{\text{DS-L}}$ and $I_{\text{DS-H}}$ rise rapidly, $V_{\text{DS-L}}$ increases, and $V_{\text{DESAT-H}}$ decreases. Under the combined effect of $I_{\text{DESAT-charge}}$ and $\mathrm{d}V_{\text{DS-L(up)}}/\mathrm{d}t$, $V_{\text{DESAT-L}}$ rises rapidly and exceeds V_{ref}, eventually being clamped by the Zener diode D_z. Due to the effect of $\mathrm{d}V_{\text{DS-L(up)}}/\mathrm{d}t$, a spike appears on $V_{\text{GS-L}}$. Q_{H}'s DESAT remains unchanged as it is in the blanking time.

(5) $t_4 - t_5$

At time t_4 , the blanking time of Q_{H}'s DESAT ends, and $V_{\text{DESAT-H}}$ gradually rises under the effect of $I_{\text{DESAT-charge}}$. It's worth noting that at this moment, the rate of decrease $\mathrm{d}V_{\text{DS-H(down)}}/\mathrm{d}t$ of $V_{\text{DESAT-H}}$ is relatively slow, insufficient to cause a decrease in $V_{\text{DESAT-H}}$. Due to the action of D_z, $V_{\text{DESAT-L}}$ remains unchanged.

(6) $t_5 - t_6$

At time t_5, the DESAT protection of Q_L activates and turns it off. As $V_{\text{GS-L}}$ decreases, $I_{\text{DS-L}}$, $I_{\text{DS-H}}$, $V_{\text{DESAT-L}}$, and $V_{\text{DESAT-H}}$ decrease, while $V_{\text{DS-L}}$ increases. Due to the relatively fast rate of decrease $\mathrm{d}V_{\text{DS-L(down)}}/\mathrm{d}t$ of $V_{\text{DESAT-H}}$ at this moment, $V_{\text{DESAT-H}}$ decreases and is clamped by the diode D_p. Simultaneously, $V_{\text{GS-H}}$ experiences a downward drop.

(7) $t_6 - t_7$

At time t_6, $I_{\text{DS-L}}$ and $I_{\text{DS-H}}$ decrease to zero. The load inductor current is freewheeling through the body diode of Q_{H}, and $V_{\text{DESAT-H}}$ stabilizes at zero. $V_{\text{DESAT-H}}$ gradually rises under the effect of $I_{\text{DESAT-charge}}$ and stabilizes at the sum of the voltage drops across $I_{\text{DESAT-charge}} \cdot R_{\text{DESAT}}$ and D_{DESAT}. $V_{\text{DS-L}}$ stabilizes at V_{Bus} after oscillation, and $V_{\text{DESAT-L}}$ stabilizes at $V_{\text{DESAT-discharge}}$.

(8) t_8-

At time t_8, Q_H turns off, and $V_{\text{DESAT-H}}$ stabilizes at $V_{\text{DESAT-discharge}}$, while other waveforms remain unchanged.

11.6.2.6 Effect of DESAT on Load Fault

The effect of DESAT on single-switch fault short circuit is shown in Fig. 11.52.

(1) $-t_1$

Q_H and Q_L are in the off-state, and Q_L withstands the bus voltage V_{Bus}. $V_{\text{DESAT-H}}$ and $V_{\text{DESAT-L}}$ are the sum $V_{\text{DESAT-discharge}}$ of the conduction voltage drop of $I_{\text{DESAT-charge}}$ across $T_{\text{DESAT-discharge}}$ and $D_{\text{DESAT-discharge}}$.

(2) $t_1 - t_2$

At time t_1, Q_L turns on, $I_{\text{DS-L}}$ linearly rises, and $V_{\text{DS-L}}$ rapidly decreases. Under the effect of $\mathrm{d}V_{\text{DS-L(down)}}/\mathrm{d}t$, $V_{\text{DESAT-L}}$ experiences a downward spike and is clamped by D_p. When $V_{\text{DS-L}}$ stabilizes at zero, $V_{\text{DESAT-L}}$ returns to $V_{\text{DESAT-discharge}}$. $V_{\text{DS-H}}$ rapidly increases and stabilizes at V_{Bus}. Under the effect of $\mathrm{d}V_{\text{DS-H(up)}}/\mathrm{d}t$, $V_{\text{GS-H}}$ experiences an upward spike, and $V_{\text{DESAT-H}}$ shows a small upward protrusion.

(3) $t_2 - t_3$

At time t_2, the blanking time of Q_L's DESAT ends, and $V_{\text{DESAT-L}}$ gradually rises under the effect of $I_{\text{DESAT-charge}}$ and exceeds V_{ref}, eventually being clamped by the Zener diode D_z. During this stage, the control signal $T_{\text{DESAT-discharge}}$ becomes low, $I_{\text{DESAT-charge}}$ charges C_{BLANK}, $I_{\text{DS-L}}$ and I_{DS} continue to linearly rise, and V_{DESAT} gradually increases until it reaches V_{ref}.

(4) $t_3 - t_4$

At time t_3, as $V_{\text{GS-L}}$ decreases, $I_{\text{DS-L}}$, $V_{\text{DESAT-L}}$, and $V_{\text{DS-H}}$ decrease, while $V_{\text{DS-L}}$ increases. Due to the relatively fast rate of decrease $\mathrm{d}V_{\text{DS-L(down)}}/\mathrm{d}t$ of $V_{\text{DS-H}}$ at this moment, $V_{\text{DESAT-H}}$ decreases and is clamped by the diode D_p. Simultaneously, $V_{\text{GS-H}}$ experiences a downward drop. Eventually, $I_{\text{DS-L}}$ and $V_{\text{DS-H}}$ decrease to zero, $V_{\text{DS-L}}$ stabilizes at $V_{\text{DESAT(th)}}$ after oscillation, and $V_{\text{DESAT-L}}$ and $V_{\text{DESAT-H}}$ stabilize at $V_{\text{DESAT-discharge}}$.

(5) t_4-

At time t_4, the drive circuit initiates the turn-off of the SiC MOSFET. The $T_{\text{DESAT-discharge}}$ control signal transitions to a high state. Consequently, I_{DS} gradually decreases to zero, and V_{DESAT} gradually decreases to $V_{\text{DESAT-discharge}}$. V_{DS} exhibits a peak and oscillations before stabilizing at V_{Bus}. During this process, due to the influence of $\mathrm{d}V_{\text{DS(up)}}/\mathrm{d}t$, V_{DESAT} experiences a transient rise during the descent.

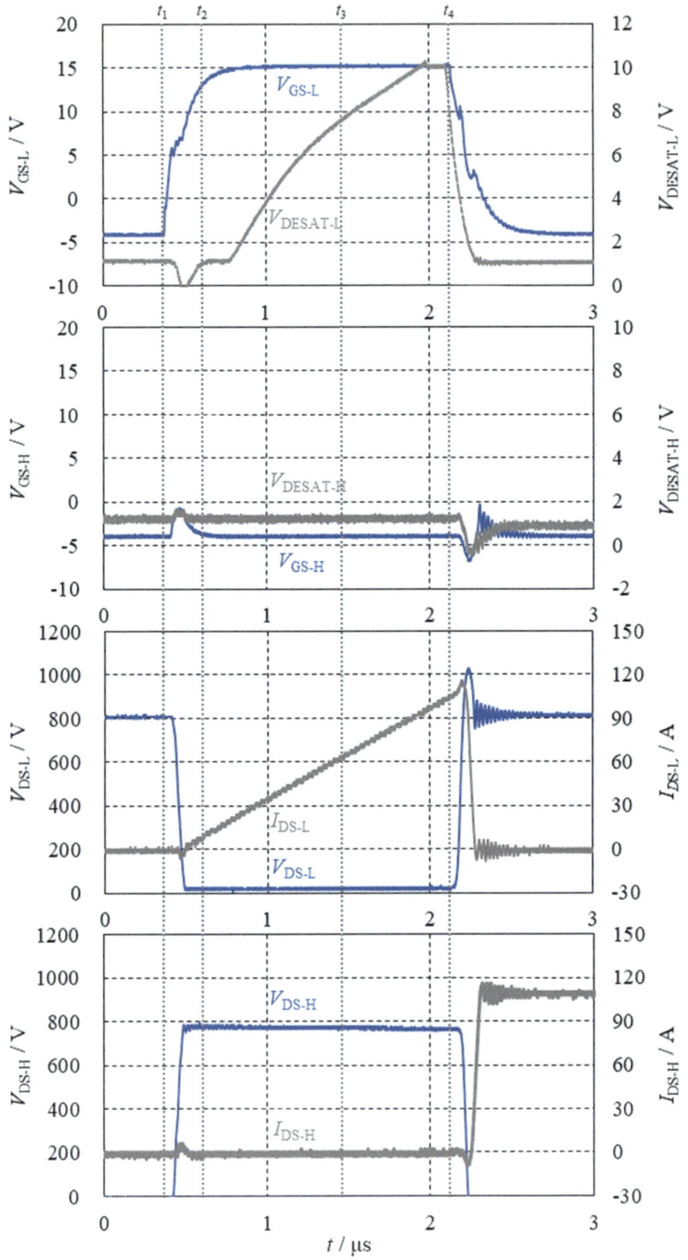

Fig. 11.52 Effectiveness of DESAT protection during load fault short-circuits

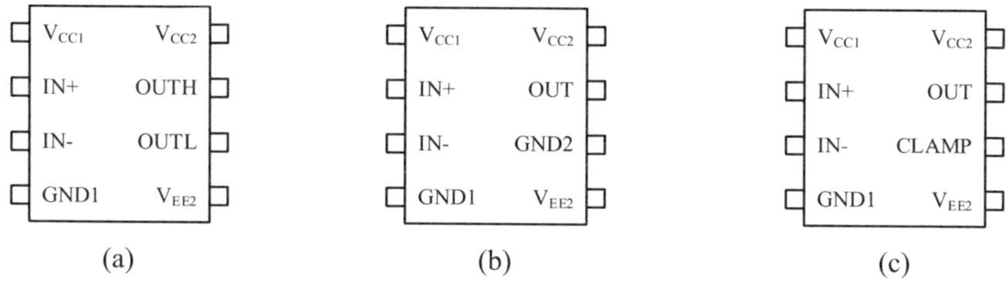

Fig. 11.53 8-pin single-channel isolation driver chips

11.7 Reference for Drive Circuit Design

11.7.1 8-Pin Single-Channel Isolated Driver Chip

Common 8-pin single-channel isolation driver chips are available in three forms, as illustrated in Fig. 11.53. The primary-side pin definitions are identical for all three types, with distinctions in the high-voltage side power supply pin, drive pin, and active Miller clamp pin. Figure 11.53a features independent drive pins for turn-on and turn-off, unipolar power supply, and no active Miller clamp function. Figure 11.53b shares drive pins for turn-on and turn-off, has bipolar power supply, and lacks an active Miller clamp function. Lastly, Fig. 11.53c shares drive pins for turn-on and turn-off, employs unipolar power supply, and incorporates an active Miller clamp function.

The design of a driving circuit based on an 8-pin single-channel isolation driver chip should consider the following points:

(1) Capacitors should be configured for isolation power input and output pins as well as the driver chip power supply pin. These capacitors serve the purpose of filtering and decoupling. Adopting a combination of high and low capacitance values covers a broader frequency range, thereby enhancing the stability and reliability of the driving circuit.
(2) The energy for charging and discharging the input capacitor during the SiC MOSFET turn-on and turn-off processes primarily comes from the capacitor on the driver chip's power supply pin. Therefore, the capacitor should be sufficiently large to avoid excessive voltage fluctuations. Generally, a capacitance of 1μC corresponding to a gate charge of 3uF is suitable.
(3) In cases where turn-on and turn-off are driven by independent pins, it is possible to set separate drive resistors for turn-on and turn-off. When turn-on and turn off share a drive pin, drive resistors for turn-on and turn-off can be separately set by incorporating diodes. It is essential to note that due to diode voltage drops, the actual drive voltage will be slightly lower than the driver chip's power supply voltage.
(4) For a driver chip with an active Miller clamp function, the active Miller clamp pin should be directly connected to the gate pin of the SiC MOSFET.

(5) Grounding for the primary side and the high-voltage side of the isolation driver chip should be separated and should not overlap. The routing of the lines for both the primary side and high-voltage side should stay within their respective grounding areas. The routing of output drive signals (G and S terminals) should ideally overlap vertically to minimize interference.

Taking the form of the driving chip in Fig. 11.53c as an example, the driving circuit based on the 8-pin single-channel isolated driving chip is illustrated in Fig. 11.54. Two modes are presented: zero-voltage turn-off and negative voltage turn-off.

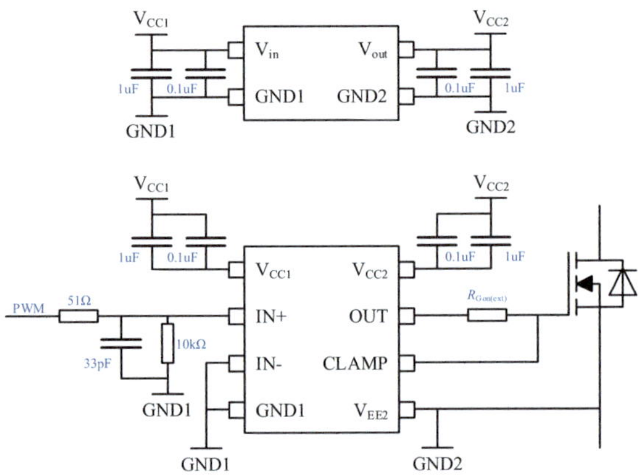

(a) Zero voltage turn off

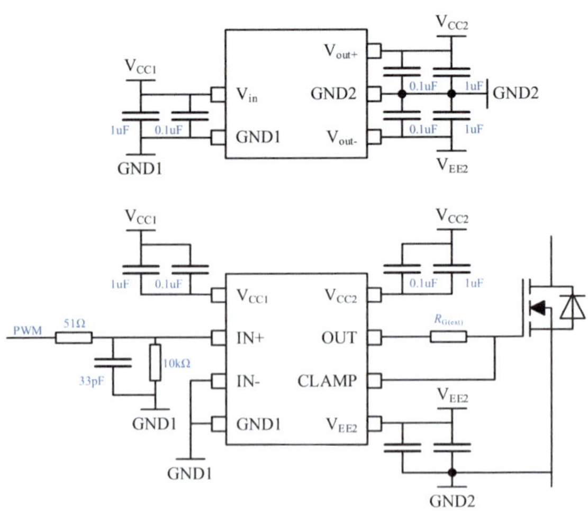

(b) Negative voltage turn off

Fig. 11.54 Driver circuit based on 8-pin single-channel isolated driver chip

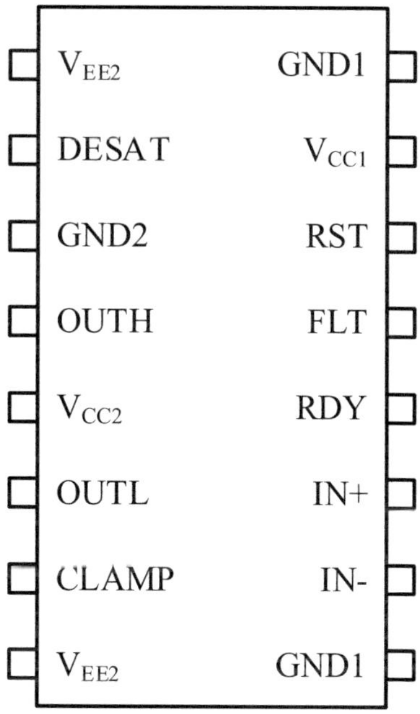

Fig. 11.55 16-pin single-channel isolated driver chip

11.7.2 16-Pin Single-Channel Isolated Driver Chip

Common 14-pin single-channel isolated driving chips, exemplified by those possessing DESAT short-circuit protection and active Miller Clamping functionalities, are prevalent. Some feature independently driven pins for activation and deactivation, alongside control and status feedback pins on the primary side, as depicted in Fig. 11.55.

(1) RDY、FLT上拉至VCC, The design of the driving circuit based on the 16-pin single-channel isolated driving chip requires attention to the following aspects:
(2) Place RC filtering circuits near the RDY, FLT, and RST pins to prevent signal interference.
(3) Pull RDY and FLT to VCC and control the LED through a transistor for convenient status observation.
(4) Design the DESAT detection circuit following the guidelines outlined in Sect. 11.6.2.2.
(5) Due to the presence of high-voltage transients, ensure that the DESAT circuit's serial resistor is not positioned above the ground on the high-voltage side.

The driving circuit based on the 16-pin single-channel isolated driving chip is depicted in Fig. 11.56.

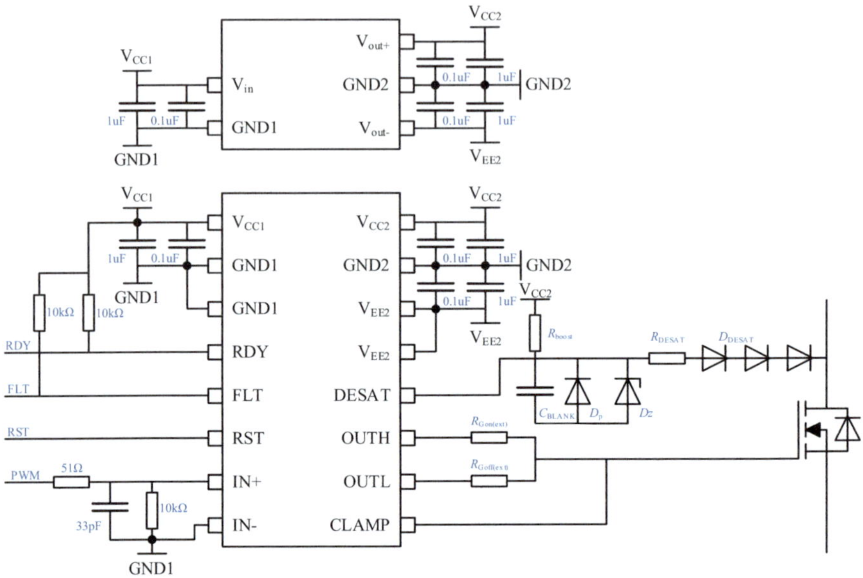

Fig. 11.56 Drive circuit based on 16-pin single-channel isolated driver chip

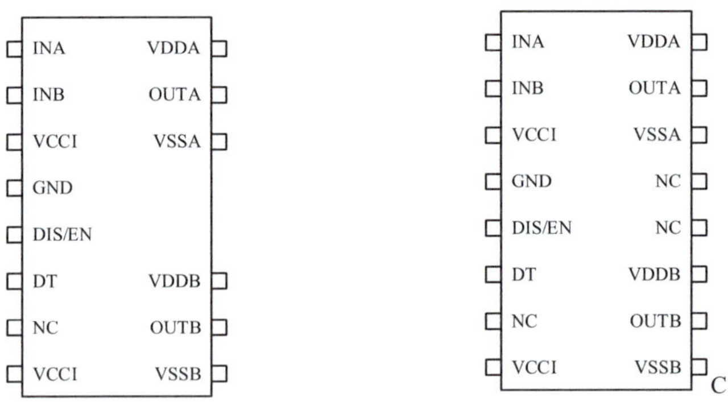

(a) 14-pin dual-channel isolated driver chip (b) 16-pin dual-channel isolated driver chip

Fig. 11.57 Dual channel isolation driver chip

11.7.3 14/16-Pin Dual-Channel Isolated Driver Chip

The commonly encountered 14-pin dual-channel isolated driving chip is illustrated in Fig. 11.57a, while the 16-pin dual-channel isolated driving chip is depicted in Fig. 11.57b. Their high-voltage side features dual-channel outputs, specifically designated for driving the upper and lower arms of a half-bridge circuit. The key distinction

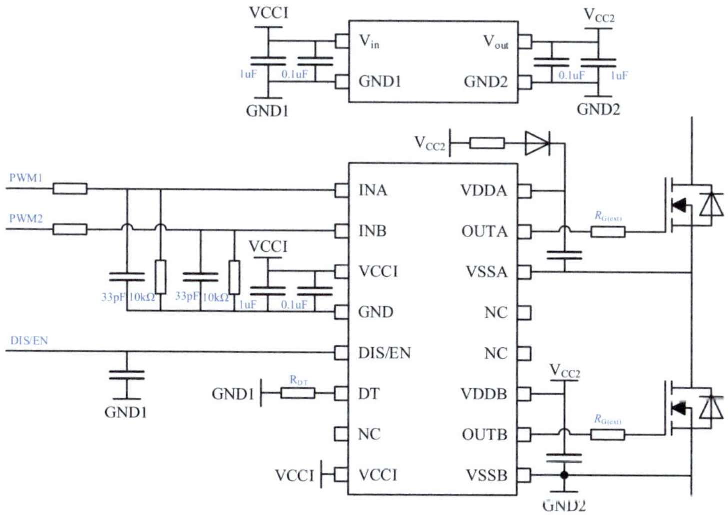

(a) High-side device driver isolated power supply – zero-voltage turn off

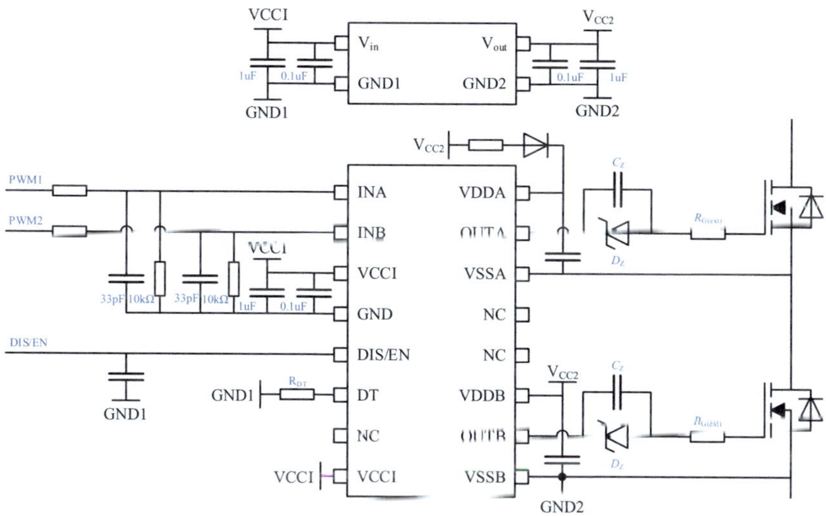

(b) High-side device drives isolated power supply – negative voltage turn off

Fig. 11.58 Drive circuit based on 14/16-pin dual-channel isolated driver chip lies in the 16-pin chip having two additional no connection (NC) pins between the dual channels.

The design of the driving circuit based on the 14/16-pin dual-channel isolated driving chip should take into consideration the following points:

(1) Set the dead time resistor and its bypass capacitor as close as possible to the DT pin.
(2) When the DIS/EN pin is connected to a remote processor, place a 1nF bypass capacitor near the DIS/EN pin.
(3) Maintain safety clearance between the primary side and the high-voltage side ground, as well as between the grounds of the dual-channel outputs on the high-voltage side, ensuring no overlapping.

The driving circuit based on the 14/16-pin dual-channel isolated driving chip is depicted in Fig. 11.58. The power supply for driving the upper bridge arm devices includes both an isolated power source and a bootstrap circuit. The turn off driving voltage is available in both zero-voltage and negative voltage variants.

References

1. Infineon Technologies AG (2017) 1EDC/1EDI compact family technical description technical description. Application Note, AN2014-06, Rev. 1.03
2. ROHM Co., Ltd. (2017) BD7F100HFN-LB BD7F100EFJ-LB datasheet. Rev.003
3. Silicon Laboratories, Inc. (2018) Si8281/82/83/84 Datasheet. Rev. 1.0
4. Infineon Technologies AG. (2019) Advantages of Infineon's high-voltage gate driver ICs (HVICs) based on its silicon-on-insulator (SOI) technology. Application Note, AN2019-12, Rev. 2.0
5. Silicon Laboratories, Inc. (2018) High-side bootstrap design using ISO drivers in power delivery systems. AN486, Rev. 0.2
6. Fairchild Semiconductor Corporation (2014) Design and application guide of bootstrap circuit for high-voltage gate-drive IC. Rev. 1.4
7. Rüedi H, Thalheim J, Garcia O (2009) Advantages of advanced active clamping. CT-Concept Technologie AG, Power Electronics Europe, Issue 8, pp 27–29
8. Dominik F, Olivier G (2014) SCALE™-2 IGBT gate drivers ease the design of optimized renewable inverter systems. CT-Concept Technologie GmbH, Bodo's Power Systems, pp 22–26
9. Infineon Technologies AG (2019) Guidelines for CoolSiC™ MOSFET gate drive voltage window. Application Note, AN2018*09, Revision 1.1
10. Infineon Technologies AG (2014) External booster for driver IC. Application Note, AN2013-10, Revision 1.6
11. Avago Technologies (2015) ACPL-339J datasheet. AV02-3784EN
12. Power Integration (2017) SID1102K datasheet. Rev. D.
13. Avago Technologies (2014) Optocouplers designer's guide. Design Guide, AV02-4387EN
14. David K (2011) Anatomy of a digital isolator. Technical Article, MS-2234, Analog Devices, Inc.
15. Silicon Laboratories, Inc. CMOS Digital Isolators supersede Optocouplers in Industrial Applications. Rev 0.2.
16. Infineon Technologies AG (2019) Infineon EiceDRIVER™ gate driver ICs selection guide 2019
17. Analog Devices, Inc. (2005) Coupler® digital isolators protect RS-232, RS-485, and CAN buses in industrial, instrumentation, and computer applications. Analog Dialogue 39-10, pp 1–4

18. Tom B (2017) Enabling high voltage signal isolation quality and reliability. White paper, Texas Instruments (2017)
19. Morgan M, Frattini G (2014) Accelerating automated manufacturing with advanced circuit isolation technology. White Paper, SSZY018, Texas Instruments Inc.
20. Kamath AS, Kannan S (2014) High-voltage reinforced isolation: definitions and test methodologies. White Paper, SLYY063, Texas Instruments Inc.
21. Luke T (2018) Considerations for selecting digital isolators. Application Note, SLLA426, Texas Instruments Inc.
22. Texas Instruments Inc. (2014) Isolation glossary. Application Note, SLLA353A
23. Silicon Laboratories Inc. Safety considerations and layout recommendations for digital isolators. Application Note, AN583, Rev. 0.3
24. IEC 61800-5-1 Ed 2.0: adjustable speed electrical power drive systems, safety requirements, electrical, thermal and energy (2007)
25. Infineon Technologies AG (2016) Control integrated power system (CIPOS™) inverter IPM reference board type 3 f or 3-shunt resistor. Application Note, AN2016-12, Revision 1.11
26. Tripathi A, Mainali K, Madhusoodhanan S et al (2016) A MV intelligent gate driver for 15kV SiC IGBT and 10kV SiC MOSFET. IEEE Appl Power Electron Conf Expo (APEC) 2016:2076–2082
27. Wiesner E, Thal E, Volke A et al (2016) Advanced protection for large current full SiC-modules. PCIM Europe
28. ON Semiconductor Corp (2017) Current sensing power MOSFETs. Application Note, AND8093/D, Rev. 6
29. Mocevic S, Wang J, Burgos R et al (2018) Phase current sensor and short-circuit detection based on Rogowski coils integrated on gate driver for 1.2 kV SiC MOSFET half-bridge module. In: 2018 IEEE energy conversion congress and exposition (ECCE). IEEE, pp 393–400
30. Wang J, Mocevic S, Xu Y et al (2018) A high-speed gate driver with PCB-embedded Rogowski switch-current sensor for a 10 kV, 240 A, SiC MOSFET module. IEEE Energy Convers Congr Expo (ECCE) 2018:5489–5494
31. Bortis D (2009) 20MW Halbleiter-Leistungsmodulator-system. Swiss Federal Institute of Technology
32. Bayarkhuu B, Bat-Ochir B O, Hasegawa K et al (2019) Analog basis, low-cost inverter output current sensing with tiny PCB coil implemented inside IPM. In: 2019 31st international symposium on power semiconductor devices and ICs (ISPSD), pp 251–254
33. Sun K, Wang J, Burgos R et al (2019) Design, analysis, and discussion of short circuit and overload gate-driver dual-protection scheme for 1.2 kV, 400 A SiC MOSFET modules. IEEE Trans Power Electron 35(3):3054–3068
34. Mocevic S, Wang J, Burgos R et al (2018) Comparison between desaturation sensing and Rogowski coil current sensing for shortcircuit protection of 1.2kV, 300A SiC MOSFET Module. In: 2018 IEEE applied power electronics conference and exposition (APEC). IEEE, pp 2666–2672

Further Reading

35. Infineon Technologies AG (2018) Advanced gate drive options for silicon-carbide (SiC) MOSFETs using EiceDRIVER™. Application Note, AN2017-04, Rev. 1.1
36. Infineon Technologies AG. How to choose gate driver for SiC MOSFETs and SiC MOSFET modules. Webinar. https://www.infineon.com/cms/en/product/power/gate-driver-ics/#!trainings
37. Zheyu Z (2019) Driving, monitoring, and protection technology for SiC devices using intelligent gate drive. Webinar, PELS. https://resourcecenter.ieee-pels.org/webinars/PELSWEB071823v.html

38. Balogh L (2017) Fundamentals of MOSFET and IGBT gate driver circuits. Application Report, SLUA618, Texas Instruments
39. ROHM Co., Ltd. (2023) Basics and design guidelines for gate drive circuits. Application Note, No.66AN032E, Rev.001
40. Toshiba Electronic Devices & Storage Corporation (2018) MOSFET gate drive circuit. Application Note
41. Markus H (2017) IGBT driver calculation. Application Note, AN-7004, Rev. 00., SEMIKRON
42. Avago Technologies (2014) Gate drive optocoupler basic design for IGBT/MOSFET applicable to all gate drive optocouplers. Application Note 5336, AV02-0421EN
43. Diogo V, Menditti MC (2020) Isolated gate driving solutions increasing power density and robustness with isolated gate driver ICs. Application Note, AN_1909_PL52_1910_201256, Rev. 1.0, Infineon Technologies AG
44. Infineon Technologies AG. How to choose gate driver for IGBT discretes and modules. Webinar. https://www.infineon.com/cms/en/product/power/gate-driver-ics/#!trainings
45. Infineon Technologies AG. Every switch needs a driver—the right driver makes a difference. Webinar. https://www.infineon.com/cms/en/product/power/gate-driver-ics/#!trainings
46. Audrey D (2019) HEV/EV traction inverter design guide using isolated IGBT and SiC gate drivers. Application Report, SLUA963, Texas Instruments Inc.
47. Texas Instruments (2019) IGBT & SiC gate driver fundamentals. SLYY169
48. Infineon Technologies AG. SOI level-shift gate driver IC in LLC half-bridge topologies. Webinar. https://www.infineon.com/cms/en/product/power/gate-driver-ics/silicon-on-insulator-soi/?redirId=104509
49. Fairchild Semiconductor Corporation (2008) Design guide for selection of bootstrap components. Rev. 1.0.0
50. Andrea M (2007) Bootstrap network analysis: focusing on the integrated bootstrap functionality. Application Note, AN-1123, Rev. 01, International Rectifier Corporation, Inc.
51. Merello A, Rugginenti A, Grasso M (2016) Using monolithic high voltage gate drivers. Application Note, Rev. 01, International Rectifier Corporation
52. Rossberg M, Vogler B, Herzer R (2008) 600V SOI gate driver IC with advanced level shifter concepts for medium and high power applications. In: IEEE European conference on power electronics & applications
53. Song J, Frank W (2015) Robustness of level shifter gate driver ICs concerning negative voltages. In: PCIM Europe, international exhibition & conference for power electronics, pp 140–146
54. Markus H (2017) Gate resistor—principles and applications. Application Note, AN-7003, Rev. 00, SEMIKRON International GmbH
55. Infineon Technologies AG (2015) EiceDRIVER™ gate resistor for power devices. Application Note, AN2015-06, Revision 1.0
56. ON Semiconductor (2020) Analysis of power dissipation and thermal considerations for high voltage gate drivers. Application Note, AND90004, Rev. 0
57. Ryan S (2018) Rarely asked questions—issue 158 driving a unipolar gate driver in a bipolar way. Analog Dialogue, 52–10
58. Silicon Laboratories Inc. Driving MOSFET and IGBT switches using the Si828x. Application Note, AN1009, Rev. 0.1
59. Longo G, Fusillo F, Scrimizzi F (2012) Power MOSFET: RG impact on applications. Application Note, AN4191, Doc ID 023815, Rev. 01, STMicroelectronics
60. Brian K (2016) Using on-off keying, digital isolators in harsh environments. Webcast, Analog Devices, Inc.
61. Texas Instruments Inc. (2018) Digital isolator design guide. Developer's Guide, SLLA284B, Rev. B.
62. Ziegler S, Woodward RC et al (2009) Current sensing techniques: a review. IEEE Sensors J 9(4):354–376
63. Yanick L (2014) Closed-loop IGBT gate drive and current balancing concepts. ETH
64. Miftakhutdinov R, Li X, Mukhopadhyay R et al (2018) How to protect SiC FETs from short circuit faults-overview. In: 2018 European conference on power electronics and applications (EPE'18 ECCE Europe)

65. Rothmund D, Bortis D, Kolar JW (2019) Highly compact isolated gate driver with ultrafast overcurrent protection for 10kV SiC MOSFETs. CPSS Trans Power Electron Appl 3(4):278–291
66. Infineon Technologies AG (2014) EICEDRIVER™ high voltage gate drive IC 1ED family technical description. Application Note, Rev. 1.4
67. Infineon Technologies AG. (2019) Using the EiceDRIVER™ 2EDi product family of dual-channel functional and reinforced isolated MOSFET gate drivers. Application Note, AN_1805_PL52_1806_095202, Rev. 2.0
68. Infineon Technologies AG (2018) EiceDRIVER™ safe high voltage gate driver IC with reinforced isolation 1EDS-SRC technical description. Application Note, AN2014–03, Rev. 1.2
69. Infineon Technologies AG (2018) PCB layout guidelines for MOSFET gate driver. Application Note, AN_1801_PL52_1801_132230, Rev. 01
70. Fairchild Semiconductor Corporation (2014) Driving and layout design for fast switching super-junction MOSFETs. Application Note, AN-9005, Rev. 1.0.1
71. ON Semiconductor (2019) NCD(V)57000/57001 gate driver design note. Application Note, AND9949/D, Rev. 0

Chapter 12
Main Applications of SiC Devices

Previous chapters have provided detailed introductions and discussions on various aspects of SiC devices. Chapter 1 comprehensively introduced the current status and future trends of the SiC industry and device development. Chapters 2–4 detailed the main characteristics of SiC diodes and SiC MOSFETs, as well as the differences between Si devices. Chapters 5 to 6 discussed the relevant technologies of parameter testing, production testing, application system testing, reliability testing, and failure analysis of SiC devices. Chapters 7–11 analyzed the challenges encountered in using SiC devices and provided feasible countermeasures. This allows engineers to have a comprehensive understanding of the characteristics, testing, and application technologies of SiC devices, thereby clearing technical barriers for practical use of SiC devices.

Compared to Si FRDs, the reverse recovery characteristics of SiC diodes can almost be neglected. Compared to Si SJ-MOSFETs, SiC MOSFETs can simultaneously meet high-voltage and high-current requirements, while their body diode reverse recovery characteristics have been significantly improved. Compared to Si IGBTs, SiC MOSFETs have significantly reduced switching losses and lower conduction losses under low-current conditions. Additionally, compared to Si devices, SiC devices can operate at higher junction temperatures. Therefore, the use of SiC devices can effectively improve the efficiency of converters, operate power converters at higher switching frequencies, reduce the usage of capacitors and magnetic components, and significantly increase the power density of power converters while reducing overall costs. These advantages are the foundation for the gradual widespread application of SiC devices in various fields, and as prices continue to decrease, the market size continues to increase.

Combining the characteristics of SiC devices and the specific requirements for power devices in practical applications, the main applications of SiC devices currently include main drive inverters, onboard chargers, onboard DC-DC converters, charging stations, photovoltaic and energy storage systems, uninterruptible power supplies (UPS), power supplies, and motor drives.

Y. Gao and Y. Zhang, *Silicon Carbide Power Devices*, CPSS Power Electronics Series,
https://doi.org/10.1007/978-981-96-3480-4_12

12.1 Main Drive Inverter

Over the past two decades, collaborative efforts between academia and industry have driven extensive research and development of SiC devices, including BJTs, JFETs, and MOSFETs. These efforts have focused primarily on high-voltage electric vehicle applications, such as main drive inverters, onboard chargers, and onboard DC-DCs, with the potential to replace traditional Si IGBTs, Si FRDs, and Si SJ-MOSFETs with SiC devices to improve efficiency, increase power density, and improve overall performance. Despite the obvious advantages of SiC devices, there is still some controversy and hesitation in the industry about their use in the automotive industry, including discussions about which type of SiC devices is best suited for automotive applications. However, with the launch of the Tesla Model 3, which uses a SiC MOSFET-based inverter, the situation began to change significantly. This major development has triggered a broader industry trend, with more and more original equipment manufacturers (OEMs) now launching SiC-based models covering 400 V and 800 V platforms, indicating that there is a growing consensus on the feasibility of SiC MOSFETs for main drive inverters.

In the main drive inverter, the circuit topology commonly used is the two-level three-phase voltage source inverter (2L-3Φ VSI), as shown in Fig. 12.1. In some cases, a non-isolated converter is incorporated between the battery and the DC capacitor of the inverter to regulate and stabilize the voltage supplied to the inverter. The two-level three-phase voltage source inverter has established a record of reliability and robustness for decades, thanks to carefully designed control strategies to ensure its effective operation. Although alternative topologies such as multiphase inverters, multilevel inverters, and certain current source inverters offer obvious advantages, this discussion focuses on power devices and will not discuss these topologies in detail for the time being.

SiC MOSFET has significant advantages over traditional Si IGBT anti-parallel Si FRD and can be fully utilized in main drive inverter applications, providing a strong reason for replacing traditional Si devices.

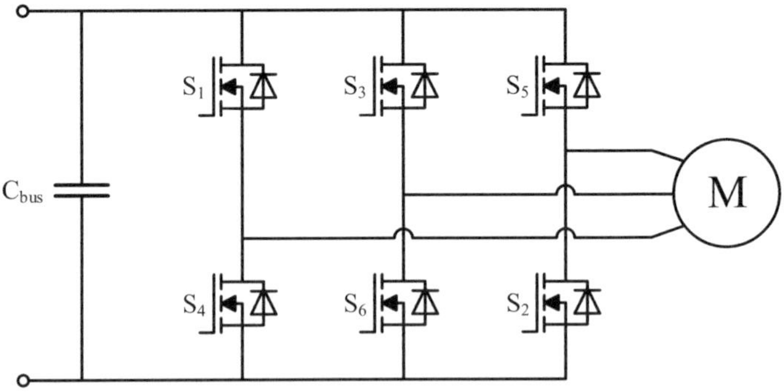

Fig. 12.1 2L-3Φ VSI inverter

First, SiC MOSFETs have significantly lower switching losses, thanks to their shorter switching delays and faster switching speeds. Compared with Si devices, this advantage is more obvious in inverter systems with higher power (>150 kW) and higher bus voltage (>600 V). High-voltage (1200 V or higher) Si IGBT requires high doping concentration through ion implantation or diffusion to offset the increase in drift region thickness and maintain low forward voltage drop during high-current conduction, which significantly reduces its switching performance, while SiC MOSFET has no such problem. Generally, the upper limit of the switching frequency of the main drive inverter based on Si IGBT is 20 kHz, and the conventional range is 8 to 12 kHz. The switching loss accounts for approximately 50% of the overall loss of the Si IGBT. The main drive inverter based on SiC MOSFET can achieve higher switching frequency, and its switching loss ratio is about 30%. Although in the main drive inverter system, the increase in switching frequency does not bring about the dramatic reduction in the size of the passive components (capacitors, inductors, and transformers) as the switching frequency increases like OBC and DC-DC. But high switching frequency can reduce motor noise and achieve low torque ripple and sufficient motor control resolution, especially for small high-speed motors.

Secondly, the on-resistance of SiC MOSFET used in the main drive inverter is only a few milliohms, and its output characteristic curve is a ray starting from zero, and the conduction loss is proportional to the conduction current. However, since Si IGBT and Si FRD contain pn junctions in their conduction structures, their output characteristic curves will not have current flowing until after their saturation voltage drop, that is, no matter how small the conduction current is, a certain forward voltage threshold is required to conduct. Based on the introduction in 4.2.2 Junction, SiC MOSFET has smaller conduction loss at light load (low output current), while Si IGBT has smaller conduction loss at heavy load (high output current). In a typical driving cycle, the main drive inverter operates under light load conditions for most of the time, so SiC MOSFET exhibits better conduction loss than Si IGBT. This is also one of the most important reasons why SiC MOSFET may still be a better choice than Si IGBT from the perspective of mileage improvement or battery pack saving even in a 400 V battery platform (such as Tesla Model 3).

The development to the 800 V platform is one of the trends in xEV technology. Its advantages include faster charging speed, higher efficiency, enhanced performance, reduced size and weight, etc. SiC MOSFETs play a key enabler role here, especially in main drive inverter systems. Figure 12.2 shows that using SiC MOSFETs on an 800 V platform can extend the drive range by 5 to 7% compared to IGBT solutions, depending on the SiC MOSFET technology. This improvement may seem small, but the savings on the battery are significant and the cost premium for SiC MOSFETs is justified. At the same time, continued improvements in SiC MOSFET technology will provide higher efficiency and lower costs.

In addition, the maximum operating temperature of SiC MOSFET is also significantly higher than that of Si IGBT, which improves the peak power capability of the

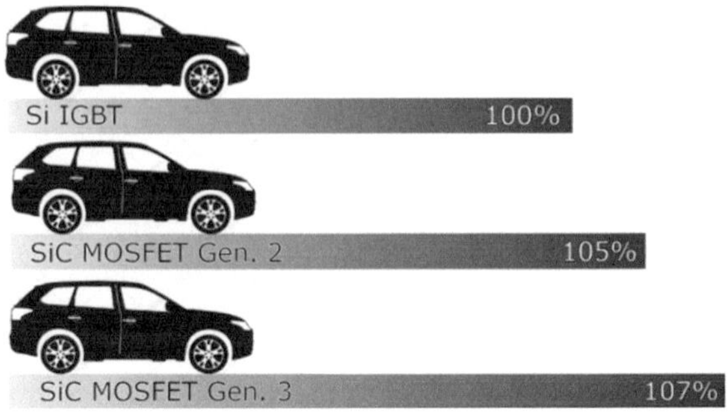

Fig. 12.2 SiC MOSFET improves drive range [1]

inverter without increasing the chip area. In order to fully utilize this advantage, innovative technologies and advanced packaging materials are needed to meet reliability requirements.

Finally, the body diode and synchronous rectification capabilities of SiC MOSFET are often overlooked advantages. When only one type of chip is being processed, especially in power modules, this facilitates the package design and assembly process, where multiple chips are bonded and packaged together. For example, when multiple SiC MOSFET chips are connected to a power module using pressure sintering technology, all chips have the same height, which reduces the trouble of comparing two chips with different heights (such as Si IGBT and Si FRD).

The hardware of the main drive inverter can generally be divided into power stage, control stage, sensing and feedback circuit, functional safety circuit, auxiliary circuit, and mechanical structure and components. Here, the focus is on the power stage. The power stage includes power semiconductors, drivers, DC and AC busbars, and DC voltage link capacitors. Replacing the traditional Si solution with SiC MOSFETs requires some major design changes, with the goal of designing a cost-effective and compact main drive inverter system to meet the application requirements.

(1) **Power Semiconductors**

Power semiconductors' conventional non-isolated discrete devices contain only one SiC MOSFET chip, such as TO-247 and D2PAK. Their performance defects and challenges in system-level design have led to their seldom use in main drive inverters. In the 2L-3Φ VSI topology, SiC MOSFET power modules have three common configurations: 1in1, 2in1, and 6in1, which indicates how many switches are integrated into one module.

1in1 modules offer the best scalability and flexibility in inverter design and are usually the lowest cost of the three solutions. In almost all cases, multiple 1in1 modules are connected in parallel to meet the power requirements of the main drive inverter application. Therefore, one of the design challenges is to balance the current of these parallel modules. The busbar and cooling structure design also require additional design. However, the successful case of using TPAK modules in Tesla Model 3 has made it a popular choice. Some power module suppliers have begun to provide 1in1 SiC MOSFET modules that are pin-compatible with TPAK.

2in1 modules are often called half-bridge topologies, which provide mechanical and electrical design flexibility and a certain degree of integration. There are various 2in1 modules available to meet various market needs, not only for different power level requirements, but also for various mounting and cooling requirements. It is feasible to connect two 2in1 modules in parallel for high currents, but this solution is less common due to the challenges of current sharing between the two modules.

The 6in1 module integrates three 2in1 modules to form a complete three-phase power module solution. Single-sided direct cooling is the most common configuration and usually requires the least effort in mechanical design.

As the core component of the power stage and cooling system in the main drive inverter, the power module converts DC power from the battery to AC power to drive the motor and dissipates its power consumption as heat to the cooling system. Therefore, it is crucial to choose the right power module. A rule of thumb is to choose the module that produces the least power loss and has the lowest thermal resistance to ensure that the chip junction temperature remains below its maximum value during operation.

In Si IGBT modules, the cost of the chip accounts for a small proportion, so increasing the total chip area is usually a good way to reduce power losses and improve module performance. In SiC MOSFET modules, chip costs dominate, and because the size of a single SiC MOSFET chip is much smaller, multiple chips must be connected in parallel to meet power requirements. Therefore, the way to improve the performance of SiC MOSFET modules is to increase the maximum allowable junction temperature and reduce thermal resistance, both of which rely heavily on advances in module packaging materials and technologies. This explains why the sintering process of SiC MOSFET modules has become a hot topic in recent years, even though the technology has been developed for decades. The use of new ceramic materials, such as Si_3N_4 and AlN, for SiC MOSFET module substrates follows the same concept.

For SiC MOSFET modules in the main drive inverter, the internal layout of the module is also key to achieve low switching losses through fast switching, maintain current balance and junction temperature uniformity between parallel chips, and avoid excessive EMI problems. In addition, due to the high cost of innovative ceramic substrates and the need for compact design, the ratio of total chip area to module area is now another indicator of module layout design.

(2) Drive Circuit

The drive circuit in the high-voltage main drive inverter has the functions of amplifying the control signal to drive the power device, providing isolation between high and low voltage for function and safety, and providing protection functions for fault detection and emergency shutdown.

Driving SiC MOSFET is to charge and discharge its junction capacitance, which can be achieved by voltage-type drive circuit or current-type drive circuit, among which voltage-type drive circuit dominates the market. At the same time, some driver IC suppliers provide current-type drive circuit solutions specifically for SiC MOSFET in the main drive inverter. This is mainly because the total charge required by SiC

MOSFET is less than that of IGBT, and the current-type drive circuit can more accurately control the SiC MOSFET switching transient for optimal performance while managing EMI and other problems caused by fast switching. With some advanced features provided by today's advanced gate driver ICs, such as gate strength control and overvoltage sensing, the switching performance of SiC MOSFET can be further improved.

It is common to integrate isolation functions into driver ICs. To ensure fast speed and long life, capacitance-based or magnetic-based isolation technology can be used. In addition to isolation voltage, another key parameter for isolated driver ICs is common-mode transient immunity (CMTI). In SiC MOSFET-based main drive inverters, the CMTI of the isolated driver IC should be at least 100 V/ns. The CMTI capability is improved to 200 V/ns and higher to accommodate not only the high dv/dt during switching transients, but also the high-frequency noise from excessive oscillations.

It is well known that SiC MOSFETs have a shorter short-circuit withstand time than Si IGBTs. However, short-circuit withstand capability is critical to safely shut down the power switch in such catastrophic situations. Although there are several alternative methods for short-circuit protection, such as chip-based current sensors and stray inductance-based overcurrent protection, the mainstream is still based on desaturation protection, usually provided by the driver IC. Typically, a lower desaturation trigger voltage can be set for faster response.

For SiC MOSFETs in most main drive inverter applications today, the manufacturer recommends a turn-on drive voltage between 15 and 20 V. As for the turn-off drive voltage, a negative voltage is recommended to avoid arm shorting due to crosstalk that causes SiC MOSFET to be falsely turned on. One of the challenges in SiC MOSFET drive circuit design is the contradiction between the required large fluctuations in drive voltage and the narrow tolerance of their absolute maximum values compared to Si IGBTs. Therefore, leaving a safety margin for the gate voltage is crucial in SiC MOSFET drive circuit design. In addition, since the on-resistance of SiC MOSFET is more sensitive to gate voltage than in the case of IGBTs, the value of undervoltage lockout (UVLO) should be set higher.

(3) **Bus Capacitors and DC/AC Busbars**

The design or selection of bus capacitors requires that the capacitance is large enough to meet the maximum voltage and current ripple requirements, and the temperature of the capacitor core is below its maximum allowable value. In addition, the low ESL of the capacitor can reduce the voltage overshoot and signal ringing caused by the fast di/dt of SiC MOSFET.

The busbar also needs to achieve a low stray inductance design by creating coupling between the positive and negative busbars and shortening the distance from the DC link capacitor to the DC power terminals of the power module. Laminated DC busbar structures can achieve both good mechanical strength and low stray inductance. The AC busbar should be designed to carry the required amount of AC current to drive the motor. In addition, the design geometry of the three-phase AC busbar should take into account the current sensor.

Currently, in designs using SiC MOSFETs, laser welding technology is widely used for terminal connections between busbar capacitors, DC/AC busbars, and power modules, replacing screw connections. Laser welding offers many key advantages such as lower resistance, enhanced electrical reliability, high mechanical strength and durability, excellent performance under vibration, better corrosion resistance, and improved heat dissipation. In addition, laser welding can achieve a more compact and simple design because it eliminates screws, washers, and additional space required for tool access. It is also a cost-effective solution because it eliminates material costs and significantly reduces labor costs on automated assembly lines.

In summary, the main drive inverter design using SiC MOSFET as the power switch has a lot in common with the design using Si IGBT anti-parallel Si FRD. The main technical challenges arise from the need for faster switching speeds, higher switching frequencies and harsher operating temperatures, as these are key features for achieving efficient, compact, cost-effective, and reliable inverter designs.

In traction inverters, the outlook for SiC MOSFETs is very promising, driven by their superior performance characteristics and increasing adoption in the electric vehicle market. SiC MOSFETs offer significant advantages over traditional silicon-based power devices, including higher efficiency, greater power density, and better thermal management. These advantages can translate into extended range, shorter charging times, and improved overall performance for electric vehicles. SiC technology is becoming a key enabler as manufacturers strive to meet increasing demands for efficiency and cost reduction.

12.2 Onboard Charger

For high-voltage traction batteries in vehicles, it is necessary to rely on power devices to convert alternating current from the grid into direct current (i.e., AC-DC conversion) for charging purposes. Specifically, DC charging stations are located outside the vehicle and use commonly known "fast charging connectors" to deliver direct current to the vehicle's battery. On the other hand, the onboard charger (OBC) inside the vehicle is another device situated within the vehicle. It acquires alternating current from the grid through what is commonly referred to as a "slow charging connector", completes the conversion to direct current, and outputs it to charge the batteries inside the vehicle. The on-board charger (OBC) is shown in Fig. 12.3.

In terms of the direction of energy transmission, onboard chargers (OBCs) can be divided into two types: unidirectional and bidirectional. Unidirectional OBCs are primarily responsible for converting alternating current from home or other AC charging stations into direct current to charge the high-voltage traction battery inside the vehicle. Bidirectional OBCs refer to OBCs equipped with vehicle-to-everything (V2X) functionality, which means that certain vehicle models have OBCs capable of inverting power through the high-voltage traction battery. This allows the vehicle to feed energy back to external power devices (off-grid) or to the grid (on-grid).

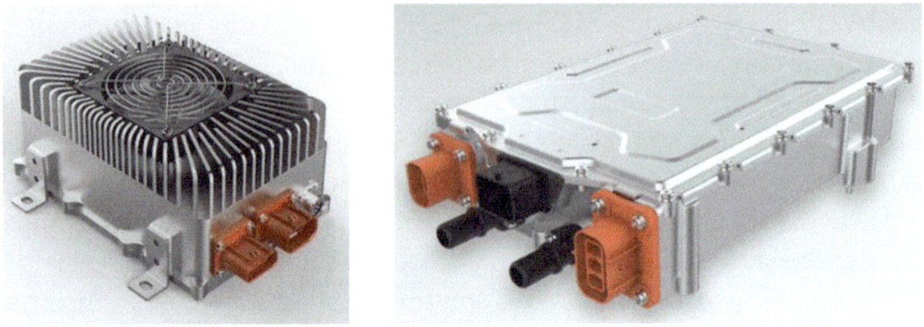

Fig. 12.3 Onboard charger

The output voltage range of the onboard charger (OBC) is influenced by the paired battery platform, which can be either 400 V or 800 V. The actual battery voltage exhibits significant variations relative to the 400 V or 800 V nominal levels. Generally, there is a significant difference in voltage between these two battery platforms, leading to the adoption of different design schemes and resulting in different product layouts for different vehicle models. In the passenger vehicle sector, the 400 V platform has been predominantly used in the past. However, in recent years, in order to further increase charging power, the 800 V platform has gradually become popular in some new vehicle models.

The power ratings of onboard chargers (OBCs) are mainly concentrated in 3.3, 6.6, and 11 kW, while some vehicle models and markets in Europe and America may also have demands for 3.6 kW, 7.2 kW, 19.2 kW, or even 22 kW. Taking China as an example, the mainstream specification currently is single-phase 6.6 kW. The reason why larger power specifications are not popular is that different countries and regions have restrictions on the maximum current and maximum power of single-phase AC input. Higher-power demands require three-phase AC input, but building dedicated three-phase AC access devices may overlap with existing DC charging stations in terms of usage scenarios, limiting the popularity of higher-power OBCs.

The technical challenges of OBC are mainly as follows:

(1) Wide voltage gain range: The input AC voltage and output DC voltage range varies widely, requiring the converter to have strong gain adjustment capability.
(2) Often, compatibility with both single-phase and three-phase AC inputs is required, along with the goal of achieving bidirectional power transmission, in conventional two-stage architectures, posing higher design difficulties for front-end AC-DC stage.
(3) In conventional two-stage architectures, the downstream DC/DC stage must handle the primary task of wide voltage range regulation while operating under high-voltage, high-power, and even bidirectional power flow conditions. Poses significant challenges for the topology and design optimization of the rear-end DC/DC stage.
(4) For electrical safety and compliance with safety regulations, OBC requires electrical isolation.

(5) Compliance with Automotive EMC Standards for AC Input: OBC must meet the mandatory requirements of automotive EMC standards under AC input conditions.

OBCs can be distinguished by single-phase and three-phase inputs, as shown in Fig. 12.4. In some cases, OBCs with three-phase inputs also need to be compatible with single-phase inputs, which can be achieved through phase loss detection and mode switching circuits (usually using relays or thyristors as switching devices).

According to the topology, OBCs have two architectures: single-stage and two-stage cascade. Currently, mature OBC products mainly adopt the two-stage cascade scheme, namely the front-end non-isolated AC-DC (PFC) stage and combinations of downstream isolated DC-DC stages.

Taking single-phase input as an example, according to the bus voltage, the traditional approach typically fixes the output voltage of the single-phase PFC (i.e., the DC bus voltage) stage at around 380–400 V. Then, the voltage is converted through the rear-end DC-DC stage to charge the 400 V battery platform. Considering the wide voltage range of the 400 V battery platform, in order to improve the efficiency of the rear-end resonant DC-DC stage (such as LLC resonant converter and CLLC resonant converter), and to achieve a nearly 1:1 regulation relationship between voltage gains, the range of bus voltage variation needs to be increased. The maximum range may

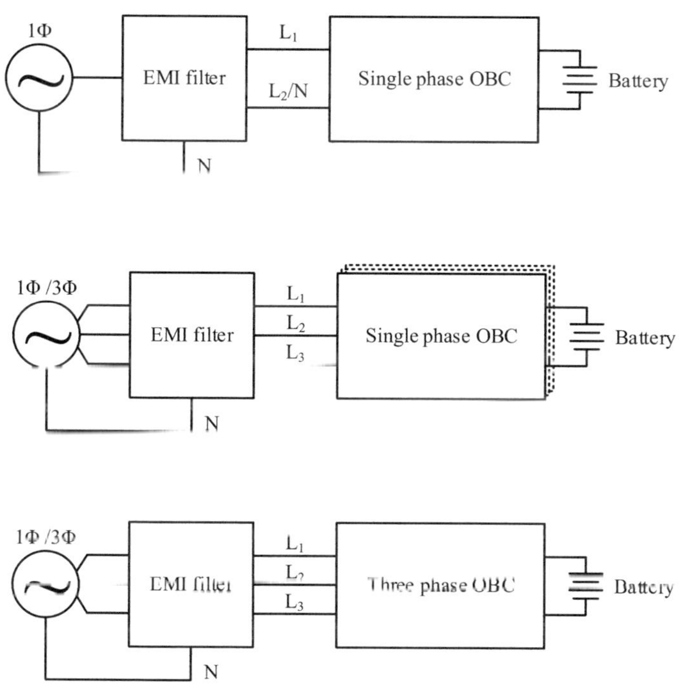

Fig. 12.4 OBC solutions for single-phase and three-phase input

approach levels close to 800 V. In the 800 V battery platform, for the same considerations of efficiency in the rear-end DC-DC conversion, the bus voltage may also be designed to approaching or exceeding 800 V, or even higher levels.

When comparing at the same voltage level, SiC MOSFETs have higher current ratings compared to Si SJ-MOSFETs, superior diode reverse recovery characteristics, and better temperature variation characteristics. Compared to Si IGBTs, SiC MOSFETs have lower switching losses.

At around 400 V bus voltage, it's quite common to replace Si diodes with 650 V SiC diodes of the same voltage rating. However, the advantage in switching losses of 650 V SiC MOSFETs over 650 V Si SJ-MOSFETs is not significant. Considering factors such as cost and maturity, SiC MOSFETs have not been widely used. In the case of around 800 V bus voltage, to avoid the complexity of using three-level topologies and maintain higher conversion efficiency, 1200 V SiC MOSFETs become the preferred choice.

For the front-end AC-DC stage, single-phase input is illustrated in Fig. 12.5a, while three-phase input is depicted in Figs. 12.6 and 12.7. In the case of single-phase input, for unidirectional OBCs employing traditional boost PFC and dual-boost PFC circuits and interleaved parallel operation, SiC diodes can be chosen for the diodes. Meanwhile, bidirectional OBCs can utilize totem-pole PFC circuits and interleaved parallel operation, with SiC MOSFETs selected for the high-frequency switching transistors. Leveraging their advantages of low switching losses and excellent reverse

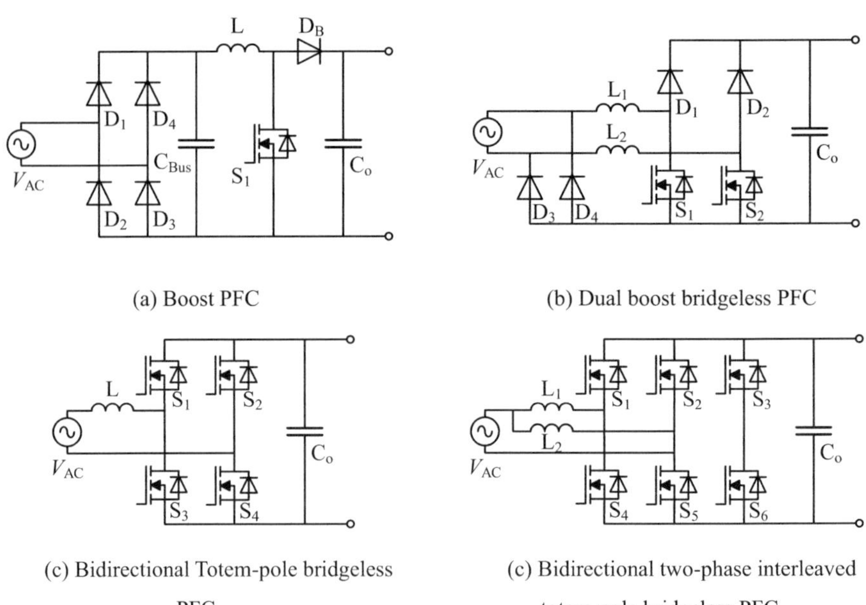

Fig. 12.5 OBC front-end AC-DC topology

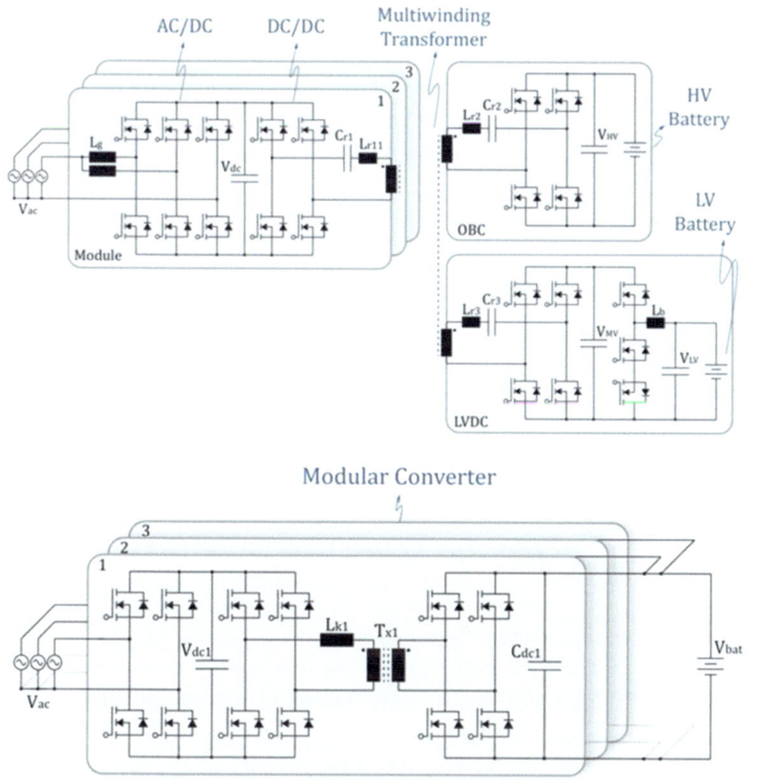

Fig. 12.6 AC-DC front-end topology for OBC (three-phase input, single-phase rectifier in parallel)

recovery characteristics of the body diode, SiC MOSFETs have become the "benchmark" topology for achieving the highest efficiency PFC in recent years. In the scenario of three-phase input, there are solutions with three single-phase modules connected in parallel, with selection considerations aligning completely with those for single-phase input. Alternatively, there are dedicated solutions for three-phase rectification, such as the three-phase six-switch topology, which can also fulfill the requirements of bidirectional OBCs.

For the rear-end DC-DC stage, which needs to handle high power, pursue high efficiency, and possess the capability to adjust to the varying voltage range of the battery, almost all designs opt for resonant topologies such as full-bridge LLC (unidirectional) and CLLC (bidirectional), and interleaved parallel operation as well as their variants. In recent years, there have also been solutions employing the dual active bridge (DAB) topology. Generally, for these topologies capable of achieving zero-voltage switching (ZVS), SiC MOSFETs exhibit lower equivalent junction capacitance, allowing for shorter ZVS times, simplified external inductor design, and minimization of dead times, thereby further enhancing the efficiency of soft-switching

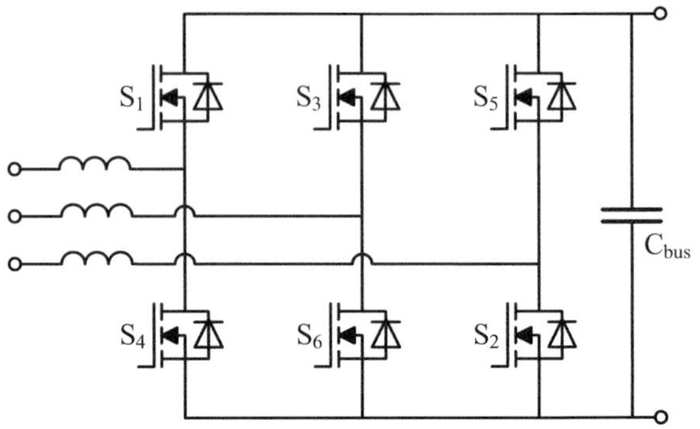

Fig. 12.7 OBC front-end AC-DC topology (three phase input, three phase rectifier)

converters (Fig. 12.8). The typical topologies for downstream DC-DC stages are shown in Fig. 12.8.

12.3 Onboard DC-DC

The onboard DC-DC converter is the power device inside electric vehicles that converts the energy from the high-voltage traction battery to supply power to the low-voltage battery pack and low-voltage electrical components inside the vehicle, such as lights, in-car entertainment, and power steering, in DC form. Similar to the development of OBCs, in recent years, bidirectional energy transfer capabilities have also emerged for onboard DC-DC converters, allowing for reverse charging of the high-voltage traction battery, offering more flexibility and intelligence for the vehicle's energy management.

It's worth noting that due to the highly integrated and modular trend in power control units within vehicles, OBCs and onboard DC-DC converters are increasingly being designed together. For example, deep integration of OBCs and onboard DC-DC converters, along with the addition of Power Distribution Unit (PDU) modules, results in "two-in-one" or "three-in-one" products. Some even propose further integration of electronic control units for main drive, motor, and gearbox to create a "multi-in-one" product. Overall, onboard power systems will continue to evolve towards modularity, integration, low cost, high efficiency, intelligence, and fully digital control.

The input of the onboard DC-DC converter comes from the high-voltage traction battery, while the output voltage range needs to adapt to the varying range of the low-voltage battery, typically within the range of 9–16 V for passenger vehicles. As intelligent driver assistance systems and more complex in-car functionalities emerge, the output power of onboard DC-DC converters has gradually increased from the earlier range of 1–2 kW to larger ranges of 2–4 kW.

Which only have the battery connected on the output side, the input and output sides of the onboard DC-DC converter are both connected to batteries, the voltage gain range also exhibits significant breadth. Common resonant converters such as

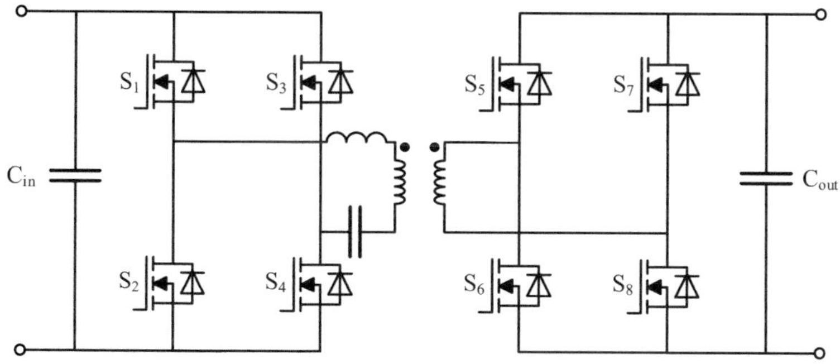

(a) Full- Bridge LLC Resonant Circuit

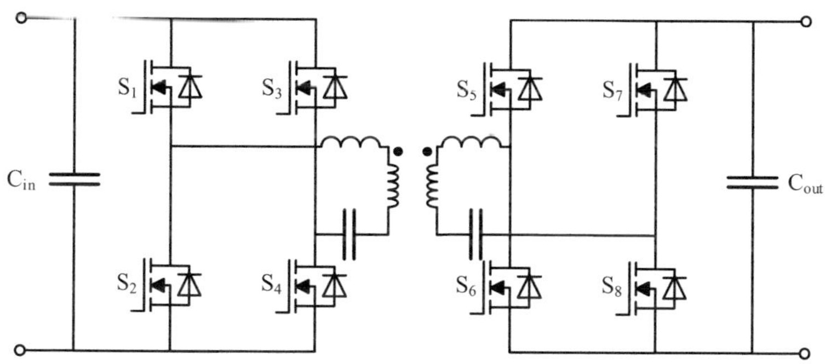

(b) CLLC Resonant Converter

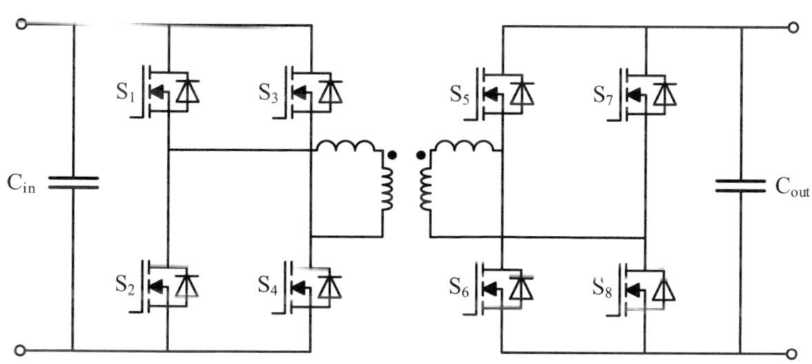

(c) DAB Converter

Fig. 12.8 Post-stage DC-DC topology of OBC (three-phase input)

LLC cannot maintain high overall efficiency within such a wide voltage range. To address this issue, other types of single-stage isolated PWM topologies and two-stage cascaded topologies are more commonly used in onboard DC-DC converters.

The technical challenges of onboard DC-DC converters mainly include:

(1) Wide gain range: This refers to a significant variation in the ratio between output voltage and input voltage, manifested by a wide input voltage range and a wide output voltage range. This requires the converter to have strong gain adjustment capabilities.
(2) At low-voltage and high-current output, to achieve higher efficiency and power density, the switching frequency needs to be further increased, posing significant challenges in terms of efficiency and thermal management. However, if soft-switching techniques are used, it is difficult for related topologies to achieve the highest efficiency across the entire operating range (full voltage and full load range).
(3) For the sake of electrical safety and compliance with safety regulations, LV DC requires electrical isolation.
(4) Compliance with automotive-related EMC standards mandated for DC inputs.

In isolated DC-DC converters for onboard applications, a half-bridge structure is commonly used on the primary side when the output power level is between 1 and 2 kW, and a half-bridge configuration is typically adopted on the primary side when the output power level is between 2 and 4 kW. To handle low-voltage high-current output, the secondary side of isolated DC-DC converters often employs full-wave rectification combined with synchronous rectification technology. Additionally, to balance the pressure on heat dissipation, the structure of multiple transformers with series and parallel windings is also common.

Single-stage solutions must have a wide gain adjustment capability while striving for maximum conversion efficiency possible. Options include: hard-switching half-bridge/full-bridge, active-clamp forward flyback, phase-shift full-bridge (PSFB, also referred to as ZVS PWM Full Bridge) designs, among others as illustrated in Fig. 12.9.

In a two-stage cascaded architecture, non-isolated topologies are typically used for gain adjustment, often employing buck or boost circuits, while efficient isolated topologies are cascaded for electrical isolation, commonly utilizing LLC circuits as illustrated in Fig. 12.10.

In vehicle-mounted DC-DC converters, SiC MOSFETs can be used on the primary side of high-voltage isolated DC-DC circuits. In hard-switching applications, they can significantly reduce switching losses, while in ZVS soft-switching applications, they can reduce dead time, thus improving efficiency. When the input voltage is from an 800 V battery platform, similar to the considerations for the 800 V bus voltage in OBCs, the design consideration for the secondary stage of the DC-DC converter would favor the use of 1200 V SiC MOSFETs.

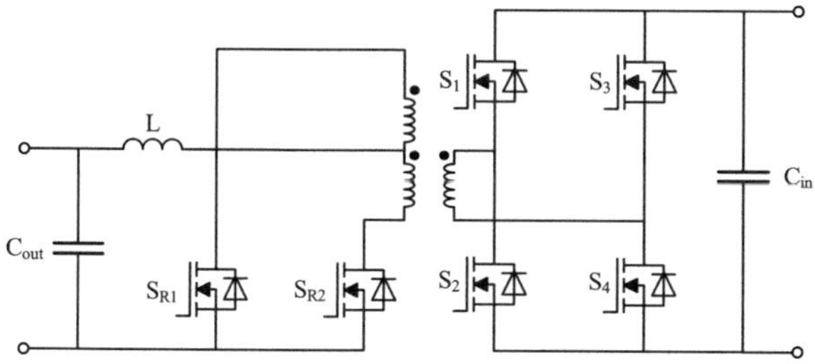

(a) Hard-Switched Full-Bridge Circuit with Full-Wave Rectification

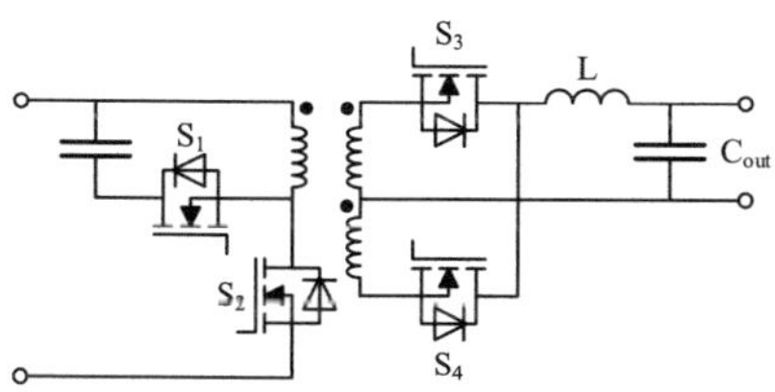

(b) Active-Clamp Forward-Flyback Converter with Full-Wave Rectification

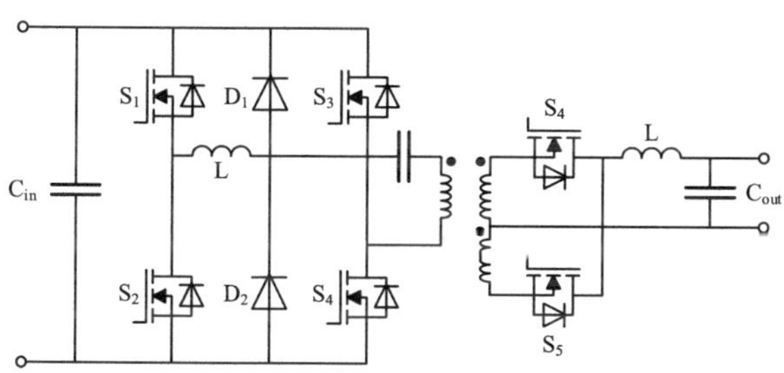

(c) Phase-Shifted Full-Bridge Circuit with Transformer-Secondary Clamp (T-C Type) and Full-Wave Rectification

Fig. 12.9 Single-stage topology solutions for automotive DC-DC converters

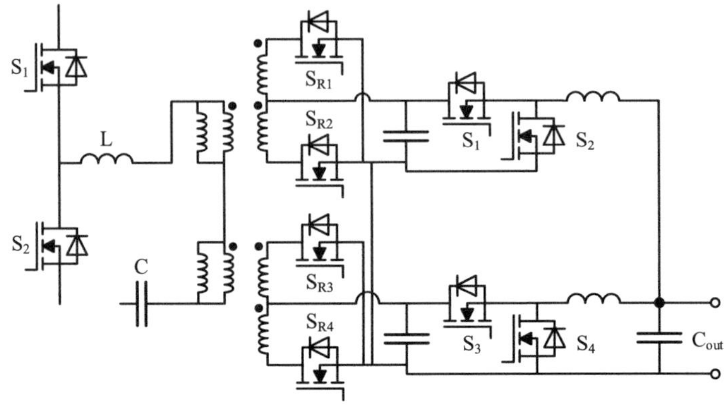

(a) Transformer series-parallel LLC + interleaved parallel buck

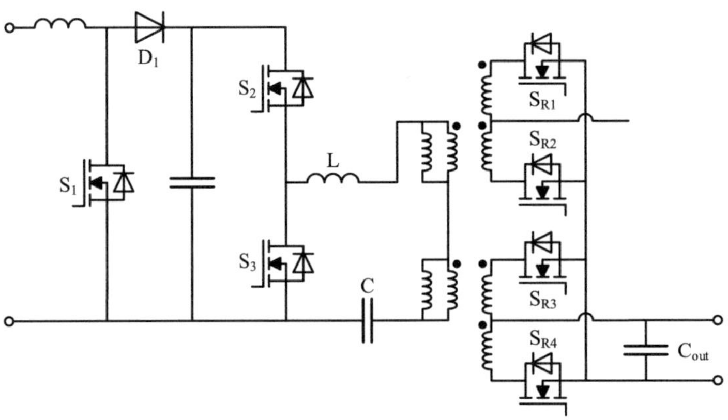

(b) Boost+ transformer series-parallel LLC

Fig. 12.10 Dual-stage topology solution used in automotive DC-DC

12.4 Charging Stations

With the advancement of the electrical energy revolution in the transportation sector, vehicles, ships, aircraft, and other means of transportation have embarked on the path of electrification. Particularly in recent years, new energy vehicles, mainly electric cars, have seen a significant increase in global ownership due to their higher energy transmission efficiency and more environmentally friendly characteristics.

Compared to traditional fuel vehicles, electric cars primarily rely on charging stations for energy supplementation. Charging stations are divided into AC charging stations and DC charging stations based on their charging methods and application scenarios. AC charging stations provide AC input to electric vehicles, and the alternating current is converted to direct current through the onboard charger (OBC) in the vehicle, thereby providing charging energy to the electric vehicle battery. Figure 12.11 shows an AC charging station located in a parking garage in Hong Kong, China, by Seng Hung Electrical. Due to power constraints imposed by the OBC in the vehicle, these stations typically have relatively low power ratings, mostly ranging from 6.6 to 11 kW, and are primarily used in residential settings or in underground parking garages of shopping malls.

DC charging stations typically range in power from 30 to 1000 kW and are mainly used in public charging stations, similar to traditional fueling stations for gasoline vehicles. These stations play a crucial role in rapidly replenishing energy for electric vehicles. Figure 12.11 shows a DC charging station located in Sweden by Shenghong Electrical (Fig. 12.12).

DC charging stations are typically classified into integrated charging stations and separated charging stations based on whether the power conversion part is integrated with the charging terminal or not, as shown in Fig. 12.13. The power of most integrated charging stations ranges from 30 to 400 kW, making them suitable for small-scale charging operations.

The power conversion unit of separated charging stations is centrally located in the main unit and is separated from the charging terminal, as shown in Fig. 12.14. When

Fig. 12.11 Shenghong electric's AC charging pile in an underground garage in Hong Kong, China

Fig. 12.12 Shenghong electric's DC charging pile in Stockholm, Sweden

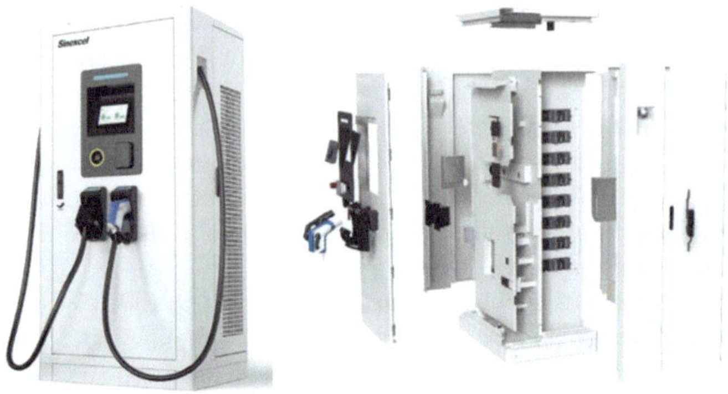

Fig. 12.13 Shenghong electric integrated charging pile

charging, the vehicle is connected to the charging terminal for charging. The power of most main units of separated charging stations ranges from 400 to 1000 kW. This configuration increases the sharing of charging modules and is suitable for medium to large-scale charging stations (Fig. 12.14).

Whether it's a DC integrated charging station or a DC separated charging station, the charging module is the core component responsible for energy conversion in DC charging stations. Due to the requirements of input grid and battery pack isolation, the typical approach is to use an AC-DC rectifier circuit with PFC in the front stage and an isolated DC-DC circuit in the back stage.

Since DC charging stations have high-power ratings, the majority of charging modules all have three-phase inputs, with a DC bus voltage ranging from 650 to 850 V. This necessitates the use of either a three-level circuit or a topology with positive and negative bus interleave, when employing traditional 650 V-rated Si MOSFETs, to lower the voltage requirements per device through device series connection. Currently, most schemes opt for a Vienna PFC circuit in the front stage

Fig. 12.14 Shenghong electric split charging pile

and a three-level or positive and negative bus interleaves LLC or phase-shift full-bridge isolated topology in the back stage. This results in more devices being used in the charging module and more complex circuit control. In this scenario, adopting higher voltage-rated SiC devices would effectively simplify the topology and enhance system reliability.

As shown in Fig. 12.15, using SiC MOSFETs in the front-end PFC allows for a reduction in the number of traditional Vienna rectifier devices, shifting from the relatively complex three-level control to a simpler two-level control. However, some power equipment also utilizes six-switch circuits based on high-voltage IGBTs as the PFC input stage. Compared to such circuits, the smaller conduction resistance parameters and excellent reverse recovery characteristics of SiC MOSFETs can reduce power device losses, thereby improving circuit efficiency.

In the rear-stage DC-DC section of the charging module circuit, traditional three-level circuits or positive-negative bus interleaved circuits often utilize multiple Si MOSFET devices in series. However, the use of high-voltage SiC MOSFETs enables the simplification of relatively complex circuits and control methods to a two-level isolation topology, as illustrated in Fig. 12.16. Moreover, since SiC MOSFETs are better suited for high-frequency applications, the circuit operating frequency

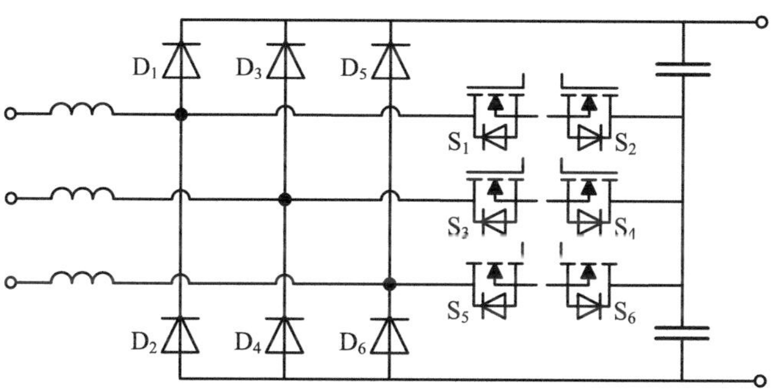

(a) Vienna Rectifier with Silicon (Si) Devices

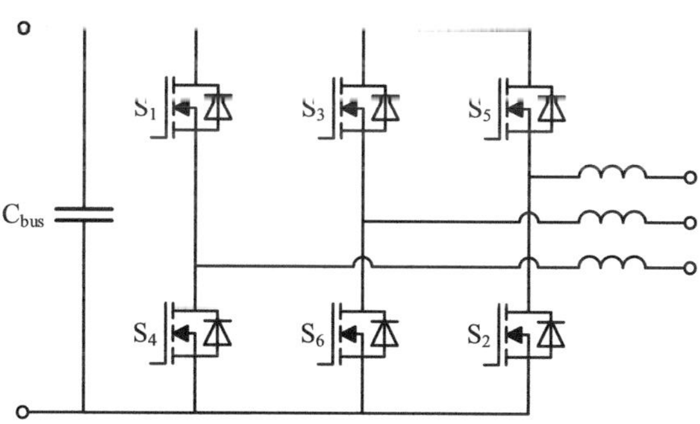

(b) SiC-Based Full-Bridge Rectifier

Fig. 12.15 Front-end AC-DC circuit of a charging module

can be further increased, significantly reducing the volume and cost of magnetic components.

Additionally, for some direct current charging stations used in vehicle-to-grid (V2G) or vehicle-to-home (V2H) applications, bidirectional circuits are employed in the charging module, as depicted in Fig. 12.17. Traditional setups utilize 1200 V IGBTs as the power semiconductor devices in the circuit. However, replacing these IGBTs with 1200 V SiC MOSFETs results in significant improvements in system

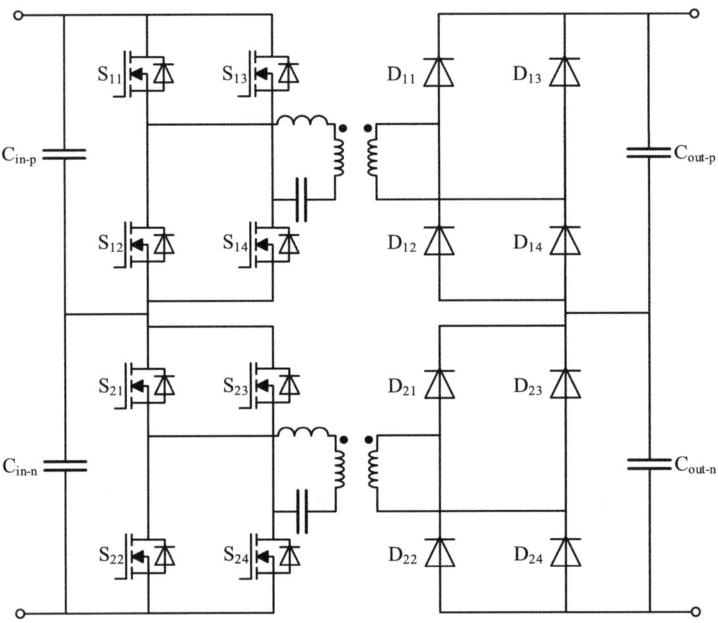

(a) Interleaved LLC Resonant Converter with Symmetrical Dual-Bus for Si-Based Devices

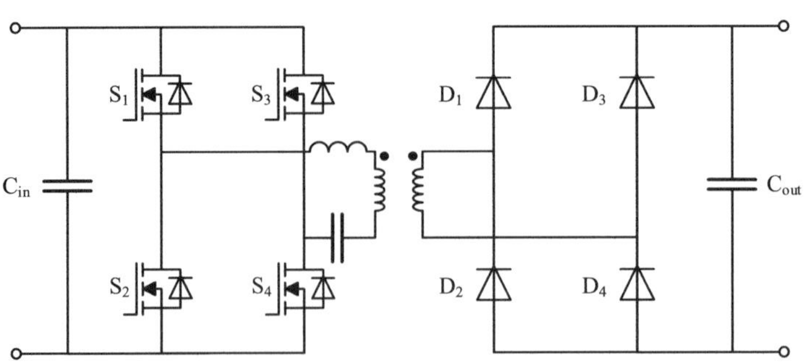

(b) SiC-Based LLC Resonant Converter

Fig. 12.16 Post-stage DC-DC circuit of charging module

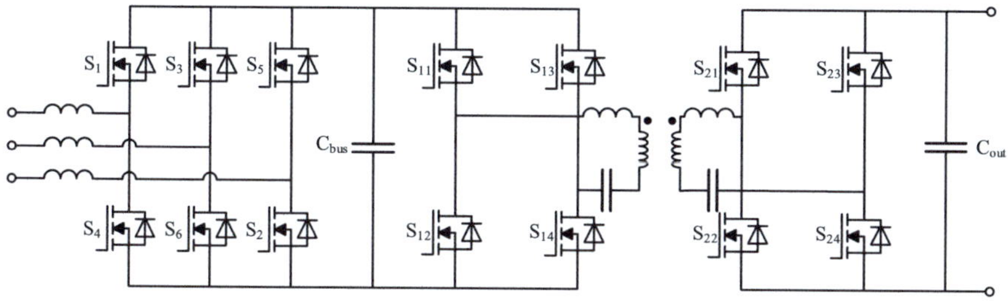

Fig. 12.17 Bidirectional DC charging module solution using SiC MOSFET

efficiency due to the excellent conduction and high-frequency characteristics of SiC. Although there are no major changes in the circuit structure, the use of SiC MOSFETs enhances system efficiency. Moreover, the high-frequency operation reduces the volume and cost of magnetic components, thereby improving the cost-effectiveness of bidirectional charging modules.

Indeed, electric vehicle charging stations represent a rapidly growing market in the global power electronics industry. With the integration of SiC MOSFET devices into direct current charging station equipment, certain parts of the charging module circuits can be made more streamlined and cost-effective. As a result, the industry has already begun widespread adoption of this technology.

12.5 Photovoltaics

1. **String Photovoltaic**

In a series-connected photovoltaic (PV) system, multiple PV panels (also known as PV modules) are connected together in a circuit according to a certain connection method, forming an electrical group. Then, multiple electrical groups are connected in parallel, as shown in Fig. 12.18. This series-parallel connection method can increase the system's voltage, thereby reducing line losses and improving system efficiency. Additionally, it can lower the risk of system failures and enhance system reliability.

The structure of the series-connected photovoltaic inverter is illustrated in Fig. 12.19, consisting of two parts: DC/DC and DC/AC. The DC/DC part is utilized

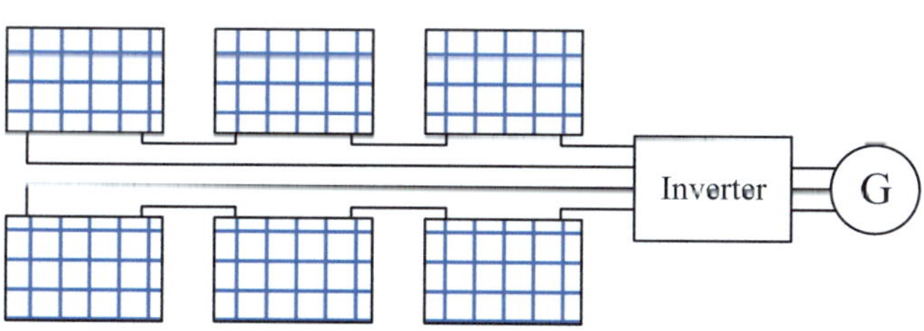

Fig. 12.18 System structure of string photovoltaic inverter

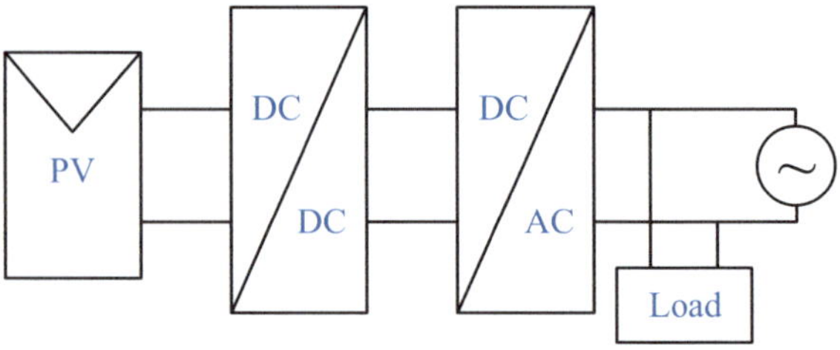

Fig. 12.19 Structure of string photovoltaic inverter

for achieving maximum power point tracking (MPPT) function, typically implemented with a boost circuit. The DC/AC part is employed for realizing inversion and grid connection functions. There are various circuit typologies for DC/AC, such as single-phase inverters including H4, H6, H6.5, Heric, and three-phase inverters including full-bridge circuits, T-type three-level, and I-type three-level circuits (Fig. 12.19).

In the DC/DC boost circuit, reverse recovery of the diode leads to reverse recovery losses, thereby resulting in increased losses in the switch as well. To effectively mitigate this issue, replacing traditional silicon freewheeling diodes (Si FRD) with silicon carbide (SiC) diodes can be employed. This substitution not only alleviates the problem without introducing any additional risk to the product, but also significantly improves product performance, resulting in considerable enhancements in volume, weight, and thermal management aspects. In single-phase systems, a common approach involves using silicon (Si) MOSFETs in conjunction with SiC diodes, which enables balancing the need for straightforward device replacement to enhance partial performance and increasing the switching frequency to improve overall performance. In three-phase systems, selecting Si IGBTs paired with SiC diodes may limit the increase in switching frequency. In such cases, further replacement of Si IGBTs with SiC MOSFETs can be considered.

In the DC/AC inverter circuit, the reverse parallel Silicon Freewheeling Diode (Si FRD) with Insulated Gate Bipolar Transistor (IGBT) can also lead to reverse recovery losses, resulting in increased losses in the IGBT. However, while silicon carbide (SiC) MOSFETs also have a reverse recovery process in their body diode, their reverse recovery characteristics are significantly superior to Si FRDs. Therefore, employing SiC MOSFETs can notably enhance efficiency. Furthermore, the switching losses of SiC MOSFETs are significantly lower than those of IGBTs. Consequently, using SiC MOSFETs allows for an increase in the switching frequency, thereby achieving comprehensive improvements in volume, weight, and thermal management aspects. Although the application of SiC MOSFETs in DC/AC inverter circuits is limited due to their insufficient pulse current performance, the outstanding characteristics of SiC MOSFETs continue to attract attention in the industry, prompting advancements in power electronics technology to compensate for this limitation. Currently, several companies are progressively integrating SiC MOSFETs into DC/AC inverter circuits. Among these, circuits such as the Heric circuit, 1000 V system three-phase full-bridge circuit, and 1500 V system I-type three-level circuit have utilized SiC MOSFETs, as depicted in Figs. 12.20, 12.21, and 12.22, respectively.

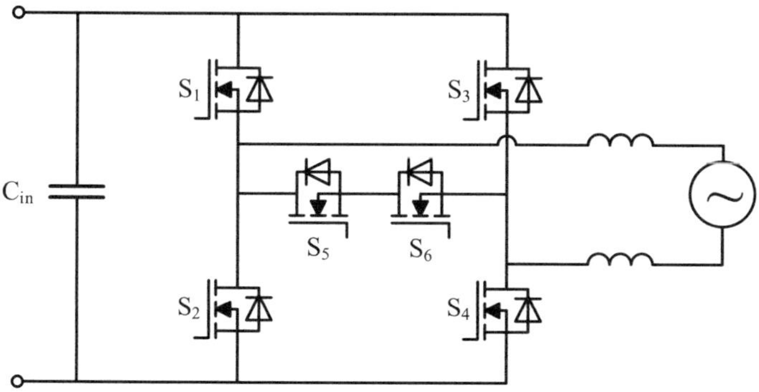

Fig. 12.20 Heric circuit

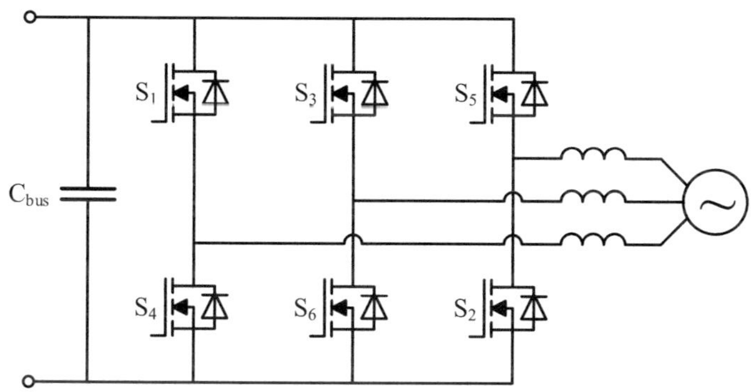

Fig. 12.21 1000 V system three-phase full-bridge circuit

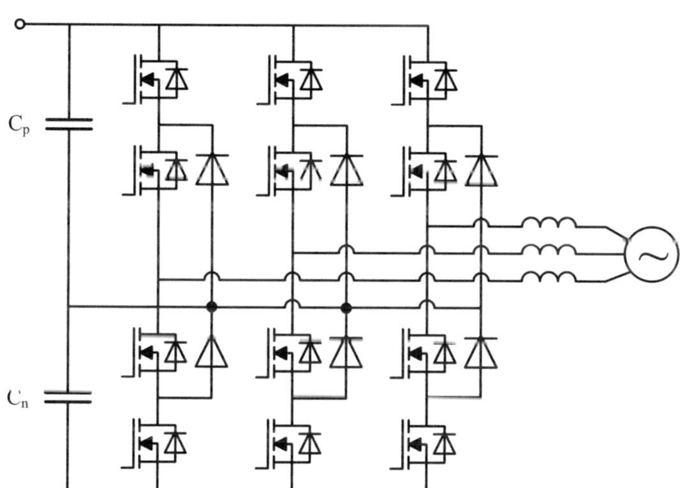

Fig. 12.22 1500 V system I-type three-level circuit

(a) MIN 2500-6000TL-XH (b) MOD 4000-10000TL3-XH (c) MAX 350KTL3-X HV

Fig. 12.23 Shenzhen Growatt NewEnergy Technology Co.,Ltd. string PV inverter

Taking the string inverter products from Shenzhen Growatt NewEnergy Technology Co.,Ltd. as an example, their MIN 2500-6000TL-XH, MOD 4000-10000TL3-XH, and MAX 350KTL3-X HV series products, through the application of the aforementioned new components and technologies, have achieved leading performance in comprehensive aspects, as illustrated in Fig. 12.23. Their peak efficiency, European efficiency, and MPPT efficiency reach 98.1%, 97.5%, and 99.5% respectively. They can operate stably in extreme temperatures ranging from − 30 to + 60 °C, at altitudes of up to 4000 m, and under humidity levels from 0 to 100%. With sleek design and user-friendly interfaces, they have garnered extremely high customer satisfaction.

2. **Microinverter**

Microinverters generally refer to photovoltaic inverters with a power rating less than 1000 W, featuring module-level maximum power point tracking (MPPT). The full name is micro photovoltaic grid-connected inverter, and "micro" is in relation to traditional centralized inverters. Unlike traditional centralized and string inverters, microinverters invert power on a per-module basis, as depicted in Fig. 12.24. Their advantage lies in their ability to independently control MPPT for each module, significantly enhancing overall efficiency. Additionally, they can mitigate issues such as high DC voltage, low-light performance, and the "Christmas light effect" commonly associated with centralized inverters.

Fig. 12.24 System architecture of microinverter

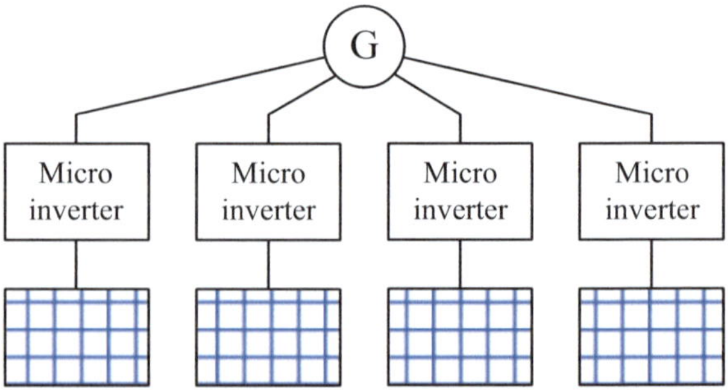

Microinverters take input from a single photovoltaic module, with their operating voltage typically ranging from 20 to 45 V. However, the maximum voltage on the inverter side is $\sqrt{2}$ times the grid voltage. Therefore, the scheme of string inverters is no longer applicable to microinverters. To address this, microinverters utilize high-frequency transformer isolation topologies to elevate the voltage of the photovoltaic module from 20 to 45 V to the corresponding output voltage after AC rectification. Simultaneously, this scheme achieves electrical isolation between the photovoltaic module and the grid.

The circuit of a microinverter mainly consists of the DC conversion section, isolation transformer, and inverter section, as depicted in Fig. 12.25. Commonly used configurations for the DC conversion section include flyback, phase-shift full bridge, and dual active bridge (DAB). For the inverter section, configurations such as full-bridge, half-bridge, three-level, novel multilevel circuits, and new circuits based on resonance technology are employed. Presently, commonly used schemes in microinverter products include flyback + full bridge, phase-shift full bridge + full bridge, and DAB + full bridge, as shown in Figs. 12.26, 12.27, and 12.28, respectively.

The circuit schemes mentioned above are relatively mature in both circuit design and control. However, to achieve further improvements in product performance, the adoption of newer and more efficient components can be pursued. For the DC conversion section, with an input range of 20 to 45 V, both flyback circuits, which subject the switching devices to higher stress, and bridge circuits, which subject the switching devices to lower stress, can utilize switching devices with a voltage rating

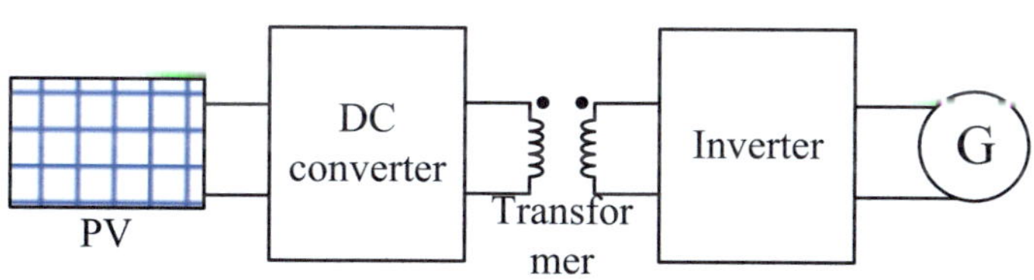

Fig. 12.25 Microinverter structure

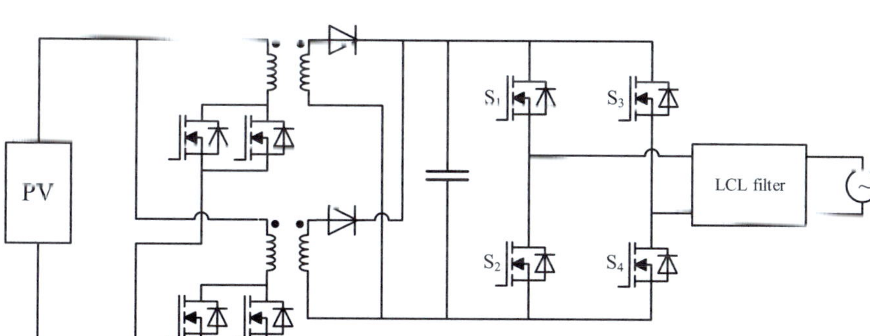

Fig. 12.26 Flyback + full-bridge solution for microinverter

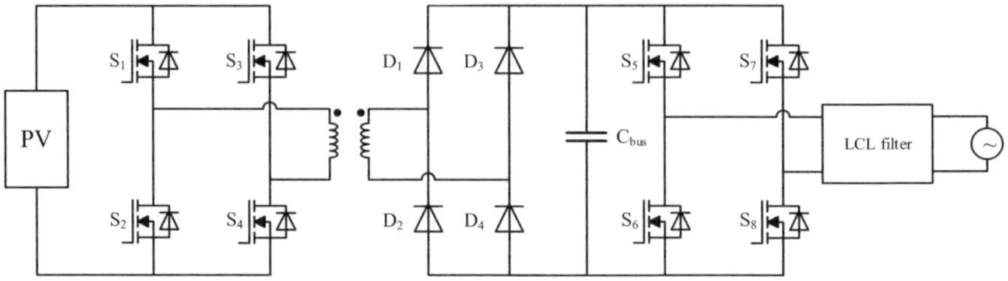

Fig. 12.27 Phase-shifted full-bridge + full-bridge solution for microinverters

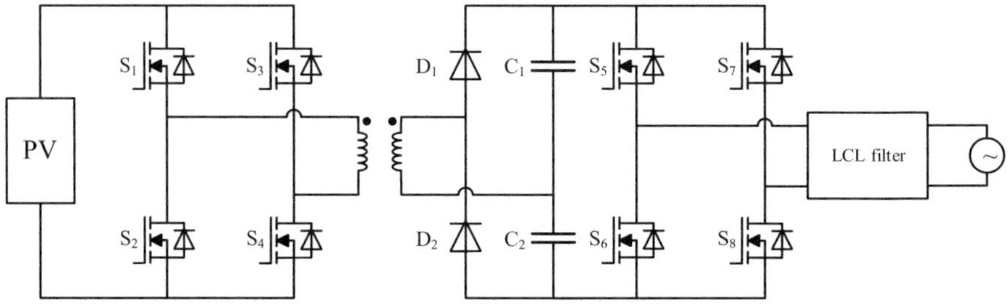

Fig. 12.28 Dab + full-bridge solution for microinverters

of at least 150 V. Currently, there is a wide selection of Si MOSFETs available with excellent performance within this voltage range, rendering SiC devices unnecessary. In the inverter section, where the maximum voltage on the inverter side is $\sqrt{2}$ times the grid voltage, the voltage rating of the switching devices should be greater than 650 V for bridge circuits and at least 800 V for flyback circuits. In this scenario, SiC devices become a viable option. In microinverter products, due to constraints on volume and product form, surface-mount packaged devices such as TO-263 and DFN5 $\times$ 6 are generally chosen. This selection meets the requirements for convenient production and high yield while also optimizing volume and thermal design.

Experimental data indicates that the efficiency of microinverters improves by 0.8% with the adoption of SiC devices. Consequently, the temperature rise decreases, leading to reduced heat dissipation requirements. Additionally, as the switching frequency increases, the volume of magnetic components decreases. Remarkably, despite these enhancements, there is no increase in the overall product cost.

Taking Shenzhen Growatt NewEnergy Technology Co.,Ltd. as an example, their microinverters NEO 2000M-X and NEO 600-1000M-X employ SiC devices, as depicted in Fig. 12.29. These products achieve leading performance in comprehensive aspects, with peak efficiency and MPPT efficiency reaching 97.3% and 99.5%, respectively. They can operate stably in extreme conditions ranging from − 40 to + 65 °C, at altitudes of up to 4000 m, and under humidity levels from 0 to 100%. With aesthetically pleasing design and user-friendly interfaces, they have garnered extremely high customer satisfaction (Fig. 12.29).

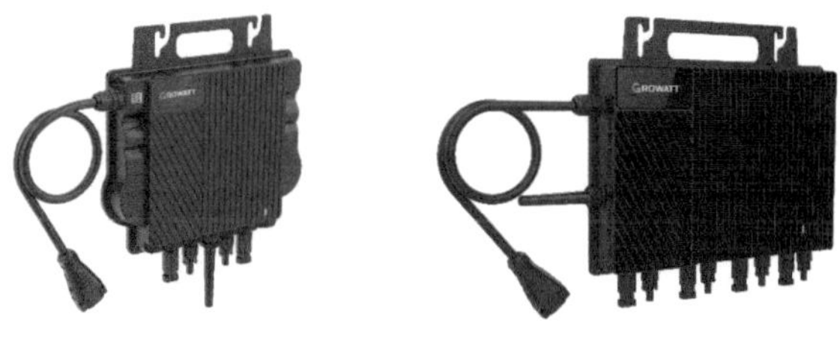

Fig. 12.29 Shenzhen Growatt NewEnergy Technology Co.,Ltd.'s microinverter

12.6 Energy Storage

The power system is increasingly incorporating renewable energy sources such as solar and wind power. However, their inherent variability and intermittency pose significant challenges to the scheduling and stability of the power system. The future energy system will be dominated by renewable energy, with various forms of energy collectively forming a diversified energy landscape. To fully utilize solar and wind energy and transform their unstable characteristics into the advantages of flexibility, energy storage technology has emerged.

Based on operational data, it is observed that the trends of grid load consumption and photovoltaic output are not consistent, leading to a significant mismatch between the two. Energy storage can effectively address this issue by storing excess solar output during periods of high photovoltaic output but low load consumption and supplying energy to the load during periods of no solar output or when solar output is lower than load consumption. Therefore, integrating storage elements, or energy storage systems, into photovoltaic systems or renewable energy systems in general can not only address the instability and intermittency of renewable energy but also maximize the utilization of renewable energy. Consequently, the widespread adoption of renewable energy through efficient energy storage technology is highly anticipated to drive the technological path toward low-carbon energy transformation.

After integrating an energy storage system into a photovoltaic system, it is referred to as a photovoltaic energy storage system. This system involves adding a battery system to the photovoltaic system, with the integration point being between the aforementioned DC/DC and DC/AC converters of the photovoltaic inverter. It forms a DC-coupled energy storage system, with its system structure and energy flow depicted in Fig. 12.30. It's important to note that the battery system mentioned above is a broad concept. It can consist of either pure chemical batteries or a combination of chemical batteries and power converters. The photovoltaic panels, batteries, loads, and grid collectively form a three-port network. The DC/DC converter on the photovoltaic side is a unidirectional boost circuit as described in Sect. 12.5, facilitating energy flow from the photovoltaic panels to the battery, load, and grid, represented by the blue and light blue energy flows. The DC/AC converter is a bidirectional circuit capable of facilitating energy flow from the photovoltaic panels and batteries to the load and grid, as depicted in Fig. 12.30 the blue and light blue energy flows,

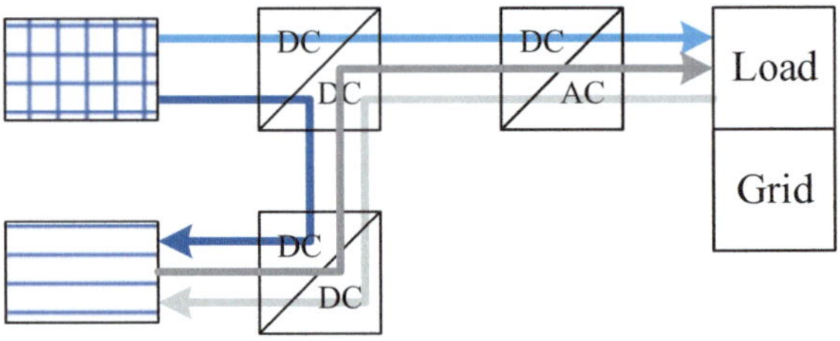

Fig. 12.30 Structure and energy flow of photovoltaic energy storage system

as well as from the grid to the batteries, as indicated by the light gray energy flow. The DC/DC converter on the battery side is also bidirectional, enabling energy flow from the photovoltaic panels, grid to the battery, as depicted by the light blue and gray energy flows, and from the battery to the load and grid, as indicated by the gray energy flow. Through power electronics control technology, it achieves flexible energy flow among the three ports, addressing the instability and intermittency issues of renewable energy and achieving the goal of flexibility in energy storage systems.

In Sect. 12.5, the DC/DC and DC/AC converters on the photovoltaic side were introduced. This section focuses on introducing the DC/DC converter on the battery side, which we will refer to as DC/DC2 for convenience. Unlike in conventional unidirectional applications, where energy is generated by the battery itself, in this scenario, the battery serves as a storage device, storing energy for later use. Therefore, from energy generation to application, the energy needs to be stored in the battery before being released, involving two energy conversions before utilization. In this process, maximizing energy utilization requires that the conversion efficiency of DC/DC2 be optimized. Its bidirectional efficiency, encompassing both charging and discharging, is referred to as the cycle efficiency and is a crucial consideration point. In this context, SiC devices can enhance the efficiency of the converter, contributing to improved overall system efficiency, which will not be elaborated further here.

The selection of the DC/DC2 topology must consider not only the requirements for battery energy conversion but also the adaptability of the battery within the system. As photovoltaic systems include both single-phase and three-phase systems, it is necessary for DC/DC2 to adapt to both. This means accommodating both the 550 V single-phase system and the 1100 V three-phase system. When using traditional Si devices, the 550 V system typically employs a relatively simple two-level circuit, both in terms of circuit design and control. In contrast, the 1100 V system requires a more complex three-level circuit. However, with SiC devices, the three-level circuit can be simplified into a two-level circuit. In non-isolated applications, DC/DC2 can utilize a bidirectional buck-boost topology, typically implemented with two or multiple interleaved parallel paths to reduce ripple current, as depicted in Fig. 12.31. Taking Shenzhen Growatt NewEnergy Technology Co.,Ltd.'s BMS controller BDC 95045-A1 as an example, it adopts the aforementioned solution. Compared to products using Si devices, it achieves a 1.9% increase in efficiency, a 0.1% decrease in cost, a reduction in weight by 1.8 kg, and a decrease in volume by 21%, as illustrated in Fig. 12.32.

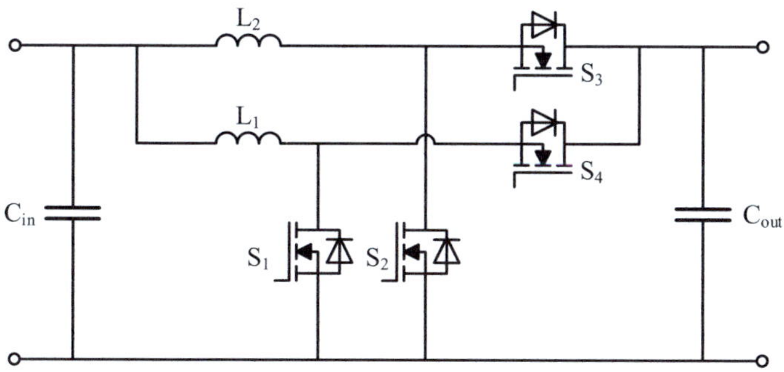

Fig. 12.31 Interleaved parallel buck-boost

Fig. 12.32 Shenzhen Growatt NewEnergy Technology Co.,Ltd. BMS controller BDC 95045-A1

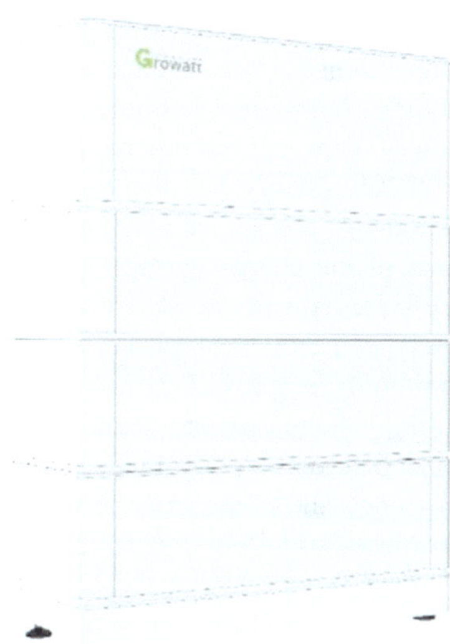

Utilizing the BDC 95045-A1 enables flexible combinations of batteries and photovoltaics, providing great freedom and adaptability in terms of time and capacity. Further integrating the above technology with photovoltaics results in integrated energy storage converters that can reduce space occupation and wiring complexity. An example of a product employing integrated technology is the residential energy storage inverter SPH 4000-10000TL3-BH from Sungrow Power Supply Co., Ltd. This product also offers high-efficiency, low-cost, compact size, and lightweight advantages. The cycle efficiency of the energy storage DC/DC2 reaches 98.2%, as shown in Fig. 12.33.

Fig. 12.33 Shenzhen Growatt NewEnergy Technology Co.,Ltd.'s residential energy storage inverter SPH 4000-10000TL3-BH

With the continuous development of energy storage converters, higher demands are placed on safety, depth of energy utilization, the use of multiple brands of energy storage cells, and the mixing of new and old batteries. Previous simple parallel connections for low-voltage batteries or simple series connections for high-voltage batteries have become bottlenecks for meeting these requirements. The concept of battery optimizer has emerged to address these needs. It not only maximizes the use of battery energy but also accommodates the use of multiple brands of cells and the on-site mixing of new and old batteries, thereby enhancing the fine management of the energy storage system. The optimizer is connected to the battery on one side, typically at 51.2 V, and outputs high voltage on the other side. To ensure safety and achieve high-voltage conversion ratios, an isolated DC/DC circuit is commonly employed. However, DC/DC circuits incur losses and thermal issues during voltage conversion and energy transfer. Therefore, while introducing an optimizer brings many advantages, it also comes with drawbacks such as reduced system efficiency and complexity in thermal design. To mitigate the disadvantages of the optimizer, soft-switching topologies like phase-shift full bridge, LLC, or DAB are often employed. Among these, the series-resonant LLC can achieve both zero-voltage switching (ZVS) and zero-current switching (ZCS) with simple control, making it widely used, as depicted in Fig. 12.34. For ease of design, the resonant tank is placed on the high-voltage side, with ZCS occurring on the low-voltage side and ZVS on the high-voltage side. However, hard switching still occurs on the high-voltage side, resulting in switching losses. SiC MOSFETs can be used to improve the efficiency of the entire optimizer and alleviate the disadvantages of reduced system efficiency and complexity in thermal design. Taking Shenzhen Growatt NewEnergy Technology Co.,Ltd.'s high-voltage battery system APX 5.0-30.0P-S as an example, it adopts the aforementioned battery optimizer technology, achieving a cycle efficiency of 92.3%, as shown in Fig. 12.35.

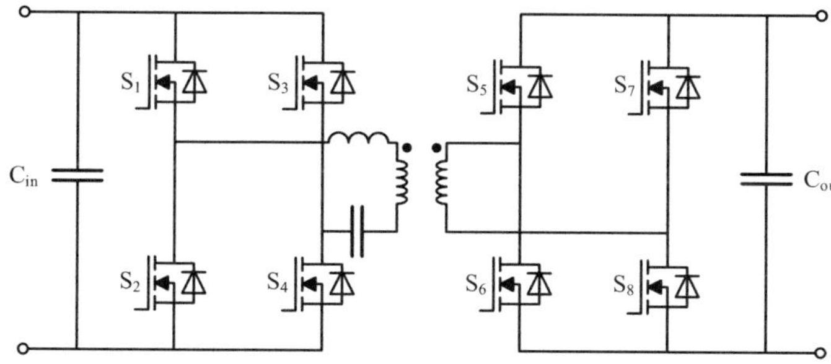

Fig. 12.34 LLC resonant circuit

Fig. 12.35 Shenzhen Growatt NewEnergy Technology Co.,Ltd. high-voltage battery system APX 5.0-30.0P-S

12.7 Uninterruptible Power Supply

Uninterruptible power supply (UPS) is a device designed to provide stable output energy during grid failures or instabilities to ensure uninterrupted power supply to equipment. Power interruptions not only result in losses due to business disruption but also lead to decreased productivity caused by power outages. With the increasing prevalence of various electrical devices in recent years, UPS systems have become increasingly important.

UPS systems feature off-grid power supply backup capabilities. With the continuous increase in equipment loads and changes in power supply architectures, there has been a rise in large-scale UPS power supply systems, becoming an essential branch of grid-type power supply systems. During the power supply and protection process, UPS systems need to calibrate and improve the quality of electrical energy through active power factor correction to the grid, isolating load-side harmonic loads to mitigate their impact on the grid and meet the IEEE 519 harmonic standards.

Currently, UPS capacities range from as low as 300VA (sufficient to support a single computer and monitoring devices) to over 2MVA (capable of supplying power to the equivalent of 175 households). The power supply capacities of cloud computing and intelligent data center projects have reached 60–80 MW. With the deepening development of AI technology and the deployment of GPU supercomputing power, project power supply capacities have exceeded 120–150 MW. For large-scale data centers, energy efficiency is becoming increasingly important in reducing operational costs. There is a growing demand for power equipment efficiency and overall data center energy consumption (PUE). Currently, large-scale UPS systems (ranging from 500 to 1.2 MW) are undergoing rapid development and technological iteration. Products like the Eaton 93PR UPS represent this trend, boasting efficiency levels exceeding 97%. The integration of SiC devices can further drive the technological development of the UPS industry.

The system structure of the currently popular online UPS is shown in Fig. 12.36.

The UPS contains multiple functional units internally, including rectifiers, inverters, DC converters, bypass modules, and maintenance bypass modules, each providing different channels for energy conversion and transmission. Among them, the rectifiers, DC converters, and inverters are power converters based on digital control. The rectifiers and inverters constitute the power supply circuit in the online AC mode, isolating the impedance and harmonic effects between the input and output

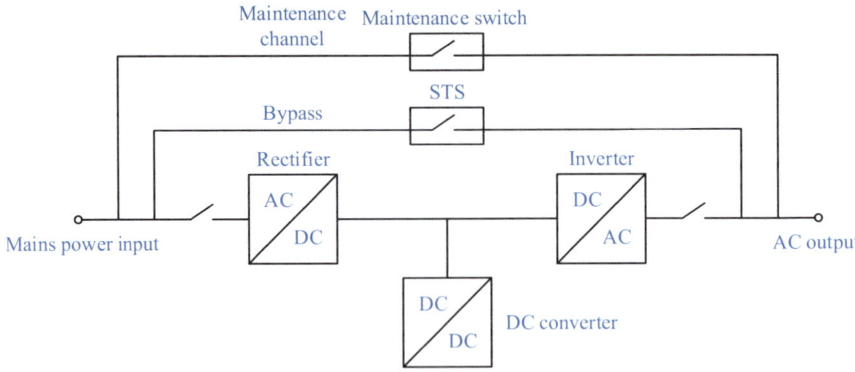

Fig. 12.36 Online double-conversion UPS system structure

terminals through AC-DC and DC-AC two-stage converters. Currently, in energy storage UPS systems, the participation in grid-side energy storage business through the grid-connected operation of rectifiers is further achieved through the design technology of bidirectional rectifiers. Representative products in this category include Eaton's 9395 and 93PR series. The DC converter is the core module inside the UPS, responsible for converting battery voltage to system bus voltage, accommodating a wider range of battery voltage. The DC converter and inverter together form the power supply circuit for battery backup mode in case of AC power abnormality. The UPS achieves uninterrupted switching between AC and battery modes, ensuring uninterrupted UPS output voltage within 0 ms of interruption.

Both double-conversion UPS systems and multimode UPS systems are equipped with rectifiers. The rectifier is a module in the UPS system that directly connects to the input grid, converting the AC voltage into the reference voltage of the internal DC bus. Currently, the majority of UPS systems use IGBTs as rectifier components, enabling sinusoidal control of the input current, thereby achieving a power factor of 0.99 or higher at the input end of the UPS system.

The typical circuits for the two common rectifiers are shown in Fig. 12.37. The three-level rectifier achieves higher efficiency by using relatively low-voltage power semiconductor devices through different pulse width modulation (PWM) control techniques. The relative disadvantage is that it requires twice the number of devices compared to the two-level rectifier, making the PCB layout and design relatively complex. In the T-type three-level rectifier, using SiC diodes for D_1 and D_2 can achieve higher converter efficiency. In the two-level rectifier, replacing Si IGBT with

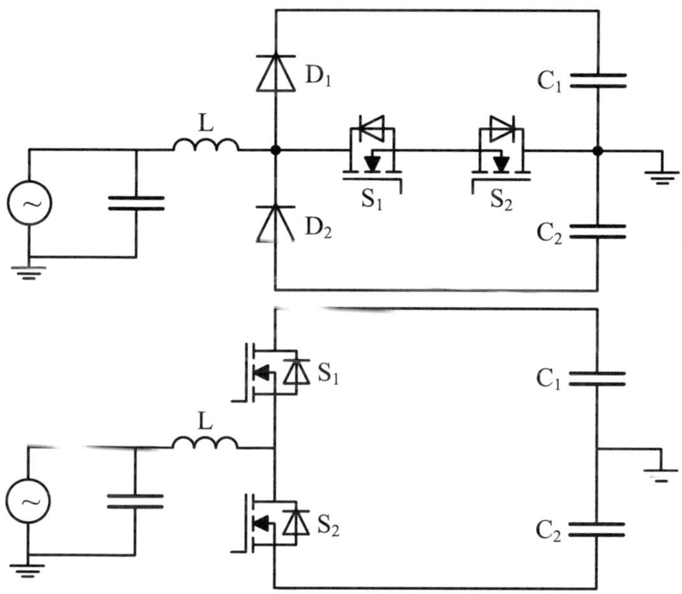

Fig. 12.37 Typical rectifier circuit

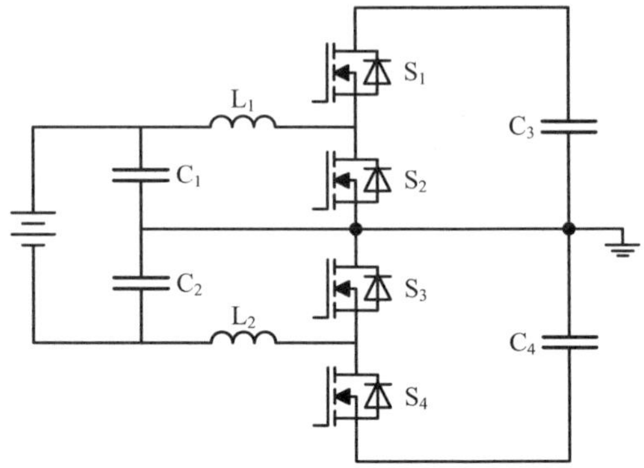

Fig. 12.38 DC converter circuit

SiC MOSFET for Q_1 and Q_2 allows for higher efficiency and power density in the converter design through higher switching frequencies. This improves the overall reliability of the converter while simplifying the device count.

The purpose of using a DC voltage converter internally in a UPS system is twofold. Firstly, it combines the high- or low-voltage internal DC bus (output voltage of the rectifier) to charge the batteries. Secondly, it performs the conversion between the voltage of the backup power supply and the DC bus voltage. When the input power from the grid is interrupted, the DC voltage converter converts the voltage from the backup energy source to match the DC bus voltage, thereby supplying power to the inverter via the DC bus voltage.

As shown in Fig. 12.38, the typical circuit of the DC voltage converter possesses bidirectional energy conversion capability. In the charging mode, the outer transistors Q_1 and Q_4 switch at high frequency, allowing energy to flow from the DC bus to the battery. In the discharging mode, the inner transistors Q_2 and Q_3 switch at high frequency, enabling energy to flow from the battery to the DC bus. Q_1 and Q_4 can be designed with diodes to form a unidirectional DC voltage converter, operating in the discharging mode. In the discharging mode, Q_1 and Q_4 contribute significantly to the device losses. By employing higher-performance and lower-switching-loss SiC diodes, the efficiency of the DC voltage converter can be enhanced, particularly suitable for scenarios with a wide range of battery voltage. Additionally, if SiC MOSFETs are used for Q_2 and Q_3, the system's operating frequency can be increased by 2 to 5 times, thereby improving the design power density through high-frequency operation and reducing the voltage ripple on the DC bus, effectively enhancing the battery's lifespan.

Inverters convert direct current voltage or DC backup power into alternating current output voltage, which is then connected to electrical devices. Inverters are typically designed based on the importance of system power supply and cost requirements. For small-power, low-cost inverters, transistors or MOSFETs are commonly used as switching components, with outputs in the form of square waves or modified sine waves. High-power inverters typically use IGBTs as switching components and employ output filters to generate high-precision sine waves. High-power UPS systems often utilize three-level inverters as the design solution. The advantage of this approach is the reduction of voltage stress on individual components and magnetic

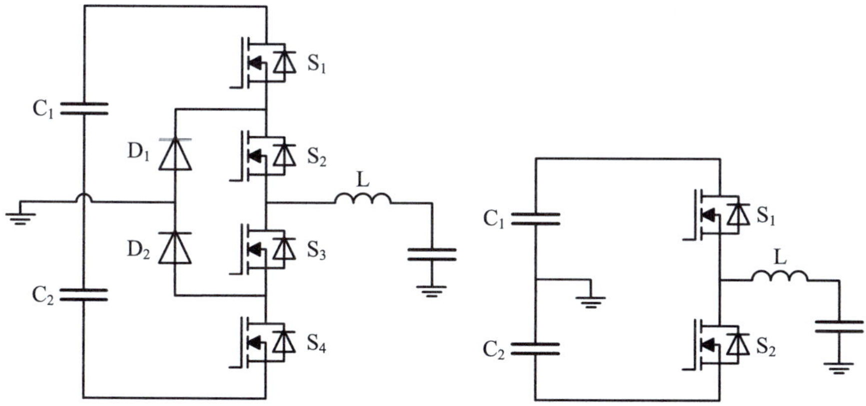

Fig. 12.39 Typical inverter circuit

elements through series connection of devices. This design allows IGBT components to operate in a mode with higher conversion efficiency compared to two-level inverters, resulting in an average efficiency improvement of around 3% for the system.

The typical circuits of the two common inverters are shown in Fig. 12.39. In inverters, the three-level architecture is widely used and is currently considered a well-balanced solution in terms of performance and economy. However, three-level inverters have relatively complex controls, especially when specific timing control mechanisms are required for current limiting protection scenarios. By using SiC MOSFETs, the inverter topology can be simplified from a three-level to a two-level half-bridge inverter. Leveraging the advantages of silicon carbide devices, such as fast switching and low losses, allows for higher switching frequencies to improve efficiency and power density, and high-frequency operation can also balance system costs. Furthermore, the simplification of the topology contributes to enhancing the overall reliability of the inverter.

Eaton's 93PR UPS product is an excellent representation of power products utilizing SiC technology. Through the hybrid module design incorporating IGBTs and SiC diodes, it elevates the efficiency of traditional UPS systems from 95–96 to 97%. Additionally, by leveraging the low conduction loss characteristics of SiC, it delivers higher operational efficiency within the 20–50% load range (Fig. 12.40).

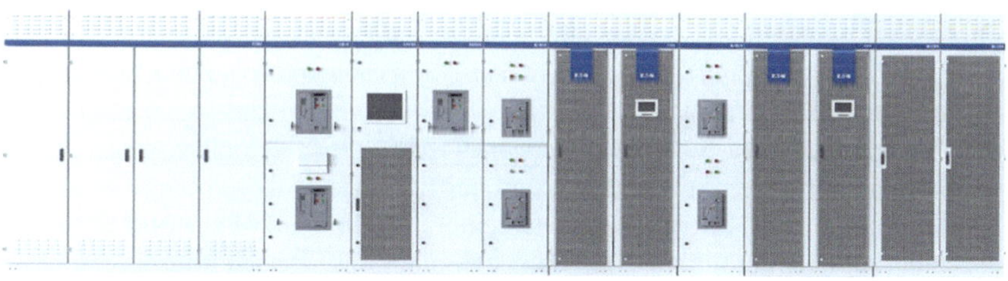

Fig. 12.40 Data center power module products based on Eaton's 93PR UPS

12.8 Power Supply

As a crucial component of people's daily lives, power supplies are widely distributed across various fields. However, with the continuous development of technology and the expansion of application scope, power supplies are facing multiple challenges. Firstly, high efficiency and energy conservation have become the primary directions for the development of the power supply industry. With increasing concerns about energy consumption and environmental impact, power supply manufacturers are striving to adopt advanced technologies and materials, optimize design and management processes to reduce energy loss and waste, and improve energy utilization efficiency. Secondly, the trend of continuous miniaturization and lightweighting of electronic products poses higher requirements for the size and weight of power supplies. Therefore, the power supply industry is actively developing more compact and lightweight power equipment, increasing its integration to meet the demands of different application scenarios. Additionally, with the widespread application of electronic products in various fields, the demand for power supply performance and reliability is also increasing. The power supply industry must continuously introduce products with stronger performance and higher reliability to meet various complex and demanding application requirements. Finally, with the emergence of new technologies such as artificial intelligence, 5G, and the Internet of Things, the power supply industry must continuously innovate and upgrade to adapt to new application scenarios and requirements, providing stable and reliable power support for these emerging technologies. Therefore, the power supply industry faces multiple challenges, but it also provides abundant opportunities for its development.

1. **Server Power Supply**

Data centers (Refer to Fig. 12.41) play an indispensable role in today's society, contributing significantly to human life and productivity, with their main demands covering several aspects. Firstly, digital transformation is a significant source of demand for data centers. Enterprises and organizations utilize data centers to store, process, and analyze vast amounts of data, enabling automation of business processes, optimization of decision-making, and provision of innovative digital services. Secondly, the emergence of big data and artificial intelligence (AI) also imposes higher requirements on data centers. Big data analytics and AI applications require immense computing power and storage resources, and data centers support these highly complex computing tasks, helping businesses extract valuable insights from massive datasets for intelligent decision-making and predictive analysis. Additionally, the rapid development of e-commerce and social media platforms drives the growth in demand for data centers. E-commerce and social media platforms rely on robust data centers to handle user transactions, search requests, content uploads, sharing, etc., and the stability and performance of these platforms directly depend on the efficient operation of data centers. In the field of scientific research, data centers support simulations, modeling computations, and scientific experiments, providing researchers with computational and storage resources, thus fostering innovation and discovery in the scientific domain. Finally, data centers play a critical role

Fig. 12.41 Data center

in supporting public services and emergency response. Governments, healthcare institutions, and non-profit organizations can utilize data centers to store and process vast amounts of critical information, enabling the delivery of more efficient services and emergency responses.

To meet such enormous demand, large-scale data centers are being rapidly constructed worldwide. These data centers typically occupy an area of hundreds of thousands of square feet and can accommodate tens of thousands of servers, providing support for large-scale cloud computing, online services, and data storage globally. As of September 2023, the USA, Germany, the UK, China, and the Netherlands respectively have over 5,300, 480, 450, 440, and 290 data centers. According to forecasts by the International Data Corporation (IDC), the data center services market in China is expected to continue growing at a compound annual growth rate (CAGR) of 18.9% over the next five years, with the market size expected to reach 307.5 billion yuan by 2027 (Fig. 12.41).

Such massive data centers consume an astonishing amount of electricity. According to relevant studies, the global power demand of data centers is approximately 416 hundred million kilowatt-hours, accounting for about 3% of all the electricity generated on Earth. Although this already represents a huge power demand, it is expected to increase over time with the growing number of operational data centers each year. Today, the issue of power shortages faced by data centers is becoming increasingly prominent, necessitating the adoption of greener and more efficient measures to address this challenge. Typically, energy-saving measures such as optimizing IT equipment, cooling systems, and power supply systems can be implemented to reduce the energy consumption of data centers.

Fig. 12.42 Server power supply

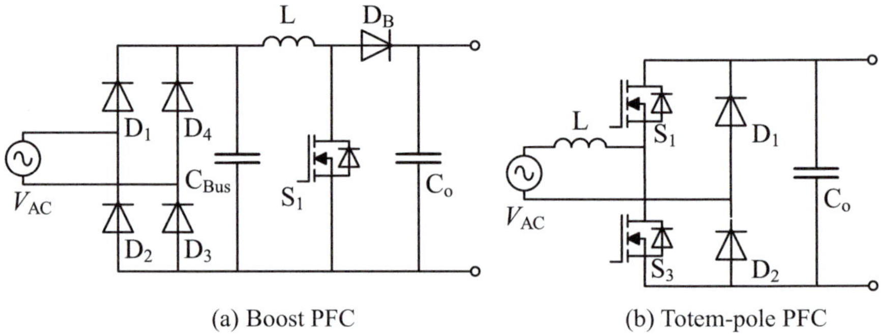

Fig. 12.43 PFC circuit topology

Among them (Refer to Fig. 12.42), server power supplies, as an integral part of the power supply system, have long been a subject of widespread attention regarding their efficiency. As early as 2004, the 80 PLUS certification program was introduced to promote the efficient use of computer power supply units. This certification requires that the energy efficiency of products should be above 80% at rated loads of 20%, 50%, and 100%, and at 100% load, the power factor should reach 0.9 or higher. Among these, the gold, platinum, and titanium levels require efficiency to be above 92%, 94%, and 96%, respectively, at 50% load. To meet these requirements, the use of SiC devices has become an inevitable choice (Fig. 12.42).

Server power supplies typically consist of a front-end PFC (Refer to Fig. 12.43) and a back-end isolated DC-DC converter. Traditional front-end PFCs use boost circuits and Si fast recovery diodes. By replacing these diodes with SiC diodes, the almost negligible reverse recovery losses can significantly improve the efficiency and power density of server power supplies. Furthermore, by transitioning to totem-pole PFC for the front-end PFC and selecting SiC MOSFETs as switching devices, component count can be reduced, EMI can be mitigated, and higher efficiency can be achieved (Fig. 12.43).

Fig. 12.44 Communication base station

2. **Power Supply for Communication Base Station**

The power supply of communication base stations (Refer to Fig. 12.44) has important value in modern society, it guarantees the stable operation of mobile communication networks and ensures the continuity and quality of communication services, providing people with the convenience of communicating and surfing the Internet anytime and anywhere (Fig. 12.44).

By the end of 2022, the total number of mobile communication base stations in my country reached 10.83 million, with a net increase of 870,000 throughout the year. Among them, there are 2.312 million 5G base stations, and 887,000 new 5G base stations were built throughout the year, accounting for 21.3% of the total number of mobile base stations. Generally speaking, the power consumption of communication base stations can range from a few kilowatts to tens of kilowatts. With the development of communication technology and the increase in network capacity, as well as the popularization of 5G technology, the power consumption of communication base stations may increase (Fig. 12.45). Figure 12.45 shows the power supply system for a communication base station.

In recent years, with the popularization of 5G, higher requirements have been put forward for communication base station power supplies. First, compared with 3G and 4G, 5G base stations require greater output power and higher efficiency. The output power of a single sector of the antenna unit (Voltage Switching) has increased from 40W ~ 80W to 200W or even higher, and the power of the BBU (baseband processing unit) has also exceeded 1000W. Due to the uneven data traffic of 5G communication, the load range of the communication power supply will change from light load to full load, so the efficiency requirements need to reach the highest value

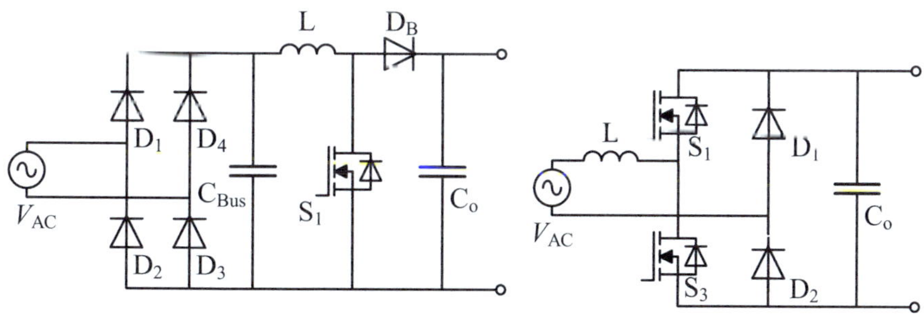

Fig. 12.45 Communication base station power supply

in a wide range, and high efficiency has become the key to reducing operating costs. Secondly, the power cabinet space of the 5G base station is limited, requiring the power module to increase the output power significantly while keeping the volume basically unchanged. At the same time, the power supply size of the 5G micro base station needs to be as small as possible, and the power density needs to be greatly improved. Finally, due to the substantial increase in the number of 5G micro base stations, and most of them are installed in confined spaces, only natural heat dissipation can be used, which poses a greater challenge to heat dissipation design.

SiC diodes have higher operating temperatures and almost negligible reverse recovery losses, which makes them widely used in PFC circuits in communication base station power supplies and play an important role.

3. **PD Power Supply**

In 1994, Intel and Microsoft initiated the International Organization for Standardization, referred to as "USB-IF". The organization has developed a series of universal serial bus specifications and specifications, which are the famous USB PD protocol. In addition, they have also developed a standard interface for adaptation based on this specification, that is, the USB interface. Nowadays, USB PD is one of the most universal and influential public fast charging protocols in the world. After several version upgrades, USB PD updated the PD 3.1 version on May 25, 2021, and the power was increased to a maximum of 240W.

As a portable electronic product, consumers require it to have high performance, high efficiency, and high power density. To meet end-user demands for lightweight and compact designs, PD fast charging must be high-frequency to reduce the volume of the magnetic core device, thereby reducing the volume of the entire machine to increase the power density. The switching loss of Si MOSFET and the reverse recovery loss of Si diode are proportional to the operating frequency. The higher the frequency, the greater the loss and the higher the temperature rise. To further improve the power density, power devices with lower losses are needed. The high-frequency and high-efficiency characteristics of gallium nitride (GaN HEMT) and silicon carbide diode (SiC-SBD) just meet this requirement.

Compared with traditional superjunction Si MOSFET, GaN HEMT with the same on-resistance has smaller parasitic capacitance, faster switching speed, and smaller switching loss, as shown in Table 12.1. Compared with Si fast recovery diodes, SiC diodes have half the reverse recovery current and lower losses, which help with the thermal management of high-power density PDs.

For high-power PD power supplies, to enhance performance and power density, they typically employ a PFC + LLC architecture, with GaN HEMTs used as power switching devices and SiC diodes utilized as PFC freewheeling diodes, as shown in Fig. 12.46.

Table 12.1 Innoscience INN650D240A (GaN HEMT) and superjunction Si MOSFET

Parameter	INN650D240A	Superjunction Si MOSFET
Rdson (mΩ)	165	159
Ciss (pF)	73	1080
Coss (pF)	20	18
Crss (pF)	0.2	4

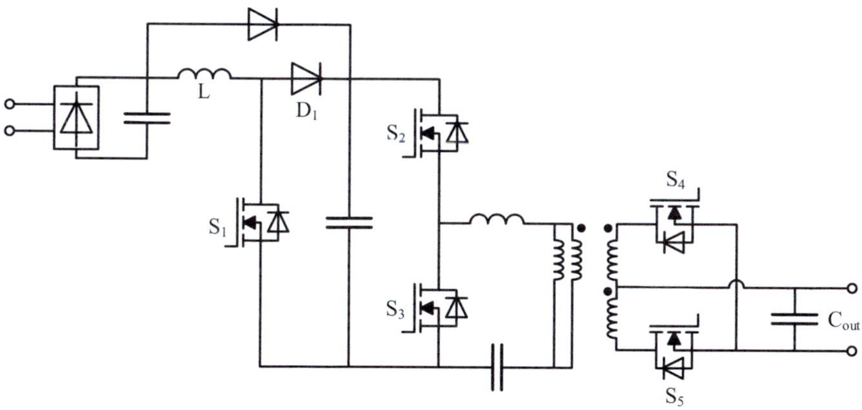

Fig. 12.46 140W PD power supply typical circuit

(a) PC Power Supply

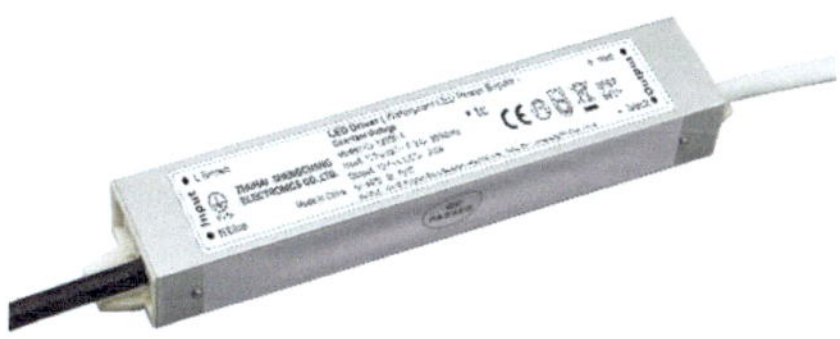

(b) LED Power Supply

(c)Edical Power Supply

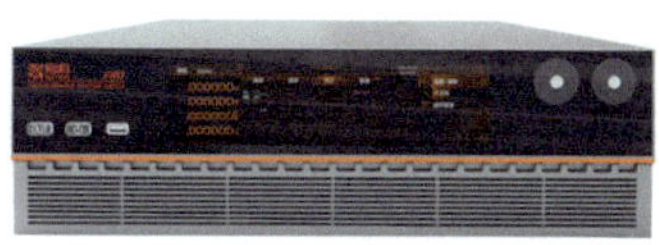

(d) Test Power Supply

Fig. 12.47 Various power supplies using SiC devices

4. **Other Power Supplies**

In addition to the server power supplies, communication power supplies, and PD power supplies mentioned above, SiC devices are also widely used in various other types of power supplies, including PC power supplies and LED power supplies using SiC diodes, as well as medical power supplies and test power supplies using both SiC diodes and MOSFETs (Fig. 12.47).

12.9 Motor Drive

According to statistics, over half of the global electricity is used to drive motors. Maximizing the efficiency of electric drives is crucial for reducing overall energy consumption and maximizing the usage time of battery-powered systems. Countries worldwide are continuously raising efficiency standards for motor drives, such as IE4 and IE5 for industrial applications, and SEER (in the USA), GB21455 (in China), and ESEER (in Europe) for air-conditioning systems.

In recent years, the development of motor drive technology has brought opportunities for the application of SiC devices. Firstly, low inductance motors can be used in applications such as axial flux machines, slotless motors, and low-leakage induction motors, which require acceptable ripple levels only at switch frequencies between 50 and 100 kHz. Therefore, IGBTs cannot meet the requirements of such applications, necessitating the use of SiC MOSFETs. Secondly, high-speed motors can be used in high-power density motors, high-pole-count high-speed motors with high torque density, and megawatt-level high-speed motors. Due to their high fundamental frequencies, these motors also require high switching frequencies, thus also requiring the use of SiC MOSFETs. Finally, power devices in applications such as electric vehicles, underwater, underground, and space applications operate in extremely harsh environments. SiC devices can operate at higher temperatures and are more suitable for such applications.

The motor drive inverter is divided into two parts: the front-end AC-DC and the back-end DC-AC. For applications with lower power requirements, the front-end AC-DC traditionally adopts a natural rectification plus boost PFC topology, where SiC diodes can be used in the boost PFC, and SiC MOSFETs can be used to replace the front-end AC-DC with totem-pole PFC. The back-end DC-AC traditionally uses a three-phase full-bridge topology composed of IGBTs, which can be replaced with SiC MOSFETs. For high-power motor drive applications, both the front-end AC-DC and the back-end DC-AC are composed of a three-phase full-bridge topology using IGBTs, which can be replaced with SiC MOSFETs (Fig. 12.48). Figure 12.48 shows the motor drive inverter circuit.

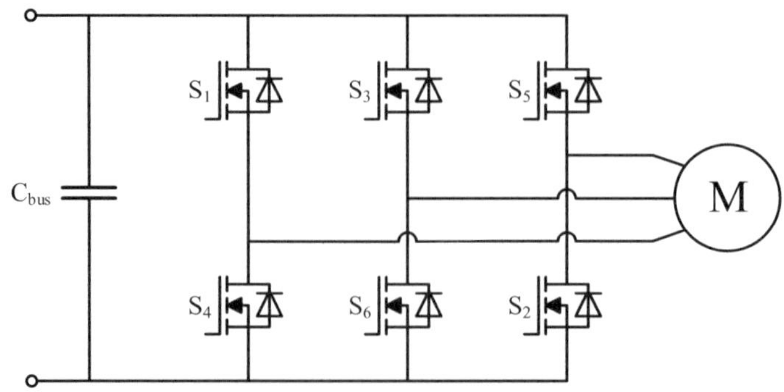

Fig. 12.48 Inverter circuit for motor drive

References

1. Hain S, Meiler M, Denk M (2019) Evaluation of 800V traction inverter with SiC-MOSFET versus Si IGBT power semiconductor technology, PCIM Europe 2019. In: International exhibition and conference for power electronics, intelligent motion, renewable energy and energy management, Nuremberg, Germany, pp 1–6.

Further Reading

2. Kampl S (2020) Topologies in on-board charging—semiconductor recommendations for reliable charging solutions, Whitepaper, Infineon Technologies AG., Rev. 1.0
3. Wang FF, Zhang Z (2016) Overview of silicon carbide technology: device, converter, system, and application. CPSS Trans Power Electron Appl 1(1):13–32
4. She X, Huang AQ, Lucía Ó, Ozpineci B (2017) Review of silicon carbide power devices and their applications. IEEE Trans Industr Electron 64(10):8193–8205
5. Do TV, Li K, Trovão JP, Boulon L (2020) Reviewing of using wide-bandgap power semiconductor devices in electric vehicle systems: from component to system. In: 2020 IEEE vehicle power and propulsion conference (VPPC), Gijon, Spain, pp 1–6
6. Morya AK et al (2019) Wide bandgap devices in AC electric drives: opportunities and challenges. IEEE Trans Transp Electrification 5(1):3–20
7. Wolfspeed, Inc. (2024) Build more efficient heat pumps and air conditioners with silicon carbide. Webinars
8. Bryson A, Albayrak M (2024) Silicon carbide for sustainable transportation. Bodo's Power Systems, P22–26
9. Mariani GB (2023) WBG switches in motor drive systems, Whitepaper, Infineon Technologies AG., Rev. 1.0

MIX
Papier aus verantwortungsvollen Quellen
Paper from responsible sources
FSC® C105338

If you have any concerns about our products,
you can contact us on
ProductSafety@springernature.com

In case Publisher is established outside the EU,
the EU authorized representative is:
Springer Nature Customer Service Center GmbH
Europaplatz 3, 69115 Heidelberg, Germany

Printed by Libri Plureos GmbH
in Hamburg, Germany